Morphology, Molecules, Evolution and Phylogeny in Polychaeta and Related Taxa

Developments in Hydrobiology 179

Series editor
K. Martens

Morphology, Molecules, Evolution and Phylogeny in Polychaeta and Related Taxa

Edited by

T. Bartolomaeus[1] & G. Purschke[2]

[1]*Freie Universität Berlin, Germany*
[2]*Universität Osnabrück, Germany*

Reprinted from Hydrobiologia, volumes 535/536 (2005)

Springer

Library of Congress Cataloging-in-Publication Data

A C.I.P. Catalogue record for this book is available from the Library of Congress.

ISBN 978-90-481-6754-8
e-ISBN 978-1-4020-3240-0

Published by Springer,
P.O. Box 17, 3300 AA Dordrecht, The Netherlands

Cover Illustration: SEM of the anterior end of *Nereis* sp. with everted pharynx and extended jaws. Dorsolateral palps show bundles of sensory cilia at their tip. Cladogram to the fore is based on molecular data and shows Clitellata and Echiura among polychaete taxa. (Original figure by Purschke, cladogram from Jördens et al. (2004): J. Zool Syst. Evol. Res 42: 270–289.)

Printed on acid-free paper

TABLE OF CONTENTS

Hydrobiologia (2005) **535/536:** vii–viii
T. Bartolomaeus & G. Purschke (eds), Morphology, Molecules, Evolution and Phylogeny in Polychaeta and Related Taxa

Preface

Polychaetes and Clitellata, which together constitute Annelida, are one of the most significant groups of metazoans, not only because they are so numerous – about 20,000 species have been described – but also because of their remarkable diversity. Recently evidence is accumulating that even some of the groups formerly regarded as independent "phyla" such as Pogonophora (Frenulata and Vestimentifera), Echiura, Myzostomida and perhaps Sipuncula, are most probably nothing else than greatly modified Annelida. The relationship of Clitellata to the other annelids has also recently become controversial. Hence these animals as a whole are highly significant to our understanding of fundamental questions about morphological and adaptive diversity, as well as clarifying evolutionary changes and phylogenetic relationships. In recent years a great deal has been learned about these animals, from morphological studies employing conventional and modern methods plus, increasingly and importantly the use of molecular markers and computer-assisted kinship analyses. This new knowledge has also raised new questions and controversies. By no means everything is known in polychaetes and related taxa and much is still to be disclosed.

So far there has been no modern, comprehensive presentation and discussion of the recent research results, nor has there been a clear statement of the open questions and current problems in the area. Such a summary also seems necessary in particular because the different research groups, as a result of the wide range of methods they employ publish in extremely diverse journals. One aspect of the topic, ultrastructure, has for many years been covered by the comprehensive and much quoted work "The Ultrastructure of Polychaeta", edited by W. Westheide & C.O. Hermans (Mikrofauna Marina 4, 1988), which resulted from a similar conference held in 1986. Somewhat later than this volume, in the series "Microscopic Anatomy of Invertebrates", Volume 7 "Annelida" (F.W. Harrsion & S.L. Gardiner, eds, Wiley-Liss. Inc., 1992) was published. Although in many respects these publications are by no means outdated, the first book is out of print now, and furthermore they treat only one aspect, morphology. During the 15 years that have been elapsed since that conference new morphological (e.g., confocal laser scanning microscopy) and molecular (e.g., DNA sequencing) methods came into use. And of course many other new results, obtained with different methods, have become available since then. However, the overall subject area indicated by the title of the symposium including evolution and phylogeny with reference to morphological and molecular characters, is not treated anywhere. The best way to achieve this goal appeared to gather specialists in a workshop and to edit a comprehensive volume with chapters written by these specialists, putting emphasis on open questions, current problems and possible directions of future research as well. We are happy that these specialists were willing to write these contributions. We are also happy that the resulting programme was attractive enough for many others who came and attended this symposium in part with short contributions and posters. We both hope that this book will be a good foundation for future work and will help to keep the topic attractive enough to encourage young scientists to work with polychaetes and related taxa. Last but not least we would like to dedicate this volume to Professor Wilfried Westheide on the occasion of his 65th birthday and retirement in recognition of his outstanding contributions in the fields of annelid systematics, evolution, phylogeny, reproductive biology and morphology.

We thank the Vice President of the University of Osnabrück, Prof. Dr P. Hertel, for his welcoming remarks at the opening of the symposium. We are extremely grateful to the Stiftung Volkswagenwerk, Hannover, Germany, for funds supporting the travel and accommodations of the participants. We are grateful to the President of the University of Osnabrück for additional financial support. Without these supports the workshop and the volume would not have been made possible. Specials thanks are due to Andrea Noël for her unceasing help and care in organizing this symposium. Scientists, staff and students from Osnabrück and Berlin contributed to the organization of the conference or have helped in many ways in editing this volume: Dr Monika C.M. Müller, Thorsten Hinken, Jens Rüchel, Sonja Raabe, Janina Jördens, Torsten Struck, Martina Biedermann, Werner Mangerich, Eva Hass-Cordes, Mechthild Krabusch, Lars Vogt, Christoph Bleidorn, Georg Mayer, Dr Markus Böggemann.

T. Bartolomaeus
Freie Universität,
Berlin, Germany

G. Purschke
Universität Osnabrück,
Osnabrück, Germany

December 2003

Hydrobiologia (2005) **535/536:** ix–xii
T. Bartolomaeus & G. Purschke (eds), Morphology, Molecules, Evolution and Phylogeny in Polychaeta and Related Taxa

List of participants

Günter ARLT, Universität Rostock, FB Biowissenschaften, Abt. für Stoffwechselphysiologie, Albert-Einstein-Straße 3, D-18059 Rostock, Germany
guenter.arlt@biologie.uni-rostock.de

Thomas BARTOLOMAEUS, Systematik und Evolution der Tiere, Institut für Biologie, Freie Universität Berlin, Königin-Luise-Straße 1-3, D-14195 Berlin, Germany
tbartol@zoosyst-berlin.de

Ralf BASTROP, Universität Rostock, FB Biowissenschaften, Abt. für Stoffwechselphysiologie, Albert-Einstein-Straße 3, D-18059 Rostock, Germany
ralf.bastrop@biologie.uni-rostock.de

Andreas BICK, Universität Rostock, FB Biowissenschaften, Allgemeine und Spezielle Zoologie, D-18051 Rostock, Germany
andreas.bick@biologie.uni-rostock.de

Miriam BLANK, Universität Rostock, FB Biowissenschaften, Abt. für Stoffwechselphysiologie, Albert-Einstein-Straße 3, D-18059 Rostock, Germany
miriam.blank@stud.uni-rostock.de

Christoph BLEIDORN, Universität Bielefeld, Fakultät für Biologie, Zoomorphologie und Systematik, Postfach 10 01 31, D-33501 Bielefeld, Germany
christoph.bleidorn@biologie.uni-bielefeld.de

Markus BÖGGEMANN, Habsburgerallee 37, D-60385 Frankfurt/Main, German
mboeggem@aol.com

Regine BÖNSCH, Institut für Angewandte Ökologie GmbH, Alte Dorfstraße 11, D-18184 Neu Broderstorf, Germany
boensch@ifaoe.de

Monika BRIGHT, Meeresbiolgie IECB, Universität Wien, Althanstraße 14, A-1090 Vienna, Austria
monika.bright@univie.ac.at

Ed B. CUTLER, Utica College of Syracuse University & Museum Associate, Department of Invertebrate Zoology, Museum of Comparative Zoology, Harvard University, Cambridge, MA 02138, U.S.A.
ebcutler@earthlink.net

Daniel DAUER, Department of Biological Sciences, Old Dominion University, Norfolk, VA 23529, U.S.A.
ddauer@odu.edu

Adriaan DORRESTEIJN, Institut für Allgemeine und Spezielle Zoologie, Universität Giessen, Stephanstrasse 24, D-35390 Giessen, Germany
adriaan.dorresteijn@zool.bio.uni-giessen.de

Nicole DUBILIER, MPI für Marine Mikrobiologie, Celsiusstraße 1, D-28359 Bremen, Germany
ndubilie@mpi-bremen.de

Kevin J. ECKELBARGER, Darling Marine Center, University of Maine, 193 Clark's Cove Road, Walpole, ME 04573, U.S.A. and School of Marine Sciences, University of Maine, Orono, ME 04469, U.S.A
kevine@maine.edu

Igor EECKHAUT, Marine Biology, Natural Sciences Building, University of Mons, Av. Champs de Mars 6, B-7000 Mons, Belgium
igor.eeckhaut@umh.ac.be

Danny EIBYE-JACOBSEN, Zoological Museum, Invertebrate Department, Universitetsparken 15, DK-2100 Copenhagen Ø, Denmark
dejacobsen@zmuc.ku.dk

Christer ERSÉUS, Department of Invertebrate Zoology, Swedish Museum of Natural History, Box 50007, S-10405 Stockholm, Sweden
christer.erseus@nrm.se

Kristian FAUCHALD, Smithsonian Institution, U.S. National Museum of Natural History, Department of Invertebrate Zoology, Washington, DC 20560, U.S.A.
fauchald.kristian@nmnh.si.edu

Marco FERRAGUTI, Dipartimento di Biologia, 26, Via Celoria, I-20133 Milano, Italy
marco.ferraguti@unimi.it

Dieter FIEGE, Forschungsinstitut und Naturmuseum Senckenberg, Sekt. Marine Evertebraten II, Senckenberganlage 25, D-60325 Frankfurt/Main, Germany
dfiege@senckenberg.de

Anna FILIPPOVA, Department of Invertebrate Zoology, Biological Faculty, State University of Moscow, 119899 Moscow, Russia
afilippowa@mail.ru

Albrecht FISCHER, Universität Mainz, Zoologisches Institut, D-55099 Mainz, Germany
afischer@mail.uni-mainz.de

Kirk FITZHUGH, Research & Collections Branch, LA County Museum of Natural History, 900 Exposition Blvd, Los Angeles, CA 90007, U.S.A.
kfitzhug@nhm.org

Olav GIERE, Universität Hamburg Zoologisches Institut and Museum, Martin-Luther-King-Platz 3, D-20146 Hamburg, Germany
olav.giere@zoologie.uni-hamburg.de

Ken HALANYCH, Department of Biological Sciences, Auburn University, 101 Life Science Building, AL 36849, USA
ken@auburn.edu

Gerhard HASZPRUNAR, Zoologische Staatssammlung München, Münchhausenstraße 21, D-81247 München, Germany
haszi@zi.biologie.uni-muenchen.de

Bilke HAUSAM, Universität Bielefeld, Fakultät für Biologie, Zoomorphologie und Systematik, Postfach 10 01 31, D-33501 Bielefeld, Germany
bhausam@biologie.uni-bielefeld.de

Harald HAUSEN, Systematik und Evolution der Tiere, Institut für Biologie, Freie Universität Berlin, Königin-Luise-Straße 1–3, 14195 Berlin, Germany

Andreas HEIJNOL, TU Braunschweig, Institut für Genetik, AG Schnabel, Spielmannstraße 7, D-38106 Braunschweig, Germany
a.heijnol@tu-bs.de

René HESSLING, Im Siek 14, D-37136 Waake, Germany
r.hessling@zeiss.de

Brigitte HILBIG, Universität Hamburg, Zoologisches Institut und Zoologisches Museum, Martin-Luther-King-Platz 3, D-20146 Hamburg, Germany
hbilbig@zoologie.uni-hamburg.de

Ulrich HOEGER, Universität Mainz, Zoologisches Institut, D-55099 Mainz, Germany
uhoeger@mail.uni-mainz.de

Dietrich K. HOFMANN, Ruhr-Universität Bochum, AG Entwicklungsphysiologie der Tiere, ND 05/779, D-44780 Bochum, Germany
dietrich.k.hofmann@ruhr-uni-bochum.de

Pat HUTCHINGS, The Australian Museum, Marine Invertebrates, 6, College Street, Sydney NSW 2000, Australia
path@austmus.gov.au

Janina JÖRDENS, Universität Osnabrück, FB 5, Spezielle Zoologie, D-49069 Osnabrück, Germany
joerdens@biologie.uni-osnabrueck.de

Reinhardt KRISTENSEN, Zoologisk Museum, University of Copenhagen, Universitetsparken 15, DK-2100 Copenhagen, Denmark
rmkristensen@zmuc.ku.dk

Otto LARINK, TU Braunschweig, Zoologisches Institut, Spielmannstraße 8, D-38106 Braunschweig, Germany
o.larink@tu-bs.de

Carsten LUETER, HU Berlin, Museum für Naturkunde, Institut für Systematische Zoologie, Invalidenstraße 43, D-10099 Berlin, Germany
carsten.lueter@museum.hu-berlin.de

Roberto MAROTTA, Dipartimento di Biologia, Sezione di Zoologia e Citologia, Via Celoria, 26, I-20133 Milano, Italy
roberto.marotta@unimi.it

Georg MAYER, Freie Universität Berlin, Institut für Biologie, Systematik und Evolution der Tiere, Königin-Luise-Str. 1–3, 14195 Berlin
gmayer@zoosyst-berlin.de

Damnhait McHUGH, Department of Biology, Colgate University, Hamilton, NY 13346, U.S.A.
dmchugh@mail.colgate.edu

Karin MEISSNER, Invertebrate Division, The Australian Museum, 6 College Street, Sydney NSW 2010, Australia
karinm.@austmus.gov.au

Joao MIGUEL DE MATOS NOGUEIRA, Departamento de Zoologia, Instituto de Biociências, Universidade de São Paulo, IB-USP, Rua do Matão, travessa 14, n° 101, CEP: 05508-900, SP, Brazil
nogueira@ib.usp.br

Monika C.M. MÜLLER, Universität Osnabrück, FB 5, Spezielle Zoologie, D-49069 Osnabrück, Germany
monika.mueller@biologie.uni-osnabrueck.de

Claus NIELSEN, Zoologisk Museum, Universitetsparken 15, DK-2100 Copenhagen, Denmark
cnielsen@zmuc.ku.dk

Lars ORRHAGE, Berzeliigatan 21, S-41253 Göteborg, Sweden

Hannelore PAXTON, Department of Biological Sciences, Macquarie University, Sydney NSW 2109, Australia
hpaxton@rna.bio.mq.edu.au

Mary E. PETERSEN, University of Copenhagen, Zoological Museum, Universitetsparken 15, DK-2100 Copenhagen, Denmark
mepetersen@zmuc.ku.dk

Günter PURSCHKE, Universität Osnabrück, FB 5, Spezielle Zoologie, D-49069 Osnabrück, Germany
purschke@biologie.uni-osnabrueck.de

Vasily RADASHEVSKY, Institute of Marine Biology, Vladivostok, 690041, Russia
radashevsky@mail.ru

Nicola REIFF, Aidenbachstraße 111a, D-81379 München, Germany
nicola_reiff@gmx.de

Reinhard M. RIEGER, Universität Innsbruck, Institut für Zoologie, Technikerstraße 25, 6020 Innsbruck, Austria
reinhard.rieger@uibk.ac.at

Greg W. ROUSE, South Australian Museum, North Terrace, Adelaide, SA 5000, Australia, School of Environmental & Earth Sciences, The University of Adelaide, Adelaide, SA 5005, Australia
gregr@saugov.sa.gov.au

Jens RÜCHEL, Universität Osnabrück, FB 5, Spezielle Zoologie, D-49069 Osnabrück, Germany
jens.ruechel@biologie.uni-osnabrueck.de

Hilke RUHBERG, Universität Hamburg, Zoologisches Institut und Zoologisches Museum, Martin-Luther-King-Platz 3, D-20146 Hamburg, Germany
ruhberg@zoologie.uni-hamburg.de

José I. SAIZ-SALINAS, Departamento de Zoologia y DCA, Univ. Pais Vasco, E-48080 Bilbao, Apdo. 644, Spain
zopsasaj@1gdx04.1g.ehu.es

Christoffer SCHANDER, University of Göteborg, Department of Zoology, Box 463, S-40530 Göteborg, Sweden
c.schander@zool.gu.se

Andreas SCHMIDT-RHAESA, Universität Bielefeld, Fakultät für Biologie, Zoomorphologie und Systematik, Postfach 100131, D-33051 Bielefeld, Germany
a.schmidt-rhaesa@biologie.uni-bielefeld.de

Paul SCHROEDER, School of Biological Sciences, Washington State University, Pullman, Washington, WA 99164-4236, U.S.A.
schroede@wsu.edu

Anja SCHULZE, Department of Invertebrate Zoology, Harvard University, MCZ, 26 Oxford Street, Cambridge, MA 02138, U.S.A.
aschulze@oeb.harvard.edu

Margarita SHABANOVA, Festivalnaya ul. 15-3-72, 125195 Moscow, Russia
shabanova@mailru.com

Mark E. SIDDALL, Division of Invertebrate Zoology, American Museum of Natural History, Central Park at 79th Street, New York, NY 10024, U.S.A.
siddall@amnh.org

Kathrin SOBJINSKI, Biologie/Zoologie, Philipps-Universität Marburg, Karl-v-Frisch-Straße, D-35032 Marburg, Germany
sobjinsk@mailer.uni-marburg.de

Eve SOUTHWARD, Marine Biological Association of the UK, Citadel Hill, Plymouth PL1 2PB, UK
100721.3720@compuserve.com

Volker STORCH, Ruprecht-Karls-Universität Heidelberg, Zool. Institut, Abt. 5 und 6, Im Neuenheimer Feld 230, D-69120 Heidelberg, Germany
volker.storch@urz.uni-heidelberg.de

Torsten STRUCK, Department of Biological Sciences, Auburn University, 101 Life Science Building, AL 36849, USA
structh@auburn.edu

Alexander B. TZETLIN, Department of Invertebrate Zoology, Biological Faculty, State University of Moscow, 119899 Moscow, Russia
tzetlin@soil.msu.ru

Lars VOGT, Systematik und Evolution der Tiere, Institut für Biologie, Freie Universität Berlin, Königin-Luise-Straße 1–3, D-14195 Berlin, Germany
lars.vogt@zoosyst-berlin.de

Thomas WEHE, Ruprecht-Karls-Universität Heidelberg, Zool. Institut, Abt, 5 und 6, Im Neuenheimer Feld 230, D-69120 Heidelberg, Germany
t.wehe@urz.uni-heidelberg.de

Wilfried WESTHEIDE, Universität Osnabrück, FB 5, Spezielle Zoologie, D-49069 Osnabrück, Germany
westheide@biologie.uni-osnabrueck.de

Katrine WORSAAE, Zoologisk Museum, Universitetsparken 15, DK-2100 Kopenhagen, Denmark
kworsaae@zmuc.ku.dk

Anna ZHADAN, Novatorov str. 4-1-27, 119421 Moscow, Russia
azhadan@mail.ru or azhadan@ecocenter.msu.ru

Sven ZÖRNER, Universität Maniz, Zoologisches Institut, D-55099 Mainz, Germany
zoersooo@mail.uni-mainz.de

Addresses of Editors:

Thomas BARTOLOMAEUS, Systematik und Evolution der Tiere, Institut für Biologie, Freie Universität Berlin, Königin-Luise-Straße 1-3, D-14195 Berlin, Germany
tbartol@zoosyst-berlin.de

Günter PURSCHKE, Universität Osnabrück, FB 5, Spezielle Zoologie, D-49069 Osnabrück, Germany
purschke@biologie.uni-osnabrueck.de

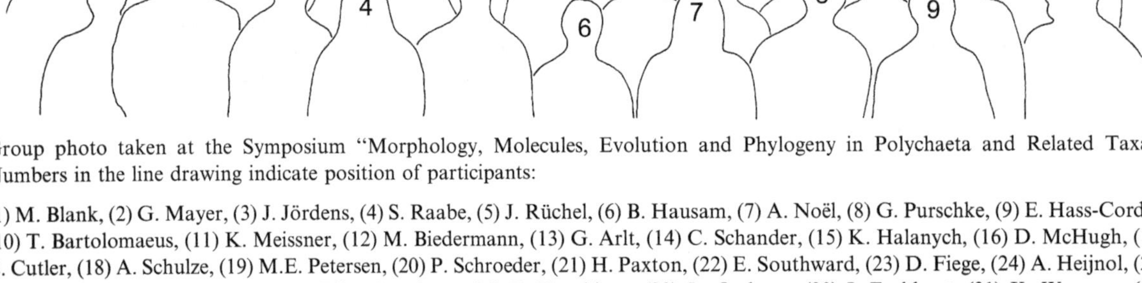

Group photo taken at the Symposium "Morphology, Molecules, Evolution and Phylogeny in Polychaeta and Related Taxa". Numbers in the line drawing indicate position of participants:

(1) M. Blank, (2) G. Mayer, (3) J. Jördens, (4) S. Raabe, (5) J. Rüchel, (6) B. Hausam, (7) A. Noël, (8) G. Purschke, (9) E. Hass-Cordes, (10) T. Bartolomaeus, (11) K. Meissner, (12) M. Biedermann, (13) G. Arlt, (14) C. Schander, (15) K. Halanych, (16) D. McHugh, (17) E. Cutler, (18) A. Schulze, (19) M.E. Petersen, (20) P. Schroeder, (21) H. Paxton, (22) E. Southward, (23) D. Fiege, (24) A. Heijnol, (25) R. Bönsch, (26) C. Nielsen, (27) D. Eibye-Jacobsen, (28) P. Hutchings, (29) L. Orrhage, (30) I. Eeckhaut, (31) K. Worsaae, (32) A. Cutler, (33) K. Fauchald, (34) W. Westheide, (35) M. Siddall, (36) E. Borda, (37) M. Bright, (38) A. Filippova, (39) G. Rouse, (40) M.C. Müller, (41) G. Haszprunar, (42) J.M.M. Nogueira, (43) T. Wehe, (44) C. Erséus, (45) M. Ferraguti, (46) A. Tzetlin, (47) G. Kristensen, (48) M. Shabanova, (49) U. Hoeger, (50) A. Fischer, (51) H. Ruhberg, (52) D. Hofmann, (53) R.M. Rieger, (54) M. Böggemann, (55) L. Vogt, (56) D. Dauer, (57) unknown, (58) C. Bleidorn, (59) K. Fitzhugh, (60) T. Struck, (61) P. Kestler, (62) O. Giere, (63) A. Schmidit-Rhaesa, (64) A. Unger, (65) A. Gruhl, (66) V. Storch, (67) C. Lueter, (68) H. Hausen, (69) A. Berentzen, (70) K. Sobjinski, (71) A. Zhadan, (72) R. Bastrop (73) A. Bick, (74) R.M. Kristensen.

Not in photo: B. Hilbig, J. Saiz-Salinas, O. Larink, V. Radashevsky, R. Hessling, N. Reiff, N. Dubilier, R. Marotta.

Hydrobiologia (2005) **535/536:** 1–22
T. Bartolomaeus & G. Purschke (eds), Morphology, Molecules, Evolution and Phylogeny in Polychaeta and Related Taxa

Cell lineage and gene expression in the development of polychaetes

Adriaan Dorresteijn
Institut für Allgemeine und Spezielle Zoologie, Universität Giessen, Stephanstrasse 24, D-35390 Giessen, Germany

Keywords: polychaete, cell lineage, determination, gene expression

Abstract

The developmental strategies of embryos within the various taxa of polychaetes are designed to set up the fate of the cell lines. Some of these traits of pattern formation are considered to be ancestral, but we also find a number of derived developmental characteristics, some of which might be useful indicators for phylogenetic relationships. A combination of ooplasmic segregation and anisotropic cleavage rapidly establishes the fate of several larval cell lines in the polychaete embryo. The setting-up of the primary trochoblasts basically concerns the same cell line ($1a^2$–$1d^2$) in polychaetes and even in molluscs. Such mechanisms may thus be regarded as ancestral. The determination of the mesoderm precursor occurs very late in both equally cleaving annelids and mollusks, indicating that an equal cleavage pattern may be regarded as an ancestral trait. Since both disproportionate cytoplasmic distribution (either by spindle shift or polar lobe formation) and cell cycle asynchronies appear to speed up the development of the mesoderm-forming cell line, these strategies represent derived traits. An analysis of these derived traits of early development is given and is discussed in the light of the phylogenetic relationships among the polychaetes. These data are extended by an analysis of some of the postlarval structures in polychaetes and the molecular developmental circuitry involved.

Introduction

Numbering approximately 9000 species, annelids belong among the larger groups of invertebrate animals. Although annelids occupy various terrestrial and freshwater habitats, it is beyond a shadow of a doubt that the ancestral life forms in this group initially lived in a marine environment. Even nowadays almost 60% of all annelid species described so far are marine and nearly all of these species are polychaetes. With only few exceptions the majority of the ancestors of the polychaetes related to the marine environment, whereas the ancestors of oligochaete and hirudinean species adopted a terrestrial and/or freshwater life style. Despite the fact that polychaete species have developed a tremendous diversity in body plan, ranging from free living to completely sedentary life styles, they never show the typical reproductive adaptations found in the clitellates. With only few exceptions, clitellates form cocoons with extremely yolky eggs which develop by a modified spiral cleavage pattern. The precursor cells for the majority of tissues of the adult body are determined at a very early developmental stage. Due to their size they can easily be identified at the posterior dorsal side of the embryo and are therefore called teloblasts (Whitman, 1878; Schleip, 1914). In the hirudinean development the N-, O/P-, O/P-, Q- teloblasts provide for the ectoderm and neural elements of the leech trunk (Stent & Weisblat, 1982). Each of the bilaterally symmetrical positioned M-teloblasts represents the precursor cell for the formation of a mesodermal germ band. The leech embryo skips a larval stage and develops all 33 segments simultaneously. Even though the number of segments in oligochaete development is not fixed by the end of embryonic development, the same type of germ band leads to the immediate formation of a small adult, thus skipping a larval stage. Shimizu (1982) showed that the isolated precursor of the teloblasts, the D-blastomere, is able to form normal young worms. Not only the adult body plans of the clitellates, but also the developmental strategies of their embryos thus share several derived characters. In contrast, most polychaete embryos develop by a type of spiral cleavage in which the formation of larval structures prevails and leads to the formation of a trochophore larva. Similar developmental strategies leading to similar larvae are encountered in the development of Mollusca, Echiurida, Sipunculida and Nemertea. The larval conditions enable the developing individual to disperse, feed and grow after the termination of embryogenesis. This developmental strategy has several advantages:

- it allows the production of a larger number of small, oligolethal eggs, thereby reducing the maternal investments in a single embryo;
- a large number of offspring secures the gene pool of the parental pair in future reproductive cycles;
- the dispersal of numerous small larvae prevents predation and supports the spreading of the species;
- the development and growth of adult tissues (in polychaetes: combined with the formation of segments) can be spread over or postponed until after the larval phase.

These, and a large number of more derived developmental characteristics, can be encountered in the various taxa of the polychaetes. This chapter focuses on the various aspects of pattern formation and the determination of larval and adult cell lines. It also provides some of the recent molecular data we have gathered on the differentiation of such cell lines. Although this chapter predominantly deals with developmental processes of polychaetes it also attempts to answer some questions of phylogenetic descent within the spiralians.

Pattern formation in early embryos

Polarity of the oocyte

In a previous review of the ultrastructural aspects of development in polychaetes, Dorresteijn & Fischer (1988) stated that pattern formation in part is prepared by the setting up of egg polarity during oogenesis. Most polychaete eggs develop within an ovary and are frequently accompanied by somatic cells (follicle or nurse cells) or they are attached to the epithelium of the ovary wall, of the peritoneum, or of a blood vessel. The diversity of polychaete oogenesis and of ovarian types has been treated by Eckelbarger (1988; this volume) and will not be repeated here. The oocytes of some polychaete groups, like the Nereididae, Phyllodocidae, Sphaerodoridae and Alciopidae, sprout from still unknown tissues and undergo previtellogenic and vitellogenic development in a free-floating condition within the coelomic fluid. But even in such cases of diffuse oogenesis the eggs show signs of polarity which must have been set up at earlier stages. In the Nereidid *Platynereis dumerilii* the mature oocyte represents a rotational ellipsoid with an equatorial radius of 75–80 μm and a rotational axis of approximately 135 μm (Fig. 1a and b). Kluge (1991) showed that the rotational axis corresponds to the future animal-vegetal axis of the embryo and the antero-posterior axis of the larva. The oocytes of *Pomatoceros triqueter* show an even more pronounced degree of animal-vegetal flattening, giving the unfertilized egg the appearance of a discus (Fig. 1c and d). Similar egg forms were also described in other polychaetes, e.g. *Arenicola marina* (Meijer, 1979) and *Podarke obscura* (Treadwell, 1901). Within the non-spherical ellipsoidal eggs, the short axis of which relate to the animal-vegetal axis, developmental factors may have been allocated in the cortex at the future polar regions. Dorresteijn & Fischer (1988) described the presence of such cortical fields with accumulations of ribosomes, ER and numerous vesicles during polar body formation of the fertilized egg of *Pomatoceros*. Other egg forms have been described by Villa (1976) for the sternaspid *Sternaspis* which produces spindle-shaped eggs with polar accumulations of mitochondria. But even in the spherical eggs of *Sabellaria alveolata* (Hatt,

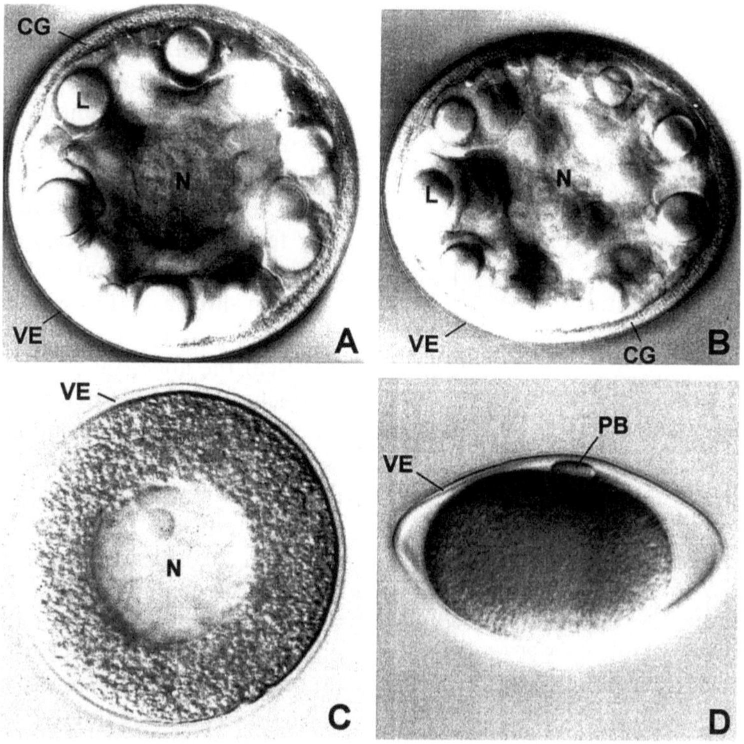

Figure 1. Oocytes of most polychaete species show signs of a preexisting polarity. The oocytes of *Platynereis dumerilii* (A, B) are shaped as rotational ellipsoids, the rotational axis of which corresponds to the animal-vegetal axis. The oocytes of the serpulid *Pomatoceros triqueter* are lense-shaped (C: top view). Although the egg slightly rounds up after fertilization the initial lense-shape of the vitellin envelope remains (D). The position of the polar body in this image labels the animal pole. The rotational axis of the oocyte thus foreshadows the future egg axis. CG: cortical granules; L: lipid droplets (surrounded by protein yolk granules); N: nucleus; VE: vitellin envelope; PB: polar body. Egg diameters: *Platynereis* (160 μm), *Pomatoceros* (65 μm).

1932; Render, 1983; Speksnijder & Dohmen, 1983; Dorresteijn & Fischer, 1988) there are experimental and ultrastructural correlates for the localization of developmental potential in the vegetal hemisphere. Jeffery (1984) demonstrated an asymmetric distribution of mRNA in the cortex of the oocytes of *Chaetopterus*. As in many other animals, the oocytes of polychaetes are liable to produce maternal factors which are allocated during oogenesis or by ooplasmic segregation during early development. The localisation of preformed factors rather than the *de novo* synthesis of such factors speeds up the processes of early development by setting up and determining the fates of some of the early blastomeres. The properties of the egg cortex, recently reviewed by Sardet et al. (2002), not only offer a chance for this precocious localization of developmental factors, it also allows a rapid reorganization of such factors upon fertilization, and, due to the interactions with mitotic spindles, set up the site of polar body formation and the various cleavage furrows. Since most polychaete eggs are not perfectly spherical and the shape relates to the animal-vegetal polarity, developmental factors may well be or get distributed in a polar fashion (Fig. 2). Unfortunately, our knowledge about the existence and the distribution of developmental factors in the polychaete egg and early embryo is still very limited. After the successful isolation of larval and postlarval developmental factors (Arendt et al., 2001; Seaver et al., 2001; Arendt et al., 2002) we should now focus on those factors involved in the earliest stages. So far, experimental studies disturbing the cortex by means of destabilizing the microfilament or microtubule network indicate that this influences both the position of the spindles as well as the determina-

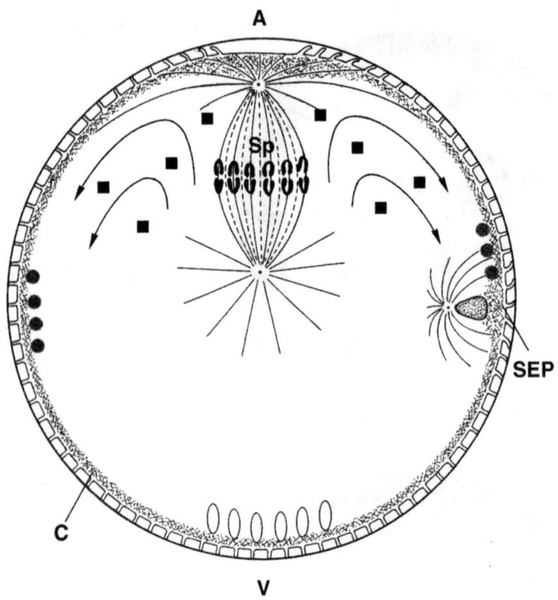

Figure 2. The animal-vegetal polarity of polychaete oocytes and eggs could well be instrumentalized for the prelocalization of developmental potential. The egg cortex (C) plays a crucial role in the allocation of determinants (indicated by the grey rounded symbols). Part of the determinants (depicted by black squares) may reside in the cytoplasm and get segregated during the cleavage process. A break of the initial rotational symmetry the axis of which gets manifest by the position of the spindle of the first meiotic division is caused by the sperm entry point (SEP). A: animal pole; V: vegetal pole.

tion of cell fates (Dorresteijn et al., 1987; Dorresteijn & Kluge, 1990)

The cleavage pattern

As in the Mollusca, Nemertea, Sipunculida and Echiurida, the embryos of the Annelida follow a quartet spiral cleavage. According to Siewing (1980), Ivanova-Kazas (1981) and several authors ever since, spiral cleavage is a autapomorphy of animal taxa grouped together as the spiralians and most likely evolved from the radial cleavage pattern of the diploblasts. The cleavage pattern of spiralian embryos is the result of the inclination of the mitotic spindles with respect to the animal-vegetal axis. In most spiralian embryos the first sign of this inclination can be encountered at second cleavage (Fig. 3F), when the spindle in either blastomere is not oriented in the horizontal plane, but stands at an angle to it. As a result, each of the two cells divides into an animal and a vegetal daughter cell. In such cases the animal cells contact each other at the animal pole, whereas the vegetal cells broadly contact each other at the so-called vegetal cross furrow. So, even at the four-cell stage the blastomeres of the spiralian egg are not in a perfect radial arrangement. The eggs of only few polychaete species, like *Pomatoceros triqueter,* lack such animal and vegetal cross furrows and then show a perfect radial arrangement of the blastomeres at the four-cell stage. This may well represent the ancestral condition.

Each of the blastomeres of the four-cell stage is the founder cell of an embryonic quadrant. We discern the A-, B-, C-, and D-quadrant. Later in their development, the cells of these quadrants participate in the formation of specific structures pertaining to dorsal, ventral or lateral fates and can then be denominated according to their fates. Conventionally, the D-quadrant is the dorsal quadrant of Mollusca and Annelida and produces the stem cell of the mesoderm, i.e. the mesentoblast 4d. The cell material within a quadrant is generated by a series of alternating dexiotropic and laeotropic spiral cleavages. For example, at third cleavage the spindles are oriented slightly tilted with respect to the animal-vegetal axis (Fig. 3H). As a result the otherwise equatorial cleavage plain becomes slanted and produces an animal daughter cell to the left and a vegetal daughter cell to the right, viewed on from lateral aspect of the embryo. Viewed from the animal pole, the animal daughter cells, the micromeres 1a–1d, are shifted clockwise (dexiotropic) with respect to the vegetal daughter cells, the micromeres 1A–1D. The fourth cleavage is characterized by the formation of a second generation of micromeres, the blastomeres 2a–2d. Almost simultaneously the first generation of micromeres divides and produces two cell tiers: $1a^1$–$1d^1$ round the animal pole and $1a^2$–$1d^2$ in a vegetal direction. During these cleavages the mitotic spindles stand at a 90° angle to the positions of spindles at the third cleavage. Consequently, this cleavage is a counter-clockwise (laeotropic) cleavage. This pattern of alternating spindle positions in embryos of polychaetes (and other organisms with spiral cleavage) is repeated at subsequent cleavages. The alternation of dexiotropic and laeotropic cleavages results in an animal-vegetal zig-zag orientation of

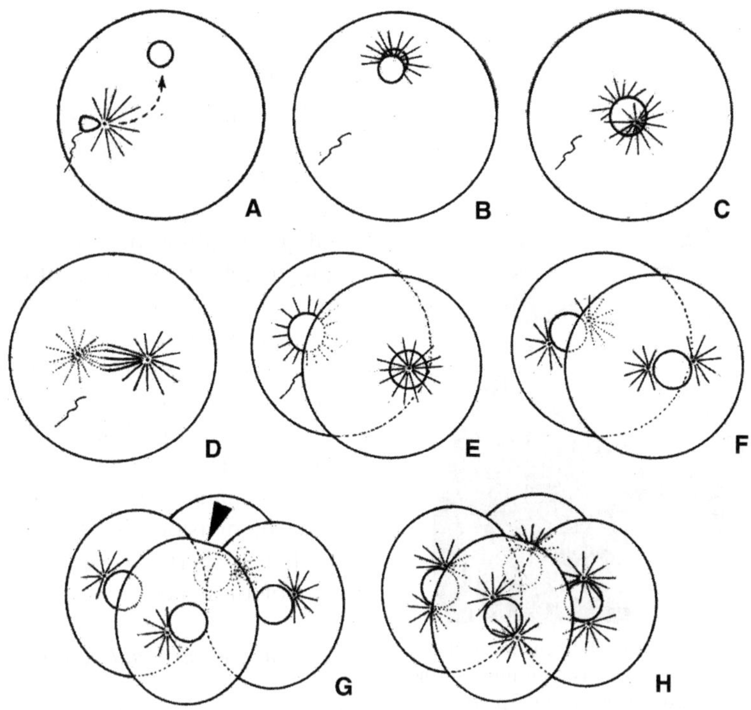

Figure 3. This simplified diagram shows the destiny of the paternal centrosome from fertilization up to the third cleavage. Initially, it forms the sperm aster which pulls the paternal pronucleus towards the maternal pronucleus (A, B). The centrosome (possibly enriched by centrosomal material of maternal origin) divides and forms the asters of the first cleavage spindle (C). The spindle stands at right angles to the path followed by the sperm aster after fertilization (D). As a result the first cleavage cuts through the sperm entry point (E). The centrosomes in either blastomere divide, and the daughter centrosomes take opposite positions on the nuclear perimeter (F). Note that one of the asters lies slightly shifted towards the animal pole and causes the spindles to take positions deviating from the equatorial plane. Due to this tilt two opposing blastomeres form a cross furrow at the animal pole (arrowhead), the other two form a cross furrow at the vegetal pole.

the blastomeres in each of the quadrants. On the basis of its cleavage history, each of these blastomeres has a unique position and set of neighbors. The constancy of spiral cleavage seems to obey certain mathematical laws. In his important study on the orientation of centrioles in embryos of the turbellarian *Polychoerus*, Costello (1961) proposed that the remnant of the previous cleavage, the so-called mid-body, might orientate the spindle at the subsequent cleavage. The same kind of cortical bias is proposed by Morris et al. (1989) in their comparison of cleavage patterns in *Nereis, Styela and Xenopus*. But can a single directional cue be sufficient to explain the alternation of dexiotropic and laeotropic spindle positions? Although we cannot rule out the role of the cortex and will discuss it later on in the setting up of dorsoventral polarity and of cell fates, we dare not forget the behavior of the centrosome. Polychaete eggs inherit a single centrosome after the mitotic divisions of the oogonium. This centrosome supports meiosis by the formation of the polar bodies. However, after polar body formation the quality of the maternal centrosome seems weakened. Studies on artificially activated eggs of *Platynereis dumerilii* (Spek, 1930; Kluge, 1991) and *Chaetopterus* (Brachet, 1937) demonstrated that the postmeiotic centrosome cannot form a bipolar mitotic spindle. The bipolar cleavage spindle must thus originate from or predominantly relate to the paternal contribution, the sperm centrosome. Let us follow the history of this centrosome, which is schematically shown in the center of the asters in Figure 3. This sperm centrosome accompanies the paternal pronucleus towards the maternal pronucleus (Fig. 3A and B). It then replicates and the sister

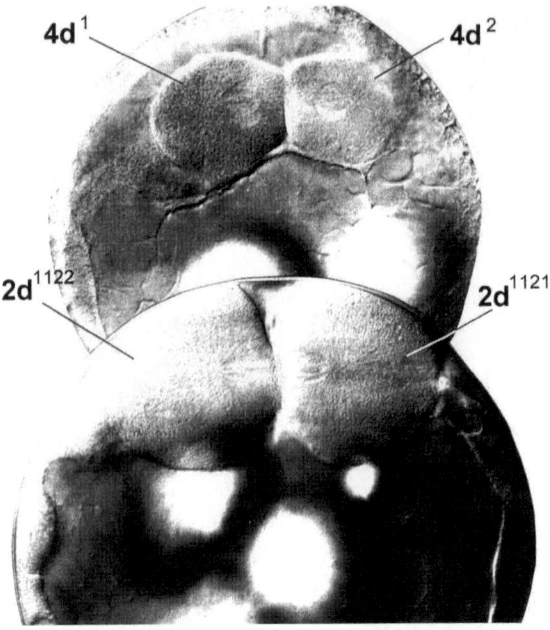

Figure 4. Photographic montage of two different focal planes within *Platynereis* embryos showing the dorsal positions of the descendents of the 2d and 4d blastomeres. As an expression of their determined state these cells no longer cleave in a spiral mode, but divide in a bilaterally symmetrical pattern. The sister blastomeres $2d^{1121}$ and $2d^{1122}$ in the equatorial level and $4d^1$ and $4d^2$ at the vegetal pole take perfect mirror positions with respect to the dorsal median.

centrosomes move and take opposite sides at the conjunction of the pronuclei (Fig. 3B and C). Although the astral microtubules interact with the cortex of the egg, the primary orientation of the spindle is nevertheless within a plane perpendicular to the direction of the movement of the sperm centrosome (Fig. 3D). This also explains why the first cleavage in polychaete eggs with high incidence cuts through the sperm entry point (Fig. 3E). If the position of a centrosome after cleavage is at the far side of the reconstituting nucleus within a blastomere and if the sister centrosomes developing from it take opposite sides of that nucleus, then the subsequent cleavage should always be at right angles with respect to the previous cleavage (Fig. 3E and F). Although the proposed mechanism does not explain the oblique position of spindles at the third cleavage, it accounts for the perpendicular spindle positions of each of the subsequent cleavages (compare Figs. 3 C, F, H). As in mollusks, the obliqueness of the spindles at second and third cleavage may be caused by the inheritance of maternal cytoplasmic and/or cortical factors (Sturtevant, 1923; Boycott et al., 1930; Freeman & Lundelius, 1982).

The number of cleavages with a spiral character is not indefinite in polychaete eggs. After the formation of the fourth generation of micromeres morphogenetic cell movements set in, and gastrulation positions the germ layers. Although cell divisions continue during gastrulation, there is no sign of spiral cleavage. In *Platynereis dumerilii* the pattern of cell divisions after sixth cleavage is bilaterally symmetrical, which is most clearly shown in the pattern of $2d^{112}$ and 4d, the descendents of the D-quadrant lying on the dorsal median (Fig. 4). The reasons for the change of a spiral mode to a bilateral mode of divisions have not been investigated so far, but are certainly indicative of the determined state of these cells.

Cellular diversity and determination of cell fate

In the previous sections we focused on the development of egg polarity and a multicellular pattern in the polychaete embryo, stressing some of the similarities studied. Basically, polychaete eggs show an architecture with animal-vegetal polarity and form quartets of cells by spiral cleavage. Although the development of egg polarity is a general feature in the development of metazoans, the spiral cleavage with its alternating dexotropic and laeotropic spindle positions is so uniquely different from radial and bilateral holoblastic cleavage that we may regard it as a sound autapomorphy for the Spiralia. This is also supported by molecular data (Peterson & Eernisse, 2001). In the following we seek explanations for the diversity of embryos and larvae found in the various groups of polychaetes. In their historic paper on the cleavage patterns and phylogenetic relations within the monophyletic gastropods, van den Biggelaar & Haszprunar (1996) describe the variations of a single ontogenetic character, i.e. the timing and mode of determination of the mesentoblast, and place this in relation to the evolution of body plans. They provide strong evidence that an equal cleavage pattern within the molluscs is to be regarded as the ancestral condition. All deviating modes of spiral development, such as polar lobe development and unequal cleavage by spindle shift (Dorresteijn &

Fischer, 1988) thus originated during the phyletic development of the gastropod taxa. If an equal cleavage pattern is the ancestral condition for the gastropods and even characterizes the embryonic development of the polyplacophorans, we must expect an equal cleavage pattern at the base in each of the spiralian phyla, including the polychaetes. If the development of ancestral polychaetes was characterized by an equal cleavage pattern, the various mechanisms to accomplish cellular diversity must have developed later in phylogeny. Should there be an evolutionary pressure to accelerate the creation of cellular diversity in a dorsoventral pattern, as we find it in the gastropods, then the eggs of more derived polychaetes should also convergently develop polar lobes, spindle shifts and asymmetrical segregation of cytoplasm to create immediate differences among the early blastomeres. Table 1 summarizes the data on the symmetry or asymmetry of the first cleavage of polychaete eggs collected from the available literature of the past 130 years. The most striking conclusion after a careful survey of the data presented must be that only very few polychaete embryos actually show an equal cleavage pattern. One can even question whether the equal first cleavage shown for *Myrianida fasciata* by Malaquin (1893) might not have been based on the fact that he missed the formation of a very small polar lobe, because other members of the Syllidae form polar lobes. Nevertheless, a limited number of small polychaete eggs, like those of the ophelids *Armandia, Ophelia,* and *Euzone,* the polynoids *Harmothoe* and *Lepidonotus*, the hesionid *Podarke,* the sigalionid *Sthenelais,* as well as the serpulids *Pomatoceros, Hydroides,* and *Serpula* really cleave symmetrically or almost so at the first and second cleavage. This results in a homoquadrantal development in which all the blastomeres have equal developmental capacities and divide in perfect synchrony. In the ophelid *Armandia brevis* even the third cleavage turns out equal, making the "micromeres" of the animal pole (1a–1d) just as large as the "macromeres" at the vegetal pole (1A–1D) (Hermans, 1964). In this case the micromeres can only be identified by the position of the polar bodies. The thorough developmental study by Hermans also shows that fourth cleavage is equal too and thus produces four quartets of equal-sized cells. Fifth cleavage is only slightly asynchronous between the four tiers of cells, but the homologous cells in each of the quadrants cleave synchronously. This cleavage leads to the first visible size difference between the blastomeres. The animal quartet $1a^{11}$–$1d^{11}$ and the vegetal macromeres 3A–3D are clearly larger than all the other cells. Nevertheless, the radial symmetry remains unperturbed. The sixth cleavage is strongly asynchronous. The largest blastomeres 3A–3D and $1a^{11}$–$1d^{11}$ divide first. As in other polychaetes, the diversity in the nuclear-cytoplasmic ratio (for explanation see Dorresteijn & Luetjens, 1994) may be the cause of this asynchrony between the various tiers of blastomeres. As shown for molluscs by van den Biggelaar (van den Biggelaar & Haszprunar, 1996), the determination of the mesentoblast would lead to asynchronous cell proliferation among the cells within the tier of macromeres or of the micromere quartets. In *Armandia brevis* this is postponed until after the 64-cell stage. Hermans (1964) describes how the macromeres and the fourth quartet of micromeres 4a–4d protrude deeply into the blastocoel. It is predictable that these cells come into contact with animal micromeres and that a single micromere of the fourth quartet is induced to become the mesentoblast 4d. So it seems that the cleavage pattern of this species represents the ancestral pattern at the base of the polychaete tree. It would be interesting to examine the cleavage patterns of other ophelids which on the whole appear to have an equal cleavage pattern. *Armandia brevis,* the other ophelids, but also the maldanids belong to the Scolecida and also the latter polychaete taxon shows an equal spiral cleavage pattern. This thus supports the phylogenetic tree proposed by Rouse & Pleijel (2001) in which the branch of the Scolecida starts at the very base of annelid phylogeny. However, we must bear in mind that even within the Scolecida we encounter groups like the orbiniids and the Arenicolidae with an unequal spiral cleavage. The groups Scolecida thus contain groups that have conserved ancestral conditions of development and others that have developed a rapid diversification among the early blastomeres. We cannot rule out that these developed from two different ancestors, one already having developed an unequal spiral cleavage.

The embryos of the species within the Eunicida seem to follow an unequal cleavage. No data

Table 1. Modes of development in annelids. The number of species listed in the various orders is arbitrary; the numerous question marks label those species whose development to my knowledge has not been published, but the data of which are essential for a sound statement on the phylogeny of cleavage patterns

Group	Genus	Mode of cleavage	Source
Scolecida	*Arenicola*	Unequal (spindle asymmetry)	Watson & Bentley (1998)
	Scoloplos	Unequal	Delsman (1916)
	Maldane	?	
	Clymenella	Unequal	Mead (1897)
	Axiothella	Unequal	Bookhout & Horn (1949)
	Heteromastus	?	
	Capitella	Unequal	Eisig (1898)
	Armandia	Equal	Hermans (1964)
	Ophelia	Equal	Bullot (1904; cited by Hermans, 1964)
	Euzonus	Equal	Dales (1952; cited by Hermans, 1964)
Eunicida	*Eunice*	?	
	Ophryotrocha	Unequal (spindle asymmetry)	Jacobsohn (1999)
	Histriobdella	?	
	Lumbrineris	Unequal	Sato et al. (1982)
	Onuphis	?	
	Hyalinoecia	?	
	Dinophilus	Unequal	Nelson (1904); Traut (1970)
	Diopatra	Unequal (spindle asymmetry)	Allen (1953)
Amphinomida	*Eurythoe*	?	
	Hermodice	?	
Phyllodocida	*Anaitides*	?	
	Aphrodite	?	
	Hermione	Unequal (spindle asymmetry)	von Drasche (1885)
	Harmothoe	Equal	Mead (1898)
	Lepidonotus	Equal	Mead (1897)
	Hesionides	Equal	
	Podarke	Equal	Treadwell (1901)
	Microphthalmus	?	
	Ophiodromus	?	
	Sthenelais	Adequal	Bumpus (1898)
	Exogone	?	
	Brania	?	
	Myrianida	Equal (small polar lobe)	Malaquin (1893)
	Odontosyllis	Unequal (polar lobe)	Fischer & Fischer (1995)
	Autolytus	Unequal (spindle shift; polar lobe)	Allen (1964)
	Nereis	Unequal (spindle asymmetry)	Wilson (1892)
	Platynereis	Unequal (spindle asymmetry)	Dorresteijn (1990)
	Tylorrhynchus	Unequal (spindle asymmetry)	Osanai (1978)
	Nephtys	?	
	Tomopteris	Unequal (spindle asymmetry)	Åkesson & Melander (1967)
	Glycera	?	
Spintherida	*Spinther*	?	
Nerillidae	*Nerilla*	?	
	Troglochaetus	?	

Continued on p. 9

Table 1 (Continued.)

Group	Genus	Mode of cleavage	Source
Aberrantida	*Aberranta*	?	
Spionida	*Cheatopterus*	Unequal (polar lobe, spindle shift)	Henry (1986)
	Spio	Unequal	Claparède & Mecznikow (1869)
	Magelona	?	
	Marenzelleria	Unequal	Bochert & Bick (1995)
	Pseudopolydora	Unequal (polar lobe)	Myohara (1979)
	Polydora	Unequal	Woodwick (1960)
	Pygospia	?	
Sabellida	*Sabella*	Unequal	McEuen et al. (1983)
	Spirographis	?	
	Fabricia	?	
	Pomatoceros	Equal/adequal	von Drasche (1884); Dorresteijn & Luetjens (1994)
	Hydroides	Equal	Costello & Henley (1971)
	Eupomatus	Equal	Hatschek (1885)
	Serpula	Equal	Stossich (1878)
	Spirorbis	Unequal	Goette (1881)
	Riftia (Pogonophora)	Unequal	Marsh et al. (2001)
	Siboglinum (Pogonophora)	Adequal	Bakke (1975), Ivanov (1975)
	Nereilinum (Pogonophora)	Unequal	Gureeva (1979)
Terebelliformia	*Amphitrite*	Unequal (spindle asymmetry)	Mead (1897)
	Alvinella	Unequal	Pradillon et al. (to be published)
	Pectinaria	Unequal (spindle asymmetry)	Costello & Henley (1971)
	Lanice	?	
	Sabellaria	Unequal (polar lobe)	von Drasche (1885)
	Phragmatopoma	?	
Sternaspidae	*Sternaspis*	Unequal	Child (1900)
Cirratuliformia	*Tharyx*	Unequal (?)	Dales (1951)
	Dodecaceria	?	
	Ctenodrilus	?	
Protodrillida	*Potodrilus*	?	
Polygordiidae	*Polygordius*	Equal	Wolterek (1904)
Aeolosomatidae	*Aeolosoma*	?	
	Rheomorpha	?	
Potamodrillidae	*Potamodrilus*	?	
Parergodrillidae	*Parergodrilus*	?	
Psammodrillidae	*Psammodrilus*	?	
Myzostomida	*Myzostoma*	Unequal (polar lobe)	Wheeler (1898)
Echiura	*Echiurus*	?	
	Bonellia	?	
	Urechis	Equal	Newby (1940)
	Ikeda	?	
Clitellata	*Helobdella (Hirudinaea)*	Unequal (teloplasm)	Weisblat et al. (1978)
	Clepsine (Hirudinea)	Unequal (teloplasm)	Whitman (1878)
	Tubifex (Oligochaeta)	Unequal (teloplasm)	Penners (1922)
	Eisenia (Oligochaeta)	Unequal (teloplasm)	De Vries (1968)
	Lumbricus (Oligochaeta)	?	

were available on the development of the Amphinomida.

Although the polynoids, the hesionids and the sigalionids have an equal spiral cleavage the phyllodocida predominantly contain species whose eggs divide unequally. One taxon within the Phyllodocida, the Syllidae, shows the largest variety of cleavage patterns. According to the figures in the study by Malaquin (1893), the egg of *Myrianida* divides equally. In contrast, the egg of *Autolytus fasciatus* divides unequally and simultaneously forms a small polar lobe. Fischer & Fischer (1995) were able to show that the cleavage pattern of *Odontosyllis enopla* also includes the formation of a polar lobe. In case Malaquin missed the formation of a (small) polar lobe in the *Myrianida* egg, then polar lobe formation might be a general characteristic of the syllids.

Again, no early developmental data are available on the Spintherida, Nerillida, and Aberrantia. Insofar as data were available, the eggs of the representatives within the order of the Spionida follow an unequal cleavage pattern. On the basis of the reliability of the developmental data we observe two different cleavage strategies within the Spionidae. *Spio* and *Marenzelleria* divide by an unequal first and second cleavage, whereas Myohara (1979) describes the formation of a large polar lobe prior to first cleavage in the egg of *Pseudopolydora kempi japonica*. This polar lobe fuses with one of the otherwise equal-sized blastomeres and creates both a size difference as well as a difference in developmental capacity. Only the blastomere receiving the material of the polar lobe forms a second polar lobe prior to the second cleavage, which then fuses with either of the sister blastomeres. This larger cell is now the D-quadrant and will form the mesentoblast in later development. The adoption of two such different developmental strategies within a single taxon is unique. It may be indicative of the possibility that the Spionidae developed from an equally cleaving ancestor and have developed two modes of unequal cleavage.

Although no data on the early development of *Sabella, Spirographis* and *Fabricia* were found in the literature, the eggs in the majority of described species in the Sabellida cleaved equally or almost so (adequally). Even for the embryos of the pogonophorans, newly introduced into this taxon, Bakke (1975) and Ivanov (1975) postulate an equal first cleavage. However, recent studies by Dorresteijn & Dorresteijn (to be published) show that the allocation of cytoplasm, lipid and yolk within the egg of *Siboglinum fiordicum* is very asymmetrical between the blastomeres of the two-cell stage, and second cleavage is always unequal, thus producing two larger blastomeres in the longitudinal axis of the future larva and two smaller lateral blastomeres. The cleavage inequality of the Pogonophora is also shown for *Nereilinum murmanicum* (Gureeva, 1979) and for *Riftia pachyptila* by Marsh et al. (2001), who recently presented micrographs of the early cleavages in that species.

With the exception of the genus *Spirorbis*, the Serpulidae show an equal spiral cleavage pattern. Not only are the first and second cleavages equal but even the third cleavage is equal or adequal. The micro- and macromeres can only be discriminated on the basis of the positions of the polar bodies. Morphometric measurements of the individual blastomeres of early cleavage stages in *Pomatoceros triqueter* (Dorresteijn & Luetjens, 1994) have demonstrated that small size differences between the blastomeres exist which finally lead to consistent size differences among the macromeres. This may well bias the possibilities for cellular interactions in later stages of this species and thereby play a role in the determination of the mesoderm stem cell (3D). Due the minor size differences among the various tiers and between blastomeres of different quadrants, the cleavages do not run perfectly synchronously throughout the embryo. A dramatic increase in this asynchrony occurs at sixth cleavage. This could be an indication that the mesoderm precursor is induced during the interval between fifth and sixth cleavage. The mechanism by which this induction takes place is still unknown, however.

The Terebellida show a similar diversity in developmental modes we have previously seen in the Phyllodocida. All embryos divide unequally either by spindle asymmetry (e.g. *Amphitrite, Pectinaria)* or by the formation of a polar lobe (e.g. *Sabellaria*). The data on the early development in the Sternaspidae and in the Cirratuliformia come from studies on the *Sternaspis scutata* (Child, 1900) and *Tharyx marioni* (Dales, 1951), respectively. The embryos of both species divide by an unequal cleavage due to spindle asymmetries.

Among the interstitial annelids the development of Polygordiidae has been documented by Woltereck (1904), who studied the cleavage pattern and larval development of *Polygordius*.This rather small egg divides equally.

As was shown for *Myzostoma glabrum* by Wheeler (1898), the Myzostomidae, a special group of ectoparasites of echinoderms, develop a polar lobe at first and second cleavage and thus cleave unequally.

This brief survey of the various cleavage patterns in polychaetes summarized in Table 1 also contains the Echiura whose representatives have been placed in close vicinity to or even within the annelids on the basis of molecular data (McHugh, 1997). Indeed, the cleavage of *Urechis caupo* is a typical equal spiral cleavage. Table 1 also includes the cleavage patterns of the oligochaetes in order to show that all species investigated divide by an unequal cleavage pattern. The eggs of the oligochaetes never cleave equally, nor do they form a polar lobe. This underlines the monophyletic origin of the clitellates. It must, however, be mentioned that some clitellates show interesting modifications of the cleavage inequality (e.g. in *Acanthobdella* (Dohle, 1999) and *Erpobdella* (Dimpker, 1918)), which clarify the true meaning of an asymmetrical distribution of cytoplasm rather than size differences of cells. This will be the issue of the following section.

Summarizing the published data of the early cleavage patterns in annelids, we find that all three modes of spiral cleavage described in Dorresteijn & Fischer (1988) are spread over the various orders of polychaetes in an almost random fashion. If we were to assume that an equal cleavage pattern is the ancestral condition and the polychaetes would be monophyletic, then unequal cleavage by spindle shift or by polar lobe formation would have developed several times independently. The equal cleavage pattern is found scattered over the entire polychaete tree and is found in species with small eggs within the Scolecida, Aciculata and Canalipalpata. If we combine this observation with the fact that the species at the roots of other spiralian trees also show an equal cleavage pattern, then the equal cleavage pattern would be ancestral. However, it may well be that these ancestral groups soon began to incorporate yolk reserves into their eggs, allowing a lecithotrophic development of the larvae. In that case the segregation of the yolk stores and "normal" cytoplasm might have been disproportional among the blastomeres and led to differences of their sizes. It cannot be ruled out that both developmental strategies were followed by a single species, e.g. in a seasonally dependent manner. Variations of developmental strategies within a single species were described for *Streblospio benedicti* by Levin et al. (1991). The development of a polar lobe may initially have been coupled to the process of unequal cleavage to secure the segregation of vegetal cytoplasm independent of the distribution of the yolk. The egg of *Chaetopterus* shows a combination of unequal cleavage and polar lobe formation. In this species the polar lobe contains cytoplasm with developmental factors for the differentiation of photocytes (Henry, 1986).

Unfortunately, the data in Table 1 are incomplete due to the fact that to my knowledge no published data exist on the development of the Aeolosomatidae, Potamodrillidae, Parergodrillidae, and Psammodrillidae. The development of the species in these taxa may well be modified to comply with the conditions of the habitat.

Differences of blastomere size and their significance for fate determination

The previous section treated the occurrence of the three basic cleavage modes among the various polychaete taxa. We have seen some of the arguments in favor of the idea that the equal cleavage pattern is the ancestral condition in every group of spiralians, including the polychaetes. We now try to elucidate the purpose of the introduction of an unequal cleavage strategy during the evolution of the polychaetes. This phylogenetic development must have a meaning since the majority of polychaete eggs divide by an unequal cleavage pattern. This section also shows some differences in the developmental strategies of eggs with spindle shift and of those forming a polar lobe.

Equally cleaving polychaete embryos, like those of *Armandia* or *Hydroides*, form four equal-sized embryonic quadrants and within each of these quadrants a number of equal-sized blastomeres. As a result the cleavages occur perfectly synchronously and the embryo is in fact radially symmetric with respect to the animal–vegetal axis.

The break of symmetry in such embryos lies in the formation of a single mesoderm-forming quadrant, the so-called D-quadrant. In equally cleaving molluscs the determination of the mesoderm mother cell 3D depends upon cellular interaction between the descendents of 1a-1d at the animal pole with 3A–3D at the vegetal pole. Although all blastomeres lie in a zig-zag configuration along the animal-vegetal axis, they have a unique position as well as a unique set of neighboring cells giving them positional information. Due to fact that the animal and vegetal cells come into contact within the central blastocoel an asymmetrical blastomere configuration is evoked. In the molluscan embryo one of the vegetal cells has long-lasting contacts to all of the animal cells and is induced by them during the interval between fifth and sixth cleavage. Developmental studies of equally cleaving polychaete embryos did not give any clues as to whether the determination of the D-quadrant occurs similarly. In *Armandia* Hermans (1964) describes the stretching of the animal and vegetal blastomeres towards the center of the blastocoel which might be indicative for such cellular interactions. Morphometric measurements of blastomeres at successive cleavage stages of *Pomatoceros* have revealed minor but consistent size differences among the blastomeres (Dorresteijn & Luetjens, 1994). These size differences correlate with minor differences in cell cycle duration up to the fifth cleavage in which one of the quadrants cleaves slightly earlier than the other three. At sixth cleavage the vegetal cell of this quadrant starts cleaving approximately 5–10 min ahead of other macromeres and subsequently dividing cells. In the case of the *Pomatoceros* egg minor size differences leading to cell cycle differences might be causal for the break of symmetry and the determination of the dorsoventral axis.

The cytological details of the break of symmetry in an unequally cleaving polychaete and an oligochaete embryo was studied by light microscopy immunocytochemical techniques. Dorresteijn & Kluge (1990) describe the process of first cleavage in *Platynereis dumerilii* using time-lapse recording with contrast enhancement. It shows that the asters of the first cleavage spindle grow at different rates. During early anaphase the larger aster takes a central position within the animal allotment of pole plasm. As a consequence the smaller aster is forced into a more peripheral position. In the *Tubifex* egg the spindle asymmetry is caused by the asymmetric distribution of centrosomal material (Shimizu et al., 1998). This may also be true for *Platynereis* and other polychaetes and would certainly explain the differences of aster growth. The asymmetric position of the spindle also leads to an asymmetric position of the actin-myosin material of the contractile ring and leads to a unequal two-cell stage with a large CD- and a small AB-cell. In most polychaete eggs observed the cleavage inequality at second cleavage is only found in the division of the CD-cell, whereas the AB-cell divides equally. The four-cell stage thus contains one significantly larger quadrant, conventionally called the D-quadrant. Morphometric analyses of blastomere size and cytoplasmic segregation in early embryos of *Platynereis* (Dorresteijn, 1990; Dorresteijn & Luetjens, 1994) have elucidated that not only the sizes of the individual quadrants differ. The larger part (60%) of cytoplasm alloted at the animal pole of the fertilized oocyte (animal pole plasm) is shunted into the D-cell. In *Nereis limbata* (Wilson, 1892) and *Platynereis dumerilii* (Dorresteijn, 1990) the cell lineage of the D-quadrant differs significantly from the cell lineages of the A-, B-, and C-cell lines. The D-cell line forms relatively small micromeres at the third and at the fifth cleavage, but particularly large micromeres, i.e. 2d and 4d, at the fourth and at the sixth cleavage respectively. These cells also inherit the cytoplasm alloted to the D-quadrant. In contrast, the sizes of the various micromere generations within the A-, B-, and C-quadrant decreases at successive stages. In *Platynereis massiliensis,* whose eggs are stuffed with yolk and measure about ten times the volume with respect to the sibling species *Platynereis dumerilii,* the same cleavage strategy of cytoplasmic segregation and cleavage pattern is found (Schneider et al., 1992). From these data, showing the differences of cleavage dynamics, we learn that the D-quadrant must already be determined at the four-cell stage.

The formation of a polar lobe is the other mechanism of unequal cleavage in some polychaete eggs and can be observed during first and second cleavage in the eggs of such species. The first polar lobe is always formed at the onset of the first cleavage as an anuclear vegetal protrusion of species-specific size. It is formed by a ring-like subequatorial constriction of the egg surface. In *Sabellaria* (Hatt, 1932) and *Pseudopolydora* (Myohara, 1979) the volume of the polar lobe is about one-third of the entire egg volume. After the fusion of such a polar lobe with one of the equal-sized blastomeres this so-called CD-cell is twice the size of the AB-cell. At second cleavage a second polar lobe forms at the vegetal pole of the CD-cell only. After the anaphase this second polar lobe fuses with the equal-sized daughter cells of CD. The largest cell is conventionally called the D-blastomere. Again, the D-quadrant of polar lobe forming eggs follows a cleavage strategy which clearly differs from that of the A-, B-, and C-quadrant, indicating that the D-cell must be determined as early as the four-cell stage.

Segregation of developmental potential

The elegant study of Wilson (1892) on the very constant cell lineage of the unequally cleaving embryo of *Nereis limbata,* as well as his studies on isolated blastomeres of another spiralian embryo, the molluscan egg of *Patella* (Wilson, 1904), have led to the assumption that the development of spiralians is a mosaic development. In such mosaic embryos the fates of the blastomeres are restricted by the differential segregation of developmental potential initially present in the ooplasm or cortex of the oocyte. As a result, each of the blastomeres has a restricted fate map and the cleavage pattern is adapted to that fate map. Ever since, embryologists have been searching for the factors responsible for the cellular diversity in the spiralian egg, predominantly in embryos of molluscs, polychaetes, and clitellates. Although the nature of diversification among the blastomeres is still unknown, these substances are conveniently called "determinants". Future analyses using molecular biological techniques will elucidate whether such factors for early developmental steps in polychaete (and other spiralian) embryos exist and if so, more about their molecular nature. Some molecules for rather late steps of polychaete development have been identified and the expression patterns of their genes were analyzed by *in situ* hybridization. This will be treated in the final section of this review.

Determinants need not necessarily be immediate gene regulatory factors. When we compare the cytological details of the unequal cleavage patterns of polychaetes and oligochaetes, the common aspect is blastomere diversification by the asymmetrical and disproportional distribution of cytoplasm, i.e. the pole plasm (in polychaetes) or teloplasm (in clitellates). The necessity for this cytoplasmic diversification for normal pattern formation has been shown in experiments. Experimental changes of normal cytoplasmic distribution at the two-cell stage in the polychaete eggs of *Nereis* (Tyler, 1930), *Chaetopterus* (Tyler, 1930; Henry & Martindale, 1987), *Platynereis* (Dorresteijn et al., 1987), and in the clitellate eggs of *Tubifex* (Penners, 1922, 1924) led to the formation of two, in most cases opposing quadrants, with D-quadrant cleavage characteristics. The embryos developed into larvae or young worms with a duplication of the dorsoventral pattern. In the leech *Helobdella* (Astrow et al., 1987) and in the polychaete *Platynereis* (Dorresteijn & Eich, 1991) an experimental equal distribution of pole plasm (teloplasm) leads to complete duplication of the D-quadrant in neighboring quadrants. For *Platynereis* this resulted in a young worm with one head and two complete trunks. This clearly demonstrates that the D-quadrant dictates the dorsoventral pattern and must therefore interact with non-D-quadrant cells. As in mollusks, (a) certain descendent(s) of the D-quadrant obtain (s) the role of an organizer of dorsoventral polarity. This requires cellular interaction and speaks against a mere mosaic development. One may still argue that it is not the inevitable change of size of the blastomeres in such experiments but rather that the cytoplasm might play the decisive role in D-quadrant determination. Indeed, all D-quadrants in normal development of polychaete embryos develop from the largest cell at the four-cell stage. However, we find an exception to this rule in the clitellates. In the leech *Acanthobdella* the A- and B-cell are the largest blastomeres, the D-cell is rather small and only twice the size of the C-cell (Dohle, 1999). Nevertheless, in this embryo the largest part of the cytoplasm is shunted into the D-cell, to a lesser degree into the C-blastomere, and the smallest portion ends up in the A- and B-blastomere. The latter two blastomeres only contribute a few cells and seem to play the role of a "yolk sac". This natural experiment thus detaches blastomere size from cytoplasmic allocation and shows that

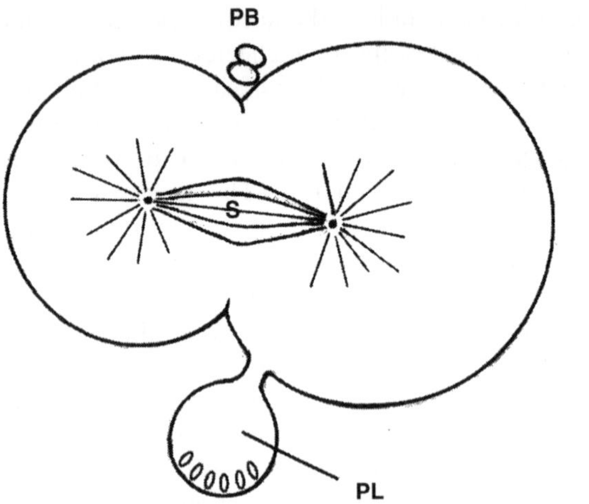

Figure 5. In the polychaete *Chaetopterus* cleavage inequality is accompanied by polar lobe formation. Determinants allocated to the vegetal cortex (see Fig. 2; represented by oval symbols) are incorporated in the polar lobe and, due to the asymmetrical position of the cleavage furrow, the determinants are automatically transferred into the largest of the cleavage cells. PB: polar bodies (at the animal pole); PL: polar lobe (at the vegetal pole); S: spindle of the first cleavage in an asymmetrical position.

the disproportional cytoplasmic segregation is relevant for D-quadrant specification.

The disproportional allocation of the yolk-free cytoplasm (pole plasm) over the various cell lines not only influences the geometry of the cleavage pattern, it also increases the cleavage rates in those cells receiving the bulk of it. Consequently, this leads to a cleavage asynchrony in which the D-quadrant descendents divide more rapidly than other cell lines. It is our hypothesis that the cytoplasm synthesizes molecular components for cell cycle control. By speeding up their cell cycle blastomeres like 2d, 4d and some of their descendents might reach a critical cell cycle in which the zygotic expression starts well ahead of all the other cell lines. After the fate of these cells has been settled, they act as organizers and induce the necessary fates of their neighboring cells. This hypothesis is currently being investigated in the embryo of *Platynereis*. Although the expression of only few of the cell cycle controlling genes has been screened and a differential expression could not be detected during early development, later stages show a differential expression of the Cyclin E gene with an elevated transcriptional level in the 2d-descendents (Dorresteijn et al., 2002).

If the polar lobe of embryos was formed in conjunction with an unequal cleavage, as is still the case in *Chaetopterus,* then the polar lobe must have served another purpose than a mere asymmetric allocation of cytoplasm. Hatt (1932) showed that developmental factors essential for setting up dorsoventral polarity in the *Sabellaria* embryo are allocated in the vegetal region of the fertilized egg before and during the formation of the polar bodies. Experiments by Render (1983) and Speksnijder & Dohmen (1983) provide evidence for the cortical binding of such developmental factors during meiotic events in *Sabellaria cementarium* and *Sabellaria alveolata,* respectively. As described in the previous section, the size of the polar lobe varies in a species-specific manner, and the lobe of *Sabellaria* certainly belongs to the larger ones. But if only the cortical binding of developmental factors (determinants) matter, then why do the polar lobes differ so dramatically in size? Of course we can only speculate on this topic, but in my opinion there can be only one logical explanation. The yolk-free cytoplasm of unequally cleaving embryos contains a homogeneous pool of determinants and gets distributed in a quantitatively asymmetric fashion. After an asynchronous cleavage process these factors become activated in those D-quadrant derivatives that reached a critical cell cycle first. This concerns the particularly large cells 2d and 4d or their descendents. In a hypothetical ancestor showing both spindle shift as well as a polar lobe, the factors for the fate determination of 2d and 4d or there descendents were confined to the cortex of the future polar lobe. In this case the cleavage inequality merely served to produce cells of adequate size within the D-quadrant, since 2d and 4d produce quite a large number of cellular offspring. The precocious formation of a polar lobe separated these factors from the subsequent cleavage process. Since the cleavage furrow bisected the egg in two unequal cells the polar lobe was always incorporated by the larger cell (Fig. 5). The same process would then be repeated at the second cleavage and possibly at the third cleavage. After the third cleavage the factor would exclusively end up in the 1D-cell, the progenitor of both the 2d and 4d cell. The precocious confinement of determinants and the secure mode of their subsequent allocation allowed the cleavage process to become less unequal in time.

Those species requiring a large cellular offspring in the 2d- and 4d-cell line secondarily shifted cytoplasm into the polar lobe, in order to enlarge the D-cell line. This hypothetical treatise on the phylogeny of polar lobe development also accounts for differences in size of polar lobes. Presumably, the ancestral polar lobes were small, like the polar lobe of *Autolytus fasciatus* (Allen, 1964) which represents only a few percent of the entire egg volume.

Development of the main body regions

Now that we have learned more about the development of the dorsoventral axis of polychaete embryos, we focus on the development of the main body regions. After embryogenesis the embryos in the majority of polychaete species develop into a trochophore larva (see Nielsen, this volume). Although the trochophore larvae of the various polychaete species differ in size, yolk content, and may follow either a planktotrophic or lecithotrophic life-style, the presumptive areas for the basic structures (apical tuft, head region, prototroch, neurogenic region, setal sac anlagen, proliferation zone, pygidium) are foreshadowed in the cellular pattern of the embryo. Such presumptive areas based on the embryonic lineages for *Podarke obscura* (Treadwell, 1901) and *Scoloplos armiger* (Delsman, 1916; Anderson, 1959) are given and compared by Anderson (1966, 1973). After comparing the cleavage patterns from the literature, Anderson (1973) came to the conclusion that the pattern is always constant, i.e. homologous structures are produced by identical lineages. On the other hand, the descriptions of the cell lines contributing to these structures in the same paper frequently show differences. This is not surprising, because the embryogenesis within the polychaetes (and other annelids) is not identical in every detail: structures formed in the larva of a certain species may not be formed in the larva of another species. The fact that most of the older studies are based on line-drawings of fixed successive stages provides an additional complication. In some cases cleavages have been overlooked, leading to an incomplete cell lineage tree or an incorrect denomination of the blastomeres (cf. Wilson, 1892, 1898; Dorresteijn & Fischer, 1988). Moreover, fixed stages do not allow one to study the dynamics of cell movements (e.g. during epiboly) and positional changes. Time-lapse videography, 3D-reconstructions and labeling techniques of embryos and larvae will elucidate some of the inconsistencies of polychaete fate maps. Ackermann & Fischer (1999) have iontophoretically injected TRITC-dextrane into the 2d- and 4d-blastomeres *of Platynereis dumerilii* and confirmed the results of cell lineage analysis for *Nereis limbata* (Wilson, 1892) and *Platynereis dumerilii* (Dorresteijn, 1990). Despite the possible mistakes in the cell lineage analyses we are able to generalize certain developmental constraints within the cell patterns of polychaete embryos. The first micromere quartet 1a-1d forms the head pattern ($1a^1$–$1d^1$) with the apical tuft ($1a^{111}$–$1d^{111}$) and the largest part of the prototroch ($1a^2$–$1d^2$). Part of the second micromere quartet 2a–2c forms the stomodaeum and its surrounding ectoderm on the ventral side, whereas 2d forms the ectoderm on the dorsal and lateral side. In unequally cleaving embryos like those of *Platynereis*, 2d is a particularly large blastomere on the dorsal median of the embryo. This blastomere proliferates in a bilaterally symmetric fashion producing numerous small cells which overgrow the large macromeres sideways until the front of these cells meet at the ventral midline. Only few of the 2d-progeny remain at the dorsal side and form a thin ectodermal layer. The lateral epithelium of the 2d-progeny thickens, invaginates and forms three pairs of setal sac anlagen. The ventral part of the 2d-progeny forms the neurogenic region which bifurcates around the stomodaeum. The fate of the micromeres 3a–3d is ectodermal. The fate of 4a–4c has not been investigated. The 4d cell divides in a bilaterally symmetric pattern and form the progenitor cells $4d^1$ and $4d^2$ of the mesodermal bands. These progenitors immigrate into the blastocoel in the equally cleaving polychaete eggs. In the unequally cleaving polychaete egg the mesodermal progenitors slide between the proliferating 2d descendents and the macromeres. The macromeres 4A–4D exclusively form the endoderm. In equally cleaving polychaete embryos these cells invaginate into the blastocoel. In the unequally cleaving embryos the macromeres are frequently contain the yolk reserves and gastrulation occurs by epiboly.

This general outline of early polychaete development does not account for the variation of larval morphology. In part this is due to the yolk content. Polychaete embryos with moderate yolk reserves develop by an equal cleavage pattern and gastrulate by invagination. Since the first micromere quartet is relatively large (e.g. in *Armandia, Hydroides,* and *Pomatoceros*) the episphere of the early larva almost equals the size of the posttrochal hyposphere. The number of differentiated cells is limited to ciliated trochal cells (apical tuft, prototroch, metatroch, neurotroch, telotroch), a pair of protonephridia and a functional larval gut. All the differentiated cell types are predominantly meant to support a longlasting planktotrophic life style. Polychaete embryos with yolky eggs cleave unequally and form a relatively small first quartet of micromeres. As a result the episphere is small compared to the posttrochal hyposphere. The largest part of the posttrochal cell material emanates from the 2d- and 4d-progeny. This type of larva follows a lecithotrophic life style. Consequently, the number of larval differentiations is limited. Although the prototroch is still present and functional, the apical tuft is barely developed, the meta- and neurotroch are missing. A continuous and thus functional gut is not formed until the yolk reserves are almost exhausted. Instead these larvae rapidly develop adult tissues, like the setal sacs and the ventral nerve chord. The egg of *Platynereis massiliensis* contains tremendous yolk reserves (Schneider et al., 1992) and the first quartet of micromeres is small compared to the macromeres. Although the embryo still forms a prototroch, other larval structures are lacking. The larva remains in brood care in the parental tube. This shows that an increase of yolk reserves correlates with a suppression of larval structures and an acceleration in the development of adult tissues (heterochrony). One can imagine how the increase of yolk could finally lead to a complete omission of larval differentiation and a direct development as found in the clitellates.

Control of differentiation of larval and adult structures

As described in the previous section, the successive micromere quartets of the polychaete egg adopt specific fates. This raises the question whether the fates of these cells are controlled by determinants distributed along the animal-vegetal axis. Some evidence for the allocation of determinants has been provided by experimental studies of Costello (1945) in which he analyzed the development of isolated blastomeres of *Nereis limbata*. An isolated micromere of the first quartet develops essentially identical to a comparable micromere of the intact embryo, thus demonstrating its determined condition. However, the only structural differentiations in such isolates were the trochoblasts and apical tuft cells. These ciliated cells are larval differentiations. However, the cells within the isolates did not differentiate any non-larval cell types. The conclusion from these experimental data should thus be that there is evidence for the localization of determinants for larval structures. The localization phenomenon was also studied in artificially activated eggs of *Chaetopterus* (Lillie, 1902; Brachet, 1937). These eggs remain uncleaved, but nevertheless form a circumferential band of longer cilia in the supra-equatorial plane of the egg. A later analysis of the same species by Eckberg & Kang (1981) using cytoskeletal destabilizing drugs demonstrated that the localization of the determinants for regional ciliation depends upon the microtubular system. In *Platynereis dumerilii* the determinants for larval gland cell differentiation in the ventral head region of the trochophore become allocated in a step-by-step process. By means of experimental cleavage arrest of successive cleavage stages of *Platynereis*, Dorresteijn & Graffy (1993) were able to show that these determinants are initially located in the ooplasm and are consequently shunted into 1a- and 1b-cell line. All the experiments described above show that at least larval cell differentiation requires the localization of determinants. The evolutionary pressure to form the proper larval structures in a correct spatial and temporal pattern must be very high.

How is the pattern formation of non-larval cells in polychaete embryos controlled? This question is currently studied by a molecular biological approach in a very small number of polychaete and clitellate embryos. This brief section presents and interprets the first published data on the genes involved in the determination and differentiation in embryonic and larval development. The majority of data collected concern the expression of orthologues to well-known developmental genes of other organisms, like *Drosophila, Xenopus* and *Mus*. Although the studies by Jeffery (1983) and Jeffery & Wilson (1983) on the uncleaved egg of *Chaetopterus* demonstrated the localization of polyA-RNA, very little attention has been paid to the distribution of maternal mRNAs in polychaetes. Pilon & Weisblat (1997) were the first to demonstrate the presence and distribution of a maternal transcript with a high level of developmental significance in the leech *Helobdella robusta*. In this species the nanos orthologue Hr-nos, initially distributed in an animal-vegetal gradient, rapidly degrades during early development, but the protein was localized by a polyclonal antiserum and due to its asymmetric distribution between ectodermal and mesodermal teloblasts seems to control a process in the mesoderm. A later study by Kang et al. (2002) revealed that this molecule is involved in the formation or further differentiation of the primordial germ cells.

Another interesting result recorded by the Weisblat group is the fact that the orthologue of *Drosophila* twist, a basic helix-loop-helix transcription factor involved in mesoderm development of the fly, is a maternal factor in *H. robusta* and present within the ooplasm of oocyte (Soto et al., 1997). This would imply that this determinant of the mesoderm is allocated by an asymmetric segregation process.

Very promising results have been collected on the expression patterns of early segmentation genes in polychaetes. Werbrock et al. (2001) have isolated the hunchback orthologue from *Capitella capitata* and were able to show that the expression of this zinc-finger transcription factor is widespread over all ectodermal and endodermal tissues, but not in the segmental precursor cells of the trunk. Similar patterns of hunchback expression were found in the leech *Helobdella triserialis* (Savage & Shankland, 1996). The overt differences in the expression patterns of hunchback between annelid and insect embryos demonstrate that the developmental significance of this gene must have been altered during the phylogeny of the arthropod tree.

Enormous efforts have been undertaken to isolate polychaete Hox genes and to study the expression pattern. The results of Dick & Buss (1994) for *Ctenodrilus serratus* suggest the presence of a single HOM/HOX cluster. Similar results were reported for *Chaetopterus variopedatus* by Irvine et al. (1997). Five of the Hox genes (CH-Hox1–CH–Hox5) within the cluster were analyzed in the same species by *in situ* hybridization techniques (Irvine & Martindale, 2000). Although the Hox genes investigated are expressed in bilateral stripes in posterior larval structures, the pattern does not explain the regionalization of the body, and the Hox genes are not involved prior to larval development. Peterson et al. (2000) suggest that Hox genes are not expressed in embryonic or larval cells, but rather in the set-aside cells for the formation of adult structures. However, a possible developmental role for a transcript of the Abdominal-B-like gene Nvi-Post1 of *Nereis virens* was found in the anlagen of the parapodium of young larvae (Kulakova et al., 2002). As in *Ctenodrilus* there seems to be only one Hox cluster in the *Nereis virens* genome (Andreeva et al., 2001).

Using the monoclonal antibody 4D9 against the engrailed protein of *Drosophila*, Dorresteijn et al. (1993) specifically labeled both the nuclei and the cytoplasm of the posterior cells in the first three larval segments of *Platynereis dumerilii*, suggesting the presence of an engrailed or engrailed-like protein. Recently, we were able to isolate the engrailed and two Wnt gene fragments and studied the role of these genes at the segmental border (Prud'homme et al., in press). The spatio-temporal expression patterns found in *Chaetopterus* do not seem to support a role in the segmentation process in that species (Seaver et al., 2001).

Recent expression studies of late developmental genes have been performed on early and late larvae of *Platynereis dumerilii*. Arendt et al. (2001) have described a strong expression of the *brachyury* gene round the blastoporus of the young trochophore. In *Platynereis* the rim of the blastoporus comprises the posterior edge of the stomodaeum, the ventral midline and the proctodaeum. The expression of *brachyury* gene in the foregut, like in echinoderm and enteropneust larvae, speaks for a common ancestry of proto- and deuterostomes. The *Platynereis* orthologue of the gene *otx* was investigated in the same study. These cells predominantly lie in the circumference of the stomodaeum and are part of cell girdles immediately anterior and posterior to the prototroch. Small cell groups expressing *otx* are also found in the cephalic epidermis and the brain. The *Platynereis* orthologue of *goosecoid* is expressed by the majority of stomodaeal cells.

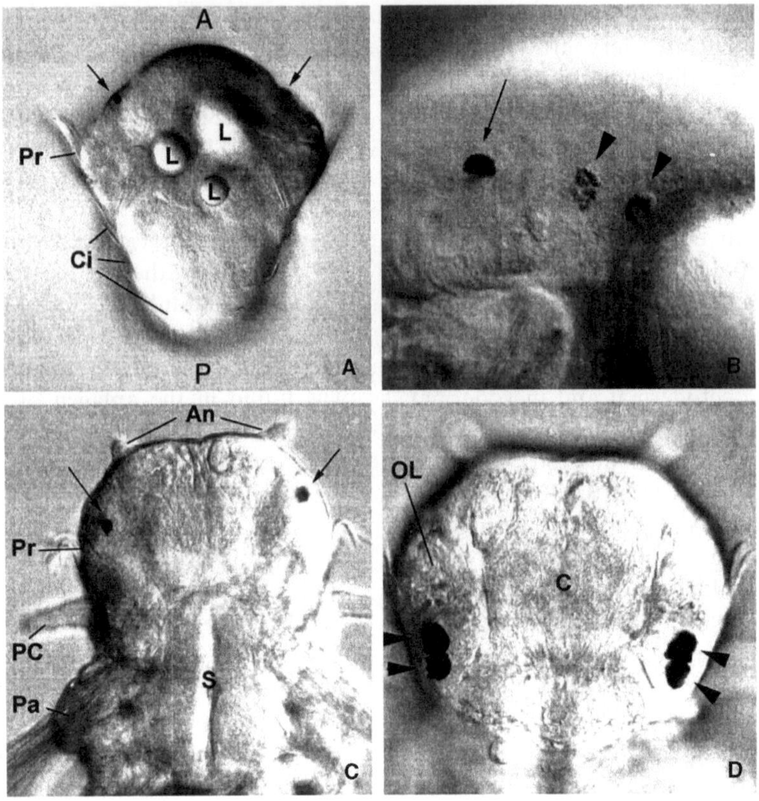

Figure 6. The development of both larval and adult eyes in *Platynereis dumerilii* starts in two lateral regions of the episphere. Although the larval eyes (simple ocelli; arrows) develop round 22 h after fertilization the adult eyes develop round 48h after fertilization (coarse arrowheads) only few cell diameters from the larval eyes. During subsequent development the larval eyes shift towards a ventral portion of the head and end at the base of the palps (6C), whereas the adult eye anlagen differentiate and shift towards the dorsal side of the head (6D). A: anterior; An: Antennae; C: central lobes of the cerebral ganglion; Ci: segmental ciliary bands; L: lipid droplets; OL: optical lobes of the cerebral ganglion; P: posterior; Pa: parapodium; PC: peristomial cirrus; Pr: prototroch; S: stomodaeum; *Platynereis* stages: 51 h (A, B); 77 h (C, D).

Another combined effort was to clarify the control of eye development in *Platynereis dumerilii* (Arendt et al., 2002). Approximately 22 h after fertilization the *Platynereis* larvae form a single pair of larval ocelli in a lateral position of the head ectoderm. The formation of two additional sets of eyes more dorsomedial of the larval eyes develop after an additional day of development (Fig. 6A, B). These are the anlagen of the adult eyes. The primordia of both larval and adult eyes lie over the lateral optical lobes of the brain and are initially only 4-5 cell diameters apart, but during further development the larval eyes move to the base of the ventral palps and the adult eyes in a dorso-lateral position on the head (Fig. 6C, D). After a screen for developmental genes involved in the formation of eyes in metazoans was performed on a 24 h-cDNA-library (Heimann, 2000) the expression patterns of *pax6, six1/2* and *ath* were described. The optic anlagen are characterized by the expression of *six1/2.* The primordia of the larval eyes express *pax6,* but surprisingly, there is hardly *pax6* expression in the adult eyes. Instead, the adult eyes transiently express *atonal (ath).* Since *pax6* is also expressed in the eyes of other bilaterians the larval eye may well have been the ancestral condition for the development of the visual system of all bilaterians.

Although the molecular data on developmental genes are still scanty, the first results in polychaetes are promising. Not only do they give us the chance to compare developmental processes and pattern formation within the polychaeta, they also provide an additional tool to study the phylogenetic relationship within the protostomia. Even if molecular data will probably not solve all questions in the field of phylogenetic research they will at least strengthen the effort of the developmental biologist for the experimental study of the process of determination long before a morphological differentiation occurs.

Acknowledgements

Part of this work was financially supported by grants of the Deutsche Forschungsgemeinschaft to the author (Do 339/3-1; Do 339/5-1).

References

Ackermann, Ch. F. & A. Fischer, 1999. A cell lineage study by microinjection of fluorochrome markers in blastomeres of *Platynereis dumerilii* (Annelida, Polychaeta) embryos. Zoology Supplement II: p. 20.

Åkesson, B. & Y. Melander, 1967. A preliminary report on the early development of the polychaete *Tomopteris helgolandica*. Arkiv för Zoology 20: 141–146.

Allen, M. J., 1953. Development of the polychaete, *Diopatra cuprea* (Bosc). The Anatomical Record 117: 572–573.

Allen, M. J., 1964. Embryological, development of the syllid, *Autolytus fasciatus* (Bosc)(Class Polychaeta). Biological Bulletin 127: 187–205.

Anderson, D. T., 1959. The embryology of the polychaete *Scoloplos armiger*. Quarterly Journal of Microscopic Science 100: 89–166.

Anderson, D. T., 1966. The comparative embryology of the polychaeta. Acta Zoologica 47: 1–42.

Anderson, D. T., 1973. Polychaetes. Embryology and Phylogeny in Annelids and Arthropods. Pergamon Press, New York, pp. 1–495.

Andreeva, T. F., C. E. Cook, N. M. Korchagina, M. Akam & A. K. Dondua, 2001. Cloning and analysis of structural organisation of Hox genes in the Polychaete *Nereis virens*. Ontogenez 32: 225–233 (in Russian).

Arendt, D., U. Technau & J. Wittbrodt, 2001. Evolution of the bilaterian larval foregut. Nature 409: 81–85.

Arendt, D., K. Tessmar, Medeiros de Campos-Baptista M-I, A. Dorresteijn & J. Wittbrodt, 2002. Development of pigment-cup eyes in the polychaete *Platynereis dumerilii* and evolutionary conservation of larval eyes in Bilateria. Development 129: 1143–1154.

Astrow, S., B. Holton & D. Weisblat, 1987. Centrifugation redistributes factors determining cleavage patterns in leech embryos. Developmental Biology 120: 270–283.

Bakke, T., 1975. Early cleavage in embryos of *Siboglinum fiordicum* Webb (Pogonophora). Zeitschrift für Zoologische systematik und Evolutionsforschung 13s: 7–9.

van den Biggelaar, J. A. M. & G. Haszprunar, 1996. Cleavage patterns and mesentoblast formation in the gastropoda: an evolutionary perspective. Evolution 50: 1520–1540.

Bochert, R. & A. Bick 1995. Reproduction and larval development of *Marenzelleria viridis* (Polychaeta: Spionida). Marine Biology 123: 763–773.

Bookhout, C. G. & E. C. Horn, 1949. The development of *Axiothella mucosa* (Andrews). Journal of Morphology 84: 145–183.

Boycott, A. E., C. Diver, S. L. Garstang & F. M. Turner, 1930. The inheritance of sinistrality in *Limnaea peregra* (Mollusca: Pulmonata). Philosophical Transactions of the Royal Society London B 219: 51–131.

Brachet, J., 1937. La différenciation sans clivage dans l'oeuf de Chaetoptère envisagée aux points de vue cytologique et metabolique. Archives of Biology 48: 561–589.

Bumpus, H. C., 1898. The breeding of animals at Woods Hole during the months June, July and August. Science 8: 850–858.

Child, C. M., 1900. The early development of *Arenicola* and *Sternaspis*. Wilhelm Roux' Arch. Entwicklungsmech. Organismen 9: 587–722.

Claparèdè, E. & E. Mecznikow, 1869. Beiträge zur Kenntnis der Entwicklungsgeschichte der Chaetopoden. Zeitschrift für Wissenschaftliche Zoologie 19: 163–205.

Costello, D. P., 1945. Experimental studies of germinal localization in Nereis. I. The development of isolated blastomeres. Journal of Experimental Zoology 100: 19–66.

Costello, D. P., 1961. On the orientation of centrioles in dividing cells, and its significance: A new contribution to spindle mechanics. Biological Bulletin 20: 285–312.

Costello, D. P. & C. Henley, 1971. Methods for Obtaining and Handling Marine Eggs and Embryos, 2nd edn. Marine Biological Laboratory, Woods Hole MA, USA.

Dales, R. P., 1951. Notes on the reproduction and early development of the cirratulid *Tharyx marioni* (St. Joseph). Journal of the Marine Biology Association, U.K. 30: 113–117.

Delsman, H. C., 1916. Eifurchung und Keimblattbildung bei *Scoloplos armiger* O.F. Müller. Tijdschrift van de Nederlandsche Dierkundige Vereeniging, Leiden. Series 2/14: 383–498.

De Vries, J., 1968. Les premières étapes de la segmentation (formation de la jeune blastule) chez le Lombricien *Eisenia foetida*.. Bulletin de la société zoologique de France 93: 87–97.

Dick, M. H., & L. W. Buss, 1994. A PCR-based survey of homeobox genes in *Ctenodrilus serratus* (Annelida: Polychaeta). Molecular Phylogenetics and Evolution 3: 146–158.

Dimpker, A. M., 1918 Die Eifurchung von *Herpobdella atomaria* Carena (*Nephelis vulgaris* Moqu. Tand.). Zoological Journal of Anatomy 40: 245–290.

Dohle, W., 1999. The ancestral cleavage pattern of the clitellates and its phylogenetic deviations. Hydrobiologia 402: 267–283.

Dorresteijn, A. W. C., 1990. Quantitative analysis of cellular differentiation during early embryogenesis of *Platynereis dumerilii*. Roux's Archives of Developmental Biology 199: 14–30.

Dorresteijn, A. W. C., H. Bornewasser & A. Fischer, 1987. A correlative study of experimentally changed first cleavage and Janus development in the trunk of *Platynereis dumerilii* (Annelida, Polychaeta). Roux's Archives of Developmental Biology 196: 51–58.

Dorresteijn, A. W. C. & P. Eich, 1991. Experimental change of cytoplasmic composition can convert determination of blastomeres in Platynereis dumerilii (Annelida, Polychaeta). Roux's Archives of Developmental Biology 200: 342–351.

Dorresteijn, A. W. C. & A. Fischer, 1988. The process of early development. Mikrofauna Marina 4: 335–352.

Dorresteijn, A. W. C. & C. Graffy, 1993. Competence of blastomeres for the expression of molecular tissue markers is acquired by diverse mechanisms in the embryo of *Platynereis* (Annelida). Roux's Archives of Developmental Biology 202: 270–275.

Dorresteijn, A. W. C. & B. Kluge, 1990. On the establishment of polarity in polychaete eggs. In H.-J.Marthy (ed.), Experimental Embryology in Aquatic Plants and Animals. Plenum Press, New York: 197–209.

Dorresteijn, A. W. C. & C. M. Luetjens, 1994. Morphometric analysis of cellular specification in *Platynereis* and *Pomatoceros* embryogenesis (Annelida, Polychaeta). Mémory Mus Nat Hist Nat 162: 45–50.

Dorresteijn, A. W. C., B. O'Grady, A. Fischer, E. Porchet-Hennerè & Y. Boilly-Marer, 1993. Molecular specification of cell lines in the embryo of *Platynereis*. Roux's Archives of Developmental Biology 202: 260–269.

Dorresteijn, A. W. C., J. Rieck, C. Schaub, D. Arendt & A. Fischer, 2002. Differential control of the cell cycle in embryos and early larvae of *Platynereis dumerilii* (Annelida, Polychaeta). Zoology 105(Suppl. V): p. 12.

von Drasche, R., 1884. Beiträge zur Entwicklung der Polychaeten. I. Entwicklung von *Pomatoceros triqueter* L. (published privately in Vienna): 1–10.

von Drasche, R., 1885. Beiträge zur Entwicklung der Polychaeten. 2.Heft. Entwickelung von *Sabellaria spinulosa* Lkt., *Hermione hystrix* Sav. und einer Phyllodocide. Verlag von Carl Gerald's Sohn (published privately): 3–16.

Eckberg, W. R. & Y-H. Kang, 1981. A cytological analysis of differentiation without cleavage in cytochalasin B- and colchicine-treated embryos of *Chaetopterus pergamentaceus*. Differentiation 19: 154–160.

Eckelbarger, K. J., 1988. Oogenesis and female gametes. Mikrofauna Marine 4: 281–307.

Eisig, H., 1898. Zur Entwicklungsgeschichte der Capitelliden. Mitteilungen aus der Zoologischen Station zu Neapel. 13: 1–292.

Fischer, A. & U. Fischer, 1995. On the life-style and life-cycle of the luminescent polychaete *Odontosyllis enopla* (Annelida: Polychaeta). Invertebrate Biology 114: 236–247.

Freeman, G. & J. W. Lundelius, 1982. The developmental genetics of dextrality and sinistrality in the gastropod *Lymnaea peregra*. Roux's Archives of Developmental Biology 191: 69–83.

Goette, A., 1881. Abhandlungen zur Entwickelungsgeschichte der Tiere. Untersuchugen zur Entwickelungsgeschichte der Wuermer: *Spirorbis nautiloides*. Zoologischer Anzeiger 4: 189–191.

Gureeva, M. A., 1979. A contribution to the study of the early development of *Nereilinum murmanicum* Ivanov 1961 (Pogonophora). Proceedings Zoology Institution Academy Science USSR 84: 63–72 (in Russian).

Hatschek, B., 1885. Entwicklung der Trochophora von *Eupomatus uncinatus*, Philippi (*Serpula uncinata*). Arbeiten aus dem Zoologischen Institut Wien 6: 121–148.

Hatt, P., 1932. Essais expérimentaux sur les localisations germinales dans l'oeuf d'un annélide (*Sabellaria alveolata* L.). Archives Anatomic Microscopy 28: 81–98.

Heimann, C., 2000. Differentielle Zellzykluskontrolle bei Spiraliern und ihre Bedeutung für die Determination der Blastomeren. Dissertation (Universität Mainz). pp. 1–155.

Henry, J. J., 1986. The role of unequal cleavage and the polar lobe in the segregation of developmental potential during first cleavage in the embryo of *Chaetopterus variopedatus*. Roux's Archvies of Developmental Biology 195: 103–116.

Henry, J. J. & M. Q. Martindale, 1987. The organizing role of the D quadrant as revealed through the phenomenon of twinning in the polychaete *Chaetopterus variopedatus*. Roux's Archives of Developmental Biology 196: 499–510.

Hermans, C. O., 1964. The reproductive and developmental biology of the ophelid polchaete , *Armandia brevis* (Moore). Thesis, University of Washington, USA.

Irvine, S. Q. & M. Q. Martindale, 2000. Expression patterns of anterior Hox genes in the polychaete *Chaetopterus:* Correlation with morphological boundaries. Developmental Biology 217: 333–351.

Irvine, S. Q. & S A. Warinner, J. D. Hunter, M. Q. 1997. A survey of homeobox genes in *Chaetopterus variopedatus* and analysis of polychaete homeodomains. Molecular Phylogenet Evolution 7: 331–345.

Ivanov, A. V., 1975. Embryonalentwicklung der Pogonophora und ihre systematische Stellung. Zeitschrift für Zoologische Systematik und Evolutionsforschung 13: 10–44.

Ivanova-Kazas, O. M., 1981. Phylogenetic significance of spiral cleavage. Biologiya Morya 5:3–14.

Jacobsohn, S., 1999. Characterization of novel F-actin envelopes surrounding nuclei during cleavage of a polychaete worm. International Journal of Developmental Biology 43: 19–26.

Jeffery, W. R., 1983. An organelle complex responsible for mRNA localization in the cortex of *Chaetopterus* eggs. Biological Bulletin 165: 515–516.

Jeffery, W. R., 1984. Patterns of maternal mRNA distribution and their role in early development. In Sawyer R. H. & R.M. Showman (eds) The Cellular and Molecular Biology of Invertebrate Development. The Belle W. Baruch Library in Marine Science Vol. 15. University of South Carolina Press, Columbia SC, USA, 1984: 125–151.

Jeffery, W. R. & Wilson, L.J. 1983.Localization of messenger RNA in the cortex of *Chaetopterus* eggs and early embryos. Journal of Embryology and Experimental Morphology 75: 225–239.

Kang, D., M. Pilon & D. A. Weisblat, 2002. Maternal and zygotic expression of a nanos-class gene in the leech *Helobdella robusta:* primordial germ cells arise from segmental mesoderm. Developmental Biology 245: 28–41.

Kluge, B., 1991. Cytologische Analyse der frühesten Entwicklungsvorgänge bei *Platynereis dumerilii* (Annelida,

Polychaeta). Doctoral Thesis (University of Mainz): pp. 1–245.

Kulakova, M. A., R P. Kostyuchenko, T F. Andreeva & A. K. Dondua, 2002. The abdominal-B-like gene expression during larval development of *Nereis virens* (polychaeta). Mechanical Development 115: 177–179.

Levin, L. A., J. Zhu & E. Creed, 1991. The genetic basis of life-history characters in a polychaete exhibiting planktotrophy and lecithotrophy. Evolution 45: 380–397.

Lillie, F. R., 1902. Differentiation without cleavage in the egg of the annelid *Chaetopterus pergamentaceus*. Archives Entw Mech Org 14: 477–499.

Malaquin, A., 1893. Recherches sur les Syllidiens. Morphologie, anatomie, reproduction, dèveloppement. Mémory Society Science Agriculture Lille. Series 4/18: 1–477.

Marsh, A. G., L. S. Mullineaux, C. M. Young D. T. Manahan, 2001. Larval dispersal potential of the tubeworm *Riftia pachyptila* at deep-sea hydrothermal vents. Nature 411: 77–80.

McEuen, F. S., B. L. Wu & F. S. Chia, 1983. Reproduction and development of *Sabella media*, a sabellid polychaete with extratubular brooding. Marine Biology 76: 301–309.

McHugh, D., 1997. Molecular evidence that echiurans and pogonophorans are derived annelids. PNAS USA 94: 8006–8009.

Mead, A. D., 1897. The early development of marine annelids. Journal of Morphology 13: 227–326.

Mead, A. D., 1898. The origin and behaviour of centrosomes in the annelid egg. Journal of Morphology 14: 181–212.

Meijer, L., 1979. Hormonal control of oocyte maturation in *Arenicola marina* L. (Annelida, Polychaeta). I. Morphological study of oocyte maturation. Development Growth of Differentiation 21: 303–314.

Morris, V. B., K. E. Dixon & R. Cowan, 1989. The topology of cleavage patterns with examples from embryos of *Nereis, Styela* and *Xenopus*. Philosophical Transactions of the Royal Society Series B325: 1–36.

Myohara, M., 1979. Reproduction and development of *Pseudopolydora kempi japonica* (Polychaeta: Spionidae), with special reference to the polar lobe formation. Journal of Fac Sci Hokka VI/21: 355–364.

Nelson, J. A., 1904. The early development of *Dinophilus:* A study in cell-lineage. Proceedings of the Academy of National Science Philadelphia 56: 687–736.

Newby, W. W., 1940. The embryology of the echiuroid worm *Urechis caupo*. Mem Am Philadelphia Soc 16: 1–219.

Osanai, K., 1978. Early development of the japanese palolo, *Tylorrhynchus heterochaetus*. Bulletin of Marine Biology 16: 59–69.

Penners, A., 1922. Über Doppelbildungen bei *Tubifex rivulorum* Lam. Verh Dtsch Zoology 27: 46–48.

Penners, A., 1924. Experimentelle Untersuchungen zum Determinationsproblem am Keim von *Tubifex rivulorum* Lam. I. Die Duplicitas cruciata und organbildende Keimbezirke. Wilhelm Roux's Archieves 102: 51–100.

Peterson, K. J., S. Q. Irvine, R. A. Cameron & E. H. Davidson, 2000. Quantitative assessment of Hox complex expression in the indirect development of the polychaete annelid *Chaetopterus*. PNAS USA 97: 4487–4492.

Peterson, K. J. & D. J. Eernisse, 2001. Animal phylogeny and the ancestry of bilaterians: inferences from morphology and 18S rDNA gene sequences. Evolution and Development 3: 170–205.

Pilon, M. & D. A. Weisblat, 1997. A nanos homolog in leech. Development 124: 1771–1780.

Prud'homme, B., R. de Rosa, D. Arendt, J. F. Julien, R. Pajaziti, A. W. C. Dorresteijn, A. Adoutte, J. Wittbrodt, G. Balavoine, 2003. Arthropod-like expression patterns of *engrailed* and *wingless* in the annelid *Platynereis dumerilii* suggest a role in segment formation. Current Biology (in press).

Render, J. A., 1983. The second polar lobe of the *Sabellaria cementarium* embryo plays an inhibitory role in apical tuft formation. Wilhelm Roux's Archieves 192: 120–129.

Rouse, G. W. & F. Pleijel, 2001. Polychaetes. Oxford University Press, London.

Sardet, C., F. Prodon, R. Dumollard, P. Chang & J. Chenevert, 2002. Structure and function of the egg cortex from oogenesis through fertilization. Developmental Biology 241: 1–23.

Sato, M., T. Makoto & N. Moritaka, 1982. Ecological aspects of the development of the polychaete, *Lumbrineris latreilli* (Audouin et Milne-Edwards): significance of direct development and non-simultaneous emergence of the young from the jelly mass. Bull Marine Biol. Stn Asamushi 17: 71–85.

Savage, R. M. & M. Shankland, 1996. Identification and characterization of a hunchback orthologue, *Lzf2*, and its expression during leech embryogenesis. Developmental Biology 175: 205–217.

Schleip, W., 1914. Die Entwicklung zentrifugierter Eier von *Clepsine sexoculata*. Verh Dtsch Zool 1914: 236–253.

Schneider, S., A. Fischer, & A. W. C. Dorresteijn, 1992. A morphometric comparison of dissimilar early development in sibling species of *Platynereis* (Annelida, Polychaeta). Roux's Archives of Developmental Biology 201: 243–256.

Seaver, E. C., D. A. Paulson, S. Q. Irvine & M. Q. Martindale, 2001. The spatial and temporal expression of Ch-en, the engrailed Gene in the Polychaete *Chaetopterus*, does not support a role in body axis segmentation. Developmental Biology 236: 195–209.

Shimizu, T., 1982. Development in the freshwater oligochaete *Tubifex*. Dev Biol Fr lnv: 283–316.

Shimizu, T., R. Ishii & H. Takahashi, 1998. Unequal cleavage in the early *Tubifex* embryo. Development Growth Differentiation 40: 257–266.

Siewing, R., 1980. Körpergliederung und Phylogenetisches System. Zoologische Jahrbücher 103: 196–210.

Soto, J. G., B. H. Nelson & D. A. Weisblat, 1997. A leech homolog of twist: evidence for its inheritance as a maternal mRNA. Gene 199: 31–37.

Spek, J., 1930. Zustandsänderungen der Plasmakolloide bei Befruchtung und Entwicklung des *Nereis*-Eies. Protoplasma 9: 370–427.

Speksnijder, J. E. & M. R. Dohmen, 1983. Local surface modulation correlated with ooplasmic segregation in eggs of *Sabellaria alveolata* (Annelida, Polychaeta). Roux's Archives Developmental of Biology 192: 248–255.

Stent, G. S. & D. A. Weisblat, 1982. The development of a simple nervous system. Scientific American 246/1: 100–110.

Stossich, M., 1878. Beiträge zur Entwicklungsgeschichte der Chaetopoden. Sitzungsberichte der Kaiserlichen Akademie der Wissenschaften, Wien 77: 533–544.

Sturtevant, A., 1923. Inheritance of direction of coiling in *Limnaea*. Science 58: 269–270.

Traut, W., 1970. Zur Sexualität von *Dinophilus gyrociliatus* (Archiannelida). III. Die Geschlechtsbestimmung. Biologisches Zentralblatt 89: 137–161.

Treadwell, A. L., 1901. The cytogeny of *Podarke obscura* Verill. Journal of Morphology 17: 399–486.

Tyler, A., 1930. Experimental production of double embryos in annelids and mollusks. Journal of Experiment Zoology 57: 347–407.

Villa, L., 1976. An ultrastructural investigation of the polar plasm of the egg of *Sternaspis* (Annelida, Polychaeta). Acta Embryologica Morphologica Experimentale 1976: 153–165.

Watson, G. J. & M. G. Bentley, 1998. Oocyte maturation and post-fertilization development of *Arenicola marina* (L.) (Annelida: Polychaeta). Invertebrate Reproductive Development 33: 35–46.

Weisblat, D. A., R. T. Sawyer & G. S. Stent, 1978. Cell lineage analysis by intracellular injection of a tracer enzyme. Science 202: 1295–1298.

Werbrock, A. H., D. A. Meiklejohn, A. Sainz, J. H. Iwasa & R. M. Savage, 2001. A polychaete hunchback ortholog. Developmental Biology 235: 476–488.

Wheeler, W. M., 1898. The maturation, fecundation and early cleavages of *Myzostoma glabrum* Leuckart. Archs Biol Liége 15: 1–17.

Whitman, C. O., 1878. The embryology of *Clepsine*. Quarterly Journal of Microscopic Science 18: 215–315.

Wilson, E. B., 1892. The cell-lineage of *Nereis*. A contribution to the cytogeny of the annelid body. Journal of Morphology 6: 361–480.

Wilson, E. B., 1898. Considerations on cell-lineage and ancestral reminiscence. Annals of the New York Academy of Science 11: 1–27.

Wilson, E. B., 1904. Experimental studies on germinal localization. II. Experiments on the cleavage-mosaic in *Patella* and *Dentalium*. Journal of Experimental Zoology 1: 197–268.

Woltereck, R., 1904. Beiträge zur praktischen Analyse der *Polygordius*-Entwicklung nach dem Nordsee- und dem Mittelmeer-Typus. Wilhelm Roux's Archieves 18: 377–403.

Woodwick, K. H., 1960. Early larval development of *Polydora nuchalis* Woodwick, a Spionid polychaete. Pacific Science 14: 122–128.

Hydrobiologia (2005) **535/536**: 23–24
T. Bartolomaeus & G. Purschke (eds), Morphology, Molecules, Evolution and Phylogeny in Polychaeta and Related Taxa

Trochophora larvae and adult body regions in annelids: some conclusions

Claus Nielsen
Zoological Museum, University of Copenhagen, Universitetsparken 15, DK-2100 Copenhagen, Denmark
E-mail: cnielsen@zmuc.ku.dk

The early development and cell-lineage of several polychaetes was studied in great detail about a century ago, for example *Neanthes* (Wilson, 1892, 1898, as *Nereis*), *Capitella* (Eisig, 1898), *Arenicola* (Child, 1900), *Amphitrite* and *Clymenella* (Mead, 1900), *Podarke* (Treadwell, 1901), *Thalassema* (Torrey, 1903), *Polygordius* (Woltereck, 1904), *Dinophilus* (Nelson, 1904), and *Scoloplos* (Delsman, 1916), and reviewed by Anderson (1966, 1973). *Polygordius* and *Thalassema* have planktotrophic larvae, whereas the other species have leicithotrophic larvae or more or less direct development.

Cleavage follows the well-known spiral pattern in all species and also the cell-lineage pattern is remarkably conservative. The prototroch is formed from the primary trochoblasts ($1a^2$–$1d^2$) in all species, with the addition of accessory trochoblasts ($1a^{1222}$–$1c^{1222}$) in some species, for example *Amphitrite* and *Podarke*, and a varying number of secondary trochoblasts (descendants of the cells $2a^1$–$2c^1$). This results in a horseshoe-shaped prototroch with a posterior gap. Descendants of the 2d-cell carry the telotroch in *Arenicola* and *Amphitrite*. In the well-known cell-lineage study of *Polygordius*, Woltereck (1904) interpreted a pair of ciliated cells behind the lateral corners of the mouth (descendants of $3c^1$ and $3d^1$) as the metatroch cells, but since their ciliation is continuous with the oral ciliation they should probably be interpreted as the first cells of the adoral ciliary zone. Two 2d-cell descendants situated between these cells and the mid-ventral band of cells appear as the most likely precursors of the metatroch cells. These observations are in complete accordance with the trochaea theory (Nielsen, 1979, 2001), which proposes that proto-, meta- and telotroch are segments of the circumblastoporal archaeotroch of the holoplanktonic protostomian ancestor; the prototroch should then by derived from the four primary prototroch cells, together with varying numbers of secondary and accessory cells from the A–C-quadrants, whereas meta- and telotroch should come from 2d-cell descendants.

Blastomeres of the first micromere quartet cover the episphere of the trochophore, but cells of the D-quadrant (1d-cell-descendants) extend posteriorly through the dorsal gap in the prototroch in some forms, such as *Amphitrite* and *Podarke*. The ephisphere becomes the prostomium at metamorphosis.

The 2d-cell, often called the somatoblast, divides prolifically and the resulting somatic plate expands from the dorsal side along the lateral sides of the embryo finally to fuse in the ventral midline covering the cells of the other quadrants in the whole region behind the metatroch. This is in full accordance with the observations of Henry & Martindale (1987) who marked individual blastomeres of 4d-cell stages of *Chaetopterus* and followed their development. The growth zone anterior to the telotroch is formed by 2d-cells, but true teloblasts, as those known from clitellates, have not been reported from any "polychaete" (although the word teloblasts has sometimes been used).

These facts can be used to refine the nomenclature used for some body regions in the polychaetes (Fig. 1).

At metamorphosis, the larval episphere becomes the prostomium, and this could perhaps with advantage be added to the definition of the prostomium. The posterior dorsal extention of 1d-cells mentioned above may contribute to the understanding of anterodorsal structures, for example the palps of *Scolelepis* (see for example Foster, 1917), which can be interpreted as prostomial if they develop from a region covered by epithelium that originated from the 1d-cell. This agrees well with the innervation of these structures (Orrhage, 1964).

The segmented body region between peristomium the pygidium, i.e., between metatroch and telotroch when present, is well-defined, and it could perhaps be advantageous to define the annelid body as consisting of three primary body regions, prostomium, peristomium and pygidium,

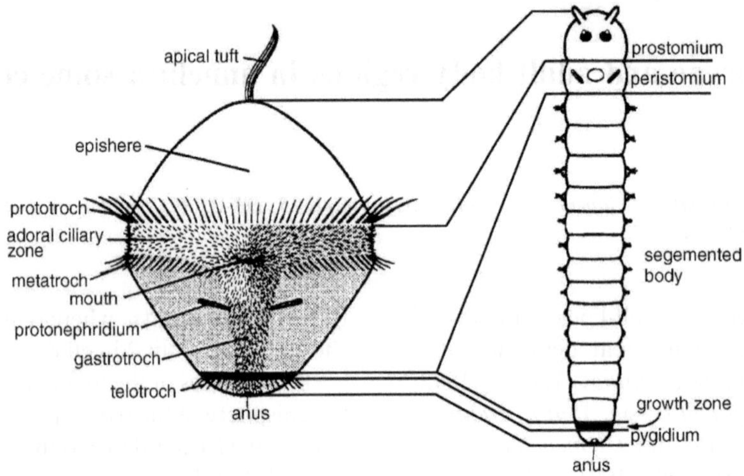

Figure 1. Diagram of corresponding body regions of a planktotrophic polychaete trochophore and an adult, both seen from the ventral side. The epithelium formed from the D-quadrant is shaded and the posterior growth zone black. The metatroch is shaded in accordance with the interpretation mentioned in the text.

and a region between peristomium and pygidium consisting of a posterior growth zone and a number of true segments (Fig. 1). With this definition, neither peristomium nor pygidium would be segments. The peristomium has a pair of larval protonephridia, whereas the true segments have various types of proto- and metanephridia. It appears that chaetae are not found on the peristomium; the true segments usually carry paired parapodia with chaetae, but these are lacking completely, for example in *Polygordius*, and achaetous segments are found in several other species, especially in the anterior body region.

References

Anderson, D. T., 1966. The comparative embryology of the Polychaeta. Acta Zoologica (Stockholm) 47: 1–42.

Anderson, D. T., 1973. Embryology and Phylogeny of Annelids and Arthropods (International Series of Monographs in Pure and Applied Biology, Zoology 50). Pergamon Press, Oxford: 1–495.

Child, C. M., 1900. The early development of *Arenicola* and *Sternaspis*. Archiv für Entwicklungsmechanik der Organismen 9: 587–723.

Delsman, H. C., 1916. Eifurchung und Keimblattbildung bei *Scoloplos armiger* O.F. Müller. Tijdschrift van de Nederlandsche Dierkundige Vereeniging, 2. Ser 14: 283–498, pls 21–26.

Eisig, H., 1898. Zur Entwicklungsgeschichte der Capitelliden. Mitteilungen aus der Zoologischen Station zu Neapel 13: 1–292, pls 1–9.

Foster, N. N., 1917. Spionidae (Polychaeta) of the Gulf of Mexico and the Caribbean Sea. Studies on the Fauna of Curaçao 36: 1–183.

Henry, J. J. & M. Q. Martindale, 1987. The organizing role of the D quadrant as revealed through the phenomenon of twinning in the polychaete *Chætopterus variopedatus*. Roux's Archives of Developmental Biology 196: 499–510.

Mead, A. D., 1897. The early development of marine annelids. Journal of Morphology 13: 227–326.

Nelson, J. A., 1904. The early development of *Dinophilus:* a study in cell-lineage. Proceedings of Academy of Natural Sciences of Philadelphia 56: 687–737, pls 43–48.

Nielson, C., 1979. Larval ciliary bands and metazoan phylogeny. Fortschritte in der zoologischen Systematik und Evolutionsforschung. 1: 178–184.

Nielsen, C., 2001. Animal Evoluiton: Interrelationships of the Living Phyla, 2nd. edn. Oxford University Press, Oxford: 1–563.

Nielsen, C., 2004. Trochophora larvae: cell-lineages, ciliary bands and body regions. 1. Annelida and Mollusca. Journal of Experimental Zoology (Molecular Development and Evolution) 302B: 35–68.

Orrhage, L., 1964. Anatomische und morphologische Studien über die Polychaetenfamilien Spionidae, Disomidae und Poecilochaetidae. Zoologiska Bidrag från Uppsala 36: 335–405.

Torrey, J. C., 1903. The early embryology of *Thalassema mellita* (Conn). Annals of the New York Academy of Sciences 14: 165–246.

Treadwell, A. L., 1901. Cytogeny of *Podarke obscura* Verrill. Journal of Morphology 17: 399–486.

Wilson, E. B., 1892. The cell lineage of *Nereis*. Journal of Morphology 6: 361–480, pls 13–20.

Wilson, E. B., 1898. Considerations on cell-lineage and ancestral reminiscence. Annals of the New York Academy of Sciences 11: 1–27.

Woltereck, R., 1904. Beiträge zur praktischen Analyse der *Polygordius*-Entwicklung nach dem 'Nordsee-' und dem 'Mittelmeer-Typus'. Archiv für Entwicklungsmechanik der Organismen 18: 377–403.

Hydrobiologia (2005) **535/536:** 25–35
T. Bartolomaeus & G. Purschke (eds), Morphology, Molecules, Evolution and Phylogeny in Polychaeta and Related Taxa

Comparative structure of the epidermis in polychaetes (Annelida)

Harald Hausen
Animal Systematics and Evolution, Free University Berlin, Königin-Luise-Str. 1–3, 14195 Berlin, Germany
E-mail: hhausen@zoosyst-berlin.de

Key words: Annelida, Echiura, Sipuncula, Myzostomida, collagen, cuticle, epidermis, ultrastructure

Abstract

The polychaete epidermis generally consists of a single layer of supportive cells, gland cells and sensory cells. Except for the latter, this paper reviews the recent literature on the annelid epidermis, focussing on the mentioned cell types and the cuticle. The annelid epidermis is compared to that of Sipuncula, Echiura and Myzostomida. Supportive cells predominate in the polychaete epidermis. They show a high structural diversity even within single specimens. Ciliated cells are usually multiciliary and only two cases of monociliary epidermis cells are known. Unambigous epithilio-muscle cells are only described in feeding palps of a *Magelona* species. Secretory cells release a large number of gland products and some of them are essential for tube secretion. Rather pecularities of the cells and its arrangement within glands than the ultrastructure of the secretions is useful for phylogenetic considerations. One of the main components of the cuticle is collagen. Recent studies indicate that annelid cuticular collagen differs in several aspects from collagen of the connective tissue and might be of interest for systematics.

Introduction

The epidermis in polychaetes is composed of a cellular monolayer that is covered by a cuticle like in many other invertebrates. The main structural characteristics are well known and were reviewed repeatedly [see Gardiner (1992b); Richards (1978); Richards (1984); Southward (1984); Southward (1993); Storch (1988); Welsch et al. (1984)]. The epidermis inhabits secretory cells, sensory cells and as its main component supportive cells. The latter term is used for all cells, which are not neuronal cells or cells specialized in secretion. Adjacent epidermis cells always are apically connected by zonulae adhaerentes and septate junctions and rest on an underlying basal matrix. Nevertheless the epidermal monolayer may have a pseudostratified appearance and especially the basal parts of secretory cells may be deeply sunken in into the underlying tissue. Cells that do not extend towards the body surface are mainly basally situated neurons, which can be abundant in species with an intraepidermal nervous system. Other basal cells only occasionally have been reported and its origin and function is not completely understood [see for discussion Gardiner (1992a); Rieger (1981)].

Sensory cells are widespread in the epidermis of polychaetes. They occur scattered between other epidermis cells or in groups or are parts of more complex sense organs. They are mainly bipolar primary receptor cells and reviewed elsewhere within this issue (see Purschke, 2005).

Supportive cells

Supportive cells show different cell morphologies even within single specimens. This presumably reflects different physiological specializations. Supportive cells may be thin and flattened, or thick and of columnar, cuboidal or of squamous shape and may be differently equipped with cell organelles. They can show pigment vesicles, vacuoles of

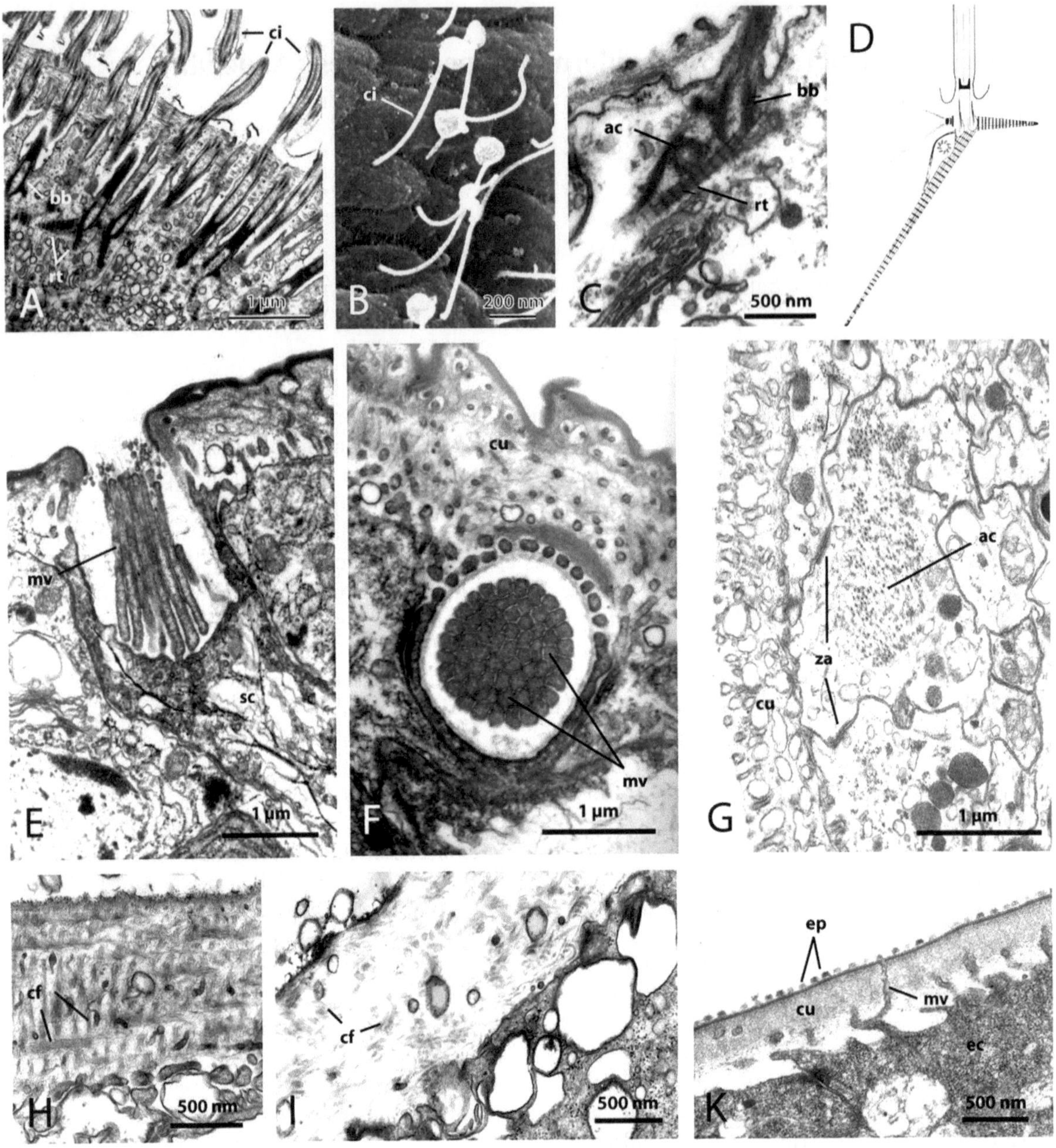

Figure 1. A: *Glycera alba*. Multiciliary epidermal cell. The cilia show a basal body (bb), two rootlets (rt) but no accessory centriol. B–D: *Magelona mirabilis*. Monociliary epidermis cells on larval feeding tentacles. B: SEM micrograph. C: basal ciliary structures with basal body (bb), two rootlets (rt) and one accessory centriol (ac). D: schematic represantation. E–F: *Pectinaria auricoma*. Circular arranged microvilli (mv) form a cone around the apical opening of a secretory cell. E: longitudinal section. F: cross section through tip of the microvillar cone. G: *Magelona mirabilis*. Epithelio-muscle cell in papilla of palps of adults with large embedded actomyosin complex (ac) H: *Eulalia viridis*. Collagenous orthogonal grid within the cuticle. I: *Ophiodromus flexuosus*. Collagenous orthogonal grid within the cuticle. K: *Poecilochaetus serpens*, larva. Cuticle devoid of collagen fibres. cf: collagen fibre; cu: cuticle; ec: epidermis cell; ep: epicuticular projection; za: zonula adhaerens.

different size, endocytotic and exocytotic vesicles, and quite different amounts of certain organelles like mitochondria, lysosoms and multivesicular bodies. Supportive cells usually bear microvilli that extend into the cuticle and – according to the above definition – all non-sensory epidermal cells with cilia are supportive cells as well. Most supportive cells contain tonofilament bundles to withstand mechanical stress. These filaments are basally connected to the underlying matrix by hemidesmosomes and often run to the apex of the cells, where they attach to hemidesmosomes at the bases of apical microvilli or project into the microvilli.

If supportive cells bear cilia like in ciliary bands, the cilia almost always occur in higher numbers per cell (Fig. 1A). The cilia lack accessory centriols, but two striated rootlets are common and usually one rootlet projects deeply into the cell while the other one runs parallel to the apical cell membrane. Only few cases of monociliary epidermal cells are known. Those on the tentacles of *Owenia fusiformis* are interspersed between unciliated and biciliated cells (Gardiner, 1978). Each cilium of the mono- and biciliated cells has one horizontal and one vertical rootlet and is accompanied by an accessory centriol. Further monociliated cells are known from *Magelona mirabilis* (Bartolomaeus, 1995). They occur on the larval feeding tentacles and show one horizontal and one vertical rootlet and an accessory centriol (Fig. 1B–D). Whereas Gardiner (1978) argues for the primary condition of monociliarity in *Owenia*, Bartolomaeus (1995) favours the hypothesis that monociliarity in both species may be caused by a truncated ciliogenesis. This process suppresses formation of further cilia in cells that formerly were multiciliated, because multiciliarity is assumed for the annelidan stem species. According to Gupta & Little (1970) a pair of centrioles is frequently found close to the apical cell surface of unspecialized epidermal cells of several pogonophoran species. One of the centriols gives rise to a cilium with complete $9 \times 2 + 2$ axonem that protrudes into a canal within the cuticle.

Epithelio-muscle cells are reported of tentacles, cirri and antennae of the syllid *Typosyllis variegata* (Storch, 1988) and in tentacles and parapodial cirri in Nereididae (Boilly-Marer, 1972). A reinvestigation of syllidan cirri, however, showed certain dissimilarities to all kinds of invertebrate muscle cells. The remarkable fibrillar structures within the epidermis cells are restricted to the apical most cell region and always are directly connected to the zonulae adhaerentes (Bartolomaeus, personal communication) instead of having a more basal position and being in contact to the plasma membrane *via* dense bodies and dense plaques. Thus, the term epithelio-muscle cells should be avoided for these cells. Nevertheless, true epithelio-muscle cells occur in the epidermis of the papilla of the palps in adult *Magelona mirabilis*. The cells contain large acto-myosin complexes, have contact to the cuticle and are interconnected to surrounding epidermis cells by zonula adhaerentes, which show no affinity to the acto-myosin (Fig. 1G).

Secretory cells

Secretory cells are abundant in the polychaete epidermis and are situated separately between the supportive cells or may be grouped within glandular fields or form complex multicellular glands. Mucous secretions assist in maintaining a mucoid film on the body surface or in feeding by forming mucus traps like in *Chaetopterus* species (Flood & Fiala-Médioni, 1982) or *Praxillura maculata* (McDaniel & Banse, 1979) or they enable the transport of food particles to the mouth. They also facilitate reproduction in producing brood chambers or egg cases in many groups. Duo-gland systems, which produce an adhesive and a releasing secretion, are used by many meiofaunal species to hold on the substrate (Martin, 1978; Gelder & Tyler, 1986). Secretions are responsible for the inner lining of burrows or are involved in the tube building process. For this purpose usually different substances produced in different cells are released (Defretin, 1971; Moermans, 1974; Vovelle et al., 1994). Least secretions also can be used to etch burrows in mineralized structures like in *Polydora* (Zottoli & Carriker, 1974).

Secretory cells release their contents *via* pores in the cuticle. The apical cell membrane may form a circle of microvilli surrounding the pore-like opening (Fig. 1E, F) and a lot of microtubules may occur in the cell apex of secretory cells. Both structures are discussed to assist in controlling the secretion process (Storch & Welsch, 1972; Richards, 1978; Hausmann, 1982; Storch, 1988). In certain secretory cells of *Spiochaetopterus typicus* the opening is surrounded by kinocilia (Storch, 1988). Singular mucous cells in several species of Pogonophora can send a short cilium into an apical lumen underneath the cuticular opening (Gupta & Little, 1970).

Several ultrastructural descriptions of secretory cells reflect the high structural diversity of the secretions from compact material to fine filamentous or granular material packed in vesicles of various sizes or in sometimes highly ordered arrangements, which seemingly may lack a surrounding membrane (Dorsett & Hyde, 1970a, b; Kryvi, 1972; Storch & Welsch, 1972; Hausmann, 1982; Welsch et al., 1984; Hilbig, 1986b; Storch, 1988; Gardiner, 1992b). Histochemical investigations revealed that epidermal cells can secrete many different substances including glycosaminoglycans like hyaluronic acid, different mucopolysaccharids, mucoproteins, proteins, enzymes, phenols, and varying inorganic components (Dorsett & Hyde, 1970a, b; Defretin, 1971; Kryvi, 1971; Moermans, 1974; Bielakoff et al., 1975; Vovelle, 1979; Vovelle & Gaill, 1986; Vovelle et al., 1994). Lectin histochemical investigations indicate the simultaneous production of different carbohydrates within different gland cells in several species (Welsch & Storch, 1986). Only few studies, however, link the histochemical knowledge to the diversity of the electron optical appearance of the secretions. Though the structure of the secretory granules or vesicles presumably largely depends on the chemical nature of its content no correlation between the structure and the chemistry of the secretions has been shown until today. Moreover, the abundance of certain organelles like free ribosomes, rough or smooth endoplasmic reticulum, and golgi stacks corresponds to the histochemical observations. Thus, ultrastructural data of the granules or vesicles of secretory cells are of minor importance for phylogenetic considerations. Other structural peculiarities of individual cells or special arrangements of secretory cells in glandular fields or multicellular glands like in the spiral glands of *Nereis* (Dorsett & Hyde, 1970b) or the pyriform glands of Pogonophora (Southward, 1993) may provide more useful characters. Furthermore standard histochemical characterization of secretions can presently not determine characters of high phylogenetic relevance. Such characters would have to include protein sequence data and exact biochemistry of the carbohydrates. Higher complexity in carbohydrate structure, however, might decrease its usefulness for systematics, because complexity of carbohydrates in secretions results from enzymatic processing within the Golgi complex so that carbohydrate structure is not as directly connected to the genes as in the case of proteins.

Ultrastructural and biochemical examination of tube building glands in Pogonophora exhibited unique differentiations within the secretory cells and clearly characterized the secreted substances. Pogonophora are the only polychaete group where chitin has been detected within the tubes (Blackwell et al., 1965; Gaill & Hunt, 1986). In *Tevnia jerichonana* and *Riftia pachyptila* many microfibrils or crystallites of β-chitin are parallel embedded in a protein matrix and form together flat ribbon like structures (Gaill et al., 1992a,b). Several crisscrossing layers of these ribbons build up the tube wall. The huge crystallites are composed of up to 6000 β-chitin chains and are secreted by specialized multicellular so called pyriform glands. In *Riftia pachyptila* the secreting cells of these glands bear a lot of cup shaped microvilli like structures, which presumably are the sites of a highly regulated microfibril formation (Southward, 1984; Shillito et al., 1993). Two proteins from the tube of *Riftia pachyptila*, which are thought to tighten the different parts of the tube by protein–protein and specific β-chitin–protein interactions, were sequenced and characterized (Chamoy et al., 2000, 2001). Its mRNA is detectable in special epidermal cells, but never within the chitin secreting pyriform glands.

Cuticle

The cuticle of adult polychaetes is composed of an amorphous or fine filamentous matrix that usually inhabits collagen fibrils. Different mucopolysaccharids and presumably also hyaluronic acid are present within the matrix (Kryvi, 1971; Manavalaramanujam & Sundara Rajulu, 1974; Moermans, 1974; Richards, 1984). Several lectins show a positive reaction to the cuticle and reveal a two layered distribution pattern of certain carbohydrates (Welsch & Storch, 1986). Cuticular proteins are thought to have a hardening effect that differs from quinone tanning (Manavalaramanujam & Sundara Rajulu, 1974). The collagen fibrils are restricted to the basal zone of the cuticle. The upper collagen free zone is termed epicuticle. This usually shows layers of different electron densities and is covered by a mostly thin surface coat varying from fine filamentous, fuzzy to electron dense granular material (Richards, 1984; Storch, 1988; Gardiner, 1992b). The collagen fibrils in the basal zone very often form a regular multilayered arrangement (Fig. 1H, I). All fibrils of one layer are parallel to each other and to the cell surface, but fibrils of adjacent layers are perpendicularly arranged. The fibrils are oriented 40°–55° to the body long axis. Such orthogonal grids are widespread within polychaetes (Gupta & Little, 1970; Storch & Welsch, 1970; Richards, 1984; Hilbig, 1986a; Gaill & Bouligand, 1987; Lepescheux, 1988; Storch, 1988; Pilato et al., 1989; Heffernan, 1990; Pilato & la Rosa, 1990; Pilato et al., 1990; Bartolomaeus, 1992; Gardiner, 1992b; Pilato & la Rosa, 1992; Tzetlin et al., 2002) and only few exceptions are known. The fibrils of several adjacent layers can be parallel to each other followed by a perpendicularly directed series of layers like in Glyceridae and Questidae (Storch, 1988) or hexagonal arrangements consisting of fibrils oriented in three different directions exist like in *Typosyllis variegata* (Storch, 1988) or in the gills of *Eunice norvegica* (Hilbig, 1986a) and other arrangements can occur [see Storch (1988)]. Gaill & Bouligand (1987) and Lepescheux (1988) showed that the collagen fibrils of the orthogonal grid in *Alvinella pompejana* and *Paralvinella grasslei* are coiled on several different levels. The collagen fibrils consist of coiled microfibrils, which are composed of triple helices containing the coiled protein chains. The undulating appearance of the fibrils in sections is due to a coiled structure of each fibril. Lepescheux (1988) could not observe endings of individual fibrils and thus assumes the fibrils to be coiled on a larger scale in running around the cylindrical bodysurface.

In young larvae collagen fibrils may lack completely (Fig. 1K) or are very thin and irregularly arranged (Holborow & Laverack, 1969; Eckelbarger, 1978; Heimler, 1981; Heimler, 1983; Schlötzer-Schrehardt, 1992). Thicker fibrils being organized in regular orthogonal grids are formed later during ontogenesis. In adults of very small polychaetes like in representatives of the interstitial fauna, however, collagen fibrils may lack or be irregularly arranged as has been shown for *Protodrilus*, *Polygordius*, *Trilobodrilus*, and *Diurodrilus* (Rieger & Rieger, 1976), *Dinophilus* species (Brandenburg, 1970), for certain hesionids (Westheide & Rieger, 1978), *Apodotrocha progenerans* (Westheide & Riser, 1983), *Ophryotrocha diadema* (Hilbig, 1986a), and *Dysponetus* spp.(Tzetlin et al., 2002). These findings can be interpreted as functional adaptions to the small body size or as the result of a progenetic or neotenic evolution. Nevertheless, the lack of a collagenous orthogonal grid is not exclusively restricted to interstitial forms. No fibers are present in the cuticle of *Sabellaria vulgaris* (Storch, 1988) and *Cossura longocirrata* (Rouse & Tzetlin, 1997). The epidermis cells in *Owenia* sp. are only covered by a loose fibrillar matrix and *Spiochaetopterus typicus* also lacks a true cuticle (Storch, 1988). The tube of *Chaetopterus variopedatus*, which shows criss-crossing layers of fibrils, is discussed to be a cuticle that has lost contact to the body wall (Brown & McGee-Russell, 1971).

Supportive cells often send microvilli into the cuticle. If an orthogonal grid is present adjacent microvilli are usually separated by one collagen fibril. The microvilli may terminate within the cuticle or cross it. In the latter case the tips of the microvilli may be somewhat inflated and lay on the apical surface of the cuticle. Those tips are often surrounded by several isolated membrane bound vesicles, which are called epicuticular projections and appear similar to the tips of the microvilli (Fig. 1K). They presumably are pinched off by the microvilli. Both, tips of the microvilli and epicuticular projections, usually are coated by fine filamentous material that may be a special glycocalyx. Many different shapes and electron densities are reported for epicuticular projections.

Comparative morphology

There is no general feature in which the unspecialized epidermis of the body wall of clitellates differs from that in polychaetes (for reviews on clitellate epidermis see Fernández et al. (1992); Jamieson

(1988); Jamieson (1992)). There are also no principle differences in the cellular components and the cuticle shows no significant deviations. The collagen fibrils usually form a regular orthogonal grid, but some modification may exist. While some members of Tubificidae and Naididae show an orthogonal grid, in a couple of oligochaete species all fibrils are parallel oriented or irregularly arranged or lack at all (Gustavsson & Erséus, 2000; Gustavsson, 2001). Fibril arrangement does not correlate with body size in tubificids and naidids and irregular patterns are described for small and large specimens in contrast to the situation in polychaetes. The different arrangements in tubifcids and naidids are thought to be useful for phylogenetic analysis.

Ultrastructural data on the epidermis of sipunculids are rare, but at least the structure of the cuticle shows striking similarities to that of annelids. Except the tentacles the entire body of *Phascolion strombus* is covered by a thick cuticle that contains collagen fibrils forming a clear orthogonal grid (Moritz & Storch, 1970). In the second larvae, the so-called pelagosphaera of *Aspidosiphon* sp. and *Paraspidosiphon* sp. the cuticle shows layers of fibrils perpendicular to one another, whereas the fibrils are irregularly arranged in larvae of *Golfingia* species (Rice, 1976). The epidermis in adult sipunculids is generally composed of cuboidal cells with deep basal infoldings (Storch, 1984). Within the trunk region the microvilli of the supporting cells do not pierce the cuticle. In *Themiste lageniformis*, however, epidermal microvilli are found on the oral surface of the tentacles, which show no collagen fibrils within their cuticle (Pilger, 1979; Pilger, 1982). Additionally the tentacles bear many ciliated cells. They are presumably sensory on the aboral surface whereas multiciliated cells on the oral surface bear kinocilia equipped with a short basal foot and only one rootlet in *Themiste lageniformis*. In contrast to this the cilia of multiciliated epidermal cells in the pelagosphera larva of *Apionsoma misakianum* represent the presumed plesiomorphic state in having two perpendicular rootlets (Lundin & Schander, 2003). The length of the horizontal rootlet secondarily is strongly reduced in the oral multiciliated cells of the tentacles in adult *Phascolion strombus*.

Echiurans have an annelid-like epidermis. In the trunk and the proboscis of *Maxmuelleria lankesteri* supportive cells of various shape and two types of secretory cells form a monolayer that partly is pseudostratified (McKenzie & Hughes, 1999). The supportive cells send many microvilli in the cuticle and the apices of the secretory cells, which release their contents *via* ducts through the cuticle, are surrounded by microvilli. Cilia are common on the proboscis though this species shows no obvious ciliated groove like other echiurans. The cilia are anchored in the cells by basal bodies and striated rootlets (McKenzie & Hughes, 1999). Monociliated cells which presumably are sensory were found in the trunk but not in the proboscis. In *Maxmuelleria lankesteri* as well as in *Urechis caupo* a lot of membrane bound epicuticular projections cover the cuticle. They are enwrapped by a glycocalyx and sometimes in contact with microvilli of the supportive cells, (Menon & Arp, 1993; McKenzie & Hughes, 1999). The cuticle consists of an outer amorphous and a basal fibrillar zone. The fibrils form no orthogonal grid but are longitudinally arranged in *Maxmuelleria lankesteri* and loosely packed in *Urechis caupo*. Both species inhabit similar looking bacteria within their cuticle. In *Urechis caupo* the secretion of metachromatic acidic mucus might be used to reduce sulphide entry from the anoxic environment (Menon & Arp, 1993).

The epidermis of myzostomids, whose membership within annelids is discussed controversely [see Eeckhaut et al. (2000); Haszprunar (1996); Mattei & Marchand (1987); Müller & Westheide (2000); Rouse & Tzetlin (1997); Zrzavý et al. (2001)] and contribution in this issue) shows some similiarities to that of polychaetes. In *Myzostoma cirriferum* flattened cells which presumably secrete the cuticular material, multiciliated cells of the same shape, and two different kinds of secretory cells are the main cellular components of the epidermis (Eeckhaut & Jangoux, 1993). Nerve cells of the ventral nerve cord send monociliated dendritic processes into the epidermis. In *Myzostoma alatum* and *Pulvinomyzostomum pulvinar* the thickness of the unciliated and multiciliated cells varies between different body parts (Kronenberger, 1997). The cuticle in all three species possesses an inner layer with a matrix of low or moderate electron density. But only in *Myzostoma cirriferum* fibrillar material within the matrix is described. The basal layer is follwed by a thin dense layer and another layer with a lucent matrix in all three species (Eeckhaut

& Jangoux, 1993) and see figures in Kronenberger (1997). The positive reaction of the cuticle to van Gieson staining in *Myzostoma cirriferum* is interpreted as a hint for collagen. Microvilli of the epidermis cells penetrate the cuticle and end on the top of the apical electron lucent layer. The tips of the microvilli are swollen, filled with electron dense granular material and covered by fine granular material. The cilia of the multiciliary cells are equipped with ciliary rootlets in all three species. Only in *Myzostoma cirriferum* myoepithelial cells are described (Eeckhaut & Jangoux, 1993). They lie basally within the epidermis and are directed perpendicular to the body long axis. The position of the nuclei of these cells remained unclear.

The collagenous cuticle of annelids, echiurids, and sipunculids is used as one argument for a common ancestry of these taxa (Ax, 2000).

Collagen ultrastructure and biochemistry

There are many observations that collagen fibrils in the cuticle of polychaetes, clitellates, pogonophorans, sipunculids and echiurans are not striated (Richards, 1984; Jamieson, 1988; Storch, 1988; Gardiner, 1992b; Jamieson, 1992; McKenzie & Hughes, 1999). This is in contrast to the collagen fibrils of the connective tissue and has recently been substantiated in further detail. A couple of biochemical studies favour the existence of at least two distinct types of collagens in annelids – a cuticular and an interstitial one. The fibrils of interstitial collagen are located within the extracellular matrix of the tissue underneath the epidermal cells. They show a cross-striation pattern *in situ* in *Alvinella pompejana* and *Riftia pachyptila* and the striation pattern of reconstituted fibrils *in vitro* is similar to that of vertebrate collagen type I (Gaill et al., 1991). Also the molecular mass in *Alvinella pompejana*, *Alvinella caudata*, *Paralvinella grasslei*, *Arenicola marina*, *Nereis diversicolor* and *Riftia pachyptila* of the individual triple helices and its length of 280–300 nm is comparable to vertebrate fibrillar collagen (Gaill et al., 1991; Gaill et al., 1995). The triple helices are composed of three identical α-chains. Such homotrimeric constitution seems not to be typical for all cuticular collagens. Though the cuticular collagen of *Riftia pachyptila* consists of only one chain type (Mann et al., 1996), in *Nereis virens*, *Nereis japonica* and *Alvinella pompejana* A- and B-chains have been described (Kimura & Tanzer, 1977; Sharma & Tanzer, 1984; Gaill et al. 1991;). The triple helices of cuticular collagen belong to the longest collagen molecules so far known. They reach between 2400 and 2600 nm in *Nereis diversicolor*, *Nereis virens*, *Alvinella pompejana*, *Alvinella caudata*, *Paralvinella grasslei* and *Arenicola marina* (Murray & Tanzer, 1985; Gaill et al., 1991; Gaill et al., 1995). Isolated triple-helices show a terminal globular domain like it is known for some vertebrate collagens like type IV and VI. The globular domains connect two or more chains in some of the preparations and may play a role in supermolecular assembly (Gaill et al., 1991; Gaill et al., 1995). The fibrils formed by this type of collagen neither *in situ* nor after reconstitution of isolated triple helices show a striation pattern.

The different types of collagen have different immunological properties. Antibodies raised against cuticular collagen of *Arenicola marina*, *Riftia pachyptila* and *Alvinella pompejana* show each a high affinity to the cuticular collagen of all three species, but a low affinity to interstitial collagen of any species (Gaill et al., 1994). Antibodies against interstitial collagen on the other hand bind effectively to interstitial collagen of all three species, but not to cuticular collagen of any species. It is possible to detect cuticular collagen fibers in *Harmothoe lunulata* and *Riftia pachyptila* immunocytochemically *in situ* with antibodies against cuticular collagen of *Arenicola marina* after freeze-fixation (Nicolas et al., 1997).

There are hints at an unusual posttranslational modification in cuticular collagen. Short carbohydrates of mainly galactose subunits are linked O-glycosidically to threonine residues in *Lumbricus terrestris*, *Nereis virens* and *Riftia pachyptila* (Muir & Lee, 1970; Spiro & Bhoyroo, 1980; Mann et al. 1996). In *Lumbricus terrestris* and *Nereis virens* also serine can be glycosylated and in the latter species an additional glucuronic acid-mannose disaccharid has been detected (Muir & Lee, 1970). In *Riftia pachyptila* the glycosylated threonine is located at the Y-position of the Gly-X-Y triplets of the amino acid sequence and presumably enhances the thermal stability of the triple helices as an adaption to hydrothermal environment of the species (Mann et al., 1996; Bann & Bächinger, 2000; Bann et al., 2000). In interstitial collagen of *Riftia pachyptila* and *Alvinella pompejana*, that also lives at hydrothermal vents, thermal stability is achieved by a high degree of hydroxylation of proline in the Y-position like it is usual for many types of collagen (Mann et al., 1992; Sicot et al., 2000).

Molecular phylogeny of annelid collagen just started. Sicot et al. (1997) and Sicot et al. (2000) got a monophyletic grouping of annelidan interstitial collagens within α-chains of fibrillar collagens of different metazoan taxa after a cladistic analysis of some cDNA sequences. No comparable phylogenetic considerations concerning the cuticular collagen have been undertaken so far. Because of the unique triple helix length, lack of striation in the fibrils, immunological properties, existence of a globular domain and the post-translational glycosylation of threonine and serine cuticular collagen might provide useful phylogenetic characters.

Acknowledgements

I would like to thank two anonymous reviewers for their helpful suggestions. The present study was supported by a grant of the Deutsche Forschungsgemeinschaft (DFG Ba 1520/2).

References

Ax, P., 2000. Multicellular Animals. Vol. 2. Berlin, New York, Springer, 396 pp.

Bann, J. G. & H. P. Bächinger, 2000. Glycosylation/hydroxylation-induced stabilization of the collagen triple helix. The Journal of Biological Chemistry 275(32): 24466–24469.

Bann, J. G., D. H. Peyton & H. P. Bächinger, 2000. Sweet is stable: glycolysation stabilizes collagen. FEBS Letters 473: 237–240.

Bartolomaeus, T., 1992. On the ultrastructure of the cuticle, the epidermis and the gills of *Sternaspis scutata* (Annelida). Microfauna Marina 7: 237–252.

Bartolomaeus, T., 1995. Secondary monociliarity in the Annelida: monociliated epidermal cells in larvae of *Magelona mirabilis* (Magelonidae). Microfauna Marina 10: 327–332.

Bielakoff, J., D. Damas & J. Vovelle, 1975. Histologie et histochimie des fromations glandulaires impliquées dans l'élaboration du tube chez *Lanice conchilega* (Annélide polychète). Archives de Zoologie Expérimentale et Génerale 116: 466–520.

Blackwell, J., K. D. Parker & K. M. Rudall, 1965. Chitin in pogonophore tubes. Journal of the Marine Biological Association of the United Kingdom 45: 659–661.

Boilly-Marer, Y., 1972. Présence de cellules de type myoépithélial chez les Nereidae (annélides polychètes). Journal de Microscopie 15: 253–256.

Brandenburg, J., 1970. Die Cuticula des *Dinophilus* (Archiannelida). Zeitschrift für Morphologie der Tiere 68: 300–307.

Brown, S. C. & S. McGee-Russell, 1971. *Chaetopterus* tubes: ultrastructural architecture. Tissue & Cell 3(1): 65–70.

Chamoy, L., M. Nicolai, B. Quennedy, F. Gaill & J. Delachambre, 2000. Characterization of a cDNA encoding RP43, a CUB-domain-containing protein from the tube of *Riftia pachyptila* (Vestimentifera) and distribution of its transcripts. The Biochemical Journal 350: 421–427.

Chamoy, L., M. Nicolai, J. Ravaux, B. Quennedy, F. Gaill & J. Delachambre, 2001. A novel chitin-binding protein from the vestimentiferan *Riftia pachyptila* interacts specifically with ß-chitin. The Journal of Biological Chemistry 276(11): 8051–8058.

Defretin, R., 1971. The tubes of polychaete annelids. In Florkin, M. H. & H. Stolz (eds), Extracellular and Supporting Structures. Comprehensive Biochemistry, Vol. 26, pp. 713–747.

Dorsett, D. A. & R. Hyde, 1970a. The epidermal glands of *Nereis*. Zeitschrift für Zellforschung 110: 219–230.

Dorsett, D. A. & R. Hyde, 1970b. The spiral glands of *Nereis*. Zeitschrift für Zellforschung 110: 204–218.

Eckelbarger, K. J., 1978. Morphogeneis of larval cuticle in the polychaete *Phragmatopoma lapidosa*. Cell and Tissue Research 186: 187–201.

Eeckhaut, I. & M. Jangoux, 1993. Integument and epidermal sensory structures of *Myzostoma cirriferum* (Myzostomida). Zoomorphology 113: 33–45.

Eeckhaut, I., D. McHugh, P. Mardulyn, R. Tiedemann, D. Monteyne, M. Jangoux & C. Milinkovitch, 2000. Myzostomida: a link between trochozoans and flatworms? Proceedings of the Royal Society of London, Series B: Biological Sciences 267: 1383–1392.

Fernández, J., V. Téllez & N. Olea, 1992. Hirudinea. In Harrison, F. W. & S. L. Gardiner (eds), Annelida. Microscopic Anatomy of Invertebrates, Vol. 7, pp. 323–394. New York, Wiley-Liss, Inc.

Flood, P. R. & A. Fiala-Médioni, 1982. Stucture of the mucous feeding filter of *Chaetopterus variopedatus* (Polychaeta). Marine Biology 72: 27–33.

Gaill, F. & Y. Bouligand, 1987. Supercoil of collagen fibrils in the integument of *Alvinella*, an abyssal annelid. Tissue & Cell 19(5): 625–642.

Gaill, F., L. Hamraoui, F. X. Sicot & R. Timpl, 1994. Immunological properties and tissue localization of two different collagen types in annelid and vestimentiferan species. European Journal of Cell Biology 65: 392–401.

Gaill, F. & S. Hunt, 1986. Tubes of deep sea hydrothermal vent worms *Riftia pachyptila* (Vestimentifera) and *Alvinella pompejana* (Annelida). Marine Ecology Progress Series 34: 267–274.

Gaill, F., K. Mann, H. Wiedemann, J. Engel & R. Timpl, 1995. Structural comparison of cuticle and interstitial collagens from annelids living in shallow sea-water and at deep-sea hydrothermal vents. Journal of Molecular Biology 246: 284–294.

Gaill, F., J. Persson, J. Sugiyama, R. Vuong & H. Chanzy, 1992a. The chitin system in the tubes of deep sea hydrothermal vent worms. Journal of Structural Biology 109: 116–128.

Gaill, F., B. Shillito, J. P. Lechaire, H. Chanzy & G. Goffinet, 1992b. The chitin secreting system from deep sea hydrothermal vent worms. Biologie Cellulaire 76: 201–204.

Gaill, F., H. Wiedemann, K. Mann, K. Kühn, R. Timpl & J. Engel, 1991. Molecular characterization of cuticle and interstitial collagens from worms collected at deep sea hydrothermal vents. Journal of Molecular Biology 221: 209–223.

Gardiner, S. L., 1978. Fine structure of the ciliated epidermis cells on the tentacles of *Owenia fusiformis* (Polychaeta, Oweniidae). Zoomorphologie 91: 37–48.

Gardiner, S. L., 1992a. Polychaeta: external anatomy. In Harrison, F. W. & S. L. Gardiner (eds), Annelida. Microscopic Anatomy of Invertebrates, Vol. 7, pp. 11–17. New York, Wiley-Liss, Inc.

Gardiner, S. L., 1992b. Polychaeta: general organization, integument, musculature, coelom, and vascular system. In Harrison, F. W. & S. L. Gardiner (eds), Annelida. Microscopic Anatomy of Invertebrates, Vol. 7, pp. 19–52. New York, Wiley-Liss, Inc.

Gelder, S. R. & S. Tyler, 1986. Anatomical and cytochemical studies on the adhesive organs of the ectosymbiont *Histriobdella homari* (Annelida: Polychaeta). Transactions of the American Microscopical Society 105(4): 348–356.

Gupta, B. L. & C. Little, 1970. Studies on Pogonophora. 4. Fine structure of the cuticle and epidermis. Tissue & Cell 2(4): 637–696.

Gustavsson, L. M., 2001. Comparative study of the cuticle in some aquatic oligochaetes (Annelida: Clitellata). Journal of Morphology 248: 185–195.

Gustavsson, L. M. & C. Erséus, 2000. Cuticular ultrastructure in some marine oligochaetes (Tubificidae). Invertebrate Biology 119(2): 152–166.

Haszprunar, G., 1996. The Mollusca: coelomate turbellarians or mesenchymate annelids? In Taylor, J. (ed.), Origin and Evolutionary Radiation of the Mollusca. Oxford University Press: 1–28.

Hausmann, K., 1982. Elektronenmikroskopische Untersuchungen an *Anaitides mucosa* (Annelida, Polychaeta). Cuticula und Cilien, Schleimzellen und Schleimextrusion. Helgoländer Meeresuntersuchungen 35: 79–96.

Heffernan, P., 1990. Ultastructural studies of the Elytra of *Pholoe minuta* (Annelida: Polychaeta) with special reference to functional morphology. Journal of the Marine Biological Association of the United Kingdom 70: 545–446.

Heimler, W., 1981. Untersuchungen zur Larvalentwicklung von *Lanice conchilega* (Pallas) 1766 (Polychaeta, Terebellomorpha). Teil I: Entwicklungsablauf. Zoologische Jahrbücher Abteilung für Anatomie und Ontogenie der Tiere 106: 12–45.

Heimler, W., 1983. Untersuchungen zur Larvalentwicklung von *Lanice conchilega* (Pallas) 1766 (Polychaeta/Terebellomorpha). Teil III: Bau und Struktur der Aulophora-Larve. Zoologische Jahrbücher Abteilung für Anatomie und Ontogenie der Tiere 110: 411–478.

Hilbig, B., 1986a. Light and electron microscopical studies of the cuticle, epidermis, and setae of some Eunicida (Annelida, Polychaeta) I.Cuticle. Zoologische Jahrbücher Abteilung für Anatomie und Ontogenie der Tiere 114: 1–41.

Hilbig, B., 1986b. Light and electron microscopical studies of the cuticle, epidermis, and setae of some Eunicida (Annelida, Polychaeta) II. Epidermal glands. Zoologische Jahrbücher Abteilung für Anatomie und Ontogenie der Tiere 114: 371–400.

Holborow, P. L. & M. S. Laverack, 1969. Cilia and other surface structures of the trochophore of *Harmothoë imbricata* (Polychaeta). Zeitschrift für Zellforschung 98: 246–261.

Jamieson, B. G. M., 1988. Oligochaete ultrastructure: some comparisons with the Polychaeta. In Westheide, W. & C. O. Hermans (eds), The Ultrastructure of Polychaeta. Microfauna Marina, Vol. 4, pp. 397–428. Stuttgart, Jena, New York, Gustav Fischer Verlag.

Jamieson, B. G. M., 1992. Oligochaeta. In Harrison, F. W. & S. L. Gardiner (eds), Annelida. Microscopic Anatomy of Invertebrates, Vol. 7, pp. 217–322. New York, Wiley-Liss, Inc.

Kimura, S. & M. L. Tanzer, 1977. Nereis cuticle collagen: isolation and characterization of two distinct subunits. Biochemistry 16(11): 2554–2560.

Kronenberger, K., 1997. Ultrastruktureller Vergleich der Epidermisstrukturen von *Pulvinomyzostomum pulvinar* (Graff) und *Myzostoma alatum* Graff (Myzostomida, Annelida). pp. 1–54. Master thesis. Ludwigs-Maximilians-Universität, München.

Kryvi, H., 1971. Histology and histochemistry of the mucous glands of *Sabella penicillum* L. (Annelida, Polychaeta). Norwegian Journal of Zoology 19: 37–44.

Kryvi, H., 1972. The fine structure of the ventral mucous cells of *Sabella penicillum* (Polychaeta). Sarsia 48: 23–32.

Lepescheux, L., 1988. Spatial organization of collagen in annelid cuticle: order and defects. Biology of the Cell 62: 17–31.

Lundin, K. & C. Schander, 2003. Epidermal ciliary ultrastructure of adult and larval sipunculids (Sipunculida). Acta Zoologica (Stockholm) 84: 113–119.

Manavalaramanujam, R. & G. Sundara Rajulu, 1974. An investigation on the chemical nature of the cuticle of a polychaete *Nereis diversicolor* (Annelida). Acta Histochemica 48: 69–81.

Mann, K., F. Gaill & R. Timpl, 1992. Amino-acid sequence and cell-adhesion activity of a fibril-forming collagen from the tube worm *Riftia pachyptila* living at deep sea hydrothermal vents. European Journal of Biochemistry 210: 839–847.

Mann, K., E. Mechling, H. P. Bächinger, C. Eckerskorn, F. Gaill & R. Timpl, 1996. Glycosylated threonine but not 4-hydroxyproline dominates the triple helix stabilizing positions in the sequence of a hydrothermal vent worm cuticle collagen. Journal of Molecular Biology 261: 255–266.

Martin, G. G., 1978. The duo-gland adhesive system of the archiannelids *Protodrilus* and *Saccocirrus* and the turbellarian *Monocelis*. Zoomorphologie 91: 63–75.

Mattei, X. & B. Marchand, 1987. Les spermatozo des Acanthocéphales et des Myzostomides. Ressemblances et conséquences Phylétiques. Comptes rendus de l'Académie des Sciences – Série III 305: 525–529.

McDaniel, N. & K. Banse, 1979. A novel method of suspension feeding by the maldanid polychaete *Praxillura maculata*. Marine Biology 55: 129–132.

McKenzie, J. D. & D. J. Hughes, 1999. Integument of *Maxmuelleria lankesteri* (Echiura), with notes on bacterial symbionts and possible evidence of viral activity. Invertebrate Biology 118(3): 296–309.

Menon, J. G. & A. J. Arp, 1993. The integument of the marine echiuran worm *Urechis caupo*. Biological Bulletin 180: 440–454.

Moermans, R., 1974. Recherches sur l'histochimie des teguments et du tube d'une annelide polychete (Eunicidae): *Hyalinoecia tubicola* (O. F. Müller). Bulletin Biologique de la France et de la Belgique 108: 41–59.

Moritz, L. & V. Storch, 1970. Über den Aufbau des Integumentes der Priapuliden und der Sipunculiden [*Priapulus caudatus* Lamarck, *Phasciolon strombi* (Montagu)]. Zeitschrift für Zellforschung 105: 55–64.

Muir, L. & Y. C. Lee, 1970. Glycopeptides from earthworm cuticle collagen. The Journal of Biological Chemistry 245(3): 10–502.

Müller, M. C. & W. Westheide, 2000. Structure of the nervous system of *Myzostoma cirriferum* (Annelida) as revealed by immunohistochemistry and cLSM analyses. Journal of Morphology 245: 87–98.

Murray, L. W. & M. L. Tanzer, 1985. The collagens of the Annelida. In Bairati, A. & R. Garrone (eds), Biology of invertebrate and lower vertebrate collagens. Vol. 93, pp. 243–258.

Nicolas, G., F. Gaill & L. Zylberberg, 1997. In situ localization of two fibrillar collagens in two compact connective tissues by electron microscopy after cryotechnical processing. The Journal of Histochemistry & Cytochemistry 45(1): 119–128.

Pilato, G. & D. la Rosa, 1990. La cuticola degli annelidi. II. Ultrastruttura della cuticola di *Marphysa sanguinea* Montagu, 1815 (Polychaeta, Eunicidae). Animalia 17: 263–271.

Pilato, G. & D. la Rosa, 1992. La cuticola degli Annelida. IV. Ultrastrattura della cuticola di *Nereis diversicolor* O.F. Muller, 1776 (Polychaeta, Nereidae). Animalia 19(1/3): 165–179.

Pilato, G., D. la Rosa & R. Avola, 1989. La cuticola degli annelidi. II. Ultrastruttura della cuticola di *Ophelia bicornis* Sav., 1818 (Polychaeta). Animalia 16: 89–102.

Pilato, G., D. la Rosa & R. Avola, 1990. La cuticola degli annelidi. III. Ultrastrutture della cuticola di *Lumbrineris crosslandi* Perkins, 1979 (Polychaeta, Lumbrineridae). Animalia 17: 301–309.

Pilger, J. F., 1979. Structure and function of a sipunculan tentacle. American Zoologist 19: 909.

Pilger, J. F., 1982. Ultrastructure of the tentacles of *Themiste lageniformis* (Sipuncula). Zoomorphology 100: 143–156.

Purschke, G., 2005. Senese organs in polychaetes (Annelida). Hydrobiologia 535/536 (Dev. Hydrobiol. 179): 53–78.

Rice, M. E., 1976. Larval development and metamorphosis in Sipuncula. American Zoologist 16: 563–571.

Richards, K. S., 1978. Epidermis and Cuticle. In Mill, P. J. (ed.), Physiology of annelids. pp. 33–61. London, New York, San Francisco, Academic Press.

Richards, K. S., 1984. Cuticle. In Bereiter-Hahn, J., A. G. Matoltsy & K. S. Richards (eds), Biology of the integument. 1. Invertebrates. pp. 311–322. Berlin, Springer.

Rieger, R. M., 1981. Fine structure of the body wall, nervous system, and digestive tract in the Lobatocerebridae Rieger and the organization of the gliointestinal system in Annelida. Journal of Morphology 167: 139–165.

Rieger, R. M. & G. E. Rieger, 1976. Fine structure of the archiannelid cuticle and remarks on the evolution of the cuticle within the Spiralia. Acta Zoologica (Stockholm) 57: 53–68.

Rouse, G. W. & A. B. Tzetlin, 1997. Ultrastructure of the body wall and gametogenesis in *Cossura* cf. *longocirrata* (Cossuridae Polychaeta). Invertebrate Reproduction and Development 32(1): 41–54.

Schlötzer-Schrehardt, U., 1992. Ultrastrukturelle Untersuchungen zur Reproduktion und Postembryonalentwicklung einschließlich Adultorganisation von *Pygospio elegans* Claparède, 1863 (Polychaeta, Annelida). Dissertation, Friedrich-Alexander-Universität Erlangen-Nürnberg.

Sharma, Y. D. & M. L. Tanzer, 1984. High-performance liquid chromatographic separation of glycopeptides from *Nereis* cuticle collagen. Analytical Biochemistry 141: 205–212.

Shillito, B., J. P. Lechaire & F. Gaill, 1993. Microvilli-like structures secreting chitin crystallites. Journal of Structural Biology 111: 59–67.

Sicot, F. X., J.-Y. Exposito, M. Masselot, R. Garrone, J. Deutsch & F. Gaill, 1997. Cloning an annelid fibrillar-collagen gene and phylogenetic analysis of vertebrate and invertebrate collagens. European Journal of Biochemistry 246: 50–58.

Sicot, F. X., M. Mesnage, M. Masselot, J.-Y. Exposito, R. Garrone, J. Deutsch & F. Gaill, 2000. Molecular adaption to an extreme environment: origin of the thermal stability of the pompeij worm collagen. Journal of Molecular Biology 302: 811–820.

Southward, E. C., 1984. Pogonophora. In Bereiter-Hahn, J., A. G. Matoltsy & K. S. Richards (eds), Biology of the integument. 1. Invertebrates. pp. 376–387. Berlin, Springer.

Southward, E. C., 1993. Pogonophora. In Harrison, F. W. & S. L. Gardiner (eds), Onychophora, Chilopoda, and Lesser Protostomata. Microscopic Anatomy of Invertebrates, Vol. 12, pp. 327–369. New York, Wiley-Liss, Inc.

Spiro, R. G. & V. D. Bhoyroo, 1980. Studies on carbohydrate of collagens. The Journal of Biological Chemistry 255(11): 5347–5354.

Storch, V., 1984. Echiura and Sipuncula. In Bereiter-Hahn, J., A. G. Matoltsy & K. S. Richards (eds), Biology of the integument. 1. Invertebrates. Berlin, Springer: 368–375.

Storch, V., 1988. Integument. In Westheide, W. & C. O. Hermans (eds), The Ultrastructure of Polychaeta. Microfauna Marina, Vol. 4 Stuttgart, Jena, New York, Gustav Fischer Verlag.

Storch, V. & U. Welsch, 1970. Über die Feinstruktur der Polychaeten-Epidermis (Annelida). Zeitschrift für Morphologie der Tiere 66: 310–322.

Storch, V. & U. Welsch, 1972. The ultrastructure of epidermal mucous cells in marine invertebrates (Nemertini, Polychaeta, Prosobranchia, Ophistobranchia). Marine Biology 13: 167–175.

Tzetlin, A. B., T. Dahlgren & G. Purschke, 2002. Ultrastructure of the body wall, body cavity, nephridia, and spermatozoa in four species of the Chrysopetalidae (Annelida, "Polychaeta"). Zoologischer Anzeiger 241: 37–55.

Vovelle, J., 1979. Les glandes cémentaires de *Petta pusilla* Malmgren, polychète tubicole Amphictenidae, et leur sécrétion organo-minérale. Archives de Zoologie Expérimentale et Génerale 120: 219–246.

Vovelle, J. & F. Gaill, 1986. Données morphologiques, histochimiques et microanalytiques sur lélaboration du tube organominéral d' *Alvinella pompejana*, Polychète des sources hydrothermales, et leurs implications phylogénétiques. Zoologica Scripta 15(1): 33–43.

Vovelle, J., M. Rusaouen-Innocent, M. Grasset & M. Truchet, 1994. Halogenation and quinone-taning of the organic tube components of some Sabellidae (Annelida Polychaeta). Cahiers de Biologie Marine 35: 441–459.

Welsch, U. & V. Storch, 1986. Carbohydrate Histochemistry of polychaetes. Zoologischer Anzeiger 217: 289–300.

Welsch, U., V. Storch & K. S. Richards, 1984. Epidermal cells. In Bereiter-Hahn, J., A. G. Matoltsy & K. S. Richards (eds), Biology of the integument. 1. Invertebrates. pp. 269–296. Berlin, Springer.

Westheide, W. & R. M. Rieger, 1978. Cuticular ultrastructure of hesionid polychaetes (Annelida). Zoomorphologie 91: 1–18.

Westheide, W. & N. W. Riser, 1983. Morphology and phylogenetic relationships of the neotenic interstitial polychaete *Apodotrocha progenerans* n.gen., n.sp. (Annelida). Zoomorphology 103: 67–87.

Zottoli, R. A. & M. R. Carriker, 1974. Burrow morphology, tube formation, and microarchitecture of shell dissolution by the spionid polychaete *Polydora websteri*. Marine Biology 27: 307–316.

Zrzavý, J., V. Hypsa & D. F. Tietz, 2001. Myzostomida are not annelids: molecular and morphological support for a clade of animals with anterior sperm flagella. Cladistics 17: 170–198.

Hydrobiologia (2005) **535/536:** 37–52
T. Bartolomaeus & G. Purschke (eds), Morphology, Molecules, Evolution and Phylogeny in Polychaeta and Related Taxa

Chaetae and chaetogenesis in polychaetes (Annelida)

Harald Hausen
Animal systematics and evolution, Free University Berlin, Koenigin-Luise-Str. 1-3, 14195 Berlin
Tel.: +30-83854884, E-mail: hhausen@zoosyst.berlin.de

Key words: Annelida, ultrastructure, chaetae, chaetogenesis

Abstract

Annelid chaetae are epidermal extracellular structures that are in general clearly visible from the exterior. Their structure is highly diverse, especially within the Polychaeta, and each species shows a specific pattern of chaetae. Chaetae have therefore gained immense significance for species determination, making them the best studied structures in polychaetes. The shape of chaetae is determined by the temporal and spatial modification of the microvilli pattern of a single cell, the chaetoblast. As chaetae are species specific, the process of their formation must be under strict control and the information needed to form certain chaetae must be highly conservative. It can be assumed that corresponding chaetogenesis is caused by commonly inherited information. Thus, comparative chaetogenesis can help to test hypotheses on the homology of certain types of chaetae and help to unravel the influence of functional constraints on the shape of chaetae. Different types of chaetae are compared here and the present state of our knowledge of their structure and formation is used to present some homology hypotheses. There are some strong arguments for a homology of uncini and certain hooks and hooded hooks. Acicula are compared to other supportive setae and the significance of the arrangement of chaeta for phylogenetic considerations is shown. Coding issues are provided in order to facilitate inclusion of information on chaetae into data matrices.

Introduction

Due to their important role in polychaete taxonomy, chaetae are possibly the epidermal structures that have been most intensively studied by light microscopy. Their main structural component is β-chitin, which is cross linked by proteins (Bobin & Mazqué, 1944; Lotmar & Picken, 1950; Picken, 1950; Jeuniaux, 1971; George & Southward, 1973; Michel & Volvelle, 1973; Rudall, 1963; Schroeder, 1984). Inorganic constituents embedded in this organic matrix may have a hardening effect (George & Southward, 1973; Bartolomaeus, 1992). The use of transmission electron microscopy has confirmed earlier light microscopic investigations on the principle structure of chaetae. They are composed of many longitudinal channels (Lippert & Gentil, 1963; Bouligand, 1966, 1967; Scherf, 1970; George & Southward, 1973; Gustus & Cloney, 1973; O'Clair & Cloney, 1974) (and see Fig. 1A). The central channels are usually larger than the peripheral ones, include an empty lumen, and the chaetal material only constitutes the walls of the channels. Peripheral channels often show very small or no lumina (Fig. 1B). The chaetal material can have a fine fibrillar appearance or a more granular structure with different densities and can from multiple layers within a channel wall. The outer surface of chaetae may be covered by a very thin, electron-dark surface coat. It is assumed that the construction of hollow channels is responsible for many of the mechanical properties of chaetae. Differences in the internal geometry, such as diameter, wall thickness and number of channels, presumably have a great influence on stiffness and flexibility (Kryvi & Sørvig, 1990; Merz & Woodin, 1991). These parameters can very to a high degree within individual chaetae. The channels in the main fang of hooks, for instance, are often completely filled with some electron-dense material and are thus very rigid structures, which rest on a more flexible shaft composed of hollow channels (Bartolomaeus, 1998).

Most chaetae project beyond the body surface and are in direct contact with the environment. Each chaetae inhabits its own open epidermal follicle (Fig. 1E). Several follicles in a group or row of chaetae form a single so-called chaetal sac, which is underlain by a common basal matrix (Bouligand, 1967). Tonofilaments cross the follicle cells and adhere to the chaeta on one face and to the basal extracellular matrix on the other face by hemides-

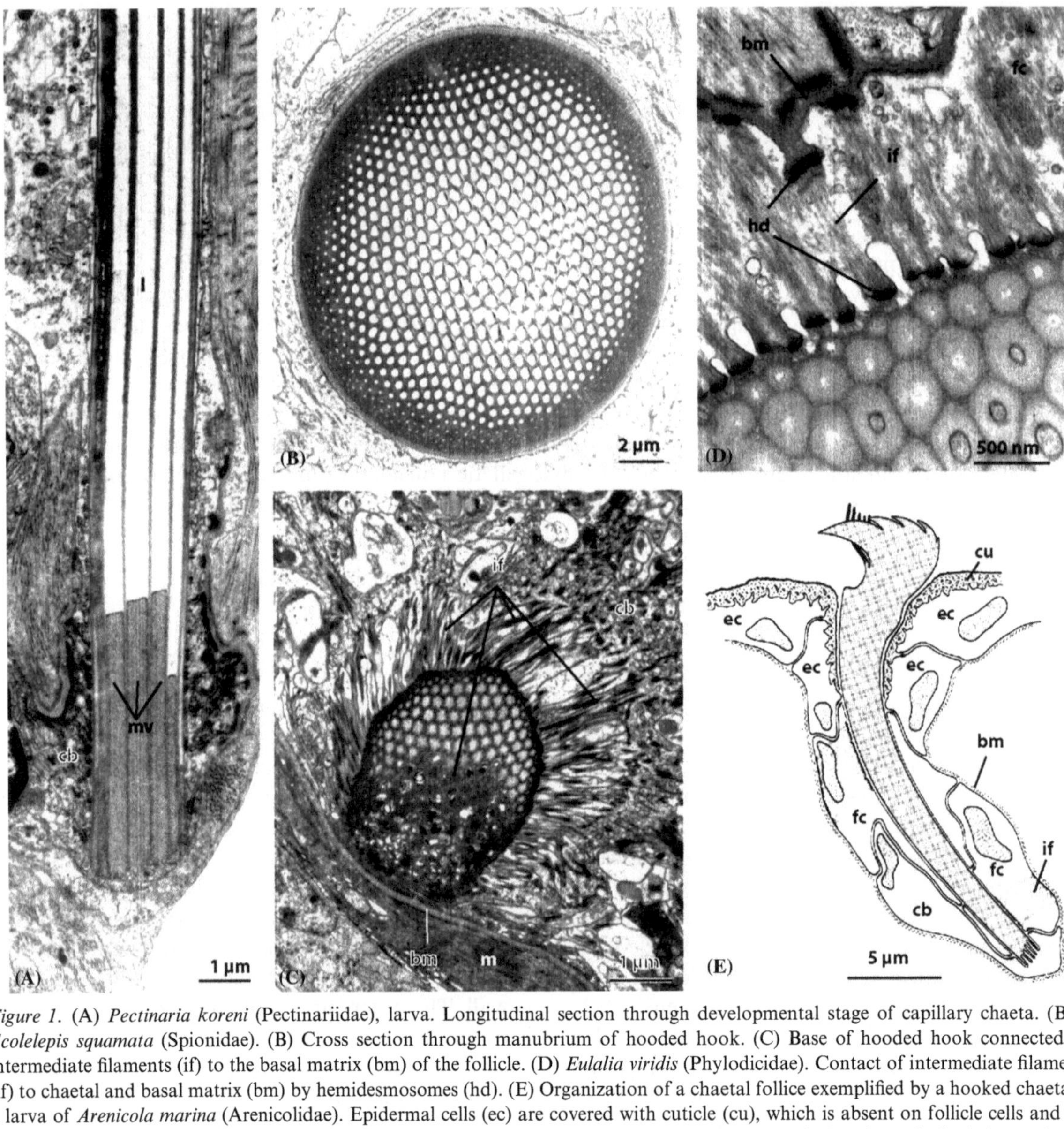

Figure 1. (A) *Pectinaria koreni* (Pectinariidae), larva. Longitudinal section through developmental stage of capillary chaeta. (B,C) *Scolelepis squamata* (Spionidae). (B) Cross section through manubrium of hooded hook. (C) Base of hooded hook connected by intermediate filaments (if) to the basal matrix (bm) of the follicle. (D) *Eulalia viridis* (Phylodicidae). Contact of intermediate filaments (if) to chaetal and basal matrix (bm) by hemidesmosomes (hd). (E) Organization of a chaetal follice exemplified by a hooked chaeta of a larva of *Arenicola marina* (Arenicolidae). Epidermal cells (ec) are covered with cuticle (cu), which is absent on follicle cells and the basal-most cell, the chaetoblast (cb). Cell crossing intermediate filaments (if) connect the chaeta to the basal matrix (bm) that underlies the follicle.

mosomes (Fig. 1C, D). Muscles that attach to the matrix are responsible for the movement of the whole group of chaetae, of few, or even of individual chaetae, as in the case of the remarkable movable spines in *Trochochaeta multisetosum*.

Chaetogenesis

Bouligand (1967) and O'Clair & Cloney (1974) presented a model of the formation process of chaetae that has been confirmed by all subsequent investigators of chaetogenesis. According to them, chaetae grow by basal apposition of new chaetal material. The most basal cell of the chaetal follicle – the chaetoblast – gives rise to a very distinct group of microvilli. New chaetal material assembles on the outer surface of these microvilli. There is no vesicular transport within the microvilli of the chaetoblast, so that the microvilli are not believed to directly supply the material for apposition on the base of the chaetae. Instead the chaetal material is secreted by the chaetoblast and by adjacent follicle cells directly into the lumen of the chaetal follicle and reaches the microvilli surface by diffusion. This material is also released between the bases of the microvilli of the chaetoblast. According to Rudall (1963) it is likely that *N*-acety-glucosamine monomers are secreted and successively added to the end of the parallel *β*-chitin chains, which afterwards are linked by proteins. Though still not confirmed by biochemical examinations, it is suggested that membrane enzymes on the surface of the microvilli are involved in the assemblage process of the chaetal material (O'Clair & Cloney, 1974). The extent to which the chaetoblast is involved in the secretion process remains unclear and no cytochemical studies have been undertaken to reveal the contents of the vesicles in the chaetoblast and the follicle cells. Autoradiographic investigations on polynoids by Zavarzin & Sergovskaya (1979) revealed that the main protein part is secreted by follicle cells during formation of capillary chaetae, but by the chaetoblast in the case of acicula. Growing by basal apposition, the chaetae are forced to move upwards during their genesis. That is the reason why chaetae consist of hollow channels: these lumina remain, where microvilli of the chaetoblast had been.

The overall structure and course of genesis of chaetae in polychaetes, clitellates and pogonophorans is identical (George & Southward, 1973; Orrhage, 1973; Schroeder, 1984; Specht & Westheide, 1988; Jamieson, 1992) and there is no doubt that these structures are homologous. Echiuran chaetae show a similar structure (Orrhage, 1971; Storch, 1984), indicating a homology as well and argue beside other characters like the nervous system (Purschke et al., 2000; Hessling & Westheide, 2002) for a position of Echiura within or as a sister group of annelids. Structures similar to annelid chaetae described in juvenile octopods (Brocco et al., 1974), a certain placophoran (Leise & Cloney, 1982), Brachiopoda (Storch & Welsch, 1972; Gustus & Cloney, 1973; Orrhage, 1973; Lüter, 2000) and Bryozoa (Gordon, 1975), however, are most probably not homologous. To assume a homology of the mollusc structures and annelid chaetae would mean arguing for a non-parsimonious reduction in at least several subgroups of molluscs and sipunculans. The chaeta-like bristles in Bryozoa are also only restricted to a subgroup of that taxon. And even if brachiopods and bryozoans are regarded as subgroups of the Spiralia (for discussion see Lüter & Bartolomaeus (1997)) a homology of its chaetae to those of annelids can only be assumed, if brachiopods and bryozoans turn out to be one of the closest relatives of annelids.

Homology of certain kinds of chaetae

Many terms are in use to manage the high structural diversity of chaetae in polychaetes, such as compound chaetae, spinigers, falcigers, spines, hooks, hooded hooks, capillaries, winged capillaries, uncini, forked chaetae, pectinate chaetae and so on. This terminology is mainly based on standard light microscopic investigations. Meanwhile, scanning electron microscopy revealed many structures of chaetae that have been beyond light microscopic resolution and this allows more detailed comparisons (see for instance Knight-Jones, 1979; Read, 1986; Imajima, 1989; Doyle, 1991; Eibye-Jacobsen, 1991). Moreover, some pitfalls of the light microscopic method became clear. For instance, capillaries in *Sabella penicillus* under the scanning electron microscope show no wings or limbations, which light microscopy suggests (Kryvi, 1989) and certain chaetae in *Poecilochaetus* show spiral palisades of fused spinules around the long axis instead of separated lateral pinnate series (Read, 1986). But though scanning electron microscopy has improved the descriptions of chaetae, it is mainly transmission electron microscopy that seems to provide the appropriate data for phylogenetics on a higher systematic level. This is because the examinations of O'Clair & Cloney (1974) consolidated our understanding of the formation process of chaetae. It became clear that the microvilli of the chaetoblast are highly dynamic structures during chaetogenesis. The entire shape of chaetae depends on the number, diameter and arrangement of the microvilli and the changes of these parameters over time. The microvilli of a single chaetoblast can be separated into different groups that may rejoin after a while. This needs variations in the adhesion of the microvilli over time. Different microvilli may fuse and microvilli may change their diameter and may be strongly bent by changes in the orientation of the surface of the chaetoblast. If polymerizing enzymes are actually present on the membranes of the microvilli, their distribution or activity along the microvilli should also be regulated. This assumption could explain why chaetal material is sometimes added on the tip of microvilli, sometimes causing completely filled channels and sometimes not, so that the channels are hollow tubes. Investigations of ultrastructure and genesis of chaetae thus provides as insight into a very complex process on a cellular or even subcellular level, i.e. the arrangement of the microvilli and possibly the distribution of enzymes on its surface. The entire course must be controlled strictly on the genetic level, as otherwise one could not observe large numbers of nearly identical chaetae within a specimen. Because the controlling mechanisms act on a low structural level, there is a good chance that these mechanisms rely on a rather basic level of the epigenetic network. To use identities in the underlying formation process between different chaetae or parts of chaetae as decisive indications for homologies thus promises very robust homology hypotheses.

Some existing reports on intraspecific variability of annelid chaetae do not necessarily contradict this view. The number of pectination teeth in chaetae of naidids and tubificids can vary with season or abiotic factors (Smith, 1985; Chapman & Brinkhurst, 1987). *Halosydna* (*Alentia*) can show varying percentages of certain chaetae due to varying temperatures (Hillger & Reish, 1970). And the number of apical teeth in hooded hooks of *Prionospio japonica* can vary with in a specimen (Ohwada & Nishino, 1991). These findings of course argue for a cautious use of fine structural features of chaetae in taxonomy on the species level. There is, however, no evidence that environmental factors influence the principle course of the formation of a certain kind of chaetae.

Hooks, unicni, and hooded hooks

Electron microscopic investigations of several sedentary taxa have led to insights in to the formation process of hooks and uncini and certain homology hypotheses (Bartolomaeus, 1995, 1998). As in all chaetae investigated, with the exception of acicula, the genesis of uncini in the terebellid *Nicolea zostericola* starts with an epidermal invagination forming the developing chaetal follicle (Bartolomaeus, 1998). The rostrum is preformed by a compact group of microvilli, on the surface of which chaetal material assembles (Fig. 2). A change in the orientation of the chaetoblast surface leads to a strong curvation of the microvilli and accordingly of the entire developing rostrum. During this process a couple of microvilli with larger diameter arise behind the rostrum. Each of these microvilli is responsible for the formation of one apical tooth behind the rostrum. As a result the rostrum of uncini is composed of many channels, whereas each apical tooth is composed of one channel. The formation proceeds with the subrostral process by addition of a large number of smaller microvilli beneath the rostrum. The short manubrium is formed by adding material on the surface of all existing microvilli. Later the formation process only proceeds on two very small groups of microvilli – one on the front and one at the back of the uncinus – leading to the anterior and posterior processes of the uncinus. When they are complete, all microvilli are replaced by small cytoplasmatic processes containing bundles of tonofilaments. These are connected via hemidesmosomes on the walls of the channels and cross the chaetoblast and adhere on that side of the cell

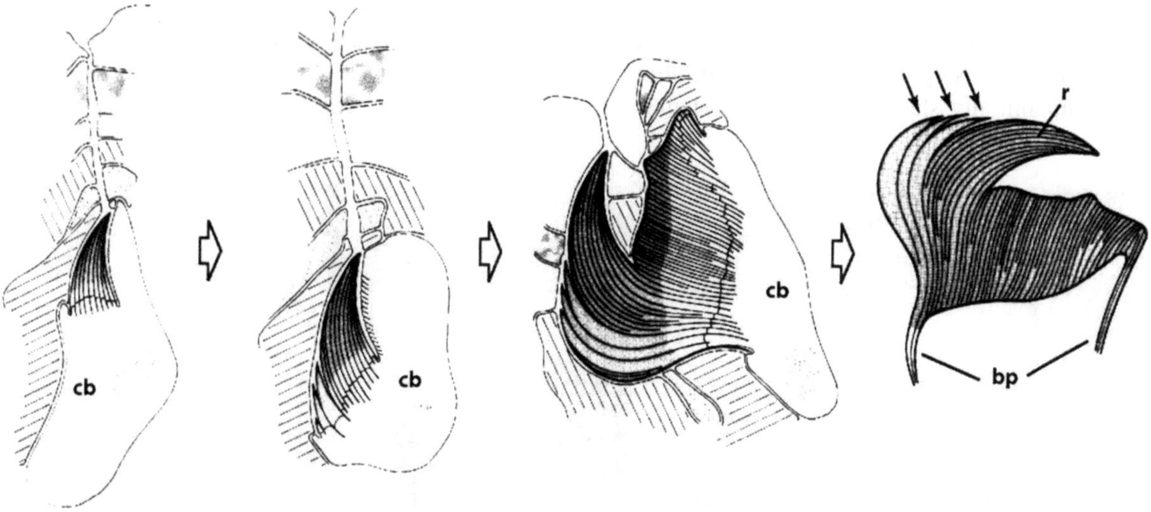

Figure 2. Nicolea zostericola (Terebellidae). Development of neuropodial uncini as reconstructed from examination of different stages. bp = basal process; cb = chaetoblast; r = rostrum; arrow = tooth of capitium.

next to the basal matrix, which underlies the chaetal sac.

The genesis of hooks and uncini in several taxa have been examined and certain identities become apparent, besides variations in form and length of hooks. A rostrum, if present, is always preformed by many microvilli and the strong bend is always the result of a reorientation of the apical membrane of the chaetoblast during genesis. In contrast, each apical tooth behind the rostrum is always preformed by only one microvillus (Fig. 3B, D, J). Thus, it seems advisable to restrict the common term capitium, used to subsume apical teeth, only to those with the described formation. The formation of a subrostral process is always realized by adding new microvilli to the anlage and the manubrium is always the product of continued assemblage of chaetal material after the tilt of the apical membrane on the microvilli so far used to preform rostrum, capitium and subrostral process. The manubrium is short in the case of uncini and long in the case of hooks.

If a rostrum is absent, the chaetogenesis starts directly by forming the capitium, as shown for the pectinariids *Pectinaria koreni* and *P. auricoma* (Fig. 3A, B) and the serpulid *Spirorbis spirorbis* (Bartolomaeus, 1995), the short shafted abdominal uncini in the sabellid *Fabricia sabella* (Bartolomaeus, 2002), the uncini in the chaetopterid *Telepsavus costarum*, and the hooks of the oweniid *Owenia fusiformis* (Meyer & Bartolomaeus, 1996). The latter contradicts Nilsen & Holthe (1985), who state that one of the two teeth is the rostrum, whereas the other belongs to the capitium. Though electron microscopic investigations in Sabellariidae have to be carried out, uncini of *Sabellaria alveolata* obviously also lack a rostrum, but show a lot of small teeth, which may form a capitium. Uncini in Siboglinidae also most probably lack a rostrum (Bartolomaeus, 1998), though Schulze (2001) interprets a group of channels in the middle of the chaetae in *Ridgeia piscesae* as rostrum, even if no curvation of these channels is described.

Independently of the existence of a rostrum, the subrostral process may be absent in certain groups as well as the basal processes, which are never present on long shafted hooks (Bartolomaeus 1995, 1998, 2002). As an additional structure, a so-called beard appears in hooks of juveniles of the arenicolid *Arenicola marina* (Fig. 3G), in maldanids (Fig. 4H) and in the psammodrilid *Psammodrilus balanoglossoides* (Meyer & Bartolomaeus, 1996; 1997). Each hair of the beard is preformed by one microvillus that originates from below the rostrum (Fig. 3I).

Despite different combinations of the chaetal parts in different taxa, the identities in structure and formation process led to a homology hypothesis of hooks and uncini in a high number of taxa. The evolution of short shafted uncini is therefore assumed to have a common stem lineage in Terebellida, Sabellariidae, Chaetopteridae, Sabellidae, Serpulidae, and Pogonophora. The uncini initially showed a strong rostrum, a capitium, a short manubrium, a subrostral process, but no basal processes. Reduction of the rostrum has to be assumed in different subgroups and the evolution of basal processes took

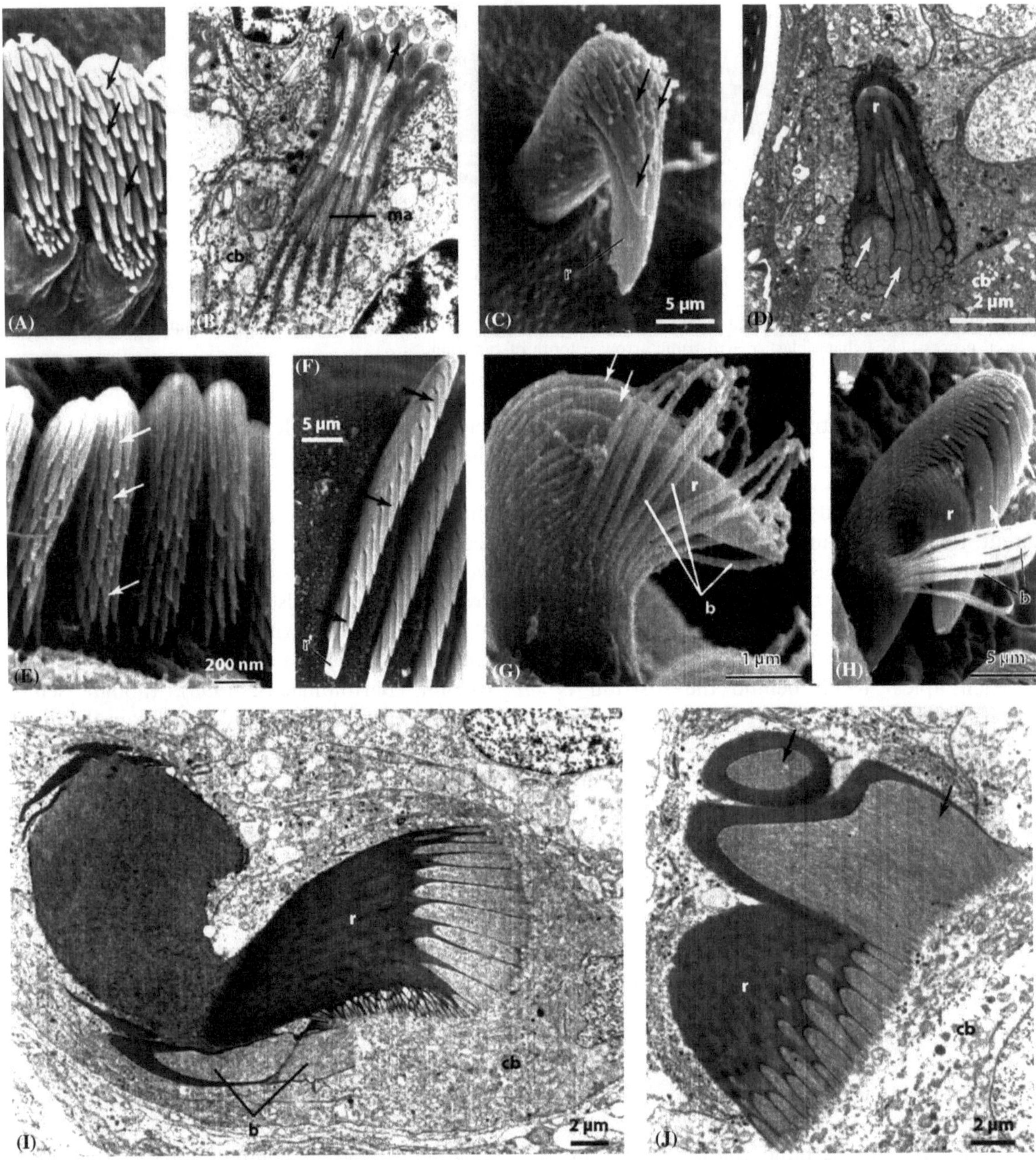

Figure 3. (A) *Pectinaria koreni* (Pectinariidae). Uncinus. (B) *Pectinaria auricoma* (Pectinariidae). Developmental stage of uncinus. (C,D) *Fabricia sabella* (Sabellidae). (C) Thoracic uncinus (D) Developmental stage of thoracic uncinus (E) *Pectinaria auricoma* (Pectinariidae). Uncini. (F) *Pomatoceros triqueter* (Serpulidae). Uncinus. (G) *Arenicola marina* (Arenicolidae). Uncinus of larva. (H) *Johnstonia duplicata* (Maldanidae). Midbody uncinus. (I,J): *Clymenura clypeata* (Maldanidae). (I) Development of rostrum and beard. (J) Development of rostrum and teeth of capitium. b: beard; cd: chaetoblast; r: rostrum; arrow: tooth of capitium

place in a common lineage leading to Terebellida, Sabellariidae and Chaetopteridae. The long manubrium of the thoracic uncini in the sabellid subgroup Fabriciinae (Fig. 3C) reflects a secondary condition. Alternatively it can be assumed that uncini evolved from long shafted hooks like those present in Oweniidae, Arenicolidae, Maldanidae and Psammodrilidae due to the identities in structure and formation of the chaetal units.

Woodin & Merz (1987) instead assume a priori that hook-bearing taxa are not closely related and argue for a numerously independent evolution of hooks in different lineages. Empirical data sug-

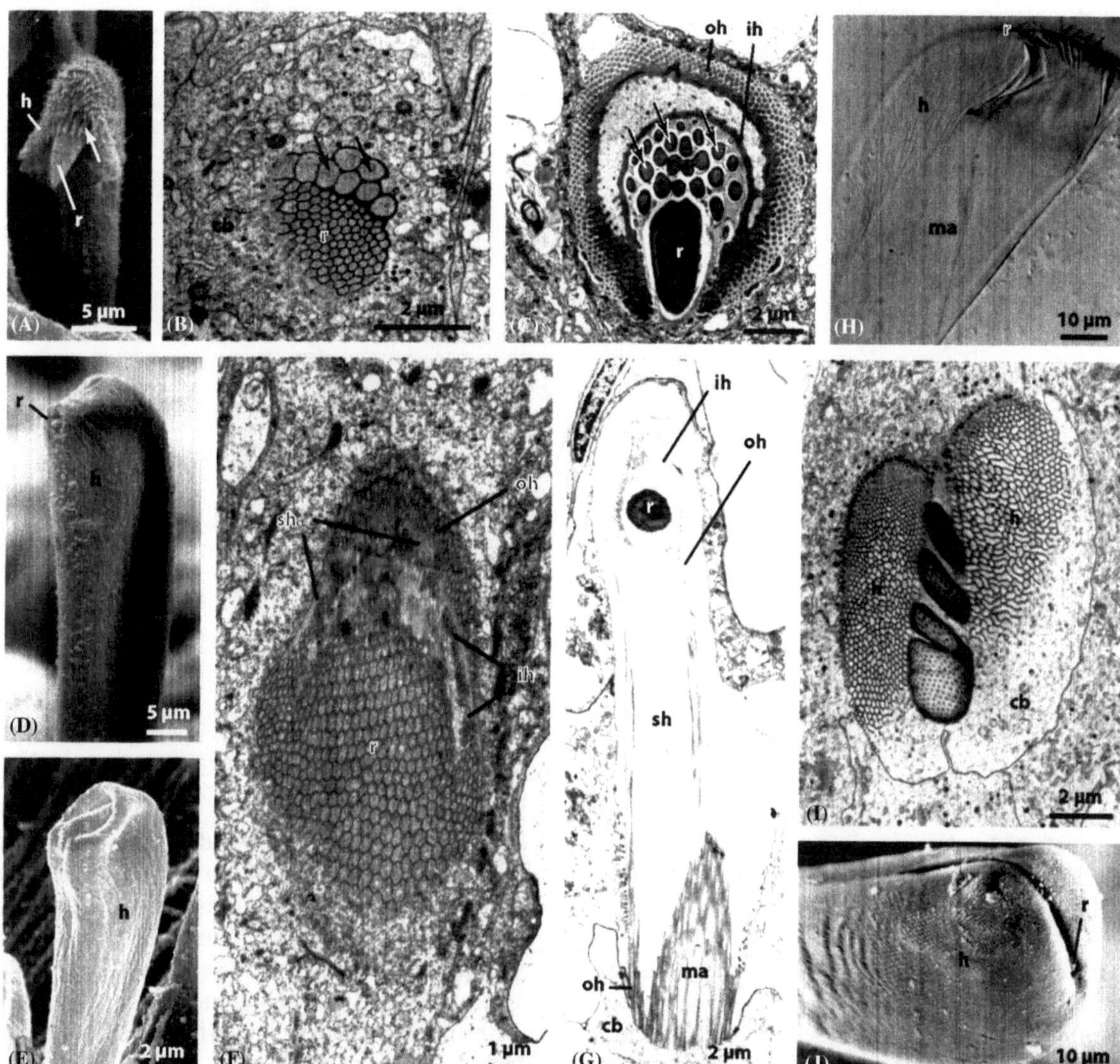

Figure 4. (A) *Notomastus latriceus* (Capitellidae). hooded hook. (B,C) *Capitella capitata* (Capitellidae). (B) development of rostrum and teeth of capitium (arrows) of a hooded hook (C) tip of a hooded hook showing rostrum, teeth of the capitium and hood structure. (D) *Malacoceros fuliginosus* (Spionidae). hooded hook. (E) *Magelona minuta* (Magelonidae). hooded hook. (F) *Scolelepis squamata* (Spionidae). Thin layers of microvilli of the chaetoblast preform the inner (ih) and the outer hood (oh) in the development of a hooded hook. (G) *Magelona minuta* (Magelonidae). Developmental stage of hooded hook. (H–K) *Lumbrineris tetraura* (Lumbrineridae). (H) hooded hook under Normarski-contrast. (I) developmental stage of hooded hook. The hood is preformed by a compact array of microvilli showing no seperate sheathes. (K) hooded hook with rostrum (r) projecting of a slit-like opening of the hood (h). cb: chaetoblast; sh: empty space within hood.

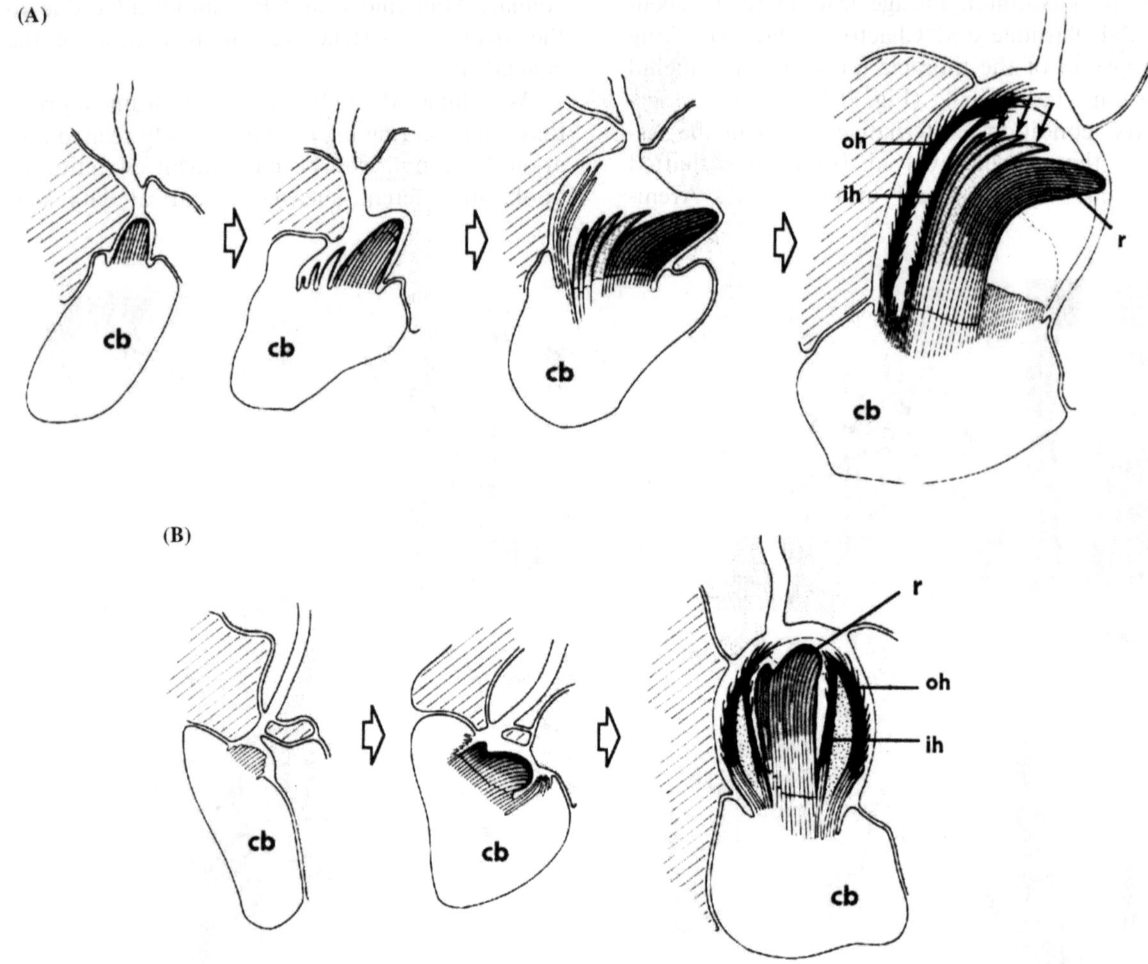

Figure 5. Chaetal development as reconstructed from examination of different stages. (A) hooded hook in *Capitella capitata* (Capitellidae). (B) hooded hook in *Scolelepis squamata* (Spionidae). cb: chaetoblast; ih: inner hood; oh: outer hood; r: rostrum; arrow: tooth of capitium.

gesting that hooks in different taxa act as anchoring devices are taken as a hint that hooks may be homoplasious (Merz & Woodin, 2000). However, as stated by Fitzhugh (1991), the only way to test a homology hypothesis based on high structural equivalence is to prove its congruence with other homology hypotheses by phylogenetic considerations. Finally, it is not surprising that a structure once evolved in a common lineage is used for the same purpose by the descendants.

Certain delicate apical structures covering the hooks are described as hoods in Lumbrineridae, Spionidae, Magelonidae, and Capitellidae (Fig. 4A, D, E, H, J). Confirming earlier light microscopic investigations on spionid chaetae by Foster (1971), ultrastructural data showed that in *Scolelepis squamata and Prionospio fallax* the hood is composed of two thin and separated sheaths, the inner one of which joins the chaetal core more apically than the outer one (Hausen & Bartolomaeus, 1998; Hausen, 2001). The two sheaths fuse on the tip, giving rise to an opening by which the core of the chaeta projects exteriorly. Two separate rings of microvilli that surround the anlage of the core are involved in the formation of the hood (Figs 4F and 5B). Almost identical in structure and formation are hoods in the capitellid *Capitella capitata* (Schweigkofler et al., 1998) (and see Figs 4C and 5A) and in the magelonid *Magelona minuta* (Hausen, 2001) (and see Fig. 4G).

There is clear evidence for a homology of the hood structure of the chaetae in Spionidae, Capitellidae and Magelonidae. Moreover, the composition of the chaetal core in *Capitella capitata* of manubrium, rostrum and capitium (Fig. 4B) favours a homology to the hooks in the other taxa (Schweigkofler et al., 1998) and argues for a close relationship of magelonids, spionids and capitellids to hook bearing annelids. A capitium may not be present at the base of the whole group since the apical teeth in the spionids *Scolelepis squamata* and *Prionospio fallax* form no capitium, because they are preformed by many channels (Hausen & Bartolomaeus, 1998; Hausen, 2001). In contrast, no evidence could be found in support of a homology hypothesis of the hooked chaetae in *Lumbrineris tetraura* and those of the other three taxa. From beginning of the hood formation a compact group of many microvilli encircles the lateral and backward part of the core (Fig. 4I), a process very dissimilar to that found in spionids, capitellids and magelonids. Only on the top and in the front of the core are the left and right group of microvilli separated, leading to a slit-like opening.

Forked chaetae

Forked chaetae apically do not taper like most other chaetae, but have two stout diverging tines, which bear spines on their inner site. All these parts in forked chaetae of *Orbinia latreillii* again are a result of the regulation of the microvilli pattern of the chaetoblast during ontogenesis (Hausam & Bartolomaeus, 2001). The investigation of different developmental stages suggests the following course of genesis. It starts with the formation of the two tines by two separated groups of microvilli. After a while individual large microvilli occur in intervals on the inner side of the tines with an angle of 20 ° to the tines. Each of the microvilli preforms one of the spines, which are composed of only one channel in outgrown chaetae. After completion of a spine the responsible microvillus changes its orientation, now running parallel to the microvilli of the tines but not fusing with them. All microvilli then converge and the chaetal shaft is built. The shaft expands in the fronto–chaudal axis by addition of further microvilli but decreases in width by merging of peripheral microvilli. Comparative studies on the formation of forked chaetae in other taxa like scalibregmatids, paraonids, and nephtyids will be worthful to decide whether or not these chaetae are homologous between the different taxa.

Chrvsopetalid paleal chaetae

The paleal notochaetae in *Chrysopetalum* sp., which cover the dorsum of the animals, exhibit a peculiar ultrastructure (Westheide & Russel, 1992). They are composed of longitudinal hollow channels of very different diameter. The centre or medulla of the flattened blade-like part of the chaetae consists of a single row of very thick channels. They are surrounded by many very small cortical channels, which contain thin lumina or none at all. However, not only the diameter differs between cortical and medullar channels. The latter are subdivided by many thin, equidistantly arranged horizontal diaphragms. The more basal shaft of the chaetae is more or less oval in cross section and all channels of the centre are of small diameter. Running upward, some of these channels broaden and become the large channels within the blade. In these chaetae each channel is also preformed by one microvillus. During genesis of the blade, the chaetoblast gives rise to very large microvilli, responsible for the formation of the medullar channels, and small peripherally arranged microvilli for the cortical channels. The cameration by the diaphragms may be due to a discontinuous rhythmic apposition of chaetal material. The systematic consequences of the similarities between the paleae of chrysopetalids and presumably camerated covering structures of some Cambrian fossils like *Wiwaxia corrugata* are still under discussion. Either *Wiwaxia* is favoured to be a closer relative of chrysopetalids or aphroditids within the Phyllodocida (Butterfield, 1990; Bartolomaeus et al., 1997), or the resemblance is taken as a hint that annelidan ancestors might have been covered by similar structures, which were lost during the evolution of annelids (Westheide, 1997).

Acicula and supportive chaetae

Acicula are chaetae with a special function. While other chaetae protrude to a large extent over the body surface and are used in its contact with the

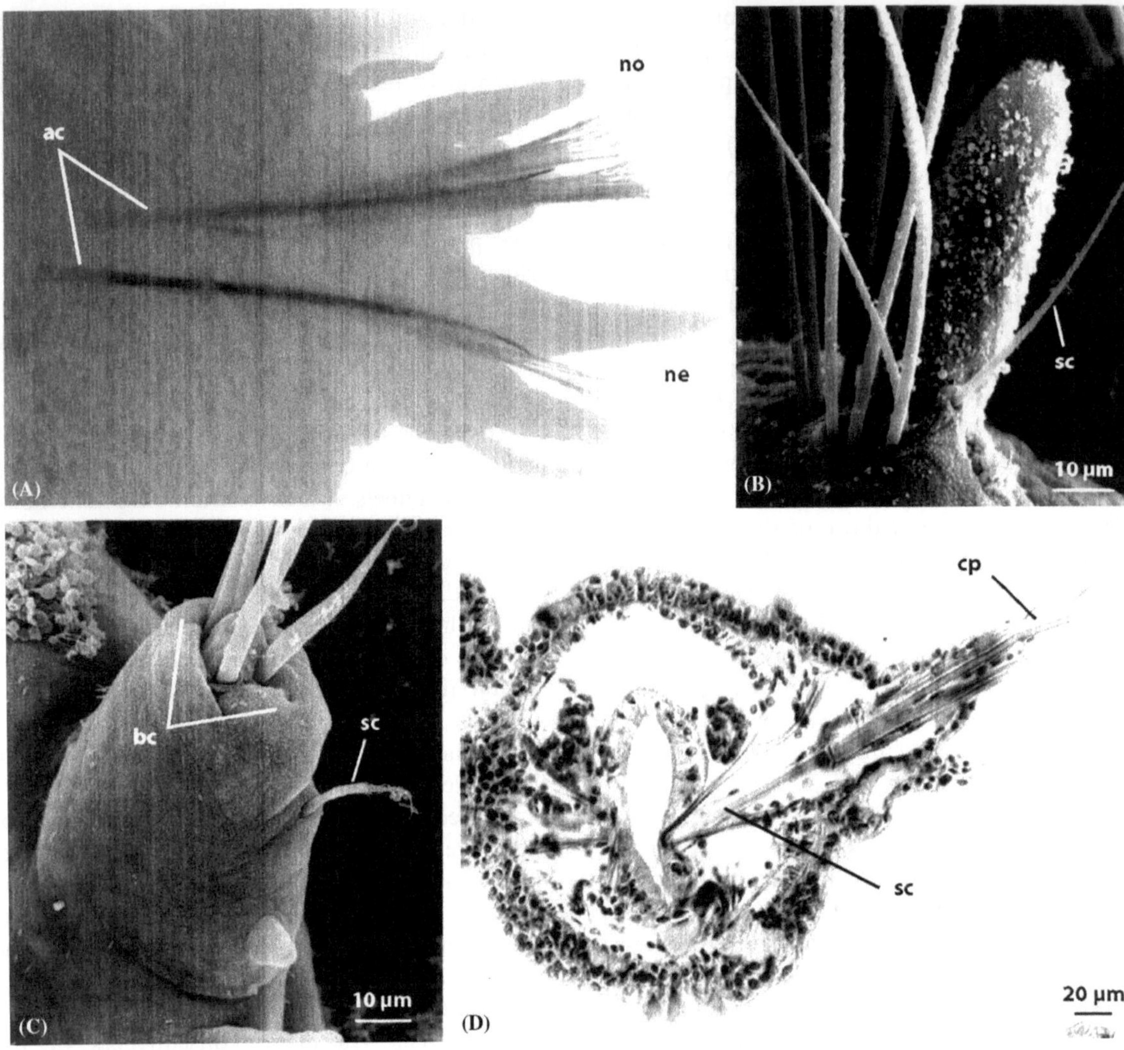

Figure 6. (A) *Nereis diversicolor*. Acicula (ac) within a parapodium. (B) *Proscoloplos cygnochaetus*. Notopodial supportive chaeta (sc) protruding far above bodysurface. (C,D) *Apistobranchus typicus*. (C) tip of neuropodial supportive chaeta (sc) lays apart of the chaetal bundle (bc). (D) Neuropodial supportive chaeta (sc) reaches far down into the body cavity. Cp: capillary chaeta; ne: otopodium; no: notopodium.

environment, acicula are deeply embedded within the tissue (Fig. 6A), often not perceivable from outside and acting as supportive structures of parapodia. They project far down into the body and are connected to strong muscles at their base and upper end via tonofilaments. Contraction of this musculature results in the movement of the whole parapodial ramus. Acicula are described for members of the Errantia sensu Ushakov (1955), such as Eunicida, Phyllodocida, and Amphinomida and there is no doubt about the homology of acicula within that group. But certain chaetae which function comparably as supportive structures are known from the sedentary groups Orbiniidae, Apistobranchidae, Psammodrilidae and Chaetopteridae and from Myzostomida. The supportive chaetae in the thoracic appendages of psammodrilids most probably evolved independently of the acicula of errant polychaetes. Investigations in *Psammodrilus aedificator* revealed a somewhat unusual structure in that they consist of Spirally wound channels, which are not connected to one another (Kristensen & Nørrevang, 1982). Furthermore, psammodrilids are in favour to be

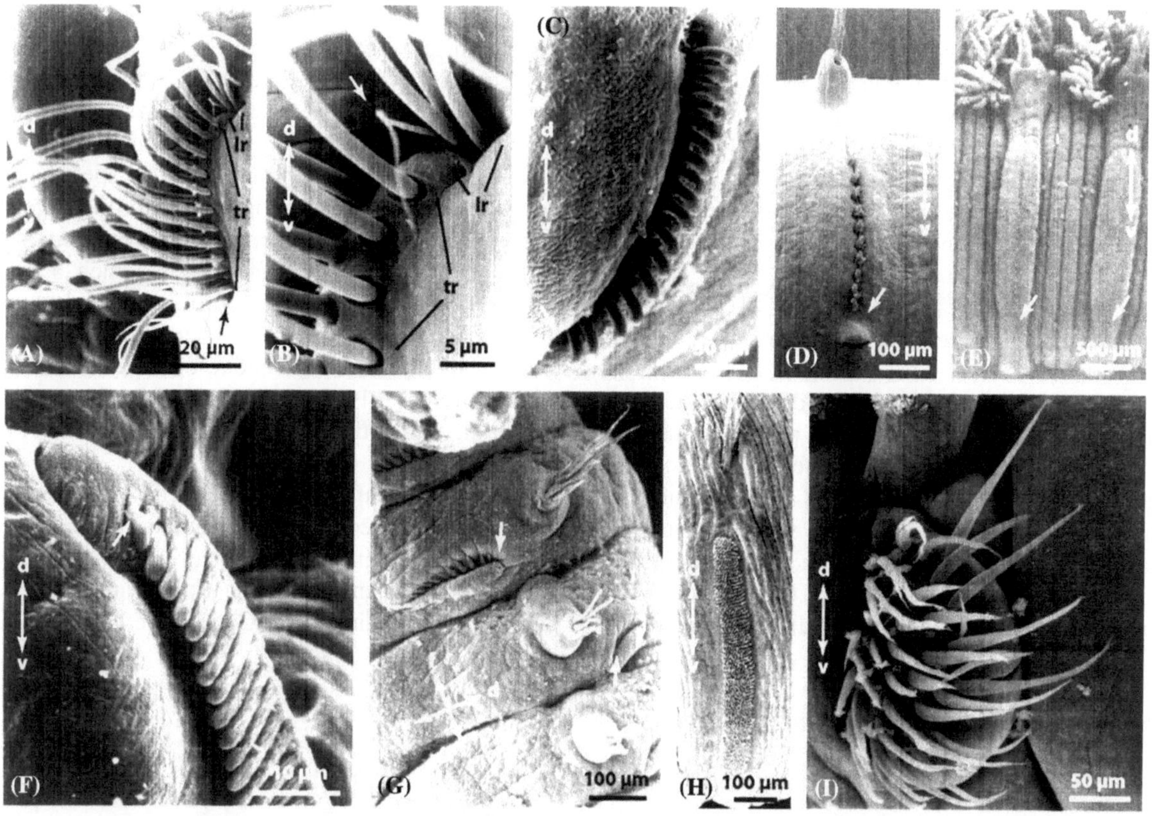

Figure 7. (A,B) *Scolelepis squamata* (Spionidae). (A) thoracacic notopodium seen from frontal with transversal and longitudinal rows of capillary chaetae. (B) Clipping from a formative site of longitudinal row. (C) *Capitella capitata* (Capitellidae). Abdominal notopodium with transversal row of hooded hooks. (D) *Axiothella isocirra* (Maldanidae). Midbody chaetiger with transversal row of hooks in neuropodium. (E) *Arenicola marina.* (Arenicolidae). Midbody chaetiger with transversal row of hooks in neuropodium. (F) *Pectinaria koreni* (Pectinariidae), dorsal part of neuropodium with transversal row of uncini. (G) *Branchiomma bombyx* (Sabellidae). Setal inversion. (H) *Owenia fusiformis* (Oweniidae). Midbody chaetiger with patch of hooks in neuropodium (I) *Apistobranchus typicus* (Apistobranchidae). Thoracic neuropodium with chaetal patch. c: caudal; f: frontal; d: dorsal; v: ventral; arrows mark formative sites.

related to several taxa without acicula or supportive chaetae (Meyer & Bartolomaeus, 1997). No matter if chaetopoterids are related to spionidan polychaetes (Rouse & Fauchald, 1997) or situated within a clade additionally composed of Sabellariidae, Terebellida, Sabellida and Pogonophora (Rouse, 2000) (and see below for a hypothesis about their position), they are also amongst groups without acicula or supportive chaetae. In the orbiniid *Proscoloplos cygnochaetus*, transmission electron microscopy revealed the ventral-most chaeta to be a supportive chaeta. Though it projects conspicuously above the body surface and resembles a short capillary, it reaches far further down than the surrounding capillary chatae (Fig. 6B) and thus differs in function. The neuropods of *Apistobranchus typicus* also show supportive chaetae with a free tip (Fig. 6C, D), whereas the notopodial supportive chatae are not visible from the outside. They are formed within a deep, apically open follicle, but growth stops before the chaetae reach the body surface (Hausen, 2001). According to Rouse & Fauchald (1997) and Rouse (2000), acicula evolved within the Polychaeta and are apomorphic for the Aciculata (same as Errantia). Westheide (1997) favours the assumption that well developed parapodia were already present in the stem species of Annelida or even in that of Articulata, but he did not indicate where aciculae evolved. It also seems conceivable that a situation like that in orbiniids and apistobranchids is an intermediate step in the evolution of acicula from normal capillary chaetae of organisms without well developed parapodia. A homology of supportive chaetae of Myzostomida to aciculae of errant polychaetes is assumed by Jägersten (1940), but the validity of this hypothesis will depend on the outcome of the discussion on the phylogenetic position of this group (see also Eeckhaut, this volume).

Arrangements of chaetae in sedentary polychaetes

Beside the ultrastructure of individual chaetae, the arrangement of chaetae also provides useful information for systematics. The assumption is that the stem species of annelids apomorphically was equipped with a dorsal and a ventral group of chaetae on the trunk segments. In many sedentary polychaetes these groups form transverse rows perpendicular to the body long axis in neuro- or notopodia or in both. Rows of chaetae in polychaetes usually have a certain turnover (Fig. 8A), because newly formed chaetae enter the row from time to time, while older chaetae are shed or resorbed (Bobin, 1935; Pilgrim, 1977; Gruet, 1986; Duchene & Bhaud, 1988). In many species the formation of new chaetae is limited to one edge of the row. Here, the epidermis repeatedly invaginates and forms a new follicle, in which a new chaeta develops while degeneration takes place on the opposite site of the row. There empty follicles occur, having lost their chaeta, or cells containing numerous large lysosomes filled with filamentous material, indicating the breakdown of an old chaeta.

Some different patterns of chaetal arrangement are known in sedentary polychaetes. Transverse rows may occur in noto- and neuropodia along the entire body, as in Spionidae, Capitellidae, Magelonidae, Cirratulidae and Cossuridae (Figs 7A, C

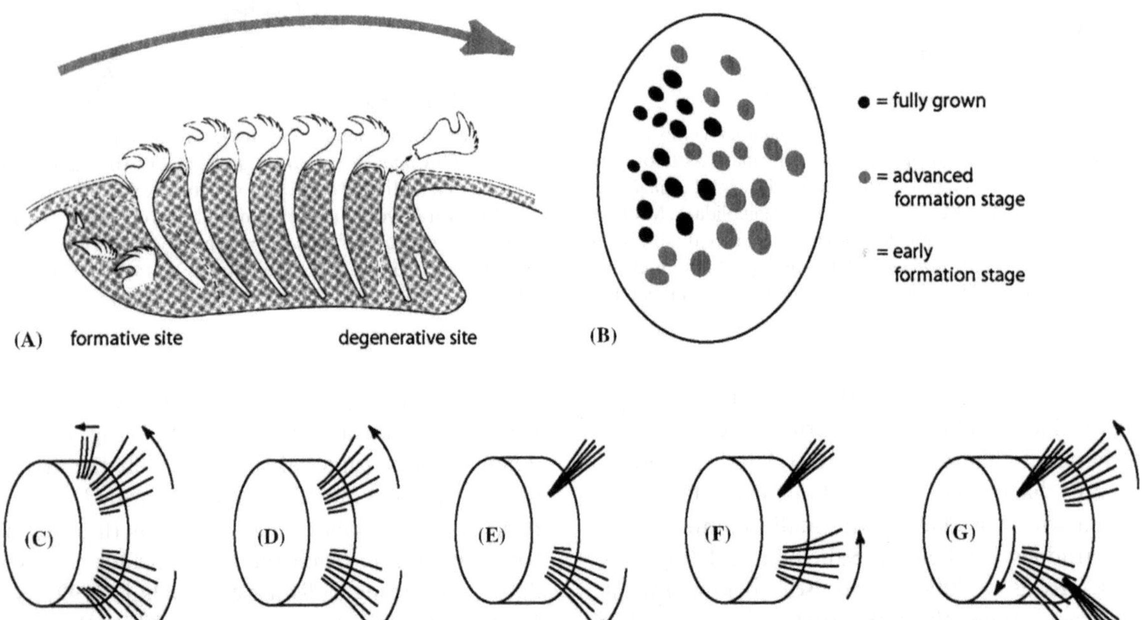

Figure 8. (A) Organization of chaetal row as present in many sedentary polychaetes with one distinct formative site. As they age, chaetae move to the degenerative site on the opposite edge of the row. (B) *Apistobranchus typicus* (Apistobranchidae). Organization of a chaetal patch of an anterior neuropodium. New chaetae are formed all along the posterior margin of the patch and move anteriorly with age. (C–F) Main types of chaetal arrangement found in sedentary polychaetes: (C) transverse rows in noto- and neuropodium with the formative sites close to the lateral midline of the body and a longitudinal row with own formative site (Spionidae, Poecilochaetidae, Trochochaetidae). (D) transverse rows in noto- and neuropodium with the formative sites close to the lateral midline of the body (Cirratulidae, Capitellidea, Magelonidae, Cossuridae). (E) Dense notopodial bundle and neuropodial transverse row with dorsal formative site (Terebellidae). (F) Dense notopodial bundle and neuropodial transverse row with ventral formative site (Arenicolidae, Maldanidae). (G) Sabellid inversion: anterior setigers as in (E) posterior setigers mirrored (Sabellidae, Serpulidae).

and 8C). The formative site of these rows lies ventrally in the notopods and dorsally in neuropods (Hausen & Bartolomaeus, 1998; Schweigkofler, et al., 1998; Radashevsky & Fauchald, 2000; Hausen, 2001). The same holds true for anterior setigers in the poecilochaetid *Poecilochaetus serpens*, anterior notopodia and at least the most anterior neuropodia in the trochochaetid *Trochochaeta multisetosum*. In other taxa, transverse rows occur in neuropodia, but notopodia bear a more condensed, bundled-like arrangement of chaetae. This situation is present in Arenicolidae, Maldanidae, Psammodrilidae, and Terebellida (Fig. 7D–F), but it is realized in two different ways. In psammodrilids and terebellids the formative site of the neuropodial rows lies at the dorsal edge. (Bhaud, 1988; Bartolomaeus, 1995; Meyer & Bartolomaeus, 1997) (and see Figs 7F and 8E), while it is at the ventral edge in arenicolids and maldanids (Bartolomaeus & Meyer, 1997) (and see Figs 7D, E and 8F). In Sabellidae and Serpulidae, thoracic notopods bear compact arrangements of chaetae, whereas thoracic neuropods bear transverse rows similar to the condition in Terebellida. Due to the sabellid inversion the abdominal notopods instead show transverse rows and abdominal neuropods resemble thoracic notopods in Sabellidae (Figs 7G and 8G). Within numerous Sabellidae and Serpulidae new chaetae are shown to enter the transverse rows dorsally in the thorax and ventrally in the abdomen (Bartolomaeus, 1995, 2002).

Oweniidae show a deviation from this pattern. They are known for their chaetal patches in the neuropods (Fig. 7H) and in adult *Owenia fusiformis* new chaetae are added dorsally along the caudal edge of the patch (Meyer & Bartolomaeus, 1996). This situation is hardly comparable with the formative site in the rows of other taxa. But newly metamorphosed stages of *Owenia fusiformis* bear single rows in neuropodia with a dorsal formative site. Thus the patches of adult specimens are interpreted as modified transverse rows.

In some other taxa the formation is not restricted to a certain edge of the transverse rows. In the chaetopterid *Telepsavus costarum* new chaetae can be formed anywhere within the rows (Bartolomaeus, personal and communication). The same is true for abdominal noto- and neuropodia in *Magelona alleni*; a derived condition, because other magelonids and groups that favoured to belong to the nearest relatives like Spioinidae show a well-defined formative site. Chaetopterids are presumably also related to groups with a well-defined formative site in the transverse rows. Obturate pogonophorans show transverse rows in the anterior segments of the opisthosoma, whereas in Perviata the girdles of the trunk are transverse rows of chaetae. Up to now there is no description of a restricted formative site in pogonophoran chaetal rows. Instead, Schulze (2001) found developmental stages of chaetae on different locations in the obturate *Ridgeia piscesae*. More investigations seem to be necessary for a decision on whether this is a derived condition within Pogonophora or apomorphic for the whole group.

In some taxa noto- and neuropodia not only bear transverse rows. So in all setigers of some spionids and in certain setigers of poecilochaetids and trochochaetids a second formative site in each ramus of the parapodia is known (Hausen & Bartolomaeus, 1998; Hausen, 2001). It generates short rows of capillary chaetae, which are aligned parallel to the body long axis and are thus longitudinal rows (Figs 7B and 8A). They are situated ventrally below the transverse rows of the neuropodium and dorsally above the transverse rows of the notopodium. New chaetae are formed caudally in the longitudinal row and push older chaetae frontally. Presumably the distinct chaetal groups dorsal of the notopodia and ventral of the neuropodia, which are described in further spionid species (Radashevsky & Fauchald, 2000), are generated by the same formative sites. Its existence in Spionidae, Trochochaetidae and Poecilochaetidae is clearly a derived condition.

The chaetal patches in frontal neuropodia of apistobranchids markedly differ from spionidan rows (Fig. 7I). New chaetae are formed all along the posterior line of the patches in *Apistobranchus typicus* and growing chaetae move anterior (Fig. 8B). Because posterior neuropodia only bear bundles of capillaries and notopodia completely lack chaetae that reach above the body surface, apistobranchids nowhere show transverse rows or longitudinal rows like spionids. Chaetation thus provides no argument for a close relationship of apistobranchids and spionids like that hypothesized by Rouse & Fauchald (1997).

Coding issues

Though the ultrastructure of individual chaetae and the arrangement of chaetae obviously provide a profound base for homology hypothesis, the different structures have to be checked for independence before use in systematics. For individual chaetae the data seem to justify 'subchaetal units' for character coding. For instance, a homology is to be assumed between hooks that can possess or lack a hood structure, a rostrum, a subrostral process, basal processes, a beard, and a long or short shaft. Despite the fact that genesis will not work without forming a shaft, no further dependencies between the substructures of hooks are to be recognized. Thus hooks can be taken as a whole character set with a large range of logical independence.

Furthermore, it has to be asked whether the type of chaetae is independent of its arrangement. Evidence for that is found in spionids. *Scolelepis squamata* exclusively bears capillary chaetae in the transverse rows of frontal noto- and neuropodia forming alternating double rows. The formative site lies dorsally in the neuropodia and ventrally in the notopodia, as it does in transverse rows of posterior parapodia. But here each second chaeta that is formed is a hooded hook and the capillary chaetae are shed very early or are resorbed (Hausen & Bartolomaeus, 1998). But the principle alignment in forming a transverse alternating double row is the same as in frontal setigers. Further in spionids hooks start more posterior in older than in young individuals (Hannerz, 1956; Mackie, 1984, 1990; Sigvaldadóttir & Mackie, 1993; Bochert & Bick, 1995). This means that from time to time the frontal-most setiger with hooded hooks loses all these chaetae and transforms into a setiger that bears only capillary chaetae, like the frontal setigers. But the alignment pattern persists. The main event is a change in the information on which chaetae have to be expressed in the formative site. This indicates an independence of alignment pattern and type of chaetae that realize the alignment. One piece of information is where to start the formation process of new chaeta, to which direction aging chaetae shift and where they degenerate. Another piece is which type of chaetae has to be formed within a certain chaetal follicle.

Further, of course, the exact elaboration of the transverse rows can change between species So transverse single, double or triple rows are realized in sedentary polychaetes. But these variations can also occur between different body parts within a specimen. Thus the principle decision to build up transverse rows seems to be independent of how exactly the rows have to appear. Last, of course, the decision where a certain type of chaetae is expressed provides a further character, though it has to be thoroughly demonstrated, whether changes during ontogeny occur, allowing only the comparison of equal developmental stages.

Investigations on chaetae in combination with other data led to certain systematic hypotheses, like a positioning of the Pogonophora within a clade also containing Terebellida, Sabellidae, and Serpulidae (Bartolomaeus, 1995). The hook-bearing Oweniidae, Maldanidae, Arenicolidae, and Psammodrilidae are thought to be the next relatives (Bartolomaeus, 1998) and a closer relationship of Capitellidae, Spionidae, and Magelonidae to all these taxa is favoured (Hausen & Bartolomaeus, 1998; Schweigkofler, et al., 1998; Hausen, 2001). The assumption is that other types of chaetae provide similar useful characteristics.

Acknowledgements

My thanks are due to T. Bartolomaeus and two anonymous referees. The study was financially supported by a grant of the Deutsche Forschungsgemeinschaft (BA 1520/2).

References

Bartolomaeus, T., 1992. On the ultrastructure of the cuticle, the epidermis and the gills of *Sternaspis scutata* (Annelida). Microfauna Marina 7: 237–252.

Bartolomaeus, T., 1995. Structure and formation of the uncini in *Pectinaria koreni, Pectinaria auricoma* (Terebellida) and *Spirorbis spriorbis* (Sabellida): implications for annelid phylogeny and the position of the Pogonophora. Zoomorphology 115: 161–177.

Bartolomaeus, T., 1998. Chaetogenesis in polychaetous Annelida – significance for annelid systematics and the position of the Pogonophora. Zoology 100: 348–64.

Bartolomaeus, T., 2002. Structure and formation of thoracic and abdominal uncini in *Fabricia stellaris* (Müller, 1774) – implication for the evolution of Sabellida (Annelida). Zoologischer Anzeiger 241: 1–17.

Bartolomaeus, T. & K. Meyer, 1997. Developmental and phylogenetic significance of hooked setae in Arenicolidae (Polychaeta, Annelida). Invertebrate Biology 116(3): 227–242.

Bartolomaeus, T., K.Meyer, H. Hausen, M. Schweigkofler & S. Schulz, 1997. Die Annelid-Borste–Morphogenese und Bedeutung für die Systematik (Teil I). Mikrokosmos 86(2): 91–98.

Bhaud, M., 1988. Change in setal pattern during early development of *Eupolymnnia nebulosa* (Polychaeta: Terebellidae) grown in simulated natural conditions. Journal of the Marine Biological Association of the United Kingdom 68: 677–687.

Bobin, G., 1935 Étude des soies simples et de leurs bulbes sétigères chez deux sabelliens.Travaux de la Station Biologique de Roscoff 13; 157–175.

Bobin, G., & H. Mazqué, 1944. Topographie, histologie, caractères physiques et chimiques des *soies d'Aphroditte aculeata* L. (Annèlide Polychète). Bulletin de la Société Zoologique de France 68: 125–134.

Bochert, R. & A. Bick, 1995. Reproduction and larval development of *Marenzelleria viridis* (Polychaeta: Spionidae). Marine Biology 123: 763–773.

Bouligand, Y., 1966. Sur l'ultrastructure des soies et des cellules associées chez une Annélde Polychète *Haplosyllis depressa*. Comptes rendus de l'Académie des Sciences Série D 263: 378–381.

Bouligand, Y., 1967. Les soies et les cellules associées chez deux Annélides Polychètes. Zeitschrift für Zellforschung 79: 332–363.

Brocco, S. L., R. M. O'Clair & R. A. Cloney, 1974. Cephalopod integument: the ultrastructure of Kölliker's organs and their relationship to setae. Cell and Tissue Research 151; 293–308.

Butterfield, N. J., 1990. A reassesment of the enigmatic Burgess Shale fossil *Wiwaxia corrugata* (Matthew) and its relationship to the polychaete *Canadia spinosa* Walcott. Paleontology 16: 287–303.

Chapman, P. M. & R. O. Brinkhurst, 1987. Hair today, gone tomorrow: induced chaetal changes in tubificid oligochaetes. Hydrobiologia 155: 45–55.

Duchene, J. & M. Bhaud, 1988. Uncinal patterns and age determination in terebellid polychaetes. Marine Ecology Progress Series 49: 267–275.

Fitzhugh, K., 1991. Uncini and the anchor-function hypothesis: empirical tests and causal explanations. Bulletin of Marine Science 48(2): 574–583.

Foster, N. M., 1971. Notes on spionid (Polychaeta) setae. Transactions of the American Microscopical Society 90(1): 34–42.

George, J. D. & E. C. Southward, 1973. A comparative study of the setae of Pogonophora and polychaetous Annelida. Journal of the Marine Biological Association of the United Kingdom 53: 403–424.

Gordo, D. P., 1975. The resemblance of bryozoan gizzard teeth to "annelid-like" setae. Acta Zoologica (Stockholm) 56: 283–289.

Gruet, Y., 1986. Premières observations sur la chute des soies operculaires ou palées chez l'Annèlide Polychète Sabellariidé *Sabellaria alveolata* (Linné). Comptes rendus de l'Academie des Sciences – Série III 302(10): 375–378.

Gustus, R. M. & R. A. Cloney, 1973. Ultrastructure of the larval compound setae of the polychaete *Nereis vexillosa* Grube. Journal of Morphology 140: 355–366.

Hannerz, L., 1956. Larval development of the polychaete families Spionidae Sars, Disomidae Mesnil, and Poecilochaetidae n. fam. in the Gullmar Fjord (Sweden). Zoologiska Bidrag från Uppsala 31: 1–204.

Hausam, B. & T. Bartolomaeus, 2001. Ultrastructure and development of forked and capillary setae in the polychaetes *Orbinia bioreti and Orbinia latreillii* (Annelida: Orbiniidae). Invertebrate Biology 120(1): 13–28.

Hausen, H., 2001. Untersuchungen zur Phylogenie "spiomorpher" Polychaeten (Annelida). Ph.D. thesis, Universität Bielefeld. Logos Verlag, Berlin, 145 pp.

Hausen, H. & T. Bartolomaeus, 1998. Setal structure and chaetogenesis in *Scolelepis squamata* and *Malacoceros fuliginosus* (Spionidae, Annelida). Acta Zoologica (Stockholm) 79(3): 149–161.

Hessling, R. & W. Westheide, 2002. Are Echiura derived from a segmented ancestor? Immunhistochemical analysis of the nervous system in developmental stages of *Bonellia viridis*. Journal of Morphology 252: 100–113.

Hillger, K. A. & D. J. Reish, 1970. The effect of temperature on the setal characteristics in Polynoidae (Annelidae: Polychaeta). Bulletin of the Southern California Academy of Sciences 69(2): 87–99.

Jägersten, G., 1940. Zur Kenntnis der Morphologie, Entwicklung und Taxonomie der Myzostomida. Nova Acta Regiae Societatis Scientarium Uppsaliensis. Ser. IV 11(8): 1–84.

Jamieson, B. G. M., 1992. Oligochaeta. In: Harrison, F. W. & S. L. Gardiner (eds), Annelida. Microscopic Anatomy of Invertebrates, Vol. 7. Wiley-Liss, New York: Inc. 217–322.

Jeuniaux, C., 1971. Chitinous structures. In: Florkin, M H. & E. H. Solz (eds), Extracellular and Supporting Structures. Comprehensive Biochemistry 26: 595–632.

Kristensen, R. M. & A. Nørrevang, 1982. Description of *Psammodrilus aedificator* sp.n. (Polychaeta), with notes on the arctic interstitial fauna of Disko Island, W. Greenland. Zoologica Scripta 11(4): 265–279.

Kryvi, H., 1989. External aspects of the setae of *Sabella penicillus* L. (Polychaeta). Sarsia 74: 269–276.

Kryvi, H. & T. Sørvig, 1990. Internal organization of limbate polychaete setae (*Sabella penicillus*), with notes on bending stiffness. Acta Zoologica (Stockholm) 71(1): 25–31.

Leise, E. M. & R. A. Cloney, (1982). Chiton integument: ultrastructure of the sensory hairs of *Mopalia mucosa* (Mollusca: Polyplacophora). Cell and Tissue Research 223: 43–59.

Lippert, W. & K. Gentil, 1963. Über den Feinbau der Schillerhaare des Polychaeten *Aphrodite aculeata* L. Zeitschrift für Morphologie und Ökologie der Tiere 53: 22–28.

Lotmar, W. & L. E. R. Picken, 1950. A new crystallographic modification of chitin and its distribution. Experientia 6(2): 58–59.

Lüter, C., 2000. Ultrastructure of larval and adult setae of Brachiopoda. Zoologischer Anzeiger 239: 75–90.

Lüter, C. & T. Bartolomaeus, 1997. The phylogenetic position of Brachiopada – a comparison of morphological and molecular data. Zoologica Scripta 26(3): 245–253.

Mackie, A. S. Y., 1984. On the identity and zoogeography of *Prionospio cirrifera* Wirén, 1883 and *Prionospio multibranchiata* Berkeley, 1927 (Polychaetae: Spionidae). In: Hutchings, P. A. (ed.), Proceedings of the First International Polychaete Conference. The Linnean Society of New South Wales, Sydney: 35–47.

Mackie, A. S. Y., 1990. *Prionospio saccifera* sp. nov. (Polychaeta: Spionidae) from Hong Kong and the Red Sea, with a redescription of *Prionospio ehlersi* Fauvel, 1928. In: Morton, B. (ed.), Proceedings of the Second International Marine Biological Workshop: The Marine Flora and Fauna of Hong Kong and Southern China, Hong Kong, 1986. Hong Kong University Press, Hong Kong: 363–375.

Merz, R. A. & S. A. Woodin 1991. The stiffness of capillary setae: a comparison among sedentariate polychaetes. Ophelia Supplement 5: 615–623.

Merz, R. A. & S. A. Woodin, 2000. Hooked setae: tests of the anchor hypothesis. Invertebrate Biology 119(1): 67–82.

Meyer, K. & T. Bartolomaeus, 1996. Ultrastructure and formation of the hooked setae in *Owenia fusiformis* delle Chiaje, 1842: implications for annelid phylogeny. Canadian Journal of Zoology 74: 2143–2153.

Meyer, R. & T. Bartolomaeus, 1997. Ultrastruktur und Morphogenese der Hakenborsten bei *Psammodrilus balanogolssiodes* – Bedeutung für die Stellung der Psammodrilida (Annelida). Microfauna Marinia 11: 87–113.

Michel, C. & J. Volvelle, 1973. Tannage quinonique des soies chez *Syllis cornuta* Rathke. Bulletin de la Société Zoologique de France 99(1): 175–181.

Nilsen, R. & T. Holthe, 1985. Arctic and scandinavian Oweniidae (Polychaeta) with a description of *Myriochele fragilis* sp.n., and comments on the phylogeny of the family. Sarsia 70: 17–32.

O'Clair, R. M. & R. A. Cloney, 1974. Patterns of morphogenesis mediated by dynamic microvilli: chaetogenesis in *Nereis vexillosa.* Cell and Tissue Research 151: 141–157.

Ohwada, T. & Y. Nishino, 1991. Morphological variations hooded hooks in *Prionospio japonica* (Polychaeta, Spionidae). Bulletin of Marine Science 48(2): 246–250.

Orrhage, L., 1971. Light and electron microscope studies of some annelid setae. Acta Zoologica (Stockholm) 52: 157–169.

Orrhage, L., 1973. Light and electron microscope studies of some brachiopd and pogonophoran setae. Zeitschrift für Morphologie der Tiere 74: 253–270.

Picken, L. E. R., 1950. Oriented protein in chitinous structures. Nature 165: 599–600.

Pilgrim, M., 1977. The functional morphology and possible taxonomic significance of the parapodia of the maldanid polychaetes *Clymenella torquata and Euclymene oerstedi.* Journal of Morphology 152: 281–302.

Purschke, G., R. Hessling & W. Westheide, 2000. The phylogenetic position of the Clitellata and the Echiura – on the problematic assessment, of absent characters. Journal of Zoological Systematics and Evolutionary Research 38: 165–173.

Radashevsky, V. I. & K. Fauchald, 2000. Chaetal arrangement and homology in spionids (Polychaeta: Spionidae). Bulletin of Marine Science 67(1): 13–23.

Read, G. B., 1986. New deep-sea Poecilochaetidae (Poecilochaeta: Spionida) from New Zealand. Journal of Natural History 20: 399–413.

Rouse, G., 2000. Bias? What bias? The evolution of downstream larval-feeding in animals. Zoologica Scripta 29: 213–236.

Rouse, G. & K. Fauchald, 1997. Cladistics and polychaetes. Zoologica Scripta 26: 139–204.

Rudall, K. M., 1963. The chitin/protein complexes of insect cuticles. Advances in Insect Physiology 1: 247–313.

Scherf, H., 1970. Struktur und Insertion der Paleen von *Pectinaria koreni* Malmgr. Zoologische Jahrbücher Abteilung für Anatomie und Ontogenie der Tiere 87: 386–401.

Schroeder, P. C., 1984. Chaetae. In: Bereiter-Hahn, J., A. G. Matoltsy & K. S. Richards (eds), Biology of the Integument. 1. Invertebrates. Springer, Berlin: 297–309.

Schulze, A., 2001. Ultrastructure of ophistosomal chaetae in Vestimentifera (Pogonophora, Obturata) and implications for phylogeny. Acta Zoologica (Stockholm) 82: 127–135.

Schweigkofler, M., T. Bartolomaeus & L. von Salvini-Plawen, 1998. Ultrastructure and formation of hooded hooks in *Capitella capitata* (Capitellida, Annelida). Zoomorphology 118: 11–28.

Sigvaldadóttir, E. & A. S. Y. Mackie, 1993. *Prionospio steenstrupi, P. fallax and P. dubia* (Polychaeta, Spionidae): re-evaluation of identity and status. Sarsia 78: 203–219.

Smith, M. E., 1985. Setal morphology and its intraspecific variation in *Dero digitata* and *Dero nivea* (Oligochaeta: Naididae). Transactions of the American Microscopical Society 104(1): 45–51.

Specht, A. & W. Westheide, 1988. Intra- and inter-specific ultrastructural character variation: the chaetation of the *Microphtalmus listensis* species group (Polychaeta, Hesionidae). Zoomorphology 107: 371–376.

Storch, V., 1984. Echiura and Sipuncula. In: Bereiter-Hahn, J., A. G. Matoltsy & K. S. Richards (eds), Biology of the Integument. 1. Invertebrates. Springer, Berlin: 368–375.

Storch, V. & U. Welsch, 1972. Über Bau und Entstehung der Mantelrandstacheln von *Lingula unguis* L. (Brachiopoda). Zeitschrift für Wissenschaftliche Zoologie 183(3/4): 181–189.

Westheide, W., 1997. The direction of evolution within the Polychaeta. Journal of Natural History 31: 1–15.

Westheide, W. & C. W. Russel, 1992. Ultrastructure of chyrsopetalid paleal chaetae (Annelida, Polychaeta). Acta Zoologica (Stockholm) 73(3): 197–202.

Woodin, S. A. & R. A. Merz., 1987. Holding on by their hooks: anchors for the worms. Evolution 41(2): 421–432.

Zavarzin, A. & T. V. Sergovskaya, 1979. The application of radioautography to analyzing the mechanisms of setal growth in polychaetes. Tsitologiya 21(4): 419–423.

Hydrobiologia (2005) **535/536**: 53–78
T. Bartolomaeus & G. Purschke (eds), Morphology, Molecules, Evolution and Phylogeny in Polychaeta and Related Taxa

Sense organs in polychaetes (Annelida)

Günter Purschke
Spezielle Zoologie, Fachbereich Biologie/Chemie, Universität Osnabrück, D-49069 Osnabrück, Germany
E-mail: Purschke@biologie.uni-osnabrueck.de

Key words: sensory cells, nuchal organs, eyes, photoreceptor-like organs, antennae, palps, dorsal organs, lateral organs

Abstract

Polychaetes possess a wide range of sensory structures. These form sense organs of several kinds, including the appendages of the head region (palps, antennae, tentacular cirri), the appendages of the trunk region and pygidium (parapodial and pygidial cirri), the nuchal organs, the dorsal organs, the lateral organs, the eyes, the photoreceptor-like sense organs, the statocysts, various kinds of pharyngeal papillae as well as structurally peculiar sensory organs of still unknown function and the apical organs of trochophore larvae. Moreover, isolated or clustered sensory cells not obviously associated with other cell types are distributed all over the body. Whereas nuchal organs are typical for polychaetes and are lacking only in a few species, all other kinds of sensory organs are restricted to certain groups of taxa or species. Some have only been described in single species till now. Sensory cells are generally bipolar sensory cells and their cell bodies are either located peripherally within the epidermis or within the central nervous system. These sensory cells are usually ciliated and different types can be disinguished. Structure, function and phylogenetic importance of the sensory structures observed in polychaetes so far are reviewed. For evaluation of the relationships of the higher taxa in Annelida palps, nuchal organs and pigmented ocelli appear to be of special importance.

Introduction

Polychaetes respond to a variety of sensory stimuli and consequently possess a wide range of sensory structures (Mill, 1978; Welsch et al., 1984). These form sense organs of several kinds, including (1) the appendages of the head region, namely palps, antennae, and tentacular cirri, (2) appendages of the trunk region and pygidium, the parapodial cirri and pygidial cirri, (3) the nuchal organs, (4) the dorsal organs, (5) the lateral organs, (6) the eyes, (7) the photoreceptor-like sense organs, (8) the statocysts, (9) various kinds of pharyngeal papillae and (10) the apical organs present in trochophore larvae. Moreover, isolated or clustered sensory cells not obviously associated with other cell types are distributed throughout the body.

Whereas nuchal organs and eyes are present in most polychaete species, all other kinds of sensory structures are restricted to certain higher taxa, to a certain group of closely related species or have, so far, only been found in single species. In polychaetes sensory structures have been reviewed by Bullock (1965), Mill (1978), Verger-Bocquet (1984, 1992), Eakin & Hermans (1988) and Storch & Schlötzer-Schrehardt (1988), those of Clitellata by Jamieson (1981, 1992), Sawyer (1986) and Fernandez et al. (1992).

A major problem is still to determine the sensory stimuli which the sensory structures mediate (Mill, 1978; Schlawny et al., 1991). Usually, responses after application of various chemical, optical or tactile stimuli have been obtained by electrophysiological recordings from nerves rather than individual receptor cells. Thus, function has mostly been deduced from structural correspondences to sensory cells for which the function has already been determined (e.g., Mill, 1978; Jouin et al., 1985). A promising approach to address these problems might be immunological labelling of specific receptor neurons (Michel et al., 1999).

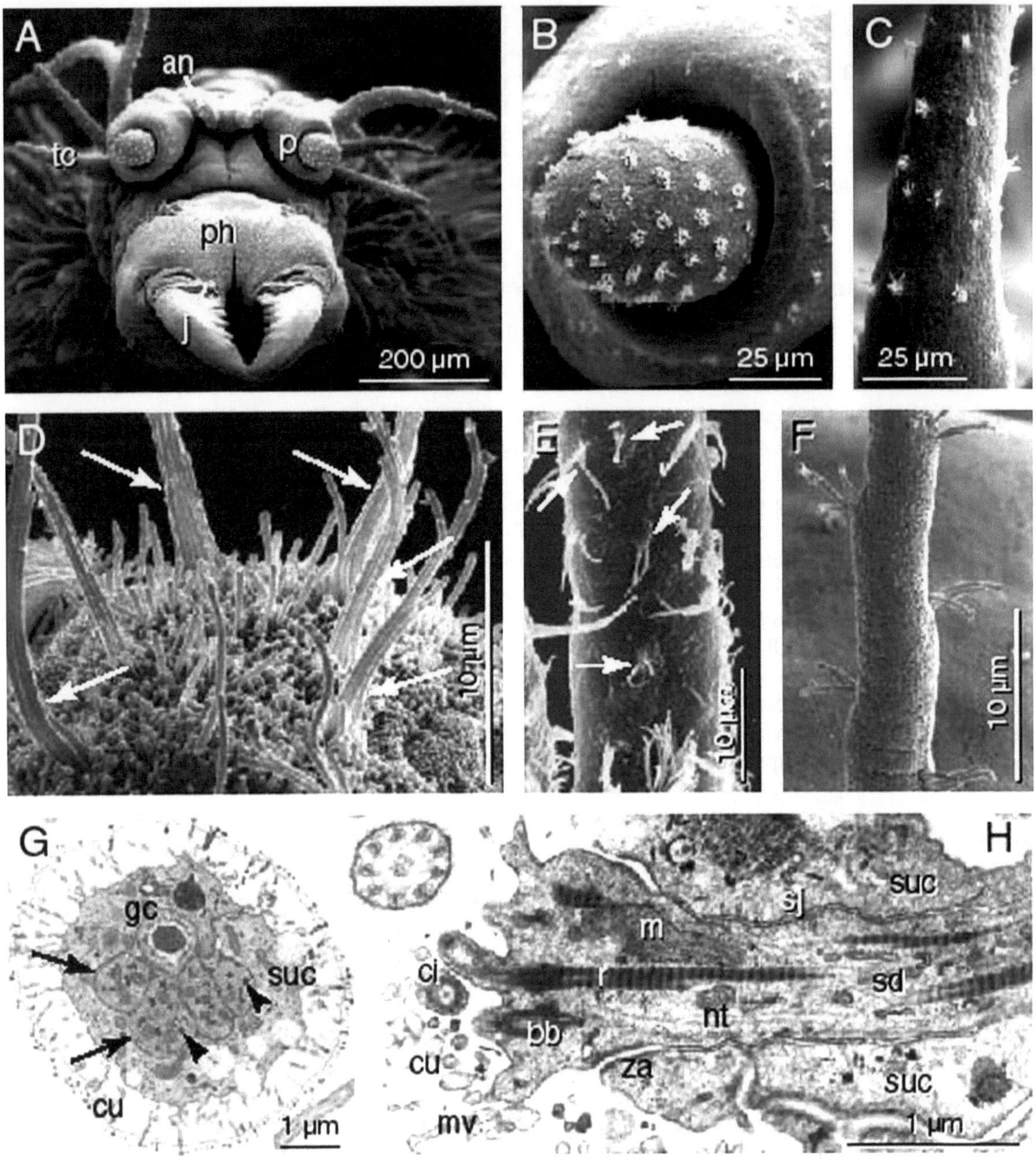

Figure 1. Multiciliate penetrative sensory cells. (A–C) *Nereis* sp. (A) Frontal view of anterior end with antennae (an), palps (p) and tentacular cirri (tc); j jaws, ph pharynx, (B) Tip of biarticulated palp with numerous sensory cilia. (C) Enlargement of first ventral tentacular cirrus. (D) *Protodrilus ciliatus*; tip of prostomium with various groups of different sensory cilia, some of which form cirri (arrows). (E) *Protodriloides symbioticus*; ventral view of palp with different kinds of sensory cells (arrows). (F) *Microphthalmus listensis*; tentacular cirrus with similar groups of sensory cilia. (G, H) *Polydora commensalis*; sensory papilla on palp. (G) Cross section through papilla with four sensory dendrites (arrows) and a gland cell neck (gc) surrounded by a single supporting cell (suc). In the dendrites 3–7 rootlets (arrowheads) are visible. (H) Longitudinal section of sensory dendrite (sd) with basal body (bb) of cilia (ci), rootlets (r), neurotubules (nt), m, mitochondrion; mv, microvillus; sj, septate junction; za, zonula adhaerens. Inset: cross section of cilium with $9 \times 2 + 2$ axoneme lacking dynein arms. D, E modified from Purschke (1993), G, H Purschke & Dauer (unpubl.). Micrographs A–C: S. Raabe.

To date the function of many sensory structures found in polychaetes is still uncertain or completely unknown.

Sensory cells

Sensory cells or receptor cells are generally bipolar primary sensory cells, the cell bodies of which are either located peripherally within the epidermis or lie within the central nervous system (Storch & Schlötzer-Schrehardt, 1988; Verger-Bocquet, 1992). As a result, there are differences in the degree of development of their dendritic processes. The peripheral dendritic processes are normally embedded in the respective epithelium, reach the epithelial surface and thus are connected to the adjoining cells by typical junctional complexes, i.e. a zonulae adhaerens followed by a septate junction (Fig. 1H). Generally the dendritic processes are ciliated and may bear a number of additional microvilli. With the exception of photoreceptor cells, the only known instance of sensory cells with microvilli alone is that reported by Dorsett & Hyde (1969) for the prostomial appendages of *Nereis diversicolor*.

A given type of sensory cell may occur in isolation, clustered, or in sense organs. Clustered sensory cells form small buds or papillae, generally composed of a few receptor cells; however, up to 16 cells have been found in sensory papillae in the caudal body region of *Arenicola marina* and up to 200 cells on the palps of *Lycastis terrestris* (see Storch, 1972; Jouin et al., 1985). Many species possess comparatively high numbers of sensory cells (Storch & Schlötzer-Schrehardt, 1988 for ref.). Since they respond to various sensory stimuli, the sensory cells differ structurally within individuals and between species. Even in the smallest polychaete known, the dwarf male of *Dinophilus gyrociliatus*, with a body length of 50 μm, no less than 40 out of a total of 68 neurons are sensory; these can be assigned morphologically to four different types (Windoffer & Westheide, 1988). This diversity is almost in the same range as has been observed in larger species (e.g., Schlawny et al., 1991; Jamieson, 1992; Purschke 1993, 1999). In spite of fine structural variations, sensory cells may be classified by the number of cilia and whether these cilia penetrate the cuticle or not, or whether they are intraepithelial (e.g., Welsch et al., 1984; Jamieson, 1992): (1) multiciliate penetrative sensory cells, (2) uniciliate penetrative sensory cells, (3) multiciliate non-penetrative sensory cells, (4) uniciliate non-penetrative sensory cells, (5) basal ciliated sensory cells. Only the first two types are externally visible (Figs 1A–F, 2A). However, care must be taken, as there are other functions of external ciliation besides detection of sensory stimuli and not every cilium, tuft of cilia or ciliary band visible in the light or scanning electron microscope is necessarily sensory or even part of a sense organ (Figs 4D, G–I, 6A, B).

Multiciliate penetrative sensory cells

Multiciliate sensory cells of this type are the most abundant of all annelid sensory cells (Fig. 1A–H). These cells differ greatly in terms of number and length of cilia; these may appear to be linked to each other to form cirri (Fig. 1D), although this could not be confirmed with transmission electron microscopy (e.g., Purschke, 1993). Sometimes different types are situated close together as, for example, on the prostomium or the palps in species of Protodrilida (Fig. 1D, E). In other species such receptor cells appear more uniform (Fig. 1B, C, F). These sensory cells are also extremely variable in other respects: length and number of additional microvilli, presence or absence of ciliary rootlets, structure of these rootlets, additional cytoskeletal elements and other fine-structural features (Fig. 1G, H). The number of cilia per cell ranges from 2 to more than 20 (Bantz & Michel, 1972; Michel, 1972; Storch & Schlötzer-Schrehardt, 1988; Schlawny et al., 1991; Verger-Bocquet, 1992; Purschke, 1993, 1999; Purschke & Jouin-Toulmond, 1994; Böggemann et al., 2000; Hessling & Purschke, 2000; Purschke & Hessling, 2002). Although mostly equipped with an axoneme showing the typical $9 \times 2 + 2$ pattern of microtubules (Fig. 1H inset), often associated with dynein arms, in many cases cilia of these cells appear to be more or less immobile (Jouin et al., 1985; Amieva et al., 1987; Purschke, 1993). It is generally assumed that these structurally different sensory cells have different functions (Toulmond et al., 1984; Jouin et al., 1985; Schlawny et al., 1991).

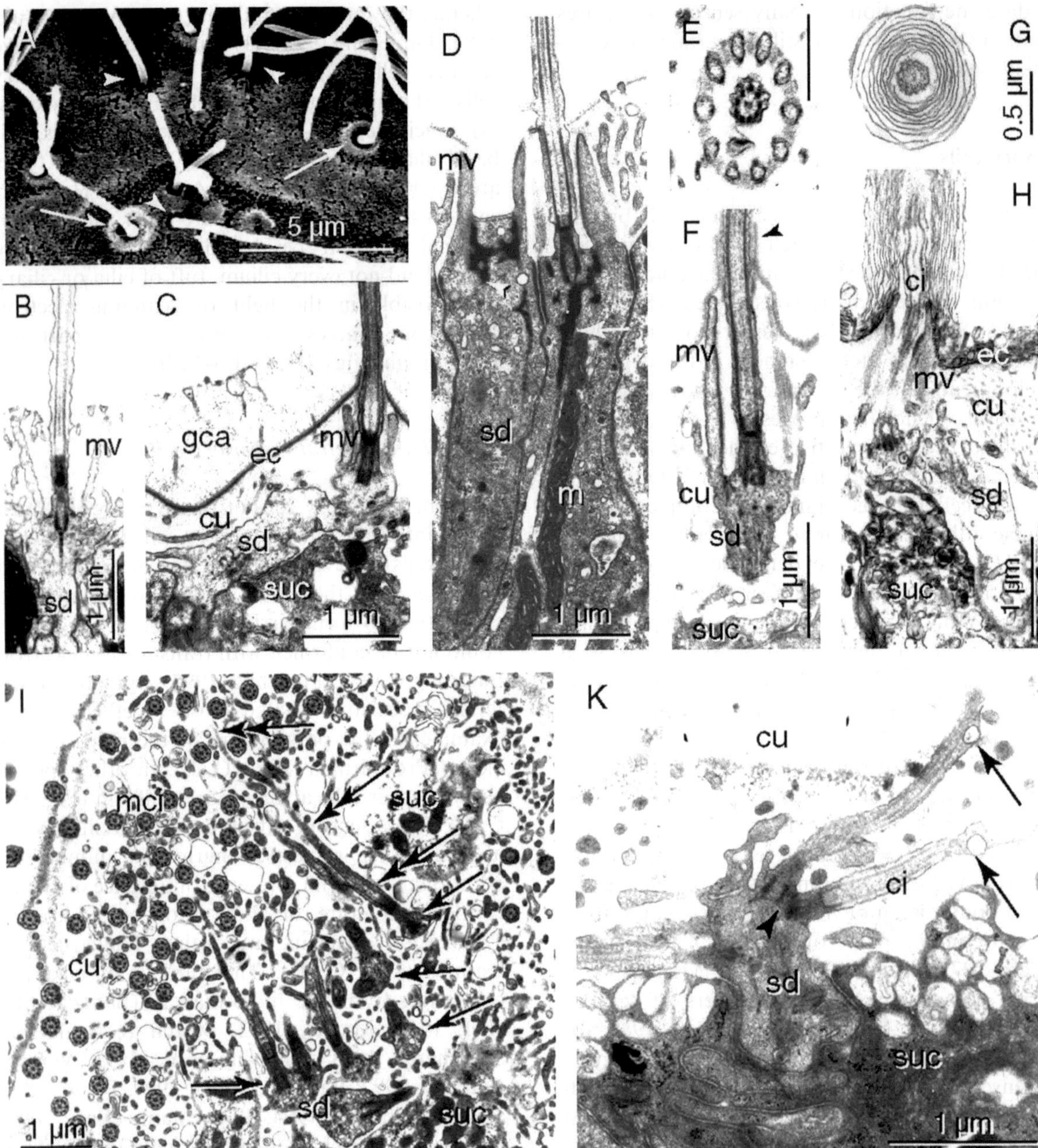

Figure 2. Sensory cells (A–H). Uniciliate penetrative sensory cells. (A) *Paranerilla limicola*; uniciliate sensory cells on the prostomium, some with collar (arrows), others without (arrowheads). SEM micrograph. (B) *Microphthalmus similis*; sensory cilium surrounded by typical microvilli (mv). (C–H) Collar receptors. (C) *Kefersteinia cirrata*; sensory dendrite (sd) projecting above surface of supporting cells (suc). Collar of microvilli not penetrating epicuticle (ec). (D) *Parenterodrilus taenioides*; collar receptors forming pore in the cuticle. Note well-developed cytoskeletal system connecting rootlet and microvilli (arrow, arrowhead). (E–F) *Polygordius appendiculatus*; cilium surrounded by 10 long microvilli, cilium basally ensheathed by epicuticle (arrowhead). Scale bar in E represents 0.5 μm. (G–H) *Glycera tridactyla*; cilium of collar receptor ensheathed by membrane-like epicuticular layers. (I–K) Non-penetrative sensory cells. (I) *Parenterodrilus taenioides*; uni- and biciliate cells (arrows) with long cilia running parallel to epithelial surface (double arrows). (K) *Stygocapitella subterranea*; sensory dendrite giving rise to numerous short cilia (ci), each with a small vacuole (arrows) and a thin process. Note absence of rootlets (arrowhead). – ci, cilium; cu, cuticle; ep, epicuticle; gca, glycocalix; m, mitochondrion; mv, microvillus; sd, sensory dendrite; suc, supporting cell. A modified from Worsaae & Kristensen (2003), D, E modified after Purschke & Jouin-Toulmond (1994), K modified from Purschke (1999).

Uniciliate penetrative sensory cells

Uniciliate sensory cells have only rarely been reported in polychaetes but most likely they are present in most if not every species. Although reviews only mention their occurrence in Nereidiidae (Dorsett & Hyde, 1969; Boilly-Marer, 1972a), they have since been found in Aeolosomatidae, Arabellidae, Dinophilidae, Dorvilleidae, Eunicidae, Glyceridae, Hesionidae, Lumbrinereidae, Lysaretidae, Nerillidae, Onuphidae, Opheliidae, Pisionidae, Polygordiidae, Protodrilidae, Sphaerodoridae, Spionidae, (Windoffer & Westheide, 1988; Schlawny et al., 1991; Purschke 1993; Hayashi & Yamane, 1994; Purschke & Jouin-Toulmond, 1994; Hessling & Purschke, 2000 and unpubl. obs.). In most cases the thin sensory dendrites (diameter about 1 μm) extend above the level of the surrounding epithelial cells (Fig 2C, F, H). Although usually equipped with a typical $9 \times 2 + 2$ axoneme (Fig 2E, G), the cilia mostly appear more or less immobile and stiff. They rest on basal bodies, and a rootlet system may be absent (Figs 2C, F, H, 6H), ill-defined (Fig. 2B) or extremely well-developed (Figs 2D, 6I). In the last case the cytoskeleton of the surrounding microvilli is often associated with the rootlet (Figs 2D, 6I). Two types of such sensory cells may be distinguished, discernable even by scanning electron microscopy (Fig. 2A): In the first type the cilia may be surrounded by typical microvilli, not reaching the cuticular surface (Fig. 2B), or microvilli are lacking (Fig. 6G, H). In the second type the cilia are surrounded by a circle of mostly 10 microvilli (Figs 2C–H, 6G, I). These microvilli are parallel to the cilium, comparatively thick and may create a pore in the cuticle open (Figs 2D, 6I). In these cases a small cuticular bulge surrounding the cilium is visible in SEM (Fig. 2A). The sensory cilia may be ensheathed

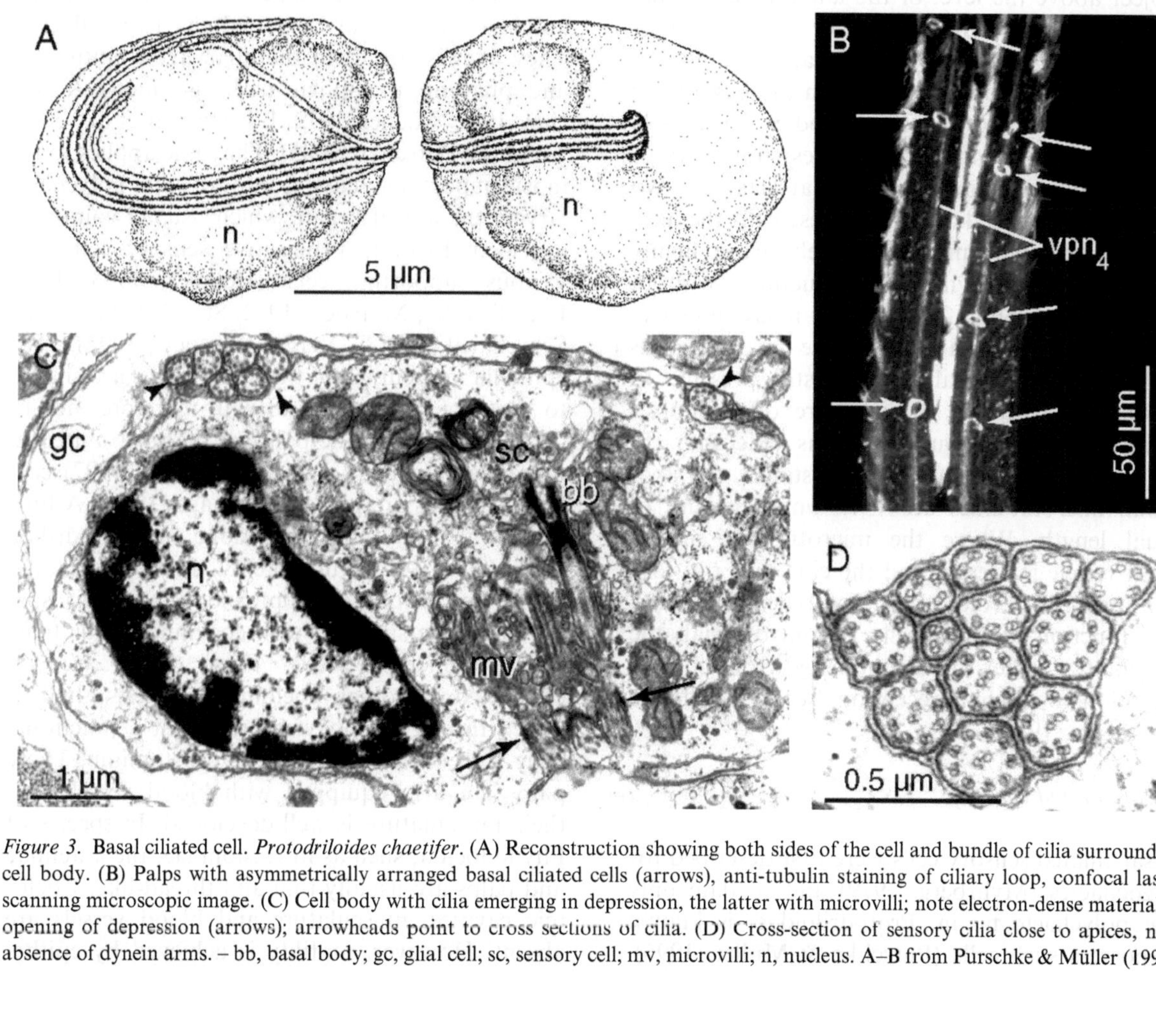

Figure 3. Basal ciliated cell. *Protodriloides chaetifer.* (A) Reconstruction showing both sides of the cell and bundle of cilia surrounding cell body. (B) Palps with asymmetrically arranged basal ciliated cells (arrows), anti-tubulin staining of ciliary loop, confocal laser-scanning microscopic image. (C) Cell body with cilia emerging in depression, the latter with microvilli; note electron-dense material at opening of depression (arrows); arrowheads point to cross sections of cilia. (D) Cross-section of sensory cilia close to apices, note absence of dynein arms. – bb, basal body; gc, glial cell; sc, sensory cell; mv, microvilli; n, nucleus. A–B from Purschke & Müller (1996).

in part or completely by the epicuticle (e.g., Polygordiidae (Fig. 2F), Nereidiidae (see Storch, 1972; Storch & Schlötzer-Schrehardt, 1988), Pisionidae). In Glyceridae they are wrapped in membrane-like epicuticular layers (Fig. 2G, H). This second type of receptor is generally called collar receptor (Schlawny et al., 1991; Purschke, 1993). It is widespread in invertebrates and thus most likely represents a plesiomorphy for Annelida (e.g., Ax, 1995). The function of these uniciliate cells is generally regarded to be mechanoreceptive.

Non-penetrative sensory cells

Non-penetrative sensory cells may either be uniciliate or multiciliate (Fig. 2I, K). They have been reported for many species (Windoffer & Westheide, 1988; Verger-Bocquet, 1992; Purschke & Jouin-Toulmond, 1994). As a rule the dendritic processes project above the level of the adjacent epidermal cells and form more or less distinct bulbs. In uni- or biciliate cells the cilia are comparatively long and run parallel to the body surface in the basal layer of the cuticle (Fig. 2I). Basal bodies are mostly without rootlets and the axonemes show various patterns of reduction. As a rule at least the two central tubules are lacking. In these cells the axonemal microtubules may successively be lost, finally resulting in microvillus-like structures without microtubules (Fig. 2I). Such microvillus-like microtubule-less cilia can still be recognized as such because a basal body is still present. In multiciliate receptors the cilia are often shorter and project in various directions (Fig. 2K). In *Stygocapitella subterranea*, for instance, the axonemal microtubulues are 1–1.5 μm long and of equal length. Where the microtubules end, a small vesicle is present and the cilia give rise to a thin process devoid of microtubules (Fig. 2K). The function of these cells is controversial; the possibilities under discussion range from chemoreception to mechanoreception (see Windoffer & Westheide, 1988).

Basal ciliated cells

Basal ciliated sensory cells were first described for oligochaetes (Myhrberg, 1979) and to date only one polychaete taxon, Protodriloidae, is known to possess such cells (Purschke & Müller, 1996). These cells are not associated with supporting cells; they are embedded between epidermal cells and do not reach the epithelial surface. A bundle of 3–15 cilia emerges from a small depression of the cell body. From this depression the cilia extend to the surface and twist once around the cell (Fig. 3A–D). This bundle of cilia can be detected with confocal LSM after labelling with anti-α-tubulin (Figs. 3B, 10B). These cells have been found in the anterior part of the brain and in the palps. The cilia have a typical $9 \times 2 + 2$ axoneme but without dynein arms (Fig. 3D). Nothing is known about the function of these sensory cells.

Sense organs

The sensory cells described above are present on the trunk and on the appendages, such as palps, antennae, tentacular cirri and the pygidial cirri. They likewise occur in the pharyngeal epithelium and pharyngeal papillae (see e.g., Bantz & Michel, 1972; Michel, 1972; Böggemann et al., 2000; Tzetlin & Purschke, 2005). The same applies to the most important sense organ of the polychaete larvae, the apical organ, which usually comprises a few multiciliate penetrative sensory cells and various supporting cells (Heimler, 1983, 1988; Lacalli, 1981; Marsden, 1982; Storch & Schlötzer-Schrehardt, 1988; Verger-Bocquet, 1992). The epithelia and cuticle of the appendages are similar to those of the trunk and will, therefore, not be described here. In the palps of the taxa united as Canalipalpata by Rouse & Fauchald (1997) coelomic cavities are present (Orrhage, 1964). A fine-structural study has been done in Protodrilida (Purschke, 1993; Purschke & Jouin-Toulmond, 1994): In *Protodrilus* spp. and *Saccocirrus* spp. the palps are supplied with coelomic canals filled with coelenchyme cells. These canals are not connected to other coelomic cavities. Fluid is added through podocytes present at a junctional zone with blood vessels in the prostomium. These palps are also equipped with blood vessels and their musculature is well developed. In species of Phyllodocida, such as in Hesionidae, the antennae and palps are usually thin and thread-like. Coelomic cavities, musculature and blood vessels are absent. However, motility is achieved by epithe-

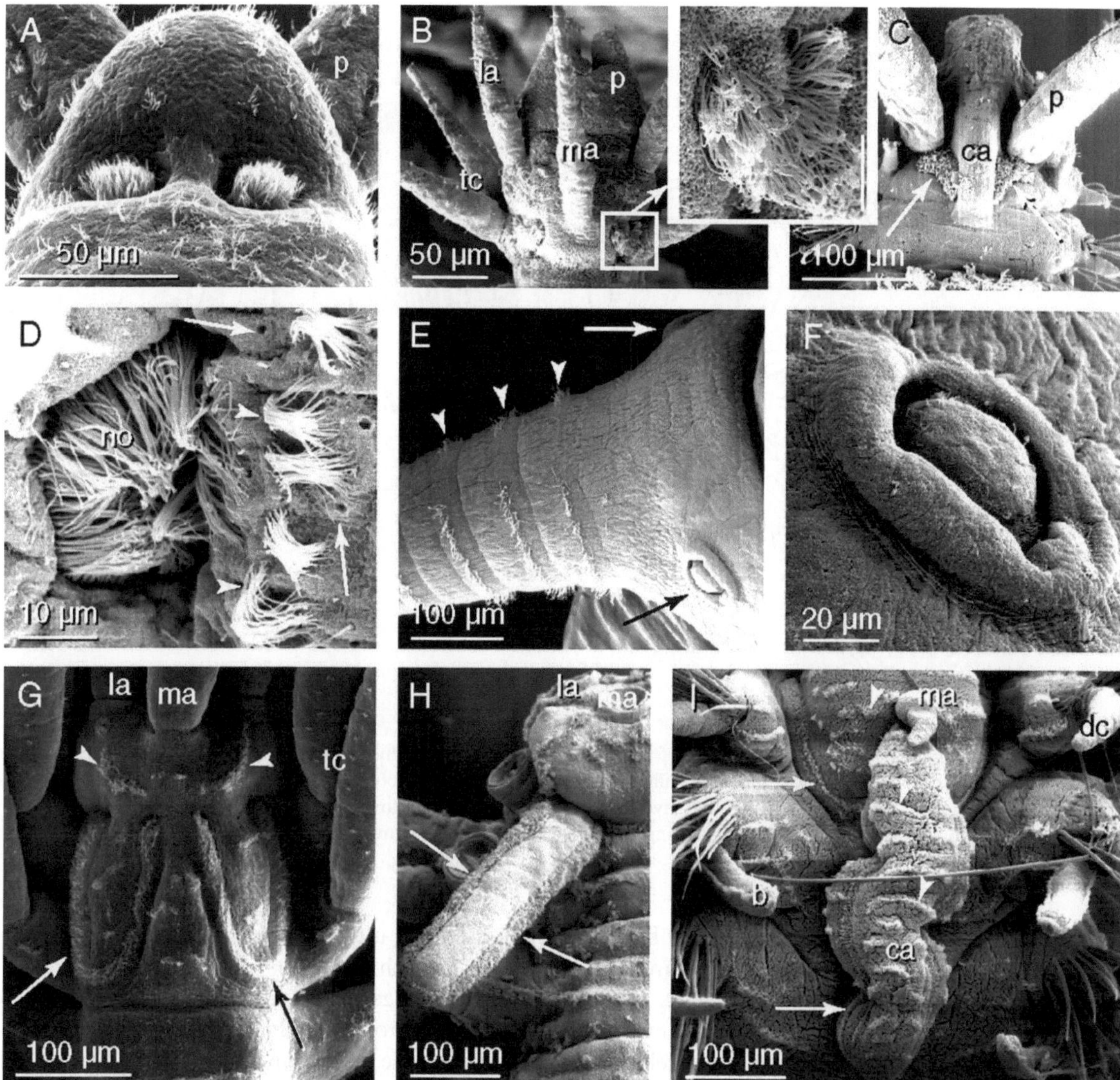

Figure 4. Nuchal organs SEM micrographs. (A–H) Typical appearance of nuchal organs as ciliated grooves or pits. (A) *Saccocirrus papillocercus*. (B) *Brania subterranea*; inset: enlargement of right nuchal organ. (C) *Polydora cornuta*; nuchal organs (arrows) situated on both sides of the caruncle (ca) behind the palps (p). (D) *Protodorvillea kefersteini*; nuchal organ (no) anterior to peristomial cilia (arrowheads); arrows point to gland cell openings. (E–F) *Glycera tridactyla* (E) Base of prostomium with nuchal organs (arrows) and rings of sensory cilia (arrowheads). (F) Detail of withdrawn nuchal organ. (G–L) Examples of hypertrophied nuchal organs. (G) *Autolytus pictus*; nuchal epaulettes (arrows) fused with peristomium; bands of cilia on prostomium (arrowheads). (H) *Amblyosyllis formosa*; appendage-like nuchal epaulettes (arrows) with ciliary band on the edges. (I) *Eurythoe complanata;* caruncle (ca) and prostomium with longitudinal ciliary bands representing nuchal organs (arrows) and transverse rows of cilia (arrowheads). – b, branchia; ca, caruncle; dc, dorsal cirrus; la, lateral antenna; ma, median antenna; no, nuchal organ; p, palp; tc, tentacular cirrus; – A modified from Purschke (1997), B, F modified from Purschke (2002a), micrographs C–E, G–I: S. Raabe.

liomuscular cells and stiffness by rootlet-like skeletal elements (Boilly-Marer, 1972b; unpubl. obs.). The innervation of the appendages is described in Orrhage & Müller (2005).

Nuchal organs

Nuchal organs are generally regarded as chemosensory (Storch & Schlötzer-Schrehardt, 1988;

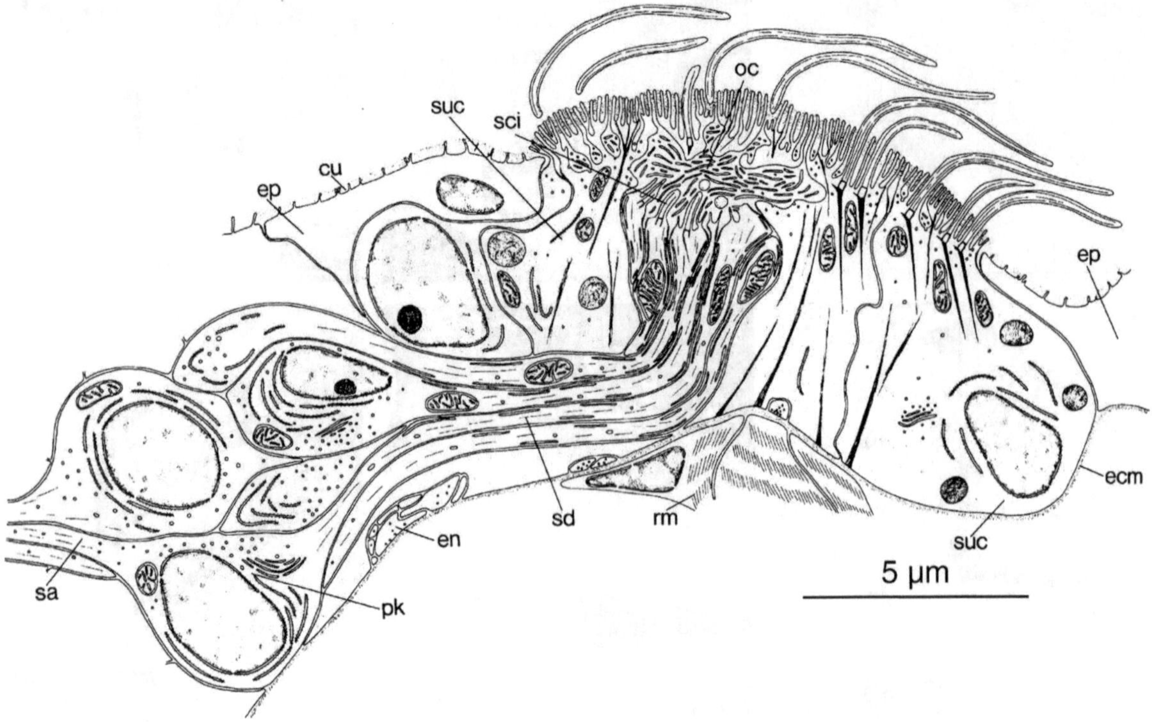

Figure 5. Nerillidium troglochaetoides. Reconstruction of nuchal organ exemplifying general organisation of nuchal organs: Ciliated supporting cells (suc) with modified cuticle and microvilli covering olfactory chamber (oc), monociliary sensory dendrites (sd), the perikarya (pk) of which form the nuchal ganglion that gives rise to axons (sa) running towards the brain; basiepithelial efferent nerve (en) innervating retractor muscle fibres (rm) and supporting cells. – cu, cuticle; ep, epidermal cell; ecm, extracellular matrix; sci, sensory cell cilium. From Purschke (1997).

Verger-Bocquet, 1992; Purschke, 1997). They have become the most important sensory structure of Annelida in terms of their phylogenetic value. The reason is that they are basically present in every species of polychaetes but absent in Clitellata without exception. There are sharply differing opinions as to whether this absence in Clitellata is primary or secondary, i.e., a loss (Fauchald & Rouse, 1997; Purschke, 1997, 1999, 2002; Rouse & Fauchald, 1997; Purschke et al., 2000; Rouse & Pleijel, 2001). In contrast to Clitellata, in the few cases of polychaetes lacking nuchal organs, such as Psammodrilidae, Pisionidae, Siboglinidae, and *Parergodrilus heideri* (see Purschke, 1997, 2000), this absence is always interpreted as a loss. Nuchal organs have been found in many polychaete species for which absence of the organs was assumed or at least considered likely (see Purschke, 1997). The light microscopical histology of these organs is well-known since Rullier's (1951, 1954) investigations; their ultrastructure has recently been reviewed (Purschke, 1997).

Nuchal organs are mostly visible as a pair of ciliated areas or spots located dorsally or dorsolaterally at the posterior edge of the prostomium (Fig. 4A–D). The cilia here are kinocilia and may be located in grooves or pits (Fig. 4A–D). These organs may be considerably larger and extend posteriorly on one or several segments (Fig. 4G–I). One of the best known examples are Amphinomidae, in many species of which a caruncle is formed as a bulging sensory area arising from the prostomium and supplied with longitudinal and transverse ciliary bands (Fig. 4I). Most likely, only the longitudinal ciliary bands represent the nuchal organ proper (unpubl. obs.). This might be the reason for different interpretations of the function of this structure (e.g., Storch & Welsch, 1969; Ameyaw-Akumfi, 1976; Storch & Schlötzer-Schrehardt, 1988). A caruncle may also be present in Spionidae, Trochochaetidae, Poecilochaetidae and Chrysopetalidae. In Spionidae the caruncle is straight with a smooth surface and the nuchal organs lie beside the caruncle, as either one pair (e.g., *Polydora cornuta*: Fig. 4C) or two pairs of ciliary bands (e.g., *Malacoceros fuliginosa*: Fig. 6A) (see Schlötzer-Schrehardt, 1986, 1987; Purschke, 1997; Jelsing, 2002a,b). These ciliary bands show considerable variation and may extend over several segments (Jelsing, 2002a,b).

Other examples of hypertrophied nuchal organs are the so-called nuchal epaulettes present in certain Syllidae (Fig. 4G, H) or Phyllodocidae (see Eibye-Jacobsen, 1991; Pleijel, 1991; Rouse & Pleijel, 2001). Whereas in *Autolytus* spp. (Syllidae), the epaulettes are completely fused with the trunk (Fig. 4G), they are appendage-like in *Amblyosyllis formosa*. A continuous ciliary band is present on the edges of all nuchal epaulettes, whereas the center is unciliated (Fig. 4G, H). These structures have not been investigated by transmission electron microscopy so far. The biological significance of these large organs is not clear since other species, often closely related to these taxa, possess nuchal organs which are not hypertrophied (Lewbart & Riser, 1996; Rhode, 1990a; Purschke, 1997).

An enlargement of these sensory organs may also be achieved by duplication: two pairs of nuchal organs are present in most (all?) Dorvilleidae and many Opheliidae (Rhode, 1989; Purschke, 1997). The size of nuchal organs may vary considerably even between closely related species.

On the other hand, especially in burrowing forms or those with a highly modified anterior end, the nuchal organs can be completely withdrawn (Fig. 4E, F) or are situated more or less internally and are invisible from the exterior; e.g., Sabellidae, *Stygocapitella subterranea*, *Hrabeiella periglandulata*, *Potamodrilus fluviatilis* (see Orrhage, 1980; Purschke, 1986, 1997, 2000; Purschke & Hessling, 2002). In Sabellidae the nuchal organs have a rather uncommon position and form a pair of pouches arising from the dorsal epithelium of the mouth cavity. This aberrant position is very likely due to the development of the branchial crown (Orrhage, 1980; Purschke, 1997). In the other species cited above, the nuchal organs are situated in deep pits and communicate with the exterior *via* small pores hardly visible even in scanning electron microscopy.

Regardless of their different external structure these organs show an overall fine-structural similarity (Fig. 5), and there is no doubt about their homology in Annelida (Rouse & Fauchald, 1997; Purschke, 1997, 2002; Rouse & Pleijel, 2001). The organs have a number of characters in common, as follows. The visible cilia are mobile and are located on the supportive cells. They are responsible for a rapid exchange of sensory stimuli by generating water currents. The sensory cells are bipolar primary sensory cells, the perikarya of which form the nuchal ganglia and their processes the nuchal nerve (Fig. 5). The nuchal nerve emanates directly from the brain as is the case for the efferent innervation, which is always separate. Usually these nerves are part of the dorsal pair of longitudinal nerves (Orrhage & Müller, 2005).

As a rule the dendrites are monociliate. Cilia, ciliary branches and true microvilli fill a subcuticular space called the olfactory chamber, which is protected by specialized cuticular or microvillar layers formed by the supporting cells. Differences between taxa include the number of ciliated supporting and sensory cells as well as the specific structure of the protective layer. These differences may be autapomorphies for certain taxa, such as the paving-stone-like microvillar cover present in Spionida and Protodrilida (Schlötzer-Schrehardt, 1986; Purschke, 1997; Jelsing, 2002b). The number of sensory cells varies from two in *Potamodrilus fluviatilis* to several hundred in *Polyophthalmus pictus* and *Armandia polyophthalma* (see Purschke, 1997; Purschke & Hessling, 2002). For many large species studied so far, e.g., *Nereis diversicolor*, *Glycera rouxi*, *Nephtys caeca* (see Whittle & Zahid, 1974), the number of sensory cells has not been determined.

Whether the absence of nuchal organs in Clitellata is primary or secondary is crucial for the systematization of Annelida (McHugh, 1997, 2000; Purschke, 1997, 2002a; Rouse & Fauchald, 1997; Westheide, 1997; Westheide et al., 1999; Purschke et al., 2000). Evidence for a primary absence would support a sister-group relationship of Polychaeta and Clitellata, the nuchal organs being the sole autapomorphy of Polychaeta, whereas evidence of a loss would support the placement of Clitellata as an ingroup of the paraphyletic 'Polychaeta'. Evidence for a loss of nuchal organs in Clitellata comes from the reduc-

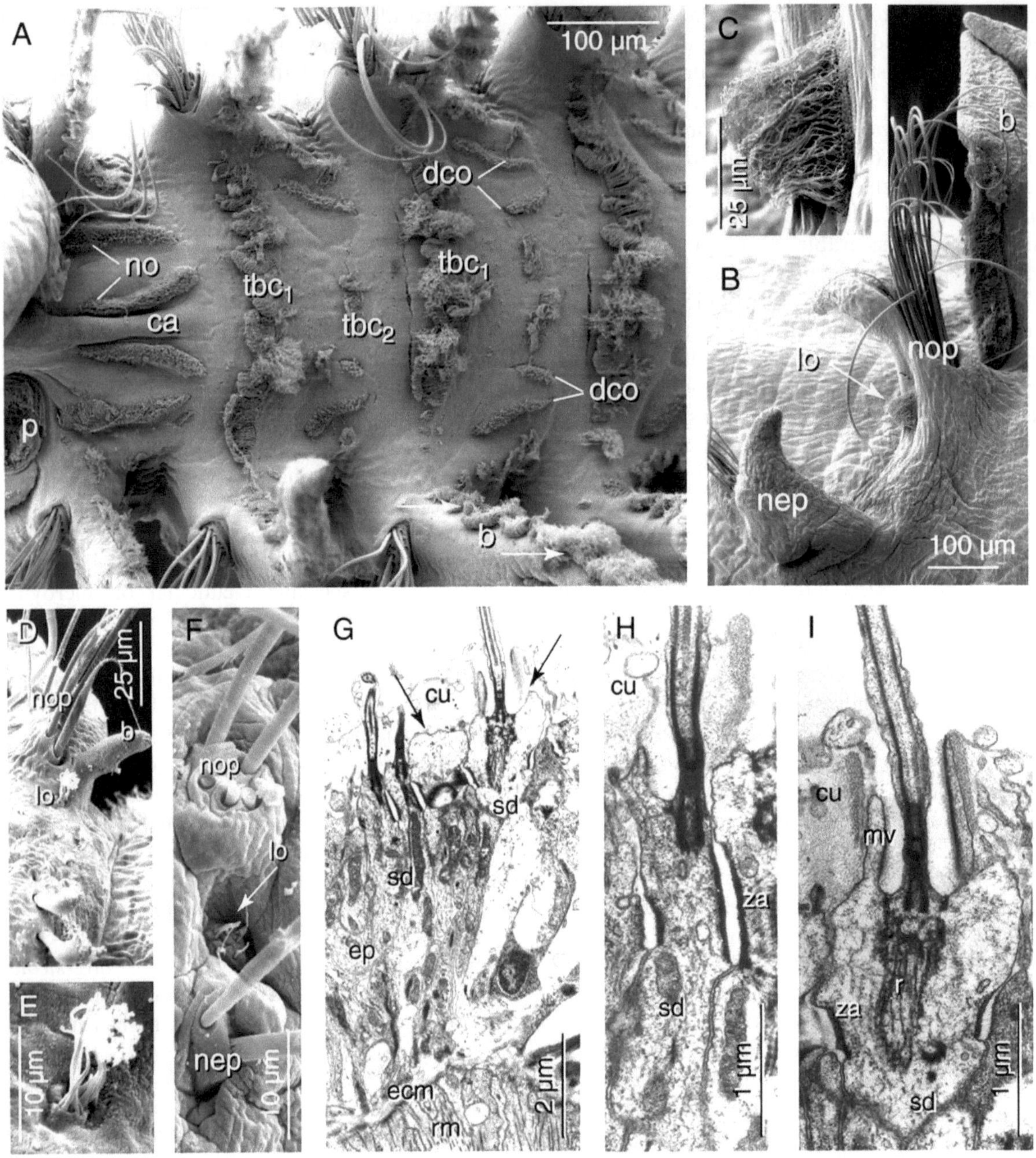

Figure 6. Dorsal and lateral ciliated organs. (A) *Malacoceros fuliginosus*; dorsal view of anterior end with nuchal organs (no) and dorsal ciliated organs (dco). Two transverse bands of cilia per segment (tbc_1, tbc_2), arrows point to ciliary band on branchia (b). (B–I) Lateral ciliated organs. (B, C) *Scoloplos armiger*; lateral organ (lo) at base of notopodium (nop). (D–E) *Polydora cornuta*; lateral organ below notopodium. (F–I) Opheliidae. (F) *Ophelia rathkei*; small lateral organ between noto- and neuropodium. (G–I) *Polyophthalmus pictus*. (G) Section through entire organ with retractor muscle (rm) and two types of monociliary sensory dendrites (sd); arrows point to collar receptors. (H) Tip of dendrite with asymmetrically situated cilium adjacent to collar receptor. (I) Collar receptor with net-like rootlet system (r). – b, branchia; ca, caruncle; cu, cuticle; do, dorsal organ; ecm, extracellular matrix; ep, epidermis; lo, lateral ciliated organ; mv, microvillus; nep, neuropodium; no, nuchal organ; nop, notopodium; p, palp; rm, retractor muscle; sd, sensory dendrite; za, zonula adhaerens. Micrographs A–F: S. Raabe, S. Göbel, B. Rohling and W. Mangerich.

tion or absence of these organs in terrestrial polychaetes, the lack of epidermal cilia in Clitellata, the backward displacement of the brain and the reduction of the prostomium (Hessling & Westheide, 1999; Purschke, 1999, 2000; Hessling et al., 1999; Purschke et al., 2000). In certain polychaetes a similar arrangement can be observed. For instance, in species of *Pisione* the anterior end is extremely modified, with a reduction of the prostomium and displacement of the brain. The large brain is situated in the first three segments, nuchal organs are obviously absent and have been lost (Siewing, 1953; Purschke, 1997; Rouse & Fauchald, 1997).

Developing nuchal organs are visible in metatrochophores (Rullier, 1951; Åkesson, 1962, 1967; Bhup & Marsden, 1982). They develop from episphere cells while the trochoblasts of the prototroch become disintegrated and are prostomial in origin (Åkesson, 1962, 1967). Ultrastructural studies on the development have not been carried out to date. There are no significant structural differences between nuchal organs of juveniles and adults in Opheliidae and Spionidae (West, 1978; Schlötzer-Schrehardt, 1986, 1987, 1991). Changes observed mainly concern the overall size and number of cells and sometimes a shift in position.

Apart from the Annelida, structures named nuchal organs are present in Sipuncula. They are situated on the introvert either between or below the tentacles (Rice, 1993). So far, ultrastructural studies have only been carried out for one species, *Onchnesoma squamatum* (see Purschke et al., 1997). This study did not favour a homology hypothesis, but additional studies in Sipuncula appear to be necessary before a final conclusion can be drawn. These findings are in contrast to the view of Åkesson (1958) derived from comprehensive light microscopic investigations. Thus, according to our present knowledge, the nuchal organ of Annelida either represents an autapomorphy of Annelida, or in case of a sister-group relationship between Polychaeta and Clitellata, an autapomorphy of the former (see Purschke, 2002a).

Dorsal organs

Dorsal ciliated organs may be of different kinds and not all of them are sensory. For instance, these structures may serve in spermatophore formation and transfer or simply generate water currents. This seemed to be true for Spionidae as well: Schlötzer-Schrehardt (1987, 1991) showed that in *Pygospio elegans* these organs are neither sensory nor have a common origin with nuchal organs but probably play a role in sperm transfer. Thus, these investigations were used to disprove earlier views (Söderström, 1927; Rullier, 1951; Orrhage, 1964). Recently, Jelsing (2002a, b) reinvestigated these organs in a comprehensive study, including several species of Spionidae. It was shown that there are various kinds of dorsal ciliation, including transverse and longitudinal bands (Fig. 6A). The longitudinal bands proved to be structurally identical to nuchal organs, so that these structures are chemosensory as well and serially homologous to nuchal organs. These structures are in fact absent in *P. elegans*. These investigations clearly demonstrate that generalisations have to be done with care and far-reaching conclusions drawn from a single species may lead to false and premature statements. However, whether the dorsal organs are innervated by the elongated nuchal nerve, as suggested by Söderström (1927) and others, remains to be proven. The dorsal organs described for Orbiniidae may also represent homonomous nuchal organs (Eisig, 1914), but this has yet to be confirmed by ultrastructural studies.

Lateral organs

Lateral ciliated organs are ciliated pits or papillae present between noto- and neuropodia in Amphinomidae, Syllidae, Eunicida, Spionidae, Opheliidae, Orbiniidae, Paraonidae, Scalibregmatidae and Pectinariidae (Fig. 6B–I; see Rouse & Pleijel, 2001; Purschke, unpubl. obs.). In the Opheliidae *Ophelia rathkei* and *Polyophthalmus pictus* these organs consist of a group of uniciliate penetrative sensory cells (Fig. 6G–I). Some of these cells are collar receptors, others are without a collar of strong microvilli. In the latter the cilium is situated eccentrically on the dendrite and is closer to the collar receptor (Fig. 6G, H). In the collar receptor a rootlet system is well-developed and connected to small hemidesmosome-like structures present between cilium and surrounding microvilli. This structure is the same as described for the dorsal cirrus organ present at the lower side of the not-

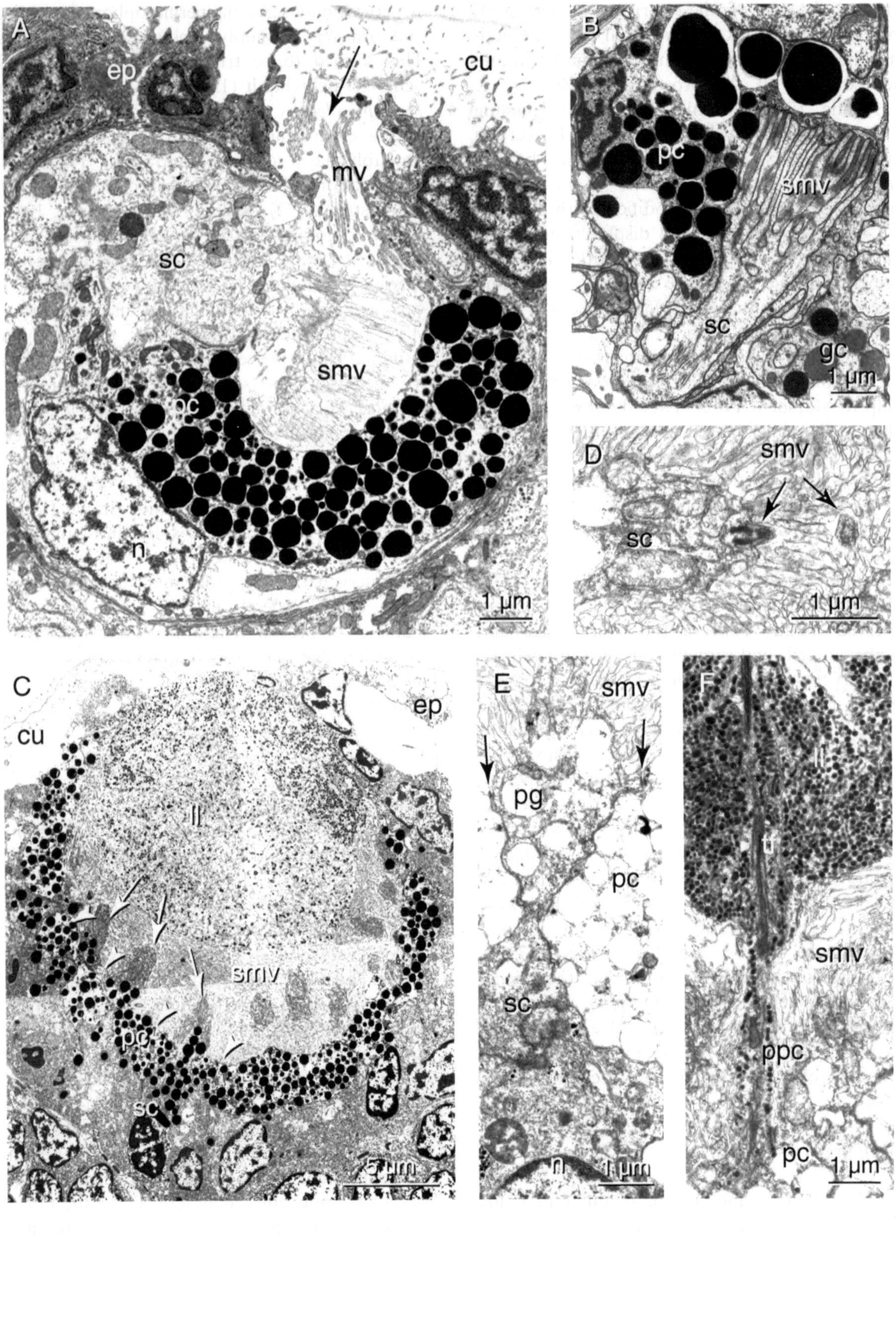
A
ep
cu
mv
sc
smv
pc
n
1 µm
B
pc
smv
sc
gc
1 µm
D
smv
sc
1 µm
C
cu
ep
ll
smv
pc
sc
5 µm
E
smv
pg
pc
sc
n
1 µm
F
ll
tr
smv
ppc
pc
1 µm

Figure 7. Cerebral eyes. (A) *Saccocirrus papillocercus*. Bicellular ocellus; pigment cell (pc) and rhabdomeric photoreceptor cell (sc), part of epidermal cell layer; eye cup open to exterior (arrow). (B) *Microphthalmus similis*. Ocellus situated deeper in prostomial tissue. (C–F) Multicellular ocelli. (C) *Gyptis propinqua*. Eye cup consisting of alternating pigment (arrowheads) and sensory cells (arrows). (D–F) *Kefersteinia cirrata*. (D) Microvilli-bearing process of sensory cell (sc) with vestigial cilium (arrows). (E) Sensory cell (sc) with pigment granules (pg) between pigmented supporting cell, arrows point to junctional complexes. (F) Process of pigment cell (ppc) extending above sensory microvilli (smv) and forming lens-like structure (ll). – cu, cuticle; ep, epidermis; gc, glial cell; ll, lens-like structure; mv, microvillus; n, nucleus; pc, pigment cell; pg, pigment granules; ppc, process of pigment cell; sc, sensory cell; smv, sensory cell microvilus; tf, tonofilaments. A modified from Purschke (1992).

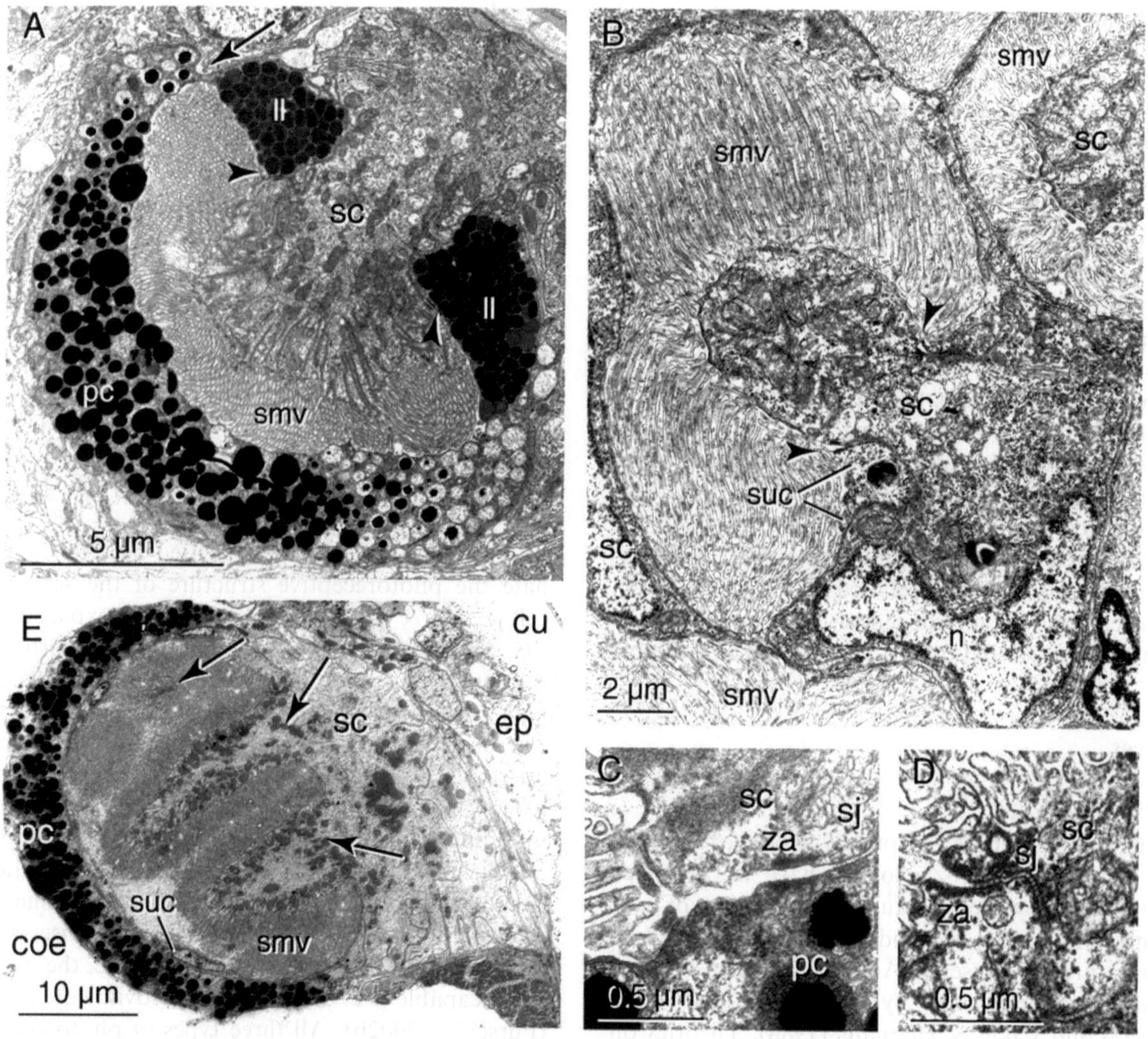

Figure 8. Eyes. (A–D) *Polyophthalmus pictus*. (A) Cerebral pigmented ocellus; note lens-like process (ll) connected to cell body of pigment cell (arrow); arrowheads point to junctional complexes enlarged in C. (B) Cerebral unpigmented rhabdomeric ocellus embedded in area made up of numerous such ocelli; arrowheads point to cell junctions shown in D. (C–D) Enlargement of junctional complexes of ocelli shown in A, B: (C) Pigmented ocellus. (D) Unpigmented ocellus. (E) *Armandia polyophthalma*: Segmental ocellus made up of large sensory cell (sc) with several rhabdomeric processes (arrows), unpigmented supporting cells (suc) and layer of mesodermal pigment cells (pc). – cu, cuticle; coe, coelom; ep, epidermis; n, nucleus; pc, pigment cell; sc, sensory cell; sj, septate junction; smv, sensory microvilli; suc, supporting cell; za, zonula adhaerens. A, B, D modified from Purschke (2003). E modified from Purschke et al. (1995).

opodia, or for the dorsal cirri, in various species of Eunicida (Hayashi & Yamane, 1994, 1997). This indicates that these organs are most likely homologus and do not represent a new type of sense organ. As argued by Rouse & Pleijel (2001), these organs should be given the same name and their phylogenetic value as an autapomorphy of Eunicida appears to be questionable. Whether this also applies to the lateral organs of Orbiniidae has still to be proven; from their appearance in SEM they seem to be different, in that they are made up of a brush of densely arranged cilia as is characteristic of multiciliate cells (Fig. 6B, C). This likewise applies to the bands of cilia present between dorsal cirrus and neuropodium in Amphinomidae as well as in *Syllides longocirrata* and other Syllidae (unpubl. obs.). An ultrastructural comprehensive investigation of these lateral organs across various taxa is needed to address the question of homology.

In contrast to these organs, the corresponding structures of Protodrilidae are clearly related to the reproductive system in being the organs of spermatophore formation (Nordheim, 1991). The lateral organs of Myzostoma have been regarded as chemoreceptive and structurally similar to nuchal organs by Eeckhaut & Jangoux (1993). However, similarities to nuchal organs have been considered superficial, and homonomy with nuchal organs has been excluded by Purschke (1997). Instead, the number of sensory cells appears to be rather low, so that other functions are conceivable as well.

Eyes

Pigmented ocelli

Most polychaetes have some type of eyes or ocelli, a diminutive eye. Zoologists have always been fascinated by the structure and function of eyes and photoreceptors, and there is an abundant literature on those of Annelida. Recent reviews have been published by Verger-Bocquet (1984, 1992) and Eakin & Hermans (1988). Theories on the evolution of eyes have been put forward by, e.g., Eakin (1963, 1982), Vanfleteren & Coomans (1976), Salvini-Plawen & Mayr (1977), Salvini-Plawen (1982), Vanfleteren (1982) and more recently using genetic and developmental data by Gehring & Ikeo (1999), Gehring (2001) and Arendt & Wittbrodt (2001).

Most eyes of polychaetes are within or adjacent to the brain and are commonly termed cerebral eyes. Others are situated on the tentacular crown in certain Sabellidae and Serpulidae, on the trunk segments, for instance in Opheliidae, Eunicidae and Sabellidae, and, finally, on the pygidium in certain Sabellidae. These eyes are called branchial ocelli, segmental ocelli or pygidial ocelli, respectively. These latter eyes are structurally much more diverse and completely different in structure compared with cerebral eyes. In addition, there is a high diversity of unpigmented ocelli and photoreceptor-like structures, often associated with or adjacent to the brain.

In size and complexity such organs range from a small ocellus composed of only two cells, one pigmented supportive and one sensory cell, which overall is only 6 μm or less in diameter, as in *Microphthalmus similis* (Fig. 7B), to large eyes of about 1 mm in diameter, made up of thousands of cells forming complex retinae, lenses and other accessory structures in Alciopidae and other planktonic polychaetes (Eakin & Hermans, 1988; Verger-Bocquet, 1992). In Alciopidae the large eyes protude laterally from the anterior end and leave a small space between them for the brain (Hermans & Eakin, 1974).

The term photoreceptor is used here to designate the photoreceptive structure of the sensory cell (Eakin & Hermans, 1988). A great expanse of the membranes bearing the light-absorbing photopigment, often in a highly ordered and regular pattern, is characteristic of photoreceptors. These membranes take the form of an array of microvilli or lamellae in rhabdomeric receptors, whereas in ciliary receptors they are outfoldings or infoldings of a ciliary membrane or, rarely, groups of cilia. A phaosome is a seemingly intracellular vacuole into which the microvillar or ciliary photoreceptors project. The phaosome arises by invagination of the apical cell membrane, the only area capable of developing microvilli or cilia (Purschke, 2002b). All three types of photoreceptors have been found in polychaetes. In addition, supportive cells are associated in some way with the photoreceptor cells. The former contain granules of shading pigment, presumably different kinds of melanins (Eakin & Hermans, 1988). These cells are also involved in the formation of lens-like structures and vitreous bodies.

In bicellular ocelli the pigment cell and the sensory cell form an extracellular cavity into which the photoreceptor projects (Figs. 7A, B, 8A). The cells are connected to one another by typical junctional complexes: a zonula adhaerens followed by a septate junction (Fig. 8C). As the photoreceptors are housed in the concavity of the pigment cell, ocelli of this type must become inverse. In many of these small bicellular ocelli, the microvilli extend as a dense brush border from the flat surface of the receptor cell (e.g., *Protodrilus* spp., *Saccocirrus* spp., *Microphthalmus listensis, Microphthalmus similis,* see Eakin et al., 1977; Pietsch & Westheide, 1985; Purschke, 1992; Fig. 7A, B). An increase in the number of receptive structures is achieved by evagination of the apical cell membrane, so that the sensory cell forms a mushroom-like process bearing densely arranged microvilli (e.g., Spionidae, Opheliidae, Polygordiidae; see Hermans & Cloney, 1966; Brandenburger & Eakin, 1981; Rhode, 1991; Bartolomaeus, 1993; Purschke, 2003; Fig. 8A).

In larger cerebral eyes, the number of cells may increase considerably, but the general structure remains similar. That is, the two cell types form a continuous epithelium around an extracellular cavity into which the sensory processes project (Fig. 7C–E; Purschke, 2003). In this epithelium pigmented supportive cells and sensory cells alternate. The cell bodies of the sensory cells are situated below the pigment layer (Fig. 7C). Usually the orientation of the sensory processes becomes everse. That is, inverse or everse design apparently reflects functional constraints and is clearly correlated with the number of cells involved in cerebral eyes of polychaetes (Purschke, 2002b). Exceptions to this rule have been described in Flabelligeridae, where the two cell types are not intermingled but separated from one another as in certain Platyhelminthes (Spies, 1975). In these eyes the photoreceptors have an inverse design. These functional constraints in orientation of photoreceptors have not been recognized in previous reviews (Eakin & Hermans, 1988; Verger-Bocquet, 1992; Arendt & Wittbrodt, 2001).

As a rule the single sensory process is cone-shaped or columnar and bears microvilli on all sides (Eakin & Hermans, 1988). These microvilli are straight and highly ordered. At the tip of the sensory process a rudimentary cilium may be present, as is the case in many bicellular ocelli (Fig. 7D). The sensory processes often contain pigment granules as well (Fig. 7C, E). In all eyes the microvilli forming the rhabdomeres are comparatively uniform in size and measure 0.07–0.1 μm in diameter and about 1–1.5 μm in length.

In addition, a lens-like structure is present in most ocelli of this type. These lenses are either formed by the supportive cells as vesicle-containing processes (Fig. 7C, F) or a secretion given off by these cells, or are formed by one or a few specialized lenticular cells (Eakin & Hermans, 1988; Verger-Bocquet, 1992). Such multicellular eyes are present in species of the Phyllodocida and have been found so far in Syllidae, Hesionidae, Nereidiidae, Phyllodocidae, Alciopidae, Polynoidae, Aphroditidae (Fischer & Böckelmann, 1965; Hermans & Eakin, 1974; Zahid & Golding, 1974; Singla, 1975; Bocquet, 1976, 1977; Verger-Bocquet, 1983a; Eakin & Brandenburger, 1985; Rhode, 1991). Very likely this type of eye is also present in Amphinomidae (unpubl. obs.). The eyes described in *Capitella* sp. I by Rhode (1993) may also belong to this type and thus may represent the only exception. Although the author regards these eyes as unique for Annelida, the adult eyes develop in the same way as other multicellular eyes but show signs of disintegration of the pigment cells after development.

This distribution indicates that this type of multicellular eye could be an antapomorphy for a taxon Aciculata sensu Rouse & Fauchald (1997). However, it is noteworthy that a similar everse type of eye is present in Sipuncula and Mollusca, indicating homology (Hermans & Eakin, 1969; Rosen et al., 1979; Land, 1984; Bartolomaeus, 1992a; Blumer, 1995; Sturrock & Baxter, 1995; Arendt & Wittbrodt, 2001).

In *Ophryotrocha puerilis* and other *Ophryotrocha* spp. a pair of ocelli is present at the posterior margin of the brain. Although situated in the peristomium, they have been regarded as cerebral by Rhode (1990b). These eyes are exceptional not only in their position in the anterior end but also in their structure (Zavarzina, 1987; Rhode, 1990b): the sensory cell bears several rhabdomeric sensory processes which are enveloped by a thin supportive cell almost devoid of pigment granules. The eye cup is formed by a layer of flat additional cells separated from the ocellus proper by an extracellular matrix. These cells contain layers of refractive crystalline platelets. The presence of an ECM indicates that two different germ layers are involved. But according to the assumption of Rhode (1990b) that the ocelli develop from the brain, the platelet-bearing cells should be mesodermal in origin instead of epidermal. Moreover, another dorvilleid, *Protodorvillea kefersteini*, possesses ocelli as described above (Purschke, unpubl. obs.). This is indicative of the presence of a secondary, newly evolved eye in *Ophryotrocha* spp. The same may apply to the eye in *Nerilla antennata*, which has an unusual structure and is composed of platelet (supportive) cells, corneal (supportive) cells and two sensory cells facing each other with their rhabdomeres (see Eakin et al., 1977; Eakin & Hermans, 1988).

In trochophore larvae of polychaetes the same type of small ocelli has been described (Eakin & Westfall, 1964; Holborow & Laverack, 1972; Brandenburger & Eakin, 1981; Verger-Bocquet, 1983a; Marsden & Hsieh, 1987; Bartolomaeus, 1987, 1992b; Rhode, 1992, 1993). Usually there is one pair of bicellular ocelli, but the number of cells as well as the number of eyes may be higher (Bartolomaeus, 1992b, 1993). These latter situations have been regarded as secondarily evolved by Bartolomaeus (1992b). These eyes may persist in the adults or may be replaced by the eyes of the adults during later development (Rhode, 1992, 1993; Bartolomaeus, 1993). During ontogenesis it is evident that the larval eyes are formed within the epithelial layer of the prostomial epidermis. After development has been completed, a pore in the eye cup may persist (Marsden & Hsieh, 1987; Bartolomaeus, 1992b; Rhode, 1992). In other species the cavity of the eye is completely closed and the ocelli are more deeply recessed into the prostomial tissues (Brandenburger & Eakin, 1981; Marsden & Hsieh, 1987; Bartolomaeus, 1992b). Such a pore may still be present in species in which the eyes of the trochophores are transferred to the adults. For instance, in *Saccocirrus* spp. (see Eakin et al., 1977; Purschke, 1992) the supportive and sensory cells are still part of the epithelium and are connected to the adjacent epidermal cells by typical apical junctional complexes (Purschke, 1992). The eye cup communicates with the subcuticular space *via* a small pore lined by an array of microvilli formed by the pigment cell, a feature unknown for other ocelli. The adult eyes develop likewise in the prostomium and in these eyes, too, a pore and a connection to the subcuticular space may still be present after differentiation has been completed (Verger-Bocquet,1992; Rhode, 1993).

Homology of cerebral eyes is generally assumed in Bilateria (e.g., Pietsch & Westheide, 1985; Bartolomaeus, 1992b; Arendt & Wittbrodt, 2001). However, a special problem appears to arise from the fact that the ocelli present in annelid larvae either degenerate and are replaced by the eyes of the adults in certain taxa or are retained in the adults in other polychaete taxa. In Mollusca the adult eyes always develop from larval eyes but are absent in basal groups (Salvini-Plawen, 1982). In spite of this difference homology of the adult eyes is conceivable since these eyes are structurally similar. Moreover, in polychaete larvae the number of cells involved and the number of eyes developed is variable between taxa investigated (see Bartolomaeus, 1992b). In these larval eyes the supportive cells have the ability to develop lens-like structures: e.g., in Syllidae, Opheliidae, or *Arenicola marina* (Verger-Bocquet, 1983b; Bartolomaeus, 1992b, 1993; Purschke, unpubl. obs.). Recent genetic work also argues for a monophyletic origin of the different types of cerebral ocelli found in extant Bilateria, all of which may have evolved from a prototypic eye by adaptive radiation (Gehring & Ikeo, 1999; Gehring, 2001). This support comes from the observation that the same master control gene *Pax6* is involved in eye development in all taxa studied so far. In addition, these cerebral eyes are formed in the *Otx*-territory (Arendt & Wittbrodt, 2001). This idea of homology of cerebral eyes throughout all Bilateria has been challenged by Arendt & Wittbrodt (2001) because in rhabdomeric invertebrate and ciliary chordate photoreceptors non-homologous cascades of phototransduction have been observed. These authors conclude that the primary rhabdomeric photoreceptor cells were lost in the stem lineage of Vertebrata and these cells were replaced by ciliary receptors.

Unpigmented ocelli

In addition to the pigmented ocelli, various kinds of ocelli without shading pigment have been

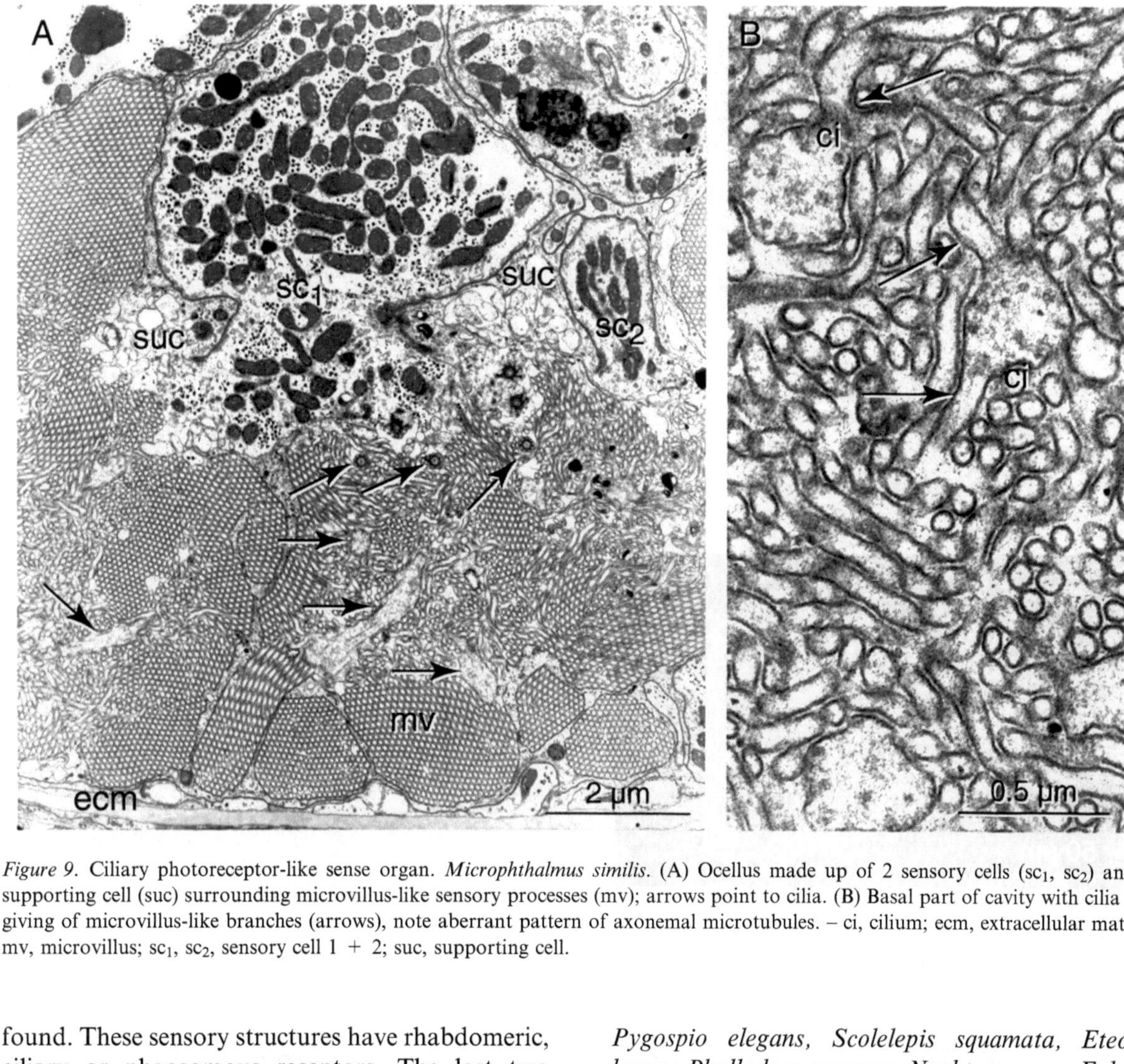

Figure 9. Ciliary photoreceptor-like sense organ. *Microphthalmus similis.* (A) Ocellus made up of 2 sensory cells (sc_1, sc_2) and 1 supporting cell (suc) surrounding microvillus-like sensory processes (mv); arrows point to cilia. (B) Basal part of cavity with cilia (ci) giving of microvillus-like branches (arrows), note aberrant pattern of axonemal microtubules. – ci, cilium; ecm, extracellular matrix; mv, microvillus; sc_1, sc_2, sensory cell 1 + 2; suc, supporting cell.

found. These sensory structures have rhabdomeric, ciliary or phaosomous receptors. The last two types will be referred to as photoreceptor-like sense organs below. Sense organs of the first type are structurally similar to pigmented ocelli, usually consisting of two cells – one supportive and one rhabdomeric sensory cell – forming an extracellular cavity (Fig. 8B, D) which houses the photoreceptive microvilli. An accessory centriole or a more or less reduced cilium may be present in the sensory cell. As a rule, the microvilli of the receptor cells are considerably longer than in pigmented ocelli and may reach up to 6 μm in length. Such ocelli have been observed in the brain or prostomium of *Ophelia rathkei, Armandia brevis, Armandia polyophthalma, Polyophthalmus pictus, Saccocirrus krusadensis, Heteromastus filiformis, Pygospio elegans, Scolelepis squamata, Eteone longa, Phyllodoce mucosa, Nephtys caeca, Eulalia viridis, and Microphthalmus* spp. (see Hermans & Cloney, 1966; Zahid & Golding, 1974; Whittle & Golding, 1974; Pietsch & Westheide, 1985; Schlötzer-Schrehardt, 1987; Rhode, 1991; Purschke, 1992, 2003; Bartolomaeus, 1993). There may only be a pair or a few ocelli of this type in individuals of a given species, but in *P. pictus* and in *A. polyophthalma* there are about 70 or 50, respectively. The main difference from pigmented ocelli is just the absence of pigment in the supportive cell, which usually becomes rather thin and may be only 40–80 nm in diamater (Purschke, 1992). These structural similarities may give some evidence for an evolution from pigmented ocelli by loss of the shading pigment, a view supported by

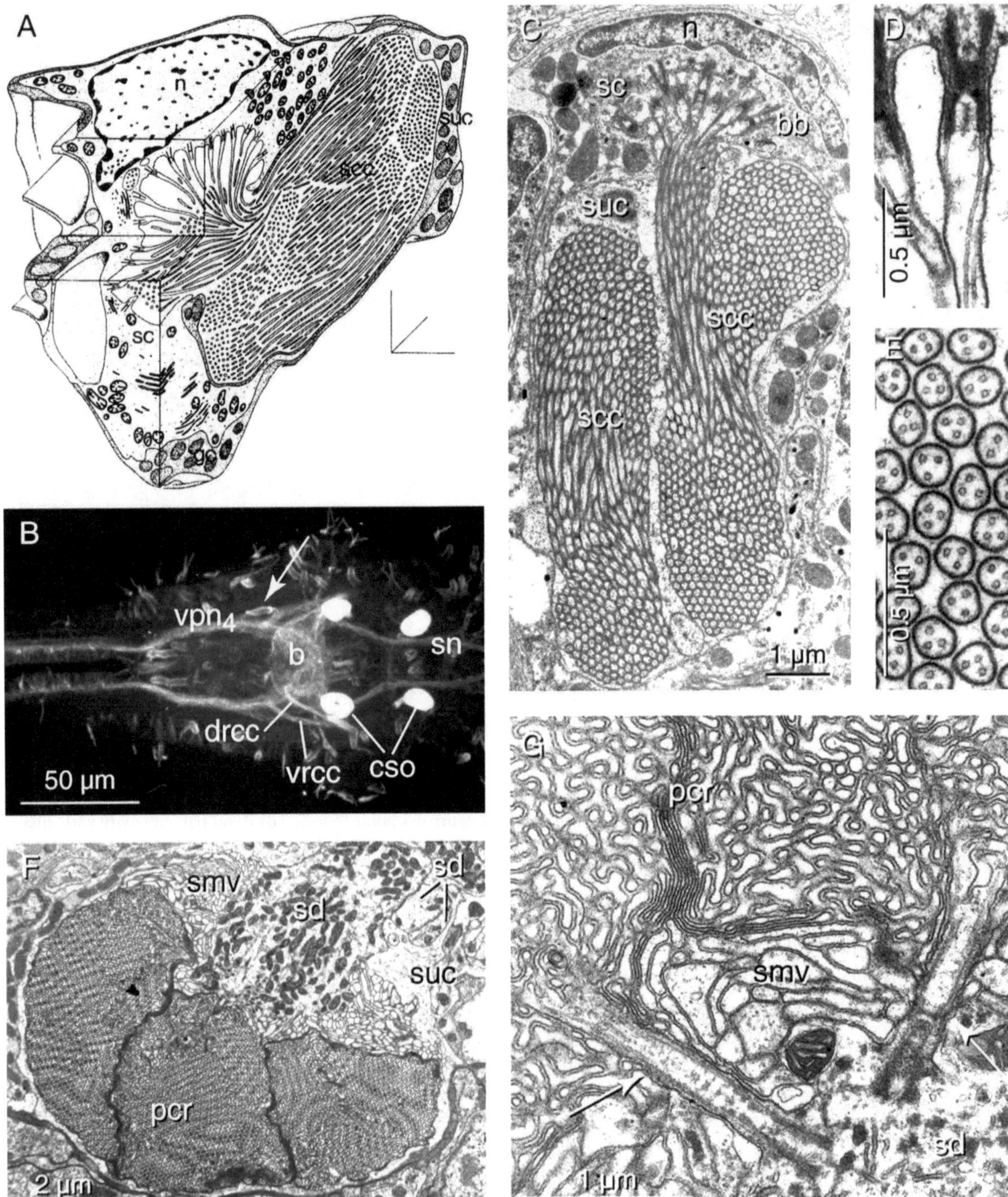

Figure 10. Ciliary photoreceptor-like sense organs. (A–E) *Protodriloides* spp. (A) *P. chaetifer*. Reconstruction of ciliary sense organ composed of sensory cell (sc) and glial supporting cell (suc). (B) *P. symbioticus*. Position of ciliary prostomial sense organs (cso) behind the neuropile of the brain (b), arrow points to basal ciliated cell; confocal laser scanning microscopic image, anti-α-acetylated-tubulin immunoreactivity. (C) *P. symbioticus*. Sensory cell (sc) with emerging sensory cilia (scc); cilia form coiled bundle. (D–E) *P. chaetifer*. Ultrastructure of cilia. (D) Base of cilia with basal body and basal region of axoneme. (E) Cross section of cilia with 3 × 1 axoneme. (F, G) *Protodrilus ciliatus*. So-called statocyst. (F) Low-power micrograph showing 3 paracrystals (pcr), sensory microvilli (smv), the three sensory dendrites (sd) and thin supporting cell (suc). (G) Cilia (arrows) without rootlets emerging from dendrite of large sensory cell splitting off and forming paracrystals. – b, brain; bb, basal body; cso, ciliary sense organ; drcc, dorsal root of circumoesophageal connective; n, nucleus; pcr, paracrystal; sc, sensory cell; scc, sensory cell cilium; sd, sensory dendrite; smv, sensory microvillus; sn, stomatogastric nerve; suc, supporting cell; vpn_4, ventral palp $nerve_4$; vrcc, ventral root of circumoesophageal connective. A–E modified from Purschke & Müller (1996), G modified from Purschke (1990b).

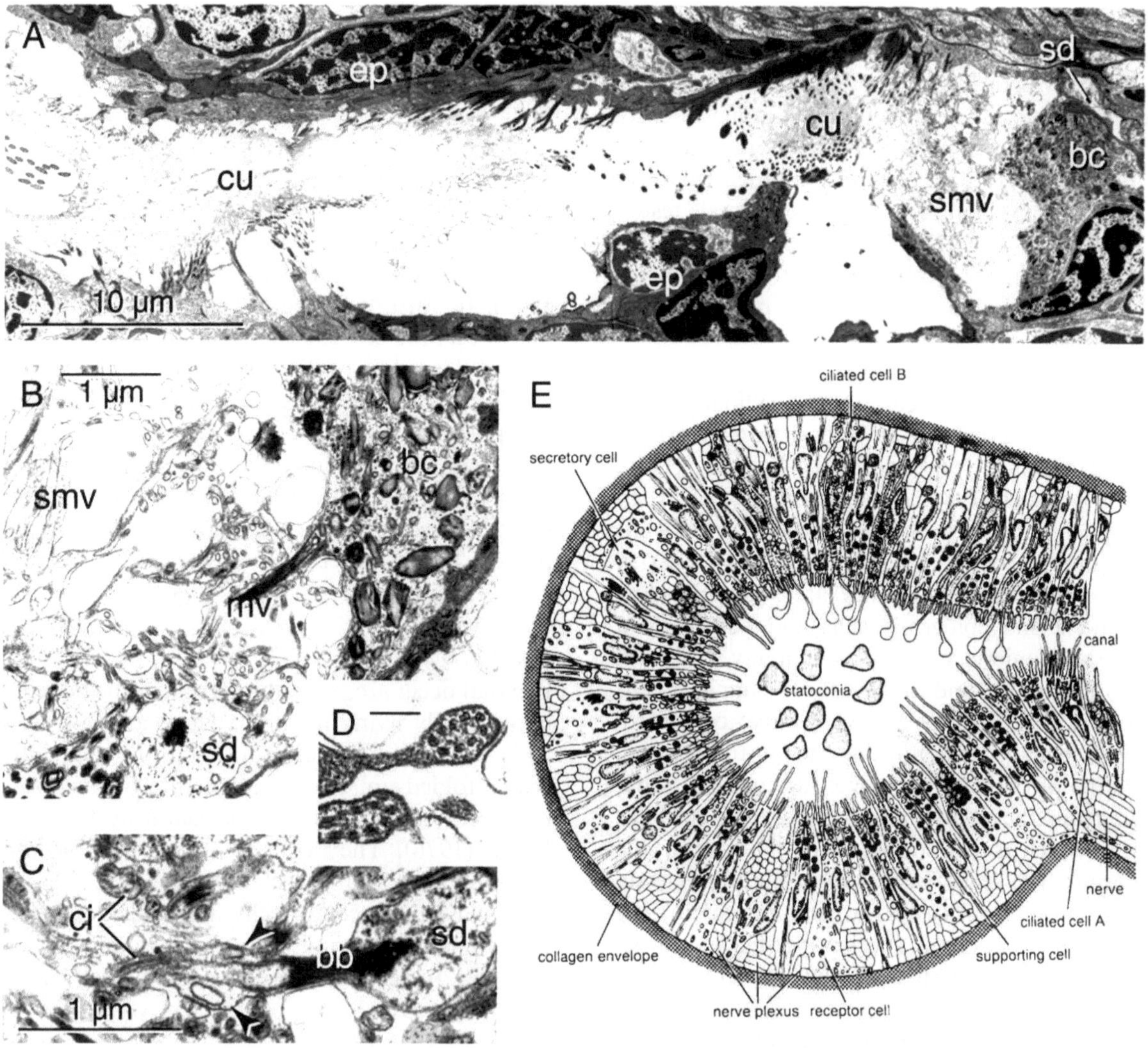

Figure 11. Other types of sense organs. (A–D) *Fauveliopsis adriatica.* Tube-like sense organ of unknown function. (A) Sense organ in longitudinal section, tube lined by epidermal invagination (ep) and filled with cuticular material (cu), at the base filled with microvillus-like sensory processes (smv). (B) Base of sense organ with sensory microvilli (smv), sensory dendrite (sd) and basal cell (bc) containing numerous vesicles. (C) Sensory dendrite with cilium, arrowheads: vesicles present in cilium. (D) Cross section of cilium with branches; scale represents 0.2 μm. (E) *Arenicola marina.* Statocyst with statoconia, epithelium made up of secretory, supporting ciliated and ciliated sensory cells. – bb, basal body; bc, basal cell; ci, cilium; cu, cuticle; ep, epidermis; mv, microvilli; sd, sensory dendrite; smv, sensory microvillus. A–D Langhage et al., unpubl., E from Storch & Schötzer-Schrehardt (1988) after Nowak (1978).

similar events observed in several platyhelminth species (Sopott-Ehlers, 1984, 1988, 1991). The functional significance of such ocelli is unknown and some hypotheses have been suggested by Bartolomaeus (1993).

Probably the so-called type-2 ocelli present in *Saccocirrus papillocercus* also belong to this type (Purschke, 1992). These are multicellular organs composed of several supportive cells and rhabdomeric sensory cells, forming a spherical structure into which the photoreceptors project. The rhabdomeres are formed by flattened cell processes already discernable in the light microcope with interference contrast.

Segmental ocelli

Segmental ocelli as well as tentacular and pygidial ocelli are structurally different from cerebral eyes and thus most likely represent independently evolved structures. When experimentally induced, expression of *Pax6* triggers development of cerebral eyes in any body region in the animals tested so far (Gehring, 2001). Thus, development of differently structured ocelli in the trunk of polychaetes thus favours the hypothesis of convergent evolution.

Segmental ocelli are present in certain Opheliidae, Eunicidae, Syllidae and Sabellidae, all of which represent different structural plans (Kernéis, 1968; Hermans, 1969; Dragesco-Kernéis, 1980; Verger-Bocquet, 1981; Purschke et al., 1995). In the opheliid taxa *Armandia* and *Polyophthalmus* the segmental ocelli are positioned somewhat in front of the parapodia. They are situated below the cuticle and epidermal cells and consist of a large rhabdomeric receptor cell, a layer of thin supportive cells and a layer of pigment cells (Fig. 8E). The sensory and supportive cells form an extracellular cavity housing the photoreceptors, while the pigment cells are mesodermal and separated from the overlain cells by an extracellular matrix which is continuous with that below the epidermis. The sensory cell is inverse and possesses several sensory processes bearing microvilli; the number of processes is species-specific and is between 10 and 25 (Hermans, 1969; Purschke et al., 1995).

In *Syllis spongicola* each chaetigerous segment of the stolon develops a pair of eyes situated below the dorsal parapodial cirri. Each compound eye consists of several units made up of supportive cells and sensory cells, being separated from one another by mucous cells. The sensory cells are characterized by an invagination of the apical photoreceptor-bearing surface and are thus phaosomous. The rhabdomeric photoreceptor has an additional vestigal cilium (Verger-Bocquet, 1981).

In the sabellid *Dasychone bombyx* these ocelli have a corresponding position between noto- and neuropodia. The ocelli are composed of several pigment and few sensory cells forming a follicle-like epidermal invagination (Kernéis, 1968; Dragesco-Kernéis, 1980). The follicle is filled with a cuticular lens. The photoreceptors are rhabdomeric. The eyes present in the eunicid *Eunice viridis* apparently represent a fourth type of such secondarily evolved eyes (Eaking & Hermans, 1988).

Branchial ocelli

Branchial ocelli, present in Sabellidae and Serpulidae, are compound eyes consisting of numerous repetitive units (reviewed by Verger-Bocquet, 1992). In various Sabellidae each unit contains a photoreceptor cell characterized by stacks of parallel and modified cilia called lamellar sacs. The cilia have a $9 \times 2 + 0$ axoneme without rootlets. A lens-like structure formed by other supportive cells than the pigment cells is also present. Apart from this common feature, there is a great diversity between species. In Serpulidae the sensory cells possess highly ordered microvilli and a stack of ciliary membranes.

Pygidial ocelli

Pygidial ocelli are present in certain Sabellidae not permanently living in tubes. These animals crawl with the pygidium in front and the tentacular crown folded up. The ultrastructure of such ocelli has been studied in *Chone ecaudata* by Ermak & Eakin (1976). They consist of a group of three types of epidermal cells: secretory cells, pigment cells and photoreceptor cells. The sensory cells are rhabdomeric, bearing numerous microvilli and two cilia lying in a depression of the cell underneath the cuticle.

Photoreceptor-like sense organs

There are various kinds of other photoreceptor-like sensory structures in polychaetes. Usually the sensory cells are ciliary, producing a great expanse of ciliary membranes as is the case in typical photoreceptors. These structures are housed in an extracellular cavity formed by sensory and supportive cells but shading pigment is normally absent in these sensory structures. Although they are structurally similar to photoreceptor structures, clear experimental evidence for light perception is lacking, Thus, other modalities are conceivable as well (see Eakin & Hermans, 1988; Rhode, 1992). Recently, presence of opsin was proven in ciliary photoreceptors of *Platynereis dumerilii*, indicating possible photoreceptive function of these ciliary receptors among polychaetes and Bilateria and thus their overall homology (Arendt et al., 2004).

Many polychaete species investigated possess at least one type of such organs, often in addition to typical pigmented eyes (e.g., Pietsch & Westheide, 1985; Rhode, 1991; Purschke, 1992; Bartolomaeus, 1993; Purschke & Jouin-Toulmond, 1994). These organs show a great diversity in number, position and structure between taxa and our knowledge about their diversity and occurrence is still rather incomplete. Thus, their importance as characters for high-level phylogenetic inference is slight, but these structures may be very useful at lower levels: species-specific differences between closely related species have repeatedly been reported (e.g., Pietsch & Westheide, 1985; Purschke, 1990a, b, 1992; Purschke & Jouin-Toulmond, 1993, 1994).

Formerly these organs were mostly called phaosomes, following the terminology of clitellate photoreceptors. However, phaosomes as defined above are extremely rare in polychaetes (Purschke, 2002b): Mostly, these structures are associated with supporting cells and the sensory cells form an extracellular cavity with them. This fact has either been overlooked because the supportive cells are relatively inconspicuous, or it has simply been neglected (e.g., Sensenbaugh & Franzén, 1987; Rhode, 1991). True phaosomes are, for instance, present in the palps of Protodrilidae (Purschke, 1993, 2003; Purschke & Jouin-Toulmond, 1993) or in Siboglinidae (Nørrevang, 1974).

The range of diversity can be demonstrated by the following examples. Species of *Microphthalmus* possess an unpaired ocellus-like structure at the posterior end of the brain (Fig. 9A, B; Pietsch & Westheide, 1985). It is characterized by a huge cavity, 10–20 μm in diameter, housing numerous microvillus-like structures. There are one or two sensory cells, each giving rise to 8–20 cilia with basal branches indistinguishable from regular microvilli (Fig. 9B). More peripherally these structures are highly ordered and resemble a rhabdomere (Fig. 9A). In addition, different paired ciliary sense organs have been found anteriorly in the prostomium of *Microphthalmus*. These anterior organs are somewhat similar to those described from the brain of *Nereis pelagica*, *Eulalia viridis*, *Anaitides mucosa*, *Eteone longa*, *Lepidonotus helotypus* and *Saccocirrus* spp (type-1 ocelli). There are usually a few cilia and ciliary branches often contain a single microtubule (Dhainaut-Courtois, 1965; Whittle & Golding, 1974; Gotow, 1976; Rhode, 1991; Purschke, 1992).

A completely different type of photoreceptor-like sense organ has been found in *Protodriloides* spp. (Fig. 10A–E; Purschke & Müller, 1996). Situated in the anterior end close to the brain, the organs can be labelled with antibodies against tubulin (Fig. 10B). Their most characteristic feature is a sensory cell giving rise to a bundle of approximately 200 unbranched cilia (Fig. 10A, C). These are strictly parallel and rolled up several times in an extracellular cavity bordered by a supportive cell. The cilia are 50–100 μm long, lack rootlets and have an extremely modified axoneme: just above the basal body the $9 \times 2 + 0$ pattern is transformed into a 3×1 pattern. Such organs have not been found in any other polychaete species.

The 'statocysts' present in most *Protodrilus* species also belong to this type (Fig. 10F, G; Purschke, 1990a,b). Regardless their size, these organs consist of a cup-shaped supportive cell and three ciliary sensory cells. The most striking feature is that the cilia of the largest sensory cell form paracrystalline structures (Fig. 10F). These paracrystals are made up of highly folded and regularly arranged ciliary membranes (Fig. 10G). The cilia of the remaining sensory cells form microvillus-like branches. There are species-specific differences in terms of size, number of cilia and structure of ciliary membranes. In view of this structure a function as statocyst appears to be questionable. Recently, rather similar organs were found in juveniles of various species of Spionida (Hausen, 2001). These structures are regarded as homologous with the 'statocysts' of Protodrilidae, giving additional support for a closer relationship of these taxa as suggested previously (see Purschke, 1993). Interestingly these organs are only present in larvae and juveniles in Spionida. They are reduced and disappear in adults. Given that there is a close relationship to the Protodrilidae, occurrence of these sense organs in the latter would speak in favour of a progenetic origin of these interstitial polychaetes.

Other types of sensory organs

A completely new type of sense organ has recently been found in *Fauveliopsis* cf. *adriatica* (Fig. 11A–D; Langhage et al., unpubl. obs.). Situated at the anterior end of the retractile prostomium of this species, these organs can be characterized as epidermal follicles extending deeply into the brain. Each is multicellular and about 40 μm deep; the lumen is filled with cuticular material except for the most posterior part (Fig. 11A). The posterior part is somewhat widened and filled with microvillus-like processes of a few sensory cells (Fig. 11B). What appears to be microvilli are mostly branches of cilia which intermingle with regular microvilli also originating from these cells. A conspicuous non-sensory cell with numerous dense vesicles forms the terminal end of the tube (Fig. 11A, B). The function of these tube-like sense organs is completely unknown. The organs appear to have some similarities with the ocellar tubes of Sipuncula (Hermans & Eakin, 1969; Rice, 1993). Further observations must show whether similar organs are present in other polychaete taxa and, for example, whether the so-called saccular apparatus in *Glycera dibranchiata* described from light microscopic observations by Simpson (1959) is a similar structure.

Statocysts

True statocysts are present in a small number of polychaete taxa such as Arenicolidae, Orbiniidae, Terebellidae and Sabellidae (Verger-Bocquet, 1992; Rouse & Pleijel, 2001). Ultrastructural studies have been carried out in only two species: *Arenicola marina* and the aulophora larva of *Lanice conchilega* (Fig. 11E; Heimler, 1983; Storch & Schlötzer-Schrehardt, 1988). Since then, no other data have been presented. The statocysts may be closed or open epidermal vesicles. The statocysts proper consist of supportive cells, gland cells and one to three types of sensory cells lining a cavity which mostly contains several statoliths. For details see Storch & Schlötzer-Schrehardt (1988) and Verger-Bocquet (1992). The scattered occurrence and variable position suggests that these organs are convergently evolved in the respective taxa, in spite of an overall similarity which might be due to functional constraints.

Conclusions

There exists a great diversity of sensory structures and sense organs in polychaetes. Whereas the structure of the sensory cells seems to be more or less completely known, this does not apply to the sense organs as a whole. Of great phylogenetic importance are palps, nuchal organs and pigmented ocelli. A remarkable aspect is the broad range of unpigmented ocelli and photoreceptor-like structures; sometimes three or four different presumed light-sensitive organs are present in one species. As demonstrated by some examples, the extent of sense organ diversity has not yet been explored and many mysteries are still waiting discovery and functional interpretation.

Acknowledgements

Thanks are expressed to Professor Wilfried Westheide for valuable discussions and support. Carola Bührmann, Sindy Göbel, Petra Langhage, Jens Koldehoff, Michael Kuper, Sonja Raabe, Birthe Rohling and Katrine Worsaae kindly supplied material or unpublished micrographs. Technical assistance by Anna Paul, Sonja Raabe, Werner Mangerich, Martina Biedermann, and Anja Ritz is gratefully acknowledged. Thanks are also due to Andrea Noël for typing the manuscript.

References

Åkesson, G., 1958. A study of the nervous system of the Sipunculoideae with some remarks on the development of the two species *Phascolion strombi* Montagu and *Golfingia minuta* Keferstein. Undersökningar över Öresund 38: 1–249.

Åkesson, G., 1962. The embryology of *Tomopteris helgolandicus* (Polychaeta). Acta Zoologica (Stockholm) 43: 135–199.

Åkesson, G., 1967. The embryology of the polychaete *Eunice kobiensis*. Acta Zoologica (Stockholm) 48: 141–192.

Ameyaw-Akumfi, C., 1976. The function of the caruncle of an amphinomid *Eurythoë* sp. – a histological investigation. Zoologischer Anzeiger 196: 318–322.

Amieva, M. R., C. G. Reed & J. R. Pawlik, 1987. Ultrastructure and behavior of the larva of *Phragmatopoma californica* (Polychaeta: Sabellariidae): identification of sensory organs potentially involved in substrate selection. Marine Biology 95: 259–266.

Arendt, D. & J. Wittbrodt, 2001. Reconstructing the eyes of Urbilateria. Philosophical Transactions of the Royal Society of London B356: 1545–1563.

Arendt, D., K. Tessmar-Raible, H. Snyman, A. W. Dorresteijn & J. Wittbrodt, 2004. Ciliary photoreceptors with a vertebrate-type opsin in an invertebrate brain. Science 306: 869–871.

Ax, P., 1995. Das System der Metazoa I. Fischer, Stuttgart, 1–226.

Bantz, M. & C. Michel, 1972. Les cellules sensorielles des papilles de la trompe chez *Glycera convoluta* Keferstein (Annélide Polychète). Zeitschrift für Zellforschung 134: 351–366.

Bartolomaeus, T., 1987. Ultrastruktur des Photorezeptors der Trophophora von *Anaitides mucosa* Oersted (Phyllodocidae, Annelida). Mikrofauna Marina 3: 411–418.

Bartolomaeus, T. 1992a. Ultrastructure of the photoreceptor in the larvae of *Lepidochiton cinereus* (Mollusca, Polyplacophora) and *Lacuna divaricata* (Mollusca, Gastropoda). Microfauna Marina 7: 215–236.

Bartolomaeus, T., 1992b. Ultrastructure of the photoreceptors in certain larvae of the Annelida. Microfauna Marina 7: 191–214.

Bartolomaeus, T., 1993. Different photoreceptors in juvenile *Ophelia rathkei* (Annelida, Opheliida). Microfauna Marina 8: 99–114.

Bhup, R. & J. R. Marsden, 1982. The development of the central nervous system in *Capitella capitata* (Polychaeta, Annelida). Canadian Journal of Zoology 60: 2284–2295.

Blumer, M., 1995. The ciliary photoreceptor in the teleplanic veliger larvae of *Smaragdia* sp. and *Strombus* sp. (Mollusca, Gastropoda). Zoomorphology 115: 73–81.

Bocquet, M., 1976. Ultrastructure de l'organe photorécepteur *d'Autolytus pictus* (Annélide Polychète). Étude chez la souche, le stolon parvenu à maturité sexuelle et la tête régénéréé. Journal de Microscopie et de Biologie Cellulaire 25: 61–66.

Bocquet, M., 1977. Étude ultrastructurale de l'organe photorécepteur *d'Odontosyllis ctenostoma* S/F: Eusyllinae (Annélide Polychète). Journal of Ultrastructure Research 58: 210–217.

Böggemann, M., D. Fiege & G. Purschke, 2000. Ultrastructure of the proboscidial papillae in some *Glycera* species (Annelida: Polychaeta: Glyceridae). Cahiers de Biologie Marine 41: 143–153.

Boilly-Marer, Y., 1972a. Étude ultrastructurale des cirres parapodiaux de Nereidiens atoques (Annélides Polychètes). Zeitschrift für Zellforschung 131: 309–327.

Boilly-Marer, Y., 1972b. Présence de cellules de type myoépithélial chez les Nereidae (Annélides Polychètes). Journal de Microscopie 15: 253–256.

Brandenburger, J. L. & R. M. Eakin, 1981. Fine structure of ocelli in larvae of an archiannelid *Polygordius cf. appendiculatus*. Zoomorphology 99: 23–36.

Bullock, T. H., 1965. Annelida. In T. H. Bullock & G. A. Horridge (eds), Structure and Function in the Nervous System of Invertebrates. Vol. I. Freeman and Company, San Francisco, U.S.A., 661–789.

Dhainaut-Courtois, N., 1965. Sur la présence d'un organe photorécepteur dans le cerveau de *Nereis pelagica* L. (Annélide Polychète). Comptes Rendus Hebdomadaires des Seances de l'Academie des Sciences Ser. D 261: 1085–1088.

Dorsett, D. A. & R. Hyde, 1969. The fine structure of the compound sense organs in the cirri of *Nereis diversicolor*. Zeitschrift für Zellforschung 85: 243–255.

Dragesco-Kernéis, A., 1980. Taches oculaires segmentaires chez *Dasychone* (Annélides Polychètes). Etude ultrastructurale. Cahiers de Biologie Marine 21: 287–302.

Eakin, R. M., 1963. Lines of evolution in photoreceptors. In D. Mazia & A Tyler (eds), General Physiology of Cell Specialization. McGraw-Hill, New York, 393–425.

Eakin, R. M., 1982. Continuity and diversity in photoreceptors. In J. A. Westfall (ed.), Visual Cells in Evolution. Raven, New York, 91–105.

Eakin, R. M. & J. L. Brandenburger, 1985. Effects of light and dark on photoreceptors in the polychaete annelid *Nereis limnicola*. Cell and Tissue Research 242: 613–622.

Eakin, R. M. & C. O. Hermans, 1988. Eyes. In W. Westheide & C.O. Hermans (eds), The Ultrastructure of Polychaeta. Microfauna Marina 4: 135–156.

Eakin, R. M. & J. A. Westfall, 1964. Further observations on the fine structure of some invertebrates eyes. Zeitschrift für Zellforschung 62: 310–322.

Eakin, R. M., G. G. Martin & C. T. Reed, 1977. Evolutionary significance of fine structure of archiannelid eyes. Zoomorphologie 88: 1–18.

Eeckhaut, I. & M. Jangoux, 1993. Integument and epidermal sensory structures of *Myzostoma cirriferum* (Myzostomida). Zoomorphology 113: 33–45.

Eibye-Jacobsen, D., 1991. A revision of *Eumida* Malmgren, 1865 (Polychaeta: Phyllodocidae). Steenstrupia 17: 81–140.

Eisig, H., 1914. Zur Systematik, Anatomie und Morphologie der Ariciiden nebst Beiträgen zur generellen Systematik. Mitteilungen aus der Zoologischen Station zu Neapel 21: 153–593.

Ermak, T. H. & R. M. Eakin, 1976. Fine structure of the cerebral and pygidial ocelli in *Chone ecaudata* (Polychaeta: Sabellidae). Journal of Ultrastructure Research 54: 243–260.

Fauchald, K. & G. W. Rouse, 1997. Polychaete systematics: past and present. Zoologica Scripta 26: 71–138.

Fernandez, J., V. Téllez & N. Olea, 1992. Hirudina. In F. W. Harrison (ed.), Microscopic Anatomy of Invertebrates. Vol. 7 Annelida. Wiley-Liss., New York, 323–394.

Fischer, A. & J. Böckelmann, 1965. Morphology and structural changes of the eye of *Platynereis dumerilii* (Polychaeta). In Rohen, J. W. (ed.), The Structure of the Eye II. Symposium Stuttgart: F.K. Schattauer, 171–174.

Gehring, W. J., 2001. The genetic control of eye development and its implications for the evolution of the various eye-types. Zoology 104: 171–183.

Gehring, W. J. & K. Ikeo, 1999. *Pax 6* mastering eye morphogenesis and eye evolution. Trends in Genetics 15: 371–377.

Gotow, T., 1976. Photoreceptor-like cells in the prostomium of a scaleworm, *Lepidonotus helotypus*. Zoological Magazine (Tokyo) 85: 265–269.

Hausen, H. 2001. Untersuchungen zur Phylogenie "spiomorpher" Polychaeten (Annelida). Logos, Berlin, 1–142.

Hayashi, I. & S. Yamane, 1994. On a probable sense organ newly found in some eunicid polychaetes. Journal of the

Marine Biological Association of the United Kingdom 74: 765–770.

Hayashi, I. & S. Yamane, 1997. Further observations of a recently found sense organ in some euniceforms, with special reference to *Lumbrinereis longifolia* (Polychaeta, Lumbrinereidae). Bulletin of Marine Science 60: 564–574.

Heimler, W., 1983. Untersuchungen zur Larvalentwicklung von *Lanice conchilega* (Pallas) 1766 (Polychaeta, Terebellomorpha). Teil III: Bau und Struktur der Aulophora-Larve. Zoologische Jahrbücher der Anatomie 111: 411–478.

Heimler, W., 1988. Larvae. In W. Westheide and C.O. Hermans (eds), The Ultrastructure of Polychaeta. Gustav Fischer Verlag, New York, pp. 353–371.

Hermans, C. O., 1969. Fine structure of the segmental ocelli of *Armandia brevis* (Polychaeta: Opheliidae). Zeitschrift für Zellforschung 96: 361–371.

Hermans, C. O. & R. A. Cloney, 1966. Fine structure of the prostomial eyes of *Armandia brevis* (Polychaeta: Opheliidae). Zeitschrift für Zellforschung 72: 583–596.

Hermans, C. O. & R. M. Eakin, 1969. Fine structure of cerebral ocelli of a sipunculid, *Phascolosoma agassizii*. Zeitschrift für Zellforschung 100: 325–339.

Hermans, C. O. & R. M. Eakin, 1974. Fine structure of the eyes of an alciopid polychaete, *Vanadis tagensis*. Zeitschrift für Morphologie der Tiere 79: 245–267.

Hessling, R. & W. Westheide, 1999. CLSM analysis of development and structure of the central nervous system in *Enchytraeus crypticus* ("Oligochaeta", Enchytraeidae). Zoomorphology 119: 37–47.

Hessling, R., Müller, M. C. & W. Westheide, 1999. CLSM analysis of serotonin-immunoreactive neurons in the central nervous system of *Nais variablis, Slavinia appendiculata* and *Stylaria lacustris* (Oligochaeta: Naididae). Hydrobiologia 406: 223–233.

Hessling, R. & G. Purschke, 2000. Immunohistochemical (cLSM) and ultrastructural analysis of the central nervous system and sense organs in *Aeolosoma hemprichi* (Annelida, Aeolosomatidae). Zoomorphology 120: 65–78.

Holborow, P. L. & M. S. Laverack, 1972. Presumptive photoreceptor structures of the trophophore *Harmothoë imbricata* (Polychaeta). Marine Behaviour and Physiology 1: 139–156.

Jamieson, B. G. M., 1981. The Ultrastructure of the Oligochaeta. Academic, London, New York, 1–462.

Jamieson, B. G. M., 1992. Oligochaeta. In F.W. Harrison (ed.), Microscopic Anatomy of Invertebrates. Vol. 7 Annelida. Wiley-Liss., New York, 217–322.

Jelsing, J., 2002a. Ultrastructural studies of dorsal ciliated organs in Spionidae (Annelida: Polychaeta). Hydrobiologia (in press).

Jelsing, J., 2002b. Ultrastructural investigations on the cephalic and metameric nuchal organs of *Spio* cf. *filicornis* (Polychaeta: Spionidae). Zoomorphology 121: 213–220.

Jouin, C., C. Tchernigovtzeff, M. F. Baucher & A. Toulmond, 1985. Fine structure of probably mechano- and chemoreceptors in the caudal epidermis of the lugworm *Arenicola marina* (Annelida, Polychaeta). Zoomorphology 105: 75–82.

Kernéis, A., 1968. Ultrastructure de photorécepteurs de *Dasychone* (Annélides Polychètes Sabellidae). Journal de Microscopie (Paris) 7:40a (Abstract).

Lacalli, T., 1981. Structure and development of the apical organ in trochophores of *Spirobranchus polycerus, Phyllodoce maculata* and *Phyllodoce mucosa* (Polychaeta). Proceedings of the Royal Society of London B212: 381–402.

Land, M. F., 1984. Molluscs. In Ali, M.A. (ed.), Photoreception and Vision in Invertebrates. Vol. 74. Plenum and NATO Sci. Aff. Div., New York, London, 699–725.

Lewbart, G. A. & N. W. Riser, 1996. Nuchal organs of the polchaete *Parapionosyllis manca* (Syllidae). Invertebrate Biology 115: 286–298.

Marsden, J. R., 1982. Morphological evidence for neural activity in prototrochal and neurotrochal cells of the trochophore larva of the serpulid polychaete, *Galeolaria caespitosa*. International Journal of Invertebrates Reproduction 5: 289–297.

Marsden, J. R. & J. Hsieh, 1987. Ultrastructure of the eyespot in three polychaete trochophore larvae (Annelida). Zoomorphology 106: 361–368.

McHugh, D., 1997. Molecular evidence that echiurans and pogonophorans are derived annelids. Proceedings of the National Academy of Sciences of the United States of America 94: 8006–8009.

McHugh, D., 2000. Molecular phylogeny of the Annelida. Canadian Journal of Zoology 78: 1873–1884.

Michel, C., 1972. Etude ultrastructurale et histochimique des papilles de la gaine de la trompe de *Notomastus latericeus* Sars (Annélide Polychète Sédentaire). Zeitschrift für Zellforschung 128: 482–503.

Michel, W. C., P. Steullet, H. S. Cate, C. J. Burns, A. B. Zhainazarov & C. D. Derby, 1999. High resolution functional labelling of vertebrate and invertebrate olfactory receptor neurons using agmatine, a channel-permeant cation. Journal of Neuroscience Methods 90: 143–156.

Mill, P. J., 1978. Sense organs and sensory pathways. In P. J. Mill (ed.), Physiology of Annelids. Academic, New York, London, 63–114.

Myhrberg, H. E., 1979. Fine structural analysis of the basal epidermal receptor cells in the earthworm (*Lumbricus terrestris*). Cell and Tissue Research 203: 257–266.

Nordheim, H. v., 1991. Ultrastructure and functional morphology of male genital organs and spermatophore formation in *Protodrilus* (Polychaeta, Annelida). Zoomorphology 111: 81–94.

Nørrevang, A., 1974. Photoreceptors of the phaosome (Hirudinean) type in a pogonophore. Zoologischer Anzeiger 193: 297–304.

Orrhage, L., 1964. Anatomische und morphologische Studien über die Polychaetenfamilien Spionidae, Disomidae and Poecilochaetidae. Zoologisk Bidrag Uppsala 36: 335–405.

Orrhage, L., 1980. On the structure and homologues of the anterior end of the polychaete families Sabellidae and Serpulidae. Zoomorphology 96: 113–168.

Orrhage, L. & M. C. M. Müller, 2005. Morphology of the nervous system of Polychaeta (Annelida). Hydrobiologia 535/536 (Dev. Hydriobol. 179): 79–111.

Pietsch, A. & W. Westheide, 1985. Ultrastructural investigations of presumed photoreceptors as a means of discrimination and identification of closely related species of the genus *Microphthalmus* (Polychaeta, Hesionidae). Zoomorphologie 105: 256–276.

Pleijel, F., 1991. Phylogeny and classification of the Phyllodocidae (Polychaeta). Zoologica Scripta 20: 225–261.

Purschke, G., 1986. Ultrastructure of the nuchal organ in the interstitial polychaete *Stygocapitella subterranea* (Parergodrilidae). Zoologica Scripta 16: 13–20.

Purschke, G., 1990a. Fine structure of the so-called statocysts in *Protodrilus adhaerens* (Protodrilidae, Polychaeta). Zoologischer Anzeiger 224: 286–296.

Purschke, G., 1990b. Ultrastructure of the "statocysts" in *Protodrilus* species (Polychaeta): reconstruction of the cellular organisation with morphometric data from receptor cells. Zoomorphology 110: 91–104.

Purschke, G., 1992. Ultrastructural investigations of presumed photoreceptive organs in two *Saccocirrus* species (Polychaeta, Saccocirridae). Journal of Morphology 211, 7–21.

Purschke, G., 1993. Structure of the prostomial appendages and the central nervous system in the Protodrilida (Polychaeta). Zoomorphology 113: 1–20.

Purschke, G., 1997. Ultrastructure of nuchal organs in polychaetes (Annelida) – new results and review. Acta Zoologica (Stockholm) 78: 123–143.

Purschke, G., 1999. Terrestrial polychaetes – models for the evolution of the Clitellata (Annelida)? Hydrobiologia 406: 87–99.

Purschke, G., 2000. Sense organs and central nervous system in an enigmantic terrestrial polychaete, *Hrabeiella periglandulata* (Annelida). Invertebrate Biology 119: 329–341.

Purschke, G., 2002. On the ground pattern of Annelida. Organisms Diversity and Evolution 2: 181–196.

Purschke, G., 2003. Ultrastructure of phaosomous photoreceptors in *Stylaria lacustris* (Naididae, "Oligochaeta", Clitellata) and their importance for the position of the Clitellata in the phylogenetic system of the Annelida. Journal of Zoological Systematics and Evolutionary Research 41: 100–108.

Purschke, G. & C. Jouin-Toulmond, 1993. Ultrastructure of presumed ocelli in *Parenterodrilus taenioides* (Polychaeta, Protodrilidae) and their phylogenetic significance. Acta Zoologica (Stockholm) 74: 247–256.

Purschke, G. & C. Jouin-Toulmond, 1994. Ultrastructure of sense organs and the central nervous system in *Parenterodrilus taenioides* and their phylogenetic significance in the taxon Protodrilida (Annelida, Polychaeta). In J.-C. Dauvin, L. Laubier & D. J. Reish (eds), Actes de la 4ème Conférence internationale des Polychètes. Memoires du Museum National d'histoire Naturelle 162: 119–128.

Purschke, G., Z. Ding & M. C. Müller, 1995. Ultrastructural differences as a taxonomic marker: the segmental ocelli of *Polyophthalmus pictus* and *Polyophthalmus qingdaoensis* sp.n. (Polychaeta, Opheliidae). Zoomorphology 115: 229–241.

Purschke, G. & M. C. Müller, 1996. Structure of prostomial photoreceptor-like sense organs in *Protodriloides* species (Polychaeta, Protodrilida). Cahiers de Biologie Marine 37: 205–219.

Purschke, G., F. Wolfrath & W. Westheide, 1997. Ultrastructure of the nuchal organ and cerebral organ in *Onchnesoma squamatum* (Sipuncula, Phascolionidae). Zoomorphology 117: 23–31.

Purschke, G., R. Hessling & W. Westheide, 2000. The phylogenetic position of the Clitellata and the Echiura – on the problematic assessment of absent characters. Journal of Zoological Systematics and Evolutionary Research 38: 165–173.

Purschke, G. & R. Hessling, 2002. Analysis of the central nervous system and sense organs in *Potamodrilus fluviatilis* (Annelida: Potamodrilidae). Zoologischer Anzeiger 241: 19–35.

Rhode, B., 1989. Ultrastructural investigations on the nuchal organ of the protandric polychaete, *Ophryotrocha puerilis* (Polychaeta, Dorvilleidae. Zoomorphology 108: 315–322.

Rhode, B., 1990a. Ultrastructure of nuchal organs in some marine polychaetes. Journal of Morphology 206: 95–107.

Rhode, B., 1990b. Eye structure of *Ophryotrocha puerilis* (Polychaeta: Dorvilleidae). Journal of Morphology 205: 147–154.

Rhode, B., 1991. Ultrastructure of prostomial receptors in four marine polychaete species (Annelida). Journal of Morphology 209: 177–188.

Rhode, B., 1992. Development and differentiation on the eye in *Platynereis dumerilii* (Annelida, Polychaeta). Journal of Morphology 212: 71–85.

Rhode, B., 1993. Larval and adult eyes in *Capitella* spec. I (Annelida, Polychaeta). Journal of Morphology 217: 327–335.

Rice, M. E., 1993. Sipuncula. In F.W. Harrison & M.E. Rice (eds), Microscopic Anatomy of Invertebrates. Vol. 12 Onychophora, Chilopoda and lesser Protostomata. Wily-Liss, New York, 237–325.

Rosen, M. D., C. R. Stasek & C. O. Hermans, 1979. The ultrastructure and evolutionary significance of the ocelli in the larva of *Katharina tunicata* (Mollusca: Polyplacophora). Veliger 22: 173–178.

Rouse, G. W. & K. Fauchald, 1997. Cladistics and polychaetes. Zoologica Scripta 26: 139–204.

Rouse, G. W. & F. Pleijel, 2001. Polychaetes. Oxford University, New York, 354 pp.

Rullier, F., 1951. Étude morphologique, histologique et physiologique de l'organe nucal chez les annélides polychètes sédentaires. Annales de l' Institut Océanographique de Monaco 25: 207–341.

Rullier, F., 1954. L'organe nucal de *Sthenelais boa* (Johnston). Comptes Rendus Hebdomadires des Séances de Ì Academie des Sciences Series D 238: 1351–1353.

Salvini-Plawen, L. von, 1982. On the polyphyletic origin of photoreceptors. In Westfall, J. A., (ed.), Visual Cells in Evolution. Raven, New York, 137–154.

Salvini-Plawen, L. von & E. Mayr, 1977. On the evolution of eyes and photoreceptors. In Hecht, M. K., W. C. Sterre & B. Wallace (eds), Evolutionary Biology, Vol. 10. Plenum, New York, 207–263.

Sawyer, R. T., 1986. Leech Biology and Behaviour, Vol. 1. Clarendon, Oxford, 418 pp.

Schlawny, A., C. Grünig & H.-D. Pfannenstiel, 1991. Sensory and secretory cells of *Ophryotrocha puerilis* (Polychaeta). Zoomorphology 110: 209–215.

Schlötzer-Schrehardt, U., 1986. Ultrastructural investigation of the nuchal organs of *Pygospio elegans* (Polychaeta). I. Larval nuchal organs. Helgoländer Meeresuntersuchungen 40: 397–417.

Schlötzer-Schrehardt, U., 1987. Ultrastructural investigation of the nuchal organs of *Pygospio elegans* (Polychaeta). II. Adult nuchal and dorsal organs. Zoomorphology 107: 169–179.

Schlötzer-Schrehardt, U., 1991. Ultrastructural differentiation of nuchal and dorsal organs during postembryonic and sexual development of *Pygospio elegans* Claparède (Polychaeta: Spionidae). Ophelia Supplement 5: 633–640.

Sensenbaugh, T. & A. Franzén, 1987. Fine structural observations of the apical organ in the larva of *Polygordius* (Annelida: Polychaeta). Scanning Microscopy 1: 181–189.

Siewing, R., 1953. Morphologische Untersuchungen am "Kopf" der Pisionidae (*Pisione puzae* nov. spec. Annelida, Polychaeta). Zoologischer Anzeiger 150: 298–313.

Simpson, M., 1959. The saccular apparatus in the brain of *Glycera dibranchiata*. Journal of Morphology 104: 561–590.

Singla, C. L., 1975. Ultrastructure of the eyes of *Arctonoë vittata* Grube (Polychaeta, Polynoidae). Journal of Ultrastructure Research 52: 333–339.

Söderström, A., 1927. Über segmental wiederholte "Nuchalorgane" bei Polychäten. Zoologisk Bidrag Uppsala 12: 1–18.

Sopott-Ehlers, B., 1984. Feinstruktur pigmentierter und unpigmentierter Photoreceptoren bei Proseriata (Plathelminthes). Zoologica Scripta 13: 9–18.

Sopott-Ehlers, B., 1988. Fine structure of photoreceptors in two species of the Prolecithophora. Fortschritte der Zoologie 36: 221–227.

Sopott-Ehlers, B., 1991. Comparative morphology of photoreceptors in free-living plathelminths – a survey. Hydrobiologia 227: 231–239.

Spies, R. B., 1975. Structure and function of the head in flabelligerid polychaetes. Journal of Morphology 147: 187–208.

Storch, V., 1972. Elektronenmikroskopische Untersuchungen an Rezeptoren von Anneliden (Polychaeta, Oligochaeta). Zeitschrift für Mikroskopisch-Anatamische Forschung 85: 55–84.

Storch, V. & U. Welsch, 1969. Zur Feinstruktur des Nuchalorgans von *Eurythoë complatana* (Pallas) (Amphinomidae, Polychaeta). – Zeitschrift für Zellforschung 100: 411–420.

Storch, V. & U. Schlötzer-Schrehardt, 1988. Sensory structures. In W. Westheide & C.O. Hermans (eds), The Ultrastructure of Polychaeta. Microfauna Marina 4: 121–133.

Sturrock, M. G. & J. M. Baxter, 1995. The fine structure of the pigment body complex in the intrapigmented aesthetes of *Callochiton achatinus* (Mollusca: Polyplacophora). Journal of Zoology London 235: 127–141.

Toulmond, A., C. Tchernigovtzeff, P. Greber & C. Jouin, 1984. Epidermal sensitivity to hypoxia in the lugworm. Experientia 40: 541–542.

Tzetlin, A. & G. Purschke, 2005. Pharynx and intestine. Hydrobiologia 535/536 (Dev. Hydrobiol. 179): 197–223.

Vanfleteren, J. R., 1982. A monophyletic line of evolution? Ciliary induced photoreceptor membranes. In J. A. Westfall (ed.), Visual Cells in Evolution. Raven, New York, 107–136.

Vanfleteren, J. R. & A. Coomans, 1976. Photoreceptor evolution and phylogeny. Zeitschrift für zoologische Systematik.und Evolutionsforschung 14: 157–169.

Verger-Bocquet, M., 1981. Etude comparative, au niveau infrastructural, entre l'œil de souche et les taches oculaires du stolon chez *Syllis spongicola* Grübe (Annélide Polychète). Archives de Zoologie Expérimentale et Génerale 122: 253–258.

Verger-Bocquet, M., 1983a. Les organes photorèceteurs des Syllidiens (Annélides Polychètes). Année Biologie 22: 169–185.

Verger-Bocquet, M., 1983b. Étude infrastructurale des organes photorécepteurs chez les larves de deux Syllidiens (Annélides, Polychètes). Journal of Ultrastructure Research 84: 67–72.

Verger-Bocquet, M., 1984. Photoréception et vision chez les Annèlides. In Ali, M. A., (ed.), Photoreception and Vision in Invertebrates. Plenum, New York, London, 289–334.

Verger-Bocquet, M., 1992. Polychaeta: sensory structures. In F. W. Harrison (ed.), Microscopic Anatomy of Invertebrates. Vol. 7 Annelida. Wiley-Liss., New York, 181–196.

Welsch, U., V. Storch & K. S. Richards, 1984. Epidermal cells. In J. Bereiter-Hahn, A. G. Matoltsy & K. S. Richards (eds), Biology of the Integument. 1. Invertebrates. Springer, Heidelberg, New York: 269–296.

West, D. L., 1978. Comparative ultrastructure of juvenile and adult nuchal organs of an annelid (Polychaeta, Opheliidae). Tissue and Cell 10: 243–257.

Westheide, W., 1997. The direction of evolution within Polychaeta. Journal of Natural History 31: 1–15.

Westheide, W., D. McHugh, G. Purschke & G. W. Rouse, 1999. Systematization of the Annelida: different approaches. Hydrobiologia 402: 291–307.

Whittle, A. C. & D. W. Golding, 1974. The fine structure of prostomial photoreceptors in *Eulalia viridis* (Polychaeta: Annelida). Cell and Tissue Research 165: 379–398.

Whittle, A. C. & Z. R. Zahid, 1974. Fine structure of nuchal organs in some errant polychaetous annelids. Journal of Morphology 144: 167–184.

Windoffer, R. & W. Westheide, 1988. The nervous system of *Dinophilus gyrociliatus* (Annelida: Polychaeta). I. Number, types and distribution pattern of sensory cells. Acta Zoologica 69: 55–64.

Worsaae, K. & R. M. Kristensen, 2003. A new species of *Paranerilla* (Polychaeta: Nerillidae) from northeast Greenland waters, arctic ocean. Cahiers de Biologie Marine 44: 23–39.

Zahid, Z. R. & D. W.Golding, 1974. Structure and ultrastructure of the central nervous system of the polychaete *Nephtys*, with special reference to photoreceptor elements. Cell and Tissue Research 149: 567–576.

Zavarzina, E. G., 1987. New data on photoreceptor structure in *Ophryotrocha dimorphica* (Polychaeta, Eunicida). Doklady Akademii NAUK SSSR 294: 224–227 (in Russian).

Hydrobiologia (2005) **535/536:** 79–111 © Springer 2005
T. Bartolomaeus & G. Purschke (eds), Morphology, Molecules, Evolution and Phylogeny in Polychaeta and Related Taxa

Morphology of the nervous system of Polychaeta (Annelida)

Lars Orrhage[1] & Monika C.M. Müller[2,*]
[1]*Berzeliigatan 21, S-412 53, Göteborg, Sweden*
[2]*Spezielle Zoologie, Fachbereich 5, Universität Osnabrück, D-49069 Osnabrück, Germany*
*(*Author for correspondence: Tel.: +49-541-969-2858/2859, Fax: +49-541-969-2587,*
E-mail: mcmueller@biologie.uni-osnabrueck.de)

Key words: Annelida, neuroanatomy, 'Polychaeta', cLSM

Abstract

The article summarizes our up to date knowledge about the morphology of the annelid, especially the polychaete, central and peripheral nervous system. Since the cephalic nervous system was in the focus of controversial discussions for decades, the structure of its neuropile, associated ganglia and nerves is reviewed in detail. The enormous variation of the ventral nerve cord and peripheral nerves is presented as well as a theory how this might have evolved. A ground pattern of the polychaete nervous system is suggested, based on developmental and regeneration studies.

Introduction

In Annelida as well as in other invertebrate taxa the nervous system is considered to be a most conservative organ system (Bullock & Horridge, 1965; Orrhage, 1974; Rouse & Fauchald, 1997; Müller, 1999b). Studies on the structure of the polychaete brain and the metameric nervous system have therefore proved to be of particular value in assessing homologies of various anterior appendages (Gustafson, 1930; Remane, 1963; Orrhage, 1974, 1980; Golding, 1992; Purschke & Jouin-Toulmond, 1994; Hessling & Purschke, 2000, Purschke, 2000) and of other structures. The morphology of the central nervous system (CNS) of Polychaeta was first investigated by dissection, then by light and electron microscopy and nowadays by combination of immunohistochemistry and confocal laserscanning microscopy (cLSM). Understanding the organization of the CNS is useful for (1) elucidating the interrelationships of the polychaete families, (2) resolving the much-debated question of the conceivable segmentation of the polychaete brain and (3) productively discussing the phylogenetic connections between Polychaeta and other invertebrate taxa (Orrhage, 1980–2001).

The brain

Orrhage (1964–2001) and Orrhage & Eibye-Jacobsen (1998) analyzed the cephalic nervous system of 32 families of Polychaeta; in 28 of them the brain and associated nerves could be analyzed in detail.

The commissures of the circumesophageal connectives

Judging from his figures, Rohde (1887) discovered that the brain of *Lepidasthenia elegans* ('*Polynoe elegans*') contains four transverse commissures. Two of them (a dorsal and a ventral one, *dc*, *vc*) were in contact with an anterior (ventral, *vr*) circumesophageal root while the other two were connected to a posterior (dorsal, *dr*) root of the connectives. In the present chapter these commissures are designated *dcvr*, *vcvr*, *dcdr* and *vcdr*, respectively. Having examined Amphinomidae, some Polynoidae, Aphroditidae and Nereididae, Gustafson (1930) certified that Rohde's observations were applicable to 'all errant polychaetes'. Although at that time this statement could have seemed insufficiently well founded, later

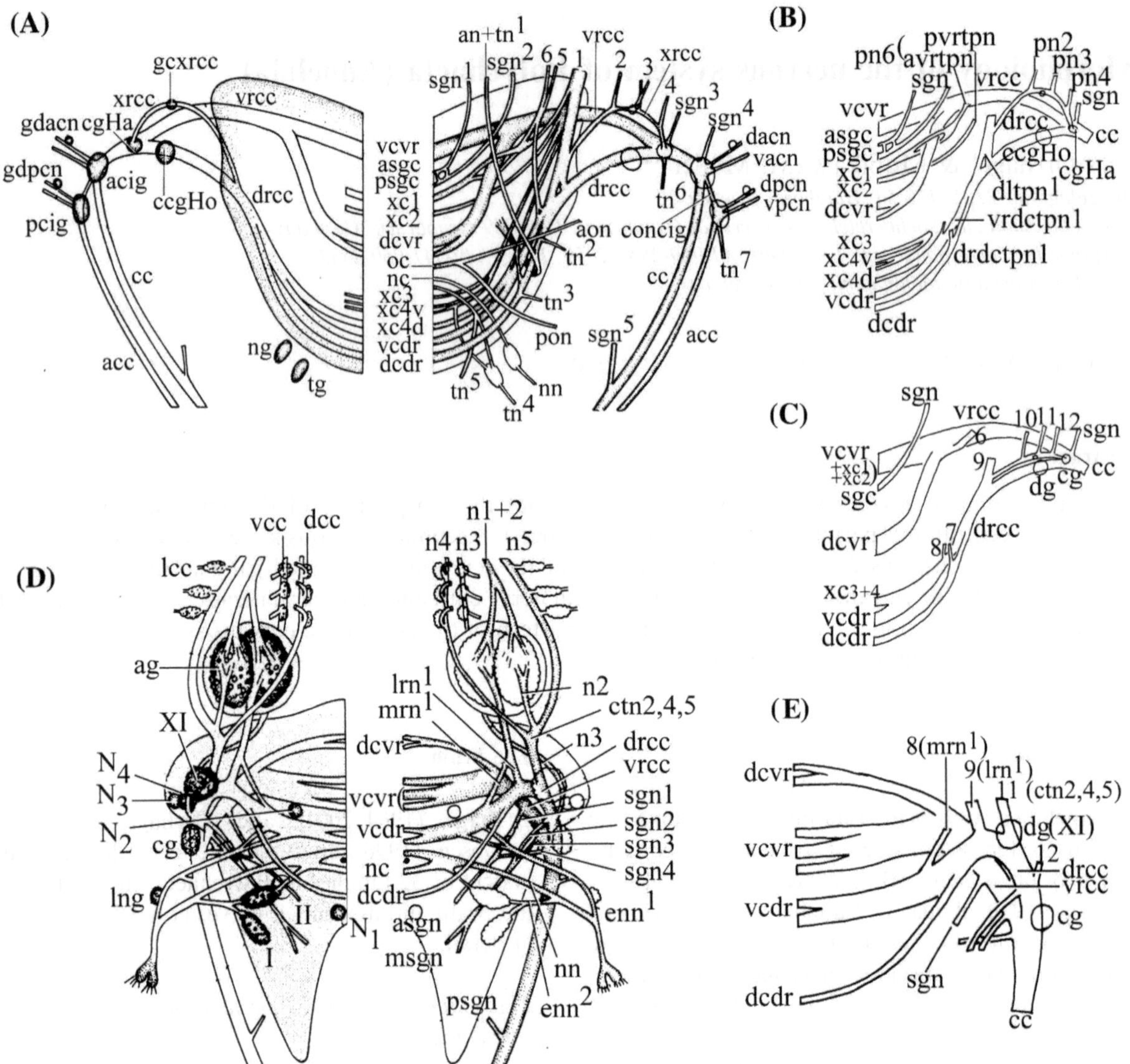

Figure 1. The cephalic nervous system of (A–C) *Neanthes virens*; (D, E) *Glycera rouxii*. Semi-schematic dorsal views. cc – circumesophageal connective; cg – cerebral ganglion; dcdr, dcvr – dorsal commissure of the dorsal (drcc) and ventral (vrcc) root of cc; dg – dorsal ganglion; vcdr, vcvr – ventral commissure of the dorsal and ventral root of cc; 1–12 – palp nerve roots. Further abbreviations see abbreviation list. Modified from Orrhage (1993) (A–C); Orrhage (1999) (D, E).

Figure 2. The anterior part of the circumesophageal connective, its dorsal and ventral roots and their brain commissures (when observed). b-brain; nla - nerves of the lateral antenna (when present, O–U, W); nma – nerves of the median antenna (when present, O–U, W); na - nerves of antennae; sgn – stomatogastric nerves; 1–12–roots of the palp nerves. The main palp nerve roots indicated by blackening. Dorsal view, right side, schematic. For further abbreviations see abbreviation list. Diagrams after or with reference to Orrhage, 1964 (I-K); Orrhage, 1966 (D, H, L, M, Y, Z, A1); Orrhage, 1974 and Purschke, 1993, combined (N); Orrhage, 1978 (A); Orrhage. 1980 (B, C); Orrhage, 1990 (O); Orrhage, 1991 (R, S); Orrhage, 1993 (V); Orrhage, 1995 (P, Q); Orrhage, 1996 (T, U); Orrhage & Eibye-Jcacobsen, 1998 (W); Orrhage, 1999 (X); Orrhage, 2001 (E–G).

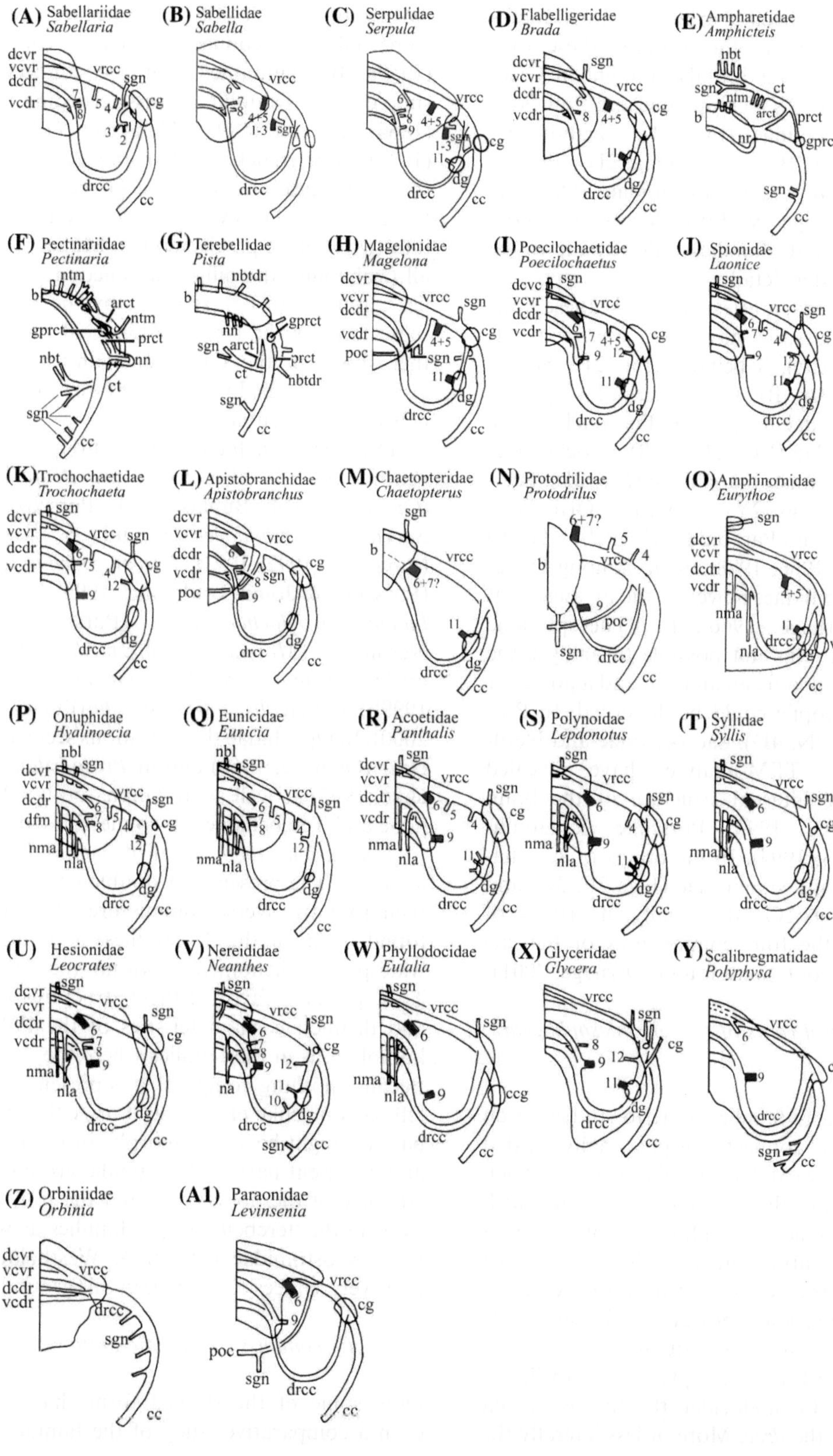
(A) Sabellariidae Sabellaria
dcvr
vcvr
dcdr
vcdr
vrcc
sgn
cg
drcc
cc
(B) Sabellidae Sabella
vrcc
4+5
1-3
sgn
drcc
cc
(C) Serpulidae Serpula
vrcc
4+5
1-3
sgn
cg
11
dg
drcc
cc
(D) Flabelligeridae Brada
dcvr
vcvr
dcdr
vcdr
sgn
vrcc
4+5
cg
11
dg
drcc
cc
(E) Ampharetidae Amphicteis
nbt
sgn
ntm
ct
arct
b
nr
prct
gprct
sgn
cc
(F) Pectinariidae Pectinaria
ntm
b
arct
ntm
gprct
prct
nn
nbt
ct
sgn
cc
(G) Terebellidae Pista
nbtdr
b
nn
gprct
sgn
arct
prct
ct
nbtdr
sgn
cc
(H) Magelonidae Magelona
dcvr
vcvr
dcdr
vcdr
poc
vrcc
sgn
4+5
cg
sgn
11
dg
drcc
cc
(I) Poecilochaetidae Poecilochaetus
dcvc
sgn
vcvr
dcdr
vcdr
vrcc
4+5
cg
12
11
dg
drcc
cc
(J) Spionidae Laonice
sgn
vrcc
sgn
cg
12
11
dg
drcc
cc
(K) Trochochaetidae Trochochaeta
dcvr
sgn
vcvr
dcdr
vcdr
vrcc
sgn
cg
12
dg
drcc
cc
(L) Apistobranchidae Apistobranchus
dcvr
vcvr
dcdr
vcdr
poc
vrcc
sgn
cg
dg
drcc
cc
(M) Chaetopteridae Chaetopterus
sgn
b
vrcc
6+7?
11
dg
drcc
cc
(N) Protodrilidae Protodrilus
6+7?
b
vrcc
poc
sgn
drcc
cc
(O) Amphinomidae Eurythoe
sgn
dcvr
vcvr
dcdr
vcdr
vrcc
4+5
nma
nla
drcc
dg
vg1
cc
(P) Onuphidae Hyalinoecia
nbl
sgn
dcvr
vcvr
dcdr
dfm
nma
nla
vrcc
sgn
cg
12
dg
drcc
cc
(Q) Eunicidae Eunicia
nbl
sgn
vrcc
sgn
nma
nla
12
dg
drcc
cc
(R) Acoetidae Panthalis
dcvr
sgn
vcvr
dcdr
vcdr
vrcc
sgn
cg
nma
nla
11
dg
drcc
cc
(S) Polynoidae Lepdonotus
sgn
vrcc
sgn
cg
nma
nla
11
dg
drcc
cc
(T) Syllidae Syllis
sgn
vrcc
sgn
cg
nma
nla
dg
drcc
cc
(U) Hesionidae Leocrates
dcvr
sgn
vcvr
dcdr
vcdr
vrcc
sgn
cg
nma
nla
dg
drcc
cc
(V) Nereididae Neanthes
sgn
vrcc
sgn
cg
12
na
11
10
dg
drcc
sgn
cc
(W) Phyllodocidae Eulalia
sgn
vrcc
sgn
nma
nla
ccg
drcc
cc
(X) Glyceridae Glycera
vrcc
sgn
cg
12
11
dg
drcc
cc
(Y) Scalibregmatidae Polyphysa
vrcc
cg
drcc
sgn
cc
(Z) Orbiniidae Orbinia
dcvr
vcvr
dcdr
vcdr
vrcc
drcc
sgn
cc
(A1) Paraonidae Levinsenia
vrcc
cg
poc
sgn
drcc
cc

investigations have been instrumental in strengthening it. Thus, among the 'errants' these four transverse commissures are also present in Glyceridae (Figs 1D–E, 2X), Onuphidae and Eunicidae (Fig. 2P–Q), Acoetidae (Fig. 2R), Sigalionidae, Syllidae (Fig. 2T), Hesionidae (Fig. 2U), Phyllodocidae (Fig. 2W) and Goniadidae (Orrhage, 1990–1999; Orrhage & Eibye-Jacobsen, 1998). Many other 'errant' families still remain to be examined in greater detail.

Among the 'sedentaries', as well, these four transverse commissures have been observed: Sabellariidae (Fig. 2A), Sabellidae (Fig. 2B), Serpulidae (Figs 2C, 3A, B), Flabelligeridae (Figs 2D, 3D,E), Magelonidae (Fig. 2H), Poecilochaetidae (Fig. 2I), Spionidae (Figs 2J, 4A,B), Trochochaetidae (Fig. 2K), Apistobranchidae (Fig. 2L), Scalibregmatidae (Figs. 2Y, 5A,B), Orbiniidae (Figs 2Z, 5C,D), and Paraonidae (Figs $2A^1$, 5E, F) (Orrhage, 1964, 1966, 1978, 1980). Among Opheliidae and Cirratulidae, however, the arrangement is unclear (Orrhage, 1966). In Chaetopteridae (Figs 2M, 4C) it was not possible to analyze the brain in any detail; only an anterior and a posterior part of the neuropile could be discerned. In Protodrilidae (Figs 2N, 4D), Saccocirridae and Nerillidae not even TEM-analyses have revealed anything of the internal structure of the brain (Purschke, 1993, 1997; Purschke & Jouin-Toulmond, 1993, 1994). In the brains of Ampharetidae (Fig. 2E), Pectinariidae (Figs 2F, 3C) and Terebellidae (Figs 2G, 3F), finally, no traces of equivalents to the four commissures of the circumesophageal roots were found (Orrhage, 2001).

Dorsal and ventral roots of the circumesophageal connectives

In most polychaete families hitherto studied, each circumesophageal connective is proximally divided into a dorsal (*drcc*) and a ventral (*vrcc*) root. Each root on one side of the animal communicates with that on the other side through two commissures, a dorsal and a ventral one (cf. above). In most families the ventral roots lie in front of the dorsal ones. In Sabellariidae, however, all four roots and their commissures are situated in approximately the same transversal plane. In Sabellidae, Serpulidae and Phyllodocidae the *vrcc* is in fact situated behind the *drcc*. More or less laterally the two roots on each side join, forming a single circumesophageal connective. Thus, von Haffner (1959a, b) maintained that the connectives are simple in the Onuphidae, and according to Ehlers (1864), Gravier (1898), Hanström (1927), Stolte (1932), Manaranche (1966) and Åkesson (1968) this is the case also with the Glyceridae. Even in these families, however, each connective is divided into two roots (e.g., Glyceridae: Müller, 1999b). In all probability the difficulties encountered by earlier authors can be explained by the fact that in these families the circumesophageal roots are rather short, in the Glyceridae almost not at all visible outside the brain proper (see also section on neuronal differentiation).

Division of each connective into a dorsal and a ventral root is also found in *Protodrilus* sp. (Orrhage, 1974; Bubko, 1981; Purschke, 1993), *Saccocirrus papillocercus* and *S. krusadensis* (Purschke, 1992, 1993), *Parenterodrilus taenioides* (Purschke & Jouin-Toulmond, 1993, 1994), *Nerillidium troglochaetoides* (Purschke, 1997), Nerillidae (*Mesonerilla intermedia*, *Nerillidium mediterraneum*, *Nerilla antennata*; Müller, 1999b) and *Aeolosoma hemprichi* (Purschke et al., 2000). In Opheliidae the conditions seem a little less clear (Orrhage, 1966) and in *Protodriloides* (Purschke, 1993) as well as in Ampharetidae, Pectinariidae and Terebellidae the connectives are simple.

Conclusions about the two roots and their commissures. The presence of double connectives and their four transverse commissures (the later were found in 26 of the 32 families) is a unique and widespread phenomenon within the Polychaeta. This speaks in favor of these structures being old and distinctive characteristics of the Polychaeta, homologous in the families where they are found. Because nothing comparable is present in the Clitellata (Purschke et al., 2000) or Arthropoda, this pattern might be an apomorphy of the Polychaeta and represent part of the cephalic ground pattern. Absence of the dorsal root and the four commissures in the 'terebellomorph' families as well as in the Myzostomidae (Müller & Westheide, 2000), however, reduces confidence in this ground pattern.

The optic commissure and the optic nerves

Only some of the studied forms have eyes and even a comparative study of the families that are

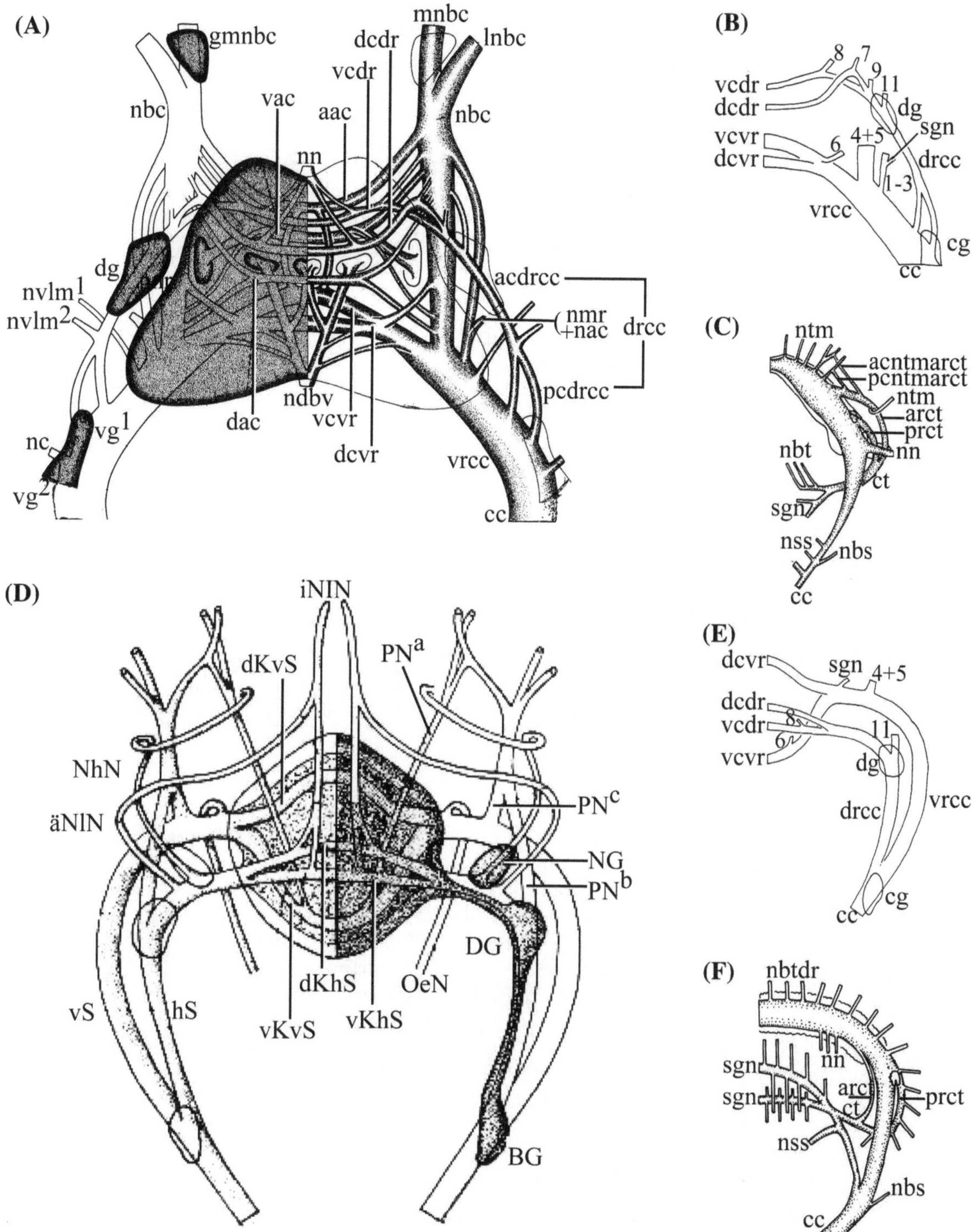

Figure 3. The cephalic nervous system of (A, B) *Serpula vermicularis*; (C) *Pectinaria* (*Amphictene*) *auricoma*; (D, E) *Brada vilosa*; (F) *Pista cristata*. Semi-schematic dorsal views. cc – circumesophageal connective; cg – cerebral ganglion; dcdr, dcvr – dorsal commissure of the dorsal (drcc) and ventral (vrcc) root of cc; dg – dorsal ganglion; sgn – stomatogastric nerve; vcdr, vcvr – ventral commissure of the dorsal and ventral root of cc; 1–11 – palp nerve roots. Further abbreviations see abbreviation list. Modified from Orrhage, 1980 (A, B); Orrhage, 2001 (C, F); Orrhage, 1966 (D, E).

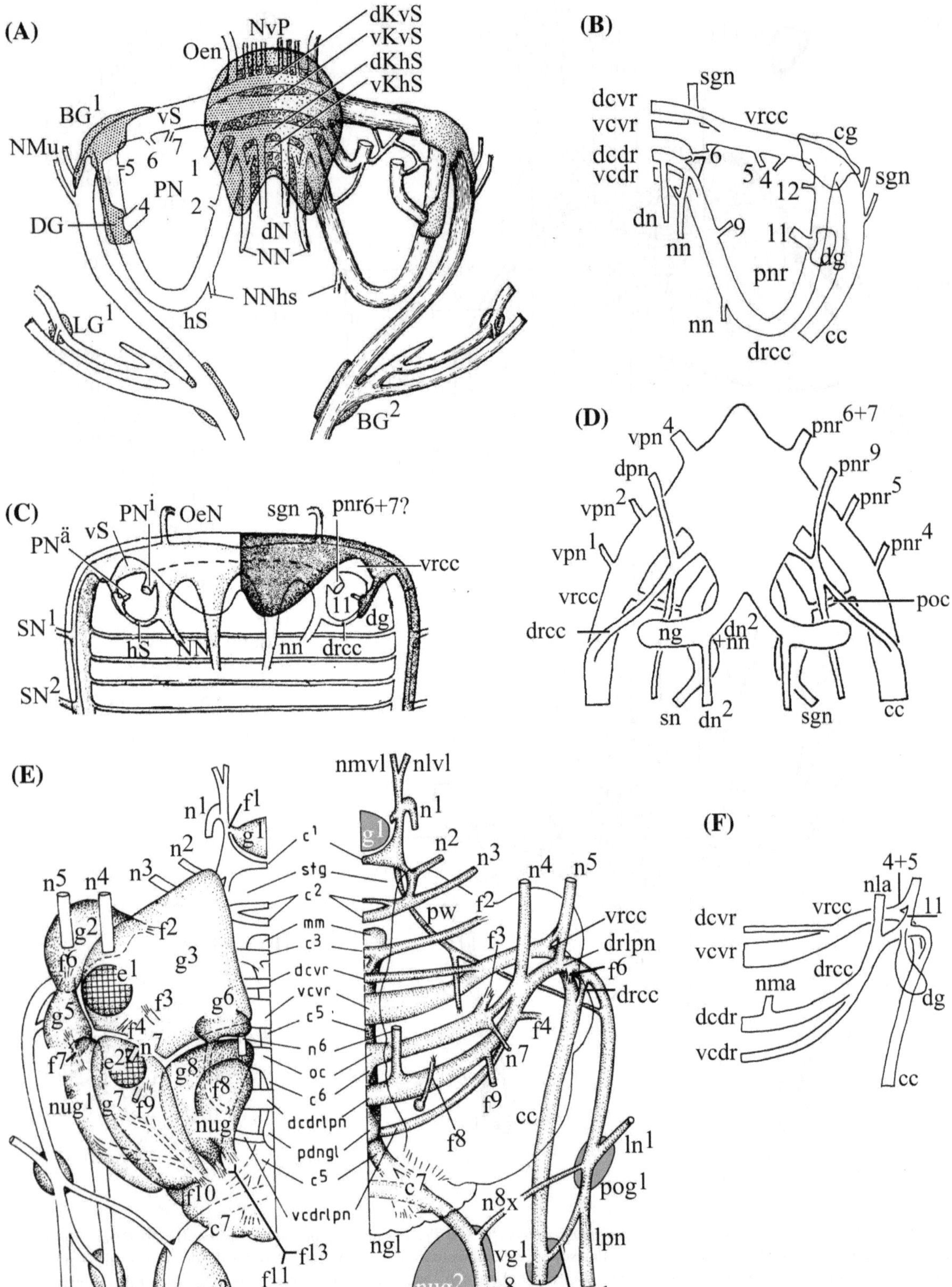

(A)
NvP
Oen
dKvS
vKvS
dKhS
vKhS
BG1
vS
NMu
5 6 7
1
PN
2
4
DG
dN
NN
NNhs
LG1
hS
BG2
(B)
sgn
dcvr
vcvr
vrcc
cg
dcdr
vcdr
6
7
5 4 12
sgn
dn
9
11
nn
dg
pnr
nn
cc
drcc
(C)
PNi
OeN
sgn
pnr6+7?
PNä
vS
vrcc
11
dg
SN1
hS
NN
nn
drcc
SN2
(D)
vpn4
pnr6+7
dpn
pnr9
vpn2
pnr5
vpn1
pnr4
vrcc
poc
drcc
ng
dn2
+nn
sn
dn2
sgn
cc
(E)
nmvl
nlvl
n1
f1
g1
n1
g1
c1
n2
n2
n3
n3
stg
n5
n4
n4
n5
c2
g2
f2
pw
f2
mm
vrcc
f6
e1
g3
c3
f3
drlpn
dcvr
f6
vcvr
drcc
g5
f3
g6
c5
f4
f4
n6
f7
e2
n7
n7
g8
oc
f8
g7
f9
c6
f9
nug1
dcdrlpn
cc
nug
f8
pdngl
ln1
c5
c7
f10
pog1
n8
x
vcdrlpn
c7
lpn
ngl
f13
vg1
f11
nug2
nug2
n8
vc
pon1
(F)
4+5
nla
11
vrcc
dcvr
vcvr
drcc
nma
dg
dcdr
vcdr
cc

Figure 4. The cephalic nervous system of (A, B) *Scolelepis cirratulus* and *S. bonnieri*; (C) *Spiochaetopterus typicus*; (D) *Protodrilus* spp.; (E, F) *Eurythoe complanata*. Semi-schematic dorsal views. cc – circumesophageal connective; cg – cerebral ganglion; dcdr, dcvr – dorsal commissure of the dorsal (drcc) and ventral (vrcc) root of cc; dg – dorsal ganglion; dpn – dorsal palp nerve; nla- nerve of lateral antenna; nma – nerve of median antenna; vcdr, vcvr – ventral commissure of the dorsal and ventral root of cc; vpn – ventral palp nerve; 1–12 – palp nerve roots. Further abbreviations see abbreviation list. Modified from Orrhage, 1964 (A); Orrhage, 1966 (C); Orrhage, 1974 and Purschke, 1993, combined (D); Orrhage, 1990 (E, F).

equipped with eyes seems rather unavailing: their optic systems are too differently modelled. Thus, in relation to the other nerve elements in the brain, the position of the optic commissure is very different. It may be found (1) between the dorsal commissure of the ventral root (*dcvr*) and the ventral commissure of the dorsal root (*vcdr*) (Nereididae), (2) between the ventral commissure of the ventral root (*vcvr*) and the dorsal commissure of the dorsal root (*dcdr*) (Amphinomidae, Polynoidae, Onuphidae, Syllidae), (3) between the ventral commissures of both roots (Hesionidae), or (4) between the dorsal commissures of both roots (Orrhage, 1990–1996, Orrhage & Eibye-Jacobsen, 1998). The families may also differ markedly from one another as concerns the course of the optic nerves: in families with only two eyes, the optic nerve may run ventral to *dcdr* (situated in the anteriormost part of the brain; Phyllodocidae) or dorsal to both roots and their commissures (Onuphidae). In families with four eyes the anterior optic nerve (*aon*) may be found (1) ventral to *drcc* (Polynoidae), (2) dorsal to *vrcc* (Syllidae), (3) dorsal to *dcdr* and *vcdr* (Nereididae), or (4) between *vcdr* and the main part of *dcdr* (actually penetrating the *dcdr*, Hesionidae). The posterior optic nerve (*pon*) of the four-eyed forms may be found between *vcdr* and *dcdr* (Polynoidae, Syllidae, Hesionidae; in the last of these penetrating the *dcdr*) or running dorsal to both *vcdr* and *dcdr* (Nereididae) (Orrhage, 1991–1996; Orrhage & Eibye-Jacobsen, 1998).

These facts indicate that it is futile to try to give any summarizing general picture of the polychaete optical system at present.

The nuchal commissure and the nuchal nerves

Among the Canalipalpata and the Scolecida hitherto analyzed, only Orbiniidae is equipped with a separate nuchal commissure (Fig. 5C: *NK*; 5D: *nc*). It is connected to both commissures of the dorsal root (*vcdr, dcdr*) and to a nuchal center (Fig. 5C, *GZNZ*) which, in turn, is associated with the central neuropile of the brain. From this nuchal commissure emanate the nerves of the nuchal as well as those of the dorsal organs. Among other 'sedentary' families studied, there is a considerable variation of the attachment site of the nuchal nerves to the brain: in Ampharetidae, Pectinariidae and Terebellidae, as well as in Protodrilidae (Fig. 4D, dn^2 + *nn*, Purschke, 1993; Purschke & Jouin-Tolmound, 1993), Saccocirridae (Purschke, 1992, 1993) and Nerillidae (Purschke, 1997) the nuchal nerves emerge diffusely from the (latero-) caudal parts of the brain. This is the case in Chaetopteridae, too, but here the *drcc* is also more directly involved. In Flabelligeridae (Fig. 3D: *iNlN, NhN, äNlN*), Spionidae (Figs. 4A: *NN, NNhS*; 4B: *nn*) and Trochochaetidae the nuchal nerves are distinctly associated with *dcdr* and *vcdr*, while in Sabellariidae, Sabellidae and Serpulidae they emanate only from them. In Scalibregmatidae the nuchal nerves are associated only with *dcdr*, whereas in Poecilochaetidae, Apistobranchidae and Paraonidae they arise from *drcc*.

With the exception of Syllidae, Hesionidae and the four aphroditacean families studied, in all aciculate forms so far analyzed, a nuchal commissure is found. Its position in relation to the other nerve elements in the brain, however, is very different (Orrhage, 1990, 1993, 1997; Orrhage & Eibye-Jacobsen, 1998). In the Amphinomidae, Onuphidae and Eunicidae the nuchal commissure is situated in the hind-most part of the brain. In the last two taxa this commissure is split into three parts; furthermore, here an additional nuchal nerve emanates from *drcc*. In Syllidae and Hesionidae the nuchal organs are innervated from two posterior ganglia (*pg*) which are associated with *dcdr* and *oc* (sic!). In Nereididae the nuchal commissure is situated in the middle of the brain (between the commissures of *vrcc* and *drcc*) and in

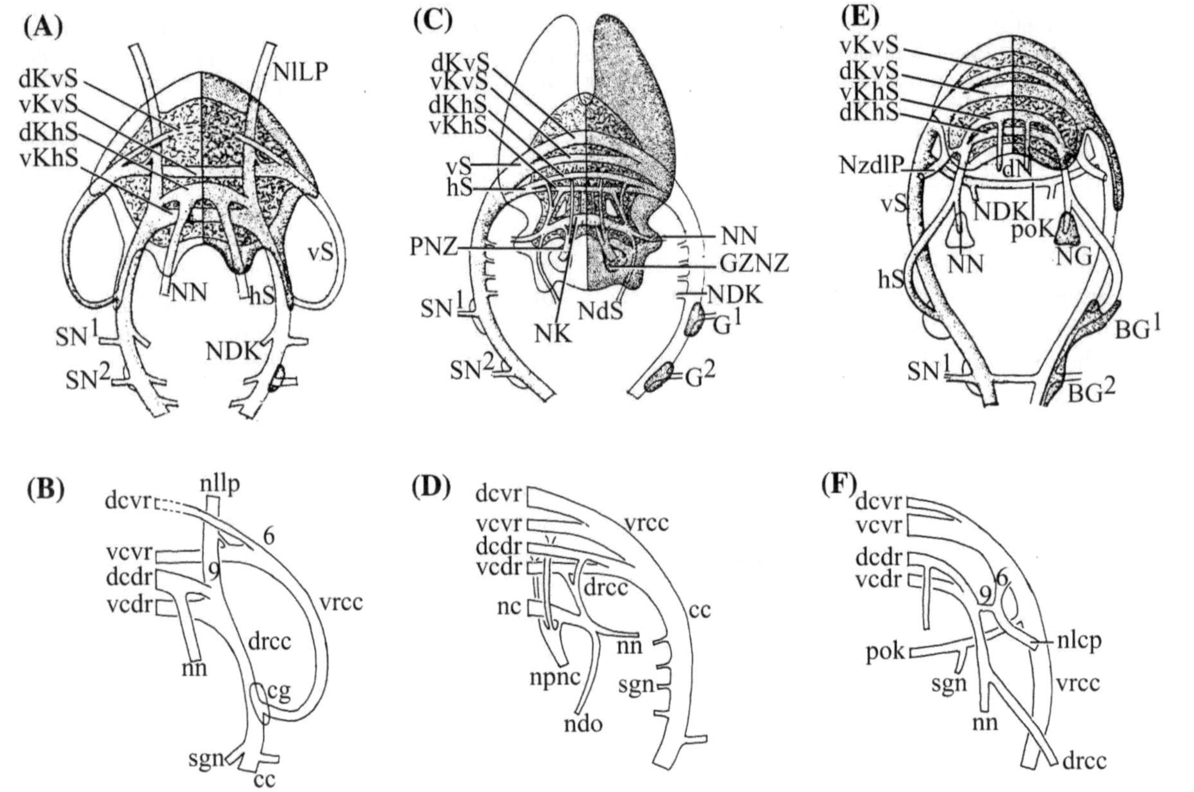

Figure 5. The cephalic nervous system of (A, B) *Polyphysa crassa*; (C, D) *Phylo norvegica* and *Orbinia sertulata*; (E, F) *Levinsenia gracilis* and *Paradoneis lyra*. Semi-schematic dorsal views. cc – circumesophageal connective; cg – cerebral ganglion; dcdr, dcvr – dorsal commissure of the dorsal (*drcc*) and ventral (*vrcc*) root of cc; vcdr, vcvr – ventral commissure of the dorsal and ventral root of cc; 6, 9 – palp nerve roots. Further abbreviations see abbreviation list. Modified from Orrhage, 1966 (A, C, E).

Phyllodocidae it is still more anteriorly located. The nuchal organs of Glyceridae and Goniadidae are innervated from a nuchal commissure and from *vcdr*. These observations indicate that at present a general pattern of the polychaete nuchal system cannot be reconstructed.

Central ganglia

Different numbers of paired ganglia are reported within the brain of Polychaeta: 26 in *Hediste diversicolor* ('*Nereis diversicolor*', Holmgren, 1916), 25 in *Nephtys* sp. (Clark, 1958), 26 in *Hermodice carunculata* (Fitzsimmons, 1964) and 17 in *Glycera convoluta* (Manaranche, 1966). Whereas Clark (1958) could homologize the posterior ganglia of *Nephtys* sp. with those of *Hediste diversicolor*, other authors emphasized great difficulties in making meaningful comparisons, e.g., between *Hediste diversicolor* and *N. irrorata* (Defretin, 1956), *Hermodice carunculata* and *Nephtys* sp. (Fitzsimmons, 1964) and *Glycera convoluta*, *Hediste diversicolor* and *Nephtys* sp. (Manaranche, 1966). Like Fitzsimmons (1964), Marsden & Galloway (1968) are skeptical of the possibility of homologizing ganglia in various polychaete brains. With some exceptions (cf. below), in many polychaete brains it has not been possible to discern any separate ganglia at all, especially not in the central neuropile of the 'sedentaries'. This, in combination with the severe obstacles to homologizing ganglia of one family with those of others, discouraged further endeavors in this field.

The commissural ganglion

In 21 of the 32 families studied a commissural ganglion (*cg*) is found at the lateral junction of *vrcc* and *drcc*. It was called 'Hamaker's commissural ganglion' (*cgHa*) by Binard & Jeener (1926, 1928) and by Orrhage (1966, 1990, 1993, 1995, 1999); it has also been referred to as 'first ventral ganglion' (Orrhage, 1964, 1966, 1974: BG^1; 1978, 1990: vg^1). The dimensions of this ganglion are different in the various families. It is quite large in Sabellariidae, Flabelligeridae, Poecilochaetidae, Spionidae, Trochochaetidae, Apistobranchidae, Acoetidae, Aphroditidae, Polynoidae, Sigalionidae and Hesionidae. In 10 other families it is of medium size or very small, especially in the Onuphidae, Nereididae, Glyceridae and Goniadidae (Fig. 2). The inconspicuous dimensions of the commissural ganglion of Nereididae and Onuphidae as well as its total absence in Eunicidae, provide one of the main arguments against Binard & Jeener's (1928) theory of a segmentation of the polychaete brain. In Cirratulidae, Ampharetidae, Pectinariidae, Terebellidae, Protodrilidae and Orbiniidae no equivalents to the commissural ganglion were found. The first ventral ganglion of Amphinomidae (vg^1) and the ganglion of the circumesophageal connective of Phyllodocidae (*ccg*) are more posteriorly located, which prevents a homologization with the commissural ganglion of other families.

The dorsal ganglion and its role in the innervation of the palps

In 20 of the 32 families studied there is a ganglion (*dg*; *DG*) on the dorsal or dorso-lateral side of the *drcc* (Fig. 3A, B, D, E). Its cells are remarkably large and plasma-rich. This is the 'optical ganglion' of many earlier authors (e.g., Retzius, 1895). Holmgren (1916) called it 'cerebrales Kommissuralganglion', a more neutral term. In many papers (e.g., Binard & Jeener, 1926, 1928) it is treated under the name of 'Holmgren's cerebral commissural ganglion'. In his comparative discussions Orrhage (1966, 1990, 1991, 1993, 1995, 1996, 1999) used this term (*ccgHo*). It is equivalent to the fifth ganglion (Fig. 4E: g^5; 4F: *dg*) of the Amphinomidae and to the eleventh ganglion (Fig. 1D: *Xl*; 1E: *dg*, (*Xl*)) of the Glyceridae and Goniadidae. In eight of these 20 families the dorsal ganglion is of great importance for the innervation of the palp (Flabelligeridae, Magelonidae, Poecilochaetidae, Spionidae, Chaetopteridae, Amphinomidae, Glyceridae and Goniadidae; Fig. 2). In Serpulidae, Acoetidae, Aphroditidae, Polynoidae and Sigalionidae the dorsal ganglia play a less prominent role in this respect. This holds true also for the Nereididae where, in fact, *Nereis pelagica* is devoid of real dorsal ganglia (Orrhage, 1993). These facts constitute further arguments against Binard & Jeener's (1928) theory of the segmentation of the polychaete brain (Orrhage, 1993, 1995). In still other families the dorsal ganglia have nothing to do at all with the innervation of the palps (Trochochaetidae, Apistobranchidae, Onuphidae, Eunicidae, Syllidae, Hesionidae). Here a comparison between the closely related Poecilochaetidae, Spionidae and Trochochaetidae is of special interest. In the two first-mentioned families the dorsal ganglia serve as a base for one of the two main palp nerve roots. In the Trochochaetidae the nerve emanating from the dorsal ganglion is instead in contact with the first lateral ganglion. For a probable explanation of these surprising conditions the reader is referred to Orrhage (1964). In Sabellariidae, Sabellidae, Cirratulidae, Ampharetidae, Pectinariidae, Terebellidae, Protodrilidae, Phyllodocidae, Opheliidae, Scalibregmatidae, Orbiniidae and Paraonidae no dorsal ganglia were observed.

Innervation of anterior appendages – homology conclusions

Median antenna

Most of the Amphinomidae, Onuphidae, Eunicidae, Acoetidae, Aphroditidae, Polynoidae, Sigalionidae, Syllidae, Hesionidae and Phyllodocidae species studied by Orrhage (1990–1996) and Orrhage & Eibye-Jacobsen (1998) are equipped with a median antenna or a nuchal papilla. The nerves (*nma*) of each of these appendages are attached to the median part of the *drcc* through two (or four: Onuphidae, Eunicidae) roots. These facts were interpreted as a proof that these appendages are homologous *inter se*. In all probability they are also equivalent to the occipital tentacle found in many Spionidae (Figs 4A,B: *dN*, *dn*) (Orrhage, 1966; Foster, 1971; Fauchald & Rouse, 1997).

Lateral antennae

The nerves (*nla*) of the lateral antennae emanate from the lateral parts of *dcdr* (Amphinomidae) or from this commissure and the central neuopile of the brain (Acoetidae, Polynoidae, Sigalionidae, Syllidae, Hesionidae, Phyllodocidae). In Nereididae the nerves of the antennae are congruent with the first tegumentary nerve (Figs 1A: $an + tn^{1}$, 2V: *na*). In Onuphidae and Eunicidae the lateral antennal nerves are also in contact with the dorsal fibril mass (*dfm*), a structure which is in all probability and at least *partim* homologous with the *vcdr* of other families. Apart from these small variations there are fundamental similarities in the innervation of the lateral antennae of the families so far studied. This indicates that these appendages are homologous *inter se*.

Palps

Out of the 32 families analyzed, 22 were found to be equipped with palps (in the sense of Binard & Jeener, 1928; Orrhage, 1964–2001; Orrhage & Eibye-Jacobsen, 1998). In Scalibregmatidae and Paraonidae palp nerves but no protruding palps were found. In Ampharetidae, Pectinariidae, Terebellidae and Orbiniidae no traces of palps or palp nerves could be found.

According to Orrhage and earlier authors, a polychaete palp is innervated through two or more nerve roots emanating from the *drcc* and the *vrcc* and their commissures. Their positions and relations to other nerve elements make it possible to homologize the palp nerve roots of one family with those present in other taxa. Until now 12 such palp nerve roots (pnr^{1-12}) were found. Additional nerve roots or nerves have been observed by Purschke (1993). In most families two of them were interpreted as main roots on account of their dimensions. The results are summarized in Figure 2 and Table 1 [excluded are Cirratulidae (not sufficiently analyzed), Opheliidae (contradictory results) as well as Ampharetidae, Pectinariidae, Terebellidae and Orbiniidae (no palps or homologous structures present)].

Evaluations. At first sight these results could give the impression of a great heterogeneity. Thus the number of palp nerve roots common in different taxa varies from eight (in Sabellidae and Serpulidae) to zero (Apistobranchidae compared with Magelonidae and Amphimomidae). The cleft between Apistobranchidae and Magelonidae/Amphinomidae can, however, be bridged by a gradually comparison from Apistobranchidae upwards (Table 1). Apistobranchidae and Trochochaetidae possess three common roots ($pnr^{6,\ 7,\ 9}$); pnr^{6} and pnr^{9} being the main roots. Trochochaetidae and Spionidae possess six common roots ($pnr^{4,\ 5\ ,\ 6,\ 7,\ 9,\ 12}$); in both families pnr^{6} is a main root. Spionidae and Poecilochaetidae possess seven common roots ($pnr^{4,\ 5,\ 6,\ 7,\ 9,\ 11,\ 12}$), pnr^{6} and pnr^{11} being the main roots. Finally, Poecilochaetidae and Magelonidae possess three common roots ($pnr^{4,\ 5,\ 11}$); pnr^{4} and pnr^{5} are fused and pnr^{11} is one of the main roots.

Apart from the heterogeneity, similar innervation patterns are also striking. In Trochochaetidae, Apistobranchidae, Acoetidae, Aphroditidae, Polynoidae, Sigalinoidae, Syllidae, Hesionidae, Nereididae and Phyllodocidae pnr^{6} and pnr^{9} are the main roots (Table 1). Palp nerve root pnr^{9} is also the main root in Protodrilidae, Scalibregmatidae, Glyceridae and Goniadidae and thus the innervation of the prostomia of the latter two families resembles the innervation of the palps in the other mentioned taxa. Equivalents to the other main root (pnr^{11}) in Glyceridae and Goniadidae are found in Flabelligeridae, Magelonidae, Poecilochaetidae, Spionidae, Chaetopteridae and Amphinomidae. The above demonstration of conformity (through intermediate states) in the innervation of palps, branchial crowns, prostomia and special areas of the prostomium of the 26 families lead to the homology conclusion, that are graphically represented in Figure 6 (Goniadidae excluded).

A summarizing picture of the twelve palp nerve roots observed by Orrhage (1964–1999) and Orrhage & Eibye-Jacobsen (1998) and their positions and relations to other cephalic nerve elements is given in Figure 7. As emphasized in earlier papers (Orrhage, 1995, 1999), such a diagram must be looked upon as a preliminary one, mirroring only the present-day knowledge. Thus, in all probability, additional nerves, nerve roots and other significant structures will be observed (cf. *Protodrilus* in Purschke, 1993) and it must be further emphasized that none of the families studied is equipped with all nerve elements.

Table 1. The palp nerve roots of 26 of the 32 families studied by Orrhage (1964–1999) and Orrhage & Eibye-Jacobsen (1998) are shown

	Nerve root number: nerve roots attached to the ventral root or its ventral commissure						nerve roots attached to the dorsal root or its commissures					
	1	2	3	4	5	6	7	8	9	10	11	12
Sabellariidae (5, 25, 36)	□	□	□	□	□		□	□				
Sabellidae (5, 25, 31, 37, 46)	■	■	■	■	■	□	□	□				
Serpulidae (5, 14, 25, 31, 37, 46)	■	■	■	■	■	□	□	□	□		□	
Flabelligeridae (5, 25, 34, 53)				■	■	□		□			■	
Magelonidae (34)				■	■						■	
Poecilochaetidae (3, 33)				□	□	■	□		□		■	□
Spionidae (5, 18, 33, 56)				□	□	■	□		□	⬚	■	□
Trochochaetidae (33, 54)				□	□	■	□		■			□
Apistobranchidae (34)						■	□	□	■			
Chaetopteridae (26, 29, 34, 51)						■ ???	■				■	
Protodrilidae (6, 30, 35, 47, 49, 50, 52)						■ ???	■		■			
Amphinomidae (15, 38, 56)				■	■						■	
Onuphidae (5, 15, 17, 20, 41)				□	□	□	□	□				□
Eunicidae (5, 15, 17, 20, 41)				□	□	□	□	□				□
Acoetidae (39)				□	□	■			■		□	
Aphroditidae (4, 5, 15, 17, 39)				□	□	■			■		□	
Polynoidae (1, 5, 15, 17, 39)				□		■			■		□	
Sigalionidae (17, 39)				□		■			■		□	
Syllidae (11, 15, 27, 42)						■			■			
Hesionidae (5, 15, 22, 42)						■	□	□	■			
Nereididae (5, 12, 15, 16, 17, 23, 40)						■	□	□	■	□	□	□
Phyllodocidae (5, 8, 13, 45)						■			■			
Glyceridae (2, 5, 14, 17, 28, 43, 55)								□	■		■	□
Goniadidae (14, 43)								□	■		■	
Scalibregmatidae (10, 17, 34)						□			■			
Paraonidae (7, 19, 34)						■			□			
	1	2	3	4	5	6	7	8	9	10	11	12

When discernible, the two main roots of each family are emphasized through dark gray. Relevant literature is given in brackets behind the family name. Not included information: Ampharetidae, Pectinariidae, Terebellidae (5, 9, 21, 24, 32, 44); Saccocirridae (47); Nerillidae (48); Orbiniidae (34).

Literature: Åkesson, 1963 (1), 1968 (2); Allen, 1904 (3); Bernert, 1926 (4); Binard & Jeener, 1928 (5); Bubko, 1981 (6); Cerruti, 1909 (7); Eibye-Jacobsen, 1993 (8); Fauvel, 1897 (9), 1927 (10); Gidholm, 1967 (11); Gilpin-Brown, 1958 (12); Gravier, 1896 (13), 1898 (14); Gustafson, 1930 (15); Hamaker, 1898 (16); Hanström, 1927 (17), 1928c (18), 1930 (19); Heider, 1925 (20); Hessle, 1917 (21), 1925 (22); Holmgren, 1916 (23); Holthe, 1986 (24); Johannson, 1927 (25); Joyeux-Laffuie, 1890 (26); Malaquin, 1893 (27); Manaranche, 1966 (28); Martin & Anctil, 1984 (29); Merker & Vaupel von Harnack, 1967 (30); Meyer 1887-1888 (31); Nilsson, 1912 (32); Orrhage, 1964 (33), 1966 (34), 1974 (35), 1978 (36), 1980 (37), 1990 (38), 1991 (39), 1993 (40), 1995 (41), 1996 (42), 1999 (43), 2001 (44); Orrhage & Eibye-Jacobsen, 1998 (45); Pruvot, 1885 (46); Purschke, 1993 (47), 1997 (48); Purschke & Jouin-Toulmond 1993 (49), 1994 (50); Racovitza, 1896 (51); Salensky, 1907 (52); Schlieper, 1927 (53); Söderström, 1920 (54); Stolte, 1932 (55); Storch, 1913 (56).

Oral filaments

The slender appendages of the lateral parts of the ventral side of the operculum of Sabellariidae were earlier quite differently interpreted: as palps homologous with those of the Spionidae (Johansson, 1927) or with those of the 'errants' (Binard & Jeener, 1928). On account of their innervation, the

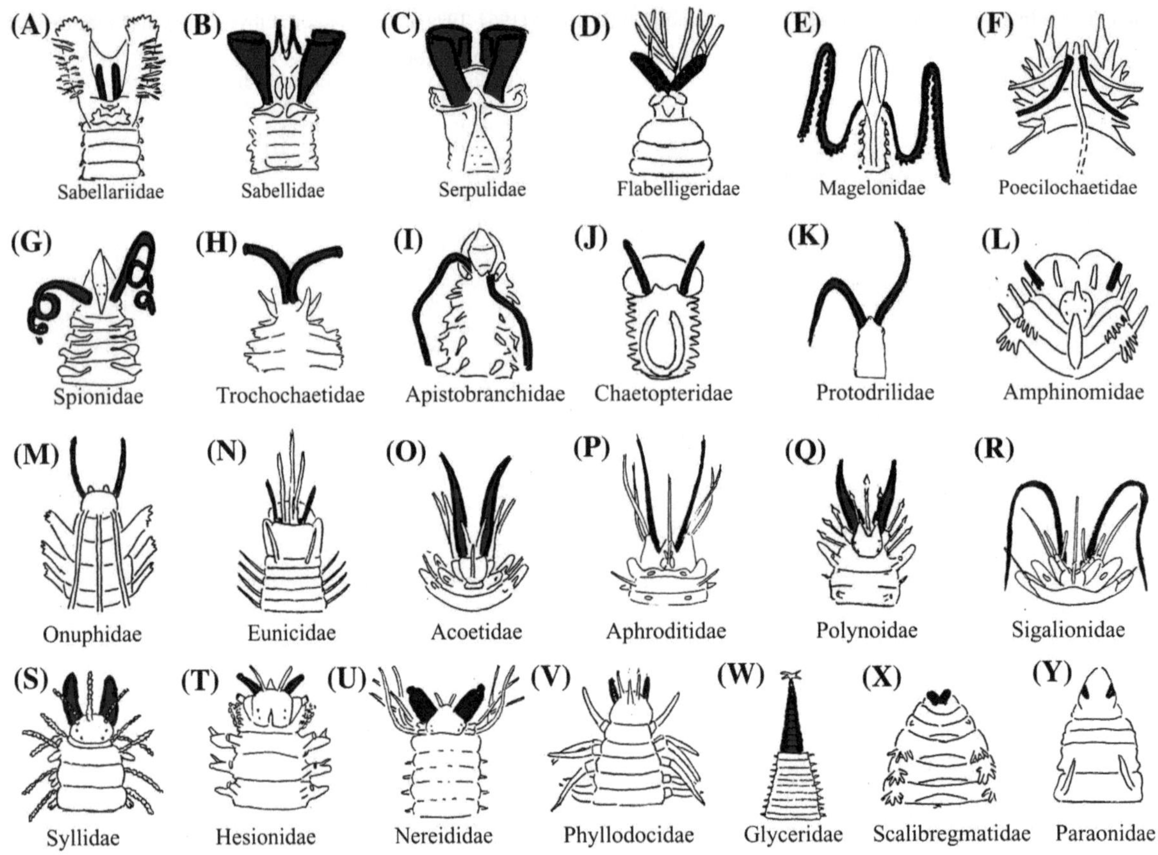

Figure 6. Anterior ends of representatives of 24 of the 32 families studied by Orrhage (1964–1999) and Orrhage & Eibye-Jacobsen, (1998). The palps of the Nereididae and their homologues in the other taxa are blackened.

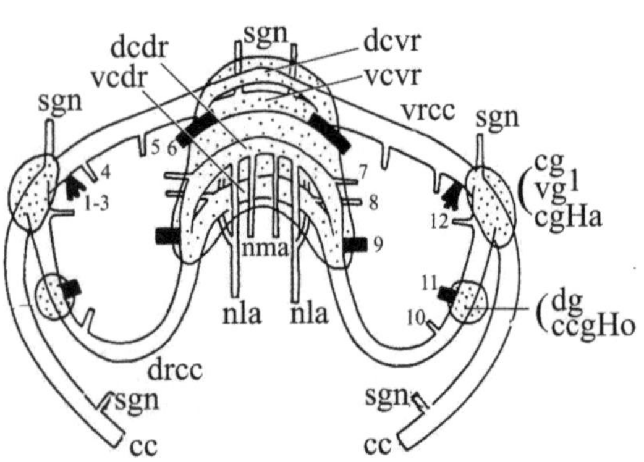

Figure 7. A synopsis of the palp nerve roots (1–12), nerves of median (nma) and lateral (nla) antennae and the stomatogastric nerves (sgn) found in families studied by Orrhage (1964–1999) and Orrhage & Eibye-Jacobsen (1998). None of these families are equipped with all nerve elements shown in the figure. ccgHo – Holmgren's cerebral commissural ganglion; cgHa – Hamaker's commissural ganglion. For further abbreviations see Figure 1 and present text. Modified after Orrhage, (1995, 1999).

two thick appendages situated in front of the mouth constitute palps. The oral filaments are nothing but extended lateral parts of the upper lip of the mouth (Orrhage, 1978).

Detached branchial radioli and pinnulae; outgrowth of the dorsal wall of the mouth cavity

The appendages of the dorsal lip of Sabellidae and Serpulidae represent various structures: in the Sabellidae they consist of radioli which have become separated from the branchial crown (*Sabella*) or they are made up of both, radioli and pinnulae (*Potamilla*, *Euchone*, *Chone*). The dorsal lip appendages of Serpulinae consist of separated pinnulae only (*Serpula*, *Pomatoceros*, *Ditrupa*, *Hydroides*, *Placostegus*). When designating these structures as palps, Johansson (1927) and Binard

& Jeener (1928) and others, were correct to some extent: these appendages are, as we know now, detached from the branchial crowns, which in themselves are equivalent to the palps. In maintaining that also the dorsal lip appendages of Filograninae are homologous with palps, however, these authors went too far: at least in *Apomatus*, *Protula* and probably *Filograna* they constitute outgrowths of the dorsal wall of the mouth cavity (cf. Orrhage, 1980).

Buccal lips

In Amphinomidae and Eunicidae, among other families, the anteriormost part of the prostomium may be equipped with outgrowths of different size and form. In the Amphinomidae these appendages were previously interpreted and designated as palps (Racovitza, 1896; Malaquin & Dehorne, 1907; Storch, 1913). On each side of an amphinomid, four nerves run to the buccal lips (Fig. 4E: *nlvl*, *nmvl*, n^2, n^3). These nerves are intimately associated with the stomatogastric ganglion (g^1) and the stomatogastric nerve (n^1). The eight nerves of the buccal lips and the stomatogastric nerves have the two anteriormost transverse brain commissures (c^1, c^2) in common. This center constitutes a complex, that is almost totally isolated from the rest of the brain. No polychaete palps are innervated like the buccal lips of the Amphinomidae. The true palps of this family are situated dorsal to the buccal lips and are innervated as described above. The buccal lips of Onuphidae and Eunicidae have also been homologized with the palps of, for instance, the Aphroditacea (Fauvel, 1923; Heider, 1925; Hanström, 1927; von Haffner, 1959a,b, 1962; Åkesson, 1967a). According to Storch (1913), Binard & Jeener (1928) and Gustafson (1930), however, the buccal lips of Eunicea are nothing but an outgrowth of the prostomium itself. The dorsal and ventral buccal lips of *Hyalinoecia* and *Nothria* and the bilobate buccal lips of *Eunice* (Orrhage, 1995, Fig. 4) are innervated through a number of nerves emanating from *vcvr*. They are rooted in the immediate vicinity of the esophageal nerve. No polychaete palps are innervated like this. The true palps of Onuphidae and Eunicidae are situated dorsal to the buccal lips and innervated as described above. The buccal lips of the Amphinomidae and those present in Eunicea are quite differently innervated. This does not speak in favor of the buccal lips of the Amphinomidae being homologous with those found in Eunicea. It is likely, therefore, that the use of a common term is misleading and inappropriate.

Buccal tentacles

The buccal tentacles are attached to the upper lip of the mouth, ventral to the tentacular membrane in Ampharetidae and Pectinariidae and to the dorsal ridge in Terebellidae (Orrhage, 2001, Fig. 3). They were earlier interpreted as palps by, e.g., Nilsson (1912), Fauvel (1927) and Binard & Jeener (1928). Rouse & Fauchald (1997) accepted this interpretation, although with hesitation for Terebellidae. In the opinion of Holthe (1986), however, these appendages are originally buccal structures. In Ampharetidae (Fig. 2E) the nerves of the buccal tentacles (*nbt*), the stomatogastric nerves (*sgn*) and the nerves running to the lateral part of the tentacular membrane (*ntm*) emanate from one and the same common tract (*ct*). The median part of the tentacular membrane is innervated by nerves coming from the brain (*ntm*). In Pectinariidae (Figs 2F, 3C) the same intimate contact is present between the nerves of the buccal tentacles (*nbt*) and the anterior stomatogastric nerves (*sgn*). Most of the nerves to the tentacular membrane (*ntm*) emanate from the brain itself. As in the Ampharetidae, however, some of the nerves running to this membrane issue from the common tract of the nerves running to the buccal tentacles and the alimentary canal. In Terebellidae (Figs 2G, 3F) the nerves supplying the buccal tentacles also run to the dorsal ridge (which is homologous with the tentacular membrane of the Ampharetidae and Pectinariidae). Most of them (*nbtdr*) emanate from the brain but quite a number are rooted in the tract leading to the anterior stomatogastric nerves.

Summing up, the conditions found in Ampharetidae, Pectinariidae and Terebellidae are variations on the same theme: the intimate contact between the nerves of the buccal tentacles and those of the intestine. It was concluded (Orrhage, 2001) that the buccal tentacles constitute parts of the alimentary canal situated outside the mouth.

In these taxa there are no neuro-anatomical indications of the presence of antennae or palps.

Outgrowths from the tip of the prostomium

As a consequence of the interpretation of the Glyceridae and Goniadidae prostomium as a pair of fused palps (Hanström, 1927; Manaranche, 1966; Orrhage, 1999), the four appendages at the tip can not be understood as either antennae (Gravier, 1898; Hanström, 1927; Binard & Jeener, 1928) or palps (Binard & Jeener, 1928).

Conclusions. Most appendages described in the last five paragraphs may be interpreted as structures sui generis. It seems probable that they are homologous *inter se*, but only within closely related taxa. Thus, for instance, no neuro-anatomical data indicate that the oral filaments of the Sabellariidae are homologous with the outgrowth of the dorsal wall of the mouth cavity of some filogranine genera.

The stomatogastric nerves

In the simplifications (Fig. 2) of most of Orrhage's (1964–2001) drawings, the attachment of the stomatogastric nerves or nerve roots (*sgn*) are shown. It is not possible to discern any general picture of the innervation of the intestinal canal in the families so far studied: the nerves or nerve roots emanate from almost any part of the CNS.

Hanström (1927, 1930) was the first to propose that during evolution, the stomatogastric nerves of the polychaetes had moved from an original position at the subesophageal ganglion (still prevailing in 'Amphictenidae') to the final position localized in the brain ('Eunicidae', 'Aphroditidae', Phyllodocidae) via the anterior part of the circumesophageal connectives and the ventral part of the brain ('Nereidae'). This idea was previously criticized by Gustafson (1930) and Snodgrass (1938), and the data presented here (Fig. 2) likewise fail to support it: the stomatogastric nerves of the Pectinariidae emanate not only from the circumesophageal connectives near the subesophageal ganglion (Fig. 3C, *nss*) but also from the neuropile of the brain (*sgn*); the fifth stomatogastric nerve of the Nereididae emerges from the posterior part of the connectives (Fig. 1A, sgn^5) and in the Eunicea a stomatogastric nerve (Fig. 2P, Q, *sgn*) arises from the lateral junction of *drcc* and *vrcc* as does the posterior root of the stomatogastric nerve of Aphroditacea and one of the stomatogastric nerves of the Phyllodocidae (Fig. 2W, *sgn*).

The composition of the polychaete brain – segmented or not?

For years students of polychaete brains have spoken in favor of two fundamentally different interpretations of the constitution of this organ. Partly at least, this was due to the investigators being supporters of one or the other of the three major theories of the origin of metamerism in the Metazoa. Most advocates of the cormen theory (cf. Haeckel, 1866) and the pseudomer theory (cf. Hatschek, 1878) interpreted the polychaete brain as non-segmented. This is in line with the fundamental idea of each of these theories. For the supporters of the cyclomer theory (cf. Sedgwick, 1884; van Beneden, 1891, 1897; Lameere, 1916, 1925), however, the polychaete brain is made up of three segments, equivalent to the medio-ventral and the two following pairs of gastral pockets of the anthozoan body organization and homodynamous with the trunk segments of the Polychaeta.

Non-segmented brain

Malaquin (1893) interpreted the brain as a single entity, homologous with a body segment. Although Racovitza (1896; supporter of the pseudomer theory) clearly emphasized that his tripartition into fore-, mid- and hind-brain is nothing but a topographic–physiological subdivision, this was misunderstood by later authors (e.g., Nilsson, 1912; Heider, 1925; von Haffner, 1959–1962; Raw, 1949; Martin & Anctil, 1984). According to Holmgren (1916) and Söderström (1920), the nuchal center should hold a somewhat different position because it appears a little later in ontogeny and is something later added to the brain, respectively. In spite of postulating a non-segmented brain in Amphinomidae, Gustafson (1930) could imagine that the palps of the 'errants' constitute segmental structures.

Segmented brain

According to Pruvot (1885) the brain consists of four centers that had migrated from the trunk in the anterior direction, along the esophageal connectives. Nilsson (1912) proposed that the brain is composed of three pairs of segmental podial ganglia pushed forwards. Storch (1913) and Hempelmann (1911) interpret 'the palp nerve' as the remnant of an earlier lateral nerve. Storch's ideas were further developed by Binard & Jeener (1928), advocates of the cyclomer theory. Gustafson (1930) and more thoroughly Orrhage (1993, 1995) criticized their theory and it turned out that the cephalic nervous system and especially the innervation of the palps is more complicated than Binard & Jeener (1928) summarized in their '*Schéma de la constitution fondamentale du système nerveux du prostomium*' (compare with Fig. 7). From the configuration and innervation of the eunicean esophagus and pharyngeal sacs Raw (1949) imagined that the mid-brain of *Eunice* sp. and other taxa is composed of three segmental ganglia which have joined, forming an 'ancient fore-brain'. He also regarded the nuchal organs as being of segmental origin. In his studies of the Eunicea Orrhage (1995) found no support for Raw's theory. Since the middle of the 20th Century no major theories concerning the segmentation of the polychaete brain seem to have been put forward. Earlier expectations have been frustrated and it may be that the problem is unsolvable.

Neuro-anatomical investigations via immunocytochemistry and cLSM

Although immunocytochemical (ICC) studies were carried out as early as 1982 (Porchet et al., 1985; Dhainaut-Curtois & Golding, 1988), until recently they were rarely applied in investigations of annelid neuroanatomy. In spite of the great variety of neuroactive substances present in annelids (acetylcholine, monoamines, neuropeptides, amino acids; see Dhainhaut-Curtois & Golding, 1988; Windoffer, 1992; Salzet & Stefano 2001), only a few of them are frequently analyzed in immunocytochemical experiments. The two most popular antibodies applied in Annelida are directed against the monoamine 5-hydroxytryptamine (5-HT, serotonin) and the tetrapeptide Phe–Met–Arg–Phe–NH_2 (FMRFamide). Their antigens are widely distributed within the nervous system (according to Miron & Anctil, 1988, 2–3% of all neurons in *Harmothoe imbricata* contain 5-HT) and therefore the immunoreactivity (IR) enables a detailed reconstruction. The FMRFamide antibody may recognize only the RFamide-motif and the staining is therefore often termed 'FMRFamide-like immunoreactivity (FMRFamide-L IR); in the following the short term 'RFamide IR' will be used. Demonstration of the entire neuronal structures can be achieved via antibodies directed against the neuronal cytoskeleton. Good results were obtained using ICC against bovine neurofilament proteins (210 kDa; Sigger & Dorsett, 1986), acetylated α-tubulin (Müller, 1999a,b; Müller & Westheide, 2000, 2002) and tyrosinated tubulin (Hessling & Westheide, 2002). The latter two can also be used to detect ciliated structures (e.g. Müller, 2002; Müller et al., 2001; Worsaae & Müller, 2003). Combination of immunocytochemistry and confocal laser-scanning microscopy (cLSM, invented by Minsky in 1984), is an extraordinarily efficient way to analyze the 3D nervous system structure in whole mounts of small and transparent animals. Furthermore, this method is also an efficient way to analyze all neuronal structures, so that the complaint of Dhainaut-Curtois & Golding (1988) '…recent investigations on the peripheral and stomatogastric nervous system have been rare…' has already been or will be corrected in the near future.

Neuronal differentiation

The development of the polychaete nervous system (NS) is still discussed controversially (Voronezhskaya et al., 2003). Investigations on neuronal differentiation contribute new arguments not only to this discussion, but also to the interpretation of the adult nervous system.

Neuronal development during ontogeny

Two contradictory hypotheses concerning the development of the annelid nervous system exist (Fig. 8). The first theory postulates that the ner-

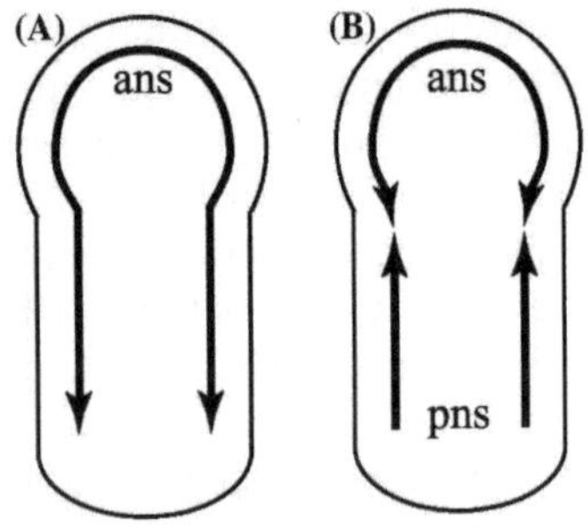

Figure 8. Theories on polychaete neuronal differentiation. (A) The nervous system solely develops from anterior ectoderm. (B) The nervous system develops from two subsystems: the anterior (ans) and the posterior (pns) nervous system.

vous system arises solely from an anterior ectoderm; the connectives grow backwards and form the ventral cord as well (Åkesson, 1968; Lacalli, 1981, 1984; Bhup & Marsden, 1982; Hay-Schmidt, 1995). According to the second theory, a posterior part of the ectoderm gives rise to additional nerves that grow towards the brain; where the anterior and posterior subsystems meet, the nerves arborize and fuse (Bullock & Horridge, 1965; Dorresteijn et al., 1993). According to Hanström (1928b) the presence of two subsystems is plesiomorphic whereas the solely anterior development is derived (as in Turbellaria and Polyplacophora). The presence of two larval types complicates the situation. The adult nervous system appears earlier in development in lecithotrophic than in planctotrophic larvae (Anderson, 1966; Heimler, 1981) and at least in the former the early primordia are retained in the adult system (Anderson, 1973). In lecitotrophic larvae, however, conflicting relations between larval and adult nervous systems are claimed: (1) the two may develop separately (Lacalli, 1984), (2) the larval system may be incorporated into the adult one (Hay-Schmidt, 1995) or (3) the larval system may form a framework along which the adult system develops (Voronezhskaya et al., 2003). While there is increasing evidence now that the planktotrophic larval nervous system develops from two subsystems (pretrochal and intratrochal (Fig. 9G), Lacalli, 1981, 1984; Voronezhskaya et al., 2003), the question of the development of the adult nervous system remains open.

In lecithotrophic larvae of *Scoloplos armiger* (Fig. 9A–C; Müller 1999b, 2003) and *Parapionosyllis minuta* (Fig. 9, E–G; Berenzen & Müller, unpublished data) the first serotonergic perikarya appear dorsolaterally in the prostomium (Fig. 9A, E). Projecting to the contralateral side, their axons form the first cerebral commissure (Fig. 9C). Next to the growth cones of these posteriorly growing neurites (*agc*) a pair of axons growing towards the anterior can be seen on the ventral side (Fig. 9B, *pgc*). In none of the species could the respective perikarya of the posterior fibers be stained. Thus, in *S. armiger* and *P. minuta* an anterior and a posterior neuronal subsystem are present. But instead of merging, the anterior fibers extend further posteriorly and medially and the posterior fibers anteriorly and laterally (Fig. 9C, E, F). In this way paired circumesophageal connectives are formed, of which the inner nerve pair represents the later ventral and the outer pair the later dorsal root. Immunreactivity to FMRFamide showed an identical developmental pattern in *Scoloplos armiger* (Müller, in press). It is assumed that during neuronal differentiation the paired esophageal connnectives fuse to some extent, as is described below.

Figure 9. Neuronal differentiation. (A–C) *Scoloplos armiger*. (A) Two spherical serotonergic perikarya (dpk) dorsally in the prostomium. (B) Neurits from anterior neurons grew caudally via anterior growth cones (agc); neurits from posterior neurons grew anteriorly via posterior growth cones (pgc). Arrows indicate growth direction. (C) In older stages the anterior neurites extend caudally and medially, the posterior ones anteriorly and laterally. cec – cerebral commissure. (D–F) *Parapionosyllis minuta*. (D) Two early, laterally located serotonergic perikarya (pk) and both growth cones are visible. (E, F) Older stage with mediad and caudad growing anterior neurites (arrow) and more lateral located nerves, formed by the posterior neurits. cc – circumesophageal connectives; sn – segmental nerve; vpk – ventral perikarya. (E) Higher magnification from F. (G) *Phyllodoce mucosa*. In the larva the first serotonergic neurites originate from one large posterior perikaryon (ppk). ptn – prototroch nerve. (H–J) *Dorvillea bermudensis*. (H) SEM image of a 3-day-old regenerate. Two old segments (os) give rise to an anterior (abl) and posterior (pbl) blastem. ac – anal cirrus; cb – ciliary band; dc – dorsal cirrus; pp – parapodium. (I) On both sides two nerves emanate from the old ventral cord (ovc). The latter ventral root of the circumesophageal connective (vrcc) already formed a commissure (cvr); the latter dorsal roots (drcc) end blind. stn – stomatogastric nerve. (J) In later stages the dorsal roots are also joined via a dorsal commissure (cdr). (A–C: Müller, 2003; D, E, F: Berenzen & Müller, unpublished; I, J: Müller & Henning, 2003).

Neuronal development during regeneration

Regeneration can be regarded as a special case of development. In contrast to embryological development, the new tissue originates from differentiated cells of the amputee (Fig. 9H). Many investigations of different taxa have demonstrated that the new nervous system of the blastema originates from the old ventral cord of the amputee (e.g., Turbellaria: Reuter et al., 1996; 'Oligochaeta': Yoshida-Noro et al., 2000 Polychaeta: Müller & Berenzen, 2002, Müller

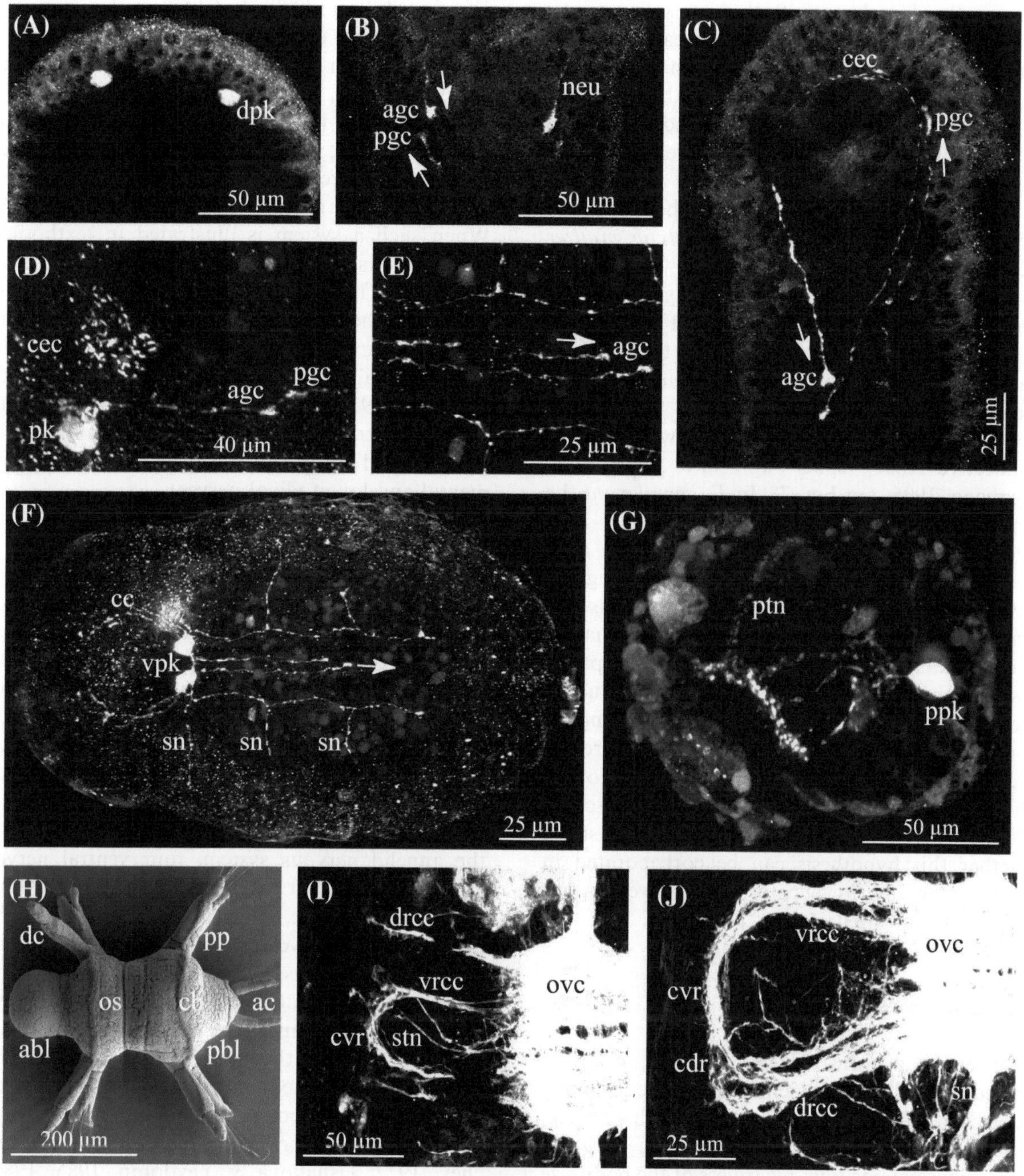

et al., 2003). Following the anterior regeneration in *Dorvillea bermudensis*, Müller & Henning (2003) reported that formation of the new nervous system starts with the outgrowth of two nerves from each side of the old ventral cord (Fig. 9I). The inner pair, the later ventral roots, fuse medially to form the ventral cerebral commissure (Fig. 10F). The outer nerve pair will become the dorsal roots. Both roots split up into two branches (Fig. 10G), of which each later on forms a commissure within the brain (Fig. 10H). The roots are still separated and thus the circumesophageal connectives are paired. The same situation was observed during regeneration in *Eurythoe complanata* and *Marphysa* sp. (Müller & Henning, 2003; Müller, unpublished data) as well as during stolonization in *Autolytus prolifer* (Kreischer & Müller, 2000). In later stages each connective pair merges, proceeding from the ventral cord towards the brain (Fig. 10I). In *D. bermudensis* and *A. prolifer* the fusion stops halfway along the connective (Müller & Henning, 2003), thus producing the typical polychaete anterior nervous system with a single connective that 'splits up' into two roots (Orrhage, 1995; see Fig. 7). In *E. complanata* the fusion is more or less complete and only relatively short roots remain (Orrhage, 1990).

From the data so far collected regarding the development and regeneration of the nervous system it can be concluded that the circumesophageal connectives are paired structures, which are partly fused in annelids possessing dorsal roots including most Polychaeta, and completely fused in 'monoconnective Polychaeta' and Clitellata where, at present dorsal roots are considered to be totally absent (e.g., Bullock & Horridge, 1965; Purschke, 2002). Further studies are needed to demonstrate whether this hypothesis can be corroborated in regenerating anterior ends or by the investigation of stolonizing Clitellata.

The ventral cord

The central nervous system (CNS) is variously embedded within the annelid body tissue. In many polychaetes the CNS retains a basiepithelial (e.g. *Owenia fusiformis*, Coulon & Bessone, 1979) or intraepidermal position, regardless of the body size (Bullock & Horridge, 1965; Hessling & Purschke, 2000; Tzetlin et al., 2002). Within Bilateria the basiepithelial position is regarded as being the plesiomorphic condition (Bullock & Horridge, 1965). Nevertheless, the polychaete CNS can also be shifted into a subepidermal position, as documented e.g., for *Nephtys* sp. (Clark, 1958), *Nereis diversicolor* (Golding, 1992) and *Myzostoma cirriferum* (Müller, 1999b). In Clitellata, however, the central nervous system is always entirely subepidermal, even in the smallest oligochaete species (Purschke et al., 2000; Purschke, 2002).

The annelid ventral nerve cord is a rope–ladder-like system, consisting of paired segmental ganglia that are connected intersegmentally by connectives and intrasegmentally by commissures. When such a system is illustrated in textbooks, most authors prefer to present an arthropod nervous system (e.g., Westheide & Rieger, 1996). This is due to the fact that 'nervous systems in the Polychaeta exhibit a surprising range of levels in organization' (also Bullock & Horridge, 1965, Golding, 1992;). This variety concerns all elements of the ventral cord (connectives, commissures, ganglia) and the peripheral nervous system (segmental and longitudinal nerves).

Ventral connectives

Commonly the presence of two separate trunks within the ventral cord is regarded as the plesiomorphic condition. The rare situation of extremely widely separated cords (Dinophilidae, Saccocirridae) as well as their fusion in the midventral line (e.g., Nerillidae) are, according to Bullock & Horridge (1965) an expression of secondary processes. According to the above-mentioned hypothesis concerning the development of the annelid nervous system, four ventral nerves are already present in early stages (Figs. 9C, F; 10C). Furthermore, in many cases a fifth, unpaired median nerve is added. It is assumed that this nerve contains neurites from ventral perikarya that are located at the transition between the esophageal connectives and the ventral cord; however, fibers of the connectives may also contribute to this median nerve (Fig. 10D). The median nerve is documented for many polychaetes (Bullock & Horridge, 1965; Bubko & Minichev, 1972; Ushakova & Yevdonin, 1985, 1987,

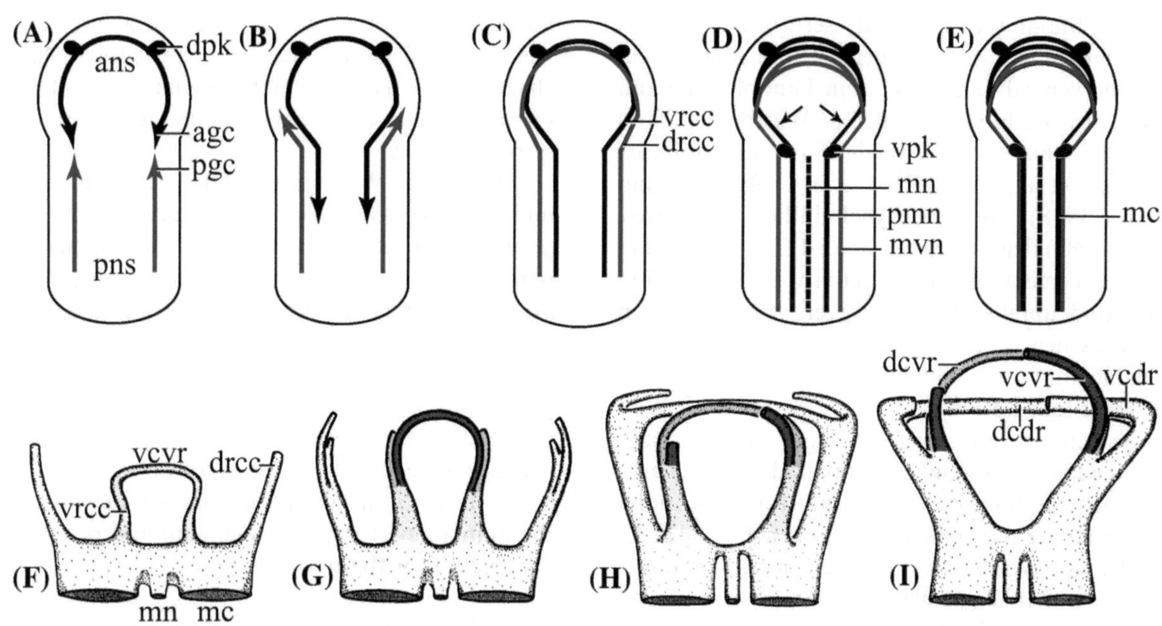

Figure 10. Neuronal differentiation; schematically demonstrated. (A–E) Development of the serotonergic nervous system during ontogeny (after *Scoloplos armiger*, *Parapionosyllis minuta*), ventral view. (A) The nervous system originates from an anterior (ans) and an posterior (pns) subsystem. The growth cones of the anterior (agc; extending caudally) and the posterior (pgc; extending anteriorly) subsystem are indicated by arrows. dpk – dorsal perikaryon. (B) The anterior system prolongs posteriorly and medially, the posterior one anteriorly and laterally. No fusion occurs. (C) The posterior system forms a second cerebral commissure. The circumesophageal connecitives are paired, possessing a dorsal (drcc) and a ventral (vrcc) root. (D) Each root splits up and forms two commissures within the brain. The roots fuse basally (arrows), leaving a dorsal root of different length behind. The first ventral perikarya (vpk) might give rise to the median nerve (mn). pmn – paramedian nerve; mvn – main ventral nerve. (E) The paramedian and the main ventral nerves of each side can fuse and this way form the main cord (mc). (F–I) Development of the cephalic nervous system during regeneration (after *Dorvillea bermudensis*, *Eurythoe complanata*, *Marphysa* sp.), ventral view. (F) From the old nerve cord (mn – median nerve; mc – main cord) two nerve pairs grow anteriorly: the latter ventral (vrcc) and dorsal (drcc) root. The inner nerves form the latter ventral commissure of the ventral root (vcvr). (G) Each root splits up into two nerves. (H) Each nerve forms a commissure. (I) The shape changes and four commissures are present in the brain: ventral (vcvr; vcdr) and dorsal (dcvr, dcdr) commissures of the ventral and the dorsal root, respectively. (A–E: Müller, 2003; F–I: Müller & Henning, 2003).

1988; Müller & Westheide, 1997, 2002) and only in seven out of 28 investigated species the nerve was absent (*Pisione remota*, *Microphthalmus listensis*, *M. sczelkowii*, *Glycera alba*, *Protodrilus* sp., *Protodriloides chaetifer*, *Saccocirrus papillocercus*; Müller, 1999b). In leeches, the median (Faivre's) nerve, which can be totally fused with one connective (Sawyer, 1986), communicates with the stomodaeal system (Bullock & Horridge, 1965). In the simple oligochaete nerve cord such a nerve is hitherto unknown. Presence of the median nerve in nearly all supraordinate polychaete taxa and Hirudinea (Bristol, 1898; Payton, 1981) indicates that it belongs to the basic annelid body plan. Furthermore, presence of the median connective in Arthropoda (Insecta: Hanström, 1928a; Decapoda: Harzsch et al., 1997; Amphipoda: Gerberding & Scholtz, 1999 suggest that it may be part of the ground pattern of Articulata. Long ago Stummer-Traunfels (1927) considered the median nerve in Myzostomidae (Nansen, 1887) an arthropod character.

Paramedian nerves have been described for Dinophilidae (Donworth, 1986; Beniash et al., 1992; Windoffer, 1992; Müller, 1999b; Müller & Westheide, 1997, 2002), Saccocirridae (Kotikova, 1973; Müller, 1999b), Protodrilidae, Protodriloidae, Ctenodrilidae and Magelonidae (Müller, 1999b). Knowing only about the presence of these nerves in Dinophilidae and Protodrilida, Windoffer (1992) proposed that the paramedian nerves coordinate the ventral cilia used for ciliary gliding. Staining of a delicate serotonergic plexus between the paramedian nerves, immediately below the ventral band of cilia in Dinophilidae and *Ophryotrocha* larvae (Müller & Westheide, 2002) supports this view. Investigations in Mollusca (Caunce et al., 1988; Syed et al., 1988; Stefano et al., 1988) and Plathyhelminthes (Sakharov et al., 1986) demonstrated that 5-HT is involved in commencement of ciliary activity in locomotion and food uptake. This, however, cannot explain the presence of corresponding nerves in Ctenodrilidae and Magelonidae, which show another mode of locomotion.

A penta-neural cord with paired main and paramedian nerves and one unpaired median nerve has thus far been described only for some polychaete larvae (*Ophryotrocha gracilis*, Fig. 11B, C, Müller & Westheide, 2002; *Scoloplos armiger*, Fig. 11A, Müller, 2003; *Myzostoma cirriferum*, Eeckhaut et al., 2003; *Capitella capitata*, *Ophryotrocha* sp., Müller, unpublished data) and adult Dinophilidae (Fig. 11E; Jägersten, 1944; Kotikova, 1973; Windoffer, 1992; Müller, 1999b; Müller & Westheide, 1997, 2002). This unique structure of the nervous system provides evidence for a progenetic origin of the dinophilids (see Westheide, 1982, 1984; Westheide & Riser, 1983), due to its common occurrence in larvae, however, it does not justify the conclusion that they are derived from Dorvilleidae or other Eunicida (Müller & Westheide, 2002). Whereas the penta-neural cord persists in Dinophilidae (autapomorphic character), it is only transient in the larvae. All variations found in the ventral polychaete cord can easily be derived from the larval nervous system. Fusion of the two peripheral nerve pairs results in a tri-neural cord (Fig. 10E). This is observable in developmental stages of *Ophryotrocha gracilis* (Fig. 11C) and *Scoloplos armiger* (not shown), and also in the posterior end of adult organisms, where the youngest, posteriormost segment possess five nerves and older, anterior ones only three (Fig. 11D), because the outer ones fuse in an anterior direction. Hypothetical additional primary or secondary absence of the median nerve would result in a di-neural cord (e.g. Glyceridae, Pisionidae). Secondary absence has been described for *Nereis virens*, in which nectochaetae possess a median nerve, whereas it is absent in adults (Ushakova & Yevdonin, 1985, 1988). Finally, midventral fusion of all nerves would result in a uni neural (simple) cord (some Nerillidae; probably 'Oligochaeta'). The latter assumption has to be tested by analyzing neuronal differentiation in the respective taxa.

Müller & Westheide (2002) suggested that midventral concentration of the nerves might have a functional explanation in the development of parapodia. In *Parapodrilus psammophilus*, for example, the nerve strands are concentrated in parapodia-bearing segments whereas they are located far apart from each other in the last, parapodia-less segment (Fig. 11G). Combination of parapodia-bearing segments with widely separated nerves can be found in *Saccocirrus*, but rather than using their stump-like parapodia for locomotion, the animals perform ciliary gliding and peristaltic contraction of the body.

Ventral ganglia

Medullary nerve cords (perikarya scattered throughout the entire length) are reported only for few polychaetes whereas most taxa possess ganglionated cords (cell-free connectives and concentrated groups of perikarya). The medullary organization might be ancestral, but embryological (Echiura, Priapulida) and phylogenetic (Onychophora) studies speak in favor of a derived character, at least in some cases (Beklemischew, 1960). The ganglia consist of a fibrous core (neuropile, if synaptic connections are present) and a peripheral rind, containing the perikarya (Golding, 1992). Typically the ganglia are located in the midventral center of the respective segment, but caudal shift in anterior body regions and anterior shift in posterior regions is common. Often ganglia span the intersegmental boundary (Smith, 1957); for instance, in *Nereis* they extend from posterior regions of one segment as far as the middle of the following one (Bullock & Horridge, 1965). For all annelid subtaxa fusion of anterior ganglia into a subesophageal ganglionic mass is known. The number of included ganglia differs in Polychaeta: e.g. two in *Ophryotrocha gracilis* (Fig. 11B, C), three in *Microphthalmus*, four in *Pisione remota* and six in *Glycera alba* (Müller, 1999b), whereas it is consistently four in the Naididae ('Oligochaeta') and Hirudinea (Hessling et al., 1999; Purschke et al., 1993). Formation of

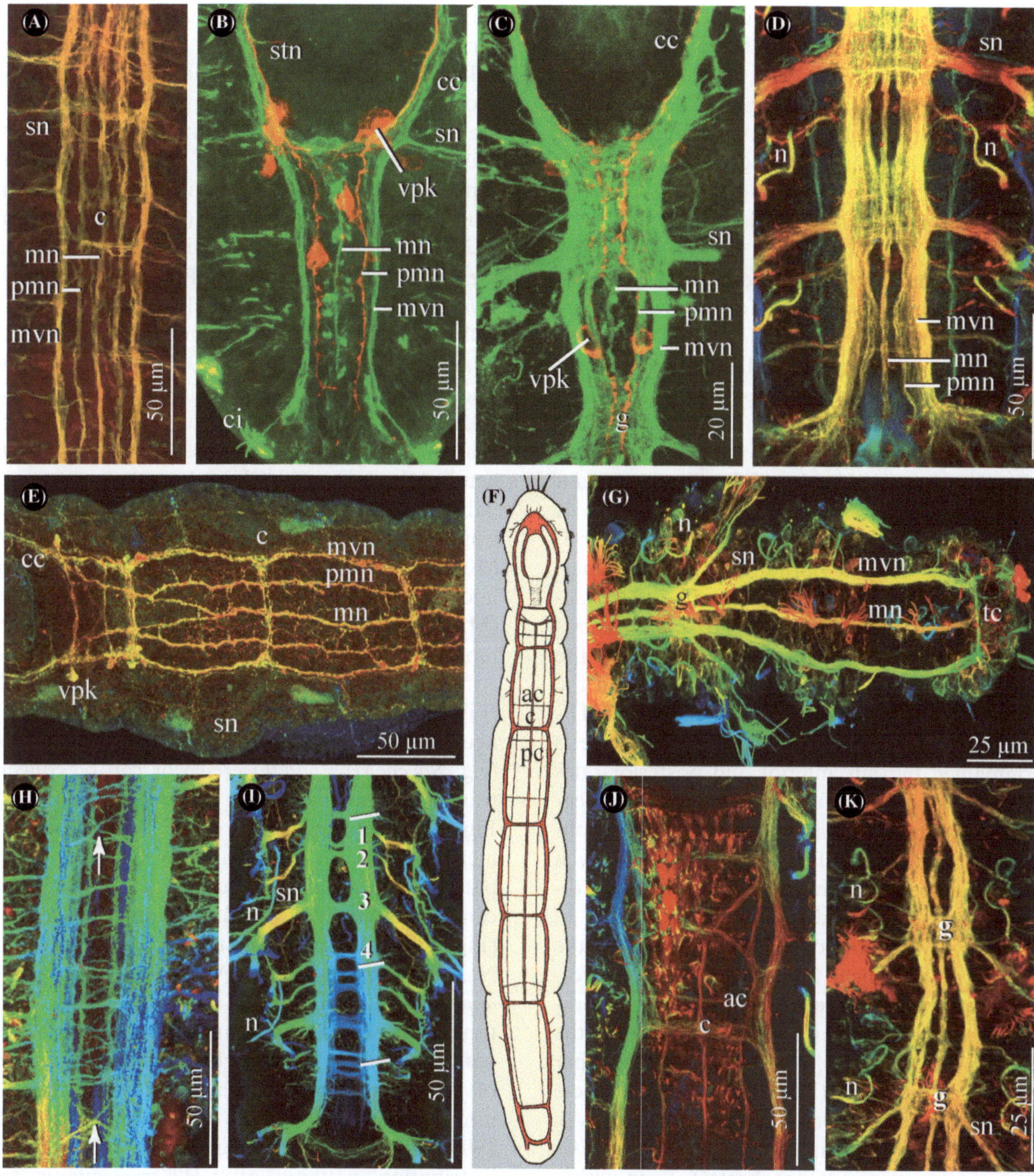

Figure 11. Confocal images (not F), 3D color-coding along the *z*-axis; red = periphery > blue = center (not B, C). (A) *Scoloplos armiger*, acetylated α-tubulin-IR (a-αT-IR); ventral cord of a larva with unpaired median (mn), paired paramedian (pmn) and main ventral (mvn) nerves. c – commissure; sn – segmental nerve. (B–D) *Ophryotrocha gracilis*. (B, C) Double staining: red = serotonergic, green = a-αT-IR. (B) Jung larva with five connectives within the ventral cord. cc – circumesophageal connectives; ci – cilia; sn – segmental nerve; stn – stomatogastric nerve; vpk – ventral perikaryon. (C) In older larvae the five connectives persist between the dense neuropile of the ganglia (g). (D) Posterior end of an adult specimen, a-αT-IR. Five connectives are visible in the posteriormost end, further anterior only three connectives persist. n – nephridia. (E) *Trilobodrilus gardineri*; serotonergic IR. Ventral cord with five connectives. (F) *Trilobodrilus axi.* Schematical drawing of the ventral cord. (G) *Parapodrilus psammophilus*, a-αT-IR, posterior end. tc – terminal commissure. (H–K) Ventral cord, tubulinergic IR. (H) *Protodrilus* sp. a-αT-IR. Arrows – repeated chiasmata. (I) *Pisione remota*, a-αT-IR, posterior end. Numbers indicate commissures per segment. (J) *Trilobodrilus hermaphroditus*, a-αT-IR, anterior end. ac – anterior commissure. (K) *Parapodrilus psammophilus*, a-αT-IR. (F: after Müller & Westheide, 2002).

more than one ganglion per segment is claimed for Sabellariidae, Serpulidae, Sabellidae (two/segment) and Pectinarridae (two or three/segment; Beklemischew, 1960). Nothing is known about how this multi-ganglionic pattern differentiates. Subdivision of a primary single ganglia seems to be likely, but this purely speculative hypothesis has to be tested and the presence of this pattern should be reinvestigated.

Commissures. The majority of nerve fibers pass to the contralateral side, thus forming well differentiated commissures (Smith, 1957; Golding, 1992). In all investigated nerillid species, *Protodrilus* sp. (Fig. 11H) and *Protodriloides chaetifer*, countless commissures interconnect the ventral cords within which, apart from chiasmata, no distinct pattern could be found. In all other taxa a definite number of commissures per segment is arranged in a segmentally repetitive formation. One (*Platynereis dumerilii*) up to seven (*Chaetopterus variopedatus*, Martin & Anctil, 1984; *Saccocirrus papillocercus*, Müller, 1999b) commissures can be counted, of which one can be termed the 'main' commissure because it is broader than the others (Müller & Westheide, 2000, 2002). In Dinophilidae, e.g., the median, main commissure is accompanied by thin anterior and posterior ones (Fig. 11F, J). Due to fusion of ganglia, the segmental pattern is often modified in the subesophageal ganglionic mass (Fig. 11J). It can vary even within one specimen: in dinophilids the subordinate commissures are differently reduced and in *Pisione remota* the four commissures present in the posteriormost segment fuse anteriorly in the following sequence: (a) median ones, (b) incorporation of the anterior, (c) incorporation of the posterior one, thus 4, 3, 2 and 1 commissures per segment can be found (Müller, 1999b). Complete fusion of the right and left hemiganglion results in so called 'unitary ganglia' (Fig. 11K, *Parapodrilus psammophilus*; also in 'Oligochaeta'). Even within these concentrations transversal 'Faserbrücken' might be identified (three in *Lumbricus terrestris*, Günther, 1971).

A tendency towards increasing neuropile concentration and parallel reduction of commissure number appears likely, but remains to be confirmed by more data.

Perikarya. The neuronal cell bodies are located dorsally and laterally in the supraesophageal (Fig. 12D,E) and ventrally and laterally in the ventral ganglia (Fig. 12A–C). In Polychaeta and 'Oligochaeta', serotonergic and RFamidergic cells occur in cerebral and ventral ganglia, whereas they are restricted to the latter in Hirudinea (Marsden & Kerkut, 1969; Wallace, 1981). Whereas 5-HT neurons, if present, are more numerous in the ventral than in the cerebral ganglia (White & Marsden, 1978; Spörhase-Eichmann et al., 1987a, b), it is just the opposite for RFamide. In Polychaeta from two (*Mesonerilla intermedia*) to 18 (*Pisione remota*, Müller, 1999b) and in 'Oligochaeta' from two (Naididae, Hessling et al., 1999) to 80–100 (*Lumbricus terrestris*, Spörhase-Eichmann et al., 1987a, b) 5-HT perikarya can be stained. With one exception (*Nerilla antennata* possesses a single, median neuron) they are arranged in bilaterally symmetrical pairs and are concentrated in two central (often a single cell) and two lateral clusters (Müller, 1999b; Müller & Westheide, 2002). At least some of the lateral neurons innervate the lateral antennae (Müller et al., 2003); therefore this cluster might represent the fifth or the eleventh ganglion in Amphinomidae and Glyceridae, respectively (Figs. 4E, 1D, E). In the ventral cord, 5-HT neurons can either have medullary distribution (e.g. *Saccocirrus papillocercus*) or can be concentrated in ganglia (e.g. *Ophryotrocha gracilis*, Fig. 12A); intersegmental clusters are also documented for Nerillidae (Müller, 1999b). Segmental patterns and, moreover, individual single neurons can hardly be identified in some Polychaeta (Fig. 12A; Müller & Westheide, 2000, 2002). In Naididae ('Oligochaeta', *Nais variabilis*, *Slavinia appendiculata*, *Stylaria lacustris*), however, serotonergic neurons are arranged in an alternating pattern in successive ganglia in posterior regions of the ventral cord, which might be an autapomorphic character for the taxon (Hessling et al., 1999). At the moment information about the distribution patterns of 5-HT neurons is too limited to allow phylogenetic comparisons, such as have been undertaken in other taxa.

The peripheral nervous system

The peripheral system consists of nerves emanating from the ventral cords (segmental or side

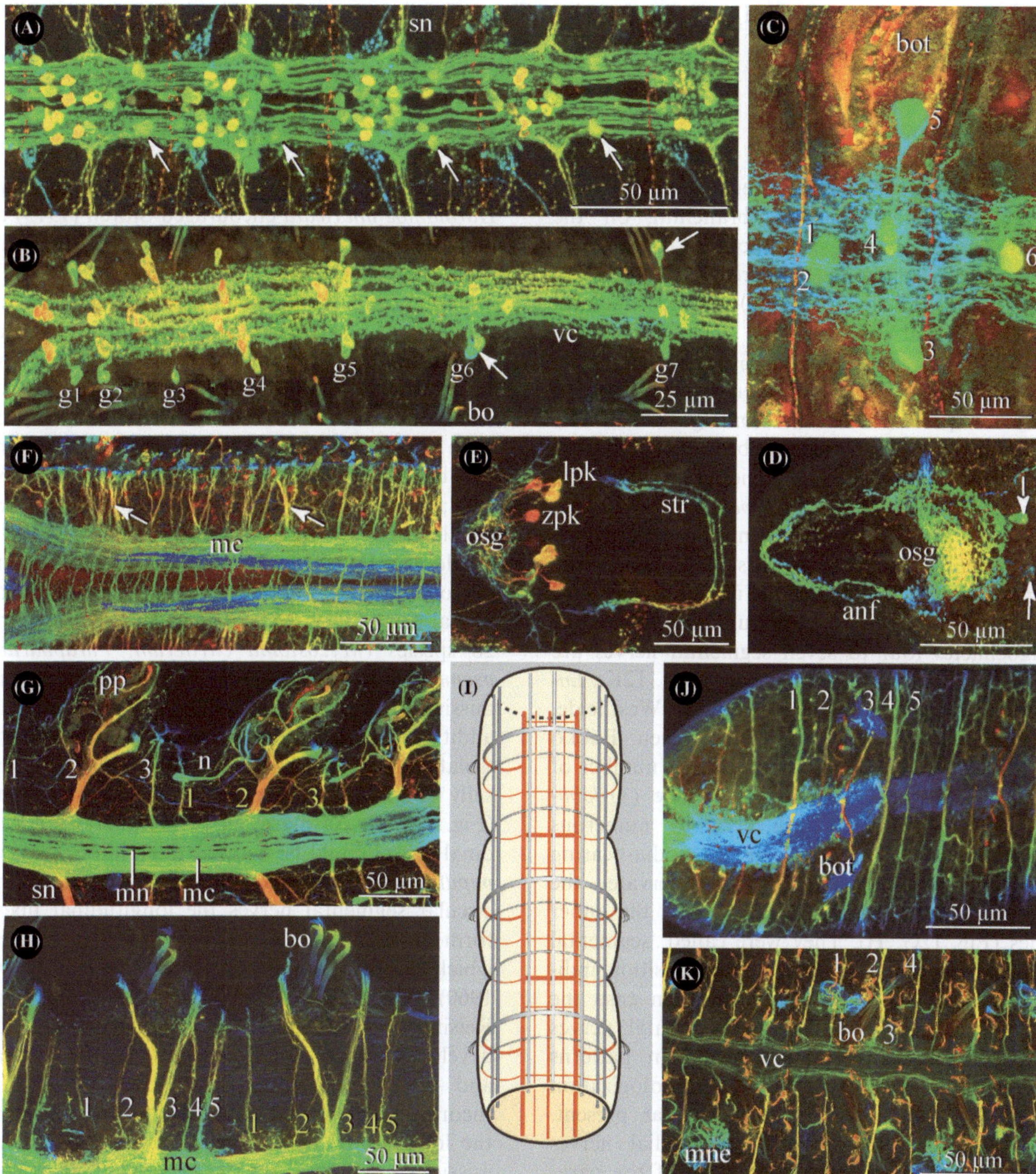

Figure 12. Confocal images (not I), 3D color-coding along the z-axis; red = periphery > blue = center. (A) *Ophryotrocha gracilis*, ventral cord, 5 HT-IR. Arrows indicate repeated perikarya. sn – segmental nerve. (B) *Pristina notopora*, ventral cord, 5 HT-IR. bo – chaetae; g1-g7 – ganglia 1–7; arrows indicate alternating perikarya. (C) *Stylaria lacustris*, single trunk ganglion, 5-HT-IR. bot – chaetal sack; number indicate single perikarya. (D) *Pristina notopora*, 5-HT-IR. Supraoesophageal ganglion (osg) with anteriorly strechtching neurites (anf) and dorso-posteriorly locacted perikarya (arrows). (E) *Trilobodrilus hermaphroditus*, 5-HT-IR. Supraoesophageal ganglion with two central perikarya (zpk) and two groups of lateral perikarya (lpk). – str – stomatogastric ring. (F-K) Lateral nerves, a-αT-IR. (F) *Portodrilus* sp., arrows – repeated thicker nerves (G) *Platynereis dumerilii.* (H) *Polydora cornuta.* (I) *Dinophilus gyrociliatus*; schematic drawing. (J) *Enchytraeus albidus*. (K) *Stylaria lacustris*. bot – chaetal sack; mc – main cord; mne – metanephridium; mn – median nerve; n – nephridium; vc – ventral cord; numbers – segmental nerves. (I: after Müller & Westheide, 2002; J: Müller & Hundsdörfer, unpublished).

nerves), additional longitudinal nerves and an intra- or subepidermal plexus.

Segmental nerves

Bullock & Horridge (1965) regard the presence of three segmental nerves, branching off the ventral cords at the ganglionic level, as the plesiomorphic condition. Whereas Beklemishew (1960) and Hanström (1928b) claimed that all annelids invariably possess three segmental nerves, some authors (e.g. Golding, 1992) reported a fourth pair and even the presence of only two pairs (Bullock & Horridge, 1965). The last condition has been described, e.g., for adult Hirudinea (Livanow, 1904; Nicholls & Van Essen, 1974). Early stages of *Erpobdella octoculata*, however, possess four segmental nerves, which subsequently become fused: first the median ones join one another and then the remaining anterior one is included (Hessling, unpublished data). For oligochaetes four to seven segmental nerves are described (*Stylaria lacustris*: four, Fig. 12K; *Enchytraeus crypticus*: five, Hessling & Westheide, 1999; *E. albidus*: five, Fig. 12J or seven, Bubko & Minichev, 1992). In Polychaeta, the number of nerves per segment ranges from none (*Trilobodrilus hermaphroditus*, Fig. 11F) to numerous (*Protodrilus* sp., Fig. 12F). In parapodia-bearing polychaetes the nerves that innervate the appendages are always thicker than the other ones (Fig. 12G, H). They were called 'segmental nerve 2' by Smith (1957), but the position indicated by this term may be wrong in case the annelid has more than three segmental nerves (Fig. 12H). Because their arborization patterns are variable or not discernible, homologization of the smaller nerves is impossible at present. However, branching of the parapodial nerves follows a fixed pattern: at the base of the parapodium the nerve splits, sending an anterior branch into the ventral cirrus while a posterior branch bends dorsal and innervates the dorsal cirrus (Dorsett, 1976; Müller & Westheide, 2002). Some segmental nerves elongate towards the dorsal side, where they fuse and form circular commissures (Fig. 12I). These nerves innervate ciliary trochs (Fig. 13A), dorsally located ciliary patches (Fig. 13D) and possibly dorsal muscle fibers.

Peripheral longitudinal nerves

The presence of paired lateral nerves in *Harmothoe imbricata* encouraged Storch (1913) to separate Amphinomidae as 'Tetraneura' from all other polychaetes, which he labeled 'Dineura'. Homologizing the respective nerves with lateral nerves of platyhelminthes and pleurovisceral connectives of Amphineura, he regarded the Amphinomidae as ancestral and the dineuralian situation as derived. Afterwards, lateral nerves were described for 'Oligochaeta' (Hanström, 1928b; Beklemischew, 1960; Bubko & Minichev, 1992; Hessling et al., 1999), Hirudinea (Beklemischew, 1960) and a few Polychaeta (Bullock & Horridge, 1965). Recent investigations demonstrated a far more common distribution of lateral nerves in polychaetes: in only three out of 28 species (*Myzostoma cirriferum, Parapodrilus psammophilus, Potamodrilus fluviatilis*) they were missing (Fig. 13E–J; Müller, 1999b). Moreover, additional longitudinal nerves were found ventrolaterally (Fig. 13A), laterally (13B) and dorsally (Fig. 13C) in larvae as well as in adults (Fig. 13E–J). As many as 17 peripheral longitudinal nerves could be demonstrated in *Saccocirrus papillocercus* (Fig. 13D, F). Together with the circular segmental nerves the longitudinal fibers form a regular grid of perpendicular nerves, which can be called an 'orthogon' (Beklemischew, 1960). From a similar neural arrangement Reisinger (1925, 1972) and Hanström (1928b) developed a theory on the evolution of the central and peripheral spiralian nervous system. This theory suggested, that – in addition to the brain – the longitudinal nerve strands and annular commissures originate from the basiepidermal plexus, become organized so that the commissures and strands are perpendicular to one another and the whole structure is displaced inward as an 'orthogon'. By reduction of the dorsal longitudinal nerve strands and concentration of perikarya into ganglia, this structure then gives rise to the nervous system of the Articulata, with its ventral ganglion chain. By providing evidence that one

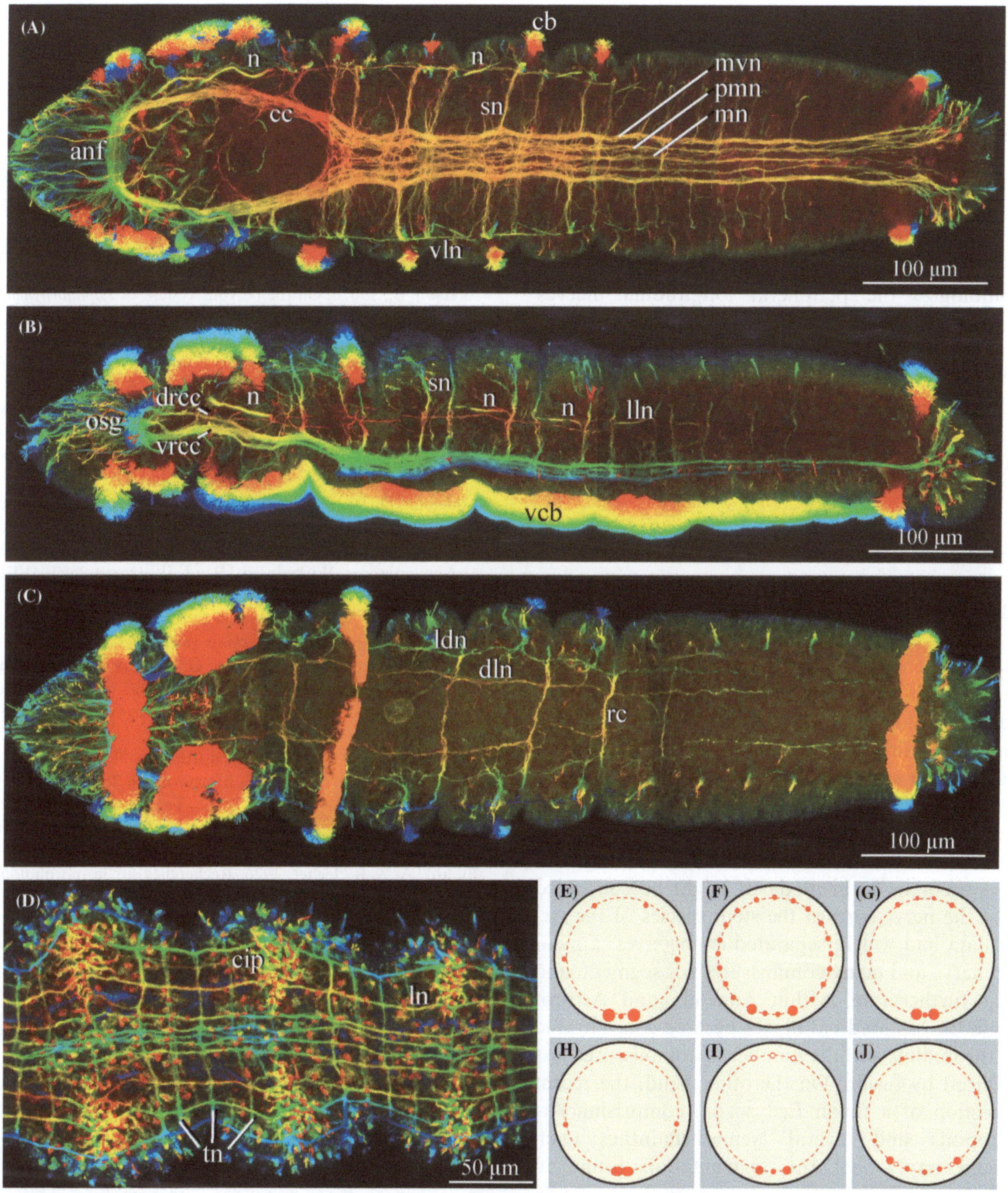

Figure 13. Peripheral nerves. (A–D) Confocal images, 3D color-coding along the z-axis; red = periphery > blue = center. (A–C) *Scoloplos armiger*, a-αT-IR. (A) Ventral view. (B) Lateral view. (C) Dorsal view. (D) *Saccocirrus papillocercus*, a-αT-IR. (E-J) Schematic cross sections with ventral nerve cord and peripheral nerves. The dotted line indicates a possible ring commissure. (E) *Scoloplos armiger*. (F) *Saccocirrus papillocercus*. (G) *Platynereis dumerilii*. (H) *Nerilla antennata*. (I) *Parapodrilus psammophilus*. (J) *Dinophilus gardineri*. anf – anterior neurites; cc – circumesophageal connective; cb – ciliary band; cip – ciliary patch; dorsal (drcc) and ventral (vrcc) root of the circumesophageal connective; dorsolateral (dln), laterodorsal, lateral (lln) and ventrolateral (vln) longitudinal nerve (ln); median (mn), main (mvn) and paramedian (pmn) nerve of the ventral cord; n - nephridium; rc – ring commissure; sn – segmental nerve; tn – transversal nerve.

of the main arguments for this scenario, an orthogon is absent in the polychaete *Lopadorhynchus*-larva (Hanström, 1928b), Åkesson (1967b) questioned this theory. In fact, homologization of the platyhelminth and annelid 'orthogon' remains difficult, because in many cases the nerves stay in an basiepithelial position in Polychaeta, whereas they are shifted inwards in Platyhelminthes. Because of their different positions and concentrations it is not even possible to homologize the peripheral longitudinal nerves within polychaetes at present. Further investigations have to clarify whether, as has been thought (Westheide & Rieger, 1996), formation or reduction of the neuronal orthogon is in fact correlated with formation and reduction of a muscular orthogon.

Annelida – Arthropoda

Most authors regard the rope-ladder-like nervous system as an apomorphic character for the Articulata. Whereas the ventral cord is highly variable in Annelida, concerning the position of the connectives, additional longitudinal nerves (unpaired median, paired paramedian) and commissure-number per segment, the Arthropod cord is much more consistent. The Onychophora, however, demonstrate that widely separated cords connected via numerous commissures per segment also occur in Arthropoda. It might be concluded that the nerve cord of the stem species of the Articulata had widely separated connectives, a median nerve and many commissures per segment, and that furthermore, within Annelida and Arthropoda the longitudinal nerves became concentrated in the midline and the number of commissures was reduced by fusion. On the other hand, the recent errection of a taxon Ecdysozoa, comprising Arthropoda and several Nemathelminthes taxa, would imply that the segmentation found in annelids and arthropods, and with it the rope-ladder-like nervous system, must be either convergent or an ancestral feature of protostomes or even bilaterians (see Scholtz, 2002 for thorough discussion). Further detailed morphological information e.g. comparison on individual single neuronal level between Annelida and Arthropoda or Cyloreuralia and Arthropoda is needed to support either the Articulata or the Ecdysozoa.

References

Åkesson, B., 1963. The comparative morphology and embryology of the head in the scale worms (Aphroditidae, Polychaeta) Arkiv för Zoologi 16: 125–163.

Åkesson, B., 1967a. The embryology of the polychaete *Eunice kobiensis*. Acta Zoologica (Stockholm) 48: 141–192.

Åkesson, B., 1967b. On the nervous system of *Lopadorhynchus* larva (Polychaeta). Arkiv för Zoologi 20: 55–78

Åkesson, B., 1968. The ontogeny of the glycerid prostomium (Annelida: Polychaeta) – Acta Zoologica (Stockholm) 49: 203–217.

Allen, E. J., 1904. The anatomy of *Poecilochaetus*, Claparède. Quarterly Journal of Microscopical Science N.S. 48: 79–151.

Anderson, D. T., 1966. The comparative morphology of the Polychaeta. Acta Zoologica (Stockholm) 47: 1–42.

Anderson, D. T., 1973. Embryology and Phylogeny in Annelids and Arthropods. Pergamon Press, Oxford, 495 pp.

Beklemischew, W. N., 1960. Grundlagen der vergleichenden Anatomie der Wirbellosen. Band II: Organologie. VEB, Berlin.

Beniash, E. A., N. N. Yerlikova & L. A. Yevdonin, 1992. Some characteristics of the *Dinophilus vorticoides* anatomy of the nervous system. In Bushinskaja, G. N. (ed.), Polychaeta and their Ecological Significance. Explorations of the Fauna of the Seas, vol. 43(51). Russian Academy of Sciences, St. Petersburg: 5–9.

Bernert, J., 1926. Untersuchungen über das Nervensystem der *Hermione hystrix* (L.M.). Z. Morph. Ökol. Tiere 6: 743–810.

Bhup, R. & J. R. Marsden, 1982. The development of the central nervous system in *Capitella capitata* (Polychaeta, Annelida). Canadian Journal of Zoology 60: 2284–2295.

Binard, A. & R. Jeener, 1926. Recherches sur la morphologie du système nerveux des Annélides. I. *Nereis*. Bull. Acad. r. Belg. Cl. Sci. 12: 437–448.

Binard, A. & R. Jeener, 1928. Morpholgie du lobe préoral des polychètes. Rec. Inst. Zool. Torley-Rousseau 2: 117–240.

Bristol, C. L., 1898. The metamerism of *Nephelis*. A contribution to the morphology of the nervous system, with a description of *Nephelis lateralis*. Journal of Morphology 15 : 17–72.

Bubko, O. V., 1981. The nervous system of *Protodrilus* (Archiannelida) (in Russian). Zoologichesky Zhurnal 60: 1749–1755.

Bubko, O. V. & Yu, S. Minichev, 1972. The nervous system in Oweniidae. Zoologichesky Zhurnal 51: 1288–1299.

Bubko, O. V. & Yu, S. Minichev, 1992. Nervous system of *Enchytraeus albidus* (Oligochaeta, Enchytraeidae). Zoologichesky Zhurnal 71: 11–17.

Bullock, T. H., 1995. Are the main grades of brains different principally in numbers of connections or also in quality? In Breidbach, O. & W. Kutsch (eds), The Nervous System of Invertebrates. An Evolutionary and Comparative Approach. Birkhäuser, Basel: 439–449.

Bullock, T. H. & G. A. Horridge, 1965. Structure and function in the nervous system of invertebrates. Vol.1, W. H. Freeman and Company, London.

Caunce, M., J. D. McKenzie, J. Tripp & W. Winlow, 1988. Serotonergic innervation of the pedal epidermis of *Lymnea*. Symposia Biologica Hungarica 36: 691–692.

Cerruti, A., 1909. Contributo all'Anatomia, biologia e sistematica delle Paraonidae (Levinsenidae) con particolare riguardo alle specie del golfo di Napoli. Mitt. Zool. Stn Neapel 19(3): 459–512.

Clark, R. B., 1958. The micromorphology of the supraoesophageal ganglion of *Nephtys*. Zool. Jb. Allgem. Zool. Phys. Tiere 68: 261–296.

Coulon, J. & R. Bessone, 1979. Autoradiographic detection of indolamine and catecholamine neurons in the nervous system of *Owenia fusiformis* (Polychaeta, Annelida). Cell and Tissue Research 198: 95–104.

Defretin, R., 1956. La neurosécrétion des polysidees et ses rapports avec l'epitoquie chez les Néréidiens. Annals Sci. Nat. Zool. 18: 209–222.

Dhainaut-Courtois, N. & D. W. Golding, 1988. Nervous System. In: Westheide, W. & C. O. Hermans (eds), The Ultrastructure of Polychaeta, 89–110.

Donworth, P. J., 1986. A reappraisal and the validation of the species *Dinophilus taeniatus* Harmer 1889 and the taxonomically significant features in monomorphic dinophilids (Annelida Polychaeta). Zoologischer Anzeiger 216: 32–38.

Dorresteijn, A. W. C., B. O'Grady, A. Fischer, E. Porchet-Henneré & Y. Boilly-Marer, 1993. Molecular specification of cell lines in the embryo of *Platynereis* (Annelida). Roux's Archives of Development Biology 202: 260–269.

Dorsett, D. A., 1976. The structure and function of proprioceptors in soft-bodied invertebrates. In: Mill, P (ed), Structure and Function of Proprioceptors in the Invertebrates. Chapman & Hall, London: 443–483.

Eeckhaut, I., L. Fievez & M. C. M. Müller, 2003. Development of *Myzostoma cirriferum* (Myzostomida). Journal of Morphology 258: 269–283.

Ehlers, E., 1864. Die Borstenwürmer nach systematischen und anatomischen Untersuchungen dargestellt. Wilhelm Engelmann, Leipzig, 479 pp.

Eibye-Jacobsen, D., 1993. On the phylogeny of the Phyllodocidae (Polychaeta, Annelida): an alternative. Zeitschrift für zoologische Systematick and Evolutions-Forschung 31: 174–197.

Fauchald, K. & G. Rouse, 1997. Polychaete systematics: Past and present. Zoologica Scripta 26(2): 71–138.

Fauvel, P., 1897. Recherches sur les Ampharétiens, Annélides polychètes sédentaires, Morphologie, Anatomie, Histologie, Physiologie. Bull. Sci. Fr. Belg. 30: 277–488.

Fauvel, P., 1923. Polychètes errantes. Faune Fr. 5: 1–488.

Fauvel, P., 1927. Polychètes sédentaires. Addenda aux Errantes. Archiannélides. Myzostomaires. Faune Fr. 16: 1–404.

Fitzsimmons, P. G., 1964. The anatomy and neurosecretory system of the supraoesophageal ganglion of *Hermodice carunculata* (Annelida: Polychaeta): M.Sc. thesis. McGill University, Monteral, Quebec, 115 pp.

Foster, N. M., 1971. Spionidae (Polychaeta) of the Gulf of Mexico and the Carribean Sea. Stud. Fauna Curacao 36: 1–183.

Gerberding, M. & G. Scholtz, 1999. Neurons and glia in the midline of the higher crustacean *Orchestia cavimana* are generated via an invariable cell lineage that comprises a median neuroblast and glial progenitors. Developmental Biology 235: 397–409.

Gidholm, L., 1967. A revision of Autolytinae (Syllidae, Polychaeta) with special reference to the Scandinavian species, and with notes on external and internal morphology, reproduction and ecology. Arkiv för Zoologie 19: 157–213.

Gilpin-Brown, J. B., 1958. The development and structure of the cephalic nerves of *Nereis*. Journal of Comparative Neurology 109: 317–348.

Golding, D. W., 1992. Polychaeta: nervous system. In Harrison, F. W. & S. L. Gardiner (eds), Microscopic Anatomy of Invertebrates 7: Annelida. Wiley–Liss, New York: 153–179.

Gravier, Ch., 1896. Recherches sur les Phyllodociens. Bull. Scient. Fr. Berg. 29: 293–381.

Gravier, Ch., 1898. Étude du prostomium des Glycériens suivie de considerations générales sur le prostomium des Annélides Polychètes. Bull. Scient. Fr. Belg. 31: 159–184.

Gustafson, G., 1930. Anatomische Studien über die Polychaeten-Familien Amphinomidae und Euphrosynidae. Zool. Bidr. Upps. 12: 305–471.

Günther, J., 1971. Mikroanatomie des Bauchmarks von *Lumbricus terrestris L.* (Annelida, Oligochaeta). Z. Morph. Ökol. Tiere 70: 141–182.

Haeckel, E., 1866. Generelle Morphologie der Organismen. Georg Reimer, Berlin.

Hamaker, J. I., 1898. The nervous system of *Nereis virens* Sars. A study in comparative neurology. Bull. Mus. Comp. Zool. Harv. 32 (6): 89–124.

Hanström, B., 1927. Das zentrale und periphere Nervensystem des Kopflappens einiger Polychäten . Z. Morph. Ökol. Tiere 7: 543–596.

Hanström, B., 1928a. Die Beziehung zwischen dem Gehirn der Polychäten und dem der Arthropoden. Z. Morph. Ökol. Tiere 11: 152–160.

Hanström, B., 1928b. Vergleichende Anatomie des Nervensystems der wirbellosen Tiere unter Berücksichtigung seiner Funktion. Julius Springer, Berlin, 628 pp.

Hanström, B., 1928c. Weitere Beiträge zur Kenntnis des Gehirns und der Sinnesorgane der Polychaeten (*Polygordius, Tomopteris, Scolecolepis*). Z. Morph. Ökol. Tiere 13: 329–358.

Hanström, B., 1930. Beitrag zur Diskussion des Kopflappens der Polychäten. K. Fysoigr. Sällsk. Lund Handl. N. F. 42(3): 1–29.

Harzsch, S. & K. Anger, R. R. Dawirs, 1997. Immunocytochemical detection of acetylated α-tubulin and *Drosophila* synapsin in the embryonic crustacean nervous system. International Journal of Devlopmental Biology 41: 477–484.

Hatschek, B., 1878. Studien über Entwicklungsgeschichte der Anneliden. Ein Beitrag zur Morphologie der Bilaterien. Arb. Zool. Inst. Univ. Wien 1: 277–404.

Hay-Schmidt, A., 1995. The larval nervous system of *Polygordius lacteus* Schneider, 1868 (Polygordiidae, Polychaetta): Immunocytochemical data. Acta Zoologica (Stockh.) 76: 121–140.

Heider, K., 1925. Über *Eunice*. Systematisches, Kiefersack, Nervensystem. Z. Wiss. Zool. 125: 55–90.

Heimler, W., 1981. Untersuchungen zur Larvenentwicklung von *Lanice conchilega* (Pallas) 1766 (Polychaeta, Terebellomorpha): Teil II. Bau und Ultrastruktur der Trochophora-Larve. Zoll. Jahrb. Anat. 106: 236–277.

Hempelmann, F., 1911. Zur Naturgeschichte von *Nereis dumerilii* Aud. et Edw. Zoologica, Stutt. 25: 1–135.

Hessle, C., 1917. Zur Kenntnis der terebellomorphen Polychaeten. Zool. Bidr. Upps. 5: 37–251.

Hessle, C. 1925. Einiges über die Hesioniden und die Stellung der Gattung *Ancistrodyllis*. Arkiv för Zoologicie 17A, 10: 1–36

Hessling, R., M. C. Müller & W. Westheide, 1999. CLSM analysis of serotonin-immunoreactive neurons in the central nervous system of *Nais variabilis, Slavinia appeniculata* and *Stylaria lacustris* (Oligochaeta: Naididae). Hydrobiologia 406: 223–233.

Hessling R. & G. Purschke, 2000. Immunohistochemical (cLSM) and ultrastructural analysis of the central nervous system and sense organs in *Aeolosoma hemprichi* (Annelida, Aeolosomatidae). Zoomorphology 120: 65–78.

Hessling, R. & W. Westheide, 1999. CLSM analysis of development and structure of the central nervous system of *Enchytraeus crypticus* ('Oligochaeta', Enchytraeidae). Zoomorphology 119: 37–47.

Hessling, R. & W. Westheide, 2002. Are Echiura derived from a segmented ancestor? Immunohistochemical analysis of the nervous system in developmental stages of *Bonellia viridis*. Journal of Morphology 252: 100–113.

Holmgren, N., 1916. Zur vergleichenden Anatomie des Gehirns von Polychaeten, Onychophoren, Xiphosuren, Arachniden, Crustaceen, Myriapoden und Insekten. K. svenska Vetensk-Akad. Handl. 56: 1–303.

Holthe, T., 1986. Evolution, systematics and distribution of the Polychaeta Terebellomorpha, with a cataogue of the taxa and a bibliography. Gunneria 55: 1–236.

Jägersten, G., 1944. Zur Kenntnis der Morphologie, Enzystierung und Taxonomie von *Dinophilus*. K. svenska Vetensk.-Akad. Handl. 21: 1–90.

Johansson, K. E., 1927. Beiträge zur Kenntnis des Polychaeten-Familien Hermellidae, Sabellidae und Serpulidae. Zool. Bidr. Upps. 11: 1–184.

Joyeux-Laffuie, J., 1890. Étude monographique du Chétoptère (*Chaetopterus variopedatus*, Rénier) suivie d'une revision des espèces du genre *Chaetopterus*. Archs Zool. Exp. Gén. 2(8): 245–360.

Kotikova, E. A., 1973. New data concerning the nervous system of Archiannelida. Zoolicheskii Zhurnal 52: 1611–1615 (in Russian).

Kreischer, S. & M. C. Müller, 2000. Development of neuronal structures in stolons of *Autolytus prolifer* (Polychaeta). Zoology 93: 1–58.

Lacalli, T. C. 1981. Structure and development of the apical organ in trochophores of *Spirobranchus polycerus*, *Phyllodoce maculata* and *Phyllodoce mucosa* (Polychaeta). Proceedings at the Royal Society of London Series B. Biological Science 212: 381–402.

Lacalli, T. C. 1984. Structure and organisation of the nervous system in the trochophore larva of *Spirobranchus*. Philosophical Transaction at Royal Society of London B. Biological Science 306: 79–135.

Lameere, A., 1916. Une theorie zoologique. Bull. Scient. Fr. Belg. 40: 378–431.

Lameere, A., 1925. Sur la signification de la trochosphère. Trav. Stn zool. Wimereux 9: 102–106.

Livanow, N. A., 1904. Untersuchungen zur Morphologie der Hirudineen. II. Das Nervensystem des vorderen Körperendes und seine Metamerie. Zool. Jb. Anat. 20: 153–226.

Malaquin, A., 1893. Recherches sur les Syllidiens. Morphologie, anatomie, reproduction, développement. Mémoires de la Société des Sciences de Lille: 1–477.

Malaquin, A. & A. Dehorne, 1907. Les Annélides Polychètes de la Baie d'Amboine. Revue suisse Zoologie 15: 335–400.

Manaranche, R., 1966. Anatomie du ganglion cérébroide de *Glycera convoluta* Keferstein (Annélide Polychète), avec quelques remarques sur certains organes prostomiaux. Cashiers de Biologie Marine 7: 259–280.

Marsden, J. R. & M. Galloway, 1968. The microanatomy of the supra-oesophageal ganglion of *Eurythoe complanata* (Polychaeta, Amphinomidae). Canadian Journal of Zoology 46: 349–367.

Marsden, J. R. & G. A. Kerkut, 1969. Fluorescent microscopy of the 5HT- and catecholamine-containing cells in the central nervous system of the leech *Hirudo medicinalis*. Comp. Biochem. and Physiol. 31: 851–862.

Martin, N. & M. Anctil, 1984. The nervous system of the tube-worm *Chaetopterus variopedatus* (Polychaeta). Journal of Morphology 181: 161–173.

Merker, G. & M. Vaupel-von Harnack, 1967. Zur Feinstruktur des „Gehirns" und der Sinnesorgane von *Protodrilus rubropharyngeus* Jaegersten (Archiannelida). Mit besonderer Berücksichtung der Neurosekretorischen Zellen. Z. Zellforsch. 81: 221–239.

Meyer, E., 1887. Studien über den Körperbau der Anneliden I-III. Mitt. Zool. Stn Neapel 7: 592–741.

Meyer, E., 1888. Studien über den Körperbau der Anneliden IV. Mitt. Zool. Stn Neapel 8: 462–662.

Miron, M.-J. & M. Anctil, 1988. Serotonin-like immunoreactivity in the central and peripheral nervous system of the scale worm *Harmothoe imbricata* (Polychaeta). Journal of Comparative Neurology 275: 429–440.

Müller, M. C., 1999a. The morphological character of the annelid peristomium – structure of the nervous system. Zoology 92: 1–43.

Müller, M. C., 1999b. Das Nervensystem der Polychaeten: immunhistochemische Untersuchungen an ausgewählten Taxa. Ph.D. Thesis, Universität Osnabrück, 402pp.

Müller, M. C. M., 2002. *Aristonerilla*: a new nerillid genus (Annelida: Polychaeta) with description of *Aristonerilla (Micronerilla) brevis* comb. nov. from a seawater aquarium. Cahiers de Biologie Marine 43: 131–139.

Müller, M. C. M., in Press. Neurogenesis in *Scoloplos armiger* (Polychaeta; Orbiniidae): evidence for paired oesophageal connectives and a pentaneural ventral cord. J. Comp. Neurol.

Müller, M. C. M. & A. Berenzen, 2002. Asexuelle Fortpflanzung bei Anneliden (Ringelwürmern). Teil III: Regeneration des Nervensystems am Beispiel von *Eurythoe complanata*. Mikrokosmos 91(2): 108–114.

Müller, M. C. M., A. Berenzen & W. Westheide, 2003. Regeneration experiments in *Eurythoe complanata* ('Polychaeta', Amphinomidae): reconfiguration of the nervous system and its function for regeneration. Zoomorphology (DOI 10.1007/s00435-003-0073-4) 122: 95–103.

Müller, M. C., J. M. Bernhard & C. Jouin-Toulmond, 2001. A new member of Nerillidae (Annelida: Polychaeta), *Xenonerilla bactericola* gen. et sp. nov., collected off California, USA. Cahiers de Biologie Marine 42: 203–217.

Müller, M. C. M. & L. Henning, 2004. Neurogenesis during regeneration in *Dorvillea bermudensis* (Dorvilleidae) – formation of the polychaete cephalic ground pattern. J. Comp. Neurol. 471: 49–58.

Müller, M. C. & W. Westheide, 1997. The nervous system of apodous polychaetes: orthogonal structures of the nervous system in juvenile stages and progenetic species. Verh. Deutsch. Zool. Ges. 90.1: 209.

Müller, M. C. M. & W. Westheide, 2000. Structure of the nervous system of *Myzostoma cirriferum* (Annelida) as revealed by immunohistochemistry and cLSM analyses. Journal of Morphology 245: 87–98.

Müller, M. C. & W. Westheide, 2002. Comparative analysis of the nervous systems in presumptive progenetic dinophilid and dorvilleid polychaetes (Annelida) by immunohistochemistry and cLSM. Acta Zoologica (Stockholm): 83: 33–48.

Nansen, F., 1887. The structure and combination of the histological elements of the central nervous system. Bergens Mus. Aarb. 1886: 27–215.

Nicholls, J. G. & D. Van Essen, 1974. The nervous system of the leech. Scientific American 230: 38–48.

Nilsson, D., 1912. Beiträge zur Kenntnis des Nervensystems der Polychaeten. Zool. Bidr. Upps. 1: 87–161.

Orrhage, L., 1964. Anatomische und morphologische Studien über die Polychaetenfamilien Spionidae, Disomidae und Poecilochaetidae. Zool. Bidr. Upps. 36: 335–405.

Orrhage, L., 1966. Über die Anatomie des zentralen Nervensystemes der sedentären Polychaeten. Ein Beitrag zur Diskussion über die Architektur des Polychaeten-Gehirns und über den Begriff Palpen bei den Borstenwürmern. Arkiv för Zoologi 19: 99–133.

Orrhage, L., 1974. Über die Anatomie, Histologie und Verwandtschaft der Apistobranchidae (Polychaeta Sedentaria) nebst Bemerkungen über die systematische Stellung der Archianneliden. Z. Morph. Ökol. Tiere 79: 1–45.

Orrhage, L., 1978. On the structure and evolution of the anterior end of the Sabellariidae (Polychaeta Sedentaria). With some remarks on the general organisation of the polychaete brain. Zool. Jb. Abt. Anat. 100: 343–374.

Orrhage, L., 1980. On the structure and homologues of the anterior end of the polychaete families Sabellidae and Serpulidae. Zoomorphology 96: 113–168.

Orrhage, L., 1987. Bristle-worms and homologies – so what? (in Swedish with English summary). Fauna Flora Upps. 82: 63–71.

Orrhage, L., 1990. On the microanatomy of the supraoesophageal ganglion of some amphinomids (Polychaeta Errantia), with further discussion of the innervation and homologues of the polychaete palps. Acta Zoologica (Stockh.) 71: 45–59.

Orrhage, L., 1991. On the innervation and homologues of the cephalic appendages of the Aphroditacea (Polychaeta). – Acta Zoologica (Stockholm) 72: 233–246.

Orrhage, L., 1993. On the microanatomy of the cephalic nervous system of Nereidae (Polychaeta), with a preliminary discussion of some earlier theories on the segmentation of the polychaete brain. Acta Zoologica (Stockholm) 74: 145–172.

Orrhage, L., 1995. On the innervation and homologues of the anterior end appendages of the Eunicea (Polychaeta), with a tentative outline of the fundamental constitution of the cephalic nervous system of the polychaetes. Acta Zoologica (Stockholm) 76: 229–248.

Orrhage, L., 1996. On the microanatomy of the brain and the innervation and homologues of the cephalic appendages of Hesionidae and Syllidae (Polychaeta). Acta Zoologica (Stockholm) 77: 137–151.

Orrhage, L., 1999. On the morphological value of the glycerid-goniadid prostomium and its appendages (Polychaeta). Acta Zoologica (Stockholm) 80: 251–264.

Orrhage, L., 2001. On the anatomy of the central nervous system and the morphological value of the anterior end appendages of Ampharetidae, Pectinariidae and Terebellidae (Polychaeta). Acta Zoologica (Stockholm) 82: 57–71.

Orrhage, L. & D. Eibye-Jacobsen, 1998. On the anatomy of the central nervous system of Phyllodocidae (Polychaeta) and the phylogeny of phyllodocid genera: a new alternative. Acta Zoologica (Stockholm) 79: 215–234.

Payton, B., 1981. Structure of the leech nervous system. In Muller, K. J., J. G. Nicholls & G. S. Stent, (eds), Neurobiology of the Leech. Cold Spring Harbor, NY: Cold Spring Harbor Laboratory Press: 35–50.

Porchet, M., N. Dhainaut-Courtois, C. Cardon & M. Bataille, 1985. Structure and functions of neuropeptides of polychaete annelids. In Kobayashi, H. H. A. Bern & A. Urano (eds), Neurosecretion and the Biology of Neuropeptides. Proceedings of the Ninth International Symposium. Japan Scientific Societies Press Tokyo and Springer-Verlag, New York: 377–385.

Pruvot, G., 1885. Recherches anatomiques et morphologiques sur le système nerveux des Annélides Polychètes. Archs Zool. Exp. Gén. 3: 211–236.

Purschke, G., 1992. Ultrastructural investigations of presumed photoreceptive organs in two *Saccocirrus* species (Polychaeta, Saccocirridae). Journal of Morphology 211: 7–21.

Purschke, G., 1993. Structure of the prostomial appendages and the central nervous system in the Protodrilida (Polychaeta). Zoomorphology 113: 1–20.

Purschke, G., 1997. Ultrastructure of nuchal organs in polychaetes (Annelida) – new results and review. Acta Zoologica (Stockholm) 78: 123–143.

Purschke, G., 2000. Sense organs and central nervous system in an enigmatic terrestrial polychaete, *Hrabeiella periglandulata* (Annelida) – implications for annelid evolution. Invertebrate Biology 119: 329–341.

Purschke, G., 2002. On the ground pattern of Annelida. Org. Divers. Evol. 2: 181–196.

Purschke, G., R. Hessling & W. Westheide, 2000. The controversial position of the Clitellata and the Echiura in the phylogenetic system of the Annelida – on the problem of the phylogenetic evaluation of absent characters. Journal of Zoological Systematics and Evolutionary Research 38: 165–173.

Purschke, G. & C. Jouin-Toulmond, 1993. Ultrastructure of presumed ocelli in *Parenterodrilus taenioides* (Polychaeta, Protodrilidae) and their phylogenetic significance. Acta Zoologica (Stockholm) 74: 247–256.

Purschke, G. & C. Jouin-Toulmond, 1994. Ultrastructure of sense organs and the central nervous system in *Parenterodrilus taenioides* and their phylogenetic significance in the taxon Protodrilida (Annelida, Polychaeta). In Dauvin, J.-C., L.Laubier, & D.J. Reish, (eds) Actes de la 4ème Conférence Internationale des Polychétes. Mém. Mus. Natn. Hist. Nat. Paris. 162: 119–128.

Purschke, G., W. Westheide, D. Rohde & R. O. Brinkhurst, 1993. Morphological reinvestigation and phylogenetic relationship of *Acanthobdella peledina* (Annelida, Clitellata). Zoomophology 113: 91–101.

Racovitza, E. G., 1896. Le lobe céphalique et l'encéphale des Annélides Polychètes (Anatomie, Morphologie, Histlogie). Arch. Zool. exp. gén. (sér. 3) 4: 133–343.

Raw, F., 1949. Some stages in the evolution of the nervous system and the foregut of the Polychaete. Smithson. Micr. Collns. 3(8): 1–35.

Reisinger, E., 1925. Untersuchungen am Nervensystem der *Bothryoplana semperi* Braun. (Zugleich ein Beitrag zur Technik der vitalen Nervenfärbung und zur vergleichenden Anatomie des Plathelminthennervensystems). Zeit. Morph. Ökol. Tiere 5: 119–149.

Reisinger, E., 1972. Zur Evolution des Orthogons der Spiralier und das Archicölomatenproblem. Z. Zool. Sys. 10: 1–43.

Remane, A., 1963. Über die Homologisierungsmöglichkeiten bei Verbindungsstrukturen (Muskeln, Blutgefäße, Nerven) und Hohlräumen. Zoologischer Anzeiger 170: 481–489.

Retzius, G., 1895. Zur Kenntnis des Gehirnganglions und des sensiblen Nervensystems der Polychaeten. Biol. Unters. 7: 6–11.

Reuter, M., I. M. Sheiman, M. K. S. Gustafsson, D. W. Halten, A. G. Maule & C. Shaw, 1996. Development of the nervous system in *Dugesia tigrina* during regeneration after fission and decapitation. Invertebrate Reproduction & Development 9: 199–211.

Rohde, E., 1887. Histologische Untersuchungen über das Nervensystem der Chaetopoden. Zool. Beitr. 2: 1–81.

Rouse, G. W. & K. Fauchald, 1997. Cladistics and polychaetes. Zoologica Scripta 26: 139–204.

Sakharov, D. A., A. I. Golubev & L. V. Malyutina, 1986. Neurociliary synapses in the planarian machinery of locomotion. Docklady Adad. Nauk SSSR 286: 1021–1024.

Salensky, W., 1907. Morphogenetische Studien an Würmern II–IV. Mém. Acad. Sci. St-Pétersb. (Sci. Math., Phys. Nat.). 19: 1–349.

Salzet, M. & G. B. Stefano, 2001. Biochemical evidence for an annelid neuroendocrine system: Evolutionarily conserved molecular mechanisms. Placebo 3(2): 54–72.

Sawyer, R. T., 1986. Leech Biology and Behaviour, Vol. 1. Clarendon Press, Oxfort; 418 pp.

Schlieper, C., 1927. *Stylarioides plumosus*, eine monographische Darstellung. Z. Morph. Ökol. Tiere. 7: 320–383.

Scholtz, G. 2002. The Articulata hypothesis – or what is a segment? Org. Divers. Evol. 2: 197–215.

Sedgwick, A., 1884. On the origin of metameric segmentation and some other morphological questions. Quarterly Journal of Microscopical Science 24: 43–82.

Sigger, J. N. & D. A. Dorsett, 1986. The morphology and physiological properties of nereid giant fibres. Comparitive Biochemistry and Physiology A – Comparitive Physiology 83: 589–595.

Smith, J. E., 1957. The nervous anatomy of the body segments of nereid polychaetes. Philosophical Transaction (B) 240: 135–196.

Snodgrass, R. E., 1938. Evolution of the Annelida, Onychophora and Arthropoda. Smithson. Misc. Collns 97(6): 1–159.

Söderström, A., 1920. Studien über die Polychätenfamilie Spionidae. Diss. Almquist & Wiksell, Uppsala, 288 pp.

Spörhase-Eichmann, U., H. Gras & F.-W. Schürmann, 1987a. Patterns of serotonin-immunoreactive neurons in the central nervous system of the earthworm *Lumbricus terrestris* L. I. Ganglia of the ventral nerve cord. Cell Tissue Research 249: 601–614.

Spörhase-Eichmann, U., H. Gras & F.-W. Schürmann, 1987b. Patterns of serotonin-immunoreactive neurons in the central nervous system of the earthworm *Lumbricus terrestris* L. II. Rostral and caudal ganglia. Cell Tissue Research 249: 625–632.

Stefano, G. B., E. Braham, J. I. Sinisterra, L. Bentvena, J. DeFiglia, E. A. Martinez & M. K. Leung, 1988. The effects of cocaine on peripheral monoaminergic regulatory mechanisms in the gill of *Mytilus edulis*. Symposia Biologica Hungarica 36: 51–65.

Stolte, H.-A., 1932. Untersuchungen über Bau und Funktion der Sinnesorgane der Polychaetengattung *Glycera* Sav.. Z. Wiss. Zool. 140: 421–538.

Storch, O., 1913. Vergleichend-anatomische Polychätenstudien. Sitz.-Ber. Akad. Wiss. Wien. (Math.-nat. Kl.) 122: 877–988.

Stummer-Traunfels, R. V., 1927. Myzostomida. In Kükenthal W, & T., Krumbach (eds). Handbuch der Zoologie.: Walter de Gruyter, Leipzig; pp. 132–210.

Syed, N. T., D. Harrison & W. Willows, 1988. Locomotion in *Lymnea* – Role of serotonergic motoneurons controlling the pedal cilia. Symposia Biologica Hungarica 36: 387–399.

Tzetlin, A. B., T. Dahlgren & G. Purschke, 2002. Ultrastructure of the body wall, body cavity, nephridia and spermatozoa in four species of the Chrysopetalidae (Annelida, 'Polychaeta'). Zoologischer Anzeiger 241: 37–55.

Ushakova, O. O. & L. A. Yevdonin, 1985. Architectonic of larval nervous system in *Nereis virens* Sars (Polychaeta, Nereidae). Untersuchungen der Meeresfauna 34: 128–131. (in Russian).

Ushakova, O. O. & L. A. Yevdonin, 1987. Development of the nervous system in larval ontogenesis of *Brania clavata* (Nereimorpha; Syllidae). Zoologizteskii Zurnal 66: 1148–1161. (in Russian).

Ushakova, O. O. & L. A. Yevdonin, 1988. Architectonic of the nervous system in larval ontogenesis of the polychaete worms. Proceedings of the Zoological Institute, Leningrad 183: 78–93. (in Russian).

Van Beneden, E., 1891. Recheres sur le dévéloppment des Arachnactis. Contribution a la morphologie de Cérianthides. Archs biol., Paris 11: 115–146.

Van Beneden, E., 1897. Les Anthozoaires de la 'Plankton-Expedition'. Lipsius & Tischer.

Von Haffner, K., 1959a. Über den Bau und den Zusammenhang der wichtigsten Organe des Kopfendes von *Hyalinoecia tubicola* Malmgren (Polychaeta, Eunicidae, Onuphidinae), mit Berücksichtigung der Gattung *Eunice*. Zool. Jahrb. Anat. 77: 133–192.

Von Haffner, K., 1959b. Untersuchungen und Gedanken über das Gehirn und Kopfende von Polychaeten und ein Vergleich mit den Arthropoden. Mitt. Hamburg. Zool. Mus. Inst. 37: 1–29.

Von Haffner, K., 1962. Über die Abhängigkeit des Gehirnbaues von den Sinnesorganen. Vergleichende Untersuchungen an Euniciden (Polychaeta). Zool. Jb. Anat. 80: 159–212.

Voronezhskaya, E. E., S. A. Tsitrin, & L. P. Nezlin. 2003. Neuronal Development in Larval Polychaete *Phyllodoce maculata* (Phyllodocidae). Journal of Comparative Neurology 455: 299–309.

Wallace, B. G., 1981. Neurotransmitter chemistry. In Muller, K.J., Nicholls, J. G. & G. S. Stent, (eds), Neurobiology of the Leech. Cold Spring Harbor Laboratory Press, Cold Spring Harbor: 147–172.

Westheide W., 1982. *Ikosipodus carolensis* gen. et sp.n., an interstitial neotenic polychaete from North Carolina, U.S.A. and its phylogenetic relationships within Dorvilleidae. Zool. Scr. 11: 117–126.

Westheide W., 1984. The concept of reproduction in polychaetes with small body size: adaptations in interstitial species. In Fischer, A, H.-D. Pfannenstiel, (eds.), Polychaete Reproduction: Progress in Comparative Reproductive Biology. Fortschritte der Zoologie, Vol. 29. G. Fischer, Stuttgart.

Westheide W. & R. Rieger, 1996. Spezielle Zoologie, Teil 1: Einzeller und Wirbellose Tiere. Gustav Fischer, Stuttgart.

Westheide W. & N. Riser, 1983. Morphology and phylogenetic relationships of the neotenic interstitial polychaete *Apodotrocha progenerans* n. gen., n. sp. (Annelida). Zoomorphology 103: 67–87.

White, D. & J. R. Marsden, 1978. Microspectrofluorometric measurements on cells containing biogenic amines in the cerebral ganglion of the polychaete *Nereis virens* (Sars). Biological Bulletin 155: 395–409.

Windoffer, R. 1992. Immunohistochemische und elektronenmikroskopische Untersuchungen am Nervensystem zweier *Dinophilus*-Arten mit unterschiedlichem Lebenszyklus. Ph.D. thesis. Universität Osnabrück, 187 pp.

Worsaae, K. & M. C. M. Müller, 2004. Nephridial and gonoduct distribution patterns in Nerillidae (Annelida: Polychaeta) examined by Tubulin Staining and cLSM. Sub. J. Morph. 261: 259–269.

Yoshida-Noro C., M. Myohara, F. Kobari & S. Tochinai, 2000. Nervous system dynamics during fragmentation and regeneration in *Enchytraeus japonensis* (Oligochaeta, Annelida). Dev. Genes Evol. 210: 311–319.

Abbreviations used in Figures 1–7

The figures are taken from papers that are published in German and in English. In the schematic drawings as well as in the list, German and English abbreviations are given in order to enable easy understanding of the original papers.

1 – 12	palp nerve roots
aac	anterior association commissure
acc	accessory connective
acig	anterior cirral ganglion
acntmarct	anterior connection between ntm (nerve of the tentacular membrane) and arct (anterior root of common tract)
ag	annexed ganglion
an + tn	antennal nerve + first tegumentary nerve
äNlN	äußerer Nerv des langgestreckten Nuchalorgans (nuchal nerve)
aon	anterior optic nerve
arct	anterior root of ct (common tract)
asgc	anterior stomatogastric commissure
asgn	anterior stomatogastric nerve
avrtpn	anterior ventro-rostral tract of palp nerve
$BG^{1,\,2}$	Bauchganglion 1, 2 (ventral ganglia 1, 2)
c^1–c^7	brain commissures
cc	circum-oesophageal connective
ccgHo	Holmgren's cerebral commissural ganglion
cg	commissural ganglion
cgHa	Hamaker's commissural ganglion
com	commissure
concig	connection between the anterior and posterior cirral ganglia
ct	common tract of sgn (*Pista*), sgn and nbt (*Pectinaria*) sgn, nbt and lntm (*Amphicteis*)
ctn	common tract nerve
dac	dorsal association commissure
dacn	dorso-anterior cirral nerve
dcc	dorsal cell cluster
dcdr	dorsal commissure of dorsal root
dcdrlpn	dorsal commissure of dorsal root and lateral palp nerve
dcvr	dorsal commissure of ventral root
dg	dorsal ganglion
DG	Dorsalganglion des hinteren Schlundkonnektivs (dosal ganglion of drcc)
dKhS	dorsale Kommissur des vorderen Schlundkonnektivs = dcdr
dKvS	dorsale Kommissur des hinteren Schlundkonnektivs = dcvr
dltpn	dorso-lateral tract of palp nerve
dn, dN	dorsal nerve

dpcn	dorso-posterior cirral nerve
dpn	dorsal palp nerv
drcc	dorsal root of circum-oesophageal connective
drdctpn	dorsal root of dctpn (dorso-caudal tract of palp nerve)
drlpn	dorsal root and longitudinal podial nerve
$e^{1,2}$	anterior (1) and posterior (2) eye
enn	extra nuchal nerve
$f^1 - f^{13}$	nerve fibres connecting different parts of the brain
$g^1 - g^9$	brain ganglia
$G^{1,2}$	ganglia 1, 2
gcxrcc	ganglion of extra root of cc
gdacn	ganglion of dorsal anterior cirral nerve
gdpcn	ganglion of dorsal posterior cirral nerve
gmnbc	ganglion of median nerve of branchial crown
GZNZ	im Anschluß an die Nuchalnerven befindliche Ganglionzellen (nuchal ganglia)
hS	hinteres Schlundkonnektiv = drcc
iNlN	innerer Nerv des langgestreckten Nuchalorgans (nuchal nerve)
lcc	lateral cell cluster
LG	Lateralganglion (lateral ganglion)
ln	lateral nerve
lnbc	lateral nerve of branchial crown
lng	lateral nuchal ganglion
lpn	longitudinal podial nerve
lrn^1	lateral root of first anterior nerve
mm	median mass
mnbc	median nerve of branchial crown
mrn^1	median root of first anterior nerve
msgn	median stomatogastric nerve
$n^1 - n^8$	brain nerves
N	giant cell
nbc	nerve of branchial crown
nbs	nerve of buccal segment
nbt	nerves of buccal tentacles
nbtdr	nerves of buccal tentacles and dorsal ridge
nc, NK	nuchal commissure
ndbv	nerve of the dorsal blood vessel
NDK, sgn	stomatogastric nerve
ndo, NdS	nerve of dorsal (sensory) organs
ng, NG	nuchal ganglion
ngl	nuchal glomeruli
NhN	Nerv zum hufeisenförmigen Nuchalorgan (nuchal nerve)
nla	nerve of lateral antenna
nlcp	nerve of lateral ciliated pad
nllp, NlLP	nerve of lateral prostomial lobe
nlvl	nerve to the latero-ventral parts of the buccal lips
nma	nerve to the median antenna ($= n^6$)
nmr	nerve of mouth region
NMu	Nerven zur Mundöffnung
nmvl	nerve to the medio-ventral parts of the buccal lips
nn, NN	nuchal nerve
NNhs	Nuchalnerv des hinteren Schlundkonnektivs (nuchal nerve of drcc)
nss	nerve of stomodeal sac
ntm	nerve of tentacular membrane
$nug^{1,2}$	nuchal ganglion 1, 2
nvlm	nerve of the ventral longitudinal muscles
NvP	Nerven zum vordersten Teil des Prostomiums (nerves extending to the anterior part of the prostomium
NzdlP	Nerven zum ziliierten dorsolateralen Teil des Prostomiums (nerves extending towards the ciliated dorsolateral part of the prostomium)
oc	optic commissure
Oen , OeN	oesophageal nerve
pcdrcc	posterior commissure of dorsal root
pcig	posterior cirral ganglion
pcntmarct	posterior connection between nerve of tentacular membrane and anterior root of cc
pdngl	posterior dorsal neuropile with globuli cells
pn, PN a-c	palp nerve
$Pn^{ä}$	äußerer Palpennerv (outer palp nerve)
Pn^{i}	innerer Palpennerv (inner palp nerve)
pnr	palp nerve root
PNZ	Pilem des Nuchalzentrums
pog^1	first podial ganglion
poK, pok	praeoral commissure
pon^1	first podial nerve
pon	posterior optic nerve
prct	posterior root of ct (common tract)
psgc	posterior stomatogastric commissure
psgn	posterior stomatogastric nerve
pvrtpn	posterior ventro-rostral tract of palp nerve
pw	nerve from lateral nerve to brain commissure
sgn	stomatogastric nerve
SN	side nerve, segmental nerve
stg	stomatogastric gap
tg	tegumentary ganglion
tn^{1-7}	tegumentary nerve
vac	ventral association commissure
vacn	ventral anterior cirral nerve
vcc	ventral cell cluster
vcdr	ventral commissure of dorsal root

vcdrlpn	ventral commissure of dorsal root and longitudinal podial nerve
vrdctpn	ventral root of dorsal-caudal tract of palp nerve
vcvr	ventral commissure of ventral root
vg	ventral ganglion
vKhS	ventrale Kommissur des hinteren Schlundkonnektivs = vcdr
vKvS	ventrale Kommissur des vorderen Schlundkonnektivs = vcvr
vpcn	ventral posterior cirral nerve
vpn	ventral palp nerve
vrcc	ventral root of cc
vrdctpn	ventral root of dorso-caudal tract of palp nerve
vS	vorderes Schlundkonnektiv = vrcc
xc	extra commissure
XI	Manaranche's Nucleus XI (dorsal ganglion)
xrcc	extra root of cc.

Hydrobiologia (2005) **535/536:** 113–126
T. Bartolomaeus & G. Purschke (eds), Morphology, Molecules, Evolution and Phylogeny in Polychaeta and Related Taxa

Muscular system in polychaetes (Annelida)

Alexander B. Tzetlin & Anna V. Filippova
Department of Invertebrate Zoology, Moscow State University, Vorobievy Gory, Moscow 119899, Russia

Key words: polychaeta, muscle system, body wall, circular muscles, longitudinal muscles, parapodia, dissepiment, septa, mesentery

Abstract

The structure of the polychaete muscular system is reviewed. The muscular system comprises the muscles of the body wall, the musculature of the parapodial complex and the muscle system of the dissepiments and mesenteries. Various types of organisation of the longitudinal and circular components of the muscular body wall are distinguished. In Opheliidae, Polygordiidae, Protodrilidae, Spionidae, Oweniidae, Aphroditidae, Acoetidae (= Polyodontidae), Polynoidae, Sigalonidae, Phyllodocidae, Nephtyidae, Pisionidae, and Nerillidae circular muscles are lacking. It is hypothesised that the absence of circular muscles represents the plesiomorphic state in Annelida. This view contradicts the widely accepted idea of an earthworm-like musculature of the body wall comprising an outer layer of circular and an inner layer of longitudinal fibres. A classification of the various types of parapodial muscle construction has been developed. Massive and less manoeuvrable parapodia composed of many components like those of *Aphrodita* are regarded to represent the plesiomorphic state in recent polychaetes. An analysis of the diversity of the muscular structure supports the hypothesis that the primary mode of life in polychaetes was epibenthic and the parapodial chaetae had a protective function.

Introduction

Polychaetes and the entire taxon Annelida are soft-bodied animals. Their mobility and the maintenance of the body shape are affected by their muscular system. The polychaete muscular system consists of several components such as circular and longitudinal fibres of the body wall, parapodial, chaetal, oblique, diagonal, dorsoventral fibres, as well as muscular structures elements of septa and mesenteries. Therefore, the knowledge of the construction of the muscle system of polychaetes is important for our understanding of the life styles observed and evolutionary trends in this group.

The ultrastructure of the muscle tissues of polychaetes was studied in detail and is well described (e.g., Mattisson, 1969; Eguileor & Valvassori, 1977; Lanzavecchia et al., 1988; Gardiner, 1992). Usually polychaete muscle fibres are double obliquely striated with the non-contractile parts located on the narrow side. But several other types of fibres including hirudinean-like fibres and cross-striated muscle cells have been described as well. These fibres mainly occur in specialised organs. Since the ultrastructure has repeatedly been reviewed (e.g., Lanzavecchia et al., 1988; Gardiner, 1992) we will only deal with available data on the anatomical structure of those muscular systems which have not been reviewed recently and which have not been considered for phylogenetic implications since the comprehensive studies of Storch (1968) and Mettam (1971).

Since the 19th century annelids and polychaetes, in particular, have been considered to possess a muscular body wall consisting of an outer layer of circular and an inner layer of longitudinal muscle fibres (e.g., Meyer, 1887, 1888). In addition diagonal fibres may also be present in the body wall. Circular muscle fibres are transversely oriented and form a cylinder, which is not even interrupted near the ventral nerve cord (Fig. 1). This pattern of organisation of the musculature has been widely accepted and has almost unchanged been adopted as characteristic for annelids in recent publications and textbooks of invertebrate zoology (Fig. 1a–c) (e.g., Storch & Welsch, 1986; Brusca & Brusca, 1990; Westheide & Rieger, 1996). However, already in the end of the 19th and the beginning of the 20th century it was noted that the so-called Archiannelida had poorly developed or even lacking circular musculature (Salensky, 1907). Absence of circular muscle fibres not only in archiannelids but also in several other polychaete species was mentioned in a number of publications (e.g., McIntosh, 1917; Hartmann-Schröder, 1958). Hartmann-Schröder (1958) and Orrhage (1962) were the first who reported lack of circular muscle fibres not only in aberrant or interstitial polychaete species but in larger species belonging to Opheliidae and Spionidae as well. These findings stimulated additional investigations which revealed that absence of circular muscle fibres occurs more often in polychaetes than generally assumed. Until today the lack of circular muscles has are recorded in macrobenthic, meiobenthic, parapodia-bearing as well as sedentary species of the following taxa: Opheliidae, Polygordiidae, Protodrilidae, Spionidae, Oweniidae, Aphroditidae, Polyodontidae, Polynoidae, Sigalonidae, Phyllodocidae, Chrysopetalidae, Nephtyidae, Pisionidae and Nerillidae (McIntosh, 1917; Hartmann-Schröder, 1958; Orrhage, 1964; Jouin & Swedmark, 1965; Mettam, 1967, 1971; Storch, 1968; Hermans, 1969; Gardiner & Rieger, 1980; Tzetlin, 1987; Ivanov & Tzetlin, 1997; Tzetlin et al., 2002a). This apparent widespread lack of circular muscle fibres raised the question whether this feature is due to convergence or represents a homologous but plesiomorphic character (Tzetlin et al., 2002a, b). The answer has far reaching consequences for our understanding of evolutionary pathways in annelids. For instance, since circular muscles are most likely important for burrowing forms but are unnecessary for animals which proceed by movements with their parapodia or cilia, this question is re-

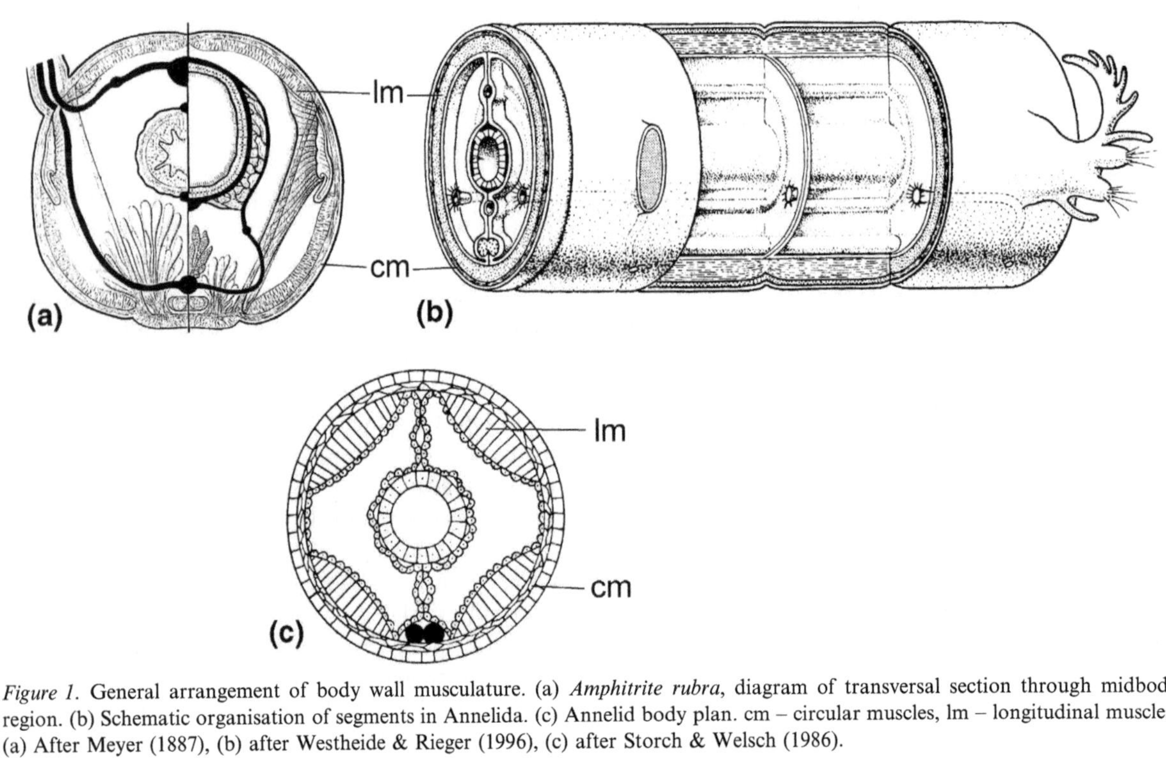

Figure 1. General arrangement of body wall musculature. (a) *Amphitrite rubra*, diagram of transversal section through midbody region. (b) Schematic organisation of segments in Annelida. (c) Annelid body plan. cm – circular muscles, lm – longitudinal muscles. (a) After Meyer (1887), (b) after Westheide & Rieger (1996), (c) after Storch & Welsch (1986).

lated to whether the polychaete stem species was epi- or endobenthic.

Until now the diversity of polychaete muscle system has usually been studied by means of routine histology, transmission as well as scanning electron microscopy (Mettam, 1967; Storch, 1968; Tzetlin et al., 2002a). At least some of these data are based on results without complete reconstruction of the muscular system of the body wall. In such cases the authors could have overlooked poorly developed muscular elements. Therefore, labelling of F-actin and subsequent confocal laser scanning microscopy is a comparatively new, excellent and accurate method for investigation of muscle fibre arrangements (Tzetlin et al., 2002b). Each muscle cell is labelled individually and, provided that the specimens do not exceed an appropriate size, can be followed along its entire

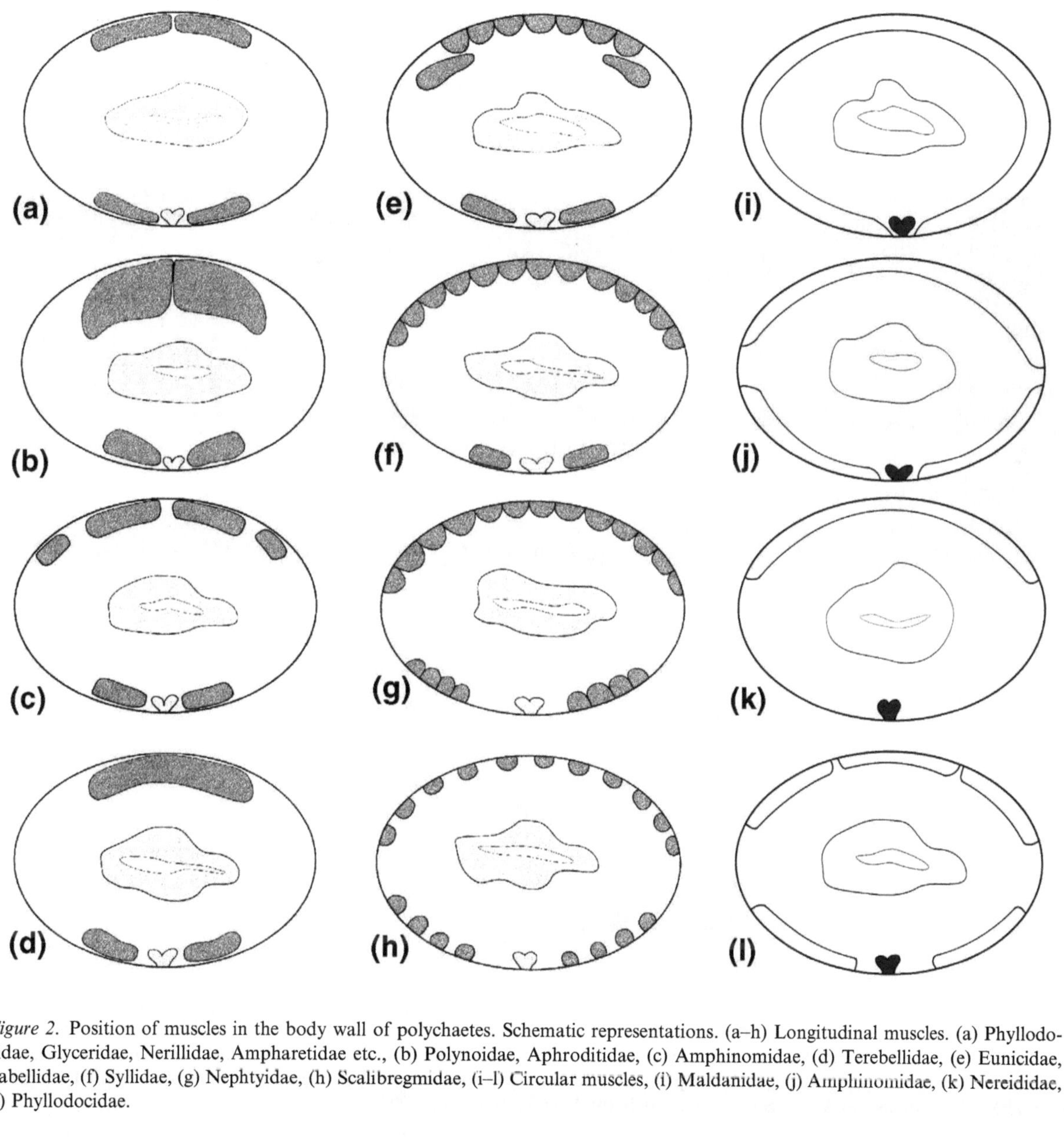

Figure 2. Position of muscles in the body wall of polychaetes. Schematic representations. (a–h) Longitudinal muscles. (a) Phyllodocidae, Glyceridae, Nerillidae, Ampharetidae etc., (b) Polynoidae, Aphroditidae, (c) Amphinomidae, (d) Terebellidae, (e) Eunicidae, Sabellidae, (f) Syllidae, (g) Nephtyidae, (h) Scalibregmidae, (i–l) Circular muscles, (i) Maldanidae, (j) Amphinomidae, (k) Nereididae, (l) Phyllodocidae.

length. Thus, this method allows investigating absence or presence of a certain type of muscle cell with a greater degree of certainty, especially if these muscles are poorly developed and hardly visible in histological sections. This method has been successfully applied in various taxa of small invertebrates as well (e.g., Rieger et al., 1994; Wanninger et al., 1999; Hochberg & Litvaitis, 2001; Müller & Schmidt-Rhaesa, 2003). To date such studies have only been carried out for a limited number of polychaete species and do not encompass the whole diversity of polychaete muscular systems. These studies will be reviewed below but these facts necessitate the need for reinvestigations of annelid musculature in a broader range of taxa.

Circular muscles

Circular and other transverse fibres usually underlay the extracellular matrix of the integument and are poorly developed compared to the longitudinal underlying the layer of circular muscles if present at all (Meyer, 1887; Westheide & Rieger, 1996). Circular muscle fibres are found in various taxa such as Amphinomidae, Nereididae, Hesionidae Glyceridae, and other Phyllodocida, Nerillidae, Capitellidae, Maldanidae, Arenicolidae, and Terbellidae. However, the structure and arrangement of these fibres vary greatly and various types may be distinguished.

In species of Glyceridae, Capitellidae, Maldanidae and Arenicolidae circular fibres are arranged

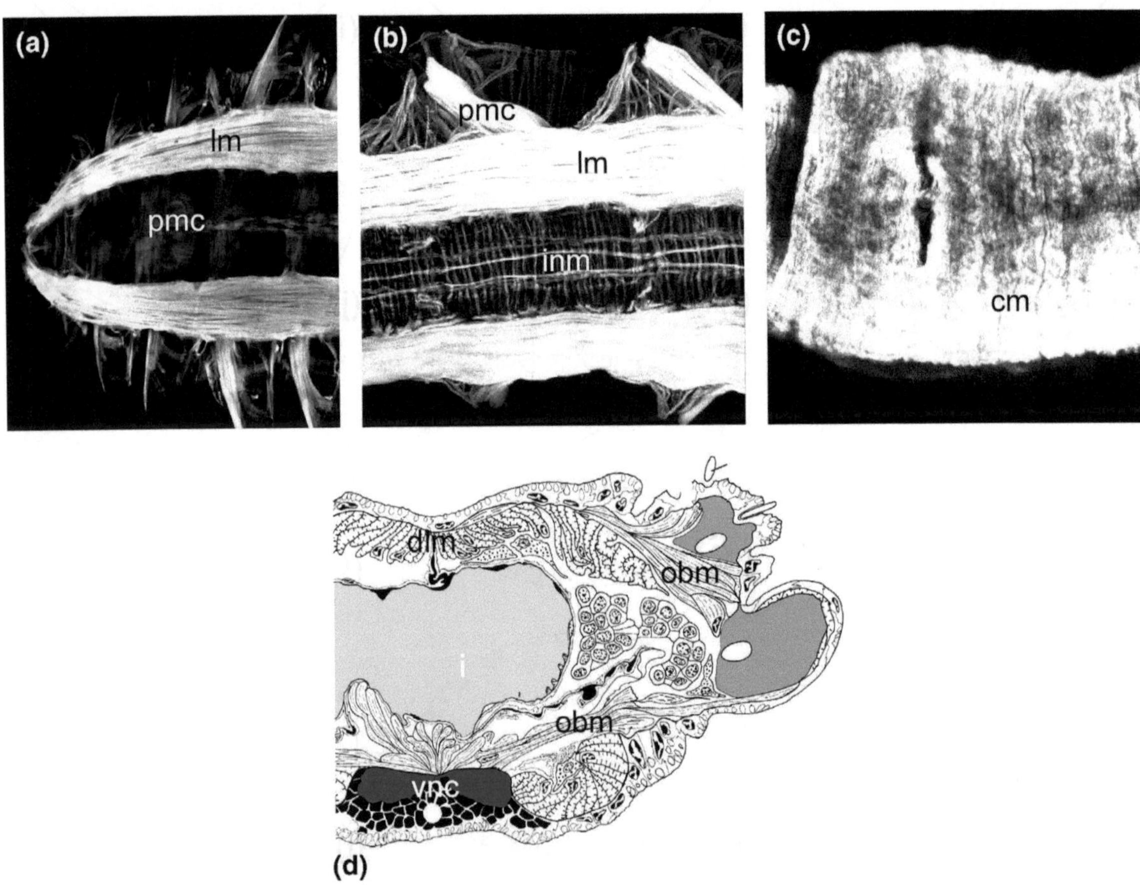

Figure 3. Arrangement of body wall muscle system. (a) *Dysponetus pygmaeus*, anterior end, ventral view. (b) *Prionospio cirrifera*, two midbody segments in dorsal view. (c) *Paraxiella praetermissa*, midbody, lateral view. (d) *D. pygmaeus*. Drawing of cross-section through midbody segment with different muscle systems. dlm – dorsal longitudinal muscle, i – intestine, lm – longitudinal muscle, obm – oblique muscle, pmc – parapodial muscle complex, vnc – ventral nerve cord. (a–c) cLSM micrographs after phalloidin-labelling, (d) after TEM observations. (a) After Tzetlin et al., 2002b, (b) After Tzetlin et al. (2002a).

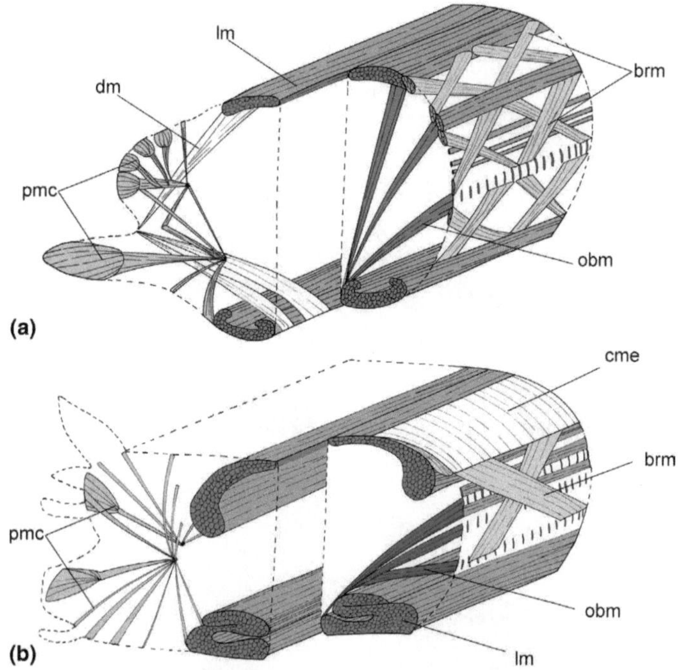

Figure 4. Schematic reconstructions of muscle systems. (a) *Aphrodite* sp., (b) *Nereis* sp. brm – bracing muscle, cme – circular muscle element, dm – diagonal muscle, lm – longitudinal muscle, obm – oblique muscle, pmc – parapodial muscle complex. After various authors.

following a pattern which corresponds to the traditional view of muscle arrangement in polychaetes (Wells, 1944, 1950; Clark, 1964). Transversal muscle fibres form almost closed circles, only interrupted at the intraepithelial nerve cord (Figs 2i and 3c). It is noticeable that even in these species the circular fibres are less developed than the longitudinal muscle fibres. The fibres form an almost complete cylinder without gaps even at the parapodia (Fig. 3c). There are, however, no circular fibres at the border of the segments. This pattern is interpreted as an adaptation to an enhanced mobility and better conjunction of the segments.

In Amphinomidae, Nerillidae and Terebellidae the circular muscles are interrupted near the parapodia (Fig. 2j and l; Storch, 1968; Marsden & Lacalli, 1978). Transverse fibres are only present in certain parts of the body and, thus, can hardly be called circular muscle cells. They are either restricted to the dorsal side (Figs 2k and 6b; e.g., Hesionidae, Nereididae) or on the ventral side as in Phyllodicidae (Storch, 1968).

In several taxa circular or transverse fibres are completely lacking. This has been observed in Aphroditidae, Chrysopetalidae, Pisionidae, Spionidae and Opheliidae (Figs 3a, b, d, 4a, b, 5a and b; see Brown, 1938; Mettam, 1971; Tzetlin, 1987; Tzetlin et al., 2002a, b, unpubl. obs.). These findings indicate that absence of circular muscle fibres is not an unusual case but a fairly common phenomenon instead (Tzetlin et al., 2002b).

In order to maintain the shape of the body it has to be expected that weak transversal muscles or entirely lacking circular muscle fibres are compensated by another system. In some taxa this requirement is achieved by so-called bracing muscles of some taxa (Fig. 4a and b). These muscles are located diagonally among the longitudinal fibres, cross each other and form a lattice. The maximal number of these groups is three: they are located ventrally, dorsally and laterally in *Aphrodita aculeata* (Fig. 4a; see Mettam, 1971). In cross-sections these muscles look like circular fibres and, most likely, were erroneously taken for them by other authors. In addition to regulating the body wall constrictions, the bracing muscles reach the parapodial muscles, so that they may be attributed to the parapodial muscle complex.

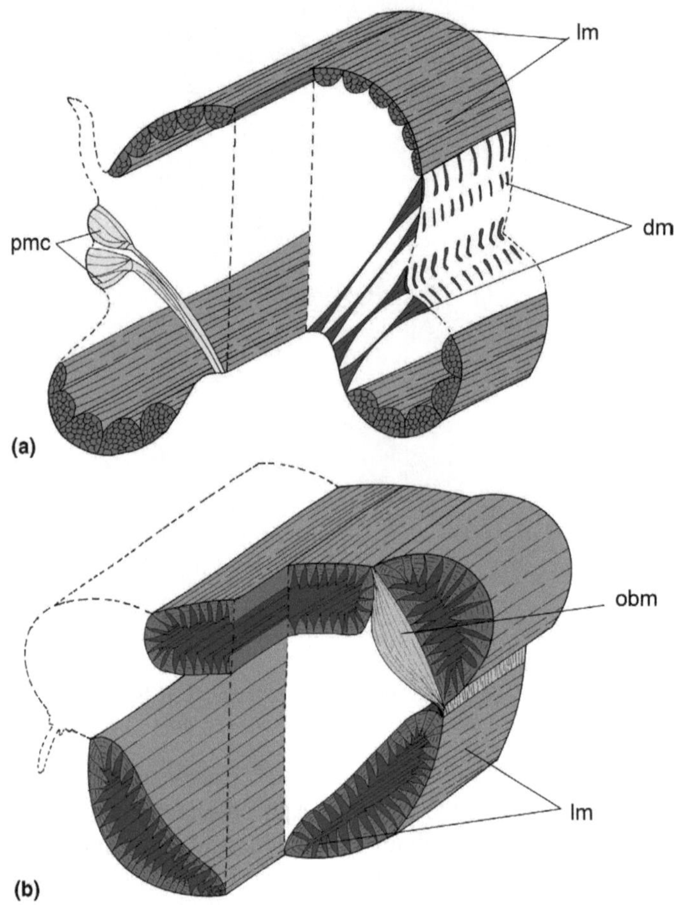

Figure 5. Schematic reconstructions of muscle systems. (a) *Ophelia* sp., (b) *Pisionidens tchesunovi*. dm – diagonal muscle, lm – longitudinal muscle, obm – oblique muscle. (a) After various authors, (b) After Tzetlin (1987).

Longitudinal muscles

Longitudinal muscles run along the whole body length and usually form discrete bands. The fibres making up the bands are arranged in different patterns (Fig. 6). A band may be formed by large flattened cells lying in a single row and not being covered by the coelomic epithelium. The nuclei of these cells are usually located on the distal part facing the coelomic cavity (Fig. 6a for Sphaerodoridae, Phyllodocidae; Tzetlin, 1987). A similar pattern is observed in Chrysopetalidae (Tzetlin et al., 2002a). However, at least a part of the bundles are covered by coelothelial cells. Sometimes the nuclei are located in the distal parts between the myofilaments or in epithelium-like processes forming a cover above the myofilaments-containing parts of the fibres (Fig. 7a and b; Phyllodocidae, see Ivanov, in press). Such processes may be misinterpreted as a coelothelium on histological sections. Each muscle cell of a longitudinal muscle band contacts the subepidermal extracellular matrix along its entire length.

In Pisionidae bands of longitudinal fibres are also formed by similar cells with their nuclei located on the distal parts devoid of myofilaments. However, the band of cells is rolled up forming a closed ellipse with a central cavity on cross-sections. The nuclei are situated in the inner part of the cavity (Fig. 6b). Such bands are covered by a coelothelium (Tzetlin, 1987).

In other taxa with well-developed longitudinal muscles these bands are rolled up differently producing a multilayered pattern of fibres. Here the

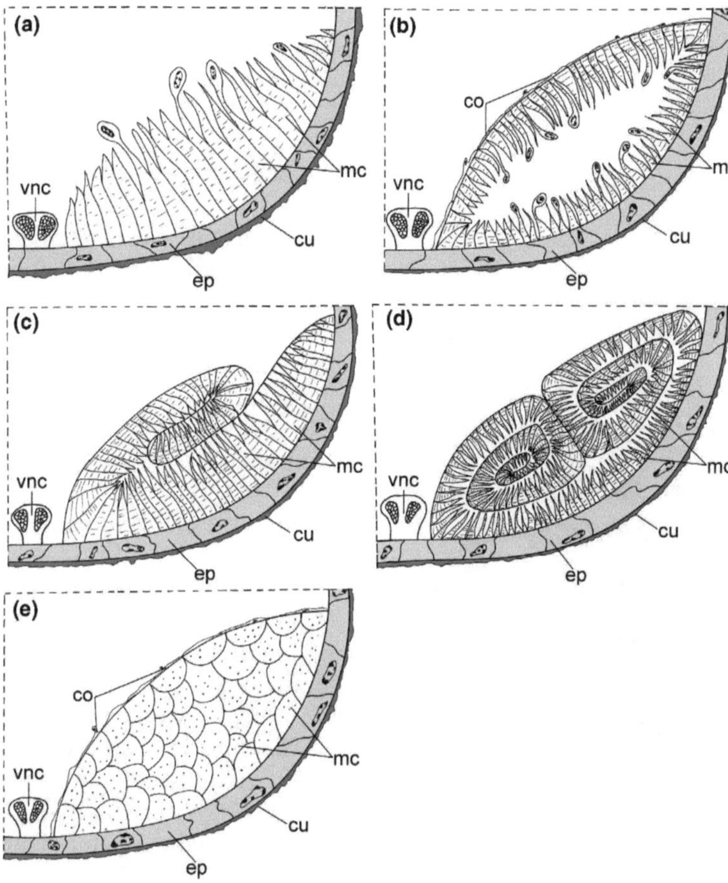

Figure 6. Patterns of arrangement of muscle fibres within longitudinal bands in polychaetes. (a) Phyllodocidae, (b) Pisionidae, (c, d) Sabellidae, Nereididae, (e) Aphroditidae. co – coelothelium, cu – cuticle, ep – epidermis, vnc – ventral nerve cord.

fibres do not form a closed cylinder but form helices with each coil tightly adjoining the previous one. The number of convolutions varies. Such bands may be s-shaped (Nereididae) or may consist of two reverse helices (Sabellidae) (Fig. 6c and d; Johansson, 1927).

Finally, a band may be formed by rounded muscle cells tightly adjoining each other. The nuclei are located centrally and are surrounded by myofilaments (hirudinean-type of muscle fibre). Each bundle is covered by a coelomic epithelium. Such bands may be rounded or flat with folded edges in cross-sections (Fig. 6e; Aphroditidae, see Storch, 1968).

In addition to these different constructions of muscle bands their number and position varies as well (Storch, 1968): (1) Four longitudinal muscle bands, two running ventrally and two running dorsally either show the same diameter as in Phyllodocidae, Glyceridae, Ampharetidae, Nerillidae, Chrysopetalidae and many other taxa (Fig. 2a; Clark, 1964; Tzetlin et al., 2002a) or the dorsal bands are more massive than the ventral ones (Fig. 2b; Eunicidae, Sabellidae). (2) Six longitudinal bands with two ventral, two dorsolateral and two dorsal bands are found in Polynoidae, Aphroditidae, Chrysopetalidae; (Fig. 2c; e.g. Tzetlin et al., 2002a). (3) In Nephtyidae and Hesionidae three longitudinal muscle bands are present, two of them running ventrally and one being located dorsally (Fig. 2e). The single dorsal band most likely corresponds to two merged dorsal bands. (4) The dorsal bands consist of up to 10 or even more smaller bundles being assembled closely to each another (Fig. 2e–g). Along with these muscles there are two dorsolateral bands and two

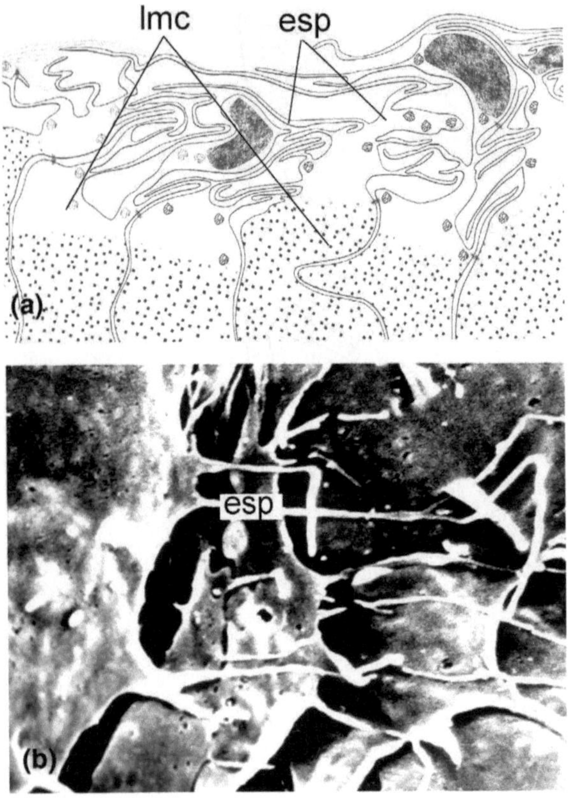

Figure 7. Phyllodoce groenlandica. Longitudinal muscle cells (lmc). A Drawaing of distal parts of muscles cells with epithelium-like processes (esp). A SEM micrograph. After Ivanov (2002).

ventral bands in Amphinomidae (Fig. 2e). In Syllidae the dorsal musculature extends laterally and there are only two ventral bands (Fig. 2f). In Terebellidae there are ventral bundles apparently comprising the ventral band being broken into bundles (Fig. 2f). (5) The longitudinal musculature is formed by small bundles which are not contacting each other and are evenly distributed (Fig. 2h).

On the ventral nerve cord a ventral median muscle may be present which may be interpreted to belong to the mesenterial musculature or represent an additional type of longitudinal muscle such as has been described for Chrysopetalidae (Tzetlin et al., 2002b).

Parapodial muscle complex

Unfortunately, data on the muscle arrangement in parapodia are scarce. Mettam (1967, 1971) described the parapodial muscle complex in species belonging to three different families of the errant type while only fragmentary data on the structure of parapodia of sedentary polychaetes are available (Brown, 1938; Dorsett, 1961; Storch, 1968). Data on spioform parapodia are lacking.

Musculature of errant parapodia

The musculature of the parapodial complex in errant polychaetes (Aciculata) consists of numerous muscles and individual muscle fibres. They will be divided in a number of functional groups according to Mettam (1967, 1971).

The muscles associated with the parapodial wall are usually only found in neuropodia (Fig. 8a and f). They start from the bases of neuropodia with two bundles, then radiate diagonally towards the upper edge of the neuropodia forming a dense lattice.

There are numerous muscles running inside the parapodium described in detail by Mettam (1967, 1971) (Fig. 8b, c, g and h). He discriminates many groups of muscles and numbers the muscles within each group. It appears more feasible only to name the main groups of muscles but not paraphrase these studies. Inside each parapodium the muscles run from the base to the tip, noto- and neuropodial intrinsic muscles are passing each other. It is noteworthy that they are neither attached to the bases of the aciculae nor connected with each other. These muscles con-

Figure 8. Schematical representation of parapodial muscle complex. (a–e) *Aphrodite* sp. (a) Parapodium wall muscles. (b, c) Different groups of intrinsic parapodial muscles. (d) Muscles associated with chaetae. (e) Complete parapodial musculature. (f–j) *Nereis* sp. (f) parapodium wall muscles. (g, h) Different groups of intrinsic parapodial muscles. (i) Muscles associated with chaetae. (j) Complete parapodial musculature. a – acicula, apr – protractor of acicula, brm – bracing muscle, chpr – chaetal protractor, chr – chaetal retractor, dm – diagonal muscle, im – intrinsic muscle, inm – intestinal muscle, levm – levator muscle, obm – oblique muscle, pwm – parapodial wall muscle. After Mettam (1971).

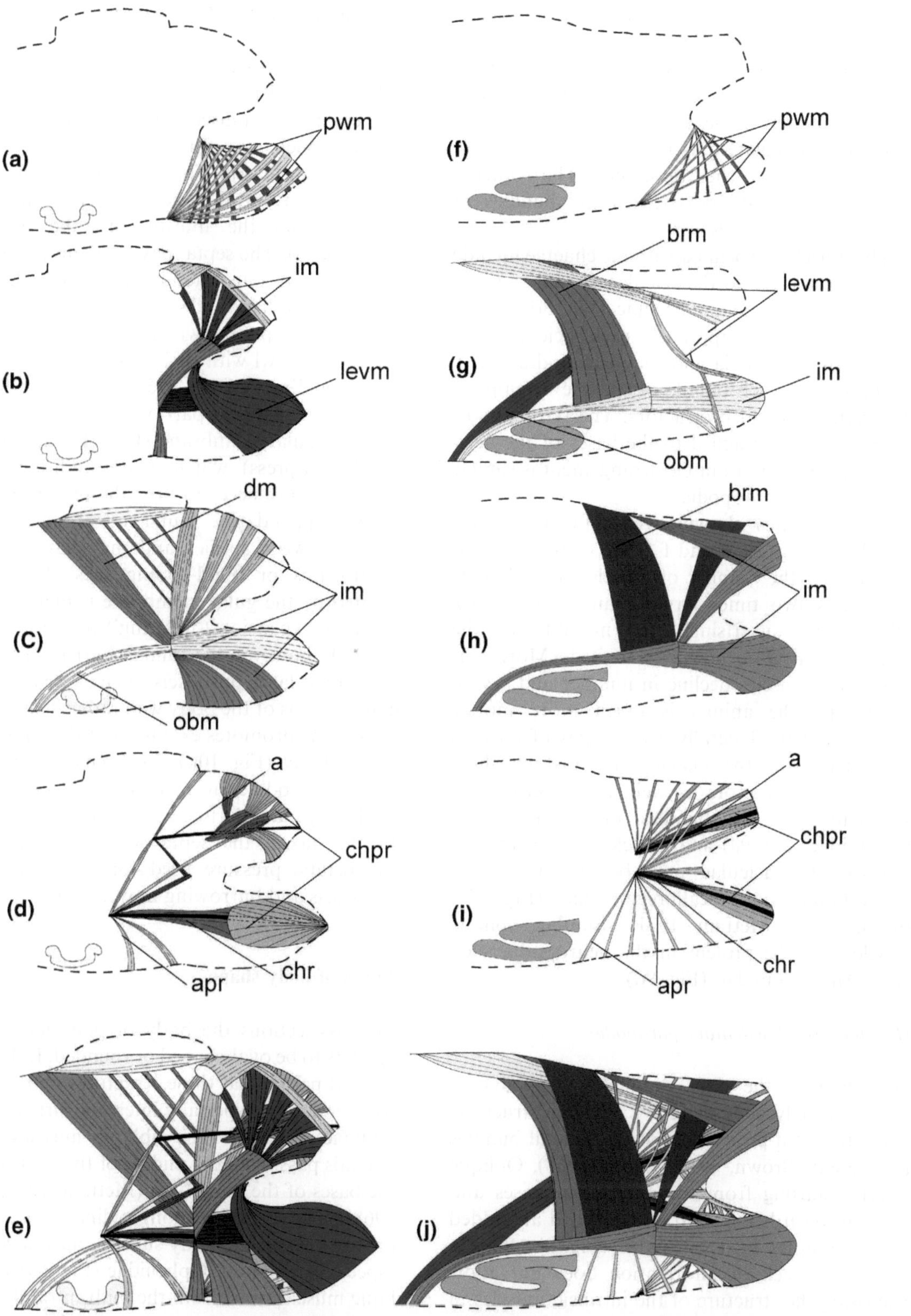
pwm
(a)
(f)
pwm
im
(b)
levm
brm
levm
(g)
im
obm
dm
im
(C)
obm
brm
im
(h)
a
chpr
(d)
apr
chr
a
chpr
(i)
chr
apr
(e)
(j)

siderably vary in number and size between species (Fig. 8b, c, g and h). In *Aphrodite aculeata* the notopodial intrinsic muscles are connected with the depressor muscles of the notopodial chaetae. Each notopdium possesses a lot of diagonal and transversal (circular) muscles. The levator muscle passes inside the neuropodium. The bracing muscles mentioned above reach the bundles of these intrinsic muscles.

The muscles associated with the chaetae include muscles connected to the aciculae as well as those attached to the regular chaetae (Fig. 8d and f). Retractor muscles approach each acicula. The muscles of the acicula and the neuropodium may be connected. Each bundle of chaetae is supplied with retractors and protractors. The retractors of the chaetae are attached to the bases of the aciculae. The pattern of muscle arrangement is similar for neuro- and notopodia.

If the figures showing the different muscle complexes (Fig. 2a–d and f–i) are combined, it is evident that the number of muscles in *Aphrodite aculeata* is two times larger than in *Nereis* sp. (39 vs. 20); surprisingly the mobility of the parapodia in *A. aculeata* is limited. Moreover, despite the muscles decline in number and size in *Nereis* sp., the animal is much more mobile (Fig. 8e and j). Generally a few types of muscles associated with the chaetae are distinguished, like acicula retractors as well as retractors and protractors of the bundles of chaetae (e.g., Nephtyidae; Fig. 9a) and muscles connecting the bases of the aciculae may be added in e.g., Aphroditidae and Scalibregmatidae (Fig. 9b; Storch 1968; Mettam 1971). Similar oblique muscles may be present in the neuropodia such as in Amphinomidae (Fig. 9d).

Musculature of sedentary parapodia

According to the data available parapodia of sedentary polychaetes are all alike in structure. Retractors and protractors of the chaetal bundles are present (Brown, 1938; Storch, 1968). Oblique muscles starting from the notopodial bases and running towards the ventral nerve cord are added to the complex in Terebellidae (Fig. 9b; Storch, 1968). The lack of information does not allow discussing the structure of the intrinsic muscles in these polychaetes.

Muscles of septa and mesenteria

The septa or dissepiments consist of the extracellular matrix situated between the adjoining coelomic epithelia or muscle cells (Fig. 10i). In addition blood vessels are formed by gaps within this extracellular matrix. The orientation of the muscles can be dorsoventral, oblique, or transverse, or they radiate from the intestine to the body wall (Fig. 10e–h). The septa may be complete as well as reduced to various extend. Accordingly the muscle cells differ in shape and size. This depends on the life style of the species, which may either dig in the sediment, crawl with their parapodia or moves by ciliary gliding (Clark, 1964). In Terebellida, the septa of the anterior part of the body are modified into the gular membrane (Fig. 10a–d; Zhadan & Tzetlin, in press), which creates additional hydrostatic pressure necessary for the protrusion of the mouth appendages. Gular membranes differ in shape between species and may have additional protrusions or blind-ending sacs. The pressure created by the gular membrane is high enough to promote expansion ('inflating') of the mouth tentacles. If, however, the animals burrow in the sludge using their large proboscis, such as *Artacama* spp., constrictions of the body wall rather than the gular membrane promotes extension of the anterior part of the body (Fig. 10k). In Arenicolidae anterior septa are modified to form a gular membrane and are highly muscularised as well (Wells, 1952, 1954). In this taxon the septa serve in creating high hydrostatic pressure used for protrusion of the proboscis and burrowing movements.

Types of body shape

In cross-sections the body of polychaetes mostly appears to be oval-shaped or rounded. If the species possess parapodia, these structures markedly protrude laterally, especially in errant forms, e.g., Nereididae; Fig. 4b). The oblique muscles in such animals pass from the middle of the ventral side to the bases of the parapodia (Mettam, 1967; Storch, 1968). The bands of longitudinal muscles form protrusions on the body surface. Such a pattern is especially visible in Opheliidae (Fig. 5a) Here oblique muscles start from the midventral line resulting in a more or less pronounced ventral ridge

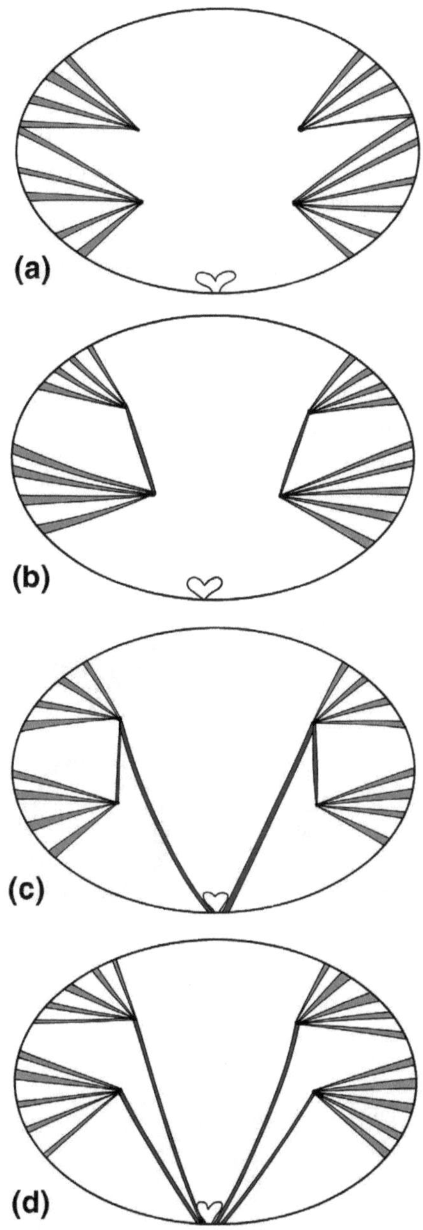

Figure 9. Different patterns of muscles associated with the chaetae. (a) Nephtyidae, (b) Aphroditidae, Scalibregmidae, (c) Terebellidae, (d) Amphinomidae.

(Brown, 1938; Hartmann-Schröder, 1958). An opposite pattern of oblique muscle arrangement resulting in corresponding grooves can be observed in certain Pisionidae as e.g. in *Pisionidens tchesunovi* (Fig. 5b; Tzetlin, 1987). Here, the oblique muscles start beneath the single dorsal band and reach the body surface below the dorsolateral bands.

Conclusions and outlook

Despite the comparatively small number of species studied in detail, it is evident that the muscular system of polychaetes is much more complex and diverse than it is described in popular zoological textbooks and manuals. The parapodia are the most vivid and typical organs in polychaetes although they may lack in certain taxa (Westheide, 1997; Purschke, 2002). However, many items remain to be studied. The discussion which type of parapodium could be regarded as the most primitive type for polychaetes, for instance, is more than 100 years old (see Ushakov, 1972). Moreover, presence or absence of parapodia in the annelid stem species is still a matter of discussion and depends on the rooting of the phylogenetic trees or in other words on the reading direction of evolutionary changes (e.g., McHugh, 1997; Rouse & Fauchald, 1997; Westheide, 1997; Westheide et al., 1999). We will not consider this discussion in detail, but would like to focus on a remarkable observation. The parapodia of Aphroditidae are used for simple movements only, although the muscular apparatus of these parapodia is very complex and massive. As is evident from Mettam's (1967, 1971) data, parapodia of Nereididae are formed by a much smaller number of muscles but possess a greater mobility including a greater diversity of movements. Although not studied in detail Tzetlin et al. (2002a) stated that the musculature of Chrysopetalidae is similar to that of Aphroditidae. The parapodia of sedentary polychaetes also consist of a small number of muscles according to the few data available. These data most likely favour the hypothesis of Westheide & Watson Russel (1992) according to which those parapodia are the most primitive ones that are noticeably located at the dorsal side. If the parapodia are divided into neuro- and notopodia the dorsal chaetae play a protection role such as in Chrysopetalidae and Aphroditidae.

As is evident from the presented data, many polychaete taxa are characterised by the absence of circular muscles in the body wall. It has been observed in Opheliidae, Polygordiidae, Protodrilidae, Spionidae, Oweniidae, Aphroditidae, Acoetidae, Polynoidae, Sigalonidae, Phyllodocidae, Chrysopetalidae, Nephtyidae, Pisionidae and Nerillidae (Salensky, 1907; McIntosh, 1917; Hartmann-

Schröder, 1958; Orrhage, 1964; Jouin & Swedmark, 1965; Mettam, 1967, 1971; Storch, 1968; Hermans, 1969; Gardiner & Rieger, 1980; Tzetlin, 1987; Ivanov & Tzetlin, 1997; Tzetlin et al., 2002a, b). Absence of these fibres has also been reported in *Jennaria pulchra* an enigmatic taxon with strong affinities to Annelida (Rieger, 1991). This suggests that the lack of circular fibres may not be rare exception but a common situation in many polychaetes.

The view that a complete muscular lining comprising outer circular and inner longitudinal fibres belongs to the ground pattern in Annelida can be traced back to the ideas of Clark (1964, 1981) regarding an oligochaete-like burrowing animal as stem species of the entire group. However, since the body cavity often lacks segmental compartments formed by complete septa, the propulsive movements caused by the antagonistic actions of circular and longitudinal fibres characteristic for larger oligochaetes are only rarely found in polychaetes (Lanzavecchia et al., 1988). In these polychaetes antagonists of the longitudinal fibres are either dorsoventral, transverse, parapodial or the remaining longitudinal fibres themselves. The ideas of Clark (1964, 1981) have recently been supported by the cladistic analyses of Rouse & Fauchald (1995, 1997), but challenged by McHugh (1997) and Westheide (1997), who, among others, consider an epibenthic parapodia-bearing and not an earthworm-like organism to be the stem species in Annelida.

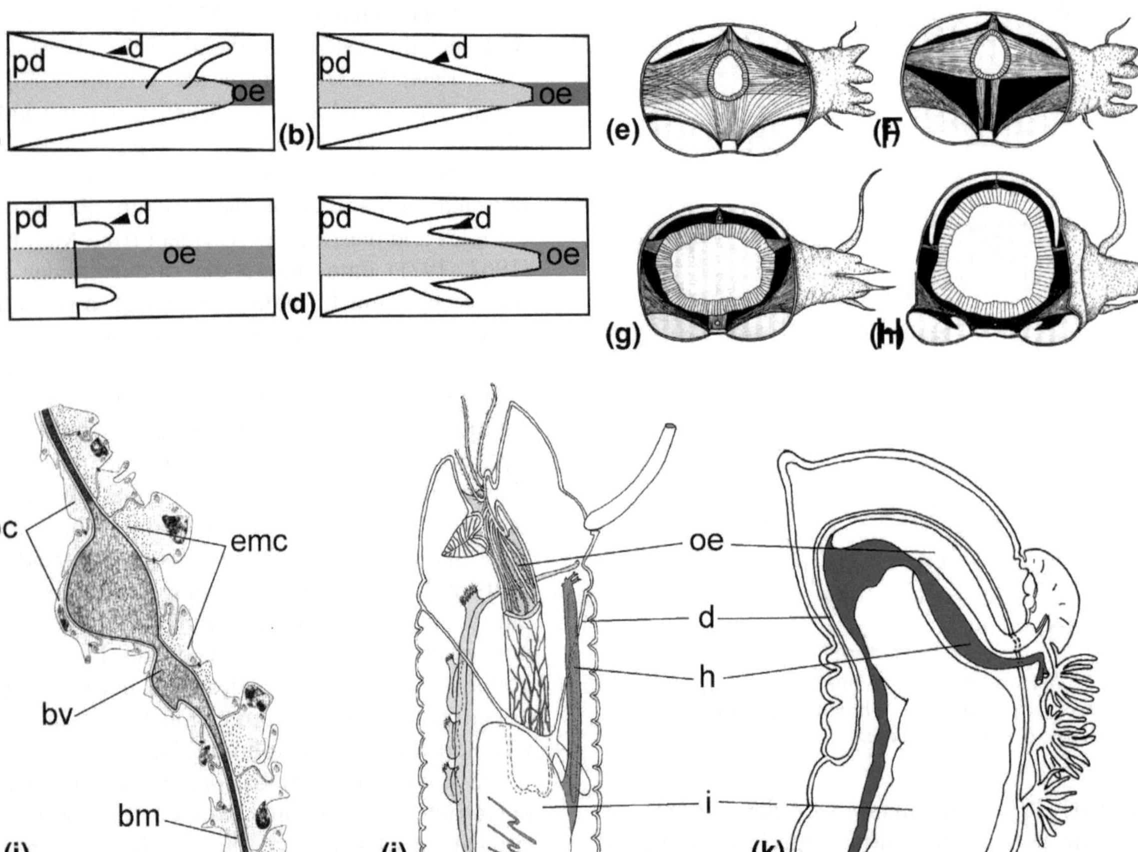

Figure 10. Structure of gular membrane (diaphragm). (a–d) Anterior part of body cavity of Terebellidae. (e–h) Supposed successive reduction of septa to form gur suspensory muscle. (i) Ultrastructure of dissepiment in Phyllodocidae. (j) Alvinellidae, sagittal section of anterior part. (k) *Artacama* sp. (Terebellidae), sagittal section of anterior end. bm – basal membrane, bv – blood vessel, cm – circular muscle, d – diaphragm, emc – epithelial muscle cell, h – heart, I – intestine, lm – longitudinal muscle, oe – oesophagus, pc – peritoneal cell, pd – prediaphragmal cavity. (a–d, j, k) After Zhadan & Tzetlin (in press), (e–h) After Clark (1964), (i) After Ivanov (2002).

Since circular muscles are especially important for burrowing forms and are not necessary for animals which proceed by movements of their parapodial appendages and chaetae (Mettam, 1971, 1985), the absence of such muscles in extant epibenthic polychaetes is related to the question, whether these muscles were present in the ancestral annelid or not. In case this stem species was in fact epibenthic and equipped with parapodia, these circular muscles do not appear to be a prerequisite for the complex movements shown by errant polychaetes. This scenario is in accordance with Mettam's (1985) hypothesis that the ancestral annelid had only longitudinal muscles used for rapid contractions of the body. Here the question arises which pattern of longitudinal muscle fibre arrangement might be the primitive. If we follow the hypothesis of Rouse & Fauchald (1997) an arrangement of longitudinal muscles in bands should be regarded as the most primitive pattern. However, which of the various patterns observed appears to be difficult to answer on our present knowledge. If parapodia-bearing taxa with protective chaetae will – with respect of their muscular system – prove to be comparatively close to the annelid stem species, a pattern of four or six bands should be the most primitive one.

In any case lack of circular fibres should now very seriously be considered in the discussion of the ground pattern of Annelida. However, further studies are required until a more definite conclusion can be drawn and how the different transverse fibres fit into this pattern, i.e. whether they represent reduced circular fibres or stages towards the development of circular muscles.

Acknowledgements

Significant parts of the data presented were gained in the department Spezielle Zoologie, University of Osnabrück and we express our deep gratitude to Professor Westheide, Dr Purschke, Dr Müller and other colleagues for their help and fruitful discussions. We are also indepted to Professor Malakhov and colleagues from the department of Invertebrate Zoology, Moscow State University supporting our work. One of the authors was partly supported by a grant from the deutsche Akademische Austauschdienst (DAAD) (Forschungsstipendium).

References

Brown, R. S., 1938. The anatomy of the polychaete *Ophelia cluthensis* McGuire, 1935. Proceedings of the Royal Society of Edinburgh 58: 135–160.

Brusca, R. C. & G. J. Brusca, 1990. Invertebrates. Sinauer Associates, Inc., Sunderland, 922 pp.

Clark, R. B., 1964. Dynamics in Metazoan Evolution. The Origin of the Coelom and Segments. Clarendon, Oxford, 313 pp.

Clark, R. B., 1981. Locomotion and the phylogeny of the Metazoa. Bolletino di Zoologica 48: 11–28.

Dorsett, D. A., 1961. The behavior of *Polydora ciliata* (Johnst.). Tube building and burrowing. Journal of the marine biological association of the United Kingdom 41: 577–590.

Eguileor, M. & R. Valvassori, 1977. Studies on the helical and paramyosinic muscles. VII. Fine structure if the body wall muscles in *Sipunculus nudus*. Journal of Submicroscopic Cytology 9: 393–372.

Gardiner, S. L., 1992. Polychaeta: General organisation, integument, musculature, coelom, and vascular system. In Harrison, F. W. & S. L. Gardiner (eds), Microscopic Anatomy of Invertebrates. Vol. 7: Annelida. Wiley-Liss, New York: 19–52.

Gardiner, S. L. & R. M. Rieger, 1980. Rudimentary cilia and muscle cells of annelids and echinoderms. Cell and Tissue Research 213: 247–252.

Hartmann-Schröder, G., 1958. Zur Morphologie der Opheliiden (Polychaeta Sedentaria). Zeitschrift für wissenschaftliche Zoologie 161: 84–113.

Hermans, C. O., 1969. The systematic position of the Archiannelida. Systematic Zoology 18: 85–102.

Hochberg, R. & M. K. Litvaitis, 2001. Functional morphology of muscles in *Tetranchyroderma papii* (Gastrotricha). Zoomorphology 121: 37–43.

Ivanov, I., in press. Ultrastructure of secondary body cavity of family Phyllodocidae (Annelida, Polychaeta).

Ivanov, I. & A. Tzetlin, 1997. Fine structure of body cavity of Phyllodocidae (Annelida, Polychaeta). Morphofunctional analysis. Doklady Akademia NAUK 354: 272–277 (in Russian).

Johansson, K. E., 1927. Beiträge zur Kenntnis der Polychaeten-Familien Hermellidae, Sabellidae uns Serpulidae. Zoologiska Bidrag fran Uppsala 11: 1–184.

Jouin, C. & B. Swedmark, 1965. *Paranerilla limnicola* n. g., n. sp., Archiannélide Nerillidae du benthos vaseux marin. Cahiers de Biologie Marine 6: 201–218.

Lanzavecchia, G., M. Eguileor & R. Valvassori, 1988. Muscles. In: Westheide, W. & C. O. Hermans (eds), The Ultrastructure of Polychaeta. Microfauna Marina 4: 71–88.

Marsden, J. R. & T. Lacalli, 1978. Morphology of the benthic larva of *Arenicola cristata* (Polychaeta). Canadian Journal of Zoology 56: 224–237.

Mattisson, A. G. M., 1969. The ultrastructure of the parapodial muscles of the spawning male of *Autolytus* (Syllidae, Polychaeta). Arkiv för Zoologi 22: 201–223.

McHugh, D., 1997. Molecular evidence that echiurans and pogonophorans are derived annelids. Proceedings of the National Academy of Sciences of the USA 94: 8006–8009.

McIntosh, W. C., 1917. On the nervous system and other points in the structure of *Owenia* and *Myriochele*. Annals and Magazine of Natural History 19: 233–265.

Mettam, C., 1967. Segmental musculature and parapodial movement of *Nereis diversicolor* and *Nephthys hombergi* (Annelida: Polychaeta). Journal of Zoology (London) 153: 245–275.

Mettam, C., 1971. Functional design and evolution of the polychaete *Aphrodite aculeata*. Journal of Zoology (London) 163: 489–514.

Mettam, C., 1985. Constraints in the evolution of the Annelida. In Conway Morris, S., D. J. George, R. Gibson & H. M. Platt (eds), The Origins and Relationships of Lower Invertebrates. Clarendon, Oxford: 297–309.

Meyer, E., 1887. Studien über den Körperbau der Anneliden. Mitteilungen aus der zoologischen Station zu Neapel 7: 592–741.

Meyer, E., 1888. Studien über den Körperbau der Anneliden. IV. Die Körperform der Serpulaceen und Hermellen. Mitteilungen aus der zoologischen Station zu Neapel 8: 462–662.

Müller, M. C. M. & A. Schmidt-Rhaesa, 2003. Reconstruction of the muscle system in *Antygomonas* sp. (Kinorhyncha, Cyclorhagia) by means of phalloidin labeling and cLSM. Journal of Morphology 256: 103–110.

Orrhage, L., 1962. Über das Vorkommen von Muskelzellen vom 'Nematoden-Typus' bei Polychaeten als phylogenetisch-systematisches Merkmal. Zoologiska Bidrag fran Uppsala 35: 321–327.

Orrhage, L., 1964. Anatomische und morphologische Studien über die Polychaetenfamilien Spionidae, Disomidae und Poecilochaetidae. Zoologiska Bidrag fran Uppsala 36: 335–405.

Purschke, G., 2002. On the ground pattern of Annelida. Organisms, Diversity and Evolution 2: 181–196.

Rieger, R. M., 1991. Neue Organisationsformen aus der Sandlückenfauna: die Lobatocerebriden und *Jennaria pulchra*. Verhandlungen der Deutschen Zoologischen Gesellschaft 84: 247–259.

Rieger, R. M., W. Salvenmoser, A. Legniti & S. Tyler, 1994. Phalloidin-rhodamine preparations of *Macrstomum hystricinum marinum* (Plathelminthes): morphology and postembryonic development of the musculature. Zoomorphology 114: 133–147.

Rouse, G. W. & K. Fauchald, 1995. The articulation of annelids. Zoologica Scripta 24: 269–301.

Rouse, G. W. & K. Fauchald, 1997. Cladistics and Polychaetes. Zoologica Scripta 26: 139–204.

Salensky, W., 1907. Morphogenetische Studien an Würmern. Mémoires de l'Académie Impériale des Sciences de St.-Pétersbourg. VIII. Serie. Classe Physico-Mathématque 19: 1–349.

Storch, V., 1968. Zur vergleichenden Anatomie der segmentalen Muskelsysteme und zur Verwandtschaft der Polychaeten-Familien. Zeitschrift für Morphologie und Ökologie der Tiere 63: 251–342.

Storch, V. & U. Welsch, 1986. Systematische Zoologie. Gustav Fischer, Stuttgart, 698 pp.

Tzetlin, A., 1987. Structural peculiarities of *Pisionidens tchesunovi* (Polychaeta) and their possible significance. Zoologisceskij Zurnal 66: 1454–1462 (in Russian).

Tzetlin, A., T. Dahlgren & G. Purschke, 2002a. Ultrastructure of the body wall, body cavity, nephridia and spermatozoa in four species of the Chrysopetalidae (Annelida, "Polychaeta"). Zoologischer Anzeiger 241: 37–55.

Tzetlin, A., A. Zhadan, I. Ivanov, M. C. M. Müller & G. Purschke, 2002b. On the absence of circular muscle elements in the body wall of *Dysponetus pygmaeus* (Chrysopetalidae, "Polychaeta", Annelida). Acta Zoologica (Stockholm) 83: 81–85.

Ushakov, P. V., 1972. Fauna of the USSR. Polychaetes. Academy of the Sciences of the USSR, Zoological Institute 102: 1–272.

Wanninger, A., B. Ruthensteiner, S. Lobenwein, W. Salvenmoser, W. J. A. G. Dictus & G. Haszprunar, 1999. Development of the musculature in the limpet *Patella* (Mollusca, Patellogastropoda). Development, Genes and Evolution 209: 226–238.

Wells, G. P., 1944. The parapodia of *Arenicola marina* L. (Polychaeta). Proceedings of the Royal Society of London 114: 100–116.

Wells, G. P., 1950. The anatomy of the body wall and appendages in *Arenicola marina* L., *Arenicola claparedii* Levinsen and *Arenicola ecaudata* Johnston. Journal of the Marine Biological Association of the United Kingdom 29: 1–44.

Wells, G. P., 1952. The proboscis apparatus of *Arenicola*. Journal of the Marine Biological Association of the United Kingdom 31: 1–28.

Wells, G. P., 1954. The mechanism of proboscis movement in *Arenicola*. Quarterly Journal of Microscopical Science 95: 251–270.

Westheide, W., 1997. The direction of evolution within Polychaeta. Journal of Natural History 31: 1–15.

Westheide, W. & R. M. Rieger, 1996. Spezielle Zoologie. Erster Teil: einzeller und Wirbellose. Gustav Fischer, Stuttgart: 1–909.

Westheide, W. & C. Watson Russel, 1992. Ultrastructure of chrysopetalid paleal chaetae (Annelida, Polychaeta). Acta Zoologica (Stockholm) 73: 197–202.

Westheide, W., D. McHugh, G. Purschke & G. W. Rouse, 1999. Systematization of the Annelida: different approaches. Hydrobiologia 402: 291–307.

Zhadan, A. & A. Tzetlin, in press. Comparative study of the diaphragm (gular membrane) in Terebellida (Polychaeta, Annelida).

Hydrobiologia (2005) **535/536:** 127–137
T. Bartolomaeus & G. Purschke (eds), Morphology, Molecules, Evolution and Phylogeny in Polychaeta and Related Taxa

The coelom and the origin of the annelid body plan

Reinhard M. Rieger[1,*] & Günter Purschke[2]
[1] *Abteilung Ultrastrukturforschung und Evolutionsbiologie, Institut für Zoologie und Limnologie, Universität Innsbruck, Technikerstrasse 25, A-6020 Innsbruck, Austria*
[2] *Spezielle Zoologie, Fachbereich Biologie/Chemie, Universität Osnabrück, D-49069 Osnabrück, Germany*
*(*Author for correspondence: E-mail: reinhard.rieger@uibk.ac.at)*

Key words: Bilateria, Annelida, Polychaeta, body cavity, coelom, acoelomate, pseudocoelomate, peritoneum, myoepithelium

Abstract

The biphasic life cycle in annelids is characterized by two completely different types of organisation, i.e. the acoelomate/pseudocoelomate larva and the coelomate adult. Based on this observation the recent literature on the different assumptions on the organisation of the bilaterian stem species with special emphasis on the evolution of the annelid body plan is reviewed. The structure of the coelomic lining ranges between a simple myoepithelium composed of epithelio-muscle cells and a non-muscular peritoneum that covers the body wall muscles. The direction of the evolution of these linings is discussed with respect to coelomogenesis. As the coelom originates from mesodermal cell bands, different assumption on the acoelomate condition in Bilateria can be substantiated. The origin of segmentation in annelids is explained by current hypothesis. Although no final decision can be made concerning the origin of the annelid body plan and the organisation of the bilaterian stem species, this paper elaborates those questions that need to be resolved to unravel the relation between the different body plans.

Introduction

The different basic designs of bilaterian body cavities, that is acoelomate/pseudocoelomate vs. coelomate constructions (Fig. 1A–E), have been employed to define body plans in several ways (see Rieger, 1985). Recently the usefulness of this concept in phylogeny has been seriously questioned both by molecular phylogenetesists (e.g., Adoutte et al., 1999, 2000) and some comparative morphologists (Willmer, 1995). This paper cannot be the place to respond in full length to this critique, it warrants, however, a general comment: irrespective of the question as to whether the different kinds of body cavities may be homologous (see Minelli, 1995 for a thorough discussion of this point), which type of body cavity may be more primitive (e.g., Rieger, 1986) or whether they may have multiple origins (e.g., Salvini-Plawen & Bartolomaeus, 1995), there is no doubt that the distinction between the a- and/or pseudocoelomate body cavity, as derivatives of the primary body cavity, and the coelomate condition as the secondary body cavity will remain a central question in discussions of the origins of the bilaterian body plans. Since Hyman's (1951) original proposal, acoelomate, pseudocoelomate and coelomate conditions have been defined further through ultrastructural work (e.g., see Rieger, 1985, 1986; Rieger & Lombardi, 1987; Fransen, 1980, 1988; Willmer, 1991; Bartolomaeus, 1994; Salvini-Plawen & Bartolomaeus, 1995; Westheide & Rieger 1996; Ax, 1996). Here it is shown for the coelomate organization of the Annelida that the concept of the three body cavity designs appears vague primarily due to the lack of sufficient ultrastructural & molecular information and that it is not simply 'imprecise and often unhelpful' (Willmer, 1995, p. 23). Until such investigations have been carried out it would seem premature to abandon the

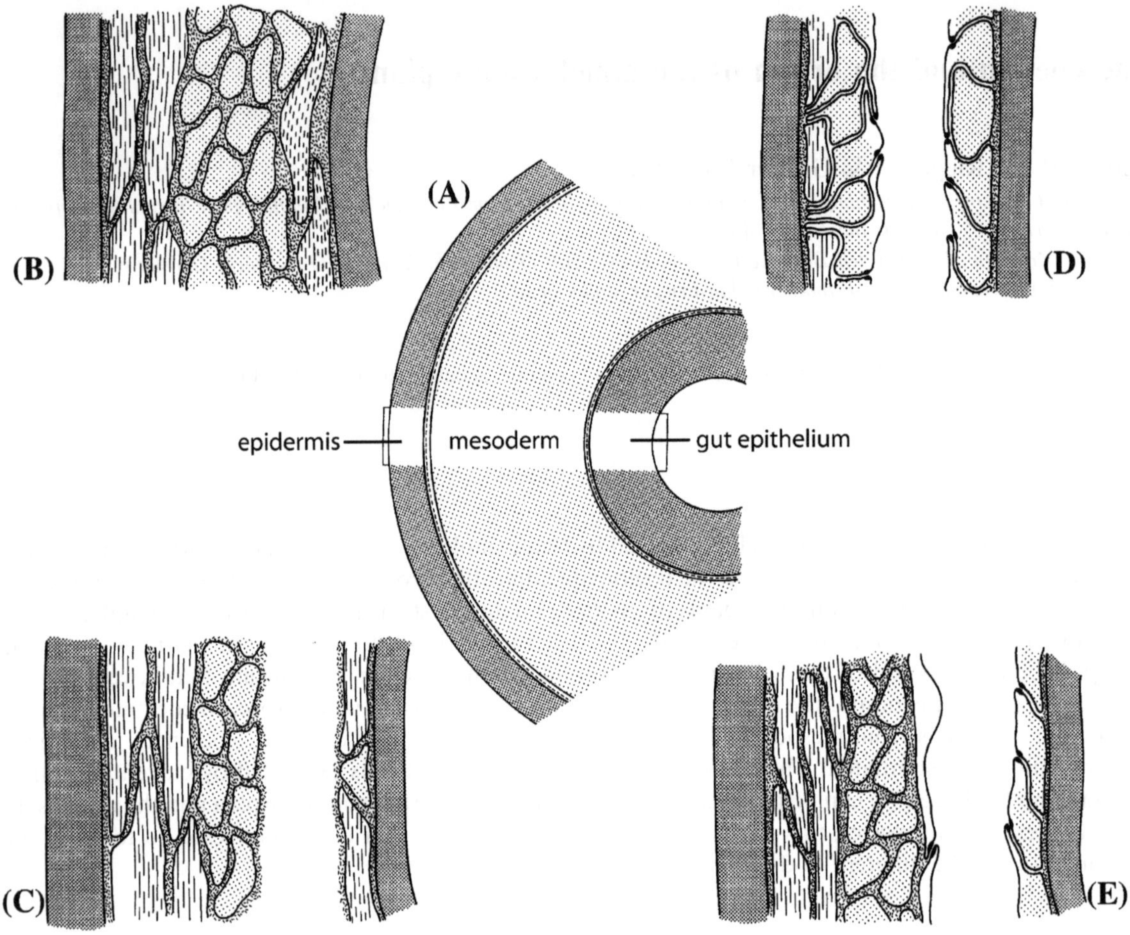

Figure 1. Diversity of mesoderm differentiation in Bilateria. (A) Mesoderm as defined as a median tissue layer situated between epidermis and gut epithelium, and primarily derived from the entoderm. (B) Acoelomate organization. (B) Pseudocoelomate organization (primary body cavity); body cavity lined by ECM only. (D–E) Coelomate organzation (secondary body cavity) as usually found in many Annelida. (D) Body cavity lined by a myoepithelium which also constitutes the muscular system of the animal. (E) Body cavity lined by a peritoneum on the somatopleure and by a visceral myoepithelium. Modified from Bartolomaeus (1994).

concept of the three body cavity designs in discussions of bilaterian phylogeny.

Adult coelomate annelids develop from pseudocoelomate larvae

If, as is widely accepted (see Nielsen, 2001), the annelid stem species exhibited a biphasic life cycle with a microscopic larva and a larger vermiform adult, it is evident that two sequential, extreme forms of body cavity design characterize the ancestral annelid body plan. It is acoelomate or pseudocoelomate in the larva and it is coelomate in adult animals (Fig. 2A, A′; Rieger, 1986, 1994). From this point of view it is not entirely correct to specify that annelids are coelomates.

As has been illustrated by Westheide (1987) progenesis is a widespread phenomenon in the evolution of interstitial annelids as well as in the interstitial fauna in general. This fact has led to the proposition that acoelomates may be secondarily derived from coelomates with schizocoelous coelom formation, without reduction of the coelom (Fig. 2B, B′; Rieger, 1986, 1991a, b; Smith et al., 1986). Among the Spiralia (but see also Tyler, 2001 for deuterostomes) the annelid example could serve as a model for the origin of acoelomate taxa

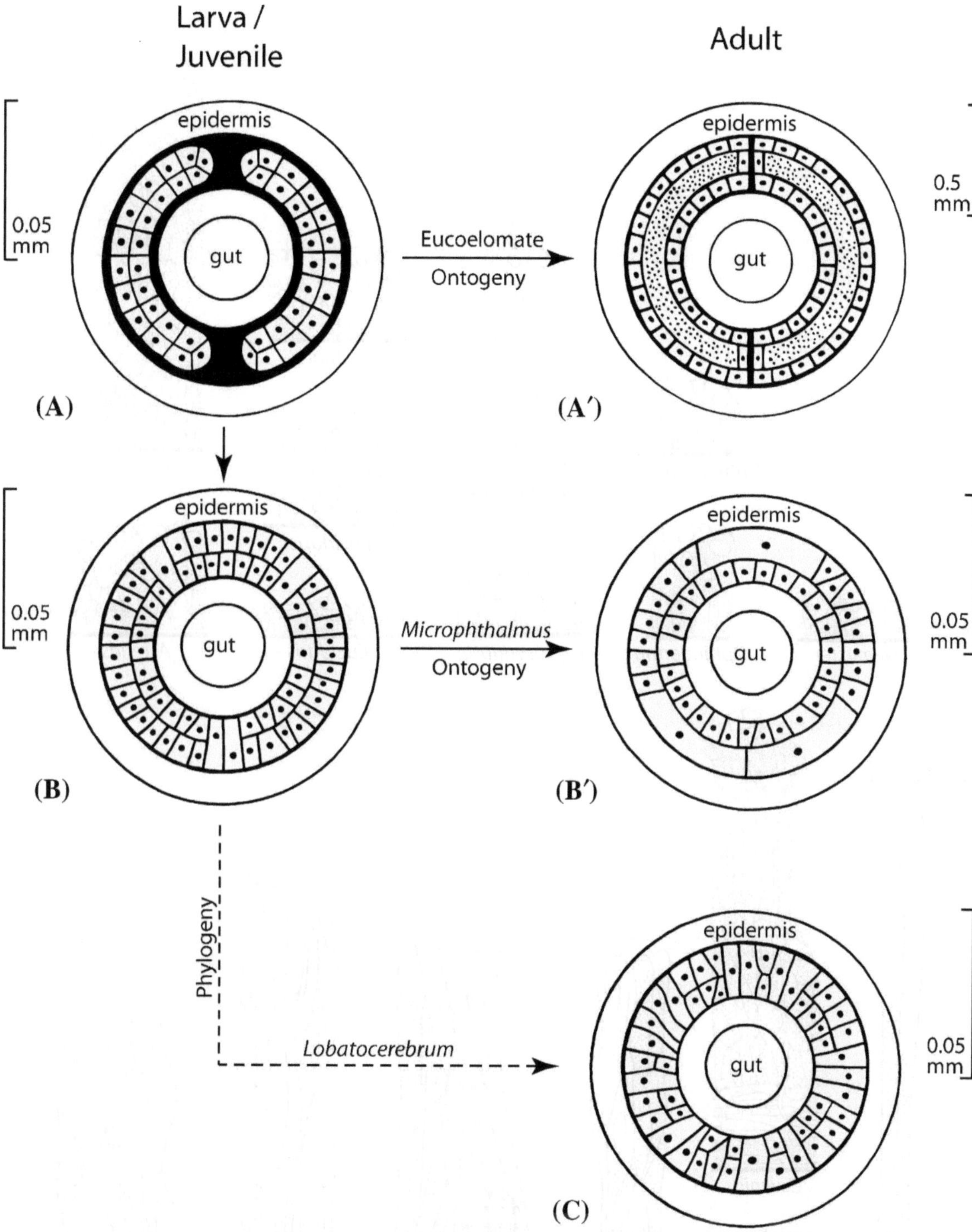

Figure 2. Body cavity designs as occurring in ontogeny and phylogeny in Annelida. (A-A′) Proposed ancestral body plan with pseudocoelomate or acoelomate larva and coelomate adult. (B–B′) Progenetic evolved interstitial species with acoelomate larva/juvenile and acoelomate adult. (C) Direct development of acoelomate adult with loss of larval stages. Original J. Lombardi, P. R. Smith & R. M. Rieger.

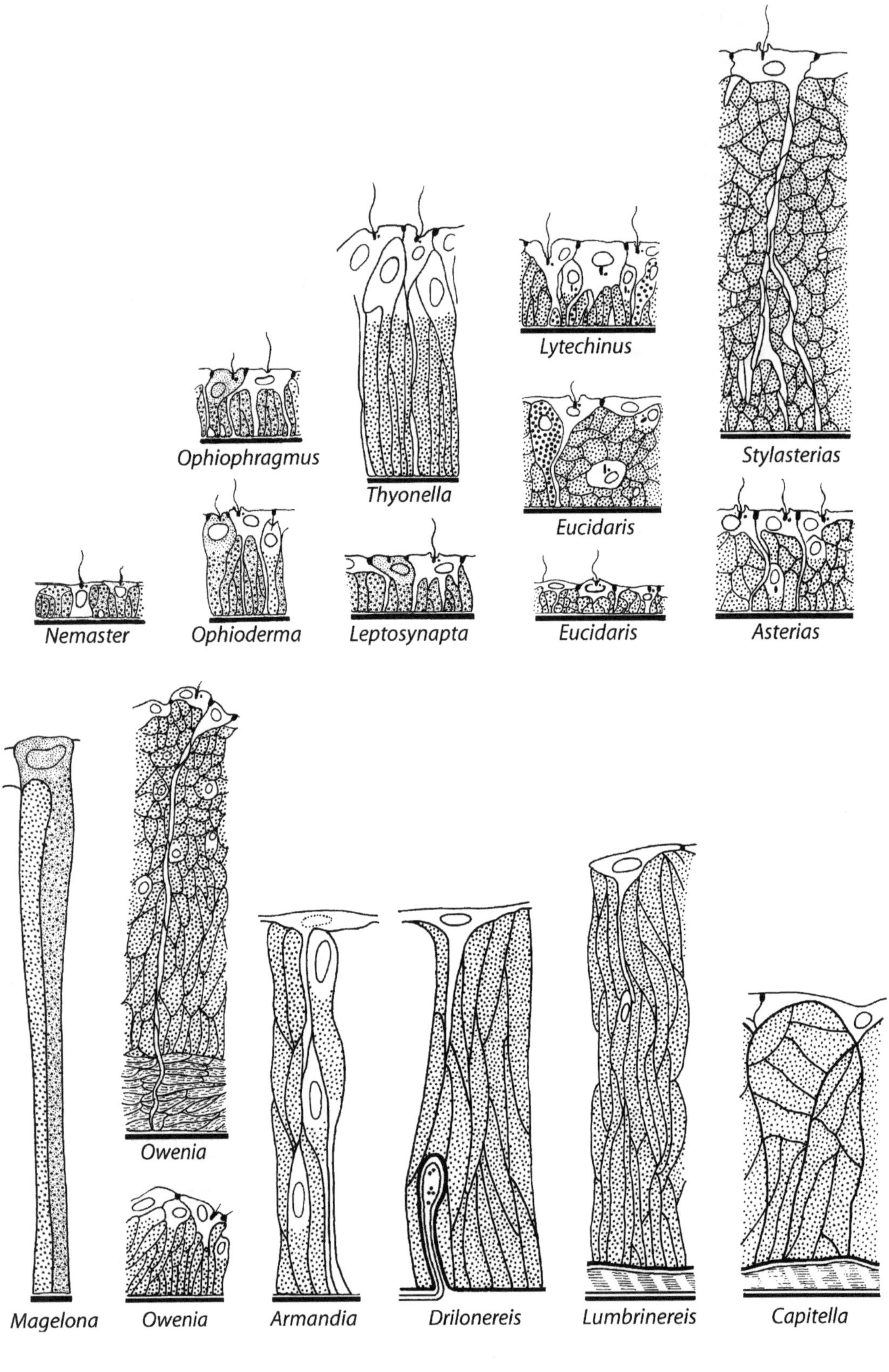
Stylasterias
Lytechinus
Ophiophragmus
Thyonella
Eucidaris
Eucidaris
Nemaster
Ophioderma
Leptosynapta
Asterias
Owenia
Magelona
Owenia
Armandia
Drilonereis
Lumbrinereis
Capitella

Figure 3. Schematic representation of different levels of histological organization within coelomic lining in Echinodermata (upper part of figure) and Annelida. Coelomic lining may constitute simple myoepithelia, pseudostratified, stratified myoepithelia and true peritoneum covering the somatic musculature. Modified from Rieger and Lombardi (1987) and Fransen (1988).

within the Bilateria from a coelomate stock (Fig. 2C).

In other proposals a microscopic organism such as a pseudocoelomate larvae of the Annelida are seen to represent the stem species of the Spiralia. The idea of a pseudocoelomate larva as ancestral body plan for the Bilateria has been recently revived by Davidson and co-workers (e.g., Davidson et al., 1995; Peterson et al., 1997, 2000; Peterson & Davidson, 2000). Adult structures such as the coelom would be a feature originating in this pseudocoelomate micrometazoan from a set of special cells (set aside cells), perhaps a kind of stem cells.

Finally, the Trochaea-theory suggests the origin of the spiralian phyla from a pseudocoelomate larva which became adapted to benthic life and developed either into an acoelomate adult ancestor for the Spiralia or a coelomate adult ancestor in the deuterostomes (Nielsen, 2001).

Histological organization of the coelomic lining in Annelida

The coelomic lining in annelids is either a myoepithelium, or a peritoneum, or a combination of these two epithelial configurations (e.g., Rieger, 1986; Fransen, 1988; Gardiner, 1992; Bartolomaeus, 1994; terminology in Rieger & Lombardi, 1987). Epithelio-muscle cells, fibre-type muscle cells, peritoneocytes and podocytes constitute the main epithelial cell types in this lining (Fig. 3). All muscle cells in coelomates may have originated in epithelia (Rieger & Ladurner, 2003).

Apical junctional complexes which are similar to zonulae adhaerentes have been described between all different epithelial cell types (e.g., Fransen, 1988; Gardiner, 1992; Bartolomaeus, 1994). Proximal to these junctions septate junctions may also occur (Fransen, 1988), but published images of these structures are lacking in Annelida (but see, e.g., Fig. 9 in Rieger & Lombardi, 1987).

A special, electron dense layer of the extracellular matrix (ECM) at the base of the coelomic lining and often an additional fibrous layer of various thickness have been found in many cases (Fransen, 1980, 1988; Gardiner, 1992; Bartolomaeus, 1994). This electron dense limiting layer of the ECM (Fransen, 1982) represents the basal lamina of the basal matrix (Rieger, 1985, 1986).

The known conformations of coelomic linings in annelids and those found in echinoderms can be aligned in a transformation sequence with a single layered myoepithelium on one end and a peritoneum with subperitoneal musculature on the other end (Rieger, 1986; Rieger & Lombardi, 1987; Stauber, 1993; Bartolomaeus, 1994). Such a transformation sequence including annelid species is shown here (Fig. 3). In annelids simple myoepithelia may occur, as well as pseudostratified and stratified myoepithelia containing all of the somatic and the splanchnic musculature (Bartolomaeus, 1994). When a non-muscular squamous peritoneum is differentiated, the somatic musculature, and in some cases also the splanchnic musculature, is found below the epithelial layer.

In echinoderms, the mechanism of the musculature sinking into the connective tissue below the epithelial level is well clarified (Stauber, 1993). It is most probable that the simple myoepithelium occurred first, followed by stepwise transformations to a non-muscular peritoneum plus subperitoneal musculature. The mechanism of myoepithelial muscle cells sinking below the level of the coelomic epithelium in annelids is less well understood (Rieger, 1986). The interpretation of a reading direction from a myoepithelium to the subperitoneal musculature is therefore less definite (but see Rieger, 1986, pp. 38–39, points A–E for general arguments in favour of that reading direction among Bilateria). A simple myoepithelial lining is currently regarded to represent the primary condition in Annelida (see Bartolomaeus, 1994 for discussion). Such transformation sequences from simple myoepithelia to separate somatic musculature and peritoneum are to be observed during development in several annelids: Coelomic cavities often arise from solid blocks of cells which become separated probably by fluid accumulation leading

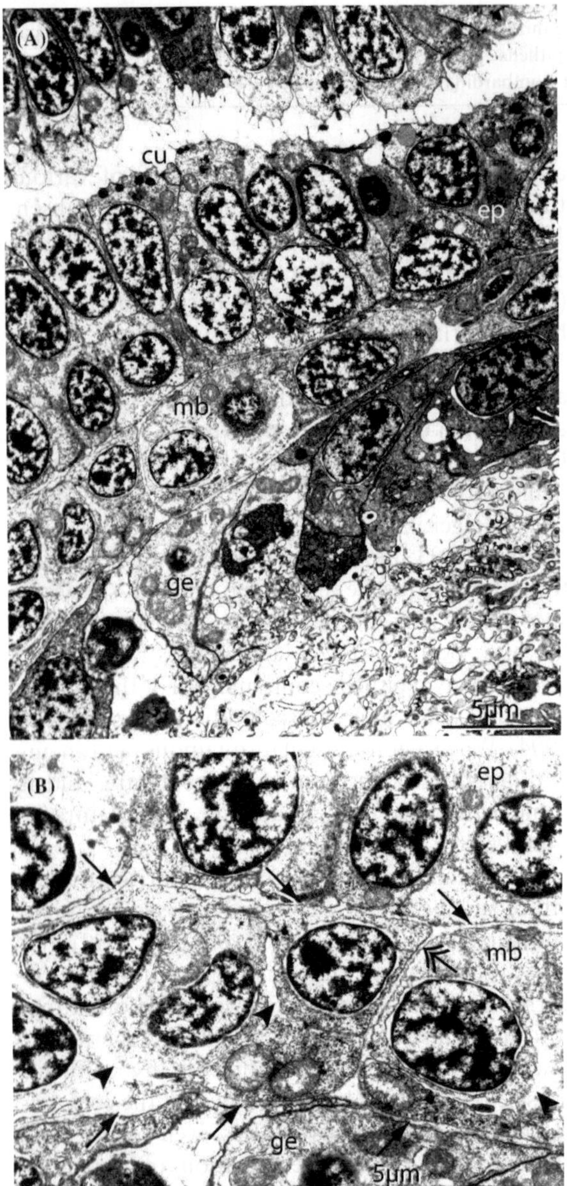

Figure 4. Ultrastructure of early mesodermal bands (mb) in a young mitraria larva of *Owenia fusiformis* (Oweniidae) in sagittal section showing their epithelial nature. (A) Low magnification to show location of mesodermal band (mb) between epidermis (ep) and gut epithelium (ge), cu cuticle. (B) Enlargement of left part of mesodermal band shown in A. Lumen of coelom represented by narrow spaces between the cells (arrowheads) which are joined by zonulae adhaerentes (double arrow). Arrows point to ECM between the different tissue layers. Micrographs: R. M. Rieger.

to an epithelial coelomic lining. Later these cells develop into peritoneal cells and muscle cells (Anderson, 1973; Gardiner, 1992; Bartolomaeus, 1994). Moreover, this pattern follows the increasing functional differentiation of cell types during evolution. Following this hypothesis, a peritoneum with underlying musculature should have evolved several times independently within Annelida.

Homology of the epithelial organization of coelomic linings in annelids

Generally the homology of the coelom in this taxon is not put into question (but see Minelli, 1995 for discussion). However, the characterization of the coelomate organization of annelids requires also strong arguments for a unique evolution of their specific epithelial differentiations. The probability of a unique evolution of myoepithelial coelomic linings in annelids, or coelomates in general, depends on the structural details defining them as 'true' epithelia, namely the apical junctional complex and the structure of the basal lamina (Rieger, 1994). Regrettably, no lanthanum preparations or freeze fracture images of apical junctional complexes in the coelomic lining of annelids have been published. These structures have been studied in other annelid epithelia (e.g., Green, 1981; Green & Bergquist, 1982). Junctions between cells in coelomic linings could now also be investigated with molecular markers (e.g., for cadherins, beta-catenin: Takeichi, 1991; Tepass et al., 2000; Tyler, 2003; B. Hobmayer, personal communication). This would be especially useful for investigating their formation during coelomogenesis. Similarly, differentiations such as basal laminas which have so far been identified with conventional TEM, could be investigated with immunocytochemical and molecular methods (see Pedersen, 1991; Kleinig & Maier, 1999; Schiebler & Schmidt, 2002; Tyler, 2003). The molecular substructure of these basal laminas could then be compared with the complex molecular networks known from basal laminas in ectodermal and entodermal epithelia of vertebrates (Fawcett, 1994; Kleinig & Maier, 1999) and other bilaterians (Tyler, 2003). Elucidating structural details of apical junctional complexes and basal laminas of the coelomic lining would yield a better understanding of coelom organization and of coelom formation. It would then be possible with increased confidence to postulate a

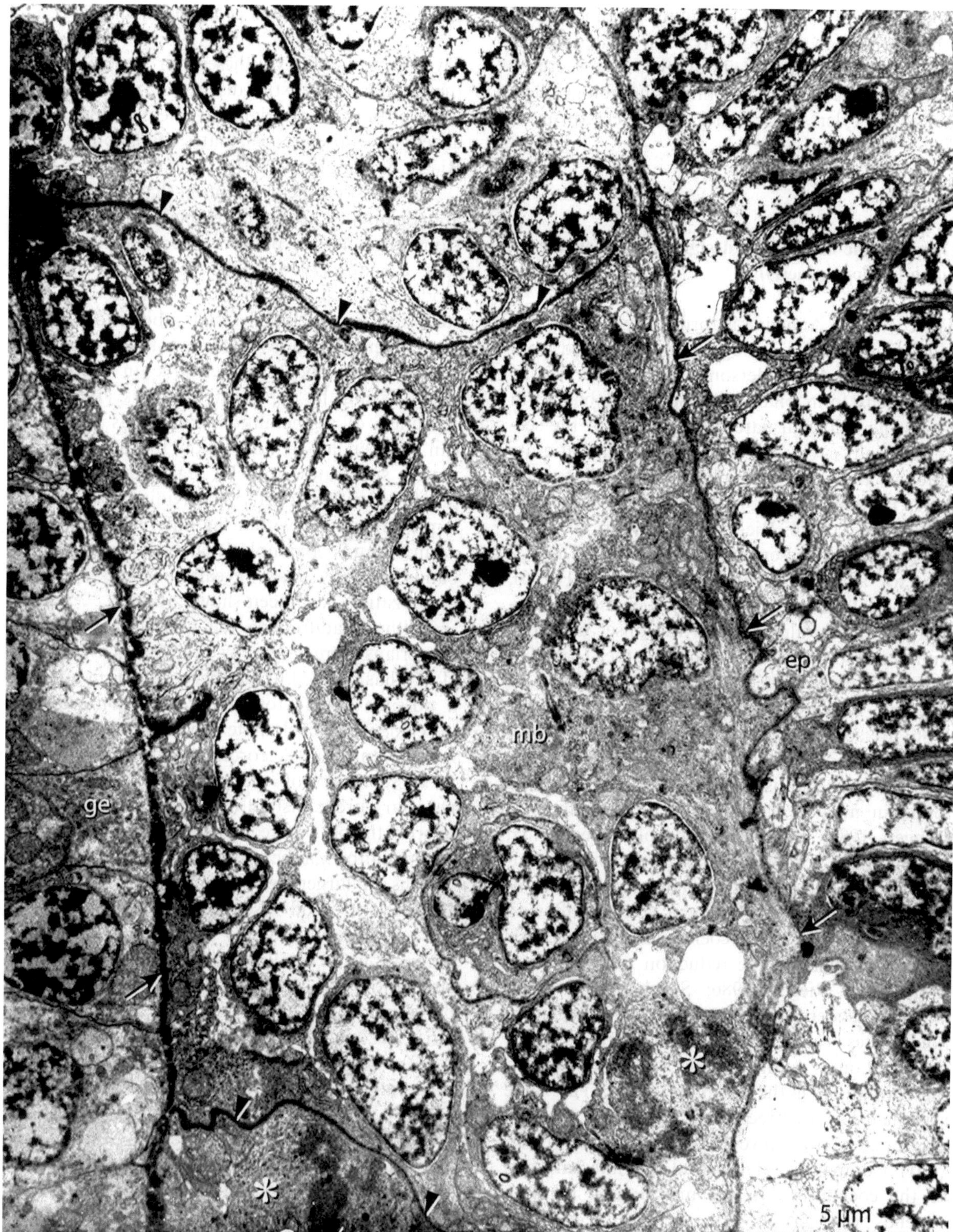

Figure 5. Ultrastructure of early mesodermal band (mb) in a early larva of an unidentified species of Oweniidae. Mesoderm shows a mesenchymal organization; arrows point to ECM between epidermis (ep), mesoderm band and gut epithelium (ge), arrowheads to ECM between adjacent epithelia in presumptive septa. Note dividing mesodermal cells (asterisks). Figure oriented perpendicular with respect to Figure 4B. Micrograph: R. M. Rieger.

monophyly of the coelomic lining in annelids, in other Spiralians and possibly even deuterostome phyla, as information about the substructures always enhances the probability of homology (e.g., Rieger & Tyler, 1979; Tyler, 1988; Haszprunar, 1996).

Origin of the histological design of the coelomic lining (coelomogenesis)

The coelom in Annelida is formed by mesodermal cells of the blastoporal region and is derived from the mesoblast 4d (Anderson, 1973; Nielsen, 2001). These cells form bilateral mesodermal bands, in which the epithelial nature of a lining of the secondary body cavity (coelom) is established at quite different times during embryogenesis (Potswald, 1981; Heimler, 1981a, b, 1983, 1988; Turbeville, 1986; Rieger, 1986; Rieger & Ladurner, 2003). In some species a mesenchymal organization is evident for a longer time in development (Fig. 5), whereas cells are arranged in epithelial configuration apparently almost from the onset of differentiation and cell proliferation in other (Fig. 4A, B), even closely related species. In the latter case coelomic cavities with collapsed lumen (Rieger, 1986) are surrounded by the epithelium (Fig. 4B). Also intermediate tissue organization, that is neither true epithelial tissue nor mesenchymal, can be found in mesodermal bands of certain annelids (Rieger, 1986). This variations of mesenchymal and epithelial conformations during coelomogenesis, and the different time points at which mesodermal bands form true coelomic epithelia allow to derive acoelomates from coelomates by progenesis without postulating the reduction of the coelom (Fig. 2A–C; Rieger, 1986; Smith et al., 1986). Other mesodermal tissues in such progenetic polychaetes can exhibit the same histological organization as do the mesodermal bands prior to the formation of the coelomic lining. The same argument has been recently proposed for deriving acoelomates from juvenile enteropneusts (Tyler, 2001). Comparative data about the differentiation of the coelom within the mesodermal band in annelids will certainly be needed for a better understanding of the evolution of the annelid body plan.

Origin of the segmented condition in annelids

Two groups of hypotheses for the origin of the Bilateria are still discussed, depending on whether bilaterians originally where coelomates or acoelomates/pseudocoelomates (Balavoine, 1998; Rieger & Ladurner, 2001):

(1) Based on the assumption that all bilaterians are coelomates having developed coelomic linings from gastrodermal pockets of coelenterate ancestors, Remane (1950, 1954, 1963a, b) proposed a model in which a vermiform coelomate bilaterian stem species developed serial subdivisions in the posteriormost region of three pairs of coelomic cavities (see model and critique in Hartmann, 1963). Using the arguments of Clark (1964), this evolution of segmentation (known as tritomery) can be seen as an adaptation for borrowing in mobile substrates. Annelid segments would have originated either from septa dividing existing coelomic cavities, or within the differentiating mesodermal bands. Examples that 'solid' mesodermal bands actually may reveal coelomic epithelial organization have been shown above.

Without addressing the question of whether bilaterians were originally coelomates, Westheide (1997) has pointed to another mechanism as being a possible key factor in the evolution from an unsegmented, coelomate ancestor to the segmented annelids. According to this model transverse septation of the coelom developed as a necessary prerequisite for transverse blood vessels to cross the body in recurring intervals and thereby ensure uniform, repeated blood supply for all body regions in larger vermiform coelomate animals. Because the blood vascular system and the coelomic organization are so intimately related (see Ruppert & Carle, 1983) the suggestion that blood circulation was a main functional factor for the development of segmentation in annelids seems most reasonable.

(2) Alternative theories propose that original bilaterians were acoelomates or pseudocoelomates, and segmentation evolved gradually (pseudometamerism hypothesis sensu Clark, 1964), together as the coelomic organization arose. Although such hypotheses have been thought to be refuted (see Remane, 1963a; Clark, 1964) they are presently discussed especially for the origin of segmentation within the arthropods (e.g., Budd,

2001). One example for pseudometamerism as origin of the segmented coelom was the gonocoel theory, which was particularly elaborated by Goodrich (1946). Pseudometamerism suggests that multiple substructures became organized in complex, sequentially arranged segments. Iterative structures such as cuticular setae, nephridia or gonads increasingly co-established iterative organs and thus segmentation, resembling the annelid body plan, arose gradually within an acoelomate or pseudocoelomate vermiform bilaterian stem species. This idea is of special significance for the Ecdysozoa-concept (e.g., Schmidt-Rhaesa et al., 1998; Budd, 2003; Garey, 2003; Schmidt-Rhaesa, 2003) which considers annelid and arthropod segmentation to have evolved in parallel. Scholtz (2003) has summarized the evidence concerning the issue of Ecdysozoan- vs. Articulata-concepts, and has argued in favour of the Articulate-hypothesis and of a homology of segmentation in annelids and arthropods.

Discussing the primary tissue organization of the mesoderm in Bilateria, Rieger and Ladurner (2003) have recently suggested that, if a model for the gradual origin of segmentation in annelids is envisioned, one driving force might be found in the strict repetitive pattern observed in the embryonic development of the circular musculature of small acoelomates such as *Convoluta pulchra* (Ladurner & Rieger, 2000). While data are lacking on myogenesis in most other spiralians (but see e.g., Reiter et al., 1996, for Macrostomorpha, Wanninger & Haszprunar, 2002, for Mollusca), the identical distances between circular muscle cells seen during early embryogenesis of *C. pulchra's* circular muscles may be due to the same or similar molecular mechanisms acting during early processes in segmentation in other protostomes and in deuterostomes (e.g., Davis & Patel, 1999; Shankland & Saever, 2000; Jouve et al., 2002).

Summary

With this paper we have tried to demonstrate the need for more detailed comparative ultrastructural and molecular analysis of the formation and of the adult organization of mesodermal tissues, in particular the histological organization and the origin of the coelomic lining in macroscopic and microscopic annelids. Although information on this subject has been accumulated during the 80s and 90s of the last century, especially detailed ultrastructural studies on apical junctional complexes and basal laminas are extremely rare and molecular information is still missing by and large. Comparisons of the formation and organization of the muscle system in spiralians usually considered as primary acoelomates (Platyhelminthes and Gnathostomulida) with that of secondary acoelomates (as so often seen among interstitial Annelida) would produce new insight into the question what makes primary acoelomates distinct from secondary ones and which of the present hypotheses might better explain the origin of the segmented coelom of Annelida. Without investigations of the features defining the 'true' epithelial organization of coelomic linings, a critical evaluation of the phylogenetic significance of the extremes of bilaterian body cavity organization (acoelomate/pseudocoelomate, coelomate) will not be possible.

Acknowledgements

Thanks are due to Wilfried Westheide and Gunde Rieger but also to Thomas Bartolomaeus for valuable comments and discussions. Supported by FWF grants (R. M. Rieger). The final preparation of the figures was carried out by Janina Jördens (University of Osnabrück).

References

Adoutte, A., G. Balavoine, N. Lartillot & R. de Rosa, 1999. Animal evolution the end of the intermediate taxa? Trends in Genetics 15: 105–108.

Adoutte, A., G. Balavoine, N. Lartillot, O. Lespinet, B. Prud'homme & R. de Rosa. 2000. The new animal phylogeny: reliability and implications. Proceedings of the National Academy of Sciences, USA 97: 4453–4456.

Anderson, D. T., 1973. Embryology and Phylogeny in Annelids and Arthropods. Pergamon, Oxford, 495 pp.

Ax, P., 1996. Multicellular Aminals. A New Approach to the Phylogenetic Order in Nature. Springer, Berlin, Heidelberg, New York, 225 pp.

Balavoine, G., 1998. Are Platyhelminthes coelomates without a coelom? An argument based on the evolution of *Hox* genes. American Zoologist 38: 843–858.

Bartolomaeus, T., 1994. On the ultrastructure of the coelomic lining in Annelida, Echiura and Sipuncula. Microfauna Marina 9: 171–220.

Budd, G. E., 2001. Why are arthropods segmented? Evolution and Development 3: 332–342.

Budd, G. E., 2003. Arthropods as ecdysozoans: the fossil evidence. In Legakis, A., S. Sfenthourakis, R. Polymeni & M. Thessalou-Legaki (eds), The new Panorama of Animal Evolution. Proceedings XVIII International Congress of Zoology, Pensoft, Sofia: 479–487.

Clark, R. B., 1964. Dynamics in Metazoan Evolution. The Origin of the Coelom and Segments. Clarendon, Oxford, 313 pp.

Davidson, E. H., K. J. Peterson & R. A. Cameron, 1995. Origin of adult bilaterian body plans: evolution of developmental regulatory mechanisms. Science 270: 1319–1325.

Davis, G. K. & N. H. Patel, 1999. The origin and evolution of segmentation. Trends in Genetics 15: M68–M72.

Fawcett, D. W., 1994. A Textbook of Histology. Chapman & Hall, New York, 964 pp.

Fransen, M. E., 1980. Ultrastructure of the coelomic organisation. I. Archiannelids and other small polychaetes. Zoomorphologie 95: 235–249

Fransen, M. E., 1982. The role of ECM in the development of invertebrates: a phylogeneticist́s view. In Hawkes S. & J. L. Wang (eds), Extracellular Matrix 478. Academic, New York: 177–181.

Fransen, M., 1988. Coelomic and vascular system. In Westheide W. & C. O. Hermans (eds), The Ultrastructure of Polychaeta. Microfauna Marina 4: 199–213.

Gardiner, S. L., 1992. Polychaeta. General organization, integument, musculature, coelom, and vascular system. In Harrison, F. W. & S. L. Gardiner (eds), Microscopic Anatomy of Invertebrates Vol. 7 Annelida. Wiley-Liss, New York: 19–52.

Garey, J. R., 2003. Ecdysozoa: the evidence for a close relationship between arthropods and nematodes. In Legakis, A., S. Sfenthourakis, R. Polymeni & M. Thessalou-Legaki (eds), The new Panorama of Animal Evolution. Proceedings XVIII International Congress of Zoology, Pensoft, Sofia: 503–509.

Goodrich, E. S., 1946. The study of nephridia and genital ducts since 1895. Quarterly Reviews of microscopical Science, New Series 86: 113–392.

Green, C. R., 1981. A clarification of the two types of invertebrate pleated septate junction. Tissue & Cell 13: 173–188.

Green, C. R. & P. R. Bergquist, 1982. Phylogenetic relationships within the invertebrata in relation to the structure of septate junctions and the development of 'occluding' junctional types. Journal of Cell Sciences 53: 279–305.

Hartmann, W., 1963. A critique of the enterocoele theory. In Dougherty, C. E. (ed.), The Lower Metazoa Vol. 5. University California, Berkeley: 55–77.

Haszprunar, G., 1996. The Mollusca: coelomate turbellarians or mesenchymate annelids? In Taylor, J. (ed.), Origin and Evolutionary Radiation of the Mollusca. Oxford University, Oxford: 1–28.

Heimler, W., 1981a. Untersuchungen zur Larvalentwicklung von *Lanice conchilega* (Pallas) 1766 (Polycheata, Terebellomorpha). Teil I: Entwicklungsablauf. Zoologische Jahrbücher, Abteilung für Anatomie und Ontogenie der Tiere 106: 12–45.

Heimler, W., 1981b. Untersuchungen zur Larvalentwicklung von *Lanice conchilega* (Pallas) 1766 (Polycheata, Terebellomorpha). Teil II: Bau und Ultrastruktur der Trochophora-Larve. Zoologische Jahrbücher, Abteilung für Anatomie und Ontogenie der Tiere 106: 236–277.

Heimler, W., 1983. Untersuchungen zur Larvenentwicklung von *Lanice conchilega* (Pallas) 1766 (Polychaeta, Terebellomorpha) Teil III: Bau und Struktur der Aulophora-Larve. Zoologische Jahrbücher, Abteilung für Anatomie und Ontogenie der Tiere 110: 411–478.

Heimler, W., 1988. Larvae. In Westheide, W. & C. O. Hermans (eds), The Ultrastructure of Polychaeta. Microfauna Marina 4: 353–371.

Hyman, L. H., 1951. The Invertebrates: Platyhelminthes and Rhynchocoela. The Acoelomate Bilateria. McGraw Hill. New York, 550 pp.

Jouve, C., T. Iimura & O. Pourquier, 2002. Onset of the segmentation clock in the chick embryo: evidence for oscillations in the somite precursors in the primitive streak. Development 129: 1107–1117.

Kleinig, H. & U. Maier, 1999. Zellbiologie. Gustav Fischer, Stuttgart, Jena, 534 pp.

Ladurner, P. & R. M. Rieger, 2000. Embryonic muscle development of *Convoluta pulchra* (Turbellaria-Acoelomorpha, Platyhelminthes). Developmental Biology 222: 359–375.

Minelli, A., 1995. Body cavities and body segmentation: problems of homology and phylogenetic reconstruction. In Lanzavecchia, G. R. Valvassori & M. D. Candia Carnevalli (eds), Body Cavities: Function and Phylogeny. Selected Symposia and Monographs, U.Z.I., 8, Mucchi, Modena: 69–73.

Nielsen, C., 2001. Animal Evolution. Interrelationships of the Living Phyla. Oxford University, New York, 563 pp.

Pedersen, K. J., 1991. Structure and composition of basement membranes and other basal matrix systems in selected invertebrates. Acta Zoologica (Stockholm) 72: 181–201.

Peterson, K. J. & E. H. Davidson, 2000. Regulatory evolution and the origin of the bilaterians. Proceedings of the National Academy of Sciences of the USA 97: 4430–4433.

Peterson, K. J., R. A. Cameron & E. H. Davidson, 1997. Set-aside cells in maximal indirect development: evolutionary and developmental significance. Bioessays 19: 623–631.

Peterson, K. J., R. A. Cameron & E. H. Davidson, 2000. Bilaterian origins: significance of new experimental observations. Developmental Biology 219: 1–17.

Potswald, H., 1981. Abdominal segment formation in *Spirorbis moerchi* (Polychaeta). Zoomorphology 97: 225–245.

Reiter, D., B. Boyer, P. Ladurner, G. Mair, W. Salvenmoser & R. Rieger, 1996. Differentiation of the body wall musculature in *Macrostomum hystricinum marinum* and *Hoploplana inquilina* (Plathelminthes), as models for muscle development in lower Spiralia. Rouxś Archives of Developmental Biology 205: 410–423.

Remane, A., 1950. Die Entstehung der Metamerie der Wirbellosen. Zoologischer Anzeiger, Supplementband 14: 16–23.

Remane, A., 1954. Die Geschichte der Tiere. In Herberer, G. (ed.), Die Evolution der Organismen. Vol. 2, 2nd edn. Fischer, Stuttgart: 340–422.

Remane, A. 1963a. The enterocelic origin of the coelom. In. Dougherty, C. E. et al. (eds), The Lower Metazoa. University of California, Berkeley, CA: 78–90.

Remane, A. 1963b. The evolution of the Metazoa from colonial flagellates vs. plasmodial ciliates. In Dougherty, C. E. et al. (eds), The Lower Metazoa. Comparative Biology and Phylogeny. University of California, Berkeley, CA: 23–32.

Rieger, R. M., 1985. The phylogenetic status of the acoelomate organization within the bilateria. a histological perspective. In Conway-Morris, S., J. D. George, R. Gibson & H. M. Platt (eds), The Origins and Relationships of Lower Invertebrates. Clarendon, Oxford: 101–122.

Rieger, R. M., 1986. Über den Ursprung der Bilateria: die Bedeutung der Ultrastrukturforschung für ein neues Verstehen der Metazoenevolution. Verhandlungen der Deutschen Zoologischen Gesellschaft 79: 31–50. (English translation at http://www.umsci.maine.edu./biology/labs/original/)

Rieger, R. M., 1994. Evolution of the 'lower' Metazoa. In Bengtson, S. (ed.), Early Life on Earth. Nobel Symposium no. 84. Columbia University, New York: 475–488 (references on 517–598).

Rieger, R. M., 1991a. Neue Organisationstypen aus der Sandlückenfauna: die Lobatocerebriden und *Jenneria pulchra*. Verhandlungen der Deutschen Zoologischen Gesellschaft 84: 247–259.

Rieger, R. M., 1991b. *Jennaria pulchra*, nov. gen. nov. spec., eine den psammobionten Anneliden nahestehende Gattung aus dem Küstengrundwasser von North Carolina. Berichte des Naturwissenschaftlich Medizinischen Vereins in Innsbruck 78: 203–215.

Rieger, R. M. & P. Ladurner, 2001. Searching for the stem species of the Bilateria. Belgian Journal of Zoology, 131 (Supplement 1): 27–34.

Rieger, R. M. & P. Ladurner, 2003. The significance of muscle cells for the origin of mesoderm in Bilateria. Integrative & Comparative Biology 43: 47–54.

Rieger, R. M. & J. Lombardi, 1987. Comparative ultrastructure of coelomic linings in echinoderm tube feet and the evolution of peritoneal linings in the Bilateria. Zoomorphology 107: 191–208.

Rieger, R. M. & S. Tyler, 1979. The homology theorem in ultrastructure research. American Zoologist 19: 654–666.

Ruppert, E. E. & K. J. Carle, 1983. Morphology of metazoan circulatory systems. Zoomorphology 103: 193–208.

Salvini-Plawen, L. v. & T. Bartolomaeus, 1995. Mollusca: mesenchymata with a coelom. In Lanzavecchia, G. R. Valvassori & M. D. Candia Carnevalli (eds), Selected Symposia and Monographs U.Z.I., 8, E. Muncchi, Modena: 75–92.

Schiebler, T. H. & W. Schmidt, 2002. Anatomie. Springer, Berlin Heidelberg, 892 pp.

Schmidt-Rhaesa, A., 2003. Integrative approaches to phylogenetic relationships of arthropods: introduction to the symposium. In Legakis, A., S. Sfenthourakis, R. Polymeni & M. Thessalou-Legaki (eds), The new Panorama of Animal Evolution. Proceedings XVIII International Congress of Zoology, Pensoft, Sofia: 461–466.

Schmidt-Rhaesa, A., T. Bartolomaeus, C. Lemburg, U. Ehlers & J. R. Garey, 1998. The position of the Arthopoda in the phylogenetic system. Journal of Morphology 238: 413–418.

Scholtz, G., 2003. Is the taxon Articulata obsolete? Arguments in favour of a cose relationship between annelids and arthropods. In Legakis, A., S. Sfenthourakis, R. Polymeni & M. Thessalou-Legaki (eds), The new Panorama of Animal Evolution. Proceedings XVIII International Congress of Zoology, Pensoft, Sofia: 489–501.

Shankland, M. & E. C. Seaver, 2000. Evolution of the bilaterian body plan: what have we learned from annelids? Proceedings of the National Academy of Sciences of the USA 97: 4434–4437.

Siewing, R., 1985. Lehrbuch der Zoologie Vol. 2, Systematik. Fischer, Stuttgart, New York, 1107 pp.

Smith, J. P. S., J. Lombardi & R. M. Rieger, 1986. Ultrastructure of the body cavity lining in a secondary acoelomate, *Microphthalmus* cf. *listensis* Westheide (Polychaeta, Hesionidae). Journal of Morphology 188: 257–271.

Stauber, M., 1993. The latern of Aristotle: organization of its coelom and origin of its muscles (Echinodermata, Echinoida). Zoomorphology 113: 137–151.

Takeichi, M., 1991. Cadherin cell adhesion receptors as a morphogenetic regulator. Science 251: 1451–1455.

Tepass, U., K. Truong, D. Goudt, M. Ikura & M. Pfeifer, 2000. Cadherins in embryonic and neural morphogenesis. Nature Reviews of Molecular Cell Biology 1: 91–100.

Turbeville J. M., 1986. An ultrastructural analyis of coelomogenesis in the hoplonemertine *Prosorhochmus americanus* and the polychaete *Magelona* sp. Journal of Morphology 187: 51–60.

Tyler, S., 1988. The role of function in determination of homology and convergence – examples from invertebrate adhesive organs. In Ax, P., U. Ehlers & B. Sopott-Ehlers (eds), Free-living and Symbiotic Plathelminthes. Progress in Zoology 36: 331–374.

Tyler, S., 2001. Origin and relationship of lower flatworms. In Littlewood, D. T. J. & R. A. Bray (eds), Interrelationsships of the Platyhelminthes. Taylor and Francis, New York: 3–12.

Tyler, S., 2003. Epithelum – the primary building block for metazoan complexity. Integrative and Comparative Biology 43: 55–63.

Wanninger, A. & G. Haszprunar, 2002. Chiton Myogenesis: Perspectives for the development and evolution of larval and adult muscle systems in molluscs. Journal of Morphology 251: 103–113.

Westheide, W., 1987. Progenesis as a principle in meiofauna evolution. Journal of Natural History 21: 843–854.

Westheide, W., 1997. The direction of evolution within the Polychaeta. Journal of Natural History 31: 1–15.

Westheide, W. & R. Rieger, 1996. Spezielle Zoologie. Teil 1: Einzeller und Wirbellose Tiere. Gustav Fischer, Stuttgart, 909 pp.

Willmer, P., 1991. Invertebrate Relationships. Patterns in Animal Evolution. Cambridge University, Melbourne: 400 pp.

Willmer, P., 1995. Modern approaches to the phylogeny of body cavities. In Lanzavecchia, G., R. Valvassori & M. D. Candia Carnevalli (eds), Body Cavities: Function and Phylogeny. Selected Symposia and Monographs, U.Z.I., 8, Mucchi, Modena: 23–39.

Hydrobiologia (2005) **535/536:** 139–165
T. Bartolomaeus & G. Purschke (eds), Morphology, Molecules, Evolution and Phylogeny in Polychaeta and Related Taxa

Structure and development of nephridia in Annelida and related taxa

Thomas Bartolomaeus* & Björn Quast
Animal Systematics and Evolution, Free University Berlin, Koenigin-Luise-Str. 1-3, 14195 Berlin
*(*Author for correspondence: E-mail: tbartol@zoosyst-berlin.de)*

Key words: protonephridia, metanephridia, ultrastructure, phylogeny, larvae, podocytes

Abstract

Two different kinds of filtration nephridia, protonephridia and metanephridia, are described in Polychaeta. During ontogenesis protonephridia generally precede metanephridia. While the latter are segmentally arranged, protonephridia are characteristic for the larva and are the first nephridial structure formed during ontogenesis. There is strong evidence that both organs depend on the same information and that their specific structure depends on the way in which the coelom is formed and which final expansion it gains. While metanephridia are regarded to be homologous throughout the polychaetes, protonephridia seem to have evolved in several lineages. Some of the protonephridia closely resemble less differentiated stages of metanephridial development, so that protonephridial evolution can be explained by truncation of the metanephridial development. Nevertheless, structural details are large enough to allow us to expect information on the polychaete evolution if the database on polychaete nephridia increases. A comparison of the polychaete metanephridia with those of the Clitellata and Sipuncula reveals some surprising details. In Clitellata the structure of the funnel is quite uniform in microdrilid oligochaetous Clitellata and resembles that of the aeolosomatids. Like the nephridia in the polychaete taxa Sabellida and Terebellida, those of the Sipunucla possess podocytes covering the coelomic side of the duct.

Introduction

Nephridia are basically defined by their function. They eliminate wastes from amino acid and nucleotide degradation as well as sometimes salt and water from the body. From the variety of such organs, filtration nephridia are assumed to belong to the ground pattern of the Bilateria (Bartolomaeus & Ax, 1992, but see Jondelius et al., 2002). In annelids these organs occur as two different structures, protonephridia and metanephridial systems (sensu Ruppert & Smith, 1988). Both organs mediate excretion in two steps: an unspecific filtration of body fluid and subsequent modification of the ultrafiltrate (Fig. 1a, b). This second step is necessary because during filtration not only metabolic wastes but also all substances with a low molecular weight, like certain anions, amino acids and different sugars, pass the filtration barrier. To utilize them again, these substances need to be reabsorbed. In both nephridia the ECM acts as filtration barrier, forming a kind of molecular sieve. This matrix needs to be stabilized to withstand the filtration pressure by special perforated cells that mediate filtration, i.e. the terminal cell or podocytes. The ultrafiltrate is modified in a simple duct. This function has been shown and experimentally confirmed in a number of papers (see Zerbst-Boroffka & Haupt, 1975; Smith & Ruppert, 1988; Bartolomaeus & Xylander, in prep.; Fig. 1c, d). The structural correlates for these functional demands, however, differ between protonephridia and metanephridial systems. In protonephridia terminal cells span the filtration barrier (Fig. 3d); these cells are directly apposed to the duct and, thus, in a protonephridium both functions are united within a single organ. In metanephridial systems the wall of transcoelomic blood vessels functions as filtration barrier, stabilized by

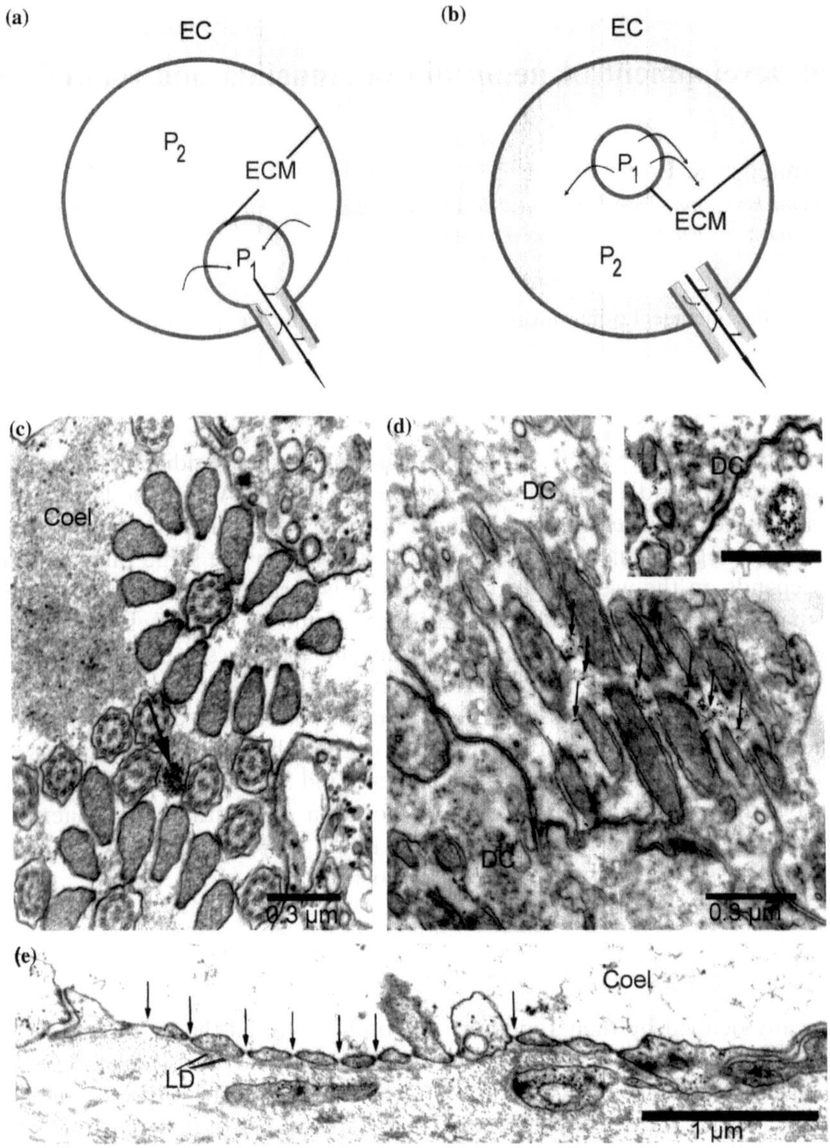

Figure 1. Function of filtration nephridia. (a) Protonephridium. $P_2 > P_1$. Pressure inside the nephridial compartment (P_1) is lower than in the surrounding interstices or body cavity (P_2). (b) Metanephridial system. $P_1 > P_2$. Pressure inside the blood vessel higher than in the coelomic cavity. (c, d) *Eulalia viridis* (Phyllodocidae). (c) Experimentally applied albumen-gold (arrow) does not pass the filtration barrier. (d) Experimentally applied iron dextrane (arrows) passes the filtration barrier and is reabsorbed in the duct (inset). (e) *Fabricia sabella* (Serpulidae). Podocyte lining the dorsal vessel in S2. Arrows mark filtration slits. *EC* epidermis, *ECM* extracellular matrix, *Coel* coelomic cavity, *DC* duct cell, *LD* lamina densa. (a), (b) modified from Ruppert & Smith (1988).

podocytes (Fig. 1e). Modification occurs spatially separated from these cells within the metanephridium. Thus, the term metanephridial system is only related to the function, because the system is formed by two structurally independent elements. The term "protonephridium", however, is related to the function as well as to the structural integrity of this organ (Fig. 2a). The term "metanephridium", finally, is related to the structural integrity alone, as it only mediates a part of the entire excretory process, i.e. modification of the ultrafiltrate (Fig. 2b). Thus, when using these different terms, we have to keep in mind that they address to completely different affinities of the nephridia.

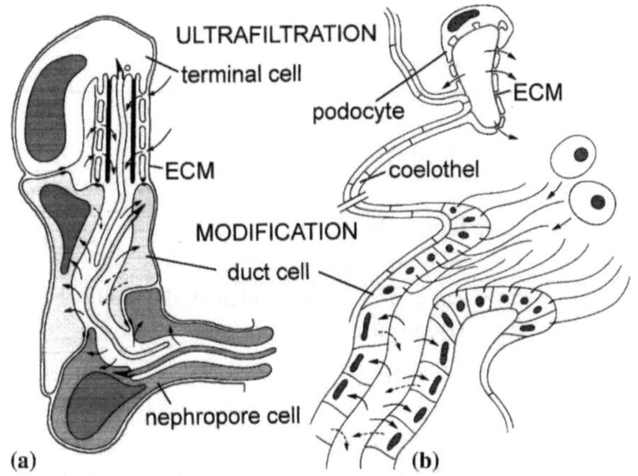

Figure 2. (a) Protonephridium and (b) metanephridial system and their substructures. The sites of ultrafiltration differ in both nephridial systems. Metanephridia may additionally serve in releasing genital products (indicated by two egg cells).

In annelids both organs occur, viz., protonephridia and metanephridia. They are related by the developmental process, because in several taxa protonephridia precede metanephridia while the animals grow (Table 1). In some annelid taxa protonephridia can be found in all stages of development, while there are metanephridia in all postlarval stages of others. Provided that annelids are monophyletic there should also be some evolutionary connection between both organs.

In this paper we summarize the present knowledge on nephridia in annelids and related taxa from a structural, developmental and comparative perspective, which leads to some evolutionary considerations.

Structure

On the structural level, organs can be described as a hierarchical system of substructures. The cell types represent the upper level of structural subsets of an organ, followed by cell surface differentiations like cilia, microvilli or structures being offset from the perikaryon. The lowest level is represented by subcellular structures, like organelles or other structures. At least one structural element should be unique to each cell type, and should define and characterise it to provide criteria for its identification. In the case of annelid nephridia, however, we shall see that these criteria may also be derived from developmental studies or from the structural context into which these cells are embedded. We are convinced that understanding organs as a hierarchical system of substructures provides essential coding issues for cladistic analyses as well as the substrate for primary homology hypotheses. On the level of organs their robustness depends on the number of identical substructures. In this section we therefore want to offer a framework for a proper description of the annelid nephridia as far as presently possible.

Protonephridia

On the cellular level, protonephridia consist of a subset of at least three different elements: terminal cell, duct cell and nephridiopore cell (Fig. 2a). They surround a compartment that is but a blindly closed cellular duct leading from the exterior into the mesodermal tissue. The entire organ is surrounded by an ECM that is continuous with the subepidermal matrix (Fig. 1a). All cells of the organ adhere to each other by cell junctions. The innermost cell is called terminal cell, flame cell or solenocyte. It mediates filtration by forming the supporting structure of the actual filtration barrier. Generally this filter is a perforated or slashed hollow cylinder that surrounds the cilium and its surrounding microvilli (Fig. 3a, b). As the filter is always connected to the adjacent duct cell by adherens junctions, these junctional complexes provide an important criterion to discriminate a circle of microvilli from a longitudinally slashed filter. Thus, the terminal cell consists of four substructures: the perikaryon, the filter, a ciliary element and a microvilli element. While the perikaryon must always be present, the remaining three substructures can be modified or may be absent. Their presence is obviously independent of each other. Distally, a duct cell is connected to the terminal cell. Like the terminal cell, the duct cell is composed of substructures: the perikaryon, a ciliary element and a microvilli element. A large number of different vesicles, some of which are coated by clathrin, are characteristic to the cytoplasm of the duct cells, indicating endo-, exo- and transcytotic processes (Fig. 4a, g). The nephridiopore or nephridiopore cell finally connects the duct and the epidermis, and its perikaryon is below the level of the epidermis. The nephridiopore cell is composed of the same substructures as a duct cell, i.e. a

Table 1. Filtration nephridia in Annelida

Taxon	Head kidney, segmental nephridia	Filtration structure	Author
Aberranta	?, ?		
Acoetidae	?, metanephridia	?	Goodrich (1945)
Acrocirridae			
Aeolosomatidae	?, metanephridia	podocytes	Bunke (1994),
Alciopidae	?, protonephridia	solenocytes	Smith & Ruppert (1988)
Alvinellidae	?, metanephridia	?	Zal et al. (1994), Zhadan et al. (2000)
Ampharetidae	?, metanephridia	?	Hessle (1917)
Amphinomidae	?, metanephridia	?	Goodrich (1900)
Aphroditidae	?, metanephridia	?	Fordham (1926), Goodrich (1945)
Apistobranchidae	?, metanephridia	?	Orrhage (1974)
Arenicolidae	?, metanephridia	podocytes	Ashworth (1912), Goodrich (1945), Wells (1959), Bartolomaeus (unpubl.)
Capitellidae	protonephridia, metanephridia	?	Eisig (1879, 1887, 1899), Goodrich (1945)
Chaetopteridae	protonephridia, metanephridia	terminal cell multic.	Bonch-Bruevich & Malakhov (1986), Joyeux-Laffuie (1890), Goodrich (1945)
Chrysopetalidae	?, metanephridia	?	Ehlers (1846), Goodrich (1945), Tzetlin et al. (2001)
Cirratulidae	?, metanephridia	?	Meyer (1887), Goodrich (1945), Olive (1970)
Cossuridae	?, ?		
Ctenodrilidae	?, metanephridia	?	Mesnil (1899), Banse (1969)
Dorvilleidae	?, metanephridia or protonephridia	podocytes or terminal cell multic., terminal cell monoc. in *Diurodrilus*	Smith & Ruppert (1988), Brandenburg (1970), Westheide (1985, 1990), Worsaae & Kristensen (this volume)
Eulepethidae	?, metanephridia	?	Goodrich (1945)
Eunicidae	?, metanephridia	?	Goodrich (1945)
Euphrosinidae	?, metanephridia	?	Gustofson (1930)
Fauveliopsidae	?, ?		
Flabelligerida	?, metanephridia	?	Goodrich (1945)
Glyceridae	?, protonephridia	solenocytes	Brandenburg (1966, 1975), Brandenburg & Kümmel (1961), Goodrich (1899, 1945), Smith & Ruppert (1988)
Goniadidae	?, protonephridia	solenocytes	Goodrich (1899, 1945), Ruppert & Smith (1988)
Hartmaniellidae	? ?		
Hesionidae	?, protonephridia or metanephridia	solenocytes multic. or terminal cells multic.	Goodrich (1898, 1945), Westheide (1986), Clausen (1986)
Heterospio	?, ?		
Histriobdellidae	?, protonephridia	terminal cell multic.	Shearer (1910), Scharnofske (1984)
Isopilidae	?, protonephridia	solenocytes	Kuper & Purschke (2001: Table 1)
Lacydonia	?, ?		
Lumbrinereidae	?, metanephridia	?	Goodrich (1945)
Magelona	protonephridia, ?	terminal cell monoc.	Bartolomaeus (1995a)
Maldanidae	?, metanephridia	podocytes	Pilgrim (1978), Bartolomaeus (unpubl.)
Nautineliidae			

Continued on p. 143

Table 1. (Continued)

Taxon	Head kidney, segmental nephridia	Filtration structure	Author
Nephtyidae	?, protonephridia	solenocytes	Brandenburg (1966), Clark (1956), Goodrich (1945)
Nereididae	?, metanephridia	podocytes	Goodrich (1945), Nakao (1974), Smith (1984)
Nerillidae	?, metanephridia or protonephridia		Jouin (1967), Saphonov & Tzetlin (1994)
Oenonidae	?, ?		
Onuphidae	?, metanephridia	?	Goodrich (1945)
Opheliidae	?, protonephridia, metanephridia	?	Hartmann-Schröder (1958), Brown (1938), McConnaughey & Fox (1945), Bartolomaeus (1993b)
Orbiniidae	protonephridia, metanephridia	terminal cell multic.	Bartolomaeus (1998), Eisig (1914), Goodrich (1945)
Oweniidae	protonephridia, metanephridia	terminal cell monoc. and podocyte-like	Smith et al. (1987), Gardiner(1979)
Paralacydonia	?, ?		
Paraonidae	?, protonephridia, metanephridia	?	Strelzov (1979)
Pectinaridae	protonephridia, metanephridia	terminal cell multic.	Bartolomaeus (1995a), Hessle (1917)
Pholoidae	?, metanephridia		Bartolomaeus (1999)
Phyllodocidae	protonephridia, protonephridia	solenocytes, solenocytes	Bartolomaeus (1989), Bartolomaeus (1999), Goodrich (1945), Smith & Ruppert (1988)
Pilargidae	? ?		
Pisionidae	?, protonephridia	solenocytes	Messenger et al. (1985), Smith & Ruppert (1988), Stecher (1968)
Poecilochaetus	metanephridia, metanephridia	podocytes	Bartolomaeus (un publ.), Allen (1905)
Poeobius	?, metanephridia		Robbins (1965)
Polygordiidae	protonephridia, metanephridia	terminal cell multic.	Smith & Ruppert (1988), Goodrich (1900, 1945)
Polynoidae	protonephridia, metanephridia	terminal cell monoc.,	Holborow (1971), Goodrich (1945), Bartolomaeus (1999)
Protodrilidae	?, protonephridia	terminal cell multic., podocytes	Nordheim & Schrader (1994)
Psammodrilidae	?, ?		
Questa	?, nephridia		Giere & Erséus (1998)
Sabellariidae	protonephridia, metanephridia	terminal cell monoc., podocytes	Smith & Ruppert (1988), Meyer (1887), Dehorne (1952), Smith (1986)
Sabellidae	?, metanephridia	podocytes	Goodrich (1945), Orrhage (1980), Bartolomaeus (1993b), Koechlin (1966), Smith & Ruppert (1988)
Scalibregmatidae	?, metanephridia	?	Dehorne & Dehorne (1913), Goodrich (1945)
Serpulidae	protonephridia, metanephridia	terminal cell multic., podocytes	Pemerl (1965), Wessing & Polenz (1974); Goodrich (1945), Bartolomaeus (1993b)
Siboglinidae (Pogonophora)	?, protonephridia or metanephridia	terminal cell monoc., ?	Southward (1993), Gardiner & Jones (1993), Schulze 2001b, Southward et al. (this volume)
Sigalionidae	?, metanephridia	?	Goodrich (1945)
Sphaerodoridae	?, protonephridia	solenocytes	Kuper & Purschke (2001)

Continued on p. 144

Table 1. (Continued)

Taxon	Head kidney, segmental nephridia	Filtration structure	Author
Sphinter	? ?		
Spionida	metanephridia, metanephridia	podocytes	Bartolomaeus (unpubl.), Goodrich (1945), Orrhage (1964), Bartolomaeus (1993b), Rice (1980)
Sternaspis	?, metanephridia		Vejdovsky (1882); Goodrich (1945)
Syllidae	protonephridia, protonephridia or metanephridia	solenocytes multic., solenocytes multic.	Bartolomaeus (1993a); Goodrich (1945), Smith (1992), Bührmann et al. (1996), Kuper & Westheide (1997), Kuper & Bührmann (1999), Kuper (2001)
Terebellidae	protonephridia, metanephridia	terminal cell multic., podocytes	Heimler (1981, 1983, 1988), Hessle (1917), Goodrich (1945), Smith (1992), Bartolomaeus (1993b)
Tomopteridae	?, metanephridia plus	solenocytes multic.	Smith & Ruppert (1988), Bartolomaeus (1997)
Trochochaeta	?, metanephridia	?	Orrhage (1964)
Typhloscolecidae	?, protonephridia	?	Smith & Ruppert (1988)
Uncispionidae	?, ?	–	

perikaryon a ciliary element and a microvillar element. A large number of different vesicles is characterstic of its cytoplasm and indicates different cytotic processes. In annelids, the nephridiopore cell is not a modified epidermal cell, as already indicated by its position. It can be discriminated from the duct cell by its delayed development; from the epidermal cells by the fact that the nephridiopore cell never sheds a cuticle. On the other hand, the nephridiopore is part of the epidermal layer as it is connected to adjacent epidermal cells by adherens junctions (Fig. 4h). Being an intermediate between epidermal cell and duct cell, the nephridiopore cell can easily be discriminated from both kinds of cells without knowing its formation in detail.

Metanephridial system

As mentioned, the term "metanephridial system" is a functional one. This excretory system is composed of two different and spatially separated substructures, the podocytes and the metanephridium. In annelids the podocytes generally rest on the coelomic side of the perivascular ECM (Fig. 2b). Podocytes consist of a perikaryon and a flat peripheral part that is perforated by meandering slits (Fig. 1e). If there are two or more podocytes they will interdigitate and will be connected by adherens junctions to each other and to adjacent peritoneal cells or epithelio-muscle cells. The meandering slits are bridged by additional extracellular material which represents the actual site of filtration. Thus, the slashed periphery of the podocytes is the supporting structure of the filtration barrier and is functionally comparable to the filter of the terminal cell. But, in contrast to the latter, it is never connected to duct cells. This is an important criterion to distinguish the podocyte from terminal cells.

The metanephridium consist of two substructures, the ciliated funnel and the nephridial duct (Fig. 2a and 6d). The ciliated funnel is composed of non-muscular ciliated cells only. These rest on the septal matrix and are composed of three different substructures: the perikaryon, a microvilli and a ciliary element; no structure offset from the perikaryon is ever found. Long, interconnected rootlets anchor the ciliary axoneme to the funnel cell (Fig. 6a). Functionally, the funnel cells also serve in releasing the genital products from the coelom during maturity. The nephridial duct consists of non-muscular cells, being composed of three substructures, the perikaryon, a ciliary and a microvilli element. A large number of different vesicles within the duct cells indicate endo-, exo-

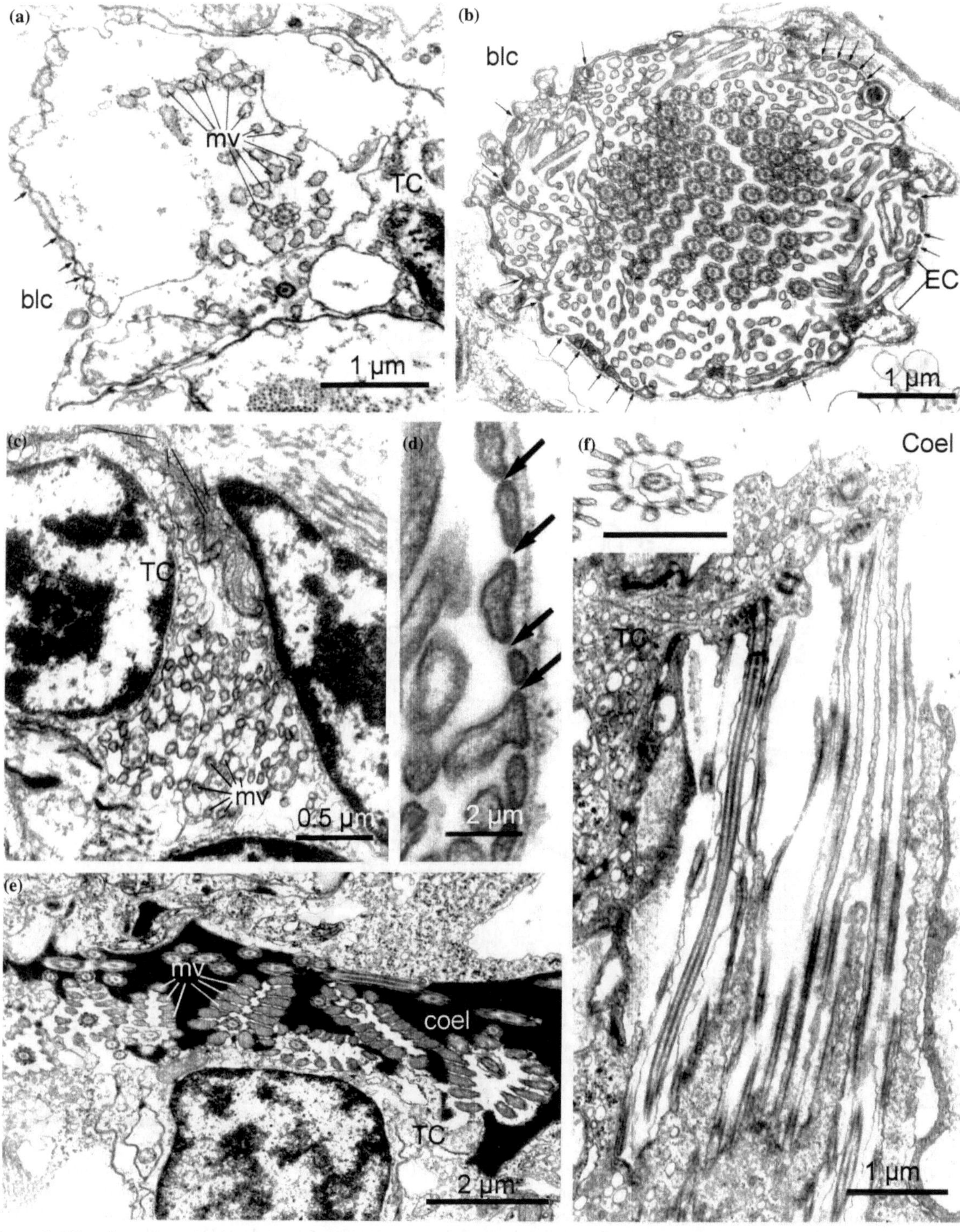

Figure 3. Filtration structures in protonephridia. (a–d) Head kidneys, (e, f) Segmental protonephridia. (a) *Magelona mirabilis,* larva. Terminal cell (TC), cross section. A filter surrounds central cilium plus its circumciliary microvilli. (b) *Spirorbis spirorbis* (Serpulidae), larva. Terminal cell, cross section. Central ciliary flame is surrounded by microvilli and the filter. (Small arrows mark ECM) (c) *Autolytus prolifer* (Syllidae), larva, Terminal cell, cross section. A filter is absent (d) *Spirorbis spirorbis* (Serpulidae), larva, Filtration barrier (arrows). (e) *Glycera alba* (Glyceridae). Solenocytes, cross section. (f) *Tomopteris helgolandica* (Tomopteridae). Multiciliated solenocyte, longitudinal section (inset: cross section). *blc* blastocoel, *coel* coelom, *ECM* extracellular matrix, *mv* microvilli.

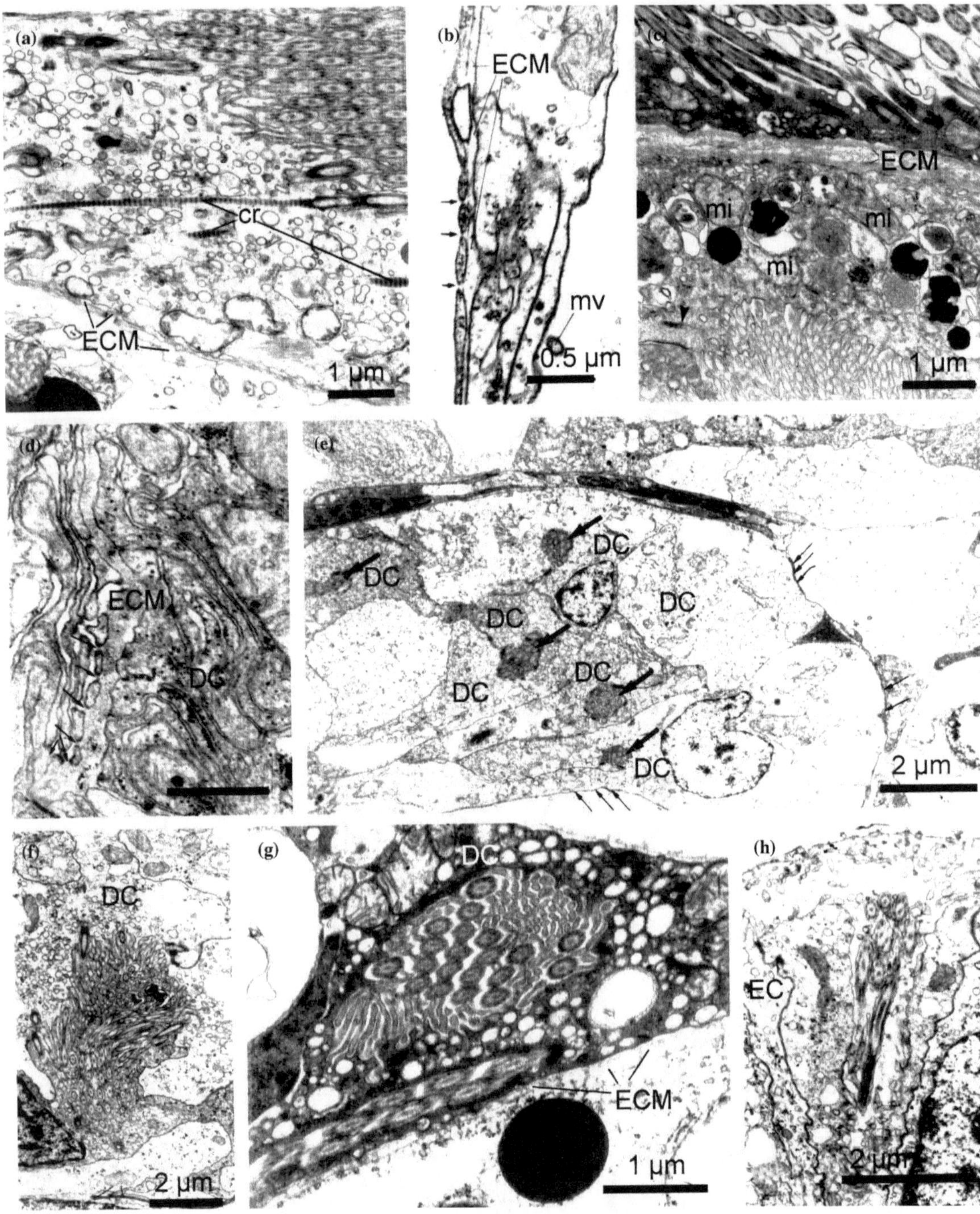

Figure 4. Modifying structures: duct and nephridiopore. (a) *Eulalia viridis* (Phyllodocidae). Ciliated duct cell. (b) *Pectinaria koreni* (Pectinariidae). Duct with podocytes resting on its coelomic face (arrows mark filtration barrier). (c, d) *Fabricia sabella* (Sabellidae). Aciliated duct and ciliated funnel cells. (d) Podocytes resting on the coelomic face of the duct. (e) *Nereis diversicolor* (Nereididae). Cross-sectioned convoluted duct (Arrows mark cilia of the duct) (f, h) *Glycera alba* (Glyceridae). Duct cells apposed to darker nephridiopore cell and (h) nephridiopore. (g) *Spirorbis spirorbis* (Serpulidae), larva, duct cell of the head kidney.

and transcytosis, as in protonephridial duct cells (Fig. 4c, e). There are a few rare reports describing the nephridiopore cells. These are almost identical to duct cells, but stain more electron-densely (Fig. 9c).

Funnel cells and duct cells can be distinguished by the apical network of interconnected ciliary rootlets, while the duct cells can be identified by their large content of vesicles (Fig. 4c and 6d). Nevertheless, identification of the funnel cells can only be done from the structural context, as a funnel is always connected to a duct. This additional criterion is needed as multiciliated peritoneal cells which are found in some annelids are structurally identical to funnel cells and also possess a strong network of interdigitating ciliary rootlets. A further criterion to distinguish funnel cells from such peritoneal cells is the way they are formed. Funnel cells always differentiate from the same anlage as the duct cells (Fig. 7a–d).

Head kidneys in Annelids

In annelids protonephridia generally precede metanephridial systems during ontogenesis and are characteristic of the annelid larva. Nevertheless, in certain species protonephridia prevail in all post-larval stages. The trochophore larva of annelids posseses a pair of nephridia called head kidneys by Hatschek (1878, 1886).

Although Hatschek (1886) used the term for the nephridia of the larva and thus restricted it to those annelids that possess a larva, we will use the term for the first pair of nephridia that is differentiated during development. This organ, however, should be situated anteriorly and closely behind the larval eyes. This term is thus defined by its course of development and its position. Today we know that head kidneys are not restricted to planktotrophic larva, which are usually translucent. Their nephridia can be observed in the living animal under the light microscope. Head kidneys are also described from lecithotrophic larvae, where they could only be detected on the electron microscopical level. They are, however, completely described in only a very few taxa (Table 2). Except in the spionids *Pygospio elegans* (Schlötzer-Schrehardt, 1992) and *P. ciliata* (Bartolomaeus, unpublished) and the poecilochaetid *Poecilochaetus serpens* (Allen, 1905; Bartolomaeus, unpublished), all species studied cell possess protonephridia that are composed of the above-mentioned substructures.

In all species studied thus far, some micrometres caudal to the eyes head kidneys have been found, generally on the level of the larval mouth. This is in accordance with the definition of the term. These organs are composed of at least three cells: one terminal cell, one duct cell and one nephridiopore cell. As such three-celled organs, they are described in *Scoloplos armiger* (Bartolomaeus, 1998), *Spirorbis spirorbis* (Bartolomaeus, 1993b), *Magelona mirabilis* (Bartolomaeus, 1995a) and *Chaetopterus variopedatus* (Bonch-Bruewich & Malakhov, 1986). In certain species the number of terminal cells and duct cell may be increased, as in *Pectinaria auricoma* (Bartolomaeus, 1995a), *Phyllodoce mucosa* (Bartolomaeus, 1989) or *Autolytus prolifer* (Bartolomaeus, 1993a). When compared between the different species, the duct cells and the nephridiopore cell merely differ in their ciliation, which may either lack or differ in number of cilia, i.e. being mono- or multiciliated (Table 2). Compared to this, a high structural diversity of terminal cells is found in the different species (Fig. 5a–e).

In the serpulid species studied thus far (Table 2) and in *Scoloplos armiger,* the terminal cell bears a bundle of cilia, sometimes referred to as a ciliary flame that extends into the small compartment formed by the cylindrical filter (Fig. 5c). Several short microvilli surround the ciliary flame. In serpulids the filter and the adjacent duct cell interdigitate by small processes (Fig. 3b). This kind of adhesion is sometimes referred to as a weir. This structure, composed of rods that alternately protrude from the terminal cell and the duct cell, is also characteristic of certain plathelminthes (Bartolomaeus & Ax, 1992) and is regarded to be a derived condition. The term weir should not be applied to the connection between terminal cell and duct cell in the studied serpulid head kidneys.

In the other annelid head kidneys studied, each cilium possesses a ring of strong, elongated microvilli surrounding it (Fig. 3a, c). This is characteristic of the protonephridia in *Magelona mirabilis*, in *Pectinaria auricoma, Polygordius* sp. as well as in the head kidneys of several phyllodocidan species. *Pectinaria auricoma* certainly

Table 2. Cellular and subsellular elements in ultrastructural studied head kidneys in Polychaeata (Annelida)

Taxon	Terminal cell				Duct cell			Nephridiopore cell			Reference
	Perikarya	Ciliary element	Microvillar element	Filter	Perikarya	Ciliary element	Microvillar element	Perikarya	Ciliary element	Microvillar element	
Oweniidae											
Owenia fusiformis	?	1	several	podocyte-like	several	1	several	?	?	?	Smith et al. (1987)
Pectinariidae											
Pectinaria auricoma	2	several	10 per cilium	+	2	several	several	1	–	several	Bartolomaeus (1995a)
Phyllodocidae											
Phyllodoce mucosa	2–3	1	15 per cilium	–	1	several	several	1	several	several	Bartolomaeus (1989)
Polygoriidae				–							
Polygordius sp.	?	several	several	+, per 1–2 cilia	?	?	?	?	?	?	Smith & Ruppert (1988)
Polynoidae											
Harmothoe imbricata	1	1	15 per cilium	+	?	several	several	?	?	?	Holborow (1971)
Sabellariidae											
Sabellaria cementarium	1–2	1	several	+	1–4	?	?	?	?	?	Smith & Ruppert (1988)
Serpulidae											
Serpula vermicularis	1	several	several	+	?	several	?	?	?	?	Pemerl (1965)
Pomatoceros triqueter	1	several	several	+	1	several	several	–	–	–	Wessing & Polenz (1974)
Spirorbis spirorbis	1	several	several	+	1	several	several	1 several	several	several	Bartolomaeus (unpubl.)
Syllidae											
Autolytus prolifer	1	several	10 per cilium	–	2	several	several	1	several	few	Bartolomaeus (1993a)
Terebellidae											
Lanice conchilega	2	several	several	?	2	several	several	1	?	?	Heimler (1981, 1983, 1988)

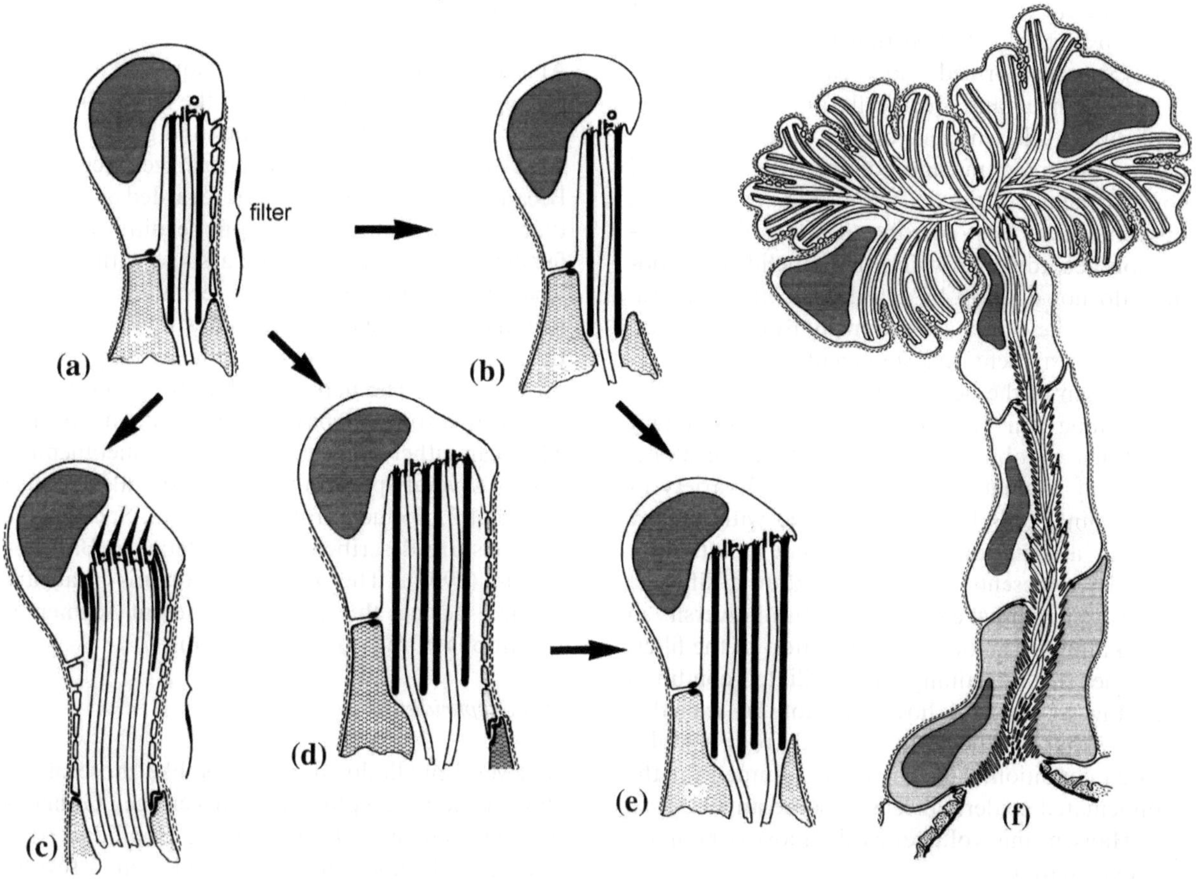

Figure 5. Terminal cells in head kidneys and segmental nephridia and their possible evolutionary relationships, if monociliarity is assumed to represent a primary condition. (a) Monociliated terminal cell as found in *Magelona mirabilis* (Bartolomaeus, 1995a), *Harmothoe imbricata* (Holborow, 1971), *Sabellaria cementarium* (Ruppert & Smith, 1988). (b) Solenocytes as found in *Phyllodoce mucosa* head kidneys (Bartolomaeus, 1989) and segmental nephridia of further Phyllodicida (Table 1). (c) Multiciliated flame cell as known from different serpulid larva (Table 2). (d) Multiciliated terminal cell with a ring of microvilli surrounding each cilium and a common filter as found in Polygordius sp. and *Pectinaria koreni* (f) head kidney (Smith & Ruppert, 1988; Bartolomaeus, 1995) and *Myzostoma cirriferum* segmental protonephridia (Pietsch & Westheide, 1987). (e) Multiciliated solenocyte like in *Autolytus prolifer* head kidney (Bartolomaeus, 1993a), *Hesionides arenaria* and *Tomopteris helgolandica* segmental protonephridia (Westheide, 1986; Bartolomaeus, 1997).

shows the most complex structure as the compartment inside the terminal cell is extremely ramified (Fig. 5f). Each ramus contains a single cilium surrounded by a ring of 10 microvilli. The lateral wall of the compartment is perforated by several slits covered by diaphragms representing the filtration barrier. Each cilium of the multiciliated terminal cell of *Polygordius* sp. is also surrounded by a ring of microvilli (Smith & Ruppert, 1988) (Fig. 5d). In certain species of the Phyllodocidae a filter is lacking, but the circumciliary ring of strong and elongated microvilli is retained and serves as supporting structure for the filtration barrier (Fig. 5b, e). Such terminal cells are called solenocytes, if they are monociliated. If these cells are multiciliated a ring of microvilli surrounds each cilium. A flame of several cilia within a ring of microvilli, but without a filter has never been described.

Only little is known of the formation of these head kidneys. According to older cell lineage studies, the head kidneys are formed by the cells $3c^{22}$ and $3d^{22}$ that sank into deeper cell layers during early development (Woltereck, 1904, 1905).

More recent electron microscopical studies of *P. mucosa* clearly showed that the protonephridium is actually formed underneath the epidermis. When terminal and duct cells are complete the nephropore cell, which is formed with some delay, pierces the epidermis and connects the nephridial compartment to the exterior (Fig. 7e, f). This observation at least corroborates the older information according to which the annelid head kidneys do not develop by ectodermal invagination. However, these data are extremely isolated and do not allow any definitive statement on the ontogenetic origin of the head kidneys.

Nothing can be said about the evolutionary directions by observing a single structure alone. Thus, any evaluation of the structural variety of the terminal cells is hardly possible without any primary assumption. If one assumes that monociliarity represents a primary condition, different pathways might have caused structural diversity in head kidneys. The one is by reduction of the filter, the other one by gaining a multiciliated condition (see Fig. 5). It can, however, not be excluded *a priori* that monociliarity evolved from a multiciliated condition. This has to be assumed for the monociliated epidermal cells in some annelid taxa (see Hausen, this volume) as the most parsimonious explanation.

Nevertheless, some concluding remarks can be made about the head kidneys. (1) These organs must be assumed for the annelid ground pattern as those excretory organs which are differentiated first during ontogenesis, irrespective of whether there was a trochophore or not. (2) Certain head kidneys are organized in ways that have been hypothesized to represent the primary condition in Bilateria (Bartolomaeus & Ax, 1992). Provided these hypotheses match the course of evolution, the structure of these head kidneys must represent the primary condition in Annelida. (3) Solenocytes and terminal cells are regarded as homologous. Reduction of the filter and adoption of its function by the microvilli is characteristic of certain Phyllodocida and also characteristic of those phyllodocidan species that possess segmental protonephridia. (4) The head kidneys of the spionid and poecilochaetid species studied thus far are metanephridia (Allen, 1905; Schlötzer-Schrehardt, 1992, own unpublished results). This observation deserves detailed analysis in the future.

Segmental nephridia

At a certain stage of annelid development a caudal, prepygidal meristematic ring of the larva, the grow zone, increases mitotic activities and gives rise to a large number of identically organized segments. Because of this, annelids can be regarded as animals that have two ontogeneses, one leading from the fertilized egg to the larva or a stage with a quiet terminal growth zone and the second leading to a metamerically organized animal with several identical segments generated by the activity of the growth zone. The nephridia that are generated by this growth zone will be called segmental nephridia. They can either be protonephridia or metanephridia. In several species these nephridia also release the genital products from the coelom. Such modifications are described in more detail in Bartolomaeus (1999). The nephridial organs that are formed during this process represent segmental repetitions of the same information.

Protonephridia

In several phyllodocidan species, like those of the Phyllodicidae, Nephtyidae, Glyceridae, Goniadidae, Pisionidae, Sphaerodoridae, Isopilidae Alciopidae, Typhloscolecidae and in several interstitially living taxa, protonephridia can be found in each body segment (see Table 1). Their terminal cells are always solenocytes (Fig. 3e). In the mentioned phyllodocidan taxa the terminal cells are monociliated. At least in those taxa, the species of which have a larger body size, ciliated cells cover the duct and close it completely towards the coelom, so that each cilium of the terminal cells plus its ring of microvilli enters the duct separately (Goodrich, 1945; Smith & Ruppert, 1988; Bartolomaeus, 1989; Smith, 1992) (Fig. 8e). According to its structure these cells must be formed in a late stage of nephridial development (Bartolomaeus, 1989 for *Phyllodoce mucosa*). This assumption was confirmed by studying the nephridial development in glycerids and nephtyids (Bartolomaeus, 1993b). Smith & Ruppert (1988) interpreted this cell as an inverted duct cell, which has not yet been confirmed by studies of the nephridial formation. Nevertheless, it seems unlikely that this cell is a modified coelothelial cell because it is directly connected to duct cells and not separated from them by a matrix, like other cells of the coelomic lining are. Further studies into nephridial formation should provide a clear solution.

There are also records from segmental protonephridia with multiciliated solenocytes. These have been described for *Tomopteris helgolandica* (Tomopteridae) and *Hesionides arenaria* (Hesionidae) (Westheide, 1986; Bartolomaeus, 1997) (Fig. 3f). Here, each protonephridium possess a single terminal cell that bears several cilia, each surrounded by a ring of 10 strong microvilli. In *T. helgolandica* a matrix connects the microvilli thus indicating ultrafiltration of coelomic fluid. Like in the taxa with monociliated solenocytes, each cilium plus its ring of microvilli enters the duct separately. After a few micrometers these small compartments join to form a larger ductule which finally merges with other such ductules to form the nephridial duct. Both lack such an inverted duct cell described above. In *T. helgolandica* a large number of cilia emanates from the solenocyte and extends into the coelom.

The multiciliated solenocyte of *T. helgolandica* is attached laterally to the funnel of a metanephridium. Although one duct cell provides a short lateral duct which seems to receive the ultrafiltrate of the terminal cell, this duct is not connected to the metanephridial duct but ends blindly. Numerous vesicles inside the duct cell indicate transcytotic processes (Bartolomaeus, 1997).

The duct of all segmental protonephridia consists of multiciliated cells that are connected by apical adherens junctions; basally to them the septate junction hinders the duct fluid from bypassing the duct cells to come into contact with the interstitial fluid. Microvilli always cover the apical surface of the duct cells. The duct cells contain numerous vesicles, lysosomes and multivesiculated bodies that indicate transcytotic processes. Coated pits and coated vesicles indicate receptor mediated endocytosis of some material from the duct. The duct is always composed of several cells. In general, ciliation, number of microvilli and content of organelles do not differ during the course of the duct. The last few duct cells that form the nephridiopore may stain more electron-densely (Bartolomaeus, 1989) (Fig. 4f, h). All ducts start in one segment and open in the preceding one to the exterior.

In some interstitial annelids the protonephridia are quite short, being composed of two to four duct cells and one or two terminal cells. These differ tremendously in structure, ranging from multiciliated solenocytes to terminal cells with a lateral tuft of cilia (Brandenburg, 1970; Clausen, 1986; Westheide, 1985). A row or double row of strengthened microvilli originating from the latter kind of terminal cell and extending deeply into the duct may separate the ciliary flame from the surrounding interstices or body cavity like a fence. These microvilli act as supporting structure for a filtration barrier. They resemble early stages of metanephridial development (Fig. 7c).

Metanephridial systems

The majority of annelid nephridia are metanephridia (Goodrich, 1945; Table 1). Starting with the funnel in one segment they pierce the septum and open to the exterior in the following segment. The internal opening is enlarged and consists of a wide funnel that rests on the septum. The cells of the funnel are generally multiciliated and rest on the septal matrix. Adjacent to the funnel epitheliomuscle cells or peritoneal cell are found. These are connected to the funnel cells by adherens junctions and also rest on the septal matrix. Generally the cilia of the funnel cell contains two ciliary rootlets, one running almost parallel to the cell surface, the other, stronger one crossing the cell obliquely while running basally. These rootlets are interconnected by intermediate filaments and form an apical network within the funnel cell (Fig. 6a, d). Apart from this there is no difference between funnel cells and adjacent duct cells. During reproduction the density of filaments apparently increases. Because the strong bundles of intermediate filaments stain intensely with certain classical colours, the difference between funnel cells and duct cells seems to be greater in histological sections than actually can be confirmed by funnel and duct cells substructures (Bartolomaeus, 1999 for discussion). The duct cells are morphologically equivalent to the protonephridial duct cells (Smith & Ruppert, 1938; Smith, 1992). Like the latter they contain numerous vesicles, lysosomes, multivesiculated body and coated vesicles. All these confirm a high endo-, exo- and transcytotic activity of duct

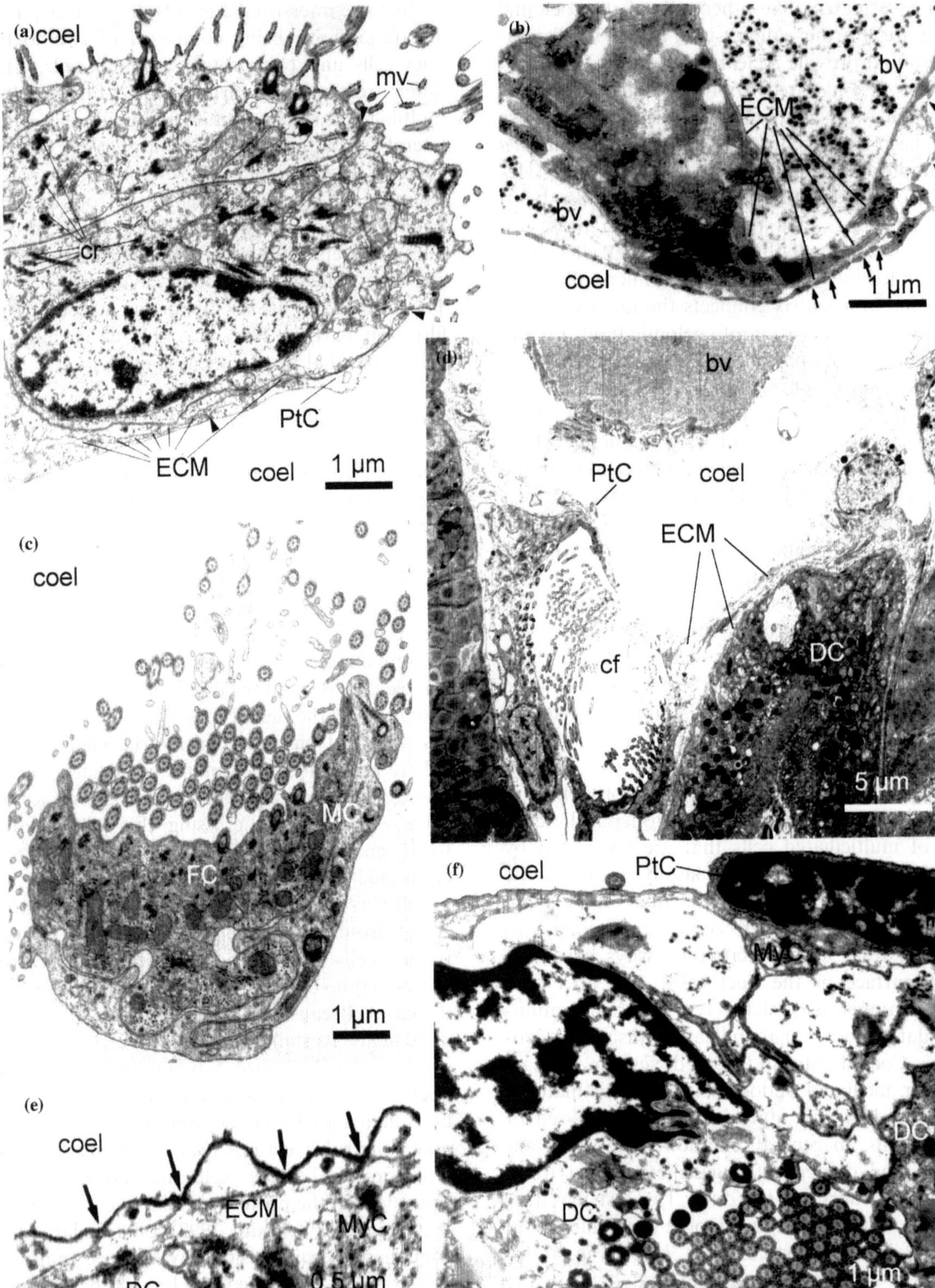

(a)
coel
mv
cr
PtC
ECM
coel
1 µm
(b)
bv
ECM
bv
coel
1 µm
(c)
coel
MC
FC
1 µm
(d)
bv
PtC
coel
ECM
cf
DC
5 µm
(e)
coel
ECM
MyC
DC
0.5 µm
(f)
coel
PtC
MyC
DC
DC
1 µm

Figure 6. Metanephridial systems. (a) *Pectinaria koreni* (Pectinariidae). Upper lip of nephridial funnel (arrowheads mark adherens junctions. (b) *Aeolosoma hemprichi* (Aeolosomatidae). Podocytes lining the perintestinal vessel (arrows mark filtration slits). (c) *Pristina longiseta* (Naididae). Nephridial funnel composed of mantel cell (*MC* and flame cell. (d) *Fabrica sabella (*Sabellidae). Ciliated funnel (*cf*), adjacent duct and blood vessel (*bv*) of the ciliated funnel. (e, f) *Golfingia minuta* (Sipuncula). Podocytes rest on the perinephridial ECM (arrows mark filtration barrier). (f) Ciliated descending branch. Muscle cells (*MyC*) of the nephridial duct are separated by an *ECM*. *coel* coelom, *cr* ciliary rootlets, *DC* duct cell, *ECM* extracellular matrix, *mv* microvilli, *PtC* peritoneocyte.

cells, as predicited by the abovementioned assumption that metanephridia merely modify coelomic fluid (Fig. 4c, e).

Nephridiopore cells have extremely seldom been studied (Bartolomaeus, 1989, 1993b; Kuper, 2001). According to studies into their development, these cells pierce the epidermis while being formed. There is some evidence that their cytoplasm stains more electron-dense than that of the adjacent duct cell, which could hint at a higher protein content and different osmotic affinities. The nucleus is always underneath the level of the epidermis cells, which obviously reflects the way in which the nephropore cell had been formed.

Some information can be obtained from a comparison of the distribution of the metanephridia within a single organism. In all species of the Serpulidae, Sabellidae, Sabellariidae and Siboglinidae (formerly Pogonophora) a single pair of nephridia opens into the first segment (Goodrich, 1945; Schulze, 2001). The duct is rather long and extends caudally for a few segments, U-turns frontally and leads to the exterior of the animal on the dorsal side. In certain Terebellida the metanephridia are restricted to a few anterior segments (Smith, 1992). Some of them may share a common duct (Goodrich, 1945). In Pectinariidae, Serpulidae and Sabellidae the metanephridium is lined by podocytes that rest on the coelomic side of the perinephridial matrix (Bartolomaeus, 1993b) (Fig. 4b, d). Here, they cannot mediate a selective filtration or bypass different fluid-filled compartments, because the perinephridial matrix never houses blood lacunae.

Podocytes surrounding the metanephridial duct are absolutely surprising, as in these taxa podocytes are also found on the perivascular ECM, as in all other species with metanephridia studied until now (Smith & Ruppert, 1988; Bartolomaeus, 1993b). Podocytes normally rest on the perivascular matrix where they span and stabilize the matrix to prevent possible rupture by the filtration pressure (Figs 1e and 6b, e). They allow a selective fluid transfer between the blood and coelom and are generally believed to adopt the function of the protonephridial terminal insofar as they eliminate metabolic wastes from the blood by ultrafiltration. Functionally, the coelomic cavity of Annelida can thus be regarded as large nephridial compartment. Podocytes are polar cells, connected by apical adherens junctions to neighbouring cells. They always possess a centriole, sometimes a small vestigial cilium and sometimes myofibrils. The latter has until now merely seen in serpulid and sabellid species (Bartolomaeus, 1993b). Podocytes were found resting on the coelomic side of the main dorsal and ventral vessel as well as on the coelomic sides of the intraseptal vessels (Fig 1e). Podocytes are generally lacking in those species that possess protonephridial excretory organs. *Protodrilus rubropharyngeus* is the only known exception in this respect. This interstitial species has segmental protonephridia; podocytes are found resting on the coelomic side of the septal matrix (von Nordheim & Schrader, 1994).

Development

Frontal to the anus a meristematic region remains until the animal reaches its final size. This region shows an enormous mitotic activity and gives rise to a certain number of identical segments. All stages of formation of these segments and their organs can be found along a caudo–frontal gradient. Most of our knowledge of the nephridial development results from ultrastructural studies into the structure of different stages of nephridial formation following a series of sections from the caudal meristematic ring to segments containing large coelomic cavities (see Goodrich, 1945; Bartolomaeus, 1993b; Bartolomaeus, 1999).

The earliest stage of nephridial formation that can be discriminated from the surrounding tissue

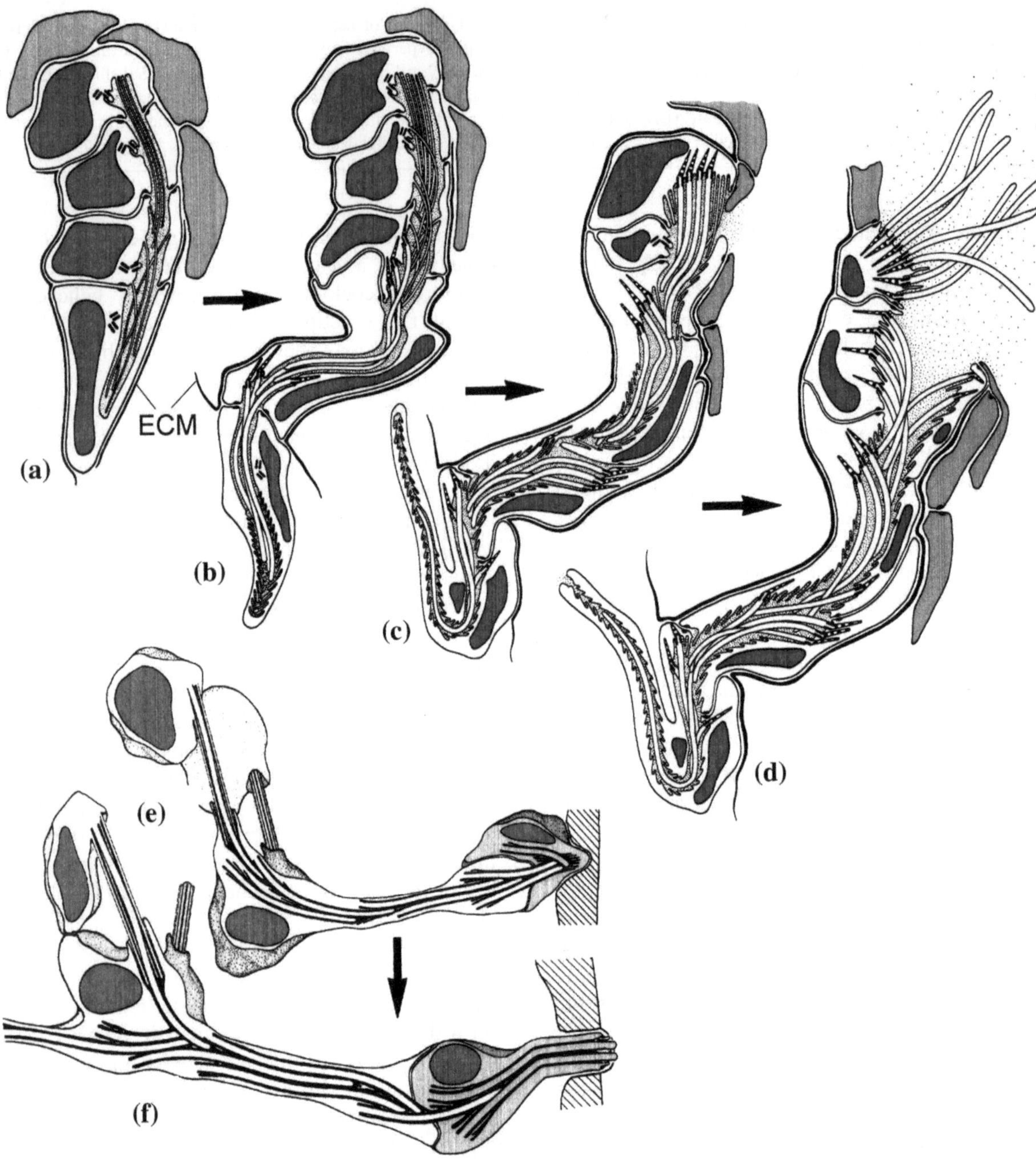

Figure 7. Nephridial development. (a–c) Metanephridia in *Ophelia rathkei* (Opheliidae), (a) Earliest recognizable anlage of the metanephridium. (b) Ciliogenesis and enlargement of the anlage, penetration of the subepidermal ECM (extracellular matrix). (c) Addition of cell, stronger ciliation in the proximal cells of the anlage, onset of fluid accumulation in the prospective coelomic cavity. (d) Movement of the surrounding cells caused by fluid accumulation. Proximal duct cell cilia face the coelom and form the funnel. (e, f) Head kidneys in *Phyllodoce mucosa* (Phyllodocidae). Nephridium is formed in deeper cell layers. Nephridopore cell forms at last (e) and finally pierces the epidermis (f).

consists of 3–4 cells that surround a small compartment (Fig. 7a). These cells rest on a very thin and incomplete matrix and show a clear polarity with apical adherens junctions and small microvilli extending into the compartment. Earlier stages do not exhibit structures that allow recognising them

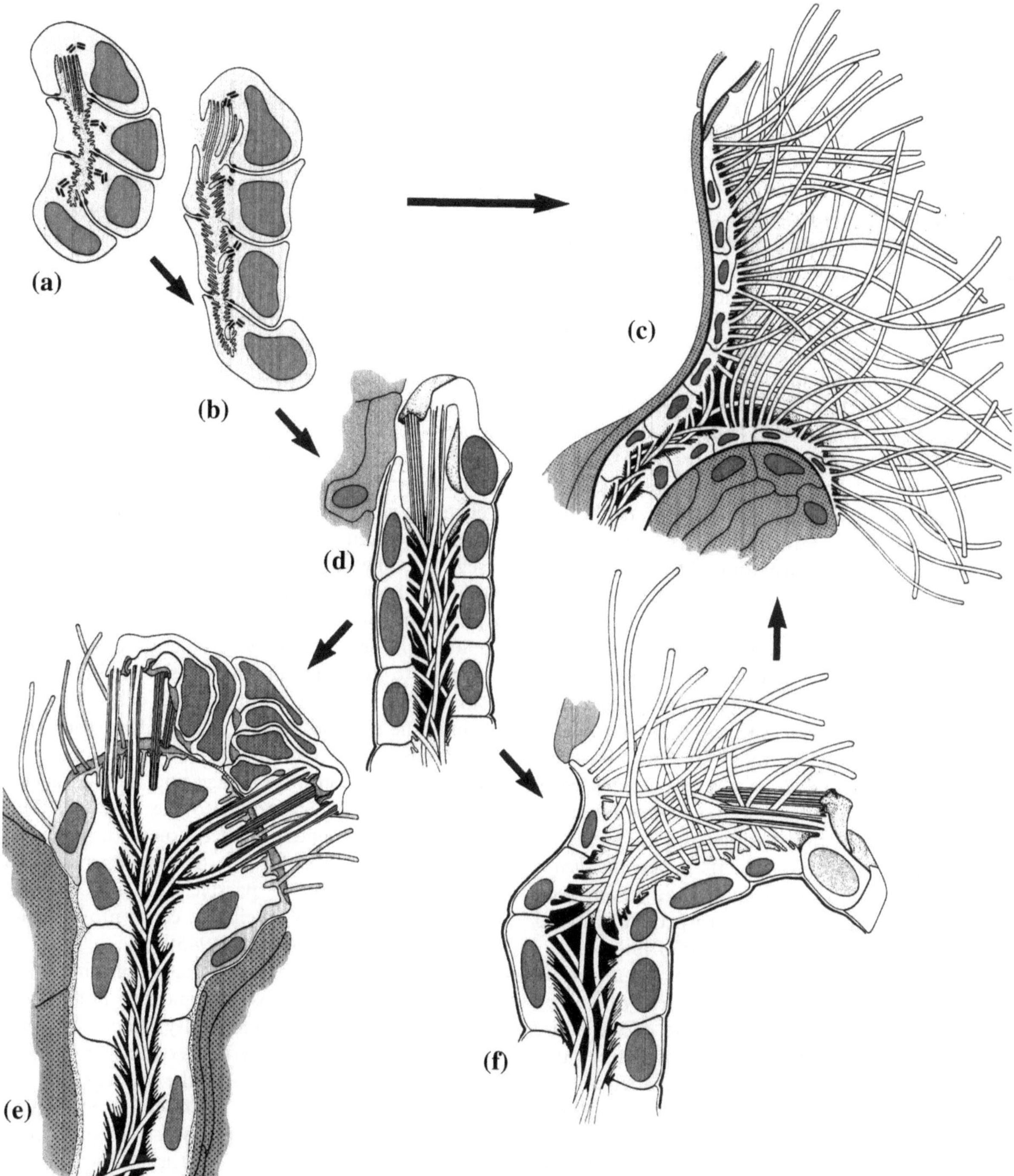

Figure 8. General scheme of developmental pathways that lead to different nephridial organs. (a) Earliest recognizable anlage. (b) Ciliogenesis starts in the proximal cells of the anlage. These cilia are lateral to the perikarya and are not completely surrounded by the cells of the anlage. These cells either become the ciliated funnels of the metanephridium during coelomogenesis (c) or they differentiate into solenocytes (d). These either move aside when the proximal section of the duct is passively opened (f) or a further cell covers the proximal section of the duct so that a protonephridium with solenocytes is formed (e). If the temporarily formed solenocytes (d) move aside (f) they degenerate during formation of the funnel. Truncation and modification cause nephridial diversity in polychaetes.

as nephridial anlage. While differentiation of the surrounding cell into muscle cells proceeds, the cells of the nephridial anlage start to form cilia (Fig. 7b). The distal duct cells elongate and the distal-most one pierces the epidermis so that the compartment the nephridial cells surrounded during development is now connected to the exterior (Fig. 7c). The number of nephridial cells increases, but the source of these cells is not clear. Although we have studied nephridial development in several taxa, there was only clear evidence in regenerating segments that additional nephridial cells result from mitosis of presumptive, but yet weakly differentiated duct cells.

Formation of cilia in the cells of the nephridial anlage always starts with a single cilium that forms a short, orimental axoneme (Fig. 8a, b). Its basal structures consist of a basal body with a short lateral basal foot and an accessory centriole lying rectangular to the basal body. In those nephridial cells that will become duct cells, this monociliated stage is passed during formation of several cilia per cell. In some species, however, the proximal nephridial cells remain monociliated. A ring of strong, elongated microvilli is formed and surrounds the cilium. These cells are typical solenocytes (Fig. 8d). This developmental path is characteristic of all species with segmental protonephridia studied thus far (see Bartolomaeus, 1999) as well as of the polynoid species *Harmothoe sarsi* and the pholoid *Pholoe inornata*. In all these species, the proximal and monociliated cells will become solenocytes. In *T. helgolandica* a monociliated intermediate stage has never been observed; here the cilia of the prospective terminal cell are each surrounded by a ring of microvilli.

Our studies into the formation of the nephridia revealed that the developmental pathway taken by the proximal cell influences the fate of the entire anlage and is responsible for its prospective function as protonephridia or as metanephridium (Bartolomaeus & Ax, 1992; Bartolomaeus, 1993b, 1997, 1999) (see in Fig. 8d to e vs. d to f and then c). We therefore will take a close look at the development of these cells. In each nephridial anlage there is some asymmetry in the organisation and attachment of the most proximal cells. In protonephridia the terminal cell is directly connected to the adjacent duct cell. The cilium plus its ring of microvilli are always lateral to the terminal cell. Surrounding muscle cells and cells that will become coelomic lining cells are apposed to both structures (Figs. 7c and 8b). The same occurs in species where multiciliated proximal cells are found in the nephridial anlage. Except for *T. helgolandica* and presumably some other species with multiciliated solenocytes not studied yet in terms of their nephridial development, these cells possess a flame of cilia extending into the duct. This bundle is lateral to the perikaryon and also bears microvilli that may surround the bundle. While the perikaryon is connected to the adjacent cells of the anlage, cilia and microvilli are next to cells of the prospective coelomic lining. Thus, irrespective of whether the anlage will differentiate into a protonephridium or a metanephridium, cilia and microvilli of the proximal cells are not surrounded by the anlage, unlike those of the adjacent cells of the nephridial anlage. Instead, these subcellular elements run between cells that will later form the coelomic lining during formation of additional segments. These proximal cells will either form the ciliated funnel or the terminal cells of the nephridium and are not completely surrounded by a matrix. At least the most proximal ones are connected to the surrounding muscle cells. Finally fluid accumulation among those cells that surround the nephridial anlage causes formation of a large coelomic cavity by schizocoely. The proximal cells of the nephridial anlage and their cilia and microvilli become exposed to the coelomic fluid and follow the general obliteration of the formerly compact tissue. In the case of a metanephridium the funnel is opened passively (Figs. 7 and 8). There is no grafting of ciliated cells onto the duct as proposed by Goodrich (1945).

In case of the protonephridia the proximal cells largely retain their position and are not moved apart while the coelomic cavity widens by fluid accumulation. In the small interstitial species this is linked to the smallness of the animals and the small coelomic cavities. In larger species with protonephridia, like nephthyids, phyllodocids, glycerids and others there is at least one cell that covers the proximal duct cells (Goodrich, 1945; Bartolomaeus, 1989) and prevents it from being opened passively. The solenocytes pierce this cell with the cilium and the circumciliary ring of microvilli to enter the duct. At least two species are known that do not possess such a covering cell,

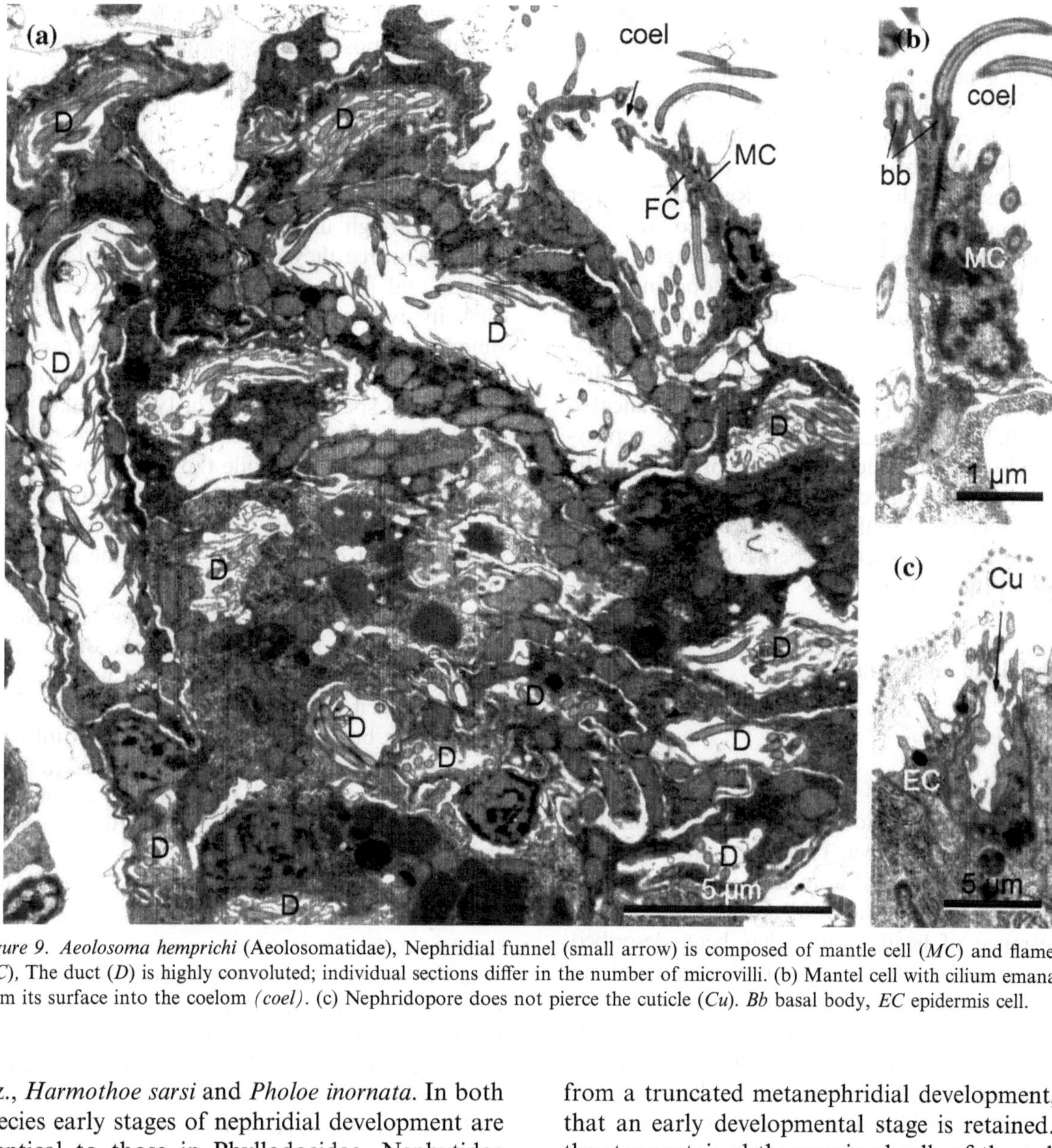

Figure 9. Aeolosoma hemprichi (Aeolosomatidae), Nephridial funnel (small arrow) is composed of mantle cell (*MC*) and flame cell (*FC*), The duct (*D*) is highly convoluted; individual sections differ in the number of microvilli. (b) Mantel cell with cilium emanating from its surface into the coelom *(coel)*. (c) Nephridopore does not pierce the cuticle (*Cu*). *Bb* basal body, *EC* epidermis cell.

viz., *Harmothoe sarsi* and *Pholoe inornata*. In both species early stages of nephridial development are identical to those in Phyllodocidae, Nephytidae and Glyceridae, but when the coelom starts to expand by fluid accumulation the solenocytes move aside so that the apices of the adjacent duct cells are exposed to the coelomic fluid by forming a funnel. During further development their solenocytes degenerate and a typical metanephridium results from this development (Fig. 8d, f, c).

Currently we assume that protonephridia in polychaetes largely depend on the size of the coelom and the size of the animals. In small animals like interstitial species protonephridia may result from a truncated metanephridial development, so that an early developmental stage is retained. In the stage retained the proximal cells of the anlage bear a lateral bundle of cilia surrounded by a ring of microvilli. Structurally, this resembles a protonephridium without a filter (see for instance Clausen (1986) for *Microphthalmus ephippiophorus* or Brandenburg (1970) for *Dinophilus gyrociliatus*). Comparative studies, however, reveal that it is identical to a developmental stage of a metanephridium. Such an interpretation is fallible, because within a single species such segmental protonephridia resulting from a truncated metanephridial development should clearly differ from the head kidney. Any evolutionary evaluation, however, has to be done in a broad comparison with the general character distribution within the polychaetes. Nevertheless, the developmental studies reviewed above clearly indicate that segmental protonephridia within the polychaetes are not necessarily homologous. Recently, Kuper (2001) showed the enormous variety of terminal sections in syllid segmental organs ranging from small funnels to slit-like terminal openings and lateral ciliary flames that extend into the duct. Several of them clearly resemble stages of metanephridial development so that variety of syllid nephridia can be explained by differently truncated nephridial development. In phyllodocids, nephtyids and glycerids, however, the protonephridia probably result from covering the proximal section of the duct by a multiciliated cells which would hinder a direct connection between nephridial lumen and the coelom.

Nephridia in related taxa

Clitellata

Head kidneys

As a result of a modified ontogenetic process, Clitellata have no larva. There are some secondary larvae in certain hirudinean taxa instead. In some lumbricid, tubificid and hirudinean species the occurrence of head kidneys has been described, which precede the formation of segmental nephridia during development (see Anderson, 1973).

As protonephridia seem to be characteristic of the annelid larva and as they are the first nephridia formed during ontogenesis, one would expect that the first nephridia formed during ontogenesis are also protonephridia in Clitellata, irrespective of whether or not they show a secondary larva. Recent studies into the ultrastructure of the secondary larva of *Erpobdella octoculata*, however, reveal that its nephridia lack a terminal cell, a characteristic essential to apply the term protonephridia to these organs (Quast & Bartolomaeus, 2001). Nevertheless, the organs are blindly closed toward the body cavity. There is some evidence that these organs are identical to the metanephridia of the adult organisms.

Segmental metanephridia and their formation

Clitellata possess segmental metanephridia which have always been described as resulting from a single anlage (Goodrich, 1945). In all non-hirudinean clitellate annelids these organs drain the coelomic cavity. As in polychaetes the organs consist of a preseptal part and a postseptal duct. Although ultrastructural data on these organs are scarce they exhibit a variety of different shapes and forms that concern the extension of duct, its exterior or interior opening and its possible fusion to form complex organs. The preseptal part consists of a small funnel or nephrostome in microdilous species (Goodrich, 1945; Bunke, 1998, 2000) or of a large and complex nephrostome in megadrilous species. While the latter are reported to possess a central ciliated cell and several marginal cells with cilia, the small nephrostomes always consist of an inner ciliary flame and some marginal cilia. Ultrastructural data of *Dero digitata*, *Nais variabilis* and *Pristina longiseta* (Naididae) show that the funnel consists of a single marginal cell called mantel cell that wraps a central flame cell (Bunke, 2000) (Fig. 6c). This cell possesses a bundle of cilia that projects into the duct, while the cilia of the mantel cell extend into the coelom. In Aeolosomatidae the metanephridial funnel is identically organized (Bunke, 1994) (Fig. 9a, b).

Development of this kind of nephridium has recently been studied by Bunke (2003). The earliest stage of development that can clearly be discriminated from the surrounding mesoderm cells consists of three cells. Their fate appears to be fixed insofar as the innermost cell that is next and ventral to the septal vessel differentiates into the mantel cells, while the median cell becomes the flame cell. Only the distalmost cell is the stem cell of the canal and generates duct cells by mitotic activities which concern the stem cells itself and its daughter cells. Ciliogenesis has not been described in detail by Bunke (2003), but his figures (Bunke, 2003: Fig. 2c) show numerous centrioles within the cytoplasm. The same is seen in developing polychaete nephridia where a monociliated stage is passed and multiplied centrioles migrate towards the adlumial cell membrane to induce formation of additional cilia. Bunke (2003) also reports that the ECM separating the duct from adjacent mesodermal cells is incomplete during early nephridial development. This is also the case during early nephridial development in polychaetes. Formation of the septa seems here to be parallel to nephridial development. There is no doubt that metanephridia in Clitellata are mesodermal. This has been shown by cell lineage studies (Weisblat et al., 1984; Kitamura & Shimizu, 2000a,b; Shimizu & Nakamoto, 2001; Shimizu et al., 2001). Due to the position of the anlage, this has also to be assumed for polychaetes.

Echiura

Head kidneys

In trochophore larva of *Echiurus echiurus* possess protonephridia with several monociliated terminal cells (Goodrich, 1945). Such head kidneys are absent in *Urechis caupo* (Newby, 1940), but can be found in fertile dwarf males of *Bonellia viridis* (Schuchert & Rieger, 1990), where they are multiciliated. It has been proposed that these protonephridia evolved secondarily (Schuchert, 1990).

Segmental nephridia and their formation

The larval protonephridia are followed by two different systems which traditionally are assumed to mediate excretion, i.e. paired segmental organs or metanephridia on the level of the ventro-medial large seta and the anal sacs. Ontogenetically the latter originate from a pair of metanephridia situated on either side of the anus. During development these metanephridia acquire additional ciliated funnels and enlarge (Baltzer, 1934). Ultrastructural data of the anal sac funnels show a grid-like arrangement of muscle cells embedded in the ECM between the duct cells and the peritoneum that rests on the coelomic side of the anal sac (Bartolomaeus, unpublished), No such muscle cells are found within the matrix between funnel cells and peritoneum. All nephridial cells are multiciliated with two rootlets per cilium. Duct cells differ from funnel cells in orientation of their rootlets and a larger number of microvilli. The rootlets in funnel cells are partly fused and form a dense intracellular meshwork.

The number of metanephridia differs among Echiura. Some species possess 1–3 pairs of such organs (Storch & Welsch, 1972; Singhal, 1982) while others like *Ikeda taenioides* are reported to possess up to 300 metanephridia (Baltzer, 1934). The number of organs varies even within a single genus (one pair in *Thalassema diaphanes* (Bock, 1942) and 27 in *T. elegans* (Baltzer, 1934)). The germ layer origin of metanephridia and anal sacs is unknown; a mesodermal origin seems likely (Baltzer, 1934).

Sipuncula

Head kidneys have never been described for Sipuncula. The first pair of nephridia that is formed is metanephridial and persists during the entire lifetime. A single pair of metanephridia seems to represent the plesiomorphic condition in Sipuncula, although their number can be increased up to three pairs (Storch & Welsch, 1972). The excretory function of the metanephridia has been shown by Pinson & Ruppert (1988). In *Golfingia minuta* these organs consist of a ciliated funnel that is directly attached to the body wall (Bartolomaeus, unpublished). The funnel cells are multiciliated and directly connected to the non-muscular, extremely flat peritoneum that lines the somatic musculature. This lining wraps the U-shaped duct and rests on the perinephridial matrix. Grid-like arrangements of longitudinal and circular muscles are found all along the course of the nephridium, except for the upper one third of the funnel. No muscle cells can be found here. The muscle cells in the region below the funnel are continuous with the body wall muscles. All muscle cells of the nephridium are embedded in the perinephridial matrix and are always true fiber muscle cells (see also Storch & Welsch, 1972 for *Phascolosoma lurco)* (Fig. 6f). The duct in *G. minuta* consists of two different sections, a narrow ciliated descending duct that is continuous with the ciliated funnel and a large, sac-like section that lacks any ciliation. This is the largest and most prominent part of the nephridium. Its diameter is up to 5% of the trunk diameter. The lining cells interdigitate and are connected by adherens and septate junctions. The cells contain numerous smaller and larger vesicles of different electron densitiy, lysosomes and endosomes. They all indicate a higher degree of cytotic processes than seen in the ciliated section. There is no matrix between the individual walls of both sections of the duct. The coelomic lining of the sac-like section contains numerous podocytes (Fig. 6e), an observation that has also been reported earlier for other sipunculan species (Serrano & Angulo, 1989). Podocytes bypass different compartments and allow for instance selective fluid transfer between the trunk coelom and the contractile vessel (Pilger & Rice, 1987). As the perinephridial matrix never contains any fluid-filled lacunae it remains enigmatic which fluid systems are selectively bypassed by these cells. The nephridioporus is below the level of the funnel and is formed by several epidermal cells that are directly connected to duct cells. The nephridopore is lined by cuticle that decreases in diameter towards the duct cells. Strong muscles of the body wall musculature underlie the nephridiopore and nerve cell processes are found in this region.

A young larval stage (according to Akesson, 1958; Gibbs, 1975) of *G. minuta* revealed some first insights into nephridial development. The nephridial anlage bulged into the coelomic cavity and was covered by flat protrusions of muscle cells and peritoneal cells. The ciliated descending part was neither connected to the coelomic cavity nor to the sac-like section of the duct. This latter was preformed by a double layer of cells. The nephridiopore could already be recognized as an cuticle-lined epidermal invagination. Although the epidermal cells directly contacted the cells of the nephridial anlage, no luminal connection was established. Those ciliated duct cells that were next to the coelomic cavity were not covered by any peritoneal or muscular cells nor by their protrusion; they were directly exposed to the coelomic fluid. Although it cannot definitively be excluded that modified peritoneal cells form the funnel, the present finding indicates that the funnel is formed by duct cells and that the anlage may be opened passively while the coelomic cavity expands.

Conclusion

It seems very likely that protonephridial head kidneys belong to the ground pattern of the Annelida (Salvini-Plawen, 1980; Bartolomaeus & Ax, 1992; Bartolomaeus, 1998). Provided that a single cilium per cell always represents a primary condition (Rieger, 1976, but see Bartolomaeus, 1995b), monociliated protonephridia consisting of three cells could represent the primary condition. In this respect the head kidneys in *Echiurus echiurus* (Echiura), the monociliarity of which is inferred from older literature (Goodrich 1945: Fig. 64B), needs to be studied urgently. However, more data are needed to elucidate the direction of evolution of the head kidneys. Although we assume that segmental nephridia and head kidneys primarily result from repetition of the information to generate a nephridium, they evolved in different directions within the Annelida. Head kidneys are characteristic of the larva, and all constraints that influence their evolution should clearly differ from those that influenced the evolution of segmental nephridia. Thus, the structure of the head kidneys and the structure of the segmental nephridia did not necessarily evolve in parallel.

Despite earlier assumptions (Bartolomaeus, 1989; Bartolomaeus & Ax, 1992), these segmental nephridia are primarily not protonephridia; instead it is more parsimonious to assume that all nephridia formed subsequently to the head kidneys primarily were metanephridia. This is also in accordance with recent cladistic analyses of polychaete phylogeny (Rouse & Fauchald, 1997; Rouse, 1999). Some of these organs, especially those in smaller annelids, closely resemble certain stages of metanephridial development. Studies of metanephridial development show that the funnel is formed when the coelomic cavity expands during coelomogenesis. Structural similarity of certain protonephridia and certain stages of the metanephridial development can be explained by a truncated formation of metanephridia due to the restricted size of the coelomic cavities. Such protonephridia should therefore primarily be found small polychaetes with large body-sized relatives. The special structure of the solenocytes in the protonephridia of certain phyllodocidans is assumed to reflect common ancestry. This is substantiated by the fact that the apical portion of the duct is covered by a multiciliated cell so that the cilia and microvilli of the terminal have to pierce to enter the nephridial duct. If this cell prohibits passive opening of the proximal section of the duct to form a typical metanephridial funnel, a different way of forming a funnel during maturity should be expected. In Phyllodocidae and Pisionidae this is actually the case (Stecher, 1968; Bartolomaeus, 1989); in Nephtyids and Glycerids the genital products are not discharged by the nephridium (see Bartolomaeus, 1999). At least in a pholoid and a polynoid the metanephridia pass a protonephridial a stage with solenocytes during development. Although it has initially (Bartolomaeus & Ax, 1992) been interpreted as recapitulation of the ancestral protonephridium, it may also represent a stage which has been preserved in those taxa with solenocytes by truncation of the development. An alternative way has been proposed by Westheide (1986), who assumed that segmental protonephridia evolved from metanephridia by reduction of the size of the funnel and its final closure by terminal cells. Kuper's (2001) comparative studies of the nephridia in Syllidae also support this assumption. A final conclusion on the direction of the nephridial evolution, however, will certainly be inferred from a sound phylogeny of the Phyllodocida. Currently neither assumption is not in accordance with the cladogram presented by Rouise & Fauchald (1997), which on the other hand has some weaknesses (see Bartolomaeus et al., 2005).

Further studies into the development of the nephridia and ultrastructural data on the head kidneys are needed for the remaining polychaete taxa. We are convinced that both structures show enough complexity and enough substructural details that their analysis could substantially increase the database to unravel polychaete evolution. The remarkable correspondence of podocyte lining the nephridia in certain Sipuncula, Sabellida and Terebellida needs further attention. It is unique and hardly explainable from a functional point of view. Further comparative studies could help to elucidate the position of the Sipuncula. There are, finally, corresponding elements in the structure of the metanephridia of Aeolosomatidae and Clitellata. Corresponding nephridial development could revive the discussion on the relationship between both groups and should urgently be studied.

Acknowledgements

We would like to thank the two referees who helped to improve the manuscript. The study was financially supported by the Deutsche Forschungsgemeinschaft (DFG Ba 1520/2 und Ba 1520/6).

References

Akesson, B., 1958. A study of the nervous system of the Sipunculidae with some remarks on the development of the two species *Phascolion strombi* Montagu and *Golfingia minuta* Keferstein. Undersökningar over Öresund. 38: 1–249.

Allen, E. J., 1905. The anatomy of *Poecilochaetus,* Claparède. Quarterly Journal of Microscopical Science 48:79–151.

Anderson, D. T., 1973. Embryology and Phylogeny in Annelids and Arthropods. Pergamon, Oxford, 495 pp.

Ashworth, J. H., 1912. Catalogue of the Chaetopoda in the British Museum (Natural History). A. Polychaeta: Part 1. – Arenicolidae. Trustees of the British Museum, London, 175 pp.

Baltzer, F., 1934. Echiurida. In Kückenthal, W. & T. Krumbach (eds), Handbuch der Zoologie 2(9). de Gruyter, Berlin: 62–186.

Banse, K., 1969. Acrocirridae n. fam. (Polychaeta Sedentaria). Journal of the Fishery Research Board of Canada 26: 2595–2620.

Bartolomaeus, T., 1989. Ultrastructure and development of the nephridia in *Anaitides mucosa* (Annelida, Polychaeta). Zoomorphology 109: 15–32.

Bartolomaeus, T., 1993a. Ultrastructure of the protonephridia in the larva of *Autolylus prolifer* (Annelida, Syllidae): implications for annelid phylogeny. Microfauna Marina 8: 55–64.

Bartolomaeus, T., 1993b. Die Leibeshöhlenverhältnisse und Nierenorgane der Bilateria –Ultrastruktur, Entwicklung und Evolution. Habilitationsschrift, Universität Göttingen.

Bartolomaeus, T., 1995a. Ultrastructure of the protonephridia in larval *Magelona mirabilis* (Spionida) and *Pectinaria auricoma* (Terebellida): head kidneys in the ground pattern of the Annelida. Microfauna Marina 10: 117–141.

Bartolomaeus, T., 1995b. Secondary monociliarity in the Annelida? Monociliated epidermal cells in larvae of *Magelona mirabilis* (Magelonida). Microfauna Marina 10: 327–332.

Bartolomaeus, T., 1997. Structure and development of the nephridia of *Tomopteris helgolandica* (Annelida). Zoomorphology 117: 1–11.

Bartolomaeus, T., 1998. Head kidneys in hatchlings of *Scoloplos armiger* (Annelida: Orbiniida): implications for the occurrence of protonephridia in lecithotrophic larvae. Journal of Marine Biology Association U.K. 78: 183–192.

Bartolomaeus, T., 1999. Ultrastructure, development and function of segmental organs in Annelida. Hydrobiologia 402: 21–37.

Bartolomaeus, T. & P. Ax, 1992. Protonephridia and metanephridia – their relation within the Bilateria. Zeitschrift für Zoologische Systematik und Evolution Forschung 30: 21–45.

Bartolomaeus, T., G. Purschke & H. Hausen 2005. Polychaete phylogeny based on morphological data – a comparison of current attempts. Hydrobiologia 535/536 (Dev. Hydrobiol. 179): 339–354.

Bonch-Bruevich, E. V. & V. V. Malakhov, 1986. Organisation of the early larva of the polychaete *Chaetopterus variopedatus* (Spiomorpha, Chaetopteridae). Doklady Akademii NAUK 292: 1013–1015.

Bührmann, C., W. Westheide & G. Pursehke, 1996. Ultrastructure of the genital organs in the interstitial syllid *Petitia amphophthalma* (Annelida, Polychaeta). Acta Zoologica 77: 201–211.

Brandenburg, J., 1966. Die Reusenformen der Cyrtocyten. Zoologische Beiträge NF 12: 345–417.

Brandenburg, J., 1970. Die Reusenzelle (Cyrtocyte) des *Dinophilus*. Zeitschrift für Morphologie und Ökologie der Tiere 68: 83–92.

Brandenburg, J., 1975. I. Organs with pressure-filtration. Morphology of the protonephridia. Fortschritte der Zoologie 23: 1–17.

Brandenburg, J. & G. Kümmel, 1961. Die Feinstruktur der Solenocyte. Journal of Ultrastructure Research 5: 437–452.

Brown, R. S., 1938. The anatomy of the polychaete Ophelia cluthensis McGuire 1935. Proceedings of the Royal Society of Edinburgh 58: 135–160.

Bunke, D., 1994. Ultrastructure of the metanephridial system in *Aeolosoma bengalense* (Annelida). Zoomorphology 114: 247–258.

Bunke, D., 1998. Ultrastructure of the nephrostome in *Enchytraeus albidus* (Annelida, Clitellata). Zoomorphology 118: 177–182.

Bunke, D., 2000. The preseptal part of metanephridia in *Nais variabilis* and *Dero digitata* (Annelida, Clitellata). Zoomorphology 120:39–46.

Bunke, D., 2003. Early development of metanephridia in the caudal budding zone of a clitellate annelid, *Dero digitata* (Naidida): an electron-microscopical study. Acta Zoologica (Stockholm) 84: 87–97.

Clark, R. B., 1956. The blood vascular system of *Nephtys* (Annelida, Polychaeta). Quarterly Journal of Microscopical Science 97: 235–249.

Clausen, C., 1986, *Microphthalmus ephippiophorus* sp. n. (Polychaeta: Hesionidae) and two other *Microphthalmus* species from the Bergen area, Western Norway. Sarsia 71: 177–191.

Daly, J. M., 1975. Reversible epitoky in the life history of the polychaete *Odontosyllis polycera* (Schmarda 1861). Journal of the Marine Biology Association U.K. 55: 327–344.

Dehorne, A., 1952, Précisions sur le type anatomique des néphridies thorciques des Sabellariides. Archives de Zoologie Experimentale et Generate 89: 37–49.

Dehorne, A.& L. Dehorne, 1913. Recherches sur *Sclerocheilus minutus* (polychète de la famille Scalibregmides). Morphologie, yeux, nèphridie et pavillon. Archives de Zoologie Experimentale et Generale 53: 61–137.

Eckelbarger, K. J. & J. P. Grassle, 1987. Spermatogenesis, sperm storage and comparative sperm morphology in nine species of *Capitella*, *Capitomastus* and *Capitellides* (Polychaeta: Capitellidae). Marine Biology 95: 415–429.

Ehlers, E., 1846. Die Borstenwürmer (Annelida, Chaetopoda) nach Systematischen und Anatomischen Untersuchungen Dargestellt, W. Engelmann, Leipzig, 268 pp.

Eisig, H., 1879. Die Segmentalorgane der Capitelliden. Mitteilungen aus der Zoologischen Station zu Neapel 1: 95–118.

Eisig, H., 1887. Monographie der Capitelliden des Golfes von Neapel und der angrenzenden Meeresabschnitte nebst Untersuchungen zur vergleichenden Anatomie und Physiologie. Fauna und Flora des Golfes von Neapel 16:1–906.

Eisig, H., 1899. Zur Entwicklungsgeschichte der Capitelliden. Mittelungen aus der Zoologischen Station Zu Neapel 13: 1–292.

Eisig, H., 1914. Zur Systematik, Anatomie und Morphologie der Ariciiden nebst Beiträgen zur generellen Systematik. Mitteilungen aus der Zoologischen Station Zu Neapel: 21: 153–600.

Fordham, M. G. C., 1926. *Aphrodite aculeata*. L.M.B.C. Memoir 27. Proceedings and Transactions of the Liverpool Biological Society 40: 121–216.

Gardiner, S. L., 1979. Fine structure of *Owenia fusiformis* (Polychaeta, Oweniidae) and its significance for annelid phylogeny. Ph.D. dissertation University North Carolina at Chapell Hill.

Gardiner, S. L. & M. Jones, 1993. Vestimentifera. In Harrison, F. W. & M. E. Rice (eds), Microscopic Anatomy of Invertebrates. Vol. 12, Onychophora, Chilopoda and Lesser Protostomata. Wiley-Liss, New York: 371–460.

Gibbs, P. E., 1975. Gametogenesis and spawning in a hermaphroditic population of *Golfingia minuta* (Sipuncula). Journal of the Marine Biology Association U.K. 55: 69–82.

Giere, O. W. & C. Erséus, 1998. A systematic account of the Questidae (Annelida, Polychaeta), with description of new taxa. Zoologica Scripta 27: 345–360.

Goodrich, E. S., 1895. On the coelom, genital ducts, and nephridia. Quarterly Journal of Microscopical Science 37: 477–508.

Goodrich, E. S., 1898. On the nephridia of the polychaeta. Part I. - On *Hesione, Tyrrhea,* and *Nephthys*. Quarterly Journal of Microscopical Science 40: 185–195.

Goodrich, E. S., 1899. On the nephridia of the polychaeta. Part II. - *Glycera* and *Goniada,* Quarterly Journal of Microscopical Science 41: 439–457.

Goodrich, E. S., 1900. On the nephridia of the Polychaeta. Part III. - The phyllodocidae, Syllidae, Amphinomidae, etc., with summary and conclusions. Quarterly Journal of Microscopical Science 43: 699–748.

Goodrich, E. S., 1945. The study of nephridia and genital ducts since 1895. Quarterly Journal of Microscopical Science 86: 113–392.

Gustofson, G., 1930. Anatomische Studien über die Polychaeten-Familien Amphinomidae and Euphrosynidae. Zoologiska Bidrag fran Uppsala 12: 305–471.

Hartmann-Schröeder, G., 1958. Zur Morphologie der Opheliiden (Polychaeta sedentaria). Zeitschrift fur Wissenchaftiche Zoologie 161: 84–143.

Hatschek, B., 1878. Studien über die Entwicklungsgeschichte der Anneliden. Ein Beitrag zur Morphologie der Bilateria. Arbeiten aus dem Zooologischen Institut der Universität Wien 1: 1–128.

Hatschek, B., 1886. Zur Entwicklung des Kopfes von *Polygordius*. Arbeiten aus dem Zoologischen Institut der Universität und der Station in Triest 6: 109–210.

Heimler, W., 1981. Untersuchungen zur Larvalentwicklung von *Lanice conchilega* (Pallas, 1766) (Polychaeta, Terebellomorpha): II. Bau und Ultrastruktur der Trochophora-Larve.

Zoologische Jahrbücher, Abteilung Anatomie und Ontogenie der Tiere 106: 236–277.

Heimler, W., 1983. Untersuchungen zur Larvalentwicklung von *Lanice conchilega* (Pallas, 1766) (Polychaeta, Terebellomorpha): teil III, Bau und Struktur der Aulophora-Larve. Zoologische Jahrbücher, Abteilung Anatomie und Ontogenie der Tiere 110: 236–277.

Heimler, W., 1988. Larvae. In Westheide, W. & C. O. Hermanns (eds), The Ultrastructure of the Polychaeta. Microfauna Marina 4: 353–371.

Hessle, C., 1917. Zur Kenntnis der terebellomorphen Polychaeten. Zoologiska Bidrag from Uppsala 5: 39–258.

Holborow, P. L., 1971. The fine structure of the trochophore of *Harmothoe imbricata*. The Fourth European Marine Biology Symposium. Cambridge University Press, Cambridge: 237–246.

Jondelius U., T. I. Ruiz, J. Baguna & M. Riutort, 2002. The Nemertodermatida are basal bilaterians and not members of the Platyhelminthes. Zoologica Scripta 31: 201–215.

Jouin, C., 1967. Étude morphologique et anatomique de *Nerillidopsis hyalina* Jouin et de quelques Nereillidium Remane (Archiannélides, Nereillidae). Archives de Zoologie Experimentale et Generale 108: 97–110.

Joyeux-Laffuie, J., 1890. Étude monographique du Chétoptère (*Chaetopterus variopedatus* Rénier) suive d'une revision des espèces du genre Chaetopterus. Archives de Zoologie experimentale et generale 8: 245–360.

Kitamura, K. & T. Shimizu, 2000a. Embryonic expression of alkaline phosphatase activity in the oligochaete annelid *Tubifex*. Invertebrate Reproduction and Development 37: 69–73.

Kitamura, K. & T. Shimizu, 2000b. Analyses of segment specific expression of alkaline phosphatase activity in the mesoderm of the oligochaete annelid *Tubifex*: implication for specification of segmental identity. Developmental Biology 219: 214–223.

Koechlin, N., 1966. Structure and function of the nephridia in *Sabella pavonina* Savigny (Polychaeta, Annelida). Comparative Biochemistry and Physiology 69A: 349–355.

Kuper, M., 2001. Ultrastrukturuntersuchungen der Segmentalorgane, der Spermien und der Brutpflegestrukturen der Syllidae (Annelida: Polychaeta). Ph.D. Thesis, University of Osnabrück, 317 pp.

Kuper, M. & C. Bührmann, 1999. Nephridial organs of interstitial Syllidae (Polychaeta). Zoology 102 (suppl. II):42.

Kuper, M. & G. Purschke, 2001. The excretory organs in *Sphaerodorum flavum* (Phyllodocida, Sphaerodoridae): a rare case of co-occurrence of protonephridia, coelom and blood vascular system in Annelida. Zoomorphology 120: 191–203.

Kuper, M. & W. Westheide, 1997. Ultrastructure of the male reproductive organs in the interstitial annelid *Sphaerosyllis hermaphrodita* ("Polychaeta", Syllidae). Zoomorphology 117: 13–22.

McConnaughey, B. H. & D. L. Fox, 1945. The anatomy and biology of the marine polychaete *Thoracophelia mucronata* (Treadwell), Opheliidae. University of California Publications in Zoology 47: 319–340.

Mesnil, F., 1899. La position systématique des Flabelligériens St. Joseph (Chlorémiens Quatrefages) et des Sternaspiens. Zoologischer Anzeiger 22: 81–85.

Messenger, E. M., P. R. Smith & E. E. Ruppert, 1985. Ultrastructure and function in the protonephridium in Nephtys (Polychaeta). American Zoologist 25: 41A.

Meyer, E., 1887. Studien über den Körperbau der Anneliden. Teil 1-3. Mitteilungen aus der Zoologischen Station zu Neapel 7: 592–741.

Nakao, T., 1974. Electron microscope study of the circulatory system in *Nereis japonica*. Journal of Morphology 144: 217–236.

Newby, W. W., 1940. The embryology of the echiuroid worm *Urechis caupo*. Memoirs of the American Philosophical Society 16: 1–213.

Olive, P. J. W., 1970. Reproduction of a Northumberland population of the polychaete *Cirratulus cirratus*. Marine Biology 5: 259–273.

Orrhage, L., 1964. Anatomische und mrophologische Studien über die Polychaeten-Familien Spionidae, Disomidae und Poecilochaetidae. Zoologiska Bidrag fran Uppsala 36: 335–446.

Orrhage, L., 1974. Über die Anatomie, Histologie und Verwandtschaft der Apistobranchidae (Polychaeta, Sedentaria) nebst Bemerkungen über die systematische Stellung der Archianneliden. Zeitschrift für Morphologie und Ökologie der Tiere 79: 1–45.

Orrhage, L., 1980. On the structure and homologues of the anterior end of the polychaete families Sabellidae and Serpulidae, Zoomorphology 96: 113–168.

Pietsch, A. & W. Westheide, 1987. Protonephridial organs in *Myzostoma cirriferum* (Myzostomida). Acta Zoologica (Stockholm) 68: 195–203.

Pilger J. F. & M. E. Rice, 1987. Ultrastructural evidence for the contractile vessel of sipunculids as a possible ultrafiltration site. American Zoologist 27: 152A.

Pilgrim, M., 1978. On the anatomy and histology of the nervous system and excretory system of the maldanid polychaetes *Clymenella torquata* and *Euclymene oerstedi*. Journal of Morphology 155: 311–326.

Pinson, D. Y. & E. E. Ruppert, 1988. The excretory role of metanephridia in *Themiste alutacea* (Sipuncula). American Zoologist 28: 752A.

Pemerl, J., 1965. Ultrastructure of the protonephridium of the trochophore larva of *Serpula vermicularis* (Annelida, Polychaeta). American Zoologist 5: 666–667.

Quast, B. & T. Bartolomaeus, 2001. Ultrastructure and significance of the transitory nephridia in *Erpobdella octoculata* (Hirudinea, Annelida). Zoomorphology 120: 205–213.

Rice, S. A., 1980. Ultrastructure of the male nephridium and its role in spermatophore formation in spionid polychaetes (Annelida). Zoomorphologie 95: 181–194.

Rieger, R. M., 1976. Monociliated epidermal cells in Gastrotricha) Significance for concepts of early metazoan evolution. Zeitschrift für zoologische Systematik und Evolution Forschung 14: 198–226.

Robbins, D. E., 1965. The biology and morphology of the pelagic annelid *Poeobius meseres* Heath. Journal of Zoology 146:197–212.

Rouse, G. & K. Fauchald, 1997. Cladistiecs and polychaetes. Zoologica Scripta 26: 139–204.

Rouse, G. W., 1999. Trochophore concepts: ciliary bands and the evolution of larvae in spiralian Metazoa. Biological Journal of the Linncan Society 66: 411–464.

Ruppert, E. E., & P. R. Smith, 1988. The functional organization of filtration nephridia. Biological Reviews 171:231–258.

Salvini-Plawen, L. v., 1980. Was ist eine Trochophora? Eine Analyse der Larventypen mariner Protostomier. Zoologische Jahrbücher Anatomie 103: 389–423.

Saphonov, M. V. & A. B. Tzetlin, 1994. An ultrastructural study of nephridial systemin two nerellids: *Nerilla jouinae* and *Trochonerilla mobilis* (Polychaeta). Mémoirs du Muséum national d' Histoire Naturelle, Paris 162: 638.

Scharnofske, P., 1984. Anatomie, Ultrastruktur und Funktion der männlichen Geschlechtsporgane der Dinophilidae und Histriobdellidae (Annelida, Polychaeta). Ph.D. Thesis, Universität. Göttingen.

Schlötzer-Schrehardt, U., 1992. Ultrastrukturelle Untersuchungen zur Reproduktion und Postembryonalentwicklung einschliesslich Adultorganisation von *Pygospio elegans* Claparede, 1863 (Polychaeta, Spionidae). 2 Bde: Text- u. Abbildungen. Dissertation, Universität Erlangen-Nürnberg.

Schuchert, P. & R. M. Rieger, 1990. Ultrastructural observations on the dwarf male of *Bonellia viridis* (Echiura). Acta Zoologica 71: 15–16.

Schuchert, P., 1990. The nephridium of the *Bonellia viridis* male (Echiura). Acta Zoologica 71: 1–4.

Schulze, A., 2001b. Comparative anatomy of excretory organs in vestimentiferan tube worms (Pogonophora, Obturata). Journal of Morphology 250: 1–11.

Serrano, T. & E. Angulo, 1989. The fine structure of the nephridial coelothelium in *Phascolosoma granulatum*. Zeitschrift für mikroskopisch-anatomische Forschung 103: 414–424.

Shearer, C., 1910. On the anatomy of *Histriobdella homari*. Quaterly Journal of Microscopic Science 55: 287–359.

Shimizu, T., K. Kitamura, A. Arai & A. Nakamoto, 2001. Pattern of formation in embryos of the oligochaete annelid *Tubifex*: cellular basis for segmentation and specification of segmental identity. Hydrobiologia 463: 123–131.

Shimizu, T. & A. Nakamoto, 2001. Segmentation in annelids: cellular and molecular basis for metameric body plan. Zoological Science 18: 285–298.

Singhal, R. N., 1982. Some morphological notes on the proboscis and trunk of four species of Echiura. Journal of Zoological Society of India 34: 96–99.

Smith, P. R., 1986. Development of the blood vascular system in *Sabellaria cementarium* (Annelida: Polychaeta): an ultrastructural investigation. Zoomorphology 106: 67–74.

Smith, P. R., 1992. Polychaeta: excretory systems. In Harrison, F. W. & S. L. Gardiner (eds), Microscopic Anatomy of Invertebrates. Vol. 7, Annelida, Wiley-Liss, New York: 71–108.

Smith, P. R. & E. E. Ruppert, 1988. Nephridia. In Westheide, W. & C.O. Hermans (eds), The Ultrastructure of the Polychaeta. Microfauna Marine 4: 231–262.

Smith, P. R., E. E. Ruppert & S. L. Gardiner, 1987. A. deuterostome-like nephridium in the mitraria-larva of Owenia fusiformis (Polychaeta, Annelida). The Biological Bulletin 173: 315–323.

Smith, R. I., 1984. The larval nephridia of the brackish-water polychaete *Nereis diversicolor*. Journal of Morphology 179: 273–289.

Smith, R. I., 1992. Three nephromixial patterns in polychaete species currently assigned to the genus *Pista* (Annelida, Terebellidae). Journal of Morphology 213: 365–393.

Southward, E. C., 1993. Vestimentifera. In Harrison, F. W. & M. E. Rice (eds), Microscopic Anatomy of Invertebrates. Vol. 12, Onychaophora, Chilopoda and lesser protostomata. Wiley-Liss, New York: 327–369.

Southward, E. C., A. Schulze & S. L. Gardiner (this volume). Pogonophora (Annelida): Form and function. Hydrobiologia.

Stecher, H.-J., 1968. Zur Organisation und Fortpflanzung von *Pisione remota* (Southern) (Polychaeta Pisionidae). Zeitschrift für Morphologie der Tiere 61: 347–410.

Storch, V. & U. Welsch, 1972. Zur Ultrastruktur der Metanephridien des landlebenden Sipunculiden *Phascolosoma lurco*. Kieler Meeresforschung 28: 277–231.

Strelzov, V. E., 1979. Polychaete worms of the family Paraonidae Cerruti, 1909. (translated from Russian) Amerind Publishing, New Delhi.

Vejdovsky, F., 1882. Untersuchungen über die Anatomie, Physiologie und Entwicklung von *Sternaspis*. Denkschrift der Mathematisch-Naturwissenschaftlichen Classe der Akademie der Wissenschaften Wien 43, 1–58.

Vejdovsky, F., 1884. System und Morphologie der Oligochaeten. Franz Rivnác, Prag.

Vejdovsky, F., 1892. Entwicklungsgeschichtliche Untersuchungen (Heft III). Die Organogenie der Oligochaeten. J. Otto, Prag, 299–360.

Von Nordheim, H. & A. Schrader, 1994. Ultrastructure and functional morphology of the protonephridia and segmental fenstrated lacunae in *Protodrilus rubropharyngeus* (Polychaeta, Protodrilidae). Helgoländer Meeresuntersuchungen 48: 467–485.

Weisblat, D. A., S. Y. Kim & G. Stent, 1984. Embryonic origins of cells in the leech *Helobdella triserialis*, Developmental Biology 104: 65–85.

Wells, G. P., 1959. The genera of the Arenicolidae (Polychaeta). Proceedings of the Zoological Society of London 133: 301–314.

Wessing, A. & A. Polenz, 1974. Structure, development and function of the protonephridium in the trochophore of *Pomatoceros triqueter* (Annelida, Polychaeta, Sedentaria). Cell Tissue Research 156: 21–33.

Westheide, W., 1985. Ultrastructure of the protonephridia in the dorvilleid polychaete *Apodotrocha progenerans* (Annelida). Zoologica Scripta 14: 273–278.

Westheide, W., 1986. The nephridia of the interstitial polychaete Hesionides arenaria and their phylogenetic significance (Polychaeta, Hesionidae). Zoomorphology 106: 35–43.

Westheide, W., 1990. Polychaetes: interstitial families. Synopses of the British Fauna, New Series 44: 1–152.

Woltereck, R., 1904. Beiträge zur praktischen Analyse der *Polygordius-Entwicklung* nach dem „Nordsee"- und dem „Mittelmeertypus". I. der für beide Typen gleichverlaufende

Entwicklungsabschnitt: vom Ei bis zum jüngsten Trochophora-Stadium. Archiv für Entwicklungsmechanik 18: 377–403.

Woltereck, R., 1905. Zur Kopffrage der Anneliden. Verhandlungen der Deutschen Zoologischen Gesellschaft 15: 154–185.

Worsaae, K. & R. M. Kristensen 2005. Evolution of interstitial Polychaeta (Annelida). Hydrobiologia 535/536 (Dev. Hydrobiol. 179): 317–338.

Zal, F., D. Desbruyères & C. Jouin-Toulmond, 1994. Sexual dimorphisms in *Paralvinella gresllei*, a polychaete annelid from deep-sea hydrothermal vents. Contes rendus de Académie Sciences, Paris, Science Vie 317: 42–48.

Zerbst-Boroffka, I. & J. Haupt, 1975. Morphology and function of the metanephridia in annelids. Fortschritte Zoologie 23: 33–47.

Zhadan, A. E., A. B. Tzetlin & M. A. Safronova, 2000. Anatomy of some representatives from the family Alvinellidae (Polychaeta, Terebellida) from the Pacific hydrothermal habitats. Zoologicheskii Zhurnal 79: 141–160.

Hydrobiologia (2005) **535/536:** 167–178
T. Bartolomaeus & G. Purschke (eds), Morphology, Molecules, Evolution and Phylogeny in Polychaeta and Related Taxa

Annelid sperm and fertilization biology

Greg W. Rouse
South Australian Museum North Terrace, Adelaide, SA 5000;
Australia & School of Environmental & Earth Sciences, The University of Adelaide, Adelaide SA 5005, Australia
E-mail: gregr@saugov.sa.gov.au

Key words: Clitellata, polychaetes, phylogeny, ultrastructure, spermathecae

Abstract

It is largely accepted that features such as parental care of young (brooding) and complex fertilization biology are derived features, having always evolved from simple reproductive methods such as external fertilization with no parental involvement in offspring survival. This is reflected in the continued use of words such as 'primitive' and 'modified' sperm. I will review cases where these terms are fundamentally misleading and reiterate calls for their abandonment. An update on the published ultrastructural descriptions of sperm in annelids since the reviews by Ferraguti (1999a) and Rouse (1999a) is provided. The structure of spermatozoa across Annelida in relation to phylogenetic hypotheses for the group is reviewed. It is concluded that until an adequate understanding of the root placement for the tree of Annelida is discovered then we cannot infer the plesiomorphic sperm and fertilization mode in the group. The fascinating fertilization mechanisms of Serpulidae are discussed. New ultrastructural data on sperm and spermathecae in the group is presented. The variability of sperm in Spirorbinae is marked and given that the fertilization mechanism is similar across the group this provides an unprecedented opportunity to correlate sperm structure with function.

Introduction

Ultrastructural studies of spermatozoa in annelids have been reviewed several times, with reference to either to polychaetes (Franzén & Rice 1988; Jamieson & Rouse, 1989; Rice, 1992; Rouse, 1999a,b) or clitellates (Fernandez et al., 1992; Ferraguti, 1999a,b). The purpose of this review is to provide an update of the most recent comprehensive reviews of sperm structure in annelids by Ferraguti (1999b) and Rouse (1999a), and to provide a brief discussion on sperm morphology and fertilization mechanisms. This is most relevant with relation to polychaetes owing to the tremendous range of reproductive modes they show. Alternative hypotheses about the phylogeny are outlined that show that we can at present not state what the plesiomorphic sperm shape or reproductive mode is for Annelida. A specific group of polychaetes will be highlighted, namely Serpulidae, and new data presented on sperm morphology and sperm storage mechanisms.

Materials and methods

Specimens were removed from their rock or algal substrate and examined while fresh and unfixed, using a Leica MZ-8 stereo microscope. For histological examination, most specimens were fixed in 3% glutaraldehyde in 0.2 M sodium phosphate buffer (pH 7.4), with 0.3 M sucrose added, for 2 h at 4 °C. Specimens were then rinsed in buffer (changed three times), post-fixed for 80 min with 1% osmium tetroxide in 0.2 M sodium phosphate buffer (pH 7.4) with 0.3 M sucrose added and then washed in buffer before dehydration. After dehydration, samples were infiltrated and embedded in Spurr's resin. For the study of the spermatheca of *Protolaeospira tricostalis* (Lamarck, 1818), serial one μm sections were cut using a Leica Reichert Ultracut S ultramicrotome, and stained with 0.3% toluidine blue in 0.1 M phosphate buffer. Micrographs were taken with a Leica DMR photomicroscope, using a 100× oil immersion objective and Kodak Tech-Pan film. For transmission electron microscopy (TEM) 50–80 nm sections were placed on uncoated 200 thin-mesh copper grids, stained with uranyl acetate and lead citrate before viewing. Electron micrographs were taken on Philips EM400 or CM12 transmission electron microscopes.

Voucher specimens of the following Serpulidae used in this study are deposited at the South Australian Museum: *Galeolaria caespitosa* Lamarck 1818, collected by G . Rouse on rocks in intertidal region North Bondi Rocks NSW, Australia, 26.11.1995; *Metalaeospira tenuis* Knight-Jones 1973, collected by G. Rouse in intertidal zone on holdfasts of brown macroalga *Ecklonia radiata* Agardh 1848 at EagleHawk Neck, Tasmania, Australia, 11.4.1996; Neodexiospira sp., collected by G. Rouse in intertidal zone on holdfasts of Ecklonia radiata in intertidal zone at EagleHawk Neck, Tasmania, Australia, 11.4.1996; *Paralaeospira* cf. *levinsoni* Caullery & Mesnil 1897, collected by G. Rouse on rocks in intertidal region North Bondi Rocks NSW, Australia, 18.10.1995; *Paraprotis dendrova* Uchida , collected from Okinawa by E. Nishi; *Pileolaria* sp., collected by G. Rouse on rocks intertidal region North Bondi rocks NSW, Australia, 26.11.1995; *Protolaeospira capensis* Day 1961, collected by G. Rouse on gastropod shells intertidal region North Bondi Rocks NSW, Australia, 26.11.1995; *Protolaeospira tricostalis* Lamarck 1818, collected by G. Rouse on rocks intertidal region North Bondi Rocks NSW, Australia, 26.11.1995; *Romanchella quadricostalis* Knight-Jones 1973, collected by G. Rouse in intertidal zone on storm-washed *Ecklonia radiata* holdfast at EagleHawk Neck, Tasmania, Australia, 11.4.1996; *Spirorbis spirorbis* Linnaeus 1758, collected by G. Rouse on *Fucus serratus* Linnaeus 1753 at low tide in pools near Tjärnö Marine Biological Station, Sweden, 28.5.1996; *Salmacina* sp., collected by G. Rouse subtidally off Carrie Bow Cay Belize.

Sperm terminology

Annelids have a great range of sperm morphologies and traditionally these have been grouped as either 'primitive' or 'modified' based on the terminology of Retzius (1904, 1905) and Franzén (1956). Polychaetes show a range of sperm shapes (Figs 1 and 2) but are traditionally referred to by these two terms only, whilst all clitellates would be classified as having modified sperm under this terminology. Franzén (1956) suggested that 'primitive' sperm with heads comprised of a simple acrosome, spherical nuclei and a small number of mitochondria and a free flagellum were associated with external fertilization (Fig. 1A). Modified

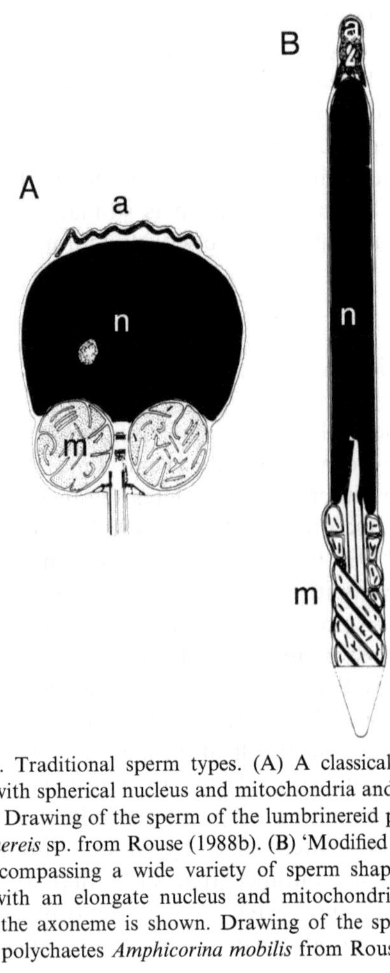

Figure 1. Traditional sperm types. (A) A classical 'primitive sperm' with spherical nucleus and mitochondria and a cap-like nucleus. Drawing of the sperm of the lumbrinereid polychaetes *Lumbrinereis* sp. from Rouse (1988b). (B) 'Modified sperm' is a term encompassing a wide variety of sperm shapes. Here a sperm with an elongate nucleus and mitochondria spiraling around the axoneme is shown. Drawing of the sperm of the sabellid polychaetes *Amphicorina mobilis* from Rouse (1992).

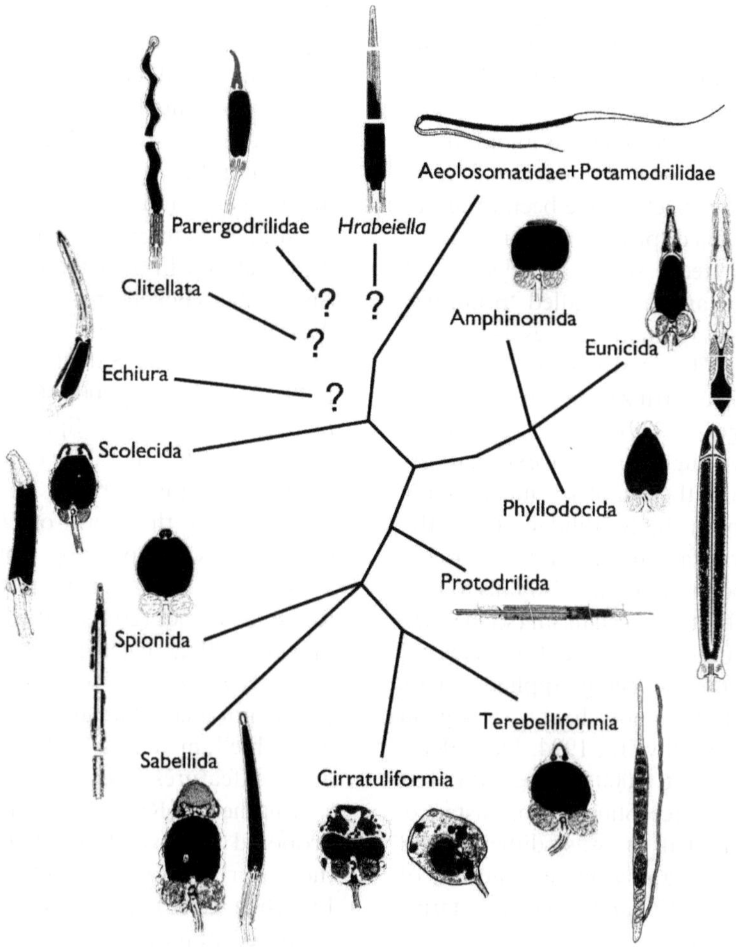

Figure 2. Unrooted version of cladogram of Annelida. Modified from that shown in Rouse & Pleijel (2003). Major taxa of Annelida are shown with no implication that all are monophyletic. Some interesting minor groups are also shown. The placement of Clitellata and Echiura within Annelida is currently unknown. Some representative sperm types from each taxon are shown in order to demonstrate that establishing the plesiomorphic sperm shape (and fertilization mode) cannot be resolved until the root placement is established. Aeolosomatidae and Potamodrilidae, sperm of *Potamodrilus fluviatilis* (modified from Bunke, 1985); Amphinomida, sperm of the amphinomid *Eurythoe complanata* (modified from Jamieson & Rouse, 1989); Cirratuliformia. Sperm of *Cirriformia* sp. (modified from Jamieson & Rouse, 1989) and *Ctenodrilus* sp. (modified from Rouse, 1999b); Clitellata, sperm of *Capilloventer australis* (modified from Ferraguti et al., 1996); Echiura, sperm of *Bonellia viridis* (modified from Franzén & Ferraguti, 1992); Eunicida, sperm of the onuphid *Hyalinoecia tubicola* sp. (modified from Cotelli & Lora Lamia Donin, 1975) and the histriobdellid *Stratiodrilus novohollandiae* (modified from Jamieson et al., 1985); Parergodrilidae + *Hrabeiella*, sperm of *Parergodrilus heideri* (modified from Purschke, 2002a) and *Hrabeiella periglandulata* (Rota & Lupetti, 1997); Phyllodocida, sperm of the polynoid *Lepidonotus* sp. (modified from Jamieson & Rouse, 1989) and the nereidid *Platynereis massiliensis* (modified from Pfannenstiel et al., 1987); Sabellida, sperm of the serpulids *Galeolaria caespitosa* (modified from Jamieson & Rouse, 1989) and *Chitinopoma serrula* (modified from Franzén, 1982); Spionida, sperm of the spionids *Tripolydora* sp. and *Prionospio* cf. *queenslandica* (modified from Rouse, 1988a); Scolecida, sperm of the capitellid *Capitella capitata* (modified from Franzén, 1982) and the arenicolid *Arenicola marina* (modified from Jamieson & Rouse, 1989); Terebelliformia, sperm of the terebellid *Streblosoma acymatum* (modified from Jamieson & Rouse, 1989) and the alvinellid *Paralvinella pandorae pandorae* (modified from Jouin-Toulmond et al., 2002).

sperm were any sperm that deviate from this pattern (e.g., Figs 1B, 2). Rouse & Jamieson (1987) subsequently proposed a new system of classifying sperm based purely on function: *Ect-aquasperm* are released into the water and fertilize similarly released eggs. This term places no phylogenetic significance on this form of fertilization. *Ent-aquasperm* are again released freely into the ambient water but differ from ect-aquasperm in being gathered by, or in some other way reaching, the female. *Introsperm* have no contact with water when passed from male to female.

This terminology was designed to avoid any *a priori* judgment of phylogenetic pattern. Additionally, a number polychaetes have been found to have sperm with so-called 'primitive' morphology, yet do not fertilize freely spawned eggs in the classical sense. For example, so-called 'primitive' sperm may be found in spermatophores as in some spionids (Richards, 1970), or stored by the female in spermathecae before fertilization e.g., alciopids (Rice & Eckelbarger, 1989). Also, Rouse & Fitzhugh (1994) have suggested that external fertilization and sperm with 'primitive' morphology are in fact secondary in the Sabellidae. Thus, the term 'primitive' sperm fails to indicate plesiomorphic status for this sperm shape and also fails to be a good guide as to fertilization mechanism (see below).

The terms 'ect-aquasperm', 'ent-aquasperm' and 'introsperm' are now being applied across various taxa in addition to polychaetes (Hodgson & Chia, 1993; Sousa & Oliveira, 1994; Degaulejac et al., 1995; Rouse & Pitt, 2000) though this has not met with universal acceptance. For instance, Rice (1992) argued that it was difficult and potentially misleading to apply the terminology of Rouse & Jamieson (1987) and used the terms 'primitive' and 'modified' in describing polychaete sperm. Since the terms of Rouse & Jamieson (1987) require knowledge of reproductive mechanisms in the species concerned before the sperm can be classified it can be difficult to classify the sperm. However, it is certainly not a misleading terminology. Given the observation by Franzén (1956) that fertilization mechanisms do tend to correlate with sperm morphology (given the caveats outlined above), it is possible to tentatively assign sperm to a category, but as pointed out above, sperm shape is not an infallible guide to reproductive mode and this really needs to be used with extreme caution.

Recent descriptions of annelid sperm ultrastructure

Since the last review on the sperm of Clitellata by Ferraguti (1999a,b), there have been two further publications on sperm of taxa in this clade (Ferraguti et al., 1999; Marotta et al., 2003). These papers have provided further evidence of the utility of sperm based characters for the phylogenetic analysis of clitellates. Since the last review of polychaete sperm by Rouse (1999a), there have been a number of interesting descriptions of polychaete sperm and fertilization mechanisms. These are studies were on Alvinellidae (Jouin-Toulmond et al., 2002), Chrysopetalidae (Tzetlin et al., 2002), Hesionidae (Eckelbarger et al., 2001), Orbiniidae (Eckelbarger & Young, 2002), Parergodrilidae (Purschke, 2002a), Sabellidae (Fitzhugh & Rouse, 1999; Gambi et al., 2000; Giangrande et al., 2000; Licciano et al., 2002), Spionidae (Williams, 2000) and Syllidae (Giangrande et al., 2002).

Also Rouse (1999a) neglected to list the description of the sperm of *Hrabeiella periglandulata* Pizl & Chalupsky by Rota & Lupetti (1997). This taxon was referred to as a polychaete by Rouse & Pleijel (2001) with its placement among this assemblage listed as *incertae sedis*. *Hrabeiella periglandulata* is a terrestrial organism, but is clearly not a clitellate annelid. Apart from the lack of a clitellum it is also lacks the various reproductive features such as seminal receptacles or spermathecae (Rota & Lupetti, 1997). It has been proposed however that *H. periglandulata* maybe the sister group to Clitellata (Purschke, 2003). Purschke (2003) based his argument on a series of proposed homologies based on the type of pharynx, cerebral sense organs and the central nervous system. The sperm of *H. periglandulata*, while filiform, shows no particular similarities with clitellates such as the presence of an acrosome tube, or interpolation of the mitochondria between the axonome and the nucleus (Rota & Lupetti, 1997). Rota & Lupetti (1997) rejected any affinity with clitellates on this basis and compared the sperm with that of a number of polychaetes but found none that were particularly similar. However, this result does not actually rule out a close, or even sister group relationship between *Hrabeiella* and Clitellata as proposed by Purschke (2003). The two sperm apomorphies above (and also a central sheath and tetragon fibres in the axonome, see Ferraguti, 1984) clearly support monophyly of Clitellata. That the sperm of *H. periglandulata* lacks these does not invalidate a close relationship with clitellates and Purschke's (2003) hypothesis deserves close attention. Of course, a real goal must also be to find the presently unknown sister group of Clitellata (and *Hrabeiella*?) among the remaining Annelida.

The recently described sperm is found in the study by Purschke (2002a) on another enigmatic terrestrial annelid, *Parergodrilus heideri* Reisinger is also of interest. This taxon is placed, along with *Stygocapitella subterranea* in Parergodrilidae , and is currently regarded as *incertae sedis* among polychaetes by Rouse & Pleijel (2001) *Parergodrilus heideri* has internal fertilization, yet the sperm presents another example where sperm shape is misleading with respect to fertilization mechanism. The sperm of *Parergodrilus heideri* is an almost classical 'primitive' sperm according to the terminology of Franzén (1977), but it is found in a terrestrial polychaetes with internal fertilization and so can hardly be regarded as a primitive annelid. While the sperm acrosome, nucleus and midpiece are very slightly elongate, the sperm is, for instance, quite similar to sperm of polychaetes such as the onuphid *Hyalinoecia tubicola* shown in Figure 2. The sperm of this particular polychaete was actually referred to as 'primitive' by Franzén (1977), so there is little doubt that that of *Parergodrilus heideri* would, completely misleadingly, also have been referred to as such. The example of *Parergodrilus heideri* provides yet more evidence for the lack of utility for the terms 'primitive' and 'modified' sperm.

Annelid sperm and systematics

Previous doubts about the use of sperm data as characters in polychaetes systematics (Jamieson & Rouse, 1989; Rice, 1992) were based on ideas of using sperm data alone for phylogenetic studies. While it is true that sperm ultrastructural characters have proved informative in clades with uniform reproductive mechanisms such as clitellates (Ferraguti, 1999b) there was thought to be a problem with polychaetes. Rice (1992, p. 150) summarized this as "variation in sperm structure within groups (such as polychaetes) may be so extensive that any number of phylogenies could be constructed depending upon which sperm type is considered to be .... plesiomorphic." Jamieson & Rouse (1989) expressed a similar view and felt that the supposed multiple origin of 'modified' forms of reproduction from external fertilization would preclude the use of sperm in systematics. These statements imply that the understanding the evolution of the tremendous variety of reproductive mechanisms among polychaetes is too problematic. However, it may be that this variability allows tests of hypotheses about the evolution of reproduction in marine invertebrates. If sperm (and other reproductive) characters are incorporated into data sets that include all morphological evidence then the problems nominated above can be eliminated. Under these circumstances the hypothesis that external fertilization is always primitive then becomes testable and homology assumptions about sperm morphology are open to reassessment. This then raises the question of what the plesiomorphic form of sperm shape and fertilization mode is for Annelida.

What is the plesiomorphic sperm type in Annelida?

Most previous influential systematizations of polychaetes (e.g., Fauchald, 1977) recognise a taxon Phyllodocida, explicitly or implicitly accepting that this is a derived annelid clade. Basal annelids, according to Rouse & Fauchald (1997), are taxa such as Clitellata and simple-bodied forms like Questidae and Paraonidae, currently regarded as part of Scolecida. As pointed out in Rouse & Pleijel (2003) this rooting of Annelida was based on outgroup choices such as Mollusca and Sipuncula, and may well be misleading. However, if it is accepted then it would suggest that the filiform sperm type and complicated reproduction seen in taxa such as most Clitellata, involving fertilization in a cocoon after sperm transfer storage in spermathecae could perhaps be the basal condition for annelids. To accept this would mean that the basal condition for polychaetes such as Scolecida would also have to be filiform sperm and some sort of 'internal' fertilization. It should be noted that most clitellates have and that this is. However, this is currently not known since a wide variety of reproduction from external fertilization to internal fertilization occurs in this group (Rouse & Pleijel, 2001). To properly understand the placement of this also requires knowledge of the plesiomorphic condition in the sister group to

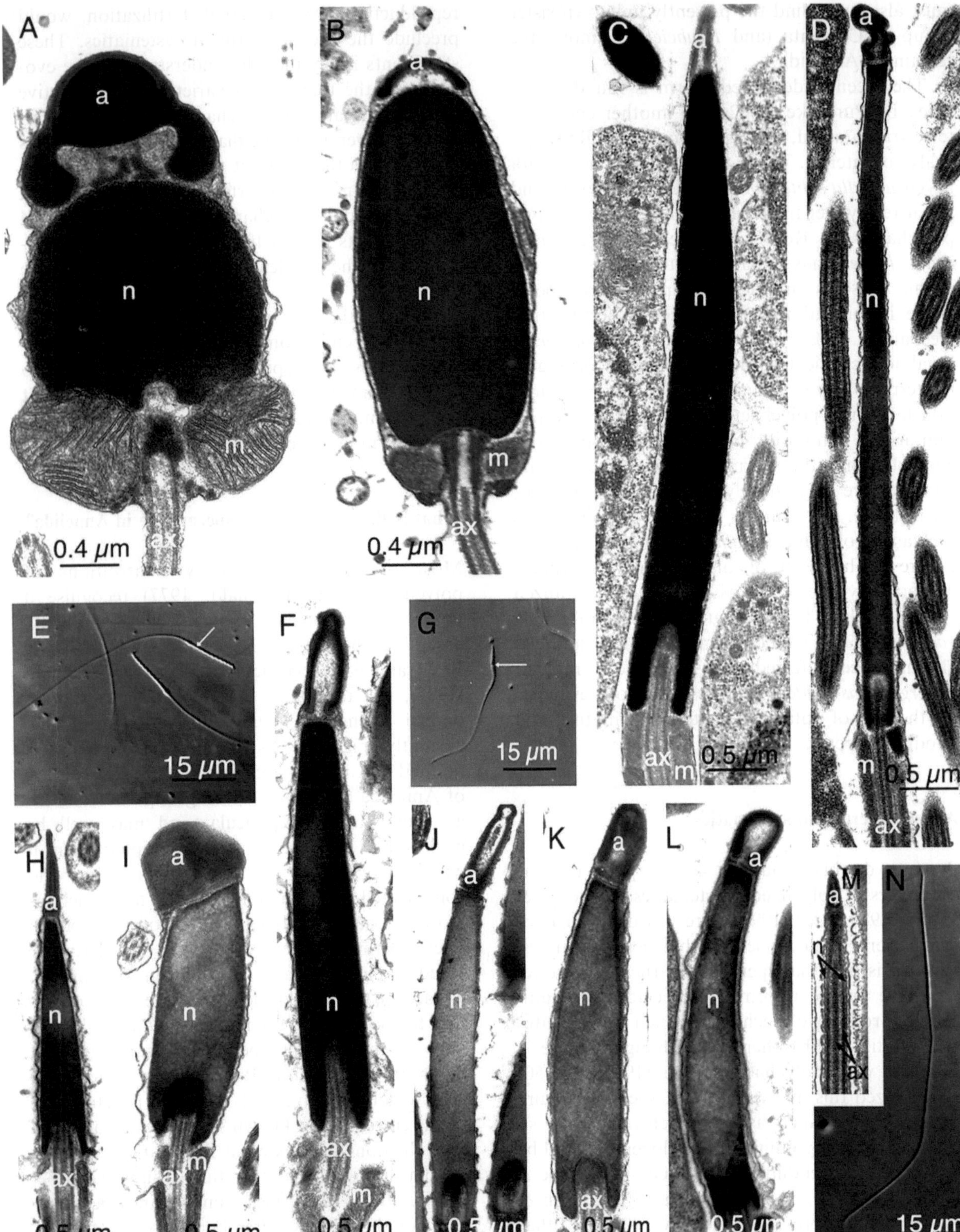
A
a
n
m
ax
0.4 μm
B
a
n
m
ax
0.4 μm
C
a
n
ax
m
0.5 μm
D
a
n
m
0.5 μm
ax
E
15 μm
F
n
ax
m
0.5 μm
G
15 μm
H
a
n
ax
0.5 μm
I
a
n
ax
m
0.5 μm
J
a
n
0.5 μm
K
a
n
ax
0.5 μm
L
a
n
0.5 μm
M
a
n
ax
N
15 μm

Figure 3. Light micrographs of whole sperm or transmission electron micrographs of longitudinal sections through sperm of various serpulid polychaetes. (A) *Galeolaria caespitosa*, (B) *Paraprotis dendrova*, (C) *Salmacina* sp. (D) *Protolaeospira tricostalis*, (E) *Paralaeospira* cf. levinsoni, (F–G) *Spirorbis spirorbis*, (H–I) *Protolaeospira capensis*, (H) narrow plane, (I) broad plane, (J) *Pileolaria* sp, (K) *Metalaeospira tenuis*, (L) *Romanchella quadricostalis*, (M) *Neodexiospira* sp. anterior part of head, (N) *Neodexiospira* sp. Abbreviations: a, acrosome; ax axoneme; m, mitochondrion; n, nucleus.

Annelida. Accepting that Echiura is within Annelida, then most authors would regard Mollusca as the sistergroup to Annelida. The plesiomorphic reproductive condition or molluscs has been suggested to be internal fertilization and filiform sperm (Buckland-Nicks & Scheltema, 1995). However, this argument was criticised by Rouse (1999b) as not being based on adequate consideration of all the data. Thus, even if we accept the topology of annelid phylogeny proposed by Rouse & Fauchald (1997) we cannot infer the basal reproductive mode and sperm shape for annelids.

There are however, several alternative hypotheses that suggest that the root for the annelid tree in a completely different location to that proposed by Rouse & Fauchald (1997) and this was recently reviewed by Purschke (2002b). Storch (1968), Westheide (1997) and Conway Morris & Peel (1995) all presented hypotheses that would root the 'crown' Annelida tree with what is here considered a part of Phyllodocida, or the more inclusive group Aciculata, though none actually explicitly suggest what the relevant basal subgroup that would be. Also the reproductive modes in these groups are also quite varied, also ranging from external to internal fertilization and our knowledge of the phylogeny within the Aciculata subgroups is not well resolved (Rouse & Pleijel, 2001). Given this controversy it becomes difficult to estimate what the plesiomorphic reproductive condition for Annelida is. If we accept that the basic topology of annelid relationships postulated in Rouse & Fauchald (1997) is correct but do no root the tree then a diagram as shown in Figure 2 is the result. This may represent the most conservative representation of our understanding of annelid relationships. On this diagram a range of sperm types have been plotted to represent the diversity of sperm and reproductive modes found in annelids. This is an extremely simplistic view since many subgroups of annelids show a wide range of reproductive modes and we have presently have no idea of how they have evolved. This will be assessed with a brief review of sperm structure in one group, Serpulidae.

Sperm and reproduction in Serpulidae

In this discussion Spirorbinae will treated as part of Serpulidae, with the group also containing Filograninae and Serpulinae (see Rouse & Pleijel, 2001). There have been studies on fertilization and larval development for around more than 30 different Serpulidae, not including Spirorbinae (Kupriyanova et al., 2001). Of these, around half are broadcast spawners with planktotrophic or lecithotrophic larvae. The rest are known to brood larvae, in a variety of fashions; free in the tube, in calcareous brooding pouches outside the tube, attached to the branchial crown, or in pouches of the thoracic membrane. Current classification would group these 30 nominal species into 20 genera. Within Spirorbinae, there is information available for reproductive methods on nearly all of the 120 nominal species (Knight-Jones & Fordy, 1979). All Spirorbinae are brooders of lecithotrophic larvae that have a swimming stage and the method of brooding has been used to classify this group for some time (Bailey, 1969). Reproduction in Filograninae is varied. *Protula tubularia* is a broadcast spawner with lecithotrophic larvae (Meyer, 1888), while other Filograninae are brooders of lecithotrophic larvae (Nishi & Yamasu, 1992a). Operculate members of Serpulinae may be broadcast spawners with planktotrophic larvae, or brooders with lecithotrophic larvae (see Nishi & Yamasu, 1992a). One unusual taxon, *Paraprotis dendrova*, broods larvae in a mass attached to a modified radiole (Nishi & Yamasu, 1992b). In all cases, it would appear that fertilization in the Serpulidae is external, either freely in the water or in the water of the tube thus making the sperm all ent-aquasperm. Where fertilization occurs in the tube then spermathecae have been

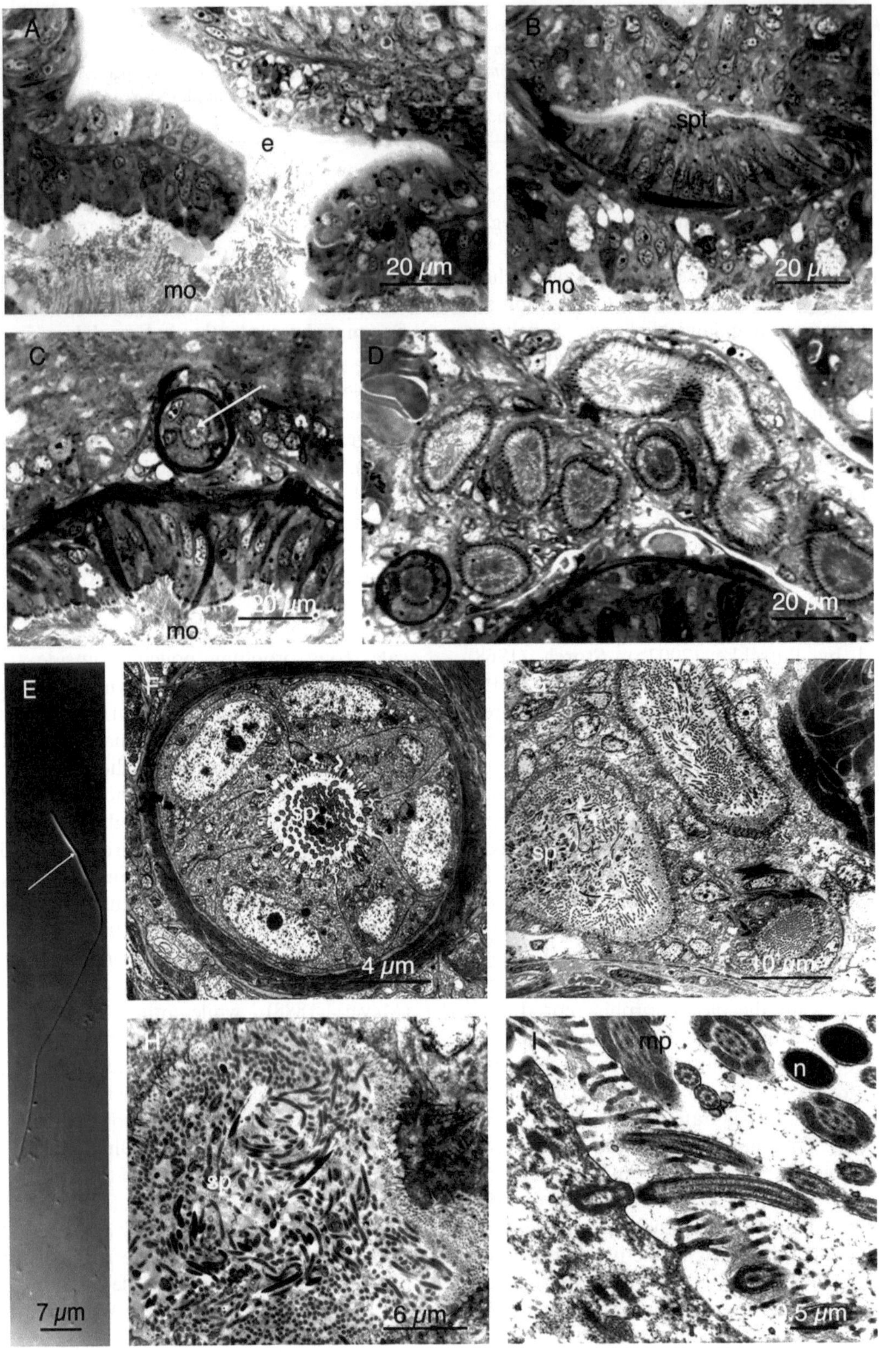
A
e
mo
20 μm
B
spt
mo
20 μm
C
mo
20 μm
D
20 μm
E
7 μm
sp
4 μm
sp
10 μm
sp
6 μm
I
mp
n
0.5 μm

Figure 4. Spermatheca of *Protolaeospira tricostalis,* (A) Mouth region showing entrance to spermatheca. (B) Mouth region showing spermatheca forming a distinct duct. (C) Anterior region of spermatheca where duct has narrowed markedly. Arrow indicates lumen. (D) Posterior region of spermatheca where duct has become very convoluted. (E) Light micrograph of whole sperm of *Protolaeospira tricostalis.* Arrow indicates junction between nucleus and midpiece. (F) Transmission electron micrograph through proximal sperm duct showing sperm free in the lumen. (G) Transmission electron micrograph through distal sperm duct showing sperm free in the lumen. Higher resolution transmission electron micrograph through distal sperm duct showing sperm free in the lumen which is heavily ciliated. (H) Transmission electron micrograph showing spermathecal wall with sperm clearly not attached to epidermal cells. Abbreviations: e, spermatheca entrance; sp, sperm; spt spermathecal duct; mo, mouth; n, nucleus. (A–D) Light micrographs of a transverse 1 μm sections.

documented, or are implicated. A comprehensive review of life histories and reproduction in Serpulidae can be found in Kupriyanova et al. (2001).

Rouse (1999a) listed four species of Serpulinae, 1 Filograninae and 2 Spirorbinae for which there was ultrastructural data of sperm structure. Illustrations of the sperm for three of these are shown here in Figure 3; the serpulin *Paraprotis dendrova* (Fig. 3B), the filogranin *Salmacina* sp. (Fig. 3C, and the spirorbin *Spirorbis spirorbis* (Fig. 3F, G). Additionally micrographs are shown for another serpulin, *Galeolaria caespitosa* (Fig. 3A) and 6 spirorbins; *Metalaeospira tenuis* (Fig. 3K), *Neodexiospira* sp. (Fig. 3M, N), *Pileolaria* sp. (Fig. 3J), *Protolaeospira capensis* (Fig. 3H, I), *Protolaeospira tricostalis* (Fig. 3D), and *Romanchella quadricostalis* (Fig. 3L). The sperm of another spirorbin *Paralaeospira* cf. *levinsoni* is also shown in a light micrograph only (Fig. 3E).

As can be seen in Figure 3, the sperm within Serpulidae vary remarkably. The sperm of *Galeolaria caespitosa* is what is typically associated with an externally fertilizing organism, and this indeed the case (Kupriyanova et al., 2001). The sperm of *Paraprotis dendrova* (Fig. 3B) also appears to be that of an animal with broadcast spawning. It has a short cylindrical nucleus capped by a simple acrosome and the midpiece is simple with a few mitochondria surrounding the anchoring apparatus. The flagellum consists of the axoneme and plasma membrane only. However, Nishi & Yamasu (1992b) described reproduction in this species, which is unusual among serpulids in that it broods larvae on a modified radiole. Some form of interaction between the sperm and female must occur (though no spermathecae are present Rouse, personal observations). Hence, in this case the sperm shape is misleading with respect to fertilization mode and the sperm are ent-aquasperm. The sperm of *Salmacina* sp. has an elongate nucleus and the mitochondria extend down the axoneme (Fig. 3C). This species is known to be a brooder of larvae and Rouse (1996) described a pair of sperm storage organs in the radiolar crown of the adults. It may that the because the sperm are stored before fertilization there has been an elongation of the nucleus and mitochondria, as discussed by Westheide (1984) and Rouse & Jamieson (1987).

As mentioned above, the various sperm of Spirorbinae shown in Figure 3 come from species that brood larvae. The method brooding varies from in the tube (*Metalaeospira, Paralaeospira, Protolaeospira, Romanchella and Spirorbis*), to special opercular capsules held outside the tube (*Neodexiospira* and *Pileolaria*), with further refinements within these broad categories (see Knight-Jones & Fordy, 1979). However, in all cases the sperm are stored prior to fertilization in a single blind spermatheca located dorsal to the mouth (Rouse, in preparation). This has previously been demonstrated for *Spirorbis spirorbis* (Daly & Golding, 1977; Picard, 1980) and is now shown here for *Protolaeospira tricostalis* (Fig. 4). Nevertheless, even though the sperm storage mechanisms are essentially similar, the sperm structures vary dramatically. The sperm of *Protolaeospira tricostalis* has a spiral acrosome (Figs 3D and 4A), while that of *Paralaeospira* cf. *levinsoni* has a spiral acrosome and nucleus (Fig. 3E). The sperm of *Spirorbis spirorbis* (Fig. 3F, G), *Pileolaria* sp. (Fig. 3J), *Metalaeospira tenuis* (Fig. 3K) and *Romanchella quadricostalis* (Fig. 3L) are basically similar, though the latter two have the sperm nucleus bent in one plane. The sperm of *Protolaeospira capensis* is rather unusual though in being flattened in one plane (Fig. 3H, I), reminiscent of the case in mammals. The sperm of *Neodexiospira* sp. is the most unusual in having an extremely elongate head 38 μm (Fig. 3N) and the nucleus is completely penetrated by the axoneme to the extent that it is merely a thin sheath (Fig. 3M).

The striking variability of sperm structure within Spirorbinae presents two opportunities. Firstly, a detailed study of the sperm storage mechanisms and fertilization biology in this group may allow for detailed inference of the actual functions of the various kinds of sperm we see in animals. To date there have been a variety of reasons proposed (see Jamieson & Rouse, 1989), but little resolution has been achieved. Spirorbinae is a small group and presents an ideal opportunity to develop a series of hypothesis on sperm shape and function that can then be assessed for other taxa. For instance the sperm are actually attached to epidermal cells in the spermatheca of Spirorbis. The duct of the spermatheca is also relatively broad and short (Daly & Golding, 1977; Picard, 1980). In contrast, the sperm lie free in the lumen of the spermatheca of *Protolaeospira tricostalis* (Fig. 4 F–I). Also, the lumen of the spermatheca is initially broad, though flattened but then becomes a very elongate narrow and twisted duct (Fig. 4A–G). There are a variety of other spermathecal ducts present in other Spirorbinae (Rouse, in preparation) and these may give valuable indicators of sperm function. Important issues in terms of sperm shape may be that how long the sperm are stored in spermathecae and how they are transferred to fertilize the eggs.

The inference of sperm storage and fertilization mechanism on sperm shape in Spirorbinae will require a good understanding of the evolution of the group. Presently, such a robust phylogenetic hypothesis is this lacking. An understanding of what the plesiomorphic sperm shape and fertilization mechanism is needed before further inferences can be drawn. The various characters present in the sperm and spermathecae will also aid in understanding the phylogeny of the group.

Acknowledgements

Thanks to Eijiroh Nishi for providing the specimen of *Paraprotis dendrova*. Brett Dicks provided valuable help with sectioning and photography. Special thanks to Thomas Bartolomaeus and Günter Purschke for inviting me to the special symposium in honour of Prof. Dr Westheide, whose work has been inspirational. This work was supported by an Australian Research Council and the South Australian Museum.

References

Bailey, J. H. 1969. Methods of brood protection as a basis for reclassification of the Spirorbinae (Serpulidae). Zoological Journal of the Linnean Society 48: 387–407.

Buckland-Nicks, J. & A. Scheltema, 1995. Was internal fertilization an innovation of early Bilateria? Evidence from sperm ultrastructure of a mollusc. Proceedings of the Royal Society of London – Series B: Biological 261: 11–18.

Bunke, D. 1985. Ultrastructure of the spermatozoon and spermiogenesis in the interstitial annelid *Potamodrilus fluviatilis*. Journal of Morphology 185: 203–216.

Conway Morris, S. & J. S. Peel, 1995. Articulated halkieriids from the lower Cambrian of north Greenland and their role in early protostome evolution. Philosophical Transactions of the Royal Society of London, Series B 347: 305–358.

Cotelli, F. & C. Lora Lamia Donin, 1975. Ultrastructural analysis of mature spermatozoa of *Hyalinoecia tubicola* (0. F Müller) (Annelida Polychaeta). Monitore Zoologico Italiano 9: 51–66.

Daly, J. M. & D. W. Golding, 1977. A description of the spermatheca of Spirorbis spirorbis (L.) (Polychaeta: Serpulidae) and evidence for a novel mode of sperm transmission. Journal of the Marine Biological Association of the United Kingdom 57: 219–227.

Degaulejac, B., M. Henry & N. Vicente, 1995. An ultrastructural study of gametogenesis of the marine bivalve *Pinna nobilis* (Linnaeus 1758). 2. Spermatogenesis. Journal of Molluscan Studies 61: 393–403.

Eckelbarger, K. J. & C. M. Young, 2002. Spermiogenesis and modified sperm morphology in the "seepworm" *Methanoaricia dendrobranchiata* (Polychaeta: Orbiniidae) from a methane seep environment in the Gulf of Mexico: Implications for fertilization biology. Biological Bulletin 203: 134–143.

Eckelbarger, K. J., C. M. Young, E. R. Llodra, S. Brooke & P. Tyler, 2001. Gametogenesis, spawning behavior, and early development in the "iceworm" *Hesiocaeca methanicola* (Polychaeta: Hesionidae) from methane hydrates in the Gulf of Mexico. Marine Biology 138: 761–775.

Fauchald, K., 1977. The polychaete worms. Definitions and keys to the orders, families and genera. Natural History Museum of Los Angeles County, Science Series 28: 1–188.

Fernández, J., V. Téllez & N. Olea, 1992. Hirudinea. In: Harrison, F. W. & S. L. Gardiner (eds), Microscopic Anatomy of Invertebrates, vol. 7. Annelida. Wiley–Liss, New York: 323–394.

Ferraguti, M., 1984. The comparative ultrastructure of sperm flagella central sheath in Clitellata reveals a new autapomorphy of the group. Zoologica Scripta 13: 201–207.

Ferraguti, M. 1999a. Euclitellata. In: Jamieson, B. G. M. (ed), Progress in male gamete ultrastructure and phylogeny,

Reproductive Biology of Invertebrates, vol. IXB. Oxford and IBH Publishing Co., New Dehli.

Ferraguti, M. 1999b. Euclitellata. Hydrobiologia 402: 225–237.

Ferraguti, M., C. Erseus, I. Kaygorodova & P. Martin, 1999. New sperm types in Naididae and Lumbriculidae (Annelida: Oligochaeta) and their possible phylogenetic implications. Hydrobiologia 406: 213–222.

Ferraguti, M., C. Erseus & A. Pinder, 1996. The spermatozoon of *Capilloventer australis* and the systematic position of the Capilloventridae (Annelida, Oligochaeta). Australian Journal of Zoology 44: 469–478.

Fitzhugh, K. & G. W. Rouse, 1999. A remarkable new genus and species of fan worm (Polychaeta: Sabellinae) associated with some marine gastropods. Invertebrate Biology 118: 357–390.

Franzén, Å., 1956. On spermiogenesis, morphology of the spermatozoon and biology of fertilization among invertebrates. Zoologiska Bidrag från Uppsala 31: 355–482.

Franzén, Å., 1977. Sperm structure with regard to fertilization biology and phylogenetics. Verhandlungen der Deutschen zoologischen Gesellschaft 1977: 123–138.

Franzén, Å., 1982. Ultrastructure of spermatids and spermatozoa in three polychaetes with modified biology of reproduction: *Autolytus* sp., *Chitinopoma serrula* and *Capitella capitata*. International Journal of Invertebrate Reproduction 5: 185–200.

Franzén, Å. & S. A. Rice, 1988. Spermatogenesis, male gametes and gamete interaction. In: Westheide, W. & C. O. Hermans (eds), The Ultrastructure of Polychaeta. Microfauna Marina 4: 309–333.

Franzén, Å. & M. Ferraguti, 1992. Ultrastructure of spermatozoa and spermatids in *Bonellia viridis* and *Hamingia arctica* (Echiura) with some phylogenetic considerations. Acta Zoologica 73: 25–31.

Gambi, M. C., A. Giangrande & F. P. Patti, 2000. Comparative observations on reproductive biology of four species of *Perkinsiana* (Polychaeta: Sabellidae: Sabellinae). Bulletin of Marine Science 67: 299–309.

Giangrande, A., M. Licciano, P. Pagliara & M. C. Gambi, 2000. Gametogenesis and larval development in *Sabella spallanzanii* (Polychaeta: Sabellidae) from the Mediterranean Sea. Marine Biology 136: 847–861.

Giangrande, A., M. Sciscioli, E. Lepore, M. Mastrodonato, P. Lupetti, & R. Dallai, 2002. Sperm ultrastructure and spermiogenesis in two *Exogone* species (Polychaeta, Syllidae, Exogoninae). Invertebrate Biology 121: 339–349.

Hodgson, A. N. & F. S. Chia, 1993. Spermatozoon structure of some North American prosobranchs from the families Lottiidae (Patellogastropoda) and Fissurellidae (Archaeogastropoda). Marine Biology 116: 97–101.

Jamieson, B. G. M., B. A. Afzelius, & A. Franzén, 1985. Ultrastructure of the acentriolar, aflagellate spermatozoa and the eggs of *Histriobdella homari* and *Stratiodrilus novaehollandiae* (Histriobdellidae, Polychaeta). Journal of Submicroscopic Cytology 17: 363–380.

Jamieson, B. G. M. & G. W. Rouse, 1989. The spermatozoa of the Polychaeta (Annelida): an ultrastructural review. Biological Reviews 64: 93–157.

Jouin-Toulmond, C., M. Mozzo & S. Hourdez, 2002. Ultrastructure of spermatozoa in four species of Alvinellidae (Annelida: Polychaeta). Cahiers de Biologie Marine 43: 391–394.

Knight-Jones, P. & M. R. Fordy, 1979. Setal structure, function and interrelationships in Spirorbidae (Polychaeta, Sedentaria). Zoologica Scripta 8: 119–138.

Kupriyanova, E. K., E. Nishi, H. A. t. Hove & A. V. Rzhavsky, 2001. A review of life history in serpulimorph polychaetes: ecological and evolutionary perspectives. 39: 1–101.

Licciano, M., A. Giangrande & M. C. Gambi, 2002. Reproduction and simultaneous hermaphroditism in *Branchiomma luctuosum* (Polychaeta, Sabellidae) from the Mediterranean Sea. Invertebrate Biology 121: 55–65.

Marotta, R., M. Ferraguti, & C. Erseus, 2003. A phylogenetic analysis of Tubificinae and Limnodriloidinae (Annelida, Clitellata, Tubificidae) using sperm and somatic characters. Zoologica Scripta 32: 255–278.

Meyer, E., 1888. Studien uber den Körperbau der Anneliden. IV. Die Körperform der Serpulacean und Hermellen. Mitteilungen aus der Zoologischen Station zu Neapel 8: 462–662.

Nishi, E. & T. Yamasu, 1992a. Brooding and larval development of a serpulid tube worm *Salmacina dysteri* (Huxley) (Sedentaria: Polychaeta). Bulletin of the College of Science, University of the Ryukyus 54: 107–121.

Nishi, E. & T. Yamasu, 1992b. Brooding habit and development of a serpulid worm *Paraprotis dendrova* Uchida (Annelida, Polychaeta, Sedentaria). Bulletin of the College of Science, University of the Ryukyus 54: 83–92.

Pfannenstiel, H.-D., C. Grünig & J. Lucht, 1987. Gametogenesis and reproduction in nereidid sibling species (*Platynereis dumerilii* and *P. massiliensis*). Bulletin of the Biological Society of Washington 7: 272–279.

Picard, A. 1980. Spermatogenesis and sperm–spermathecae relations in *Spirorbis spirorbis* (L.). International Journal of Invertebrate Reproduction 2: 73–83.

Purschke, G., 2002a. Male genital organs, spermatogenesis and spermatozoa in the enigmatic terrestrial polychaete *Parergodrilus heideri* (Annelida, Parergodrilidae). Zoomorphology 121: 125–137.

Purschke, G., 2002b. On the ground pattern of Annelida. Organisms Diversity & Evolution 2: 181–196.

Purschke, G., 2003. Is *Hrabeiella periglandulata* (Annelida, "Polychaeta") the sister group of Clitellata? Evidence from an ultrastructural analysis of the dorsal pharynx in *H. periglandulata and Enchytraeus minutus* (Annelida, Clitellata). Zoomorphology 122: 55–66.

Retzius, G., 1904. Zur Kenntnis der Spermien der Evertebraten. I. Taf. 1-X11. Biologische Untersuchungen von Gustaf Retzius, Neue Folge 11: 1–32.

Retzius, G., 1905. Zur Kenntnis der Spermien der Evertebraten. II. Biologische Untersuchungen von Gustaf Retzius, Neue Folge 11: 79–102.

Rice, S. A., 1992. Polychaeta: Spermatogenesis and spermiogenesis. In: Harrison F.W. & S.L. Gardiner (eds.), Microscopic Anatomy of Invertebrates Annelida, Vol. 7. Wiley–Liss, New York: 129–151.

Rice, S. A. & K. J. Eckelbarger, 1989. An ultrastructural investigation of spermatogenesis in the holopelagic polychaetes *Vanadis formosa and Krohnia lepidota* (Polychaeta: Alciopidae). Biological Bulletin. Marine Biological Laboratory, Woods Hole, MA. 176: 123–134.

Richards, S. L., 1970. Spawning and reproductive morphology of *Scolelepis squamata* (Spionidae: Polychaeta). Canadian Journal of Zoology 48: 1369–1379.

Rota, E. & P. Lupetti, 1997. An ultrastructural investigation of *Hrabeiella* Pizl & Chalupsky, 1984 (Annelida). 2. The spermatozoon. Tissue and Cell 29: 603–609.

Rouse, G. W., 1988a. An ultrastructural study of the spermatozoa from *Prionospio* cf. *queenslandica* and *Tripolydora* sp.: two spionid polychaetes with different reproductive methods. Acta Zoologica 69: 205–216.

Rouse, G. W., 1988b. An ultrastructural study of the spermatozoa of *Eulalia* sp. (Phyllodocidae), *Lepidonotus* (Polynoidae), *Lumbrinereis* sp. (Lumbrinereidae) and *Owenia fusiformis* (Oweniidae). Helgoländer Meeresuntersuchungen 42: 67–78.

Rouse, G. W., 1992. Ultrastructure of spermiogenesis and spermatozoa of four *Oriopsis* species (Sabellinae; Sabellidae; Polychaeta). Zoologica Scripta 21: 363–379.

Rouse, G. W., 1996. Variability of sperm storage by females in the Sabellidae and Serpulidae (Polychaeta). Zoomorphology 116: 179–193.

Rouse, G. W., 1999a. Polychaeta, including Pogonophora and Myzostomida. In: Jamieson, B. G. M. (ed.), Progress in Male gamete ultrastructure and phylogeny. Reproductive Biology of Invertebrates, Vol. IXB. Oxford and IBH Publishing Co., New Dehli: 81–124.

Rouse, G. W., 1999b. Polychaete sperm: phylogenetic and functional considerations. Hydrobiologia 402: 215–224.

Rouse, G. W. & K. Fauchald, 1997. Cladistics and polychaetes. zoologica Scripta 26: 139–204.

Rouse, G. W. & K. Fitzhugh, 1994. Broadcasting fables: is external fertilization really primitive? Sex, size and larvae in sabellid polychaetes. Zoologica Scripta 23: 271–312.

Rouse, G. W. & B. G. M. Jamieson, 1987. An ultrastructural study of the spermatozoa of the polychaetes *Eurythoe complanata* (Amphinomidae), *Clymenella* sp. and *Micromaldane* sp. (Maldanidae), with definition of sperm types in relation to reproductive biology. Journal of Submicroscopic Cytology 19: 573–584.

Rouse, G. W. & K. Pitt, 2000. Ultrastructure of the sperm of *Catostylus mosaicus* and *Phyllorhiza punctata* (Scyphozoa, Cnidaria): implications for sperm terminology and the inference of reproductive mechanisms. Invertebrate Reproduction & Development 38: 23–34.

Rouse, G. W. & F. Pleijel, 2001. Polychaetes. Oxford University Press, London, 354 pp.

Rouse, G. W. & F. Pleijel, in press. Current problems in polychaete systematics. Hydrobiologia.

Sousa, M. & E. Oliveira, 1994. An ultrastructural study of *Crassostrea angulata* (Mollusca, Bivalvia) spermatogenesis. Marine Biology 120: 545–551.

Storch, V., 1968. Zur vergleichenden Anatomie der segmentalen Muskelsysteme und zur Verwandtschaft der Polychaeten-Familien. Zeitschrift für Morphologie und Ökologie der Tiere 63: 251–342.

Tzetlin, A. B., T. Dahlgren, & G. Purschke, 2002. Ultrastructure of the body wall, body cavity, nephridia and spermatozoa in four species of the Chrysopetalidae (Annelida, "Polychaeta"). Zoologischer Anzeiger 241: 37–55.

Westheide, W., 1984. The concept of reproduction in polychaetes with small body size: adaptations in interstitial species. Fortschritte der Zoologie 29: 265–287.

Westheide, W., 1997. The direction of evolution within the Polychaeta. Journal of Natural History 31: 1–15.

Williams, J. D., 2000. Spermiogenesis and spermatozoon ultrastructure of *Polydora neocaeca* (Polychaeta: Spionidae) from Rhode Island. Invertebrate Reproduction & Development 38: 123–129.

Hydrobiologia (2005) **535/536:** 179–198
T. Bartolomaeus & G. Purschke (eds), Morphology, Molecules, Evolution and Phylogeny in Polychaeta and Related Taxa

Oogenesis and oocytes

Kevin J. Eckelbarger
Darling Marine Center, University of Maine, 193 Clark's Cove Road, Walpole, ME 04573, U.S.A.
School of Marine Sciences, University of Maine, Orono, ME 04469, U.S.A.
E-mail: kevine@maine.edu

Key words: oogenesis, ultrastructure, gametogenesis

Abstract

The morphological features of polychaete ovarian morphology and oogenesis are reviewed. Some basic information on ovarian structure and/or oogenesis is known for slightly more than half of recognized polychaete families although comprehensive studies of oogenesis have been conducted on ~0.1% of described species. Relative to other major metazoan groups, ovarian morphology is highly variable in the Polychaeta. While some species appear to lack a defined ovary, most have paired organs that are segmentally repeated to varying degrees depending on the family. Ovaries vary widely in their location but are most frequently associated with the coelomic peritoneum, parapodial connective tissue, or elements of the circulatory system. The structural complexity of the ovary is correlated with the type of oogenesis expressed by the species. In some polychaetes, extraovarian oogenesis occurs in which previtellogenic oocytes are released into the coelom from a simple ovary where differentiation occurs in a solitary fashion or in association with nurse cells or follicle cells. In other species, intraovarian oogenesis occurs in which oocytes undergo vitellogenesis within the ovary, often in association with follicle cells that may provide nutrition. Vitellogenesis probably includes both autosynthetic and heterosynthetic processes; autosynthesis involves the manufacture of yolk bodies via the proteosynthetic organelles of the oocyte whereas heterosynthesis involves the extraovarian production of female-specific yolk proteins that are incorporated into the oocyte through a receptor-mediated process of endocytosis. Variation in the speed of egg production varies widely and appears to be correlated with the vitellogenic mechanism employed. Mature ova display a wide range of egg envelope morphologies that often show some intrafamilial similarities.

Introduction

Polychaetes are well represented in marine benthic communities and currently comprise 9000 recognized species (Rouse & Pleijel, 2001). They also show spectacular diversity with respect to life history features and reproductive habits (Schroeder & Hermans, 1975; Wilson, 1991; McHugh & Fong, 2002) although the entire life history has been described for only 3% of described species. Comprehensive studies of oogenesis have been conducted in only ˜0.1% of described species (reviewed by Eckelbarger, 1983, 1984, 1988, 1992) and our knowledge of oogenesis is almost exclusively directed at shallow water species although deep sea species are gaining new attention (reviewed in Eckelbarger et al., 2001). A number of earlier reviews are available including Schroeder & Hermans (1975), Eckelbarger (1983, 1984, 1986, 1988, 1992), and Fischer & Pfannenstiel (1984).

Structure of the ovary

Polychaetes show far greater ovarian morphological diversity than most major metazoan groups including their close relatives the Mollusca and Sipuncula (Eckelbarger, 1984, 1988, 1992). Most polychaetes have well-defined ovaries but their location, number, and structural complexity are too variable to generalize. However, ovarian morphology is usually highly conserved within a family although a few exceptions have been noted. For example, in some serpulimorphs (e.g. *Hydroides dianthus*, *Ficopomatus enigmaticus*), true gonads are absent and only small proliferative germinal epithelia are observed (Kupriyanova et al., 2001). However, in other serpulimorphs (e.g. *Salmacina/Filograna* complex) a distinct gonad is present (Kupriyanova et al., 2001). A few families, notably the Nereididae (Olive, 1983), Alciopidae (Eckelbarger & Rice, 1988), Sphaerodoridae (Christie, 1984), and Phyllodocidae (Olive, 1975), appear to lack discrete ovaries altogether. In all known examples, ovaries are retroperitoneal (Potswald, 1969) and they are often segmentally repeated the length of the female or they may be restricted to a few anterior segments as in the Terebellidae, Ampharetidae, Arenicolidae, and Pectinariidae. The position of the ovary varies considerably from one family to another but they are most frequently encountered in the parapodia or ventral region of the body. Some are associated with parapodial connective tissue or the coelomic peritoneum, while others are attached to some element of the circulatory system, especially those of the intersegmental septa, the ventral or nephridial blood vessels, or various blind-ended capillaries.

Patterns of oogenesis

Two patterns of oogenesis have been identified in polychaetes including intraovarian and extraovarian (Eckelbarger, 1983) with approximately half the families falling into one pattern or the other. In species having extraovarian oogenesis, oocytes leave the ovary prior to vitellogenesis and complete differentiation in the coelom (Fig. 1B–D). In this instance, the ovary serves as a proliferative organ and generally remains rather small in size (Olive, 1971) (Fig. 1A). Oocyte development may be solitary whereby oocytes float freely in the coelom (Fig. 1B) or they may be associated with follicle or nurse cells (Fig. 1C, D) during all or a portion of the coelomic phase (Eckelbarger, 1988). The majority of polychaete families are characterized by either intraovarian or extraovarian oogenesis although there are examples of both types in the Spionidae and Arenicolidae. In the Spionidae, *Marenzelleria viridis* (Bochert, 1996) and *Streblospio benedicti* (Eckelbarger, 1980) undergo intraovarian oogenesis, whereas extraovarian oogenesis is described in *Polydora ligni* and *Spio setosa* (Eckelbarger, 1988, 1992). In the Arenicolidae, *Arenicola marina* (Newell, 1948), *A. brasiliensis* (Okada, 1941), and *Branchiomaldane vincenti* (Ashworth, 1912) have extraovarian oogenesis, while *Arenicolides ecaudata* undergoes intraovarian oogenesis (Southward & Southward, 1958).

Extraovarian oogenesis

In many families (e.g. Glyceridae, Pectinariidae, Sabellidae), individual oocytes leave the ovary as small, previtellogenic cells (Fig. 1A, B) and enter the coelom where they undergo growth and differentiation solitarily. However, in a few families, for example, Terebellidae (Eckelbarger, 1975), Cirratulidae (Olive, 1971; Christie, 1985), Amparetidae (Hutchings, 1973), Pectinariidae (Tweedell, 1966), clusters of small, previtellogenic oocytes are initially released into the coelom. These clusters are enveloped by a thin layer of follicle cells (=sheath cells) derived from the peritoneum or ovarian epithelium and the oocytes lack intercellular connections. In other families, for example, Nereididae (Fischer, 1975) and Alciopidae (Eckelbarger & Rice, 1988) (Fig. 2A–E), previtellogenic oocyte clusters are observed (Fig. 2A) but the oocytes communicate via intercellular bridges (Fig. 2D, E). In all of the above examples, oocytes rupture through the epithelium prior to vitellogenesis and disperse independently in the coloem except in *Pholoe minuta* (Pholoidae) where they remain as clusters through late vitellogenesis (Heffernan & Keegan, 1988).

Intraovarian oogenesis

In some families (e.g. Capitellidae, Sabellariidae, Orbiniidae), oocytes remain confined to the ovary until near the end of vitellogenesis (Fig. 1E–H). Developing oocytes are often closely associated with blood vessels that extend into the coelom from the body wall as in *Hesiocaeca methanicola* (Fig. 1F) or intersegmental septa as in *Phragmatopoma lapidosa* (Fig. 1G). In capitellids, oocytes develop within the ovaries while encircled with follicle cells (Fig. 1H). Eggs are ovulated into the coelom for immediate spawning via coelomoducts or they may undergo a period of coelomic storage prior to spawning.

Stages of Oogenesis

Oogenesis consists of a proliferative phase followed by an often prolonged growth phase resulting in the production of the mature ovum. The proliferative phase involves gametogonial and premeiotic (spireme) stages while the growth phase involves a previtellogenic and vitellogenic stage. Few studies have examined the origin of germ cells in polychaetes and a confusing terminology has been used to described early stages of oogenesis (reviewed in Eckelbarger, 1988, 1992).

Oogenesis – previtellogenesis

The morphological events that occur during this phase are very similar in most species. The smallest recognizable germ cells are best referred to as oogonia (Clark & Olive, 1973) and they show a consistent morphology from one species to another. These early cells usually display a disproportionately large nucleus with peripherally displaced heterochromatin (Fig. 3A) and ooplasm devoid of organelles except perhaps a few mitochondria (Fig. 3B). Oocytes in the zygotene/pachytene stage of meiotic prophase are relatively easily to identify due to their lack of nucleoli and the presence of distinctive synaptonemal complexes of synapsed chromosomes (Fig. 2B; Fig. 3A, insert). Following this premeiotic phase, the nucleolus and diffuse chromatin reappears, and the nucleus enlarges to form what is now referred to as the germinal vesicle (Fig. 3B, C).

Ultrastructural changes in the nucleus during the previtellogenic phase correlate with ooplasmic metabolic events but our knowledge is largely confined to studies of nereidid species (Dhainaut, 1972, 1984; Bertout et al., 1981). A prominent nucleolus appears during the diplotene stage and assumes a central position as oocyte growth begins (Fig. 3B, C). The nucleolus undergoes a number of ultrastructural changes and may divide into a number of accessory micronucleoli that assume positions against the inner nuclear envelope. Groups of nucleocytoplasmic granules migrate to the periphery of the nucleus and enter the perinuclear ooplasm via the nuclear pores (Fig. 3D) where they form fibrogranular clusters ('nuage') (Fig. 3E). Nuage is believed to be ribonucleoprotein of nuclear origin and it often associated with groups of pleomorphic mitochondria that greatly increase in number during this phase (Fig. 3B, D).

The release of early oocytes into the coelomic fluid usually stimulates the rapid formation of surface microvilli (Fig. 3B) that probably function in absorbing low-molecular weight yolk precursors and vital metabolites during oocyte differentiation. In most oocytes, vesicular and lamellar forms of rough endoplasmic reticulum appear (Fig. 4A, B), often in two morphological forms. For example, in the oocytes of *Platynereis dumerilii* (Fischer, 1975), *Harmothoe imbricata*, (Garwood, 1978), *Capitella jonesi* (Eckelbarger & Grassle, 1982) (Fig. 4A), and *Hesiocaeca methanicola* (Eckelbarger et al., 2001), the RER characteristically forms large, parallel arrays, while in *Chaetopterus pergamentaceus* (Villa, 1970), *Phragmatopoma lapidosa* (Eckelbarger, 1979), and *Marenzelleria viridis* (Bochert, 1996), concentric configurations are typical. Golgi complexes also appear in great abundance in the oocytes of many species (Fig. 4B) but they do not undergo product secretion until the vitellogenic phase.

Oogenesis – vitellogenesis

Vitellogenesis is often a prolonged process during the diplotene stage of the first meiotic prophase. It results in a significant increase in the volume of the oocyte largely due to the accumulation of nutritive material in the ooplasm. Nutritive stores vary but may consist of lipid droplets and glycogen granules but more often of yolk granules. Yolk granules are membrane-bounded, electron dense organelles of variable morphology often consisting of lipoglycoproteins (vitellin). The type and quantity of yolk stores appears to be species-specific with differences being noted in even closely related species (Allen, 1961; Eckelbarger & Grassle, 1983).

Vitellogenesis has been extensively documented in polychaetes but almost exclusively on the morphological level. Unfortunately, this singular

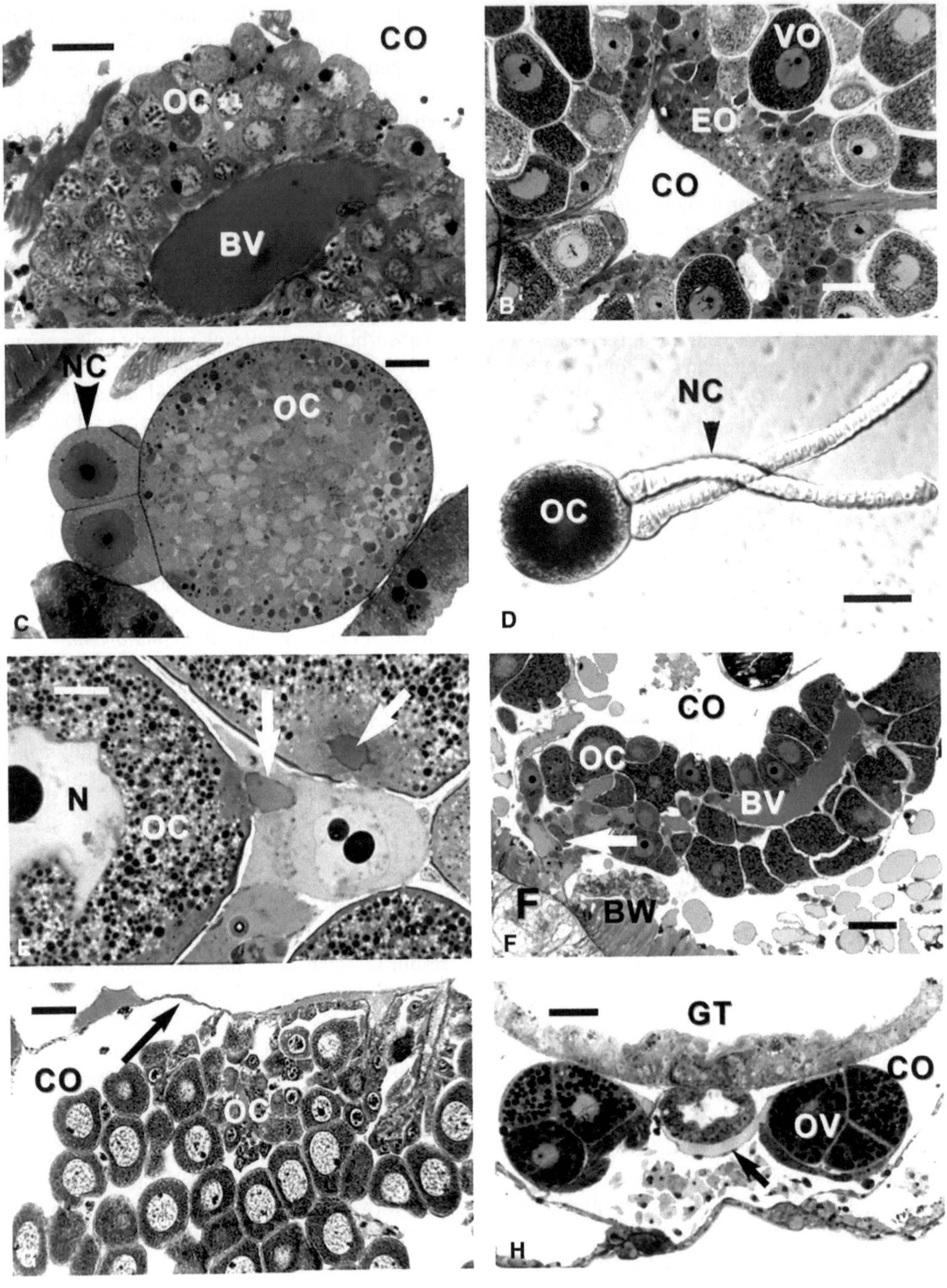
CO
OC
BV
A
VO
EO
CO
B
NC
OC
C
NC
OC
D
N
OC
E
CO
OC
BV
BW
F
F
CO
OC
GT
CO
OV
H

Figure 1. (A–D) species with extraovarian oogenesis. (E–H) species with intraovarian oogenesis. (A) Early oocytes (EO) budding from a coelomic blood vessel (BV) in *Sabella* sp. CO = coelom. Scale = 10 μm. (B) Early oocytes (EO) and vitellogenic oocytes (VO) proliferating into the coelom (CO) in *Owenia* sp. Scale = 50 μm. (C) Large vitellogenic oocyte (OC) with two attached nurse cells (arrowhead) in *Tomopteris pacifica.* Scale = 15 μm. (D) Oocyte (OC) and two chains of nurse cells (NC) in *Diopatra cuprea.* Scale = 50 μm. (E) Oocytes (OC) attached to blood vessels (arrowheads) in *Leitoscoloplos fragilis.* Scale = 10 μm. (F) Genital blood vessels (BV) proliferating from body wall (BW) (arrowhead) while surrounded with developing oocytes (OC) in *Hesiocaeca methanicola.* Scale = 90 μm. (G) Intersegmental blood vessel (arrow) with attached ovary (OV) in *Phragmatopoma lapidosa.* Scale = 90 μm. (H) Paired ovaries (OV) attached to blood vessels (arrowhead) of ventral gut (GT) in *Capitella jonesi.* CO = coelom. Scale = 90 μm.

approach can be misleading. The yolk proteins (vitellins) stored during vitellogenesis are formed from yolk protein precursors, vitellogenins, a specific class of exogenous proteins that usually comprise 60–90% of the soluble egg yolk proteins within the egg (Hagedorn & Kunkel, 1979). Two synthetic processes are believed to be involved in vitellogenesis in all metazoan oocytes: (1) autosynthesis, in which vitellin is produced by the oocyte itself using its proteosynthetic organelles, and (2) heterosynthesis, the extraovarian production of vitellogenin and transport into the oocyte via endocytosis for assembly into vitellin. In all likelihood, a combination of these processes probably occurs in most metazoans to varying degrees. In the majority of species, vitellogenic mechanisms are surmised by describing the activities of various proteosynthetic organelles during yolk synthesis. Nearly all polychaete oocytes contain Golgi complexes and RER (reviewed in Eckelbarger, 1984, 1988, 1992) and both organelles appear to collaborate in the synthesis of yolk bodies (Fig. 4B, C). The RER cisternae often swell and become distended with product and transition vesicles from the RER fuse with the forming face of the Golgi complex. Progressively larger yolk bodies result from the fusion of secretory vesicles released from the maturing face of the Golgi complex (Fig. 4C). Mature yolk bodies have been described that represent a wide variety of morphological forms (reviewed in Eckelbarger, 1988).

Based on morphology alone, it has been proposed that in a few species, notably *Dinophilus ciliatus* (Grün, 1972) and *Harmothoe imbricata* (Garwood, 1981), yolk synthesis occurs entirely within the RER cisternae, while in *Mercierella enigmatica,* Sichel (1966) proposed that yolk is derived from transformed oocyte mitochondria. In studies of oogenesis in *Ophryotrocha labronica* (Emanuelsson, 1969) and *Chaetopterus pergamentaceus* (Villa, 1970), 'multivesicular bodies' were implicated in the formation of yolk bodies. Other studies of oogenesis have described multivesicular-like bodies in oocytes including those of *Ophryotrocha puerilis* (Pfannenstiel & Grünig, 1982) and *Phragmatopoma lapidosa* (Eckelbarger, 1979) but interpreted them as immature yolk bodies.

The exclusive use of ultrastructural evidence as a means of determining the mechanism of yolk synthesis can result in misleading conclusions as demonstrated by extensive studies of nereidid oocytes. Early morphological and radioactive tracer studies suggested that oocytes manufactured yolk via autosynthetic processes whereas later studies indicated that heterosynthetic uptake of yolk precursors was the predominate mechanism (reviewed in Fischer & Hoeger, 1993; Lawrence & Olive, 1995; Fischer, 1999). There are suggestions that the length of vitellogenesis is closely related to the mechanism of yolk synthesis (auto- vs. heterosynthesis) with relatively slow yolk synthesis being associated with autosynthesis, while rapid yolk synthesis is correlated with heterosynthesis (Eckelbarger, 1983, 1988, 1992, 1994). For example, oocyte differentiation exceeds a year in the nereidids *Nereis virens* (Brafield & Chapman, 1967) and 7–8 months in *N. grubei* (Schroeder, 1968). Low endocytotic activity has been reported from these species despite documented evidence that female specific protein is incorporated into oocytes during the extended growth period (Fischer & Dhainaut, 1985) and a similar pattern is shown for other semelparous nereidids, *Nereis diversicolor*, *N. pelagica*, *Perinereis cultrifera*, and *Playtnereis dumerili* (reviewed in Fischer & Rabien, 1986). In contrast, oogenesis occurs in less than 3 months in *Neanthes arenaceodentata* (Davis, 1969) and high levels of endocytosis have been observed during vitellogenesis (Eckelbarger, unpublished).

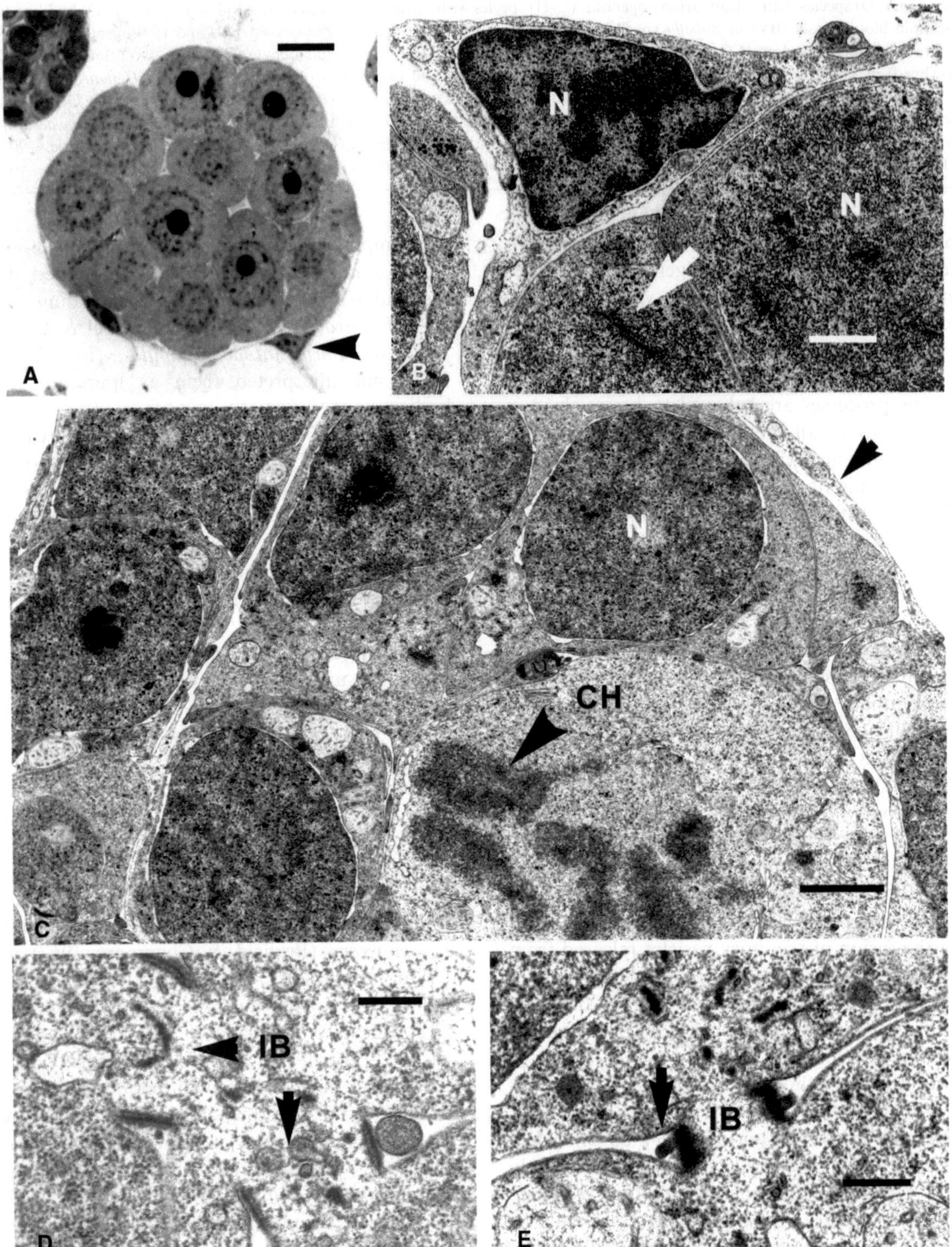

Figure 2. (A) Packet of previtellogenic oocytes floating in coelom of *Rhynchonerella angelini* while surrounded by thin layer of sheath cells (arrowhead). Scale = 50 μm. (B) Nucleus (N) of sheath cell surrounding oocytes of *Alciopa reynaudii* in zygotene/pachytene stage of meiosis. Arrowhead indicates synapsing chromosomes in nucleus. Scale = 8 μm. (C) Coelomic cluster of oogonia in *Alciopa reynaudii* showing one in mitotic division with prominent chromosomes (CH). Arrowhead indicates surrounding sheath cell. N = nucleus. Scale = 8 μm. (D) Two intercellular bridges (IB) connecting three oocytes in coelomic cluster in *Alciopa reynaudii*. Scale = 5 μm. (E) Intercellular bridge (IB) between two oocytes in *Alciopa reynaudii*. Arrowhead indicates dense external ring around bridge. Scale = 5 μm.

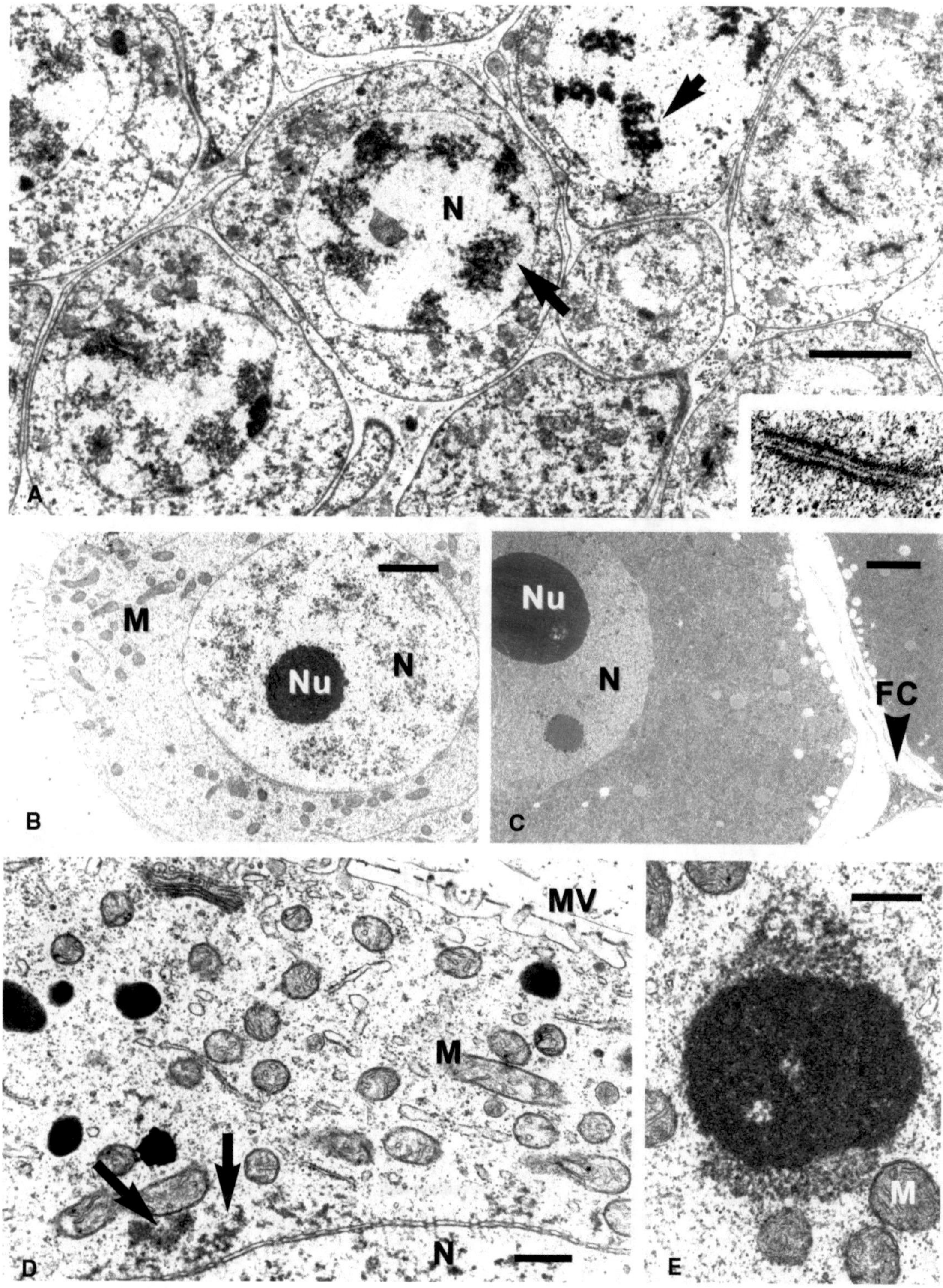

Figure 3. (A) Early oogonia in ovary of *Sabella* sp. showing patches of heterochromatin (arrows) in nuclei (N). Insert: synaptonemal complex in nucleus of *Rhynchonerella moebii*. Scale = 8 μm. (B) Previtellogenic oocyte of *Polydora ligni* with prominent nucleolus (Nu) in nucleus (N) and clusters of perinuclear mitochondria (M). Scale = 5 μm. (C) Previtellogenic oocyte of *Biremis blandi* with nucleus (N) and large nucleolus (Nu). Note surrounding follicle cells (FC). Scale = 10 μm. (D) Ooplasm of *Streblospio benedicti* showing nuage (arrows) adjacent to nucleus (N). M = mitochondria MV = microvilli. Scale = 0.5 μm. E. Nuage from previtellogenic oocyte of *Capitella* sp. II. M = mitochondrion. Scale = 0.5 μm.

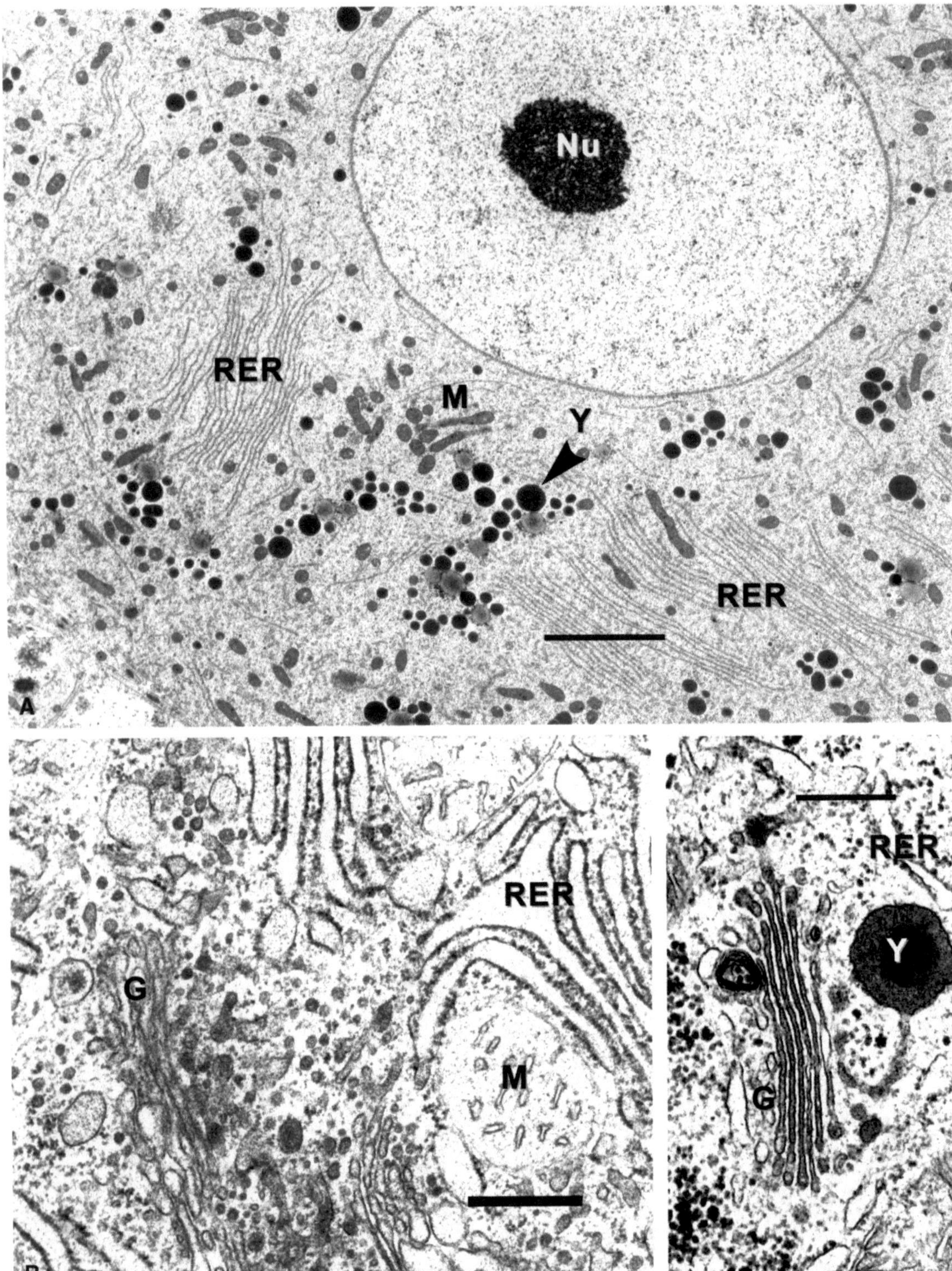

Figure 4. (A) Early vitellogenic oocyte of *Capitella jonesi* showing nucleus (N), nucleolus (Nu), parallel arrays of rough endoplasmic reticulum (RER), mitochondria (M) and yolk bodies (Y). Scale = 5 μm. (B). Ooplasm of *Alciopa reynaudii* showing Golgi complex (G), RER, and mitochondria (M). Scale = 0.5 μm. (C). Ooplasm of *Capitella jonesi* showing RER cisternae and Golgi complex (G) secreting nascent yolk body (Y). Scale = 3 μm.

In addition to yolk bodies, the mature oocytes of many species contain annulate lamellae, parallel arrays of fenestrated cytomembranes that closely resemble fragments of nuclear envelope (Fig. 5A, C). Ooplasmic and internuclear annulate lamellae have been described from at least a dozen species but their function is unknown (reviewed in Eckelbarger 1988, 1992). Many mature ova also contain cortical granules, membrane-bounded organelles of Golgi origin that are involved in the fertilization reaction in many species (Fig. 5B, D).

Blood vessel–oocyte associations

Ovaries are often associated with blood vessels but the degree of intimacy between the oocyte and blood vessel lumen and the duration of this association vary greatly. In species exhibiting extraovarian oogenesis, previtellogenic oocytes are released into the coelom from the blood vessels (Fig. 1A) while those with intraovarian oogenesis retain a blood vessel association throughout all or most of vitellogenesis (Fig. 1E–H). In some species having intraovarian oogenesis such as *Aricidea fragilis* (Eckelbarger, 1988, 1992) and *Novaquesta* sp., direct oocyte contact with blood vessel lumina is prevented by intervening myoepithelial cells (Fig. 6A,B). Similarly, in *Marenzelleria viridis* (Bochert, 1996), perivasal cells prevent direct contact with the blood vessels. In *Hesiocaeca methanicola* (Eckelbarger et al., 2001), follicle-like cells surround each oocyte so they cannot directly contact the vessel lumina. In other species such as *Flabelliderma commensalis* (Spies, 1977), *Harmothoe imbricata* (Garwood, 1981), *Diplocirrus glaucus* (Olive, 1983), *Nephtys hombergii* and *N. caeca* (Olive, 1978), and *Kefersteinia cirrata* (Olive & Pillai, 1983), oocytes contact the blood vessel lumina but there is no ultrastructural evidence of nutrient transfer. Strong ultrastructural evidence for the direct transfer of nutrients from the circulatory system to oocytes has been documented in *Phragmatopoma lapidosa* (Eckelbarger, 1979), *Streblospio benedicti* (Eckelbarger, 1980), *Spio setosa* (Eckelbarger, unpublished), and *Leitoscoloplos fragilis* (Eckelbarger, unpublished) where the oocytes have an intimate association with the vessel lumen (Fig. 6C). In these examples, high levels of endocytotic activity in the perivasal zones of the oocytes suggest heterosynthetic uptake of female-specific vitellogenins from the circulatory system (Fig. 7A–C).

Follicle cell–oocyte associations

Follicle cells are somatic, mesodermally derived cells that are associated with developing oocytes in all species studied. They are usually thin, squamous cells (Fig. 8A, B) but they can undergo hypertrophy during vitellogenesis in some species. They lack cytoplasmic continuity with oocytes but may form junctional complexes with the underlying oocytes. In an unusual variation, follicle cells in the ovary of *Cossura* cf. *longocirrata* contact the underlying oocytes via 'cytoplasmic bridges' (Rouse & Tzetlin, 1997) although they are better described as foot processes that contact the oocyte surface but lack cytoplasmic continuity. These follicle cells are strikingly similar to those documented in gastropod limpets (Hodgson & Eckelbarger, 2000).

Follicle cells have been hypothesized to play a variety of supportive roles during oogenesis including: (1) mechanical support for oocyte clusters, (2) synthesis and/or transfer of yolk precursors and metabolites, and (3) the digestion and resorption of moribund eggs in atretic follicles. Follicle cells originate from the peritoneum and they show great variation in their degree of temporal association with oocytes. In many species, perhaps the majority, follicle cells are small, they contain few proteosynthetic organelles suggesting little biosynthetic activity, and they do not undergo any significant morphological changes while associated with developing oocytes. In *Platynereis dumerilii* (Fischer, 1975), *Nicolea zostericola* (Eckelbarger, 1975), *Biremis blandi* (Fig. 3C), and *Rhynchonerella angelini*, *Alciopa reynaudii*, and *Vanadis formosa* (Eckelbarger & Rice, 1988) (Fig. 2A–E), for example, previtellogenic oocyte clusters are released into the coelom initially encapsulated in follicle cells that later disintegrate. Their only role appears to be maintaining the temporary structural integrity of the clusters. In other species undergoing extraovarian oogenesis but lacking initial oocyte clusters, oocytes develop solitarily in the coelom

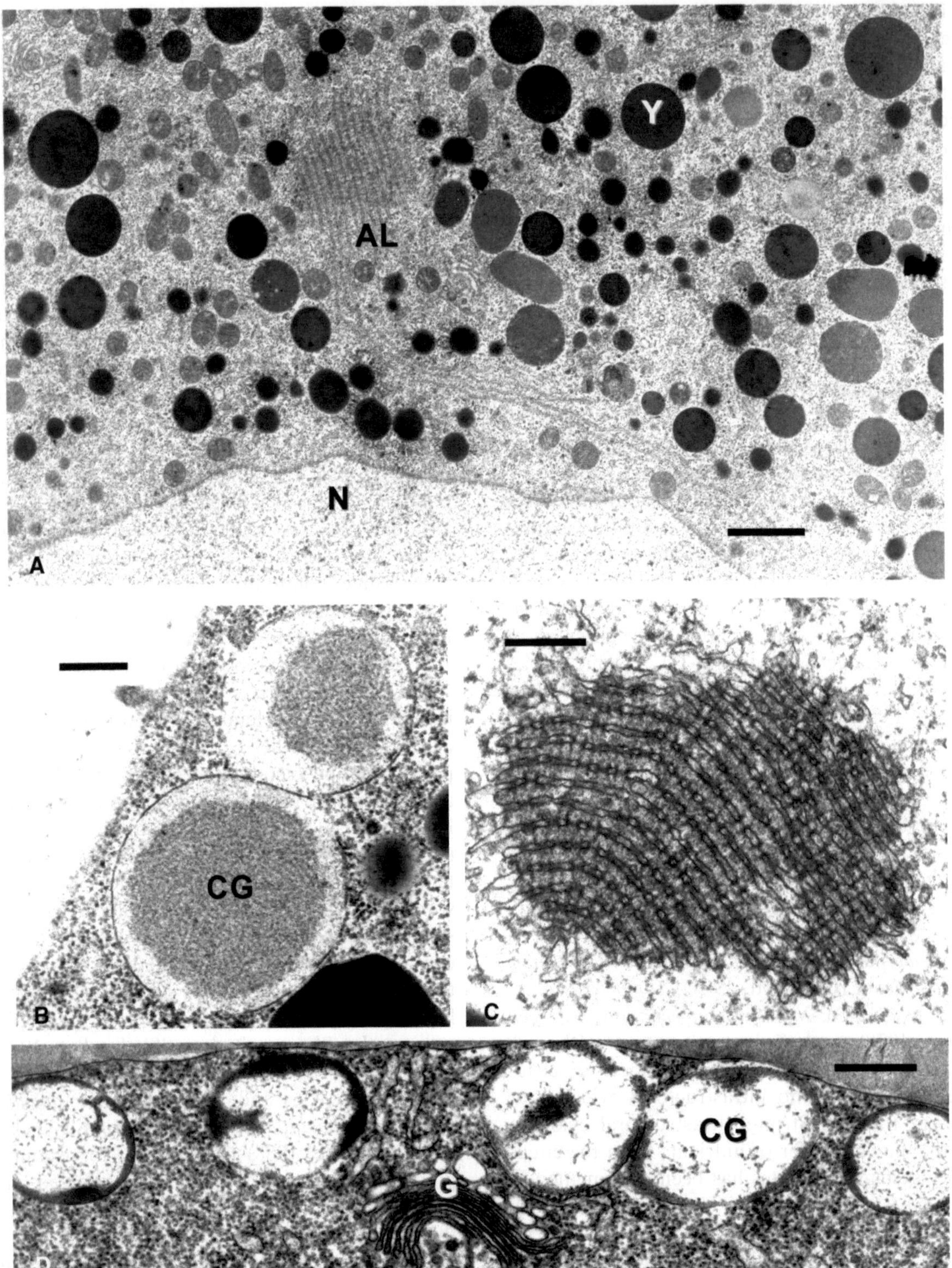

Figure 5. (A) Arrays of annulate lamellae (AL) in the perinuclear ooplasm of *Hydroides dianthus.* Y = yolk body; N = nucleus. Scale = 5 μm. (B) Cortical ooplasm in an oocyte of *Hydroides dianthus* showing two cortical granules (CG). Scale = 1 μm. (C) Annulate lamellae in the oocyte of *Pygospio elegans.* Scale = 1 μm. (D) Golgi complex (G) synthesizing cortical granules (CG) in ooplasm of *Phragmatopoma lapidosa.* Scale = 4 μm.

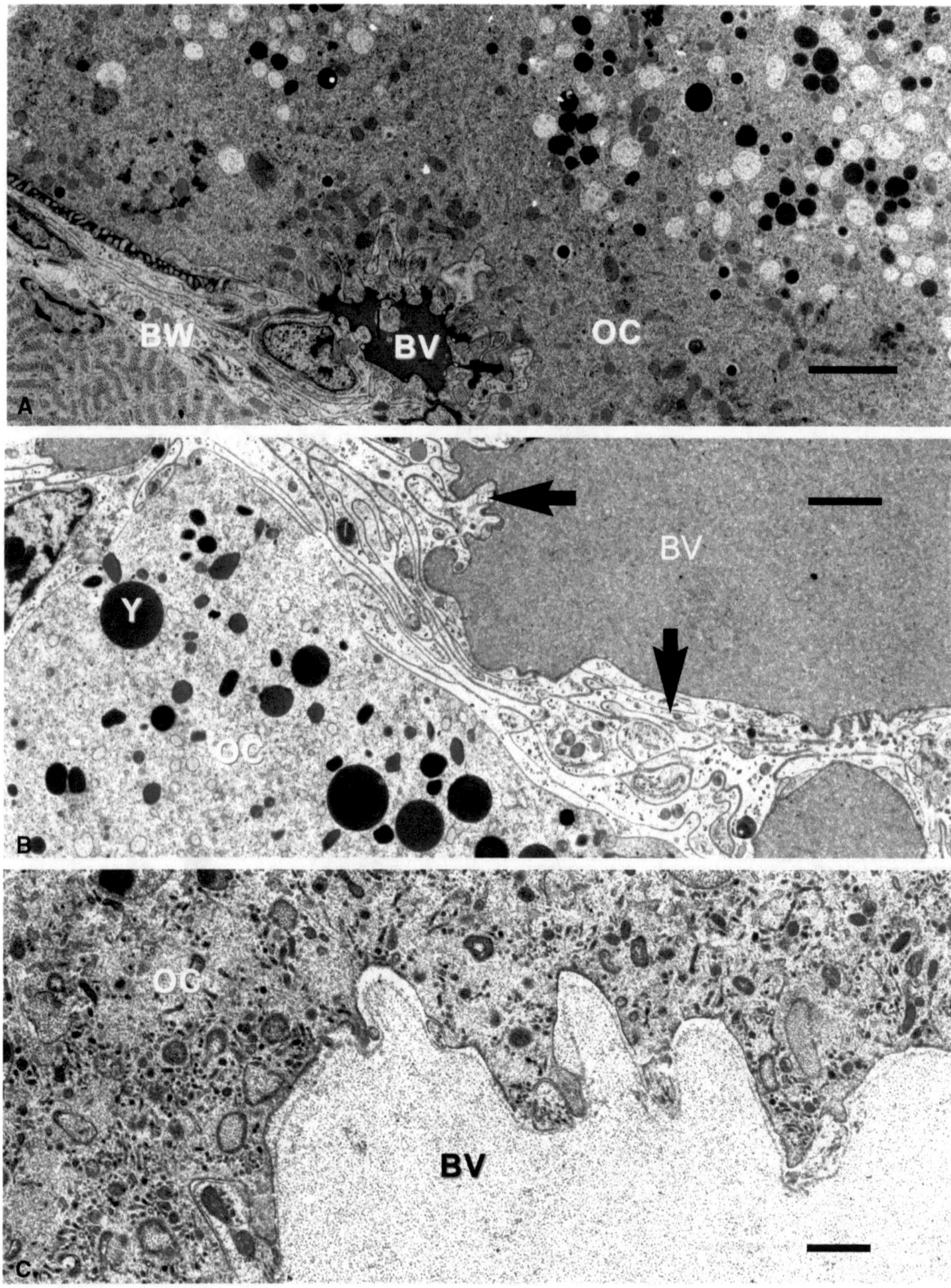

Figure 6. (A) Early vitellogenic oocyte (OC) attached to genital blood vessel (BV) in *Novaquesta* sp. Scale = 15 μm. (B) Perivasal cells (arrows) positioned between genital blood vessel (BV) and early vitellogenic oocyte (OC) in *Aricidea fragilis*. Y = yolk body. Scale = 5 μm. (C) Early vitellogenic oocyte (OC) in direct contact with genital blood vessel (BV) in *Streblospio benedicti*. Scale = 1.5 μm.

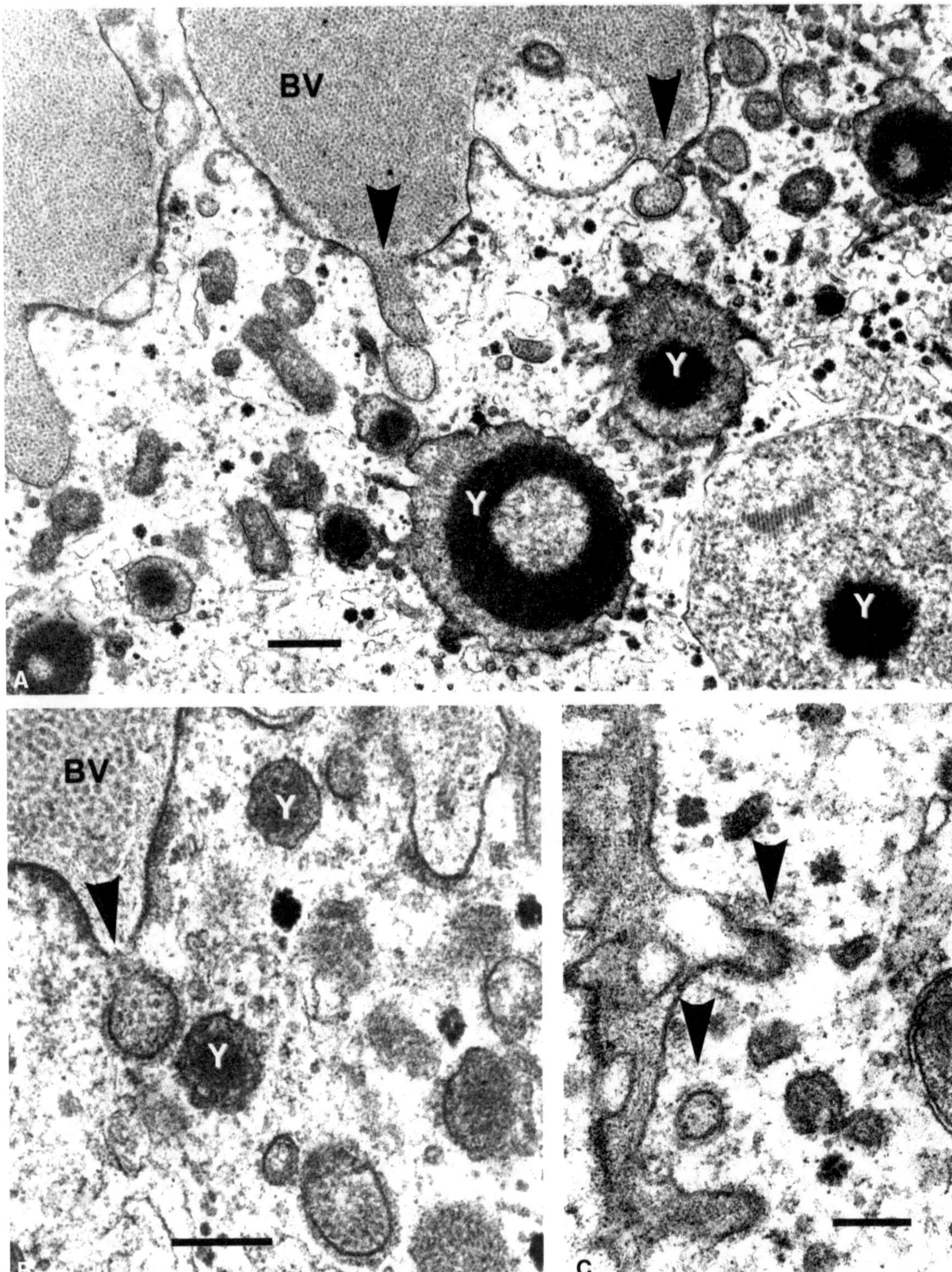

Figure 7. A–C. Perivasal region of early vitellogenic oocyte in *Streblospio benedicti*. (A). Endocytotic pit formation (arrowheads) and heterosynthetic formation of yolk bodies (Y) through fusion of endosomes. Scale = 0.3 μm. (B) Higher magnification of perivasal region showing pit formation (arrowhead) along oocyte–blood vessel interface. Y = nascent yolk body. Scale = 0.3 μm. (C) Endocytotic activity (arrowheads) along surface of oocyte. Scale = 0.5 μm.

so follicle cells play no further role. The only exception known is that of *Cossura* cf. *longocirrata* (Rouse & Tzetlin, 1997) in which follicle cells remain attached to coelomic oocytes until late vitellogenesis. In contrast, follicle cells are universally associated with oocytes throughout vitellogenesis in species undergoing intraovarian oogenesis. In *Harmothoe imbricata* (Garwood, 1978), *Nephtys hombergii* and *N. caeca* (Olive, 1978), *Naineris laevigata* (Giangrande & Petraroli, 1991), *Pholoe minuta* (Heffernan & Keegan, 1988), and *Hesiocaeca methanicola* (Eckelbarger et al., 2001), follicle cells appear to be relatively inactive. In *Kefersteinia cirrata* (Olive & Pillai, 1983), follicle cells contain abundant RER and Golgi complexes. Follicle cells in the ovary of *Phragmatopoma lapidosa* store glycogen and lipids (Eckelbarger, 1979), those of *Streblospio benedicti* contain extensive arrays of RER and active Golgi complexes (Eckelbarger, 1980) (Fig. 8C), while those of *Capitella jonesi* (Eckelbarger & Grassle, 1982) and *Capitella* sp. I (Fig. 8B, D) have highly proteosynthetic follicle cells that hypertrophy during vitellogenesis. There is also evidence that follicle cells are involved in oosorption in the ovaries of *Nicolea zostericola* (Eckelbarger, 1975), *Streblospio benedicti* (Eckelbarger, 1980), and *Capitella jonesi* (Eckelbarger & Grassle, 1984).

Nurse cell–oocyte associations

Nurse cells are accessory cells that connect with the oocytes by confluent cytoplasmic, intercellular bridges (fusomes) (Fig. 9B, C). Since they originate from the germ cell line they are homologous and therefore of potential use in phylogenic analyses. Nurse cell–oocyte associations in polychaetes are relatively rare and they occur sporadically, even within closely related species (Rouse, 1992). They have been described within the Phyllodocida in the syllid *Typosyllis pulchra* in which groups of nurse cells surround individual oocytes (Heacox & Schroeder, 1981), in some exogonine syllids in which nurse cells, oocytes, and follicle cells are associated (Cognetti-Varriale, 1965), and in the tomopterid *Tomopteris helgolandica* (Åkesson, 1962) where a cluster of seven nurse cells is attached to one end of the oocyte (Figs. 1C; 9B). Within the Eunicida, chains of nurse cells are associated with the oocytes in several species in the Onuphidae including *Diopatra cuprea* (Anderson & Huebner, 1968) (Figs. 1D; 9A, C) and *Onuphis* spp. (Paxton, 1979; Hsieh, 1984; Eckelbarger, 1988). A single polyploid nurse cell is associated with every oocyte in the dorvilleid, *Ophryotrocha* spp. (Korschelt, 1893; Ruthmann, 1964; Emanuelsson, 1969; Pfannenstiel, 1978). In the first record within the Capitellida, Rouse (1992) described clusters of nurse cells attached to individual oocytes in *Micromaldane nutricula* (Maldanidae). Some small-bodied species have also been reported to have nurse cells including some myzostomids (Schroeder & Hermans, 1975), *Dinophilus gyrociliatus* (Nachtsheim, 1919; Traut, 1969; Grün, 1972), *Mesonerilla biantennata* (Jouin, 1968), and *Trilobodrilus hermaphroditus* (Riser, 1999).

Egg envelopes and microvilli

Mature eggs are surrounded by egg envelopes that consist of an extracellular matrix penetrated by oocyte microvilli of various lengths and configurations. They show wide variation in thickness and ultrastructural complexity (Fig. 10A–O) and while there are usually morphological similarities within a family (Eckelbarger, 1984) (Fig. 10A–C), even closely related species differ (Eckelbarger & Grassle, 1983). In most species, the egg envelope begins to form with the proliferation of microvilli from the oocyte surface. The microvilli usually develop a filamentous glycocalyx along their external surfaces and the perivitelline space between microvilli gradually fills in with one or more layers of extracellular material. Egg envelopes are commonly thin (< 1 μm) but they reach 4 μm in thickness in the amphinomid, *Benthoscolex* sp. (Eckelbarger, unpublished) (Fig. 10K). In some spioniform species, thick, highly ornamented or honeycombed egg envelopes have been described (reviewed in Blake & Arnofsky, 1999), while egg envelopes with a reticulated surface pattern has been reported (reviewed in Kupriyanova et al., 2001). In mature eggs, the egg envelope is invariably penetrated by the tips of the oocyte microvilli, putting the egg in direct contact with its environment. In many instances, the microvilli have simple rounded tips while others

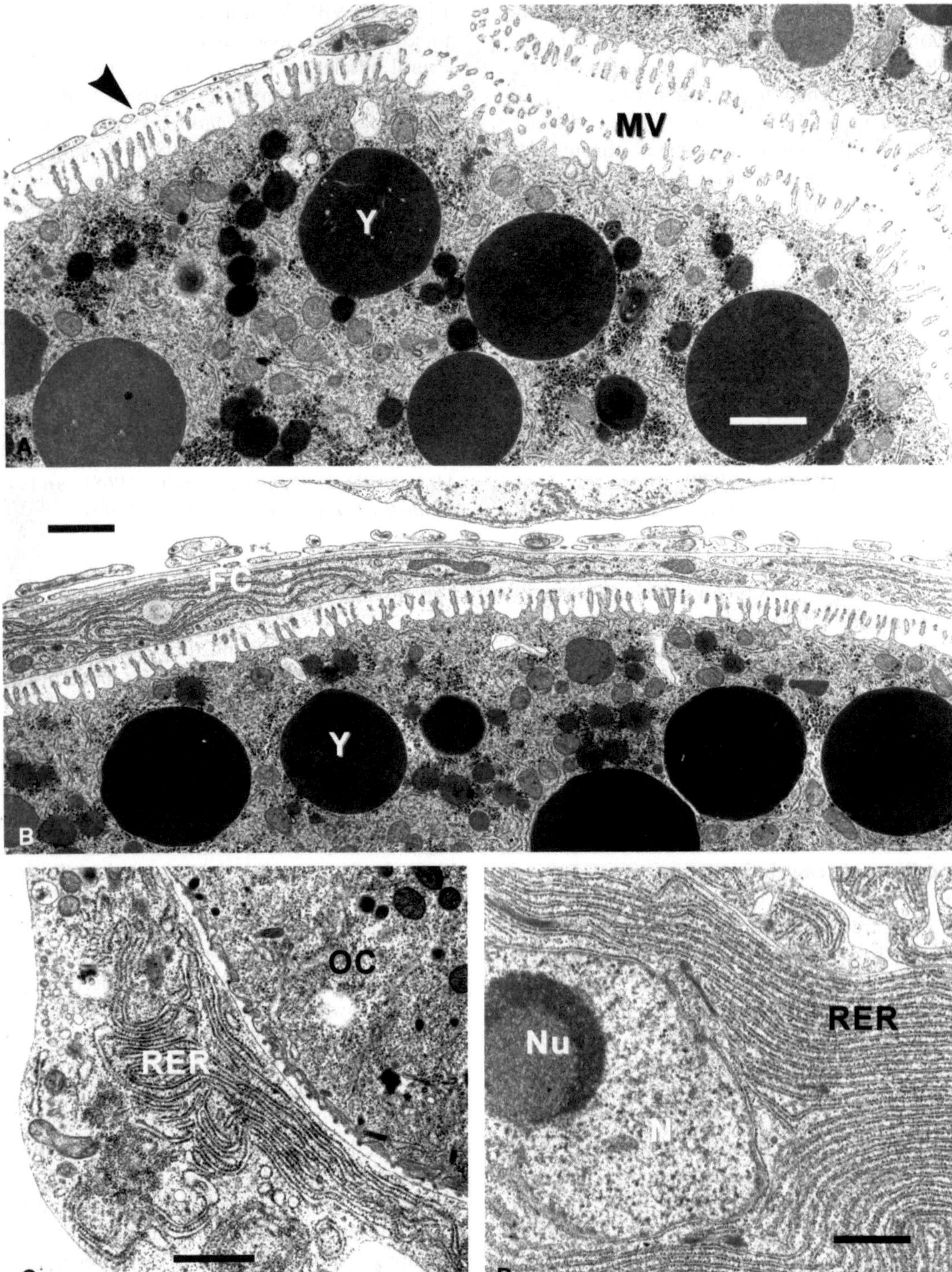

Figure 8. (A) Early vitellogenic oocyte of *Capitella jonesi* showing large yolk granules (Y), surface microvilli (MV), and thin, enveloping layer of follicle cells (arrowhead). Scale = 2 μm. (B) Follicle cell layer (FC) surrounding vitellogenic oocytes of *Capitella* sp. I. Y = yolk body. Scale = 2 μm. (C) Oocyte (OC) of *Streblospio benedicti* surrounded by follicle cell containing RER cisternae. Scale = 1.5 μm. (D) Follicle cell containing parallel arrays of RER in ovary of *Capitella* sp. I. Scale = 2 μm.

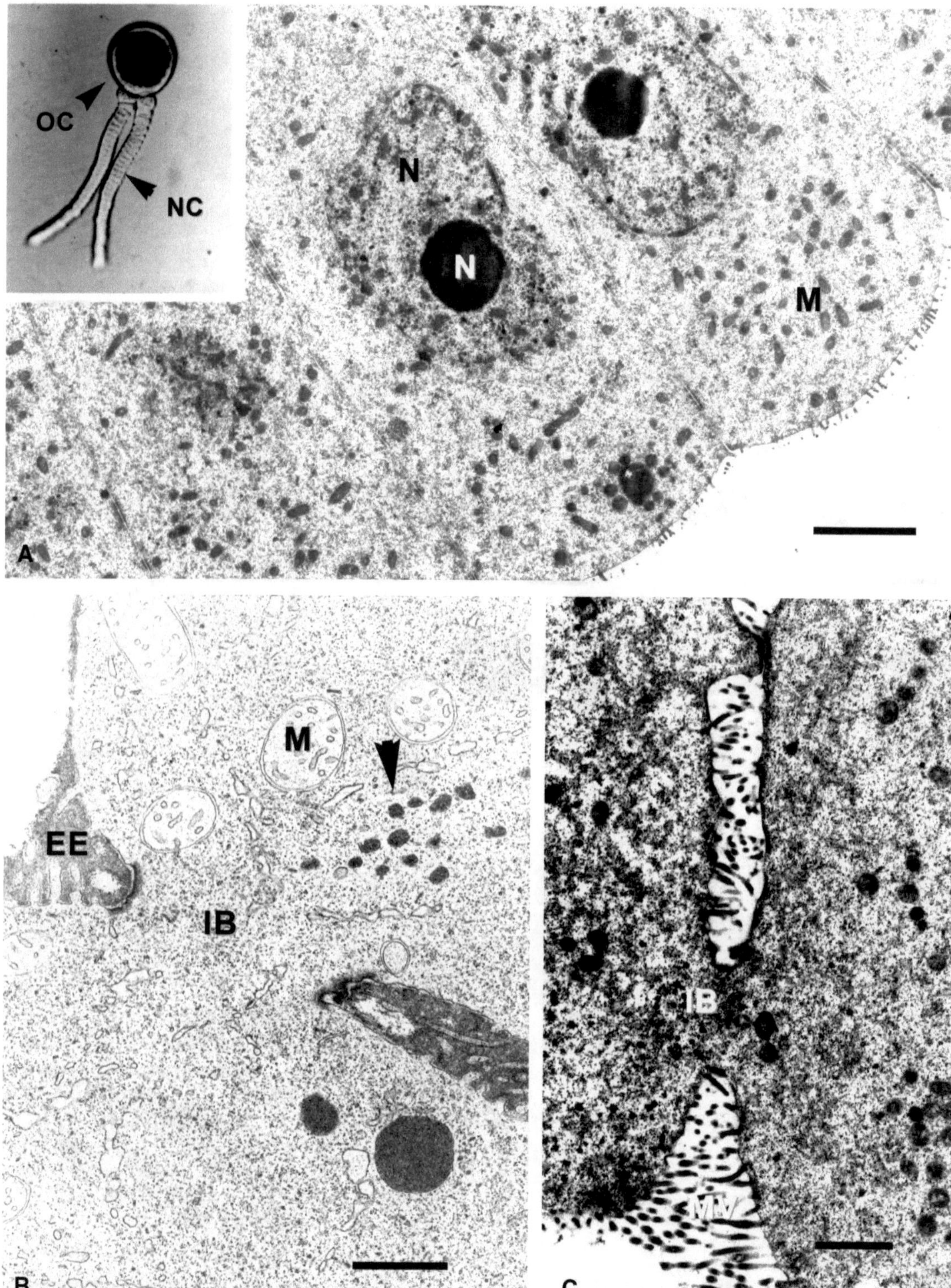

Figure 9. (A) Nurse cells of *Diopatra cuprea* with prominent nuclei (N) and nucleoli (Nu) and clusters of mitochondria (M). Scale = 5 μm. (B) Intercellular bridges (IB) between two oocytes in *Tomopteris pacifica*. Arrowhead indicates unknown dense bodies. M = mitochondrion; EE = egg envelope. Scale = 3 μm. (C) Intercellular bridge (IB) between oocytes in *Diopatra cuprea*. MV = microvilli. Scale = 2 μm.

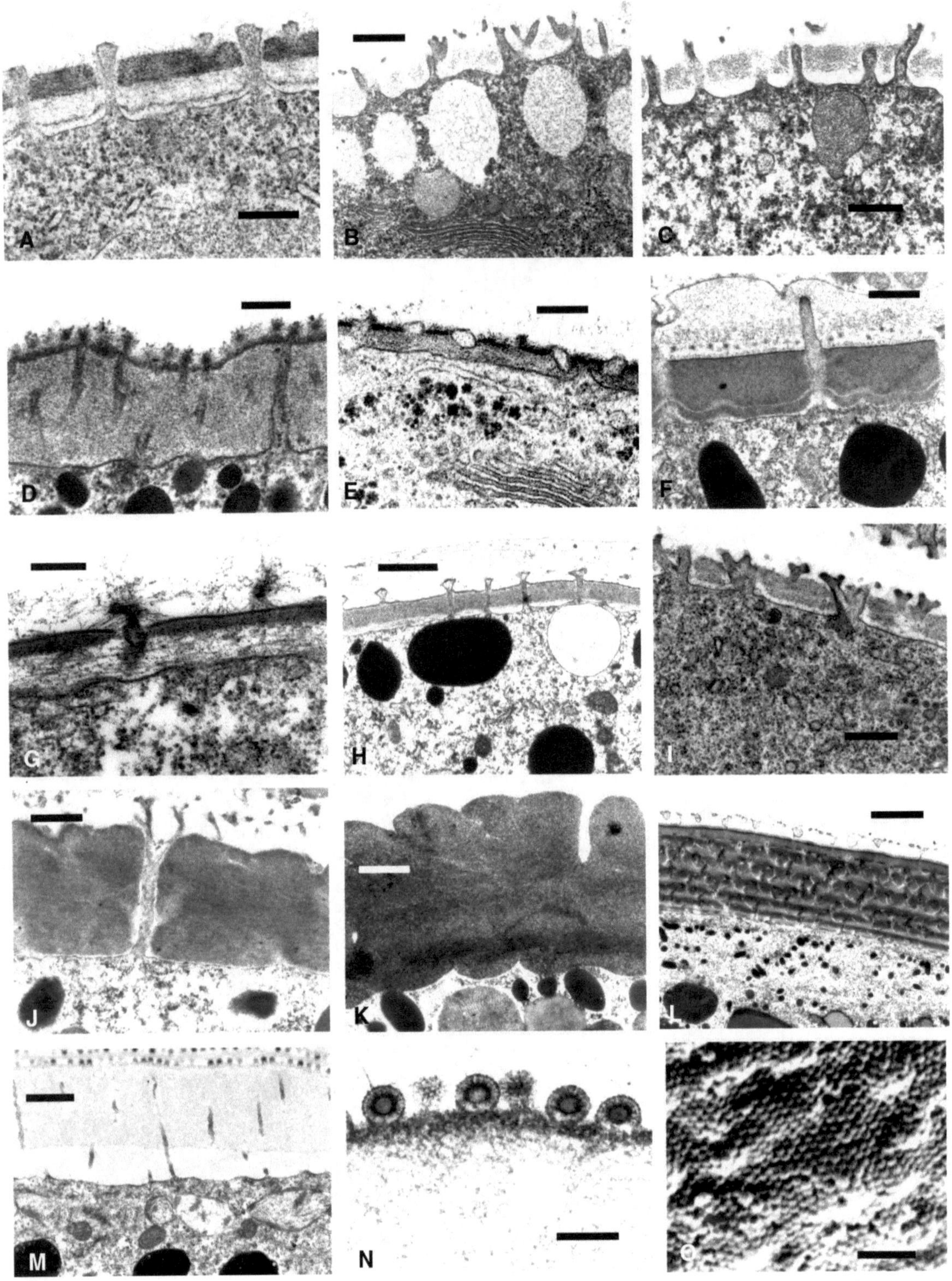

Figure 10. Egg envelopes of selected polychaete species. Scales for A–M = 1 μm. (A) *Vanadis formosa*; (B) *Rhynchonerella angelini*; (C) *Alciopa reynaudii*; (D) *Spio setosa*; (E) *Streblospio benedicti*; (F) *Lepidasthenia varia*; (G) *Proceraea fasciata;* (H) *Notomastus lobatus*; (I) *Phyllodoce fragilis*; (J) *Pseudoeurythoe ambigua*; (K) *Benthoscolex* sp.; (L) *Diopatra cuprea*; (M) *Phragmatopoma lapidosa*; (N) Surface granules from egg envelope of *P. lapidosa*. Scale = 0.1 μm. (O). Scanning electron micrograph of granules covering surface of egg envelope of *P. lapidosa*. Scale = 1 μm.

bifurcate or bear extensive filamentous adornments. In *Hydroides* spp. (Colwin & Colwin, 1961), *Pomatoceros triqueter* (Ap Gwynn & Jones, 1971), *Typosyllis pulchra* (Heacox & Schroeder, 1981) *Terebella rubra* (Eckelbarger, 1984), *Sabellaria alveolata* (Pasteels, 1965a, b; Franklin, 1966), *Phragmatopoma lapidosa* (Eckelbarger, 1979), and *Marenzelleria viridis* (Bochert, 1996), the microvilli terminate in a monolayer of surface granules. These granules attain their greatest complexity in the Sabellariidae (Fig. 10M–O) and are retained during early development through the trochophore stage (Eckelbarger & Chia, 1978). In some nereids, a specialized structure called the microvillous tip vesicle is situated at the tip of each microvillous and reportedly serves as a sperm receptor (Sato, 1999).

Conclusion

Polychaetes show a wide array of life histories and this diversity is reflected in a high degree of variability in ovarian morphology and mechanisms of oocyte differentiation unmatched in other metazoans. Differences in ovarian structure and mechanisms of vitellogenesis appear to be correlated with elements of their life history such as the speed of egg formation, the frequency of spawning, the size and energy content of the egg and resultant consequences for larval dispersal. Some aspects of oogenesis appear to have been conserved at the family level including the structure and position of the ovary, the type of oogenesis (e.g. extraovarian vs. intraovarian), and the ultrastructure of the egg envelope. However, our knowledge of oogenesis is still limited to a very small number of species in selected families. Whereas polychaete sperm structure has been very useful in phylogenetic studies, the use of characters based on oogenesis in phylogenetic analyses are very limited, at present. Nevertheless, comparative studies of polychaete oogenesis have great value because the developmental pathways established during oogenesis have a direct effect on subsequent life histories.

Acknowledgements

The author wishes to thank N.W. Riser, Edward Ruppert, Craig Young, Judith Grassle, Lisa Levin, and Claudia Mills for assistance in collecting specimens used to illustrate this chapter. Thanks also to Linda Healy and Tim Miller for providing technical assistance with the illustrations.

References

Åkesson, B.,1962. The embryology of *Tomopteris helgolandica* (Polychaeta). Acta Zoologica 43: 135–199.

Allen, M. J., 1961. A cytochemical study of the developing oocytes and attached nurse cells of the polychaetous annelid *Diopatra cuprea* (Bosc). Acta Embryol. Morph. Exp. 4: 219–238.

Ap Gwynn, I. & P. C. T. Jones, 1971. On the egg investments and fertilization reaction in *Pomatoceros triqueter* L. An ultrastructural study. Zeitschrift für Zellforschung 112: 390–396.

Ashworth, J. H., 1912. Observations on the structure and affinities of *Branchiomaldane vincenti* Langerhans. Proceedings of the Royal Society of Edinburgh 32: 62–72.

Bertout, M., F. Caner, A. Dhainaut & J. Krembel, 1981. Development of the nuclear structure and metabolism during oogenesis of *Perinereis cultrifera* (Annelida, Polychaeta). International Journal of Invertebrate Reproduction 3: 121–132.

Blake, J. A. & P. L. Arnofsky, 1999. Reproduction and larval development of the spioniform Polychaeta with application to systematics and phylogeny. Hydrobiologia 402: 57–106.

Bochert, R., 1996. An electron microscopic study of oogenesis in *Marenzelleria viridis* (Verrill, 1873) (Polychaeta: Spionidae) with special reference to large cortical alveoli. Invertebrate Reproduction and Development 29: 57–69.

Brafield, A. E. & G. Chapman, 1967. Gametogenesis and breeding in a natural population of *Nereis virens*. J. Mar. Biol. Ass. U.K. 47: 619–627.

Clark, R. B. & P. J. W. Olive, 1973. Recent advances in polychaete endocrinology and reproductive biology. Oceanography and Marine Biology Annual Review 11: 175–222.

Christie, G., 1984. The reproductive biology of a Northumberland population of *Sphaerodorum gracilis* (Rathke, 1843) (Polychaeta: Sphaerodoridae). Sarsia 69: 117–121.

Christie, G., 1985. A comparative study of the reproductive cycles of three Northumberland populations of *Chaetozone setosa* (Polychaeta: Cirratulidae). J. Mar. Biol. Ass. UK 65: 239–254.

Cognetti-Varriale, A. M., 1965. Richerche sulla biologia riproductiva dei Policheti: I. Gli ovari dell Exogoninae. Arch. Zool. Ital. 50: 26–28.

Colwin, L.H. & A.L. Colwin, 1961. Changes in the spermatozoon during fertilization in *Hydroides hexagonus* (Annelida). I. Passage of the acrosomal region through the vitelline membrane. Journal of Biophysical and Biochemical Cytology 10: 231–254.

Davis, W. R., 1969. Oogenesis and its relationship to dissolved oxygen suppression in *Neanthes arenaceodentata* (Polychaeta: Annelida). M.A. thesis, California State University, Long Beach, 62 pp.

Dhainaut, A., 1972. Évolution nucléolaire au cours de l'ovogenèse de *Nereis pelagica*. 1. Étude morphologique. Journal of Microscopy 13: 67–84.

Dhainaut, A., 1984. Oogenesis in polychaetes. Ultrastructural differentiation and metabolism of Nereid oocytes. Fortschr. Zool. 29: 183–205.

Eckelbarger, K. J., 1975. A light and electron microscope investigation of gametogenesis in *Nicolea zostericola* (Polychaeta: Terebellidae). Marine Biology 30: 353–370.

Eckelbarger, K. J., 1979. Ultrastructural evidence for both autosynthetic and heterosynthetic yolk formation in the oocytes of an annelid (*Phragmatopoma lapidosa*: Polychaeta). Tissue and Cell 11: 425 – 443.

Eckelbarger, K. J., 1980. An ultrastructural study of oogenesis in *Streblospio benedicti* (Spionidae), with remarks on diversity of vitellogenic mechanisms in Polychaeta. Zoomorphology 94: 241–263.

Eckelbarger, K. J., 1983. Evolutionary radiation in polychaete ovaries and vitellogenic mechanisms and their role in life history patterns. Canadian Journal of Zoology 61: 487–504.

Eckelbarger, K. J., 1984. Comparative aspects of oogenesis in polychaetes. In Fischer, A & H.-D. Pfannenstiel (eds), Polychaete Reproduction. Progress in Comparative Reproductive Biology, Fortschr. Zool. 29: 123–148.

Eckelbarger, K. J., 1986. Vitellogenic mechanisms and the allocation of energy to offspring in polychaetes. Bull. Mar. Sci. 39: 426–443.

Eckelbarger K. J., 1988. Oogenesis and female gametes. In Westheide, W. & C. O. Hermans (eds), The Ultrastructure of Polychaeta, Microfauna Mar. 4: 281–307.

Eckelbarger, K. J., 1992. Oogenesis. In Harrison, F.W. & S. L. Gardiner (eds), Microscopic Anatomy of Invertebrates, Vol. 7, Chapter 2 (Polychaeta). Wiley-Liss Inc., New York: 109–127.

Eckelbarger, K. J., 1994. Diversity of metazoan ovaries and vitellogenic mechanisms: implications for life history theory. Proceedings of the Biological Society of Washington 107: 193–218.

Eckelbarger, K. J. & F-S. Chia, 1978. Morphogenesis of larval cuticle in the polychaete *Phragmatopoma lapidosa*: a correlated scanning and transmission electron microscope study from egg envelope formation to larval metamorphosis. Cell and Tissue Research 186: 187–201.

Eckelbarger, K. J. & J. P. Grassle, 1982. Ultrastructure of the ovary and oogenesis in the polychaete *Capitella jonesi* (Hartman, 1959). Journal of Morphology 171: 305–320.

Eckelbarger, K. J. & J. P. Grassle, 1983. Ultrastructural differences in the eggs and ovarian follicle cells of *Capitella* (Polychaeta) sibling species. Biological Bulletin 165: 379–393.

Eckelbarger, K. J. & J. P. Grassle, 1984. Role of ovarian follicle cells in vitellogenesis and oocyte resorption in *Capitella* sp. I (Polychaeta). Marine Biology 79: 133–144.

Eckelbarger, K. J. & S. A. Rice, 1988. Ultrastructure of oogenesis in the holopelagic polychaetes *Rhynchonerella angelini* and *Alciopa reynaudii* (Polychaeta: Alciopidae). Marine Biology 98: 427–439.

Eckelbarger, K. J., C. M. Young, E. Ramirez Llodra, S. Brooke & P. A. Tyler, 2001. Gametogenesis, spawning behavior, and early development in the 'iceworm' *Hesiocaeca methanicola* (Polychaeta: Hesionidae) from methane hydrates in the Gulf of Mexico. Marine Biology 138: 761–775.

Emanuelsson, H., 1969. Electron microscope observations on yolk and yolk formation in *Ophryotrocha labronica* La Greca and Bacci. Z. Zellforsch. Mikrosk. Anat. 95: 19–36.

Fischer, A., 1975. The structure of symplasmic early oocytes and their enveloping sheath cells in the polychaete *Platynereis dumerilii*. Cell and Tissue Research 160: 327–343.

Fischer, A., 1999. Reproductive and developmental phenomena in annelids: a source of exemplary research problems. Hydrobiologia 402: 1–20.

Fischer, A. & A. Dhainaut, 1985. The origin of yolk in the oocytes of *Nereis virens* (Annelida: Polychaeta). Electron microscopic and autoradiographic studies by use of unspecific and yolk-specific markers. Cell and Tissue Research 240: 67–76.

Fischer, A. & U. Hoeger, 1993. Metabolic links between somatic sexual maturation and oogenesis in nereid annelids – a brief review. Invertebrate Reproduction and Development 23: 131–138.

Fischer, A. & H. D. Pfannenstiel (eds), 1984. Polychaete reproduction. Progress in comparative reproductive biology. Fortschr. Zool. 29: 341pp.

Fischer, A. & H. Rabien, 1986. Molecules and cellular functions driving oocyte growth in nereid annelids. In Porchet, M., J.-C. Andries & A. Dhainaut (eds), Advances in Invertebrate Reproduction, 4. Elsevier Science Publishers, New York: 195–205.

Franklin, L. E., 1966. An egg cell-membrane derivative on the vitelline membrane of annelid eggs. Expermental Cell Research 43: 673–675.

Garwood, P. R., 1978. The role of environmental factors in the control of reproduction in *Harmothoe imbricata* (Polychaeta: Polynoidae). PhD Thesis, University of Newcastle upon Tyne, UK.

Garwood, P. R., 1981. Observations on the cytology of the developing female germ cell in the polychaete *Harmothoe imbricata* (L.). International Journal of Invertebrate Reproduction and Development 3: 333–345.

Giangrande, A. & A. Petraroli, 1991. Reproduction, larval development and post-larval growth of *Nainereis laevigata* (Polychaeta, Orbiniidae) in the Mediterranean Sea. Marine Biology 111: 129–137.

Grün, G., 1972. Über den Eidimorphismus und die Oogenese von *Dinophilus gyrociliatus* (Archiannelida) Z. Zellforsch. Mikrosk. Anat. 130: 70–92.

Hagedorn, H. H. & J. G. Kunkel, 1979. Vitellogenin and vitellin in insects. Annual Review of Entomology 24: 475–505.

Heacox, A.E. & P.C. Schroeder, 1981. A light and electron microscopic investigation of gametogenesis in *Typosyllis pulchra* (Berkely & Berkeley) (Polychaeta: Syllidae) II. Oogenesis. Cell and Tissue Research 218: 641–658.

Heffernan, P. & B. F. Keegan, 1988. Quantitative and ultrastructural studies on the reproductive biology of the polychaete *Pholoe minuta* in Galway Bay. Marine Biology 99: 203–214.

Hodgson, A. N. & K. J. Eckelbarger, 2000. Ultrastructure of the ovary and oogenesis in six species of patellid limpets (Gastropoda: Patellogastropoda) from South Africa. Invertebrate Biology 119: 265–277.

Hsieh, H. -L., 1984. Morphological studies on the reproduction and larval development of *Kinbergonuphis simoni* (Polychaeta: Onuphidae). MSc Thesis, University of South Florida, Tampa.

Huebner, E. & E. Anderson, 1976. Comparative spiralian oogenesis – structural aspects: an overview. American Zoologist 16: 315–343.

Hutchings, P. A., 1973. Gametogenesis in a Northumberland population of the polychaete *Melinna cristata*. Marine Biology 18: 199–211.

Jouin, C., 1968. Sexualité et biologie de la réproduction chez *Mesonerilla* Boaden (Archiannélides Nerillidae). Cah. Biol. Mar. 9: 31–52.

Korschelt, E., 1893. Ueber *Ophryotrocha puerilis* Clap.-Metsch. und die polytrochen Larven eines andern Anneliden (*Harpochaeta cingulata*, nov. gen., nov. spec.). Z. wiss. Zool. 57: 224–289.

Kupriyanova, E. K., E. Nishi, H.A. Ten Hove & A.V. Rzhavsky, 2001. Life history patterns in serpulimorph polychaetes: ecological and evolutionary perspectives. Oceanography and Marine Biology Annual Review 39: 1–101.

Lawrence, A. & P. J. W. Olive, 1995. Gonadotrophic hormone in *Eulalia viridis* (Polychaeta, Annelida): stimulation of vitellogenesis. Invertebrate Reproduction and Development 28: 43–52.

McHugh, D. & P. P. Fong, 2002. Do life history traits account for diversity of polychaete annelids? Invertebrate Biology 121: 325–338.

Nachtsheim, H., 1919. Cytologische Untersuchungen über die Geschlechtsbestimmung bei *Dinophilus apatris* Korschelt. Arch. Mikrosk. Anat. 39: 17–140.

Newell, G. E., 1948. A contribution to the knowledge of the life history of *Arenicola marina*. J. Mar. Biol. Ass. UK 27: 554–580.

Okada, K., 1941. The gametogenesis, the breeding habits, and the early development of *Arenicola cristata* Stimpson, a tubiculous polychaete. Sci. Rep. Tohoku Imp. Univ., Biol. 16: 99–145.

Olive, P. J. W., 1971. Ovary structure and oogenesis in *Cirratulus cirratus* (Polychaeta Cirratulidae). Marine Biology 8: 243–260.

Olive, P. J. W., 1975. Reproductive biology of *Eulalia viridis* (Müller) (Polychaeta: Phyllodocidae) in the northeastern UK J. Mar. Biol. Ass. UK 55: 313–326.

Olive, P. J. W., 1978. Reproduction and annual gametogenic cycle in *Nephtys hombergii* and *N. caeca* (Polychaeta: Nephtyidae). Marine Biology 46: 83–90.

Olive, P. J. W., 1983. Oogenesis in Annelida: Polychaeta. In Adiyodi, K. G. & R. G. Adiyodi (eds), Reproductive Biology of Invertebrates, Vol. 1, Oogenesis, Oviposition, and Oosorption, Wiley, New York: 357–422.

Olive, P. J. W. & G. Pillai, 1983. Reproductive biology of the polychaete *Kefersteinia cirrata* Keferstein (Hesionidae). II. The gametogenic cycle and evidence for photoperiodic control of oogenesis. International Journal of Invertebrate Reproduction 6: 307–315.

Pasteels, J.J., 1965a. Étude au microscope électronique de la réaction corticale. II. La réaction corticale de l'oeuf vierge de *Sabellaria alveolata*. Journal of Embryological Expermental Morphogy 13: 327–339.

Pasteels, J. J., 1965b. Aspects structuraux de la fécondation vus au microscope électronique. Archives of Biology (Liège) 76: 463–509.

Paxton, H., 1979. Taxonomy and aspects of the life history of Australian beachworms (Polychaeta: Onuphidae). Australian Journal of Marine and Freshwater Research 30: 265–294.

Pfannenstiel, H. D., 1978. Die Entwicklung der Kontaktstruktur von Ei und Nährzelle im Zuge der Oogenese von *Ophryotrocha puerilis* Claparede & Mecznikow (Polychaeta, Dorvilleidae). Zoomorphologie 90: 181–196.

Pfannenstiel, H. D. & C. Grünig, 1982. Yolk formation in an annelid (*Ophryotrocha puerilis*, Polychaeta). Tissue and Cell 14: 669–680.

Potswald, H. E., 1965. Reproductive biology and development of *Spirorbis* (Serpulidae, Polychaeta). PhD Thesis, University of Washington, Seattle.

Potswald, H. E., 1969. Cytological observations on the so-called neoblasts in the serpulid *Spirorbis*. Journal of Morphology 128: 241–259.

Riser, N. W., 1999. Description of a new species of dinophilid polychaete, with observations on other dinophilids and interstitial polychaetes in New England. Northeast. Nat. 6: 211–220.

Rouse, G. W., 1991. Spermathecae, sperm and sperm ducts in small sabellids (Sabellidae, Polychaeta). American Zoologist 31: 129A.

Rouse, G. W., 1992. Oogenesis and larval development in *Micromaldane* spp. (Polychaeta: Capitellida: Maldanidae). Invertebrate Reproduction and Development 21: 215–230.

Rouse, G. W. & F. Pleijel, 2001. Polychaetes. Oxford University Press, New York, 354 pp.

Rouse, G. W. & A. B. Tzetlin, 1997. Ultrastructure of the body wall and gametogenesis in *Cossura* cf. *longocirrata* (Cossuridae Polychaeta). Invertebrate Reproduction and Development 32: 41–54.

Ruthmann, A., 1964. Zellwachstum und RNS-Synthese im Ei-Nährzellverband von *Ophryotrocha puerilis*. Z. Zellforsch. 63: 816–829.

Sato, M., 1999. Divergence of reproductive and developmental characteristics in *Hediste* (Polychaeta: Nereididae). Hydrobiologia 402: 129–143.

Schroeder, P. C., 1968. On the life history of *Nereis grubei* (Kinberg), a polychaete annelid from California. Pac. Sci. 22: 476–481.

Schroeder, P. C. & C. O. Hermans, 1975. Annelida: Polychaeta. In Giese, A.C. & J.S. Pearse (eds), Reproduction of Marine Invertebrates, Vol. III, Academic Press, New York, 213 pp.

Sichel, G. 1966. Modificazioni ultrastructurali dell ooplasma in rapporta alla vitellogenese in *Mercierella enigmatica* Fauvel (Annelida: Polychaeta). Atti Accad. Gioenia Sci. Nat. 18: 21–23.

Southward, E. C. & A. J. Southward, 1958. The breeding of *Arenicola ecaudata* Johnston and *A. branchialis* Aud. & Edw. at Plymouth. J. Mar Biol. Ass. UK 37: 267–286.

Spies, R. B., 1977. Reproduction and larval development of *Flabelliderma commensalis* (Moore). In Reish, D. J. & K. Fauchald (eds), Essays on polychaetous annelids, in memory of Dr. Olga Hartmann, Univervisty of Southern California, Los Angeles: 323–345.

Traut, W., 1969. Zur Sexualität von *Dinophilus gyrociliatus*. II. Der Aufbau des Ovars und die Oogenese. Biol. Zentralbl. 88: 695–714.

Tweedell, K. S., 1966. Oocyte development and incorporation of ^{3}H-thymidine and ^{3}H-uridine in *Pectinaria* (*Cistenides*) *gouldii*. Biological Bulletin 131: 516–538.

Villa, L., 1970. Dati preiliminari sull'ultrastruttura dell uovo mature di *Chaetopterus pergamentaceus* C. Acta Embryologiae Experimentals 1: 79–94.

Wilson, W.H. 1991. Sexual reproductive modes in polychaetes: classification and diversity. Bulletin of Marine Science 48: 500–516.

Hydrobiologia (2005) **535/536:** 199–225
T. Bartolomaeus & G. Purschke (eds), Morphology, Molecules, Evolution and Phylogeny in Polychaeta and Related Taxa

Pharynx and intestine

Alexander Tzetlin[1] & Günter Purschke[2,*]
[1]*Department of Invertebrate Zoology, Moscow State University, 119899 Moscow, Russia*
[2]*Spezielle Zoologie, Fachbereich Biologie/Chemie, Universität Osnabrück, D-49069 Osnabrück, Germany*
*(*Author for correspondence: E-mail: purschke@biologie.uni-osnabrueck.de)*

Key words: pharynx, axial pharynx, ventral pharynx, dorsolateral folds, dorsal pharynx, intestine, enteronephridia, jaws

Abstract

The alimentary canal of polychaetes consists of a foregut, midgut, and hindgut. The alimentary canal shows different specializations even in homonomously segmented polychaetes. The foregut gives rise to the buccal cavity, pharnyx and oesophagus, the midgut may be divided into a stomach and the intestine proper. Since polychaetes use a wide spectrum of food sources, structures involved in feeding vary as well and show numerous specializations. In the foregut these specializations may be classified as one of the following types: dorsolateral folds, ventral pharynx, axial muscular pharynx, axial non-muscular proboscis and dorsal pharynx. The latter, typical of oligochaetous Clitellata, occurs rarely in polychaetes. The structure, evolution and phylogenetic importance of these different types are described and discussed. Axial muscular and ventral pharynges may be armed with jaws, sclerotized parts of the pharyngeal cuticle. Terminology, structure, occurrence and development of the jaws are briefly reviewed. Special attention has been paid to the jaws of Eunicida including extinct and extant forms. Conflicting theories about the evolution of the jaws in Eunicida are discussed. The epithelia of the intestine may form a pseudostratified epithelium composed of glandular cells, absorptive cells and ciliated cells or only one cell type having similar functions. A conspicuous feature in the intestine of certain polychaetes is the occurrence of unicellular tubular structures, called enteronephridia. So far these enteronephridia are only known in a few meiofauna species.

Introduction

The alimentary canal of polychaetes consists of three parts: foregut, midgut and hindgut. Foregut and hindgut are of ectodermal origin, being formed by stomatodeal and proctodeal invaginations of the ectoderm and thus usually bearing a cuticle. The midgut is derived from the endoderm. The structure of the gut is correlated with adaptations to feeding and life style of polychaetes. Usually the alimentary canal shows different specialized parts even in homonomously segmented taxa (Fig. 1A–C). The foregut gives rise to the buccal cavity, pharynx and oesophagus, the midgut may be divided into a stomach and the intestine proper (Fig. 1B and C), and only the hindgut normally forms a single part. The length of the different regions varies among species according to their specific adaptive needs. Usually the intestine is no longer than the body, but in taxa with comparatively short or stout bodies the intestine may be coiled (Fig. 1B) or bear caeca to provide sufficient digestive surfaces. The gut system is attached to the body wall by means of septa and mesenteries which, however, may be more or less reduced or absent when the intestine is coiled or involved in extensive longitudinal movements. In contrast to Clitellata, no polychaete taxon is known to date in which a digestive tract is completely absent. However, in Siboglinidae (formerly

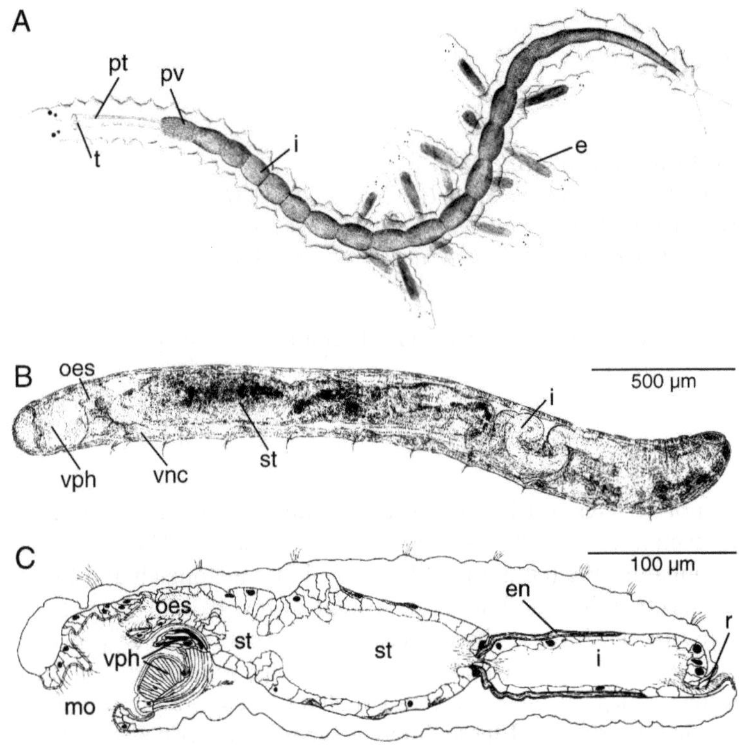

Figure 1. Different patterns of gut structure in polychaetes. (A) '*Exogone rubescens*' (Syllidae). Intestine (i) forms a straight tube. Pharynx differentiated into pharyngeal tube (pt) and proventricle (pv); e developing embryos. (B) *Stygocapitella subterranea* (Parergodrilidae). Gut divided into ventral pharynx (vph), oesophagus (oes), stomach (st), coiled intestine (i), rectum (r), and ventral nerve cord (vnc). (C) *Trochonerilla mobilis* (Nerillidae). Digestive tract comprises ventral pharynx (vph), oesophagus (oes), anterior and posterior stomach (st), enteronephridia (e), intestine (i) and rectum (r). Enteronephridia open into the gut at the border between stomach and intestine. (mo) mouth. A after Ørsted (unpubl.) from Wolff & Petersen (1991), B from Struck et al. (2002), C from Tzetlin et al. (1992).

Pogonophora, now considered to belong to Polychaeta) it is highly modified to form the trophosome housing symbiotic bacteria (Southward et al., 2005) and in *Parenterodrilus taenioides* the gut is residual and non-functional (Jouin, 1979, 1992).

The literature on the anatomy, morphology and physiology of the polychaete digestive system and feeding biology is copious and has been reviewed by Dales (1963), Jeuniaux (1969), Michel & DeVillez (1978), Fauchald & Jumars (1979), Michel (1988), Purschke (1988a), and Saulnier-Michel (1992).

Pharynx and foregut

Polychaetes use a wide spectrum of food sources and show a great diversity of feeding habits (Fauchald & Jumars, 1979). Accordingly, structures involved in feeding vary as well and show numerous specializations (for recent reviews see Purschke, 1988a; Saulnier-Michel, 1992). These specializations include the structural differentiation of the foregut proper and presence or absence of accessory structures, which can include ciliated tracts, ciliated fields or tentacles. Although feeding tentacles differ considerably between taxa, they mostly represent specialized palps and, thus, can be regarded as homologous (Orrhage & Müller, 2005).

However, in Ampharetidae, Pectinariidae and Terbellidae the feeding tentacles are considered not be modified palps (Orrhage, 2001). Moreover several other tentacle-like structures may be present such as the buccal tentacles of Cossuridae, which originate from the stomodeal epithelium (Tzetlin, 1994).

The mouth region of the larvae may give rise to a variety of adult structures often together termed the buccal organ (Rouse, 2000). In many species the foregut is more or less protrusile and thus called a proboscis, which in turn may be composed of several parts (e.g., Purschke, 1988a; Saulnier-Michel, 1992). The term pharynx is generally restricted to the muscular regions of the foregut. Since the foregut is derived from the ectoderm, its epithelium is usually covered by a cuticle. In several taxa, areas of this cuticle have become thickened, sclerotized and mineralized to form distinct jaws or tooth-like structures (p. 209). Although the structure of the foregut varies widely among the polychaete taxa, these different structures have until recently been classified as only three types of buccal organs (Fauvel, 1959; Dales, 1962; Saulnier-Michel, 1992): (1) axial muscular proboscis (pharynx), (2) ventral pharyngeal organ and (3) axial non-muscular proboscis. However, Purschke & Tzetlin (1996) described a previously unrecognized, comparatively simply structured foregut as another distinct plan of organization, which has been called dorsolateral ciliary folds. Rouse (2000) subdivided the ventral pharyngeal organs into simple ventral buccal organs and ventral buccal organs with well-developed musculature, the latter only occurring in Eunicida and Amphinomida. Moreover, the simple, tube-like foregut present in Sabellidae and Serpulidae (as well as Spirorbidae and Sabellariidae?) has been classified as absence of a buccal organ by Rouse (2000). Recent investigations of the foregut in the enigmatic terrestrial polychaete *Hrabeiella periglandulata* confirmed the presence of a dorsal pharynx in this species, elsewhere only known in and characteristic of oligochaetous Clitellata (Rota, 1998; Purschke, 2002, 2003).

Dorsolateral ciliary folds

In many polychaete species the dorsolateral walls of the foregut are differentiated into protrusible ciliated folds (Fig. 2A–H). Although sometimes briefly mentioned in the literature (e.g., Orrhage, 1964; Purschke & Jouin, 1988; Tzetlin, 1989), their importance as a distinct type of feeding structure in polychaetes has not been recognized for a long time. One reason might be their simple structure and another reason, that these folds most often occur together with a ventral pharyngeal organ (Purschke & Tzetlin, 1996). Since there are species in which these folds are the only differentiation of the foregut used for feeding, they are described separately.

These folds are always positioned dorsolaterally and are usually heavily ciliated (about 5–8 cilia per μm^2; Fig. 2D and H). As a rule the ciliated cells are associated with numerous gland cells, which may either be randomly distributed over the folds or are clustered to form distinct pharyngeal or salivary glands. The ciliation of these folds may either be restricted to the foregut or represent a continuation of ciliary fields present on the ventral side of the prostomium (Fig. 2F–H). Usually these folds take the form of a pair of inner lips, separated by an unciliated pouch from the lips surrounding the mouth (Fig. 2B–E). Musculature is only weakly developed and consists of a thin layer of transverse and longitudinal fibres. A few retractor fibres are present as well.

For feeding, the folds can be everted through the mouth opening and brought into contact with the substrate (Fig. 2B, C and E). This is mainly achieved by contraction of the musculature of the body wall without changes of the hydrostatic pressure in the body cavity (Purschke & Tzetlin, 1996). Food particles not adhering firmly to the substrate are kept and retained with mucus, collected by means of the action of the cilia of the folds, transported and sorted along the folds into the mouth and further to the oesophagus. Clearly, these structures are primarily adapted to microphagy and the organisms may be classified as surface deposit feeders (Fauchald & Jumars, 1979). With this mode of feeding, food particles attached to the substrate cannot be ingested and the area which can be grazed is comparatively small, not exceeding the width of the body. Only in species also possessing a ventral pharyngeal organ is it possible for food particles adhering to the substrate to be abraded by the action of the bulb and then to be ingested by the beating of the ciliary folds (Fig. 3B). Both modes of feeding may occur in one species (Jennings & Gelder, 1969; Schmidt & Westheide, 1972; Gelder & Uglow, 1973; Westheide & Schmidt, 1974). Given these considerations, it is not surprising that dorsolateral folds are present only in small epibenthic or interstitial species, usually not exceeding a few millimeters in length, or in juveniles of larger species. In the latter, other types of microphagous feeding structures such as feeding tentacles or a non-muscular

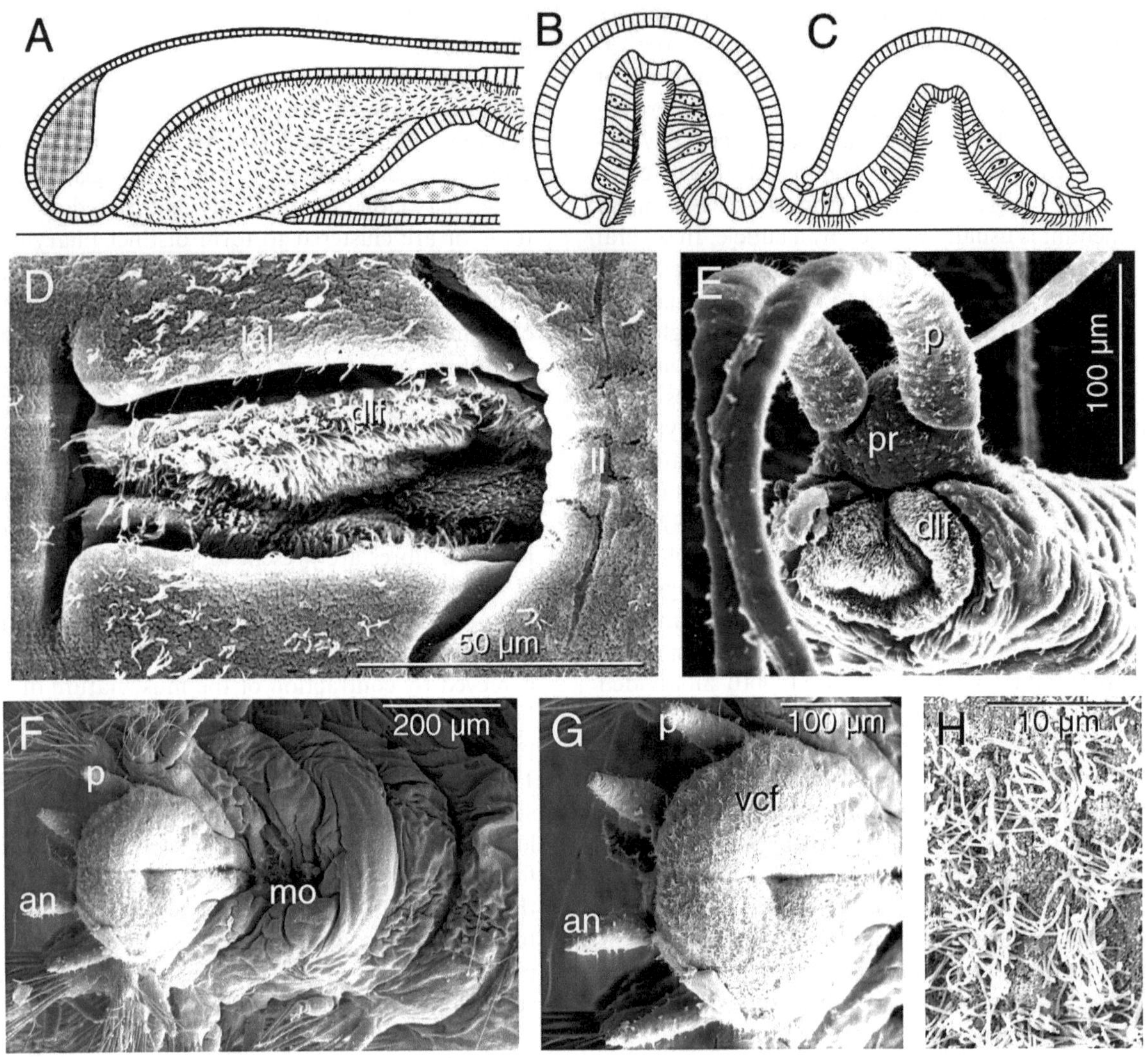

Figure 2. Foregut. Dorsolateral folds. (A–C) Schematic representation. (A) Sagittal section. (B) Cross-section in the resting position. (C) Cross-section in the feeding position. (D, E) *Saccocirrus papillocercus*. (D) Mouth opening with lateral and lower lips (lal, ll) and dorsolateral folds (dlf) in the resting position. (E) Dorsolateral folds everted for feeding. p palp, pr prostomium. (F–H) *Eurythoe complanata*. (F) Ventral view of anterior end, mo mouth. (G) Enlargement of prostomium with ciliary fields (vcf) leading into mouth; an antenna, p palp. (H) Ciliation of prostomium. A–E modified from Tzetlin et al. (1992), micrographs F–H S. Raabe.

axial proboscis develop later during ontogeny (Eisig, 1914; Dales, 1957; Heimler, 1983; Purschke, 1988b).

Ventral pharynx

Ventral pharyngeal organs are widespread within polychaetes. They exhibit a high degree of diversity among the different taxa and several types may be distinguished (Purschke & Tzetlin, 1996; Purschke, 1988a,b, 2002). Situated below the oesophagus, these organs are made up of different parts divided by a few blind-ending pouches. These are situated in the ventroposterior region of the foregut (Fig. 3A and B). The invaginations are lined by specialized epithelia and are covered by structurally differentiated cuticles. Most characteristic is the presence of transverse plate-like muscle fibers underneath one of these invaginations (Figs 3C and 4A). These transverse fibers usually form a compact muscle structure, called muscle bulb. Above this bulb a tongue-like organ may be present. A group of longitudinal fibres runs semicircularly around the ventral pharynx from the lower lip to the dorsalmost parts underneath the oesophagus. These fibers are either called sagittal muscle or, since they envelope the entire ventral pharynx, investing muscle (Purschke, 1988a). The ultrastructure of ventral pharyngeal organs is comparatively well known (Rieger & Rieger, 1975; Jouin, 1978a; Michel, 1978; Purschke, 1985a,b, 1987a,b, 1988a,b; Tzetlin, 1987, 1989, 1991; Tzetlin et al., 1987, 1992; Purschke & Jouin, 1988; Saulnier-Michel et al., 1990; Müller et al., 2001).

Ventral pharynges are used for scraping or licking off food particles from the substrate or they function as devices for processing food particles inside the oesophagus. Usually parts of the cuticle in the ventral pharynges are differentiated for this purpose. In these areas it is comparatively thick, frequently forming tooth-like structures, and microvilli do not reach the surface of the cuticle (Fig. 3B). In spite of this function the cuticle is not sclerotized in most taxa; the only exception is Eunicida which possess a characteristic jaw apparatus described below (p. 211). The stylets typical of Nerillidae are actually intracellular skeletal elements and are not cuticular jaws (Purschke, 1985b; Tzetlin & Larionov, 1988; Tzetlin et al., 1992; Müller et al., 2001). The bulb and/or the tongue-like organ are everted through the mouth, dislodging food particles; in contrast to axial pharynges these structures usually cannot be protruded very far.

The bulbus muscle serves as a firm but elastic cushion supporting the epithelial areas used for scraping. In the bulbus muscle only transverse fibers are present and antagonistic fibers do not exist. The bulb is usually surrounded by a comparatively thick extracellular matrix and several types may be distinguished. Currently six different types of ventral pharynges may be distinguished on the basis of the structure of the bulb. Amphinomida, Eunicida, Dinophilidae, Diurodrilidae and Nerillidae each possess their own type of ventral pharynx.

Type 1

In the most common type, plate-like muscle cells and interstitial cells alternate in the bulb (Fig. 3C–E). In the interstitial cells there is a prominent system of intermediate filaments oriented dorsoventrally, i.e., these filaments are perpendicular with respect to the myofilaments. In certain species these interstitial cells have voluminous cell bodies containing only the nucleus and a few organelles (Fig. 3C and D; Heimler, 1983; Tzetlin, 1987; Purschke, 1988b). Histological observations indicate that this type of bulbus muscle occurs in many species and may thus be the most frequent type. In Protodriloidae, Protodrilidae and Saccocirridae interstitial cells with prominent bundles of tonofilaments are present as well, but the cell bodies are small (Fig. 3E; Purschke & Jouin, 1988). In this type the muscle fibers contract against the extracellular matrix surrounding the bulb, opposing the internal pressure of the interstitial cells and their dorsoventral tonofilament system (Purschke, 1988b). Because these filaments do not allow a dorsoventral extension of the bulb, cushions of varying firmness and elasticity are achieved. In Parergodrilidae a similar bulbus muscle is present (Purschke, 1987a, 1988b). However, in *Stygocapitella subterranea* the bulb is small and more or less replaced by a voluminous multicellular gland. In the terrestrial *Parergodrilus heideri* the muscle bulb is lacking, completely replaced by this gland and the investing muscle fibres are transformed to make up the complex musculature of the tongue-like organ.

Type 2

In a second type of bulbus muscle a similar function is realized by bulbus muscle fibers in which the two myofilament systems are perpendicular to each other and run from right ventrolateral to left dorsolateral and vice versa (Fig. 3F and G). This bulb is found in Dinophilidae (Rieger & Rieger, 1975; Purschke, 1985a, 1988a).

Type 3

In Nerillidae a dorsoventral tonofilament system is present in each muscle fiber, crossing the myofilaments at a right angle (Fig. 4A and B; Purschke, 1985b, 1988a; Tzetlin et al., 1992; Müller et al., 2001). Very often the muscle fibers making up the bulbus are of the circomyarian type, with a central core of sarcoplasm housing mitochondria and nucleus; this fibre type is usually found in Hirudinea (Lanzavecchia et al., 1988).

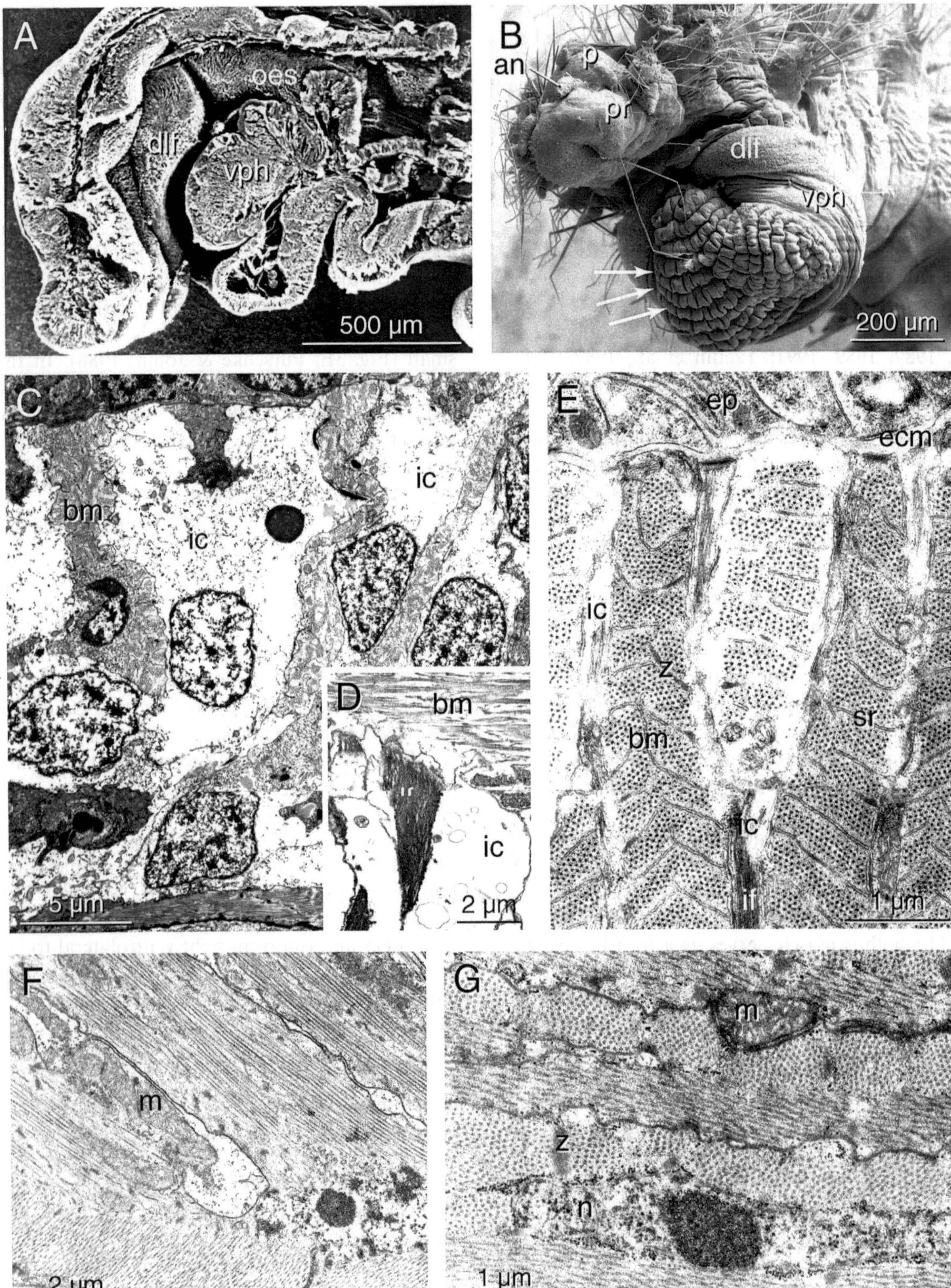
A
oes
dlf
vph
500 µm
B
an
p
pr
dlf
vph
200 µm
C
bm
ic
ic
5 µm
D
bm
ic
2 µm
E
ep
ecm
ic
z
bm
sr
ic
if
1 µm
F
m
2 µm
G
m
z
n
1 µm

Figure 3. Foregut. Ventral pharyngeal organ. (A) *Nicomache minor* (Maldanidae). Sagitally dissected specimen showing ventral pharynx (vph), dorsolateral folds (dlf) and oesophagus (oes). (B) *Eurythoe complanata* (Amphinomidae). Everted ventral pharynx (vph), surface of the bulb supplied with transverse ridges (arrows) used for scraping off food particles; an antenna, dlf dorsolateral folds, p palp, pr prostomium. (C) *Scoloplos armiger* (Orbiniidae). Longitudinal section of muscle bulb composed of large voluminous interstitial cells (ic) and bulbus muscle fibers (bm). (D) *Ctenodrilus serratus* (Ctenodrilidae). Muscle bulb (bm), large interstitial cells (ic) with prominent bundles of intermediate filaments (if) crossing myofilaments at a right angle. (E) *Saccocirrus papillocercus* (Saccocirridae). Bulb made up of small interstitial cells (ic) and plate-like muscle fibers (bm); ecm extracellular matrix, ep epithelium, if intermediate filaments, sr sarcoplasmic reticulum, z z-rod. (F, G) Dinophilidae, muscle bulb consisting of muscle fibers only; myofilaments perpendicular in both halves of the fibers. (F) *Trilobodrilus axi.* (G) *Dinophilius gyrociliatus*; m mitochondrion, n nucleus, z z-rod. A modified from Tzetlin et al. (1992); B micrograph S. Raabe, C modified from Purschke (1988a).

Type 4

In Dorvilleidae the bulbus muscle fibers are of this type as well but without tonofilaments or interstitial cells in between them (Purschke, 1987b).

Type 5

Most likely another type of pharyngeal bulb is present in Diurodrilidae, characterized by a large ventral pharyngeal gland cell and poorly developed musculature (Kristensen & Niilonen, 1982). Unfortunately, from the brief description it cannot be concluded whether this type may have been evolved from one of those mentioned above and has be included in one of the types mentioned above.

Type 6

The pharynx of Amphinomida appears to be completely different from all structures described above. Unfortunately, the highly muscularized structure has not yet been described at the ultrastructural level. It ventrally has a series of distinct transverse lamellae with a thickend and transformed cuticle (Dales, 1962; Purschke & Tzetlin, 1996).

The investing muscle fibres are not that diverse and typically are obliquely striated. In certain species the myofilament system is not helically arranged so that in sections only one myofilament system appears to be present, e.g., Nerillidae (Purschke, 1985b; Müller et al., 2001). Usually nuclei are located in posterior extensions where the fibers turn but in some species their positioning can also be of the Hirudinean type (see above).

Additional muscle fibers may be present in the tongue-like organs, which are pistil-shaped, lateral folds or flat structures attached to the lateral walls of the foregut. Intracellular skeletal elements are present in many (all?) Nerillidae. Situated in epithelial cells, they measure up to 55 μm in length and up to 3 μm in diameter. These skeletal rods are initially made up of tonofilaments which fuse to form striations with a period of 65–70 nm (Purschke, 1985b; Tzetlin et al., 1992; Müller et al., 2001). Depending on the diameter they are recognizable in the light microscope or not. In Parergodrilidae the cuticle of the tongue-like organ forms a rod-like posterior extension serving as attachment area of the musculature (Purschke, 1987a). In other ventral pharynges such specializations are unknown.

Axial non-muscular pharynx

An axial non-muscular proboscis is present in many of the sand- or mud-swallowing species of Arenicolidae, Maldanidae, Capitellidae, Opheliidae, Orbiniidae and Paraonidae (Fig. 4D and E; Purschke & Tzetlin, 1996; Rouse, 2000). These pharynges are composed of epithelial, glandular and sensory cells but musculature is weakly developed (Purschke, 1988a; Saulnier-Michel, 1992). The epithelium is unciliated and forms papillae in Arenicolidae, Maldanidae and Capitellidae. Ultrastructural studies are rare and have only been made in one species, *Notomastus latericeus* (see Michel, 1972; Saulnier-Michel, 1992). Everting movements of the proboscis are achieved by changes in hydrostatic pressure in the anteriormost compartments of the body cavity. The pressure increase effected by contractions of the musculature of the body wall is limited to the anterior part of the body cavity by the presence of a strong muscular septum in Capitellidae and Arenicolidae (Purschke & Tzetlin, 1996). The proboscis is withdrawn by the activity of retractor muscles.

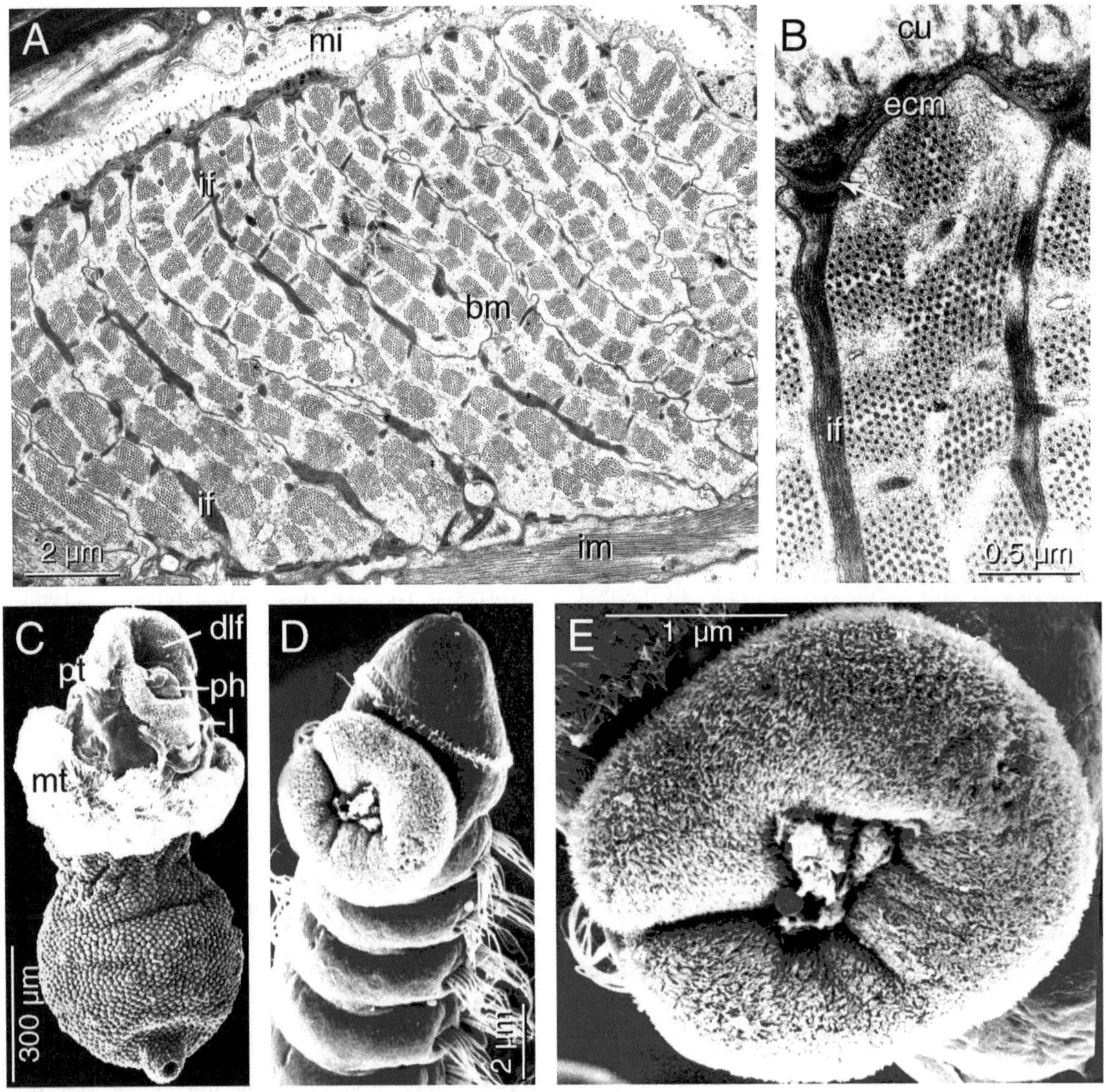

Figure 4. Foregut. (A, B) *Nerillidium troglochaetoides* (Nerillidae). (A) Muscle bulb composed of plate-like muscle cells (bm); myofilaments oriented transversally and parallel; im investing muscle, mi median invagination, if intermediate filaments. (B) Enlargement of bulbus muscle fibre with bundles of intermediate filaments (if) perpendicular to myofilaments. Note myopithelial junction (Arrow); cu cuticle, ecm extracellular matrix. (C) *Phascolosoma agassizii* (Sipuncula). Pelagosphera larva with ventral pharynx (ph); dlf dorsolateral folds, l lower lip, mt metatroch, pt prototroch. (D, E) *Paraonella nordica*. (D) Everted non-muscular axial proboscis. (E) Ciliated epithelium of proboscis. C modified from Rice, 1985.

During ontogeny, individuals of *Arenicola marina*, *Ophelia* spp., and *Scoloplos armiger* (as well as other Orbiniidae) pass through a juvenile stage possessing dorsolateral ciliary folds and a ventral pharyngeal organ (Eisig, 1914; Anderson, 1959; Purschke & Tzetlin, 1996). Later in development these structures are replaced by axial non-muscular proboscides. Interestingly, in some species of Maldanidae the adults have a foregut with a typically developed ventral pharynx and dorsolateral folds (e.g., *Nicomache minor*), whereas other species exhibit various stages of development of axial proboscis-like structures in addition to more or less reduced ventral pharynges (e.g., *Nicomache lumbricalis*, *Praxillela praetermissa*) or even possess only a non-muscular axial proboscis (*Axiothella rubrocincta*) (Fig. 5; Tzetlin, 1991). A similar situation has been observed in Orbiniidae by Eisig (1914).

Dorsal pharynx

A dorsal pharynx is typical of oligochaetous Clitellata, and *Hrabeiella periglandulata* is the only non-clitellate annelid possessing this type of pharynx (Jamieson, 1992; Rota, 1998; Purschke, 1999, 2002, 2003). It is characterized by a conspicuous thickening of the dorsal wall of the foregut owing to epithelial cells, gland cells and muscle fibers (Purschke, 2003). This dorsal pad is densely ciliated whereas the buccal cavity is always unciliated and, further posterior, only the oesophagus has an epithelium with motile cilia (Fig. 6). In *H. periglandulata* a complex system of muscle fibers serving as protractors and retractors is attached to the dorsal ciliated pad. The cell bodies of the gland cells form four pairs of lobes situated on the dorsal side of the foregut between the muscle fibers. They send long, thin processes ventrally, the openings of which are situated exclusively between the ciliated cells of the pad. The epithelial cells are characterized by a well-developed cytoskeleton consisting of bundles of intermediate filaments running in the apical-basal direction. The ciliary rootlets are connected to this filament system. The ciliary pad is surrounded by a cuticular fold and is made up of comparatively short cilia (about 5 μm long, approximately 12 cilia/μm^2) This type of pharynx is obviously adapted to ingestion of decaying plant material and detritus in terrestrial habitats (Westheide & Müller, 1996; Purschke, 2003). This pharynx is structurally rather similar to dorsolateral ciliary folds in that the ciliated and glandular area is restricted to a dorsal pad and additionally supplied with a well-developed system of pro- and retractor fibers.

Axial muscular pharynx

An axial pharynx with a well-developed muscular region is usually present in Phyllodocida (Fig. 1A; see Purschke, 1988a; Saulnier-Michel, 1992). The mouth gives rise to the buccal cavity, called proboscidian sheath (see Purschke, 1988a), which houses the muscular region of the pharynx in the resting position (Fig. 8D). These organs can often be protruded to a great distance, and at the junc-

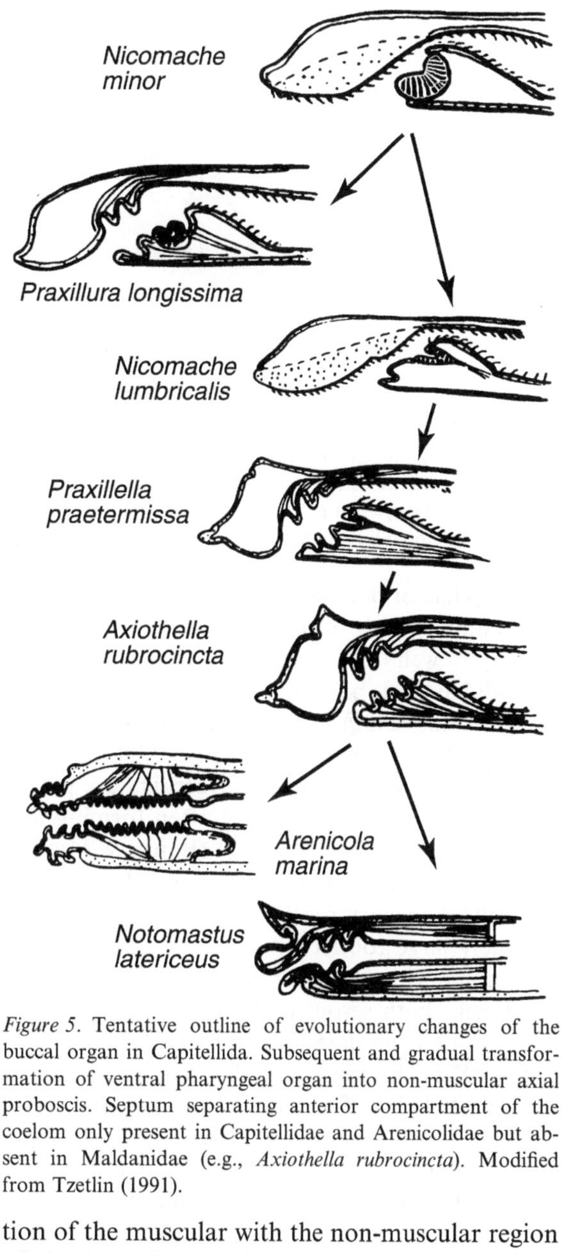

Figure 5. Tentative outline of evolutionary changes of the buccal organ in Capitellida. Subsequent and gradual transformation of ventral pharyngeal organ into non-muscular axial proboscis. Septum separating anterior compartment of the coelom only present in Capitellidae and Arenicolidae but absent in Maldanidae (e.g., *Axiothella rubrocincta*). Modified from Tzetlin (1991).

tion of the muscular with the non-muscular region of the buccal organ jaws may be present, as in Nereididae, certain Hesionidae, Pisionidae, Glyceridae or Polynoidae (Figs 7A and C, 10A–D). This region represents the physiological mouth opening in these taxa. In many species of Pisionidae, Hesionidae, Pilargidae and Syllidae the tip of the muscular region is differentiated into a number of finger-like papillae, each bearing a set of sensory cells (Figs 7C–H, 8A–F and 10A, C). The proboscidian sheath covers the pharynx externally when everted. It often bears soft papillae, sclero-

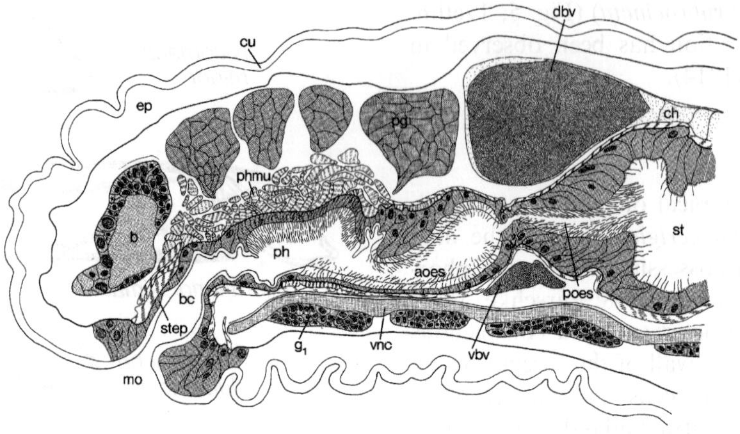

Figure 6. Foregut, dorsal pharynx. *Hrabeiella periglandulata.* Reconstruction of foregut with dorsal pharynx (ph) composed of ciliary pad, pharyngeal musculature (phmu) and pharyngeal glands (pg). Note unciliated buccal cavity (bc); aoes anterior oesophagus, b brain, ch chloragocyte, cu cuticle, dbv dorsal blood vessel, ep epidermis, g1 ganglion 1, mo mouth, poes posterior oesophagus, st stomach, step stomodeal epithelium, vbv ventral blood vessel, vnc ventral nerve cord. From Purschke (2003).

tized papillae or paragnaths (Figs 7A and B, 10D). Such papillae with sclerotized parts (onglets) are especially well developed in Glyceridae and Goniadidae and bear species-specific characters (Böggemann, 2002). Comprising two to three sensory cells with a few penetrative cilia and supportive cells, the papillae of Glyceridae are partly covered by a sclerotized cuticle often forming fingernail-like structures (Fig. 7B) (Bantz & Michel, 1971, 1972; Böggemann et al., 2000). In the sensory cells the ciliary rootlets fuse to form a single, extremely large rootlet extending through the entire papilla. This type also incorporates numerous gland cells, both serous and mucous, which assist adhesion and the digestion of food (Saulnier-Michel, 1992). The most complex axial pharynx is found in Syllidae. The anterior part of the digestive tract consists of a buccal cavity (proboscidian sheath), pharyngeal tube, proventricle and ventricle (Figs 1A, 7G and H; see Purschke, 1988a). The ventricle, sometimes supplied with caeca, is regarded as part of the intestine. The axial proboscis is divided into a long non-muscular part, the pharyngeal tube, followed by a shorter muscular part with prominent radial musculature, the proventricle. At the opening of the pharyngeal tube a circle of soft papillae is often present (Fig. 10A). This region is followed by differently arranged jaws or teeth (Fig. 10A and see below) or may be unarmed (e.g., *Syllide* sp., Fig. 7D). The pharyngeal tube is of different length and thus may be straight or coiled in the resting position.

In axial pharynges, the proboscidian sheath and the muscular region are covered by the cuticle, which may be modified in comparison to that present on the trunk. Especially regions subjected to mechanical stress possess protective structures such as the lamellar layers in Hesionidae (Fig. 8B, C, E and F; Westheide & Rieger, 1978; Purschke, 1988a). In Glyceridae such a lamellar layer is also present on the trunk (Bantz & Michel, 1971, 1972; G. Purschke & Koldehoff, unpublished data).

The musculature of this type of buccal organ is complex: circular, longitudinal and radial fibers are present, allowing sucking and swallowing movements (Fig. 8D). Among these fibers the radial ones predominant. Although usually a component of obliquely striated musculature as is typical for Annelida, these fibers exhibit several specializations: nucleus and mitochondria are often located centrally as in Hirudinea and usually a T-system is present (Purschke, 1988a). In *Nephtys* and Syllidae membrane-bounded granules containing crystalline calcium phosphate are present in the central cytoplasm (Briggs et al., 1985; Bryan & Gibbs, 1986; Purschke, 1988a). In Syllidae the radial fibres of the proventricle are not oblique-striated but cross-striated. The number of sarcomeres varies from one to about 20 depending on species. Sarcomeres may be as long as 60 μm, the longest sarcomeres found in Metazoa so far (Del Castillo et al., 1972; Smith et al., 1973; Wissocq, 1974).

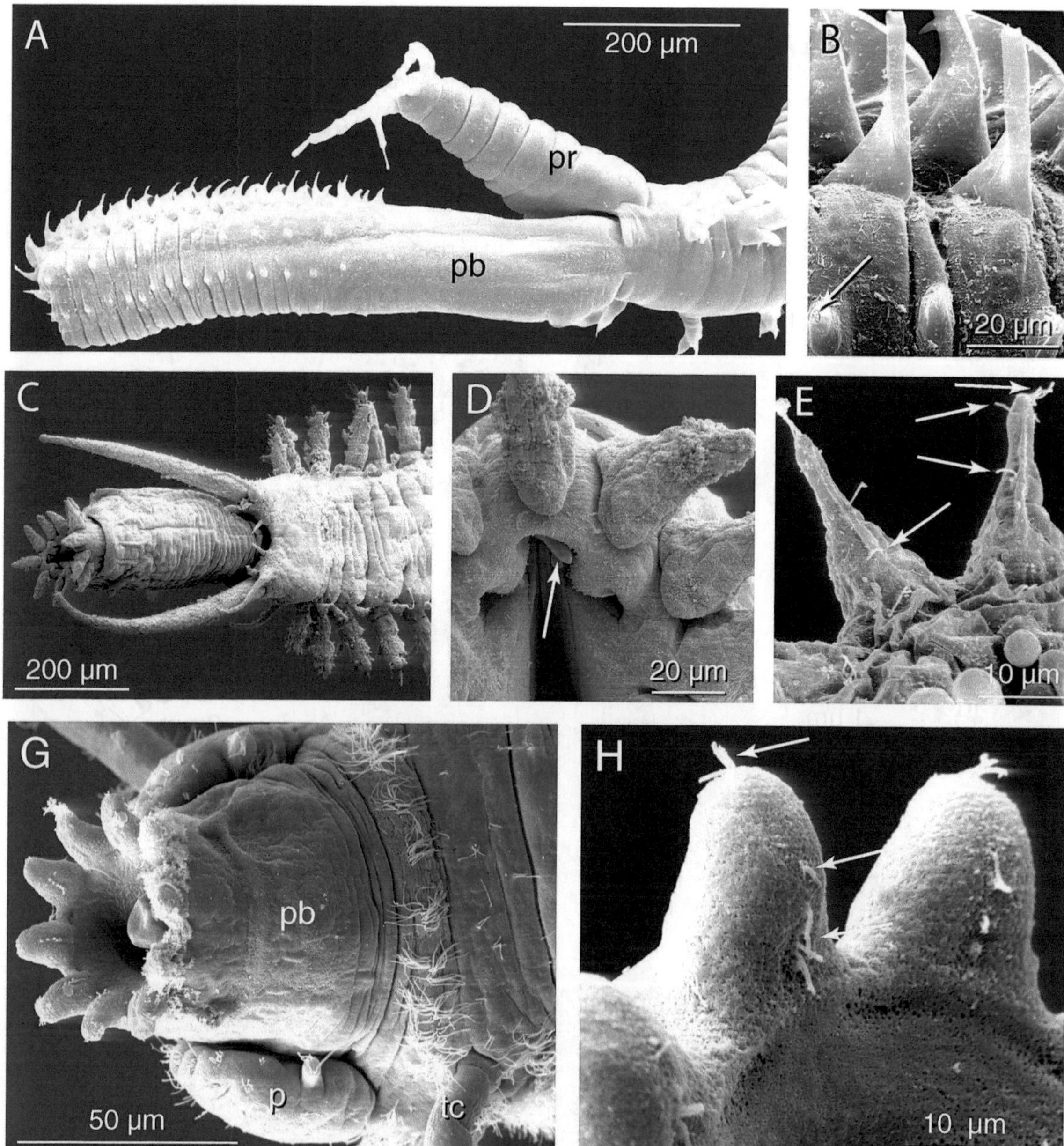

Figure 7. Foregut; axial muscular pharynx. (A, B) *Goniadides falcigera* (Goniadidae). (A) Proboscis (pb) partly everted, densely covered with finger-like papillae; pr prostomium. (B) Enlargement of papillae. Note sensory cilia on top of papillae (arrow). (C, D) *Pisione remota* (Pisionidae). (C) Specimen with everted pharynx bearing 14 terminal papillae. (D) Enlargement of papillae and pharyngeal opening with tip of jaw (arrow). (E) *Microphthalmus listensis* (Hesionidae). Two of ten pharyngeal papillae with cilia of sensory cells (arrows). (G, H) *Syllides caribica* (Syllidae); unarmed pharynx. (G) Everted proboscis with 10 soft papillae; p palp, tc tentacular cirrus. (H) Enlargement of two pharyngeal papillae with cilia of sensory cells (arrows). Micrographs A,B M. Böggemann, C, D S. Raabe, G, H M. Kuper.

Phylogenetic remarks

The phylogenetic importance of the different foregut structures is still under discussion. Dales (1962) in his comprehensive study was among the first to use the structure of the foregut as a basic criterion for grouping the polychaete families. Whereas an axial muscular proboscis is present

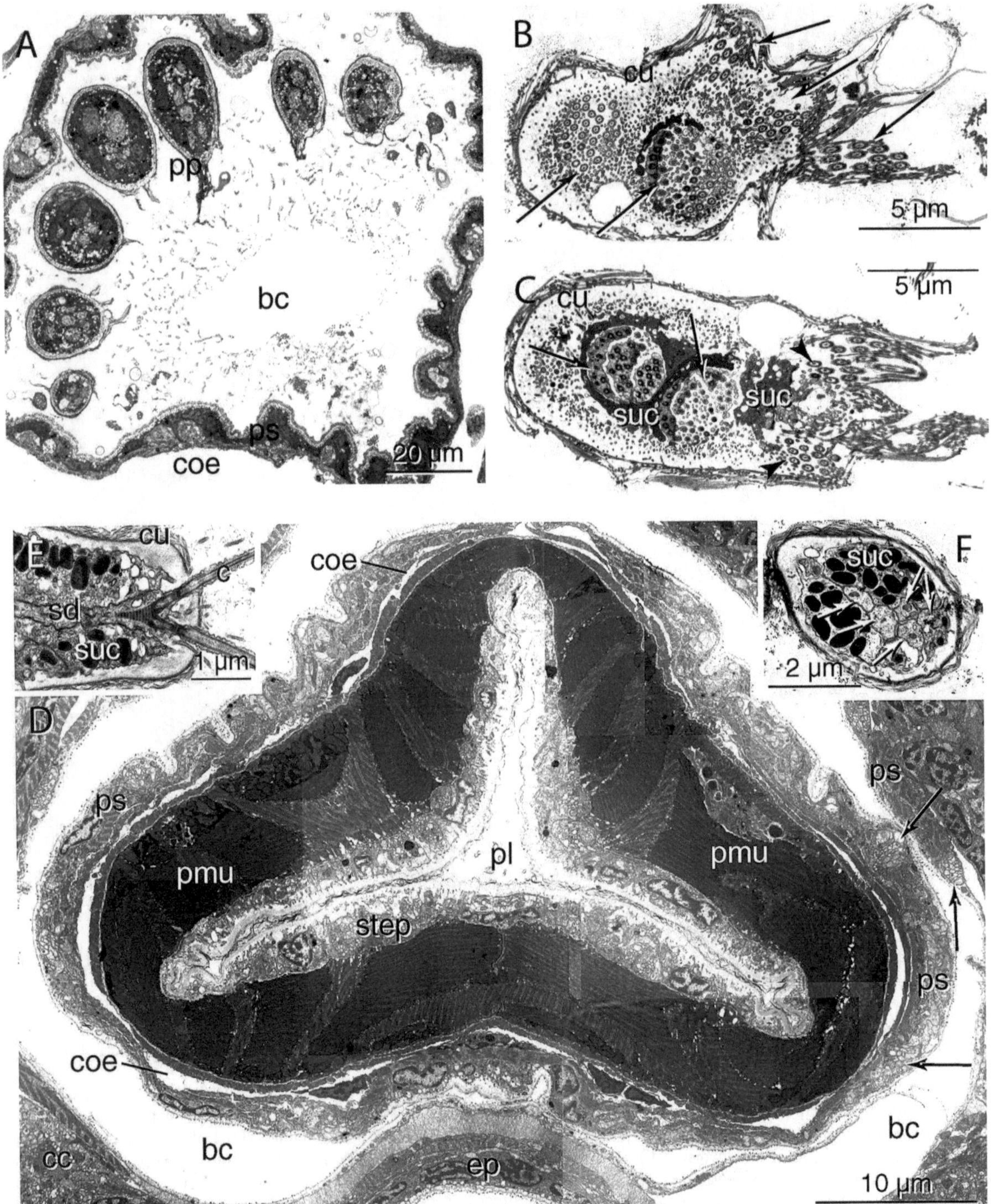

Figure 8. Foregut; axial muscular pharynx. TEM. (A–C) *Kefersteinia cirrata* (Hesionidae). (A) Buccal cavity (bc) with transverse sectioned pharyngeal papillae (pp); coe coelom, ps pharyngeal sheath. (B) Transverse section of papilla with numerous sensory cilia (arrows), pharyngeal cuticle (cu) with several epicuticular lamellar layers. (C) Somewhat deeper section with supportive cells (suc), two central groups of sensory cells (arrows) and a few peripheral sensory cells (arrowheads). (D) *Hesionides arenaria* (Hesionidae). Muscular part of pharynx; bc buccal cavity, cc circumoesphageal connective, coe coelom, ep epidermis, pl pharyngeal lumen, ps pharyngeal sheath, pmu pharyngeal musculature, step stomodeal epithelium. (E, F) *Microphthalmus listensis* (Hesionidae). (E) Tip of pharyngeal papilla, sensory dendrite (sd), two sensory cilia (c) attached to one common rootlet, cuticle (cu) with lamellar cover, suc supportive cell. (F) Cross section of pharyngeal papilla with 5 sensory dendrites (arrows), suc supportive cell.

only in Phyllodocida (and Myzostomida, a taxon of questionable affinities; see Eeckhaut & Lanterbecq, 2005), the other higher taxa of polychaetes do not exhibit such a uniform distribution of foregut structures. Even in a given family taxon different types of buccal organs may be present (Purschke & Tzetlin, 1996). Due to their structural diversity, the homology assumption of ventral pharyngeal organs (e.g., Dales, 1962, 1977) was challenged. Therefore, fine-structural studies have been conducted to elucidate whether ventral pharyngeal organs are in fact homologous or evolved convergently (Rieger & Rieger, 1975; Jouin, 1978a,b; Michel, 1978; Purschke, 1985a,b, 1987a,b, 1988b; Tzetlin, 1987, 1989, 1991; Tzetlin et al., 1987, 1992; Purschke & Jouin, 1988; Tzetlin & Larionov, 1988; Saulnier-Michel et al., 1990). These investigations confirmed great structural differences among the various polychaete taxa examined, making an overall homology unlikely. However, as more was learned about these organs, it became possible to demonstrate a plausible origin of these pharyngeal structures from only a few, convergently evolved, different types (Purschke, 1988b; Purschke & Tzetlin, 1996). Moreover, there is strong evidence that non-muscular axial buccal organs represent derived structures, most likely to have been evolved from foreguts with a ventral pharyngeal organ (Tzetlin et al., 1987; Tzetlin, 1991). Evidence is drawn from the fact that in most cases a ventral pharynx present in larvae is subsequently retained only if the adult of the species is small; it is replaced by a non-muscular proboscis in larger species. In the clade formed by Arenicolidae; Maldanidae and Capitellidae the proboscis is developed from the region in front of the ventral pharynx, in Orbiniidae behind it (Eisig, 1914; Tzetlin, 1991). In the former the epithelium of the proboscis is unciliated, whereas in Opheliidae and Paraonidae it is ciliated (Hartmann-Schröder, 1958; A.B. Tzetlin, unpublished data; Fig. 4D and E). Moreover, in taxa with appendages used for feeding (e.g., Spionidae, Ampharetidae, Terebellidae), a ventral pharynx is present in juveniles as well and may be retained by the adults.

The existence of a comparatively simple feeding structure in polychaetes was recognized by Purschke & Tzetlin (1996), namely the so-called dorsolateral ciliary folds, which are widespread among polychaetes. These microphagous feeding structures are most often found together with a ventral pharynx but may also occur alone. They are present in larvae, juveniles and adults of small species but are usually replaced by other feeding structures in larger species. In connection with their widespread occurrence, these observations strongly indicate that dorsolateral ciliary folds and a ventral pharynx represent the plesiomorphic condition for Annelida (Purschke & Tzetlin, 1996). All other types of feeding structures may have been evolved from these folds, as is indicated by structural and developmental data (Fig. 9; see Purschke & Tzetlin, 1996). Since these folds are obviously restricted to individuals of small body size in extant Annelida, it follows that the annelid stem species either was comparatively small and microphagous, or at least had a life cycle that included a developmental stage showing these characteristics (Purschke & Tzetlin, 1996; Purschke, 2002). Although it seems conceivable that dorsolateral ciliary folds might already have been present in the stem species of Annelida, it is questionable whether they are an autapomorphy of Annelida or, more probably, a plesiomorphy (Ax, 1999). The latter interpretation is supported by the presence of such ciliary folds and a ventral pharynx (buccal organ) in pelagosphera larvae of Sipuncula (Fig. 4C; Rice, 1973, 1976). Homology of these structures in Sipuncula and Annelida remains to be proven but appears to be likely. However, the position of Sipuncula is still uncertain: in analyses using morphological data they usually fall outside Annelida (Rouse & Fauchald, 1997; Ax, 1999), whereas molecular data often suggest placement of Sipuncula within Annelida (e.g., Winnepenninckx et al., 1998; Martin, 2001).

Jaws

General

Polychaete jaws are cuticular structures formed on the surface of specialized epithelial cells, often called gnathoblasts. Usually jaws are sclerotized parts of the pharyngeal cuticle which may be highly mineralized in some taxa, but chitin is absent (Saulnier-Michel, 1992). Jaws occur in Phyllodocida and Eunicida (Wolf, 1980; Paxton, 2000, Rouse, 2000; Rouse & Pleijel, 2001) and

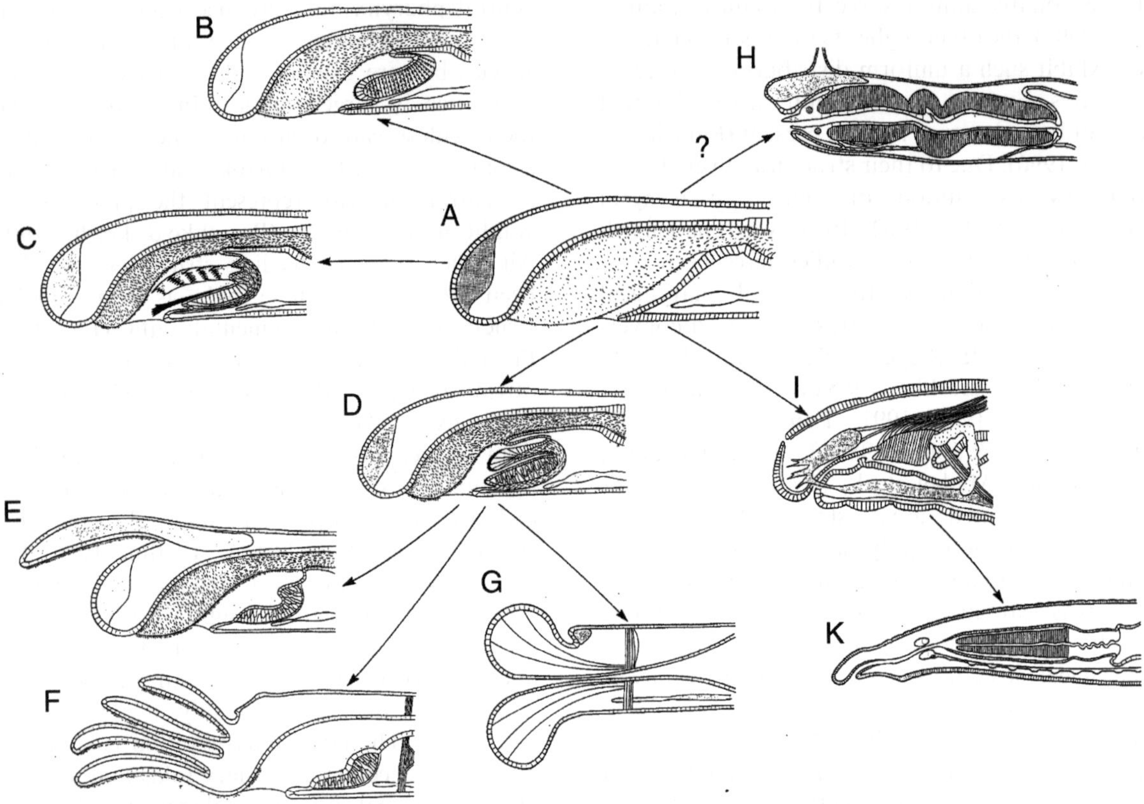

Figure 9. Schematic representation of different foregut structures in polychaetes related to dorsolateral folds (DLF). Arrows indicate probable phylogenetic pathways, some of which can be followed during ontogenesis. (A) DLF only feeding structure present (e.g., Polygordiidae). (B–E) DLF and different ventral pharyngeal organs (VPO). (B) VPO composed of muscle cells only (e.g., Dinophilidae). (C) VPO with jaw apparatus (Eunicida). (D) VPO consisting of muscle bulb, tongue-like organ and investing muscle (e.g., Orbiniidae, Ctenodrilidae, Protodrilidae). (E) VPO and additional feeding appendages (palps) (e.g., Spionidae). (F) VPO Ventral organ and numerous feeding appendages. DFG absent in adults, anterior compartment of coelomic cavity separated by muscular septum (e.g., Terebellidae). (G) DLF and VPO replaced by non-muscular axial proboscis, anterior part of the body separated by muscular septum (e.g., Maldanidae, Capitellidae, Arenicolidae). (H) DLF replaced by axial muscular pharynx (Phyllodocida). (I) DLF modified to dorsal ciliated glandular pad (oligochaetous Clitellata, *Hrabeiella periglandulata*). (K) Dorsal pharynx replaced by axial muscular pharynx (Hirudinea). Modified from Purschke (2002).

more recently jaws were found in certain Ampharetidae (Desbruyères, 1978; Mackie, 1994).

Various aspects of jaw structure are traditionally used as taxonomic and phylogenetic characters. Special attention has been paid to the jaws of Eunicida ever since the fossil scolecodonts were identified as jaws belonging to extinct Eunicida (Pander, 1856; Ehlers, 1868). First Kozlowski (1956) presented a reconstruction of a fossil jaw apparatus directly comparable with extant ones. So far, the palaeozoic jaws of Eunicida are the most abundant fossil material in Annelida (Croneis & Scott, 1933; Kielan-Jaworowska, 1966; Kozur, 1970; Mierzejewski & Mierzejewska, 1975; Mierzejewski, 1978, 1984; Colbath, 1986, 1988; Zaslavskaya, 1989). Hence the comparative analysis of extant and fossil jaw apparatuses is highly important for our understanding of evolutionary trends among polychaetes. Fossil jaws that can be referred to extant polychaete genera are Mesozoic and belong to Glyceridae, Goniadidae and Dorvilleidae (*Ophryotrocha*) (Szaniawaski, 1974), or later fossils from the lower Miocene (Szaniawaski & Wrona, 1987).

Phyllodocida

In Phyllodocida the jaws are represented as more or less numerous tooth-like structures located on the axial muscular proboscis, usually at the beginning of the muscular part (Fig. 10A–D). Jaws have been described for Acoetidae, Aphroditidae, Eulepethidae, Polynoidae, Pholoidae, Sigalonidae, Pisionidae, Chrysopetalidae, Hesionidae, Nereididae, Syllidae, Goniadidae, Glyceridae, and Nephtyidae. In most cases they are used to capture and hold the prey, and in tearing off pieces of algae or decaying matter. The number of jaws may be one, two, four or rather more (Rouse & Pleijel, 2001). Scale worms such as Polynoidae, Sigalonidae, Eulepethidae and Aphroditidae possess two pairs of dorso-ventrally oriented jaws which are more or less well-developed; ultrastructural studies of the jaws are lacking (Saulnier-Michel, 1992; Rouse, 2000). In Syllidae there may be single mid-dorsal jaw (tooth; e.g., *Exogone*, *Sphaerosyllis*, *Syllis*), a series of teeth in a ventrolateral arc (sometimes combined with the large mid-dorsal tooth, e.g., *Odontosyllis, Eusyllis*; Fig. 10A) or a complete ring of teeth called trepan (e.g., most *Autolytus* spp.) situated somewhat behind the pharyngeal papillae (Fig. 10A). In several syllid species the pharynx may be unarmed (e.g., *Syllides*, Fig. 7G); hence pharyngeal structures are highly important for taxonomy in Syllidae (e.g., Licher, 1999; Glasby, 2000). The syllid tooth is scarcely sclerotized and primarily consists of a thick cuticle (Purschke, 1988a; Saulnier-Michel, 1992). There is no experimental evidence that these teeth may be chitinized as sometimes found in the literature. In Hesionidae some taxa possess a pair of jaws (Fig. 10C), the fine structure of which is unknown. One pair of lateral jaws is also present in Nereididae, Nephtyidae, and Pisionidae and all possess heavily sclerotized jaws (Fig. 10B–D; Michel et al., 1973; Purschke, 1988a; Saulnier-Michel, 1992). In Nereididae additional hard structures, the paragnaths, are present (Fig. 10B). Other types of buccal pieces are found Goniadidae: The chevrons are additional, serially arranged v-shaped hard structures situated more basally on the muscular proboscis (Fig. 10E).

A more conspicuous jaw apparatus is found in Glyceridae (Fig. 10D): the two pairs of jaws are always associated with venom glands and each of the venom ducts opens at the tip of the jaw, with an additional series of pores on the ventral side. The jaws are situated at the end of the eversible proboscis and each bears a supporting structure called the aileron (Böggemann et al., 2000). In addition to scleroproteins the jaws are highly mineralized (see Saulnier-Michel, 1992; Böggemann et al., 2000). In the closely related Goniadidae the jaws are completely different and comprise a pair of larger jaws (macrognaths) associated with smaller ones (micrognaths) forming a complete circle (e.g., Rouse, 2000).

Ampharetidae

In a few, small species of Ampharetidae jaws were found: *Gnathampharete paradoxa*, *Ampharete* sp., and *Adercodon pleijeli* (Desbruyères, 1978; Uebelacker & Johnston, 1984; Mackie, 1994). In all species the jaws consist of a transversal row of denticles located on the posterior edge of the ventral bulb (Fig. 10I). Nothing is known about how Ampharetidae use their jaws.

Eunicida

In Eunicida the jaw apparatus consists of a pair of mandibles, located on the ventral muscle bulb of the ventral pharynx, and one or two paired rows of maxillary plates located on lateral folds of the ventral pharyngeal organ (Fig. 10F and G; Wolf, 1980; Purschke, 1987b; Paxton, 2000). In all known extant and fossil Eunicida the mandibles are paired longitudinal structures usually with denticulated frontal edges (Fig. 10F). They are anchored in deep epidermal follicles. There is an articulation in the anterior part between left and right mandible, sometimes forming a ligament-like structure or a symphysis (Hartmann-Schröder, 1967; Purschke, 1987b). Although such an articulation is present, the main movements are in the anterior–posterior direction.

The maxillary apparatus varies between different Eunicida and several types may be distinguished (Figs 12 and 13). In general, one or two pairs of longitudinally arranged rows of maxillary plates are present, which may be connected caudally by carriers or carrier-like structures (Fig. 12D–H). The individual maxillary plates are usually denticulated or, rarely, single toothed. The number of maxillary plates in each row varies from

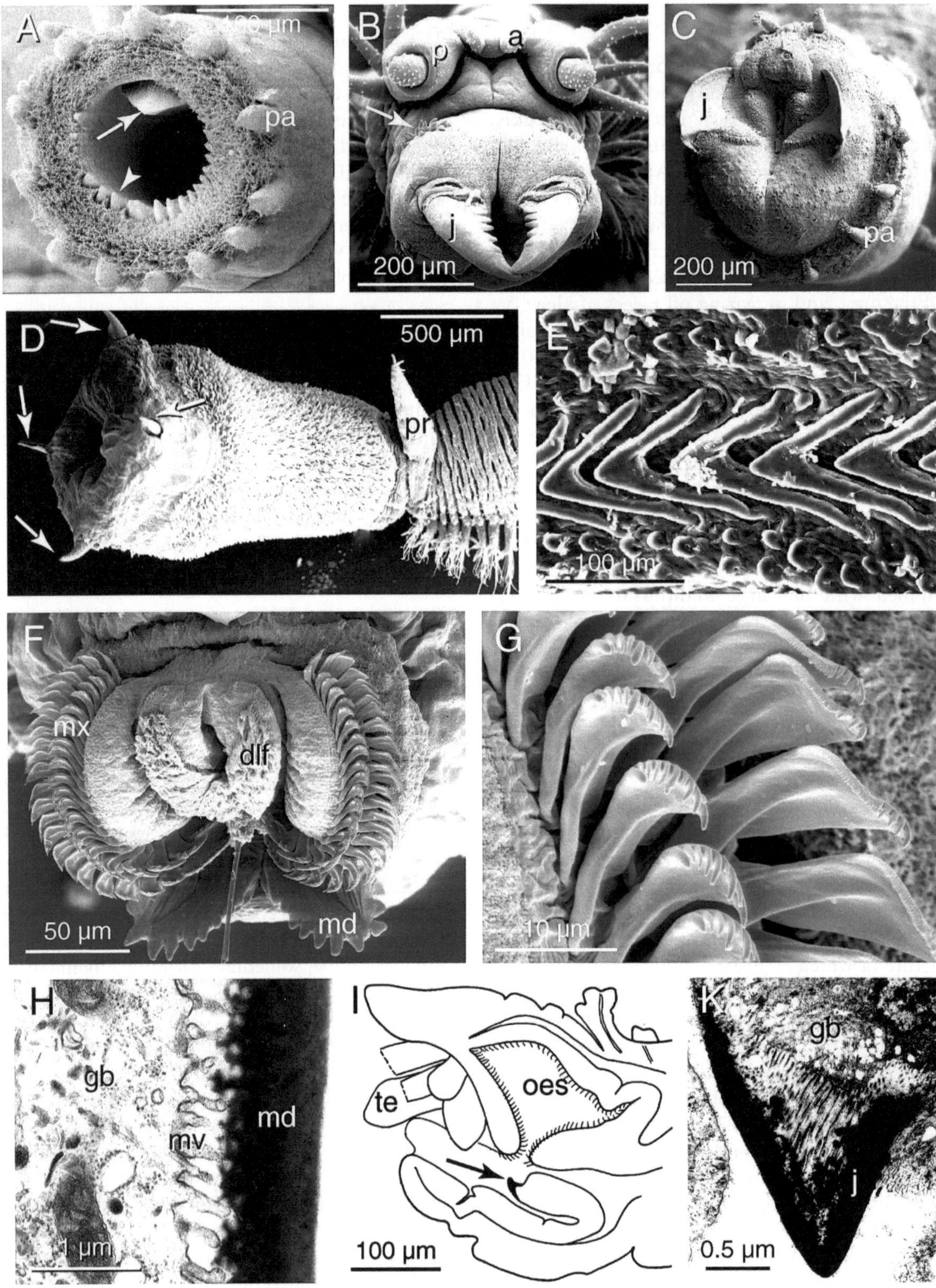
A
400 µm
pa
B
p
a
j
200 µm
C
j
pa
200 µm
D
500 µm
pr
E
100 µm
F
mx
dlf
md
50 µm
G
10 µm
H
gb
mv
md
1 µm
I
te
oes
100 µm
K
gb
j
0.5 µm

Figure 10. Jaws. (A) *Eusyllis blomstrandi* (Syllidae), pharyngeal tube with circle of papillae (pa), cilia and dorsal tooth (arrow) and row of smaller teeth (arrowhead). (B) *Nereis* sp. (Nereididae), jaws (j) and first group of paragnaths (arrow). a antenna, p palp. (C) *Syllidia armata* (Hesionidae), shovel-like jaws at the tip of everted pharynx, note unequal distribution of pharyngeal papillae (pa). (D) *Glycera alba* (Glyceridae), specimen with completely everted pharynx showing the four jaws (arrows) and numerous external papillae. pr prostomium. (E) *Goniada maculata* (Goniadidae), chevron. (F–H) *Protodorvillea kefersteini* (Dorvilleidae), jaw apparatus consisting of mandibles (md) and two rows of maxillary plates (mx). dlf dorsolateral folds. (G) Close up of maxillary plates. (H) TEM-micrograph of mandible (md), note short microvilli (mv) of gnathoblast (gb). (I–K) *Adercodon pleijeli* (Ampharetidae). (I) Position of jaws in the foregut (arrow), oes oesophagus, te tentacle. (K) TEM-micrograph of jaw (j), gnathoblast (gb) with long microvilli. Micrographs D, E M. Böggemann, I modified from Mackie (1994).

four in Lumbrineridae to 50 and even more in Dorvilleidae (Figs 10F and G, 12A–H). Mandibles and maxillae can be moved independently. The movements of the jaws are complex (Hartmann-Schröder, 1967; Wolf, 1980). In most cases Dorvilleidae use their jaws, especially the mandibles, for scraping off food particles from hard substrates and, predominantly the maxillae, for capturing and holding these particles (Tzetlin et al., 1987; A.B. Tzetlin & G. Purschke, unpubl. obs.). Larger extant Eunicida (Lumbrineridae, Onuphidae, Eunicidae) are macrophagous animals using their jaws for capturing and holding food (Jumars, 1974).

Since the mandibles are very similar among the different taxa in Eunicida, classification of eunicidan jaws is based on the structure of the maxillae. The first classification of jaw structures in Eunicida was introduced by Ehlers (1864–1868). A more detailed classification was done by Kielan-Jaworowska (1966) in her outstanding monograph on fossil polychaete jaws: (1) placognath jaws, (2) ctenognath jaws, (3) labidognath jaws and (4) prionognath jaws (Figs 12A–H, 13A and C–E). Later a fifth type, xenognath jaws, was introduced by Mierzejewski & Mierzejewska (1975) (Fig. 13B).

(1) The placognath maxillary apparatus is characterized by asymmetrically arranged denticulated plates in the posterior part and two symmetric anterior rows of apparently free maxillary plates (Fig. 13A). Carriers are lacking. Placognath jaws are only known from Ordovician until the Upper Devonian. Kielan-Jaworowska (1966) and Mierzejewski (1978) described moulting of the maxillae in Mochtyellidae. Jaws of this type appear to be unique within Eunicida and are difficult to compare with other jaws.

(2) Ctenognath jaws consist of small, symmetrically arranged basal plates in the posterior part and four rows of numerous, presumably free, so-called denticles anteriorly (Figs 11, 12A–C and

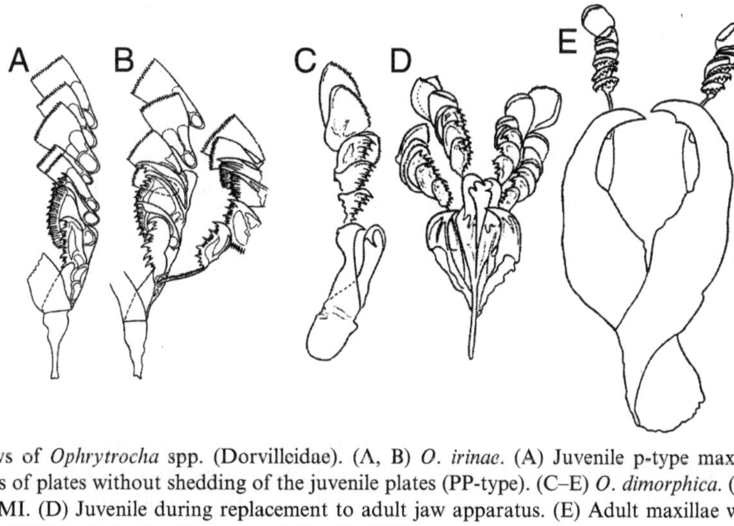

Figure 11. Different jaws of *Ophrytrocha* spp. (Dorvilleidae). (A, B) *O. irinae*. (A) Juvenile p-type maxillae. (B) Adult maxillary apparatus with two rows of plates without shedding of the juvenile plates (PP-type). (C–E) *O. dimorphica*. (C) Maxillae of 7 chaetiger juvenile with bidentate MI. (D) Juvenile during replacement to adult jaw apparatus. (E) Adult maxillae with large MI (forceps; K-type). A, B after Tzetlin (1980a), C–E after Zavarzina & Tzetlin (1986).

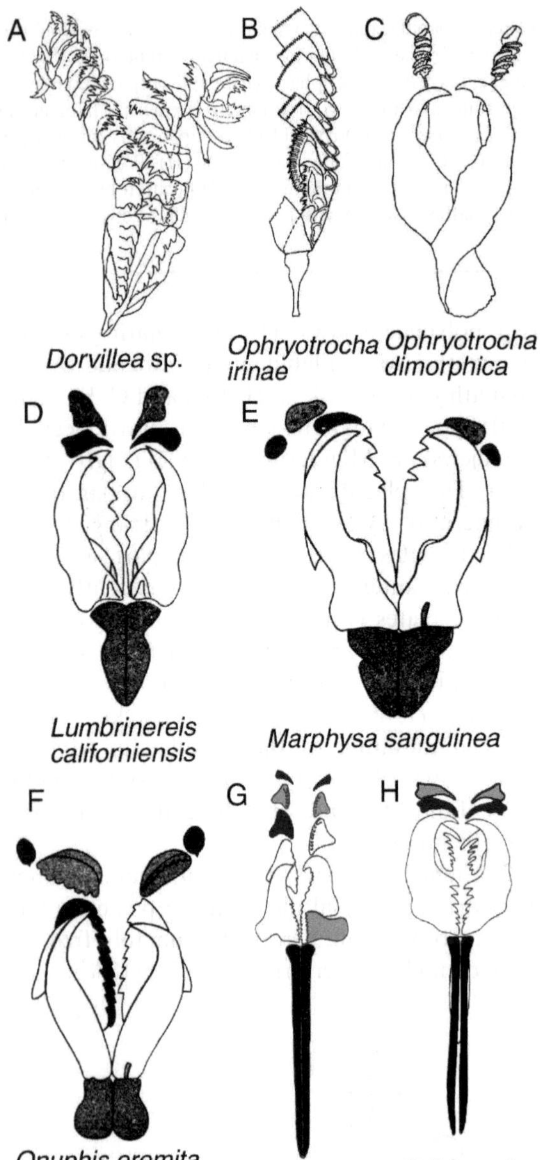

Figure 12. Diagramatic outline of different types of maxillary apparatuses present in extant and extinct Eunicida. (A–C) Ctenognath maxillae (Dorvilleidae). (A) Polymeric maxillae arranged in four rows (B) Oligomeric maxillae arranged in two rows. (C) K-type maxillae of *Ohryotrocha*. (D–F) Labidognath maxillae (D) Lumbrineridae. (E) Eunicidae. (F) Onuphidae. (G, H) Prionognath maxillae; Oenonidae (formerly classified in Lyseratidae and Arabellidae). Redrawn after Kilan-Jaworoswska (1966), Tzetlin (1980a, b), Zavarzina & Tzetlin (1986).

13C). These rows form longitudinal series extending for more than half of the length of the maxillary apparatus. Typical carriers are lacking. Ultrastructural studies of extant material have shown that all maxillary plates are fused and form one piece (Purschke, 1987b). Thus a distinction between basal plates and denticles appears to be difficult and uncertain. According to, e.g., Åkesson (1976) the neutral term 'maxillary plate' should be favored. Posterior jaw pieces (denticles and base plates) possess narrow open pulp cavities. These are present in extant (Dorvilleidae, Histriobdellidae; see Jennings & Gelder, 1976; Tzetlin, 1980b; Purschke, 1987b) and fossil forms (Tetraprionidae from Ordovicium; Kielan-Jaworowska, 1966). According to Tzetlin (1980b) the jaws of Polychaetaspidae and Paulinitidae (Fig. 13D) also belong to this type, but see Kielan-Jaworowska (1966) who classified these jaws as labidognath.

(3) Jaws of the labidognath type are arranged in a semicircle when retracted (Ehlers, 1864–1868). Labidognath maxillae have short, broad carriers embedded in the pharyngeal tissues (Kielan-Jaworowska, 1966). There are 4–5 right and 4–6 left maxillary plates (Figs 12D–F and 13D). Jaws of this type are present in Eunicidae, Hartmaniellidae, most Lumbrineridae and Onuphidae but are known from extinct forms as well (Hartman, 1944; Paxton, 2000).

(4) Prionognath maxillae (extant and extinct) are arranged in two parallel rows of up to 5 maxillary plates and possess long slender carriers, slightly longer than or as long as the group of maxillary plates (Figs 12G, H and 13E). This type is found in Oenonidae among the extant Eunicida. Sometimes the jaw apparatus of Histriobdellidae is also assigned to this type (see Paxton, 2000).

In recent material prionognath and labidognath maxillae differ not only in the maxillary arrangement but also in the structure of the posterior maxillary plate (MI; numbering from posterior to anterior). In the prionognath type it is denticulated on the inner surface, like the anterior ones, whereas in labidognath maxillae MI is forceps-like and lacks teeth. However, if the classification of Kielan-Jaworowska (1966) of the maxillae as labidognath in Polychaetapsidae and Paulinitapsidae is correct, this distinction does only apply for recent taxa. In both extant and fossil labidognaths and prionognaths, the maxillae are two-rowed jaw apparatuses each consisting of only a few denticulated or non-denticulated cuticular maxillary plates.

(5) Xenognath maxillae were described for *Archeoprion quaristata* by Mierzejewski & Mierzejewska (1975). Maxillae of this type consist of two symmetrical plates with a few transverse rows of numerous denticles (Fig. 13B). This jaw apparatus has comparatively long and slender carriers, sometimes regarded as pseudocarriers.

Maxillae in Ophryotrocha

Another classification of the maxillae was introduced to describe the pharyngeal armature in *Ophryotrocha* (Dorvilleidae). Species of this genus like other dorvilleids replace maxillae sequentially during ontogeny (maturation, changes of sex, etc.). When leaving the egg mass, juveniles possess only a few achaetigerous segments. The pharyngeal armature in this stage consists of rudiments of paired mandibles and two rows of maxillary plates (Åkesson, 1967, 1976; Tzetlin, 1980b). Initially there are only 3–4 maxillary plates in every row, but soon 7–8 pairs of denticulated maxillary plates have been produced (Fig. 11A; Åkesson, 1973a, 1976; Tzetlin, 1980b). Carriers are very weakly developed at this stage. This type of maxillae was named P-type (Düsing, 1961; Müller, 1962). Periodically, maxillae of the P-type are replaced by new ones of the same type. A number of species have such jaws throughout life, such as *Ophryotrocha geryonicola, O. gracilis* and *O. cosmetandra.*

In *Ophryotrocha irinae* juveniles have a two-rowed P-type maxillary apparatus, and later a second pair of P-type maxillae appears, while the old ones are not shed (Fig. 11A and B; Tzetlin, 1980b). As a result the maxillary armature consists of four almost equal rows of maxillary plates. Later this four-rowed maxillary apparatus may be replaced by a new one also consisting of four rows of maxillary plates (PP-type). This jaw type has been

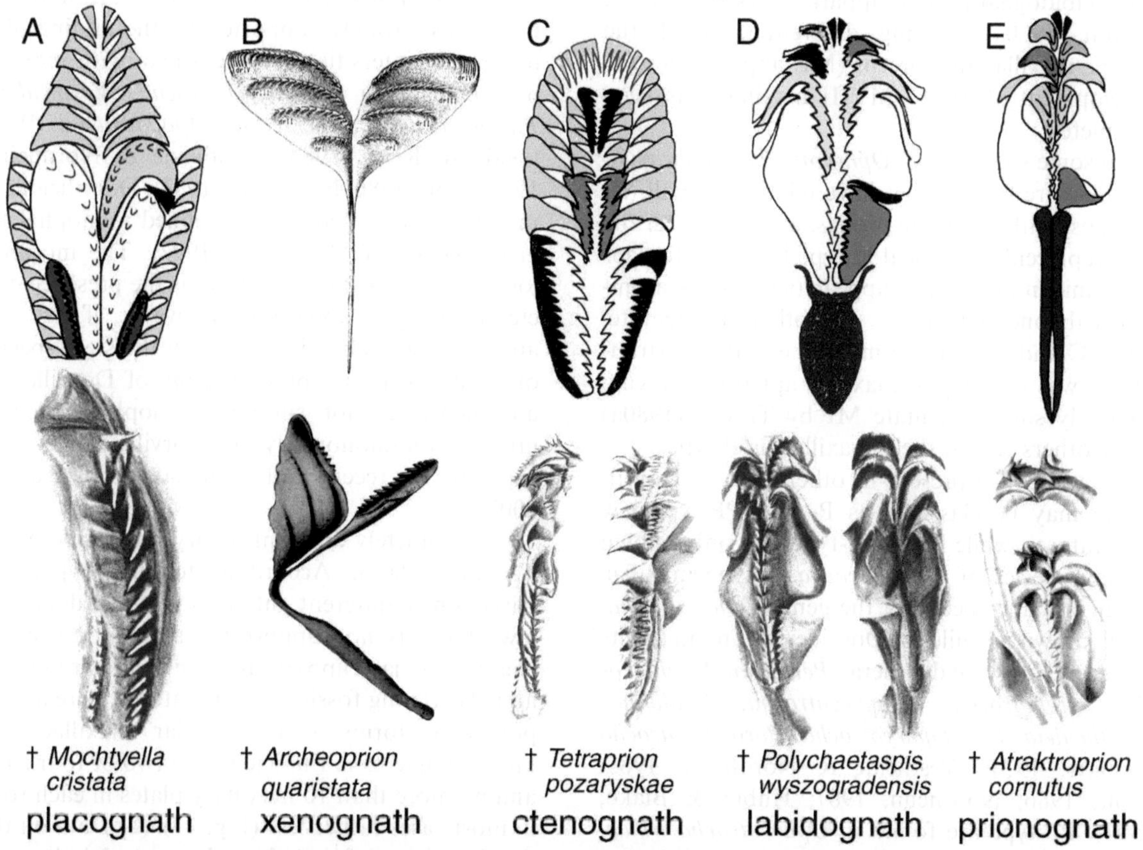

Figure 13. Types of maxillary plates found in fossil Eunicida. Upper row: reconstructions of jaws, lower row: drawings of specimens. Modified from Kielan-Jaworowska (1966) and Mierzejewski & Mierzejewska (1975).

regarded as homologous to the four rows of maxillary plates present in the maxillary apparatus of the other genera of Dorvilleidae by Tzetlin (1980a).

In other species of *Ophryotrocha*, such as *O. dimorphica* and *O. puerilis*, juveniles bear P-type maxillae and then change to another type of jaws – K-type – upon replacement (Düsing, 1961; Müller, 1962). This type of maxillae consists of a pair of large forceps (MI) and two rows of maxillary plates (MII–MVIII) identical with the juvenile P-type (Fig. 11C–E). The MI maxillary plates are rooted in carriers, and are connected with MII – MVIII by a narrow strip of sclerotized cuticle. The MI plates have unidentate or asymmetrical prongs. After K-jaws have formed they are not replaced any more. Very often development of the big MI forceps is connected with sexual maturation or change of sex in sequentially hermaphrodite species (Düsing, 1961; Müller, 1962; Åkesson, 1976; Zavarzina & Tzetlin, 1986). The general appearance of the K-type maxillae is consistent with labidognath jaws apparatus (Fig. 12C–F; Tzetlin, 1980a). During differentiation of the K-type maxillae the forceps (MI) appear after the development of MII – MVIII maxillae has been completed.

In some species (e.g., *Ophryotrocha schubravyi*), maxillae are formed that have relatively small MI (forceps) with bidentate prongs, and then after the next replacement, typical K-type large MI forceps with unidentate prongs appear (sometimes asymmetrical: one branch bi-, the other uni-dentate; Fig.11C and D). The term T-type maxillae (transition) was created for maxillae apparatus having relatively small bidentate MI by Tzetlin (1980a) while others regard these maxillae as P-type.

The maxillae present in other taxa of Dorvilleidae may be classified as P- and PP-type jaw apparatuses, while K- and T-types are only known among species of *Ophryotrocha*. P-type maxillae occur in many species of the genus *Ophryotrocha*, in all known juveniles of Dorvilleidae and in adults of several dorvilleid genera: *Petrocha*, *Pusillotrocha*, *Arenotrocha*, *Apophryotrocha*, *Exallopus*, *Westheideia*, *Pseudophryotrocha*, *Parophryotrocha* (Jumars, 1974; Westheide & Nordheim, 1985; Wolf, 1986; Nordheim, 1987; Hilbig & Blake, 1991). PP-type are found in *Ophryotrocha irinae*, and in *Dorvillea, Schistomeringos, Pettibonea, Protodorvillea*. Relatively large species of *Dorvillea* and *Schistomeringos* have more then 30 maxillary plates in each of the four rows (Hilbig & Blake, 1991).

Evolution of jaws in Eunicida

It is beyond discussion that Dorvilleidae represent the most ancient living members of Eunicida (Orensanz, 1990; Paxton, 2000). Usually polymeric maxillary apparatuses of large Dorvilleidae, such as are present in *Dorvillea* and *Schistomeringos*, were considered as plesiomorphic for Dorvilleidae and all Eunicida (Kielan-Jaworowska, 1966; Jumars, 1974; Orensanz, 1990; Eibye-Jacobsen & Kristensen, 1994; Paxton, 2000). According to Jumars (1974) the most primitive jaw apparatus consisted of numerous maxillary plates which were arranged in six rows. The evolutionary trends of the jaws in dorvilleids were (1) decrease in the number of maxillary plates and (2) subsequent reduction of number of rows from four to finally two rows of maxillary plates. This evolutionary trend was strongly supported by the finding of a number of intersitial dorvilleid taxa with reduced or even lacking jaws (*Arenotrocha*, *Westheideia*, *Pusillotrocha*, *Petrocha* and *Apodotrocha*) (Westheide & Riser, 1983; Westheide & Nordheim, 1985; Wolf, 1986; Nordheim, 1987). Dorvilleidae is thought to represent a well-defined monophyletic group (Rouse & Fauchald, 1997). The morphological basis for this hypothesis is the presence of a ctenognath jaw apparatus. However, if this jaw apparatus was already present in the stem species of Eunicida, it is a plesiomorphy of Dorvilleidae and hence cannot support monophyly of this group. Non-monophyly of Dorvilleidae is also suggested by recent molecular data (Struck et al., 2002).

A completely different theory was proposed by Tzetlin (1980b). According to this hypothesis based on a different interpretation of dorvilleid jaws there is no support for a basal polymeric ctenognath jaw apparatus. The main arguments are: (1) Among fossil jaw apparatuses there are no polymeric forms with irregular maxillae. All known Palaeozoic Eunicida have two or four rows and no more than 10 maxillary plates in each row, in most cases rather less (Fig. 13). (2) Data on the development of ctenognath and labidognath apparatuses show that always oligomeric ctenognath-like jaws are formed in the first ontogenetic stages (Åkesson, 1967, 1973a, 1976; Hsieh & Simon, 1987; Tzetlin, 1980a). Due to the resemblance of K-type maxillae of Ophryotrocha with labidognath jaws it was concluded that they most likely evolved from oligomeric ctenognath jaws (Fig. 12B–F). (3) Four rows of maxillae in Dorvilleidae most likely evolved from two-rowed jaws by prevention of shedding of the old maxillae during the processes of replacement (Fig. 12A and B). Oligomeric organization with two rows of maxillae of the P-type should thus be suggested as the plesiomorphic maxillary apparatus of Eunicida. (4) The life style of extinct Paleozoic Eunicida very likely was similar to that of extant *Ophryotrocha* sp.: i.e., small animals that feed by scraping off microfouling and other debris from hard substrates. However, this theory has not been accepted and was criticized by e.g., Orensanz (1990). Additional investigations appear to be necessary to resolve Eunicidan relationships and thus evolution of the different types of jaws.

Jaw histogenesis

Despite the great diversity in number and composition of the jaws in the different taxa there is only little information about their formation. Jaws of Phyllodocida, Eunicida and Ampharetidae are basically different and probably evolved independently. In Eunicida mandibles and maxillae are different as well. However, growth appears to be similar and only two types of jaw histogenesis and growth are known: in the first the gnathoblasts possess comparatively long microvilli and the jaws continue to grow throughout the animal's life (Fig. 10K). The gnathoblasts continuously produce new portions of the cuticular collagen matrix, which later becomes sclerotized by formation of scleroproteins. They may be calcified by calcite (Lumbrinereidae), aragonite (Onuphidae, Eunicidae) (Colbath, 1986) or otherwise mineralized by heavy metals such as iron, copper and zinc in Phyllodocida (Michel et al., 1973; Bryan & Gibbs, 1979, 1980; Gibbs & Bryan, 1980a, b; Böggemann et al., 2000). In Phyllodocida jaw growth is restricted to the basal parts (Olive, 1977, 1980). Since these jaws have a lifelong growth period, growth-lines can usually be observed (Kirkegaard, 1970; Olive, 1977, 1980; Tzetlin, 1990; Britaev & Belov, 1993; Britaev et al., 2002). Jaws of this type are the only ones found in Phyllodocida and applies for the mandibles in Lumbrinereidae, Onuphidae, Eunicidae and Ampharetidae as well (Wolf, 1980; Colbath, 1986; A.B. Tzetlin, unpublished observations).

A different type is typical for the maxillary apparatus in Dorvilleidae. It is characterized by gnathoblasts with very short microvilli or without microvilli (Fig. 10H; Damas, 1987; Purschke, 1987b). Jaws of this type do not grow after they have been formed. In many cases species with this type of jaws replace them regularly. The new ones are formed by newly developed groups of gnathoblasts (Damas, 1987; Purschke, 1987b). The old ones are shed and eliminated from the pharynx via the digestive tract (Tzetlin, 1980a; Hsieh & Simon, 1987). Whether maxillae are replaced in the other Eunicida is seen controversially and appears not to be resolved yet. Kielan-Jaworowska (1966) suggested replacement of labidognath maxillae which was refuted by Paxton (1980) who found evidence for growth throughout life without replacement in Onuphidae. Colbath (1987) suggested that Oenonidae have a fine structure of the jaws that is inconsistent with continuous growth and probably undergo regular shedding and replacement of maxillae. Replacement of maxillae has been described in Onuphidae by Hsieh & Simon (1987). So there is certain evidence indicating that shedding of the maxillae is typical of all Eunicida (Colbath, 1987; Hsieh & Simon, 1987) but further experimental evidence and ultrastructural investigations are required.

Intestine

In polychaetes the midgut may be differentiated into a stomach and intestine. Structure and function have been reviewed by Michel (1988) and Saulnier-Michel (1992) and only a few new results have since been obtained. Therefore, only a short summary is given here. The endodermal epithelia making up this part may form a pseudostratified epithelium, and rest on an extracellular matrix followed by the peritoneal lining. This lining comprises longitudinal and circular muscle fibers as well. Blood spaces are frequently found in the extracellular matrix between the two epithelia to form the blood sinus of the gut.

A distinct stomach representing a glandular or muscular organ is present in many microphagous polychaetes traditionally referred to as sedentary polychaetes (Fig 1B and C). This part of the alimentary canal is the site of extra- and intracellular digestion. It may be composed of glandular cells, absorptive cells and ciliated cells or only one cell type having similar functions. In Phyllodocida an anatomically separated stomach cannot be distinguished (Fig. 1A).

The stomach is followed by the intestine, which mostly forms a straight tube. In certain polychaetes it may be coiled (e.g., Parergodrilidae; Fig. 1B) or bear lateral caeca (e.g., Aphroditidae). The epithelium often comprises absorptive cells and gland cells but in certain species only one cell type may be present (Heffernan, 1988; Michel, 1988; Saulnier-Michel, 1992; Tzetlin et al., 1992). Usually the intestinal cells are equipped with cilia and a well-developed brush border of microvilli. Additional excretory cells filled with spherocrystals have been described for a few species (see Michel, 1988). A ciliated gutter separated from the intestine by a phalanx of straight microvilli has been described ultrastructurally in the meiofaunal polychaetes *Dinophilus gyrociliatus* and *Nerillidium troglochaetoides* (see Oster, 1986; Tzetlin et al., 1992) but a similar structure is present in many macrofaunal species as well (Saulnier-Michel, 1992). In *Notomastus latericeus* this gutter gives rise to an accessory intestine of uncertain function (Saulnier-Michel, 1992). Sometimes a posterior intestine characterized by a progressive numerical increase in absorptive cells and decrease in gland cells can be distinguished.

A conspicuous feature in the intestine of most Nerillidae is the occurrence of unique tubular structures, discovered and called 'entéronéphridies' by Jouin (1967). These enteronephridia have been reported for the taxa *Meganerilla*, *Mesone-*

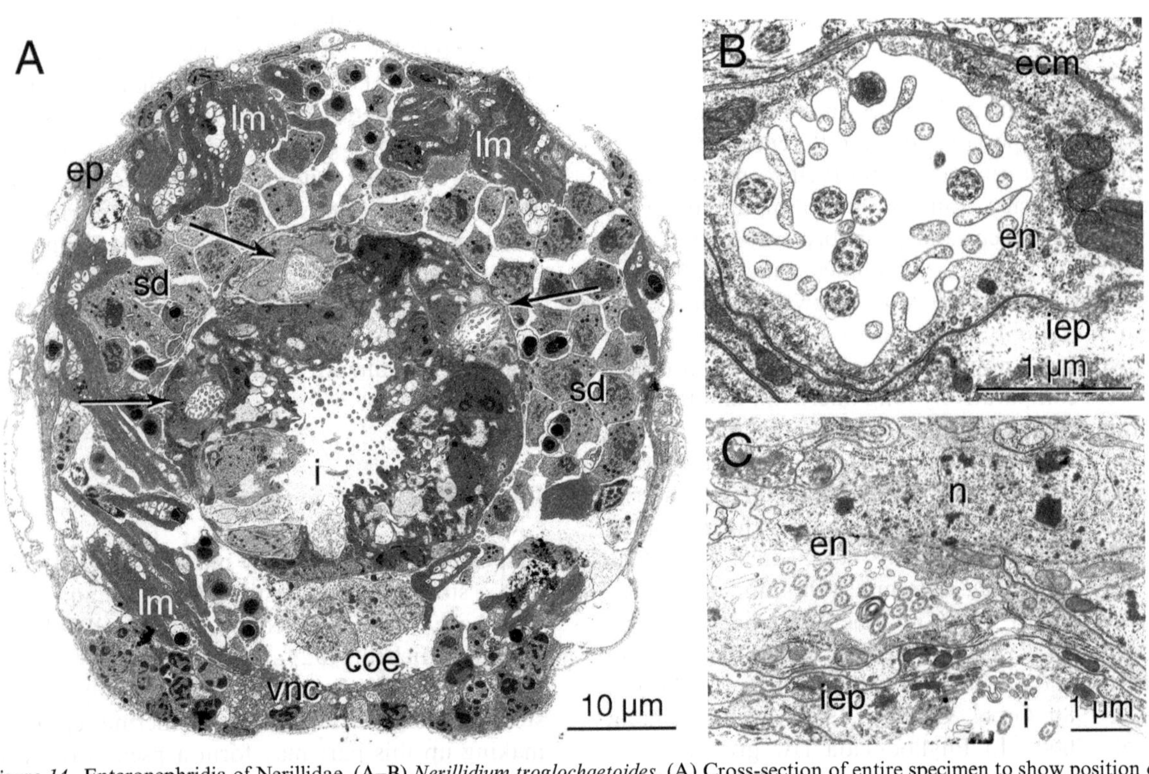

Figure 14. Enteronephridia of Nerillidae. (A–B) *Nerillidium troglochaetoides.* (A) Cross-section of entire specimen to show position of the three enteronephridia (arrows) in the epithelium of the intestine (i); coe coelom, ep epidermis lm longitudinal musculature, sd spermatids, vnc ventral nerve cord. (B) Enlargement of entronephridium (en) with microvilli and cilia, cross section; ecm extracellular matrix, iep intestinal epithelium. (C) *Troglochaetus beranecki*, limnetic species. Section with nucleus (n) of enteronephridium (en) close to opening into intestine (i).

rilla, *Nerillidium*, *Akessoniella*, *Trochonerilla*, *Aristonerilla* and *Troglochaetus* (Jouin, 1967, 1968; Tzetlin & Larionov, 1988; Tzetlin & Saphonov, 1992; Tzetlin et al., 1992; Müller, 2002). In larger species, such as *Nerilla antennata,* enteronephridia are lacking. These structures are blind-ending ciliated intestinal canals, which open into the posterior stomach and run posteriorly all along the intestine (Figs. 1C and 14A–C). Situated in the periphery of the gut, they are embedded between the regular epithelial cells (Tzetlin et al., 1992). The number of enteronephridial canals differs between species and ranges from 3 to 13. Ultrastructural investigations revealed that they are unicellular structures, each cell measuring up to 130 μm in length and approximately 5 μm across (Tzetlin et al., 1992). The canal is formed by the invaginated cell apex and is present throughout the length of the enteronephridia. The luminal cell surface bears a well-developed brush border of microvilli and a few cilia. The apical membrane is characterized by endo- or exocytic vesicles but there is no increase of the basal surface by means of basal folds or a basal labyrinth. These cells are attached to the adjacent cells of the stomach by typical junctional complexes, indicating that the enteronephridia belong to this part of the gut. The enteronephridia resemble protonephridial or metanephridial ducts (Fig. 14B and C), although all species investigated possess protonephridia (Smith, 1992; see Chapter 8). The function of these structures is presumed to be excretory, but experimental evidence is lacking. So far these organs have not been found in any other taxon of Annelida: The enteronephridia described for Megascolecida are structurally different: they are true metanephridia opening into the intestine (see Tzetlin et al., 1992). The only known species with structurally similar organs is *Jennaria pulchra* (see Tzetlin et al., 1992), a taxon with annelid affinities but uncertain systematic position (Rieger, 1991).

Rectum

The rectum is the ectodermal posterior part of the gut (Saulnier-Michel, 1992). It may be short, not exceeding the length of the pygidium, or may extend over several segments. Generally the epithelium bears a cuticle of varying thickness and may be ciliated as well. The anus usually opens terminally, dorsally or ventrally and often possesses a sphincter formed by circular fibres.

Acknowledgements

Thanks are expressed to Dr Hannelore Paxton and Prof. Dr Valadimir Malkhow for fruitful discussions and suggestions. Helpful comments of two anonymous referees are gratefully acknowledged. Part of the studies of the senior author (A.T.) was supported by the Russian fund for basic researches (N-03-04-48598 and 01-04-49093-a). We thank Dr M. Böggemann, Dr M. Kuper and S. Raabe for material or micrographs. Technical assistance by Anna Paul, Werner Mangerich, Martina Biedermann, Anja Ritz and Andrea Noel is gratefully acknowledged.

References

Åkesson, B., 1967. On the biology and larval morphology of *Ophryotrocha puerilis* Claparède & Metschnikov (Polychaeta). Ophelia 4: 111–119.

Åkesson, B., 1973a. Morphology and life history of *Ophryotrocha maculata* sp.n. (Polychaeta, Dorvilleidae). Zoologica Scripta 2: 141–144.

Åkesson, B., 1973b. Reproduction and larval morphology of five *Ophryotrocha* species (Polychaeta, Dorvilleidae). Zoologica Scripta 2: 145–155.

Åkesson, B., 1976. Morphology and life cycle of *Ophryotrocha diadema*, a new polychaete species from California. Ophelia 15: 23–35

Anderson, D. T., 1959. The embryology of the polychaete *Scoloplos armiger*. Quarterly Journal of microscopical Science 100: 89–166.

Ax, P., 1999. Das System der Metazoa II. Fischer, Stuttgart, 383 pp.

Bantz, M. & C. Michel, 1971. Revètement cuticulaire de la trompe chez *Glycera convoluta* Keferstein (Annélide Polychète). Zeitschrift für Zellforschung 118: 221–242.

Bantz, M. & C. Michel, 1972. Les cellules sensorielles des papilles de la trompe chez *Glycera convoluta* Keferstein (Annélide Polychète). Anatomie microscopique et ultrastructure. Zeitschrift für Zellforschung 134: 351–366.

Böggemann, M., 2002. Revision of the Glyceridae Grube 1850 (Annelida: Polychaeta). Abhandlungen der Senckenbergischen Naturforschenden Gesellschaft 555: 1–249.

Böggemann, M., D. Fiege & G. Purschke, 2000. Ultrastructure of the proboscidial papillae in some *Glycera* species (Annelida: Polychaeta: Glyceridae). Cahiers de Biologie Marine 41: 143–153.

Briggs, R. T., J. E. Chaffee & M. Anderson, 1985. Calcium containing granules in myoepithelial cells of the polychaete

Syllis spongiphila: possible tonic modulators. Tissue and Cell 17: 923–928.

Britaev, T. A. & V. Belov, 1993. Determination of the age of polychaetes fam. Polynoidae by the jaws growth-lines. Zoologicheskii Zhurnal 72: 15–21 (in Russian).

Britaev, T. A., M. Pluscheva & A. Buyanovski, 2002. Size-age structure of populations of *Lepidonotus squamatus* and *Harmothoe imbricata* (Polychaeta, Polynoidae) in the White Sea. Zoologicheskii Zhurnal 81: 285–291 (in Russian).

Bryan, G. W. & P. E. Gibbs, 1979. Zinc – a major inorganic component of nereid polychaete jaws. Journal of the Marine Biological Association of the United Kingdom 59: 969–973.

Bryan, G. W. & P. E. Gibbs, 1980. Metals in nereid polychaetes: the contribution of metals in the jaws to the total body burden. Journal of the Marine Biological Association of the United Kingdom 60: 641–654.

Bryan, G. W. & P. E. Gibbs, 1986. Occurrence, composition and function of intracellular calcium phosphate granules in the musculature of nephtyid polychaetes (Annelida). Journal of the Marine Biological Association of the United Kingdom 66: 343–365.

Colbath, G. K., 1986. Jaw mineralogy in eunicean polychaetes (Annelida). Micropaleontology 32: 186–189.

Colbath, G. K., 1987. Evidence for shedding of maxillary jaws in eunicoid polychaetes. Journal of Natural History 21: 443–447.

Colbath, G. K., 1988. Taphonomy of recent polychaete jaws from Florida and Belize. Micropaleontology 34: 83–89.

Croneis, C. & H. W. Scott, 1933. Scolecodonts. Bulletin of the Geological Society of America 44: 207.

Dales, R. P. 1957. The feeding mechanism and structure of the gut of *Owenia fusiformis* D.Ch. Journal of the Marine Biological Association of the United Kingdom 36: 81–89.

Dales, R. P. 1962. The polychaete stomodeum and the interrelationship of the families of Polychaeta. Proceedings of the Zoological Society of London 139: 389–428.

Dales, R. P. 1963. Annelids. Hutchison University Library, London, 200 pp.

Dales, R. P., 1977. The Polychaete stomodeum and phylogeny. In Reish, D. J. & K. Fauchald (eds), Essays on Polychaetous Annelids in Memory of Dr. Olga Hartman. Allan Hancock Foundation, Los Angeles: 525–546.

Damas, D., 1987. Étude cytologique, histochimique et microanalytique des pieces maxillaires d'*Ophryotrocha puerilis* (Annélide, Polychète, Eunicidae). Bulletin of the Zoological Society of France 111: 61–73.

Del Castello, J., M. Anderson & D. S. Smith, 1972. Proventriculus of a marine annelid: muscle preparation with the longest recorded sarcomere. Proceedings of the National Academy of Sciences of the United States of America. 69: 1669–1672.

Desbruyères, D., 1978. Un Ampharetidae (Annélides Polychètes sédentaires) à structure buccale aberrante: *Gnathampharete paradoxa* gen., sp. n. Comptes Rendus de l'Académie des Sciences Paris, D 286: 281–284.

Düsing, H., 1961. Die Umwandlung des Kauapparates in Abhängigkeit vom sexuellen Zustand bei *Ophryotrocha puerilis* Claparde et Metschnikoff. Naturwissenschaften 3: 532–533.

Eeckhaut, I. & D. Lanterbec, 2005. Myzostomida. A review of the phylogeny and ultrastructure. Hydrobiologia 535/536 (Dev. Hydrobiol. 179): 251–273.

Ehlers, E., 1864–1868. Die Borstenwürmer nach systematischen und anatomischen Untersuchungen dargestellt. Wilhelm Engelmann, Leipzig: 1–748.

Ehlers, E., 1868. Über eine fossile *Eunice* aus Solnhofen (*Eunicites avitus*) nebst Bemerkungen über fossile Würmer überhaupt. Zeitschrift für wissenschaftliche Zoologie 18: 421–443.

Eisig, H., 1914. Zur Systematik, Anatomie und Morphologie der Ariciiden nebst Beiträgen zur generellen Systematik. Mitteilungen aus der zoologischen Station zu Neapel 21: 153–593.

Eibye-Jacobsen, D. & R. M. Kristensen, 1994. A new species and genus of Dorvilleidae (Annelida, Polychaeta) from Bermuda, with a phylogenetic analysis of Dorvilleidae, Iphitimidae and Dinophilidae. Zoologica Scripta 23: 107–131.

Fauchald, K. & P. A. Jumars, 1979. The diet of worms: a study of polychaete feeding guilds. Oceanography and Marine Biology Annual Review 17: 193–284.

Fauvel, P., 1959. Classe des Annélides Polychètes. In Grassé, P.-P. (ed.), Traité de Zoologie, Vol. 5. Masson, Paris: 13–196.

Gelder, S. R. & R. F. Uglow, 1973. Feeding and gut structure in *Nerilla antennata* (Annelida: Archiannelida). Journal of Zoology 171: 225–237.

Gibbs, P. E. & G. W. Bryan, 1980a. Copper – the major metal component of glycerid polychaete jaws. Journal of the Marine Biological Association of the United Kingdom 60: 205–214.

Gibbs, P. E. & G. W. Bryan, 1980b. A note on the element composition of the jaws of *Goniada maculata* (Polychaeta: Goniadidae). Journal of the Marine Biological Association of the United Kingdom 60: 541–542.

Hartman, O., 1944. Polychaetous annelids. Part V. Eunicea. Allan Hancock Pacific Expeditions 10: 1–238.

Hartmann-Schröder, G., 1958. Zur Morphologie der Opheliiden (Polychaeta sedentaria). Zeitschrift für wissenschaftliche Zoologie 161: 84–143.

Hartmann-Schröder, G., 1967. Feinbau und Funktion des Kieferapparates der Euniciden am Beispiel von *Eunice (Palola) siciliensis* Grube (Polychaeta). Mitteilungen aus dem Hamburgischen Zoologischen Museum und Institut 64: 5–27.

Heffernan, P., 1988. Ultrastructural and histochemical studies of the digestive system of *Pholoë minuta* (Polychaeta: Sigalionidae). Journal of the Marine Biological Association of the United Kingdom 68: 447–464.

Heimler, W., 1983. Untersuchungen zur Larvalentwicklung von *Lanice conchilega* (Pallas) 1766 (Polychaeta/Terebellomorpha) Teil III. Bau und Struktur der Aulophora-Larve. Zoologische Jahrbücher. Anatomie 110: 411–478.

Hilbig, B. & J. A. Blake, 1991. Dorvilleidae (Annelida: Polychaeta) from the U.S. Atlantic slope and rise. Description of two new genera and 14 new species, with a genric revision of *Ophryotrocha*. Zoologica Scripta 20: 147–183.

Hsieh, H.-L. & J. L. Simon, 1987. Larval development of *Kinbergonuphis simoni*, with a summary of development patterns in the family Onuphidae (Polychaeta). Bulletin of the Biological Society of Washington 7: 194–210.

Jamieson, B. G. M., 1992. Oligochaeta. In Harrison, F. W. & S. L. Gardiner (eds), Microscopic Anatomy of Invertebrates. Vol. 7 Annelida. Wiley-Liss, New York: 217–322.

Jennings, J. B. & S. R. Gelder, 1969. Feeding and digestion in *Dinophilus gyrociliatus* (Annelida, Archiannelida). Journal of Zoology 158: 441–457.

Jennings, J. B. & S. R. Gelder, 1976. Observations on the feeding mechanism, diet and digestive physiology of *Histriobdella homari* van Veneden 1958: an aberrant polychaete symbiotic with North American and European lobsters. Biological Bulletin 151: 489–517.

Jeuniaux, C., 1969. Nutrition and Digestion. In Florkin, M. & B. T. Scheer (eds), Chemical Zoology. Vol. 4. Academic Press, London, New York: 69–91.

Jouin, C., 1967. Étude morphologique et anatomique de *Nerillidopsis hyalina* Jouin et de quelques *Nerillidium* Remane (Archiannélides Nerillidae). Archives de Zoologie Expérimentale et Genérale 108: 97–110.

Jouin, C., 1968. Sexualité et biologie de la reproduction chez *Mesonerilla* Remane and *Meganerilla* Boaden (Archiannélides Nerillidae). Cahiers de Biologie Marine 9: 31–52.

Jouin, C., 1978a. Anatomical and ultrastructural study of the pharyngeal bulb in *Protodrilus* (Polychaeta, Archiannelida). I. Muscles and myo-epithelial junctions. Tissue and Cell 10: 269–287.

Jouin, C., 1978b. Anatomical and ultrastructural study of the pharyngeal bulb in *Protodrilus* (Polychaeta, Archiannelida). II. The stomodeal epithelium and its cuticle. Tissue and Cell 10: 289–301.

Jouin, C., 1979. Description of a free-living polychaete without gut: *Astomus taenioides* n.gen., n. sp. (Protodrilidae, Archiannelida). Canadian Journal of Zoology 57: 2448–2456.

Jouin, C., 1992. The ultrastructure of a gutless annelid, *Parenterodrilus* gen. nov. *taenioides* (= *Astomus taenioides*) (Polychaeta, Protodrilidae). Canadian Journal of Zoology 70: 1833–1848.

Jumars, P., 1974. A generic revision of the Dorvilleidae (Polychaeta), with six new species from the deep North Pacific. Zoological Journal of the Linnean Society of London 54: 101–135.

Kielan-Jaworowska, S., 1966. Polychaete jaw apparatuses from the Ordovician and Silurian of Poland and a comparison with modern forms. Acta Palaeontologica Polonica 16: 1–125.

Kirkegaard, J. B., 1970. Age determination of *Nephtys* (Polychaeta: Nephtydae). Ophelia 7: 447–496.

Kozlowski, R., 1956. Sur quelques appareils masticateurs des Annélides Polychètes Ordoviciens. Acta Paleontologica Polonica 1: 3–16.

Kozur, H., 1970. Zur Klassifikation und phylogenetischen Entwicklung der fossilen Phyllodocida und Eunicida (Polychaeta). Freiburger Forschungsheft C 260: 35–81.

Kristensen, R. M. & T. Niilonen, 1982. Structural studies on *Diurodrilus* Remane (Diurodrilidae fam. n.), with description of *Diurodrilus westheidei* sp.n. from the arctic interstitial meiobenthos, W. Greenland. Zoologica Scripta 11: 1–12.

Lanzavecchia, G., M. de Eguileor & R. Valvassori, 1988. Muscles. In Westheide, W. & C. O. Hermans (eds), The Ultrastructure of Polychaeta. Microfauna Marina 4: 71–88.

Licher, F., 1999. Revision der Gattung *Typosyllis* Langerhans, 1879 (Polychaeta: Syllidae). Morphologie, Taxonomie und Phylogenie. Abhandlungen der Senckenbergischen Naturforschenden Gesellschaft 551: 1–336.

Mackie, A. S. Y., 1994. *Adercodon pleijeli* gen. et.sp. nov. (Polychaeta, Ampharetidae) from the Mediterranean Sea. In Dauvin, J. C., L. Laubier & D. Reish (eds), Actes de la 4ème Conferénce internationale des Polychètes. Mémoires du Muséum National d'histoire Naturelle 162: 243–250.

Martin, P., 2001. On the origin of the Hirudinea and the demise of the Oligochaeta. Proceedings of the Royal Society of London B 268: 1089–1098.

Michel, C., 1972. Étude ultrastructurale et histochimique des papilles de la gaine de la trompe de *Notomastus latericus* Sars (Annélide Polychète Sédentaire). Zeitschrift für Zellforschung 128: 428–503.

Michel, C., 1978. Étude histologique, histochimique et cytologique du stomodeum chez le Polychète Sédentaire, *Audouinia tentaculata* (Montagu). Cahiers de Biologie Marine 19: 433–446.

Michel, C., 1988. Intestine. In Westheide, W., & C. O. Hermans (eds), The Ultrastructure of the Polychaeta. Microfauna Marina 4: 157–175.

Michel, C. & E. J. DeVillez, 1978. Digestion. In Mill, P. J. (ed.), Physiology of Annelids. Academic Press, London, New York: 509–554.

Michel, C., M.-T. Fonze-Vignaux & M.-F. Voss-Foucart, 1973. Données nouvelles sur la morphologie, l'histochimie et la composition chimique des mâchoires de *Glycera convoluta* Keferstein (Annélide, Polychète). Bulletin biologique de France et de la Belgique 107: 301–321.

Mierzejewski, P., 1978. Molting of the jaws of the early Paleozoic Eunicida (Annelida: Polychaeta). Acta Palaeontologica Polonica 23: 73–88.

Mierzejewski, P., 1984. *Sycnlinophora synclinalis* Eisenack – the oldest arabellid polychaete. Review of Paleobotany and Palynology 43: 285–292.

Mierzejewski, P. & G. Mierzejewska, 1975. Xenognath type of polychaete jaw apparatuses. Acta Palaeontologica Polonica 20: 437–443.

Müller, H., 1962. Über die Sexualität des Polychaeten *Ophryotrocha puerilis*, ihre Determination und Kauapparatentwicklung. Zeitschrift für Morphologie und Ökologie der Tiere 52: 1–32.

Müller, M. C. M., 2002. *Aristonerilla*: a new nerillid genus (Annelida: Polychaeta) with description of *Aristonerilla* (*Micronerilla*) *brevis* comb. nov. from a seawater aquarium. Cahiers de Biologie Marine 43: 131–139.

Müller, M., J. M. Bernhard & C. Jouin-Toulmond, 2001. A new member of Nerillidae (Annelida: Polychaeta), *Xenonerilla bactericola* gen. et sp. nov., collected off California. USA. Cahiers de Biologie Marine 42: 203–217.

Nordheim, H. v., 1987. *Petrotrocha notogaea* gen. et sp. n., a neotenic interstitial polychaete (Dorvilleidae) from the Hauraki Gulf, New Zealand. Zoologica Scripta 16: 33 38.

Olive, P. J. W., 1977. The life-history and population structure of the polychaetes *Nephtys caeca* and *Nephtys hombergii* with special reference to growth rings in the teeth. Journal of the Marine Biological Association of the United Kingdom 57: 133–150.

Olive, P. J. W., 1980. Growth lines in Polychaeta jaws. In Rhoads, D. C. & R. A. Luts (eds), Skeletal Growth of Aquatic Organisms, Plenum, New York: 561–592.

Orensanz, J. M., 1990. The eunicemorph polychaete annelids from antarctic and subantarctic seas. With addenda to the Eunicemorpha of Argentina, Chile, New Zealand, Australia, and the Southern Indian Ocean. Biology of the Antarctic Seas 21: 1–183.

Orrhage, L., 1964. Anatomische und morphologische Studien über die Polychaetenfamilien Spionidae, Disomidae und Poecilochaetidae. Zoologiska Bidrag fran Uppsala 36: 335–405.

Orrhage, L., 2001. On the anatomy of the central nervous system and the morphological value of the anterior appendages of Ampharetidae, Pectinariidae and Terebellidae (Polychaeta). Acta Zoologica (Stockholm) 82: 57–71.

Orrhage, L. & M. C. M. Müller, 2005. Morphology of the nervous system of Polychaeta. Hydrobiologia 535/536 (Dev. Hydrobiol. 179): 79–111.

Oster, U., 1986. Microfilament-supported macrovilli in the hindgut of the polychaete *Dinophilus gyrociliatus*. Zeitschrift für Naturforschung C41: 1139–1143.

Pander, C. H., 1856. Monographie der fossilen Fische des Silurischen Systems des Russisch-Baltischen Gouvernements. St. Petersbourg: 1–91.

Paxton, H., 1980. Jaw growth and replacement in Polychaeta. Journal of Natural History 14: 543–546.

Paxton, H., 2000. Eunicida. In Beesly, P. L., G. J. B. Ross & C. J. Glasby (eds), Polychaetes and Allies: The Southern Synthesis. Fauna of Australia. Vol. 4A. Polychaeta, Myzostomida, Pogonophora, Echiura, Sipuncula. CSIRO, Melbourne: 89–104.

Purschke, G., 1985a. Anatomy and ultrastructure of the ventral pharyngeal organs and their phylogenetic importance in Polychaeta (Annelida). I. The pharynx of the Dinophilidae. Zoomorphology 105: 223–239.

Purschke, G., 1985b. Anatomy and ultrastructure of the ventral pharyngeal organs and their phylogenetic importance in Polychaeta (Annelida). II. The pharynx of the Nerillidae. Microfauna Marina 2: 23–60.

Purschke, G., 1987a. Anatomy and ultrastructure of the ventral pharyngeal organs and their phylogenetic importance in Polychaeta (Annelida). III. The pharynx of the Parergodrilidae. Zoologische Jahrbücher für Anatomie 115: 331–362.

Purschke, G., 1987b. Anatomy and ultrastructure of the ventral pharyngeal organs and their phylogenetic importance in Polychaeta (Annelida). IV. The pharynx and jaws of the Dorvilleidae. Acta Zoologica (Stockholm) 68: 83–105.

Purschke, G., 1988a. Pharynx. In Westheide, W. & C. O. Hermans (eds), The ultrastructure of Polychaeta. Microfauna Marina 4: 177–197.

Purschke, G., 1988b. Anatomy and ultrastructure of the ventral pharyngeal organs and their phylogenetic importance in Polychaeta (Annelida). V. The pharynges of the Ctenodrilidae and Orbiniidae. Zoomorphology 108: 119–135.

Purschke, G., 1999. Terrestrial polychaetes – models for the evolution of the Clitellata (Annelida)? Hydrobiologia 406: 87–99.

Purschke, G., 2002. On the ground pattern of Annelida. Organisms Diversity and Evolution 2: 181–196.

Purschke, G., 2003. Is *Hrabeiella periglandulata* (Annelida, 'Polychaeta') the sister-group of Clitellata? Evidence from an ultrastructural analysis of the dorsal pharynx in *H. periglandulata* and *Enchytraeus minutus* (Annelida: Clitellata). Zoomorphology 122: 55–66.

Purschke, G. & C. Jouin, 1988. Anatomy and ultrastructure of the ventral pharyngeal organs of *Saccocirrus* (Saccocirridae) and *Protodriloides* (Protodriloidae fam. n.) with remarks on the phylogenetic relationships within the Protodrilida (Annelida: Polychaeta). Journal of Zoology, London 215: 1–28.

Purschke, G. & A. B. Tzetlin, 1996. Dorsolateral ciliary folds in the polychaete foregut: structure, prevalence and phylogenetic significance. Acta Zoologica (Stockholm) 77: 33–49.

Rice, M. E., 1973. Morphology, behaviour, and histogenesis of the pelagosphera larva of *Phascolosoma agassizii* (Sipuncula). Smithsonian Contributions to Zoology 132: 1–51.

Rice, M. E., 1976. Larval development and metamorphosis in Sipuncula. American Zoologist 16: 563–571.

Rice, M. E., 1985. Sipuncula. Developmental evidence for phylogenetic interference. In Conway Morris, S., J. D. George, R. Gibson & H. M. Platt (eds), The origins and relationships of lower invertebrates. Clarendon, Oxford: 274–296.

Rieger, R. M., 1991. Neue Organisationstypen aus der Sandlückenfauna: die Lobatocerebriden und *Jennaria pulchra*. Verhandlungen der Deutschen Zoologischen Gesellschaft 84: 247–259.

Rieger, R. M. & G. E. Rieger, 1975. Fine structure of the pharyngeal bulb in *Trilobodrilus* and its phylogenetic significance within Archiannelida. Tissue and Cell 7: 267–279.

Rota, E., 1998. Morphology and adaptations of *Parergodrilus* Reisinger and *Hrabeiella* Pizl & Chalupsky, two enigmatic soil-dwelling annelids. Italian Journal of Zoology 65: 75–84.

Rouse, G. W., 2000. Morphology and Physiology. In Beesly, P. L. Ross, G. J. B. & C. J. Glasby (eds), Polychaetes & Allies: The Southern Synthesis. Fauna of Australia. Vol. 4A. Polychaeta, Myzostomida, Pogonophora, Echiura, Sipuncula. CSIRO, Melbourne: 9–32.

Rouse, G. W. & F. Pleijel, 2001. Polychaetes. Oxford University Press. New York: 1–354.

Rouse, G. W. & K. Fauchald, 1997. Cladistics and polychaetes. Zoologica Scripta 26: 139–204.

Saulnier-Michel, C., 1992. Polychaeta: Digestive System. In Harrison, F. W. & S. L. Gardiner (eds), Microscopic Anatomy of Invertebrates. Vol. 7. Annelida. Wiley-Liss, New York: 53–69.

Saulnier-Michel, C., G. Gaill, A. Hily, P. Alberic & M. A. Cosson-Mannevy, 1990. Structure and functions of the digestive tract of *Alvinella pompejana*, a hydrothermal vent polychaete. Canadian Journal of Zoology 68: 722–732.

Schmidt, P. & W. Westheide, 1972. *Dinophilus gyrociliatus* (Polychaeta). Nahrungsaufnahme und Fortpflanzung. Begleitveröffentlichung zum Film E1750. Encyclopaedia Cinematographica: 1–16.

Smith, D. S., J. Del Castillo & M. Anderson, 1973. Fine structure and innervation of an annelid with the longest recorded sarcomere. Tissue and Cell 5: 281–302.

Smith, P. R., 1992. Polychaeta: Excretory system. In Harrison, F. W. & S. L. Gardiner (eds), Microscopic Anatomy of Invertebrates. Vol. 7. Annelida. Wiley-Liss, New York: 71–108.

Southward, E.C., A. Schulze & S. L. Gardiner, 2005. Pogonophora (Annelida): Form and function. Hydrobiologia 535/536 (Dev. Hydrobiol. 179): 225–249.

Struck, T. H., W. Westheide & G. Purschke, 2002. Progenesis in Eunicida ('Polychaeta', Annelida) separate evolutionary events? Evidence from molecular data. Molecular Phylogenetics and Evolution 25: 190–199.

Szaniawski, H., 1974. Some mesozoic scolecodonts congeneric with recent forms. Acta Palaeontologica Polonica 19: 179–199.

Szaniawski, H., & R. Wrona, 1987. Polychaete jaws from the Cape Melville formation (Lower Miocene) of King George Island, West Antarctica. Acta Palaeontologica Polonica 49: 105–125.

Tzetlin, A. B., 1980a. Two new species of the fam. Dorvilleidae from the White and the Barents Seas. Zoologicheskii Zhurnal 59: 1817–1822 (in Russian).

Tzetlin, A. B., 1980b. *Ophryotrocha schubravyi* sp.n. and the problem of evolution of the mouth parts in the Eunicemorpha (Polychaeta). Zoologicheskii Zhurnal 59: 666–670 (in Russian).

Tzetlin, A. B., 1987. Fine morphology of the pharynx apparatus in the larvae of *Nicolea zostericola* (Polychaeta, Terebellidae). Doklady Akademia Nauk SSSR 293: 1505–1509 (in Russian).

Tzetlin, A. B., 1989. Morphological analysis of the structure of the feeding apparatus in polychaetes of the genus *Praxillura* (Maldanidae). Doklady Academia Nauk SSSR 309: 757–760 (in Russian).

Tzetlin, A. B., 1990. To the taxonomical status of the *Lumbrineris* Blainville (Polychaeta: Lumbrineridae) from the White Sea. In: Biological Resource of the White Sea, Izdatel'stvo Moskovskogo Universiteta, Moscow: 99–107 (in Russian).

Tzetlin, A. B., 1991. Evolution of feeding apparatus in the polychaetes of the order Capitellida. Zoologicheskii Zhurnal 70: 10–22 (in Russian).

Tzetlin, A. B., 1994. Fine morphology of the feeding apparatus of *Cossura* sp. (Polychaeta: Cossuridae) from the White Sea. In Dauvin, J. C., L. Labier & D. Reish (eds), Actes de la 4ème Conferénce internationale des Polychètes. Mémoires du Muséum national dHistoire naturelle, Paris. 162: 137–143.

Tzetlin, A. B. & V. V. Larionov, 1988. Morphology of a new archiannelid *Akessoniella orientalis* gen. et. sp.n. (Nerillidae). Zoologicheskii Zhurnal 67: 846–857 (in Russian).

Tzetlin, A. B. & M. V. Saphonov, 1992. *Trochonerilla mobilis* gen. et sp. n. – a meiofaunal nerillid (Annelida: Polychaeta) from a marine aquarium in Moscow. Zoologica Scripta 21: 251–254.

Tzetlin, A. B., E. G. Zavarzina & M. V. Saphonov, 1987. Morphofunctional analysis of the pharyngeal structures of the several annelids. Doklady Academia Nauk SSSR 294: 1008–1011 (in Russian).

Tzetlin, A. B., G. Purschke, W. Westheide & M. V. Saphonov, 1992. Ultrastructure of enteronephridia and general description of the alimentary canal in *Trochonerilla mobilis* and *Nerillidium troglochaetoides* (Polychaeta: Nerillidae). Acta Zoologica (Stockholm) 73: 163–176.

Uebelacker, J. M. & P. G. Johnson, 1984. Taxonomic guide to the polychaetes of the Northern Gulf of Mexico. Vol. VI. Barry A Vittor & Associates, Mobile: 39-1–44-37.

Westheide, W. & H. v. Nordheim, 1985. Interstital Dorvilleidae (Annelida: Polychaeta) from Europe, Australia and New Zealand. Zoologica Scripta 14: 183–199.

Westheide, W. & P. Schmidt, 1974. *Trilobodrilus axi* (Polychaeta). Nahrungsaufnahme und Fortpflanzung. Begleitveröffentlichung zum Film E1955. Encyclopaedia Cinematographica 1–12.

Westheide, W. & R. M. Rieger, 1978.Cuticle ultrastructure of hesionid polychaetes (Annelida). Zoomorphologie 91: 1–18.

Westheide, W. & N. W. Riser, 1983. Morphology and phylogenetic relationships of the neotenic interstitial polychaete *Apodotrocha progenerans* n. gen.,n. sp. (Annelida). Zoomorphology 103: 67–87.

Westheide, W. & M. C. Müller, 1996. Cinematographic documentation of enchytraeid morphology and reproductive biology. Hydrobiologia 334: 263–267.

Winnepenninckx, B. M. H., Y. van de Peer & T. Backeljau, 1998. Metazoa relationships on the basis of 18S rRNA sequences: a few years later ... American Zoologist 38: 888–906.

Wissocq, J. C., 1974. Étude ultrastructurale d'un organe musculaire constitué de fibres possédant les plus longs sarcomères du règne animal: Le proventricule des Syllidiens (Annélides Polychètes). Journal de Microscopie 19: 285–306.

Wolf, G. 1980. Morphologische Untersuchungen an den Kieferapparaten einiger rezenter und fossiler Eunicoidea (Polychaeta). Senckenbergiana maritima 12: 1–182.

Wolf, P. S. 1986. Four new genera of Dorvilleidae (Annelida: Polychaeta) from the Gulf of Mexico. Proceedings of the Biological Society of Washington 99: 616–626.

Wolff, T. & M. E. Petersen, 1991. A brief biography of A. S. Ørsted, with notes on his travels in the West Indies and Central America and illustrations of collected polychaetes. Ophelia Supplement 5: 669–685.

Zaslavskaya, N. M., 1989. Reconstruction of the phylogenesis of the family Hartmaniellidae (Polychaeta). In: Phylogenetic aspects of Paleontology, 35th session of the USSR Paleontological Society, Moscow. Abstracts: 18–19.

Zavarzina, E. G. & A. B. Tzetlin. 1986. Biology of *Ophryotrocha dimorphica* sp.n. (Polychaeta, Eunincida) from Peter the Great Bay (the Japan Sea). Zoologicheskii Zhurnal 65: 1808–1817 (in Russian).

Hydrobiologia (2005) **535/536:** 227–251
T. Bartolomaeus & G. Purschke (eds), Morphology, Molecules, Evolution and Phylogeny in Polychaeta and Related Taxa

Pogonophora (Annelida): form and function

Eve C. Southward[1,*], Anja Schulze[2] & Stephen L. Gardiner[3]
[1]*Marine Biological Association of the U.K., Citadel Hill, Plymouth PL1 2PB, U.K.*
[2]*Smithsonian Marine Station, 701 Seaway Drive, Fort Pierce, FL 34949, U.S.A.*
[3]*Department of Biology, Bryn Mawr College, Bryn Mawr, PA 19010, U.S.A.*
*(*Author for correspondence: E-mail: 100721.3720@compuserve.com)*

Key words: Pogonophora, Siboglinidae, Frenulata, Vestimentifera, anatomy, ultrastructure

Abstract

Pogonophora, also known as Siboglinidae, are tube-dwelling marine annelids. They rely on endosymbiotic chemoautotrophic bacteria for nutrition and their anatomy and physiology are adapted to their need to obtain both oxygen and reduced sulphur compounds. Frenulate pogonophores are generally long and slender, sediment-living tubeworms; vestimentiferans are stouter, inhabitants of hydrothermal vents and cool seeps; and moniliferans or sclerolinids are very slender inhabitants of decaying wood and sulphidic sediments. The anatomy and ultrastructure of the three groups are compared and recent publications are reviewed. Annelid characters are the presence of chaetae and septa, concentrated at the hind end. The adaptations to a specialised way of life include, in particular, the chitinous tube; the anterior appendages that function as gills; the internal tissue called the trophosome, where the endosymbiotic bacteria live; and the blood vascular system that transports oxygen, sulphide and carbon dioxide to the trophosome.

Introduction

Pogonophora are tube-dwelling marine worms that depend on internal symbiotic bacteria for nutrition. Their systematic position has been argued over for years. Because the adults have no mouth or functional gut, their dorsoventral orientation has been controversial. Furthermore, the segmented posterior end was unknown for a long time. Their habitats range from reducing sediments to decaying wood, cool sulphidic seeps and warm hydrothermal vents.

Within the Pogonophora, three major groups are recognized today: the Frenulata, the Vestimentifera and the Monilifera, the latter represented by the single genus *Sclerolinum*. Caullery (1914) described the first pogonophoran genus, *Siboglinum*, and erected the family Siboglinidae with unknown affinity to any existing phylum. Uschakov (1933), describing *Lamellisabella zachsi*, assigned his species to the sabellid polychaetes. The name Pogonophora has been in use since Johansson (1937, 1939) designated a class Pogonofora in the 'Vermes Oligomera', to include *Lamellisabella zachsi*. Ivanov (1951) united the previously described frenulates when he included Siboglinidae in the class Pogonophora. His monograph (Ivanov, 1963) in which he elevated Pogonophora to phylum level has formed the foundation for subsequent work on the group. He still rejected annelidan affinities (Ivanov, 1975, 1988) after the segmented posterior end was described (Webb 1964). The enterocoelic origin of the coelom (Ivanov, 1957, 1975, 1988) was a critical character in his opinion and that of his successors, who place Pogonophora close to Archicoelomata and Deuterostomia (Malakhov et al., 1996c). When the Vestimentifera were discovered, Ivanov (1994) accepted them as Pogonophora, though Jones (1985a) proposed a separate phylum. A review of the Vestimentifera (Malakhov & Galkin, 1998) divides the phylum Pogonophora into two subphyla, Perviata and Obturata, using the names proposed by Jones (1981a); Perviata contains two classes, Frenulata and Monilifera (Ivanov, 1991); Obturata contains only one class, Vestimentifera.

The alternative hypothesis, of a close relationship to the Annelida, developed gradually, following the discovery of the multisegmented opisthosoma, with septa between the segments and serially arranged chaetae (Webb, 1964; Southward, 1975a). The original dorso-ventral orientation of the animal was reversed. Protostomian affinity was supported by discovery of the larval mouth and interim gut in both vestimentiferans and frenulates, which confirmed the change of dorsoventral orientation (Jones & Gardiner, 1988, 1989; Southward, 1988; Callsen-Cencic & Flügel, 1995). Both molecular and morphological cladistic analyses indicate that Pogonophora are annelids (e.g., Rouse & Fauchald, 1995, 1997; McHugh, 1997; Schulze, 2003; Halanych, 2005). The name Siboglinidae has since been used to include Vestimentifera, Frenulata and *Sclerolinum* (see also Rouse, 2001). Specialists on the molecular evolution of extracellular haemoglobins consider vestimentiferans and frenulates to form a class of Annelida for which they coined the name Opisthochaeta (Zal et al., 1997; Negrisolo et al., 2001). On the other hand, Salvini-Plawen (2000) argues that the present state of knowledge is insufficient and that we need more detailed investigation before reaching any conclusion about the origin of Pogonophora. Almeida et al. (2003) place Pogonophora in the Enterocoela, which they regard as a transitional taxon between protostomes and deuterostomes.

The taxonomists among us (ECS and SLG) prefer to retain the name Pogonophora for the group and to use a Linnean system of classification, with three subclasses Frenulata, Vestimentifera and Monilifera – and the accepted families within them (Table 1). We need to keep the freedom to move genera between families, to merge families, and to propose new ones. As we continue to describe more new species, some revision of the families is certainly going to be necessary, since there are now about 135 named species of frenulates, 13 vestimentiferans and 8 moniliferans.

Our aim in this paper is to review recent work, our own and other authors', emphasizing function as well as structure. More has been published lately on vestimentiferans than frenulates because of the great geological and biogeographical interest in hydrothermal vents and cold seeps (Tunnicliffe et al., 1998), where vestimentiferans are important members of the fauna. They are large and easily collected with the aid of submersibles and experimental physiology has flourished, through use of new pressurized aquaria.

Table 1. Classification

Annelida			
Pogonophora: Subclasses	Families	Genera	Species
Frenulata	Oligobrachiidae	6	20
	Siboglinidae	2	73
	Polybrachiidae	8	29
	Lamellisabellidae	2	9
	Spirobrachiidae	2	5
Vestimentifera			
	Lamellibrachiidae	1	5
	Escarpiidae	3	4
	Riftiidae	1	1
	Ridgeiidae	1	1
	Tevniidae	2	2
	Arcovestiidae	1	1
	Alaysiidae	1	1
Monilifera			
	Sclerolinidae	1	7

Morphology

The anatomy and fine structure of Pogonophora have been documented in comprehensive publications including Ivanov's (1963) monograph, Gupta & Little (1969, 1970, 1975) and two chapters in 'Microscopic Anatomy of Invertebrates' (Gardiner & Jones, 1993; Southward, 1993).

The body plan of *Siboglinum* is typical of frenulates. Figure 1 compares the regions of the adult body with those of the larva at the time of settlement and shows the position of the first septum. The diaphragm develops later, as a muscular partition. The frenulum, or bridle, is a pair of hardened cuticular ridges around the forepart, close behind the tentacle bases. *Siboglinum* species have a single tentacle but most other genera have more (from two to over 200). The trunk is very long and the postannular region of the trunk

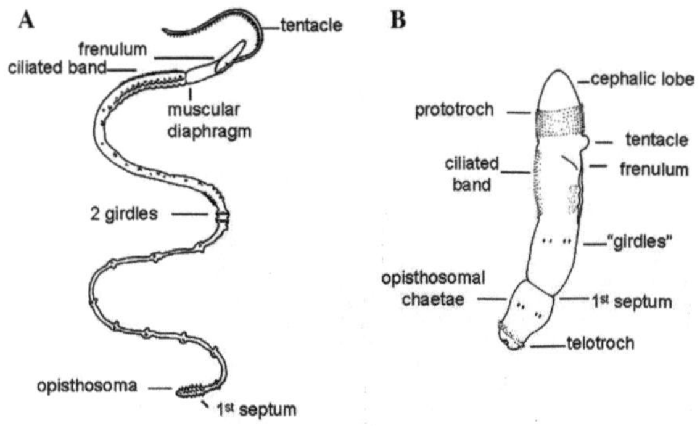

Figure 1. Frenulate body regions. (A) *Siboglinum* adult, diagrammatic. (B) Late larva of *Siboglinum fiordicum.*

contains the trophosome, visible in life as a dark streak between the red blood vessels.

Comparison of the body plans of a frenulate, a vestimentiferan and a moniliferan (Fig. 2) indicates an anterior region, a diaphragm, a trunk region and a segmented opisthosoma in all three. Whereas the anterior region of frenulates and moniliferans includes a cephalic lobe and dorsal tentacles, vestimentiferans possess a unique structure called the obturaculum which supports an extensively developed branchial plume (Fig. 3A) and which acts as an operculum to block the entrance to the tube when the animal withdraws (Jones, 1981a; Gardiner & Jones, 1993). Together, the obturaculum and associated branchial plume comprise the obturacular region of vestimentiferans.

The vestimental region is very muscular and has wide vestimental folds that normally curl over the dorsal side of the body, enclosing a space into which the gonopores open posteriorly. The outer side of the vestimental folds is covered with small papillae bearing cuticular plaques. Among them are the openings of numerous tube-secreting glands. Moniliferans have a muscular anterior region, also rich in tube-secreting glands. A ring of papillae, carrying plaques, lies just behind the tentacle bases (Fig. 3B). The trunk carries more papillae in all three groups. A ventral ciliated band is always present, but in frenulates it is on the anterior trunk, in vestimentiferans on the vestimental region and in moniliferans on the anterior region. The papillae of frenulates are varied in size and structure on different regions of the trunk. The girdle region lies at mid-trunk level and consists of four half-hoops of numerous chaetae, developed from four groups of chaetae on the second larval segment. In moniliferans trunk chaetae are just in front of the opisthosoma (absent in some species), and they are not present in adult vestimentiferans.

The opisthosoma is divided by septa into coelomate segments, with regularly arranged chaetae. Most of the features shared with annelids are

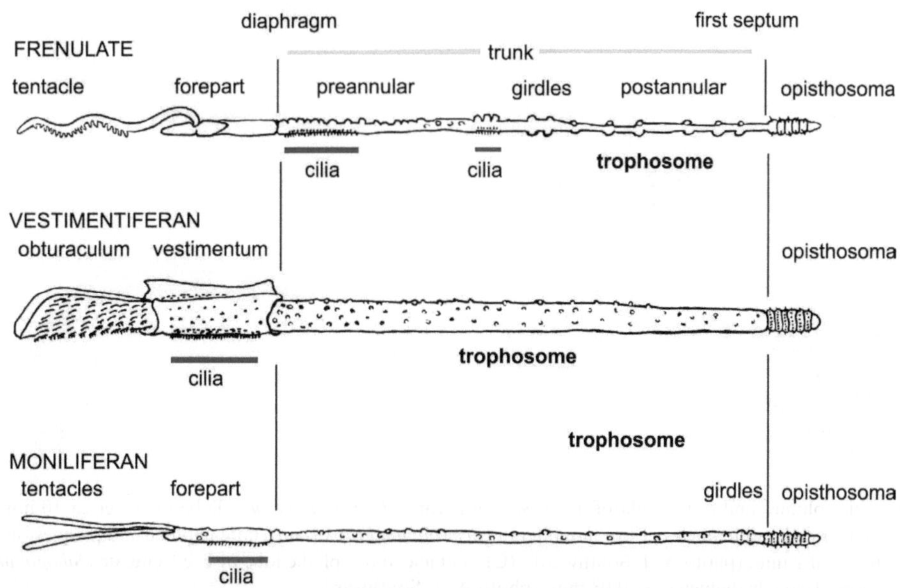

Figure 2. Comparison of body regions of frenulate, vestimentiferan and moniliferan, diagrammatic. Shading shows the extent of the trunk, trophosome and ventral ciliated bands.

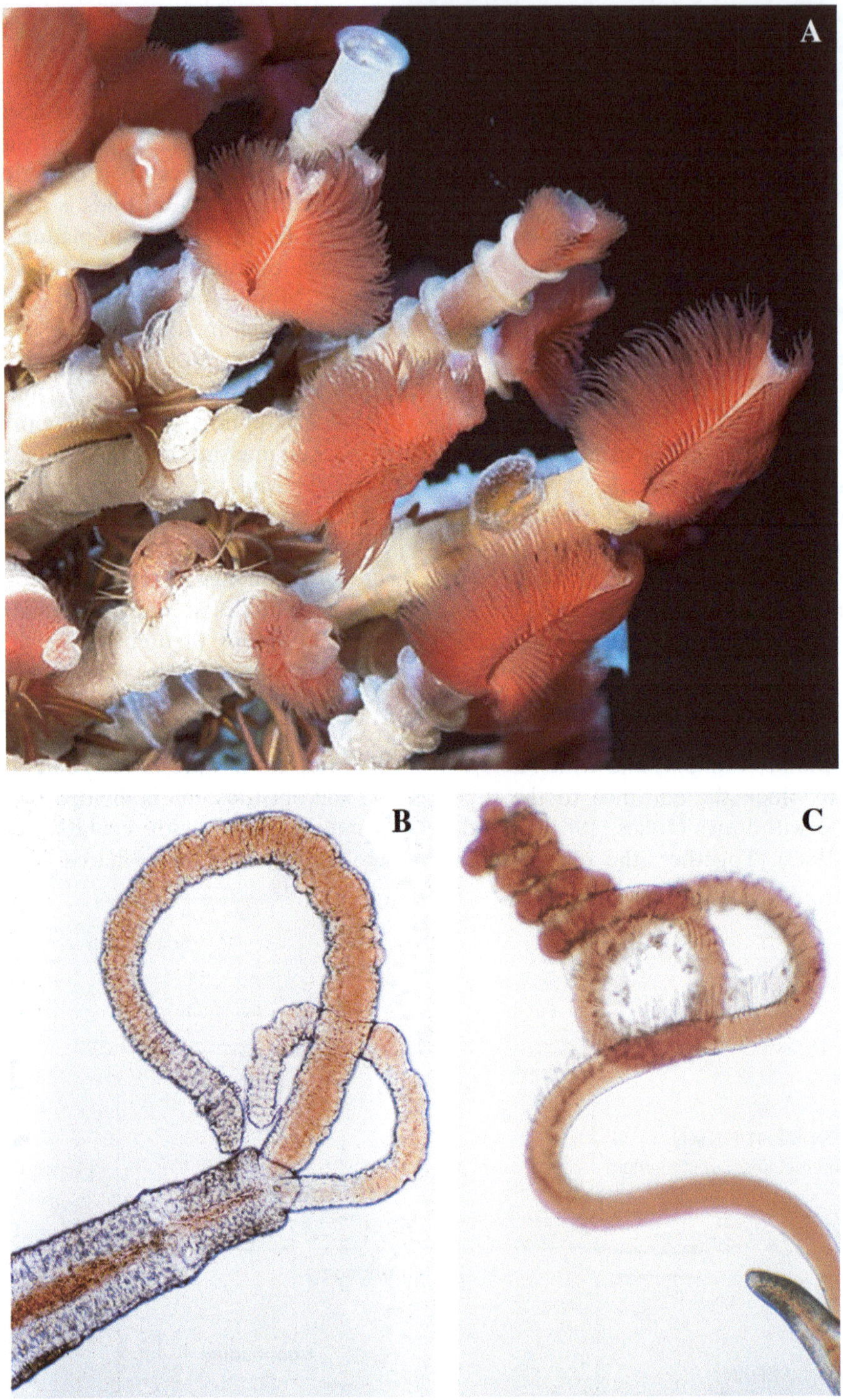

Figure 3. (A) Branchial plumes and obturacula of live vestimentiferans (*Ridgeia piscesae*). Tube diameter ca 10 mm (photo Verena Tunnicliffe). (B) Anterior end and tentacles of live moniliferan (*Sclerolinum brattstromi*), note papillae and plaques on anterior part of body. Body diameter ca. 0.1 mm. (photo A. J. Southward). (C) Tentacle and cephalic lobe of live frenulate (*Siboglinum ekmani*), note double row of pinnules. Tentacle diameter ca. 0.08 mm. (photo A. J. Southward).

concentrated in the opisthosoma, including muscular septa, segmentally arranged chitinous chaetae, ganglia and blood vessels (Southward, 1975a,b). In post-larval and juvenile vestimentiferans (up to ca. 1.7 mm long) the second segment (trunk) has chaetae of two types, capillary and claw-like (Jones & Gardiner, 1989). These stay in place while several rows of adult-type chaetae develop on the opisthosome (Fig. 4A). The claw-like larval chaetae have a group of small teeth that point anteriorly. In larger juveniles the larval chaetae are lost, and no more develop on the trunk. The adult-type chaetae of pogonophores of all three groups have long shafts and finely dentate heads, with the teeth usually in two groups, the anterior teeth pointing backward and the posterior group pointing forward (Fig. 4B, C).

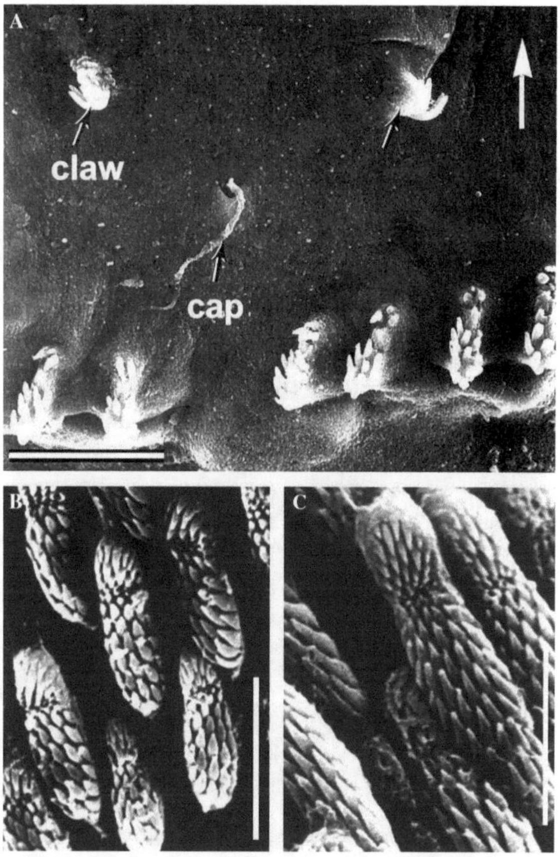

Figure 4. Chaetae. (A) Larval chaetae of trunk and adult chaetae of first opisthosomal segment of vestimentiferan post-larva (*Riftia* or *Oasisia*) (Gardiner & Jones, 1993). (B) Adult chaetae of frenulate *Siboglinum atlanticum.* (C) Adult chaetae of frenulate *Lamellisabella coronata* (B & C, From George & Southward, 1973). Scale bars = 10 μm.

Tube and tube secretion

The pogonophoran tube provides both support and protection. Webb (1971) recognized two major styles of forward growth, one continuous and flexible, the second a stiffer series of overlapping funnels, but both may be found in one species at different stages in the life history. Folding of the flexible anterior end, as found in many frenulates, can protect the animal, whereas a funnelled tube is open at the top to predators and parasites. Vestimentiferans can plug the entrance with their obturaculum, but frenulates such as *Polybrachia* and *Lamellisabella* must leave the tips of the tentacles exposed.

Chitin was originally reported from the tubes of three frenulate species (Brunet & Carlisle, 1958), and later confirmed by X-ray diffraction analysis (Blackwell et al., 1965). There are three crystallographic forms of chitin in living organisms, differing in the alignment of their chains of chitin molecules, designated α-, β- and γ-chitin. The chitin in frenulate and vestimentiferan tubes is crystalline β-chitin, which also occurs in annelid chaetae and in the 'pen' of the squid *Loligo* but is otherwise rare in the animal kingdom (Jeuniaux, 1963; Rudall & Kenchington, 1973). The use of chitin as a tube-building material is unknown in other annelids (Vovelle, 1982).

A chitin-protein complex forms a substantial proportion of the tube of the frenulate *Siboglinum fiordicum* (Foucart et al., 1965). Similarly, vestimentiferan tubes, exemplified by *Riftia pachyptila*, are composed mainly of chitin and protein (Webb, 1971; Gaill & Hunt, 1986; Shillito et al., 1995a, 1997; Chamoy et al., 2000). Chitin-containing Cambrian fossil tubes may indicate the antiquity of Pogonophora (Carlisle, 1964; Southward & Southward, 1967).

The tubes of frenulates and vestimentiferans are built up from many layers of fibrils secreted by multicellular epidermal glands (tubiparous glands or pyriform glands), with additional material from unicellular glands. A multicellular gland consists of a pear-shaped cluster of secretory cells, surrounding a central lumen, in which the fibrillar secretion of individual cells accumulates (Southward & Southward, 1966; Gupta & Little, 1970,

1975; Southward, 1984, 1993; Gardiner & Jones, 1993). Discharge is through a narrow multicellular duct. Strands of secretion emerge from the openings of such ducts in living animals and are visible in SEM preparations (Webb, 1965; Gardiner & Jones, 1993). The secretory surface of each cell forms a pocket lined with microvilli. There are simple digitate microvilli over much of the surface, but cup-shaped ones in the deeper part of the pocket or ampulla (Fig. 5B, D).

Recent studies have confirmed the suggestion made by Gupta & Little (1975) that the cup-shaped

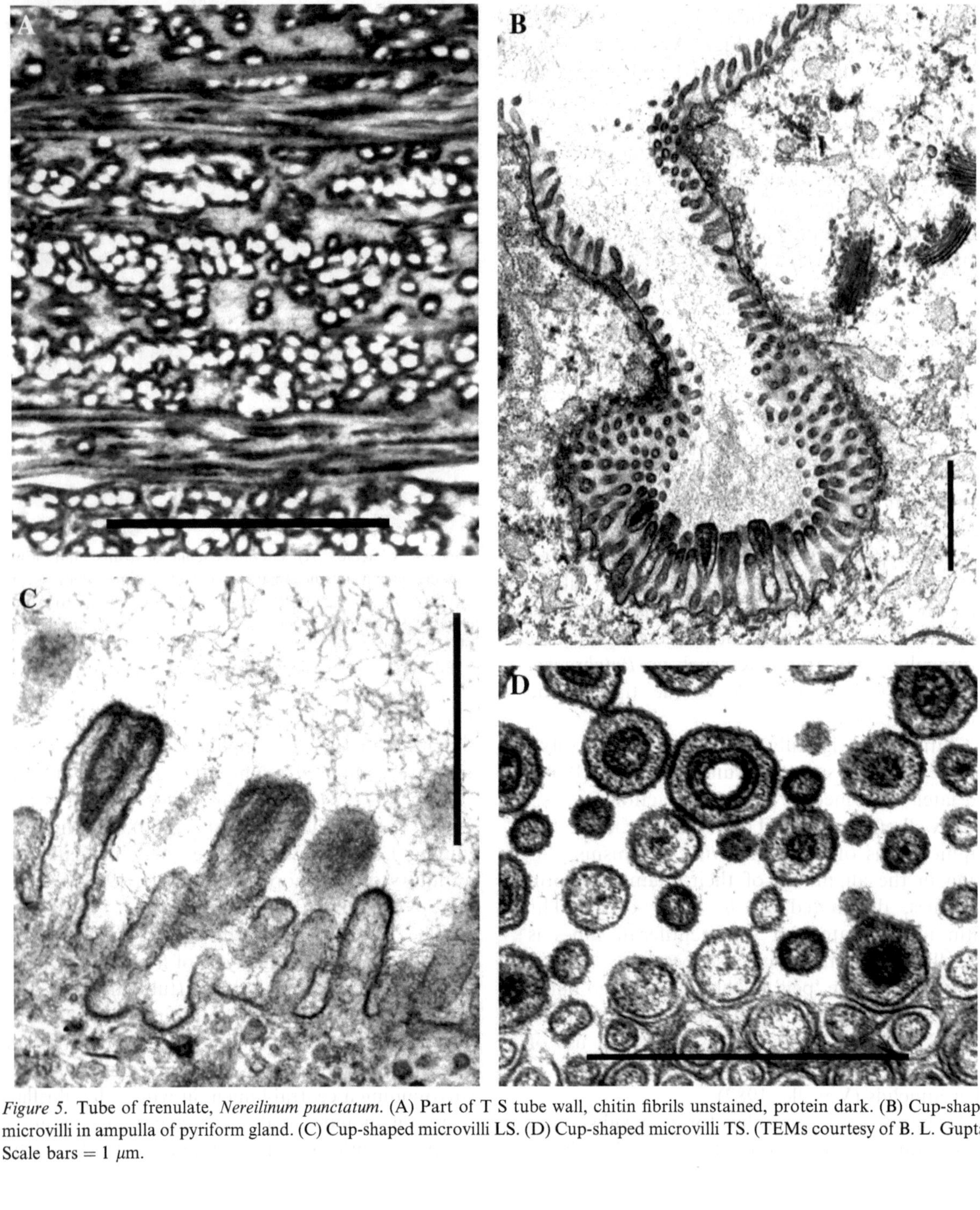

Figure 5. Tube of frenulate, *Nereilinum punctatum*. (A) Part of T S tube wall, chitin fibrils unstained, protein dark. (B) Cup-shaped microvilli in ampulla of pyriform gland. (C) Cup-shaped microvilli LS. (D) Cup-shaped microvilli TS. (TEMs courtesy of B. L. Gupta). Scale bars = 1 μm.

microvilli are the source of chitin for tube building. The site of secretion of chitin microcrystallites has been located very precisely in *Riftia pachyptila* with an immuno-gold labelling technique (Shillito et al., 1993, 1995b). There is labelling inside the hollow of the cup but not in the intracellular compartment of the gland cell, therefore the chitin molecules must be assembled in the cups. Chitin synthase activity has been identified in the tissue and the synthase is suspected to be in the cup membrane (Ravaux et al., 1998). Cells of the outer epidermis produce the major protein of the tube (Chamoy et al., 2000, 2001).

Shillito et al. (1997) describe the laying down of fibrils in plywood-like layers, 15 to 20 layers in 5 μm thickness of tube of *Riftia pachyptila*. There is a very similar fine structure in the tubes of frenulates (Fig. 5A), likened to a molecular sieve by Gupta & Little (1975). The closely packed fibrils form a physical barrier to even the smallest bacteria (0.1 μm), but water and small molecules such as sodium chloride can pass through, as shown in the tubes of the frenulates *Siboglinum atlanticum* and *S. ekmani* (Southward & Southward, 1963). The use of ^{14}C-labelled compounds showed that amino acids, glucose and fatty acids could be taken up by the animals, via their epidermis (Little & Gupta, 1968, 1969; Southward & Southward, 1970). The tube wall slowed the rate of uptake, but did not block it. After more experiments it was concluded that *Siboglinum* species could take up and metabolise dissolved organic compounds from known environmental concentrations at a rate sufficient to sustain respiration but not enough for growth and reproduction (Southward & Southward, 1981).

The symbiotic bacteria in the trophosome require oxygen and reduced sulphur compounds as a source of energy and CO_2 for carbon fixation. Dissolved gases can enter the tubes of frenulates that are buried in reducing sediments. Dissolved sulphide and other reduced sulphur compounds diffuse from the sediment through the tube wall to the trunk epidermis and thence to the trophosome and the bacteria. Oxygen from the overlying water diffuses into the top part of the tube and to the tentacles, where it is bound to haemoglobin in the blood and carried to the trophosome (Terwilliger et al., 1987). Hydrothermal vent tubeworms such as *Riftia pachyptila* are able to absorb both oxygen and sulphide from the mixed vent and ambient water bathing their branchial plumes (Childress & Fisher, 1992) but *Lamellibrachia* sp., living at cool seeps in the Gulf of Mexico, probably obtain much of their sulphide via a thin-walled basal extension of the tube that penetrates into sulphidic sediment and has been shown to be permeable to dissolved sulphide (Julian et al., 1999). Sulphide uptake by the posterior region of the body can be enough to fuel total inorganic carbon uptake via the plume (Freytag et al., 2001). Studies of *Lamellibrachia satsuma* at a shallow site in southern Japan used sulphur isotope ratios to show that the major source of sulphide to these animals was from the sediment, through a similar thin-walled extension of the tube (Miura et al., 2002).

Cuticle and body wall

Two distinct types of collagen are present in pogonophores, as shown by immuno-labelling: a cuticular and an interstitial type. Recent analysis of the two types of collagen in the vestimentiferan *Riftia pachyptila* has shown that cuticle collagen has a great fibre length (1.5 μm), is non-striated, has high thermal stability (37 °C) and high threonine content. The cuticle consists largely of layers of collagen fibrils, often arranged helically, and penetrated by microvilli from the cell surface (reviewed by Southward, 1993; Gardiner & Jones, 1993). It forms thickened ridges and plates, particularly the frenulum or bridle of frenulates, the scale-like adhesive plaques typical of all pogonophores, and the rectangular scales on the tentacles of the frenulate *Lamellisabella coronata*. Compared with this, the interstitial collagen of the extracellular matrix (ECM) is formed of shorter, striated fibres, has a melting point of 29 °C and contains less threonine (Gaill et al., 1995; Mann et al., 1996). It is present under the epidermis, between muscle layers and in the matrix of the obturaculum (Gaill et al., 1994).

The epidermis is rich in secretory cells and contains the completely intraepidermal nervous system (Gupta & Little, 1970; Southward, 1984, 1993; Gardiner & Jones, 1993). It is underlain by basal lamina, ECM of varying thickness, and the basal lamina of the body wall muscle (Gupta & Little, 1975; Jensen & Myklebust, 1975; Matsumo

& Sasayama, 2002). The myoepithelial cells of the frenulates *Siboglinum fiordicum* and *Oligobrachia mashikoi* show a relative lack of striation as compared to the tubificid *Branchiura sowerbyi*. From this observation, Matsumo & Sasayama (2002) concluded that *Oligobrachia mashikoi* could not be an annelid. However, the difference in body wall musculature is more probably the result of an inactive mode of life in pogonophores than an indication of phylogenetic affinities. Studies of muscle systems of about 25 polychaete species have shown that there is so much variation that it is difficult to draw phylogenetic conclusions from ultrastructural details (Lanzavecchia et al., 1988).

Anterior appendages

The anterior appendages of pogonophores have been termed tentacles in frenulates and moniliferans and branchial filaments in vestimentiferans. Rouse & Fauchald (1997) and Rouse (2001) coded 'peristomial grooved palps' or 'palps' as present in siboglinids/pogonophores, in their cladistic analyses of polychaetes. That is, they assume for the purpose of analysis that the anterior appendages are homologous with the peristomial grooved palps of polychaetes. The grooved palps of spionids have a longitudinal ciliated groove and a blind-ending blood vessel in a muscle-lined coelomic cavity; they receive nerves from two different parts of the forebrain (refs in Müller & Orrhage, *this volume*). They are both sensory and used for food collection. The homology with the 'tentacles' of frenulates, vestimentiferans and moniliferans is doubtful because the latter lack a ciliated groove, and are not used for food collection. Their blood system consists of two longitudinal blood vessels that are dilations of a blood sinus lying in the ECM between the epidermis and the muscle layer surrounding the central coelom. Narrow sinuses running through the pinnules where these are present (Fig. 6A, B), or under the epidermis, link the longitudinal vessels and complete the circulation (Gupta & Little, 1969; Van der Land & Nørrevang, 1977; Gardiner & Jones, 1993; Southward, 1993).

The pinnules are unicellular in frenulates and multicellular in vestimentiferans, but the cuticle is always extremely thin over the pinnules and the

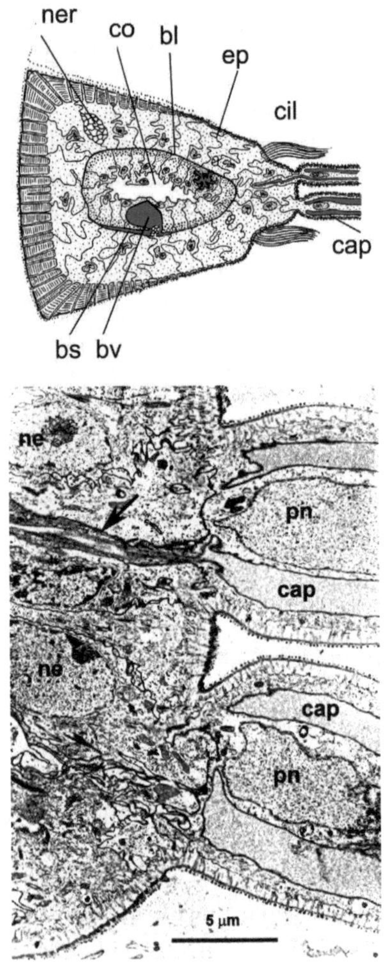

Figure 6. Frenulate tentacles, *Siphonobrachia ilyophora*. (A) TS tentacle and base of pinnules; bl – basal lamina; bs – blood sinus; bv – blood vessel; cap – capillary in pinnule; cil – cilia; co – coelom; ep – epidermis; ner – nerve. (Modified from Little & Gupta, 1969). (B) TEM of pinnule bases; cap – capillary: ne – nucleus of epidermal cell; pn – nucleus of pinnule cell; arrow – linking vessel. (Courtesy of B. L. Gupta).

blood circulates close to the surface. If ciliated cells are present, they form two narrow tracts along the tentacle or filament, on either side of the row of pinnules, but the pinnules themselves are not ciliated. Pinnules are not present in *Sclerolinum* species or some of the smaller frenulates, where the tentacles are extremely slender and their epidermal cuticle is thin. In all species the 'tentacles' seem to function as gills, and their structure is similar to that of many polychaete gills (e.g., Storch & Alberti, 1978). They have some sensory function, being provided with nerves from the sides of the intraepidermal brain.

Sensory cells are generally rather few, but *Riftia pachyptila* has specialised sensory filaments scattered among the respiratory ones (Gardiner & Jones, 1993).

Frenulates may possess over 200 tentacles/filaments, emerging from the body immediately behind the cephalic lobe (Southward, 1993). In their more complex branchial plume, vestimentiferans typically possess from several hundred to many thousands of filaments (Webb, 1969; Jones, 1981a).

In the branchial plume of vestimentiferans, the filaments are organized into right and left groups of lamellae (Fig. 3A). Depending on the species and, to some degree, where the lamellae occur along the length of the obturaculum, the filaments within a lamella are fused by their adherent cuticles along some portion of their length (Gardiner & Jones, 1993). In most vestimentiferans, the lamellae emerge directly from the anterior vestimental region and there is a basibranchial system of blood vessels connecting to the main dorsal and ventral vessels; each lamella is fused to its more medial lamella, with the two most medial lamellae (=earliest lamellae formed) fused to the lateral surface of the obturaculum (see Jones, 1985a; Gardiner & Jones, 1993; Miura et al., 1997).

In *Riftia pachyptila*, axial extensions of the main dorsal and ventral vessels run in an extended layer of vestimental tissue adhering closely to each side of the lower end of the obturaculum proper; from these axial blood vessels transverse vessels arise to supply the branchial vessels in the overlapping lamellae, from which the distal ends of separate filaments project at right angles (Jones, 1981a, Gardiner & Jones, 1993; Andersen et al., 2002; ECS, personal observation). The total branchial surface area per unit wet mass averages 22 cm^2 g^{-1} (Andersen et al., 2002), which is second among aquatic organisms only to another hydrothermal vent inhabitant, the polychaete *Paralvinella grasslei* (Jouin & Gaill, 1990). This average value translates to nine times the surface area of the remainder of the body (Andersen et al., 2002). Diffusion distances are very short, ranging from slightly more than 25 μm from the filament surface to the central coelomic cavity and its paired blood vessels to as short as 1 μm in certain regions of pinnules associated with the filaments. Their observations led Andersen et al. (2002) to conclude that the branchial plume is the predominant exchange organ in *Riftia pachyptila.*

Obturaculum

In *Riftia pachyptila*, the obturacular region occupies 11–35% of the length of the body (Jones 1981a). In other species where complete specimens are known, it comprises about 3.5% (juvenile specimen of *Alaysia spiralis*; Southward, 1991) and 2.8% (*Paraescarpia echinospica*; Southward et al., 2002) of the length of the body.

The anterior face of the obturaculum may be lined by a simple cuticle, as in *Alaysia spiralis* and *Lamellibrachia* species (see Southward, 1991, among others) and *Arcovestia ivanovi* (Southward & Galkin, 1997), or be covered by a crust-like material as observed for species of *Escarpia* (Jones, 1985a), *Seepiophila jonesi* (Gardiner et al., 2001) and *Paraescarpia echinospica* (Southward et al., 2002). In addition, some type of medial structure, which is secreted by the obturacular epithelium, may project from between the apical halves of the obturaculum. This medial structure takes the form of a small fin in *Tevnia jerichonana* and *Escarpia laminata*, an elongated spike (Jones, 1985a) in *Escarpia spicata*, a longer spike in *Paraescarpia echinospica* (Southward et al., 2002), but a series of saucer-like structures in *Oasisia alvinae* and *Ridgeia piscesae* (Jones, 1985a).

The obturaculum consists of paired obturacular halves, each surrounded by an epithelium with an overlying cuticle (see Webb, 1969; Van der Land & Nørrevang, 1977; Jones, 1981a,b, 1985b; Gardiner & Jones, 1993; Malakhov et al., 1996a). Numerous bundles of longitudinal muscle fibres are situated immediately internal to the epithelium. Most of the internal volume of the obturacular halves is occupied by an extensively developed extracellular matrix (ECM), consisting mainly of collagen fibres, among which are spindle-shaped cells believed to be involved in collagen secretion (Andersen et al., 2001). In the centre of each obturacular half is a slender perivascular coelom that surrounds a blind-ending obturacular vessel.

Andersen et al. (2001) demonstrated that the muscle bundles consist of smooth muscle fibres, which is an uncommon feature of muscles in annelids, and that the ECM displays an organization different from a typical annelid cartilage. Based on an assessment of the nature of the muscle fibres and the relationship between the muscle bundles and the ECM, Andersen et al. (2001) suggested that the muscle fibres produce a tension that allows the system to act as a catch connective tissue capable of changing its softness and stiffness. The large volume of cartilaginous matrix provides a force opposing the contraction of the surrounding muscle bands and thus maintains the shape and solidity of the obturaculum. The obturaculum supports the filaments when outside the tube (Fig. 3A) and acts as a plug when the animal retreats.

The development of the obturaculum begins later than the filaments. In a juvenile 1.8 mm long of a species of *Ridgeia*, Southward (1988) documented the first appearance of the obturaculum as a pair of D-shaped bulges of cuticle-covered epithelium projecting above the anterior end of the brain, but buried in a depression between two semicircles of seven filaments. Each bulge contained a mesodermal core with a central blood vessel. This observation suggests that the obturaculum may be an outgrowth of the vestimental region, rather than the first body segment (Jones, 1985b) or the first pair of filaments (Ivanov, 1989).

In the light of this possibility, one of us (SLG) examined serial sections of the obturaculum of a juvenile specimen (45-filament stage) of *Oasisia alvinae*. The obturaculum displays the typical organization discussed above (Fig. 7A) with the exception of the most apical region. Here the obturacular vessel is absent and each obturacular half is completely filled with cells, i.e., a layer of ECM is not apparent (Fig. 7B). Near the base of the obturaculum, the branchial lamellae lose their structural integrity, and the branchial epithelium becomes continuous with the epithelium apical to the brain. Also, a layer of epithelium, which is derived from the branchial lamellae, encircles the obturaculum providing it with a double layer of epithelium (Fig. 7C). This second epithelial layer is bounded externally by a cuticle. At the region where the obturaculum merges with the vestimentum, both epithelial layers become continuous with the epithelium surrounding the brain (Fig. 7D). These observations support Southward's (1988) observations and strongly suggest that the obturaculum is an apical outgrowth of the vestimental region. Such a developmental pattern may have significance for future discussions of patterns of segmentation in vestimentiferans.

Anatomy of the blood vascular system

The major blood vessels in Pogonophora, as in other annelids, are the dorsal and ventral vessels (Webb, 1969; Van der Land & Nørrevang, 1977; Gardiner & Jones, 1993; Malakhov et al., 1996b). The dorsal vessel is contractile throughout most of its length, but especially in the heart region, located in the anterior vestimental region in Vestimentifera (Fig. 8A) (Gardiner & Jones, 1993) and in the forepart in Frenulata (Southward, 1993). The dorsal vessel pumps blood forward into the branchiae. The flow is reversed at the tips of the branchiae and continues from there into the ventral vessel. More posteriorly, the dorsal and ventral vessels are connected by trophosomal vessels and by the septal vessels in the opisthosome.

Throughout most of its length, the dorsal vessel is suspended in paired coelomic cavities by a dorsal and a ventral mesentery. The extracellular matrix between the two cell layers of the mesenteries is continuous with the vascular lamina (Fig. 8B, C). There is no cellular endothelium. In the heart region, the perivascular cavities disappear and the lumen of the blood vessel is constricted by a thick musculature. Behind the heart, an intravasal body occupies the lumen of the blood vessel in Vestimentifera (Schulze, 2002) and Frenulata (Ivanov, 1963; Southward, 1993). An intravasal or heart-body is a strand of tissue located inside the dorsal contractile blood vessel. It is separated from the blood by a basal lamina. An intravasal body is also present in several taxa in the Terebellida (Dales & Pell, 1970; Spies, 1973; Jouin-Toulmond et al., 1996). In Vestimentifera, it extends through the trunk and into the opisthosome and adheres to the ventral side of the lumen of the dorsal vessel where the mesenteries split (Schulze, 2002) (Fig. 8B). The basal lamina of the intravasal body is thicker than the vascular lamina but constructed more loosely (Fig. 8C). The cytoplasm of the epithelial cells contains numerous mitochondria, rough endoplasmic reticulum, some Golgi vesicles and electron-dense inclusions. The electron-dense inclusions resemble haematin-bodies with their heterogeneous, lamellar substructure (Fig. 8C).

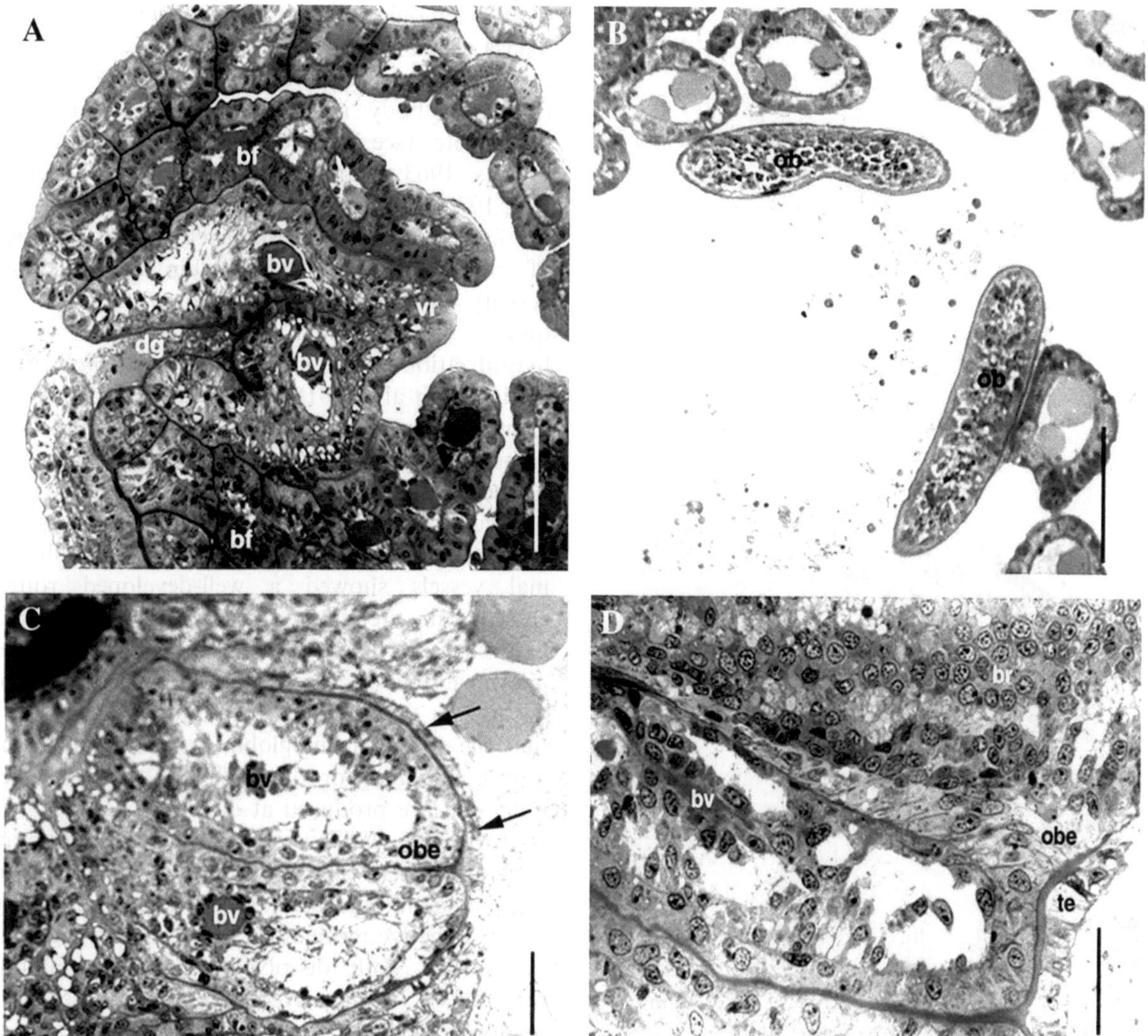

Figure 7. Obturaculum of juvenile *Oasisia alvinae*. Light microscopy. (A) cross section in mid-region showing typical shape of obturaculum with dorsal groove (dg) and prominent ventral ridge (vr). (B) cross section in apical region showing obturacular halves (ob) separated and lacking obturacular associated blood vessels. (C) cross section near base of obturaculum. Note tentacular epithelium (arrows) surrounding obturacular epithelium (obe). (D) cross section at region of merging of obturaculum with vestimental region. Note obturacular epithelium (obe) continuous with epithelium surrounding brain (br). bf – branchial filaments; te – tentacular epithelium. Scale bars: A–C = 75 μm; D = 25 μm.

Haemoglobin function and synthesis

An essential characteristic of pogonophoran blood is extracellular haemoglobin (Hb). Three types of Hb have been identified in the vestimentiferan *Riftia pachyptila* (Zal et al., 1996): V1 (~3500 kDa), V2 (~400 kDa) and C1 (~400 kDa). Whereas V1 and V2 occur in the blood, C1 is restricted to the coelom. V1 is a hexagonal bilayer hemoglobin (HBL), common in many annelid groups (Green et al., 2001). It consists of two monomeric globin chains, four linker units and one disulphide bonded dimer, whereas V2 and C1 have one monomeric globin chain and one disulphide bonded dimer, but no linker units (Zal et al., 1996). Two types of HBL exist in annelids (Jouan et al., 2001): type I occurs in oligochaete, hirudinean and vestimentiferan Hbs as well as sabellid chlorocruorins; type II occurs in other polychaetes. HBL appears to be absent in frenulates; they have V2

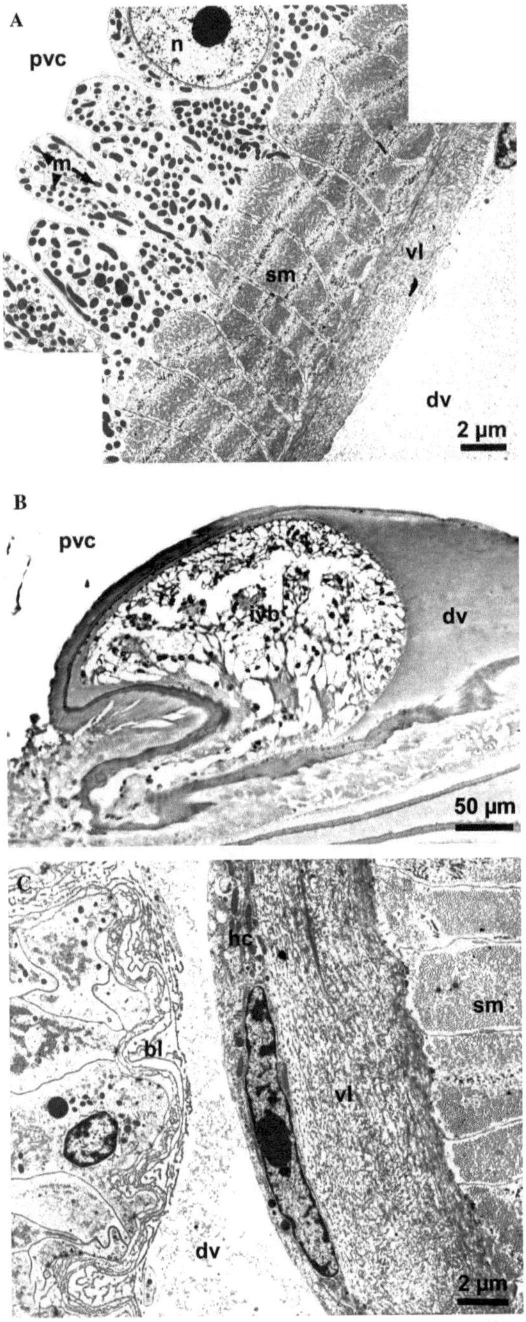

Figure 8. Dorsal vessel and intravasal body in Vestimentifera. (A) *Ridgeia piscesae*, TEM, wall of dorsal vessel in vestimental region, posterior to the heart. (B) *Paraescarpia echinospica*, LM, intravasal body in vestimentum. (C) Intravasal body (left), blood haemocyte and wall of dorsal blood vessel in vestimentum. Density and abundance of striated collagen fibres is greater in vascular lamina of dorsal vessel than in basal lamina of intravasal body. dv – dorsal vessel; hc – haemocyte; ivb – intravasal body; m – mitochondria; n – nucleus; pvc – perivascular cavity; sm – striated muscle of dorsal vessel wall; vl – vascular lamina.

(light haemoglobin) only (Terwilliger et al., 1987; Yuasa et al., 1996).

Haemoglobin synthesis may take place in the intravasal body in pogonophores as in other polychaetes (see Kennedy & Dales, 1958; Breton-Gorius, 1963; Mangum & Dales, 1965; Dales & Pell, 1970). In Pogonophora, more direct evidence for haemoglobin synthesis, such as the presence of globin mRNA, haemoglobin molecules or activity of enzymes for haeme synthesis is required. On the other hand, based on *in situ* hybridizations of juvenile *Riftia pachyptila*, Andersen et al. (2001) showed that globin A2, a haemoglobin component common to all three haemoglobin types, is present in the branchial plume, the vestimental region and trunk, and particularly in the afferent trophosomal vessels. Ultrastructural studies of the afferent trophosomal vessels showed a well-developed rough endoplasmic reticulum, making them another likely site for haemoglobin synthesis. The juveniles examined by Andersen et al. (2001) did not yet have a distinct intravasal body. It is possible that the site of haemoglobin production changes during ontogeny or that different types of haemoglobin are produced at different sites.

Arp et al. (1987) showed that vestimentiferan haemoglobin is capable of binding oxygen and sulphide simultaneously and reversibly, providing a very effective mechanism to satisfy the high demands for both dissolved gases in symbiont metabolism (Childress & Fisher, 1992) (see next section). Vestimentiferan haemoglobins vary in their affinities for oxygen and sulphide, with C1 having the highest oxygen affinity and V1 the highest sulphide affinity. Zal et al. (1997, 1998) identified two mechanisms of sulphide binding in *Riftia pachyptila*:

1. All three Hbs can form *S*-sulphohaemoglobin by binding HS^- to cysteine residues according to the following equilibrium:

$$R{-}SH + HS^- + H^+ \leftrightarrow R{-}SSH$$

2. Only V1 Hb is able to form persulphide bonds:

$$R{-}SS{-}R + 2HS^- + 2H^+ \leftrightarrow 2R{-}SSH$$

Persulphide bonds are formed near neutral pH but with an optimum at pH 7.5. The lower pH in the trophosome might trigger sulphide release to the symbionts.

Frenulates appear to have only a lower molecular weight vascular Hb. In *Siboglinum fiordicum* and *Galathealinum* sp. the Hb has a very high affinity for oxygen, and apparently a very low ability to bind sulphide (Terwilliger et al., 1987; Childress & Fisher, 1992) but the amino acid sequence of the globin-like chains of *Oligobrachia mashikoi* Hb indicates a sulphide-binding potential (Zal et al., 1997). The very long slender bodies of frenulates penetrate deeply into sediments containing low concentrations of reduced sulphur compounds. Sulphide may reach the trophosome mainly by diffusion and the carrying capacity of the haemoglobin may be less important than in vestimentiferans.

Sulphide binding might have evolved in vestimentiferans as a mechanism to detoxify sulphide, as well as to transport and store large quantities to fuel symbiont metabolism (Schulze & Halanych, 2003). On the other hand, Bailly et al. (2003) suggest that sulphide binding is a plesiomorphic feature in annelids and that secondary loss occurred repeatedly as a result of constraint relaxation.

Trophosome

The tissue that hosts symbiotic bacteria in pogonophores lies between the dorsal and ventral longitudinal blood vessels in the trunk region. It occupies most of the length of the trunk in vestimentiferans (Fig. 2) but is restricted to the posterior two thirds of the trunk in frenulates and moniliferans (Southward, 1993; Gardiner & Jones, 1993). The intracellular bacteria in the trophosome of *Riftia pachyptila* were discovered by Cavanaugh et al. (1981), who suggested that the large, globular bacteria were sulphur-oxidizing chemoautotrophs, a new concept of symbiosis at the time, but one which solved the long-standing problem of how these 'gutless' animals feed. Felbeck et al. (1981) and Rau (1981) found evidence of autotrophy. A similar type of symbiosis was found in frenulates and moniliferans, with thin rod-shaped bacteria in the trophosome of frenulates, and thick rods in *Sclerolinum brattstromi* (Southward, 1982). One frenulate, *Siboglinum poseidoni*, contains methanotrophic symbionts (Schmaljohann & Flügel, 1987), but other species examined appear to be sulphide oxidizers (Southward et al., 1986). The position and structure of the frenulate trophosome suggest that it replaces the alimentary canal. Outer peritoneal cells surround a core of bacteriocytes, with a central lumen in some species. Blood-filled sinuses between the basal laminae of the two epithelia are connected to the longitudinal blood vessels. Ultrastructural studies of juvenile *S. poseidoni* indicate that the bacteriocytes are endodermal in origin (Callsen-Cencic & Flügel, 1995).

The bulky vestimentiferan trophosome has a more complicated blood vascular system than that of frenulates, but it seems to have basically the same two-layered structure. This can be seen during the post-larval development as described by Jones & Gardiner (1988, 1989), Southward (1988) and Gardiner & Jones (1993). In the juveniles of *Ridgeia piscesae* and *Riftia pachyptila* it appears that the initial symbiotic bacteria are engulfed from the ciliated gut lumen by the endodermal cells. These bacteria remain undigested in vacuoles, while other types of bacteria are digested in separate phagocytotic vacuoles. Later, the multiplication of the symbiotic bacteria and the bacteriocytes that contain them occurs in the central region of each lobule, around an axial blood vessel. Growing bacteriocytes with larger bacteria are in the median layer while cells with degenerating bacteria lie towards the periphery, where there is a network of afferent blood vessels (Bosch & Grassé, 1984a,b; Gardiner & Jones, 1993; Bright et al., 2000; Bright & Sorgo, 2003). Blood circulates through sinuses among the bacteriocytes, from the peripheral vessels to the axial vessels.

Experimental work on uptake, transport and metabolism of sulphide and oxygen and fixation of inorganic carbon in *Riftia pachyptila* has been reviewed by Childress & Fisher (1992) and Goffredi et al. (1998). The nitrogen requirements of the symbiosis appear to be met by nitrate, absorbed by the host, reduced to ammonia by the symbionts and transferred to the host as amino acids (Girguis et al., 2000). Bright et al. (2000) used ^{14}C bicarbonate and autoradiography to show that carbon fixation and release of organic carbon from the symbionts to the host occur rapidly (within 15 min of exposure) in small *Riftia pachyptila*. The released carbon is incorporated into fast-growing tissues and into the tube-secreting glands, as might be expected from the speed of tube secretion observed by Gaill et al. (1997). The trophosome of *Riftia* contains much glycogen, both in the bacteria and the host cells (Sorgo et al., 2002). Before the trophosome of frenulates was known, glycogen was found to be abundant in the central core of the postannular trunk of the frenulate *Siboglinum atlanticum* (Southward, 1973). Later studies of the fine structure of the trophosome of *Siboglinum* and *Oligobrachia* species found that the outer, peritoneal epithelium was the chief storage region for glycogen particles and large droplets, probably of lipid and protein, (Southward, 1982, 1993). The trophosome also accumulates metals. In older animals it contains so many dark granules that it becomes brown or greenish black. In frenulates these mineral granules are multilayered and present mainly in the peritoneal cells (Southward, 1982). Granules in the peritoneal cells of *Riftia pachyptila* contain S, Fe, Cu and Ag, while granules in bacteriocytes contain Mg, P, S, Ca, Fe, Cu and Ag (Truchet et al., 1998).

Excretory organs

The role of nephridia is the removal of waste products, in particular nitrogenous wastes, by filtration of body fluids and subsequent modification of the primary urine by secretion and/or selective re-absorption. In Pogonophora, the trophosome probably partially takes over these functions; mineral particles are stored as insoluble granules and bacterial symbionts might recycle nitrogen compounds (Southward, 1993). In addition, excretory organs are present in the anterior forepart in frenulates (Ivanov, 1963; Southward, 1993) and the anterior vestimentum in vestimentiferans (Van der Land & Nørrevang, 1977; Gardiner & Jones, 1993; Schulze, 2001b). To date, no observations have been made on excretory organs of *Sclerolinum*.

Ivanov (1963) distinguished two orders within the Frenulata, based on the anatomy of excretory organs: the Thecanephria and the Athecanephria. The major difference is that nephridial ducts are longer in Thecanephria and enclosed in a 'renal sac'. Southward (1993), however, using ultrastructural methods found the distinction unclear and could not observe a renal sac in any of the species she examined.

In general, excretory organs occur in two basic designs: metanephridia and protonephridia. In metanephridia, filtration is driven by muscular contraction of blood vessels and is accomplished through the basal lamina of podocytes, representing a barrier between a blood vessel and a coelomic compartment. According to the functional model of nephridial design by Ruppert & Smith (1988) Pogonophora are expected to have metanephridia because their blood vascular system is well developed (see previous section). However, in no pogonophoran species have all three components of a metanephridial design been observed:

1. A blood vessel surrounded by a coelomic compartment: The dorsal vessel in all Pogonophora and the obturacular vessels in Vestimentifera are suspended in perivascular cavities.
2. Podocytes: not detected in any pogonophoran. In Vestimentifera, the dorsal vessel is thickly muscularized, not only in the heart region, but also along the entire length (Fig. 8A, B), making it unlikely that podocytes are present.
3. Nephrostome: Ivanov (1963) presents anatomical diagrams of the thecanephrian *Lamellisabella zachsi* and the athecanephrians *Oligobrachia dogieli* and *Siboglinum caulleryi* and shows coelomoducts in all of them. However, Southward (1993) found them only in one representative of the Thecanephria, *Siphonobrachia lauensis*, but not in several *Siboglinum* species or *Oligobrachia gracilis*. Gardiner & Jones (1993) observed nephrostomes in the vestimentiferans *Tevnia jerichonana* and *Oasisia alvinae*. Despite careful examination of serial sections of eight vestimentiferan species, including the species examined by Gardiner & Jones, Schulze (2001b) did not observe nephrostomes in any vestimentiferans.

In protonephridia, fluids are filtered through a filtration weir of a terminal cell driven by ciliary action. The only species of pogonophoran in which terminal cells have been observed are *Oligobrachia gracilis* and *Siboglinum ekmani* (Southward, 1993: Fig. 20A, B). They seem to filter blood from the ventral blood vessel (Fig. 9A). Primary urine is then transported into a median section, consisting of convoluted ducts that eventually connect to ectodermal ducts leading to the nephropores. A median section of convoluted ducts is also present in Vestimentifera (Schulze, 2001b: Figs 1–3). It is located behind the brain and in the area of the sinus valvatus, where the efferent plume vessels join to form the ventral vessel. The sinus valvatus is a valve-like structure that seems to prevent back-flow of blood into the tentacles. It is possible that the blood pressure in this area is increased due to the small opening to the ventral vessel. The convoluted, ciliated ducts of the excretory organ are closely apposed to the blood vessels in this region (Fig. 9B). Although no terminal cells have been observed, it is the most likely site for their presence.

The apparent absence of nephrostomes, at least in some species, might be correlated with the presence of extracellular coelomic hemoglobin. Coelomic proteins might be lost through coelomoducts if these were present, a case parallel to nephtyid polychaetes (Smith & Ruppert, 1988).

In conclusion: No uniform picture of the nature of nephridia in pogonophores is emerging. Considering that Pogonophora show close affinities to polychaetes that clearly have a metanephridial design (Rouse & Fauchald, 1997), it is possible that more basal groups have metanephridia, whereas more derived groups, such as vent-inhabiting vestimentiferans might have a protonephridial design and show only rudimentary nephrostomes if any.

Opisthosome

The opisthosome is the segmented posterior section of the pogonophoran body. As in other annelids, the segmentation arises from a posterior growth zone (Webb, 1964; Nørrevang, 1970; Southward, 1975a). The septa between adjacent segments consist of double mesodermal layers with a thin layer of extracellular matrix between them.

In general, the vestimentiferan opisthosome has more segments than the frenulate and moniliferan opisthosome. In frenulates, four to eight peg-like chaetae are present in each segment, arranged dorsolaterally and ventrolaterally on both sides. In vestimentiferans, the anterior opisthosomal segments bear rows of uncini (Schulze, 2001a) similar to uncini in the frenulate girdles and parapodial uncini in some Sabellidae (Knight-Jones & Fordy, 1979; Knight-Jones, 1981), Arenicolidae (Bartolomaeus & Meyer, 1997) and Oweniidae (Meyer & Bartolomaeus, 1996). Rows of uncini are also present in the opisthosomal segments of *Sclerolinum* (Southward, 1972, 2000; Smirnov, 2000). The different designs in the three groups may be correlated with different lifestyles: whereas the opisthosome of Vestimentifera and Monilifera is enclosed inside the tube and probably serves as an anchor, the opisthosome in the Frenulata extends outside the tube and is mainly used as a digging organ.

On the cellular level, chaetogenesis in Pogonophora is similar to chaetogenesis in related polychaetes (Gupta & Little, 1970; George & Southward, 1973; Bartolomaeus, 1995; Schulze, 2001a). The chaetoblast and follicle cells produce chitinous chaetal material that accumulates around the apical microvilli of the chaetoblast. As the chaeta grows, the microvilli retract and the chaetal material forms hollow cylinders. In *Ridgeia piscesae*, chaetae are formed in chaetal follicles, consisting of a chaetoblast, a follicle cell and an epidermis cell. Sometimes two follicle cells may be present. There are two follicle cells in the frenulate *Siboglinum ekmani* (Gupta & Little, 1970). No particular formative zone has been detected for chaetal formation in any pogonophoran.

When Jones (1985a) erected Obturata (=Vestimentifera) and Perviata (=Frenulata) as separate phyla he based his decision partly on the difference in opisthosomal segmentation between the two groups. While in Vestimentifera each opisthosomal segment comprises two symmetrical coelomic cavities, the opisthosomal coeloms in adult Frenulata are unpaired. However, studying early juveniles, Southward (1975a) showed that in frenulates the opisthosomal segments arise from paired mesodermal blocks and that the mesenteries are lost in later development. Furthermore, contrary to Jones (1985a), we found that in the vestimentiferan *Ridgeia piscesae* only the posterior

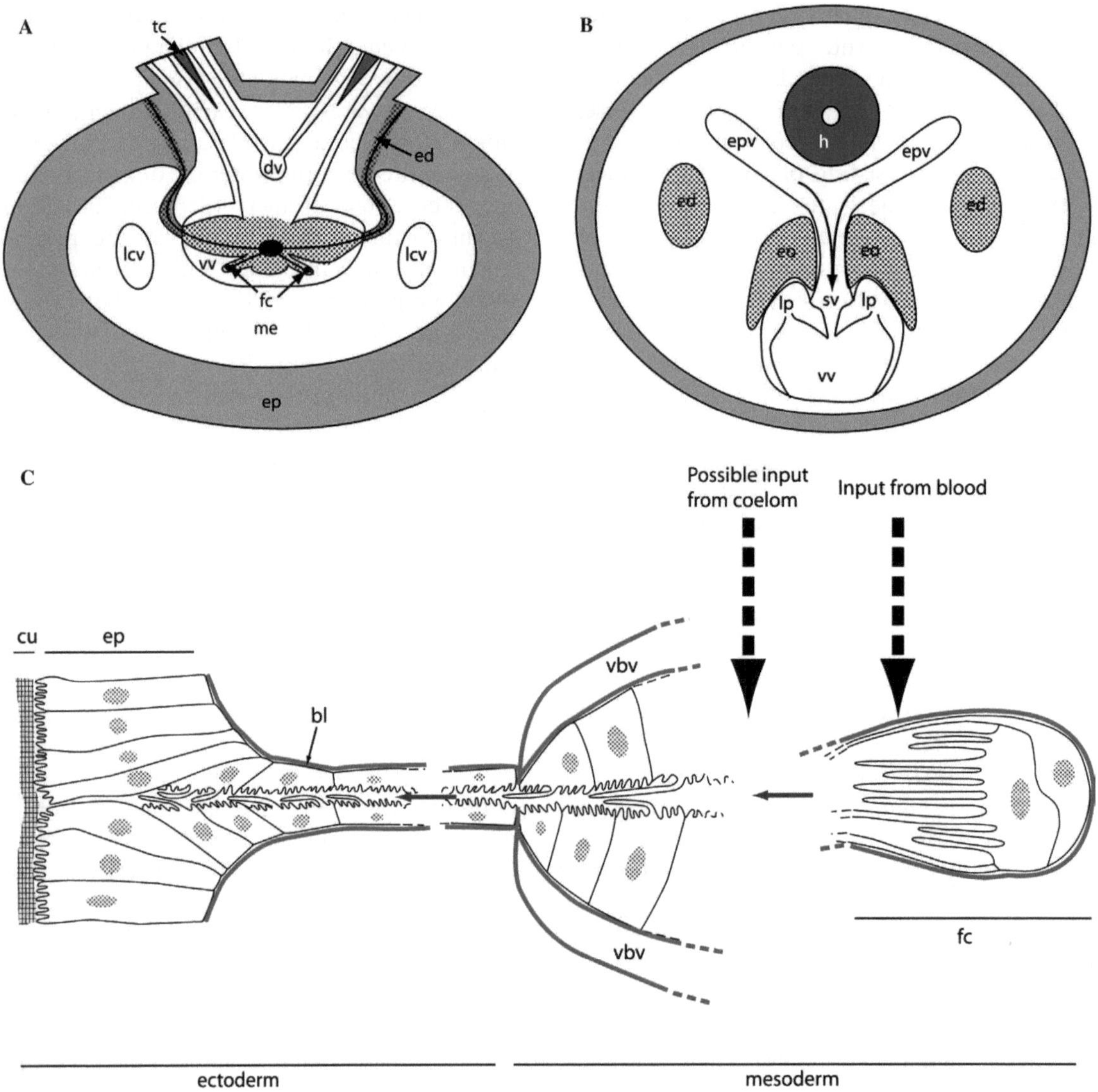

Figure 9. Association of excretory organs with blood vessels in Frenulata and Vestimentifera. (A) *Oligobrachia gracilis*, cross section in region of tentacle origin. The convoluted ducts of the medial part of the excretory organ are completely enclosed in the ventral vessel. (B) *Ridgeia piscesae*, cross section in region of heart. The convoluted ducts in the excretory organ surround the junction of the efferent plume vessels and the sinus valvatus. Arrows indicate direction of blood flow. (C) Cellular arrangement in excretory organs of *Oligobrachia gracilis*. Gray arrows indicate direction of flow of excretory fluids. Whether additional influx through coelomoducts happens in addition to filtration through filtration cells could not be detected. bl – basal lamina; cu – cuticle; dv – dorsal vessel; ed – excretory duct; eo – excretory organ; ep – epidermis; epv – efferent plume vessels; fc – filtration cell; h – heart; lcv – lateral cephalic vessels; lp – lateral pockets of sinus valvatus; sv – sinus valvatus; tc – tentacular coelom; vv – ventral vessel.

faces of the septa are muscular instead of both faces (Fig. 10). This is also the case in frenulates, suggesting that the opisthosomes are more similar in the two groups than previously assumed.

In both Vestimentifera and Frenulata, the septa are vascularised by septal vessels. These originate from the ventral vessel and run between the two mesodermal layers of the septa without a cellular lining (Fig. 10B). Glands that resemble the tube-secreting pyriform glands in the vestimentum and trunk are associated with the septa. Gardiner & Jones (1993) distinguished in *Riftia pachyptila* long glands that are linked to the blood vascular system and short glands that are not. This distinction was not clear in histological sections of *Ridgeia piscesae* or *Tevnia jerichonana*. In *Ridgeia piscesae*, the peripheral glands have a pore to the exterior and resemble the pyriform glands of the vestimentum and trunk whereas the more central glands have no direct connection to the exterior. They are formed by one of the mesodermal layers of the septa and are usually surrounded by blood spaces (Fig. 10A).

Unlike the nerve cord in the trunk, the ventral nerve cord in the opisthosome has no giant axons in either Frenulata or Vestimentifera. Southward (1975a) describes three nerve trunks in the opisthosome of the frenulate *Siboglinum fiordicum* and 'nerve bulges' that may be ganglia, probably associated with chaetal movement. No 'nerve bulges' are present in Vestimentifera but nerve tracts are often present between the chaetal sacs.

Although adult pogonophores have no continuous gut, a residual gut can be detected in the opisthosomes even in adult vestimentiferans. An anus was detected in a *Tevnia jerichonana* specimen with an opisthosome of 7 × 4 mm size (Fig. 11B). The gut tissue is closely associated with the dorsal vessel and enclosed in musculature. In a small *Ridgeia piscesae* specimen (5 mm) the gut wall in cross section consists of approximately seven multiciliated cells and is completely enclosed in a basal lamina (Fig. 11A). In another small specimen bacterial cells with diameters up to 4.3 μm were observed around the gut (Fig. 11C). Bacterial cells were absent from the lumen of the ciliated duct. Previously, a rudimentary gut has only been demonstrated in early developmental stages (Jones & Gardiner, 1988; Southward, 1988) of a maximum length of 2.6 mm. As it is not continuous in adult specimens, it is obviously non-functional as a digestive tract. Whether it has taken over another function in the adult needs to be examined.

Reproduction, embryos and larvae

In pogonophores the sexes are separate, except for the hermaphrodite *Siboglinum poseidoni*. The reproductive organs are normally paired and when mature, occupy quite a lot of space in the trunk, often severely restricting the trophosome. The male gonopores are always at the anterior end of the trunk, as are the female gonopores of vestimentiferans, but in frenulates the oviducts open more posteriorly (Ivanov, 1963). In *Sclerolinum* species paired sperm ducts open at the anterior end of the trunk, but female gonopores have not yet been found (ECS, personal observation). The long-headed sperm of frenulates and vestimentiferans have mitochondria wrapped around the nucleus (Gardiner & Jones, 1985, 1993; Southward, 1993). Frenulates produce spermatophores wrapped in a mucopolysaccharide coat (Flügel, 1978), secreted in a special region of the sperm duct. No spermatophores have been observed in *Sclerolinum* species. The early report of spermatophores in *Sclerolinum sibogae* (Southward, 1961) was erroneous; recent re-examination of the type material (ECS) found sperm in the gonoducts but no spermatophores, while examination of living *Sclerolinum brattstromi* found small clusters of long-headed sperm able to swim in concert (personal observation ECS). In vestimentiferans, sticky sperm masses are transferred between males and females of *Ridgeia piscesae* (Southward & Coates, 1989; MacDonald et al., 2003) and *Tevnia jerichonana* (personal observation ECS). *Riftia pachyptila* in natural populations release puffs of eggs and sperm into the surrounding water (Van Dover, 1994), but internal fertilization remains a strong possibility in this species because sperm have been detected inside the oviducts (Gardiner & Jones, 1985; P. A. Tyler, personal communication).

Some frenulates incubate large eggs in their tubes until the settlement stage, whereas vestimentiferans disperse their eggs and have lecithotrophic planktonic development. Frenulates with small eggs possibly have planktonic larvae but these have never been reared (Southward, 1999). The known larvae have been described as trochophore-like.

Marsh et al. (2001) reared *Riftia pachyptila* at *in situ* temperature and pressure (2 °C and 250 atm). Cleavage was slow but a swimming ciliated larva was observed after 34 days, when it had two equatorial bands of simple cilia, but no mouth, apical tuft or telotroch. The larval metabolic lifespan was estimated to be 34–44 days, based on respiration rate and energy reserves. Study of the current regime showed that there is a high probability that larvae will remain within a few tens of km of their source but the potential along-ridge dispersal could be as much as 100 km. Cool seep vestimentiferans studied

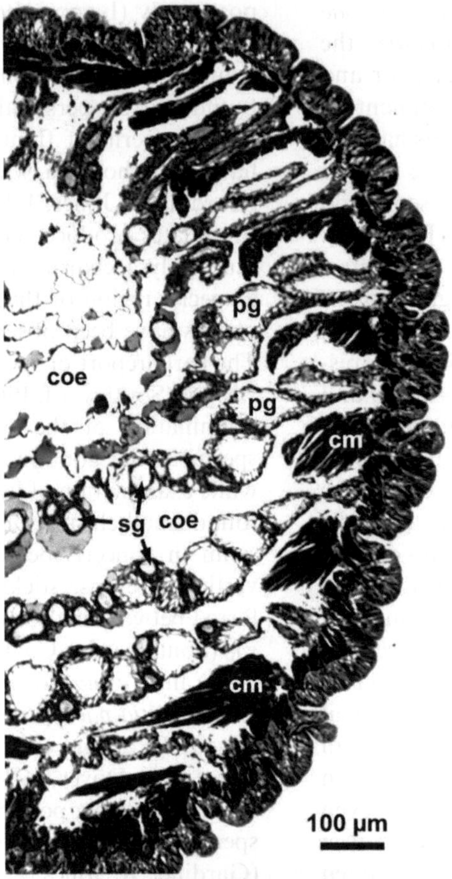
A
pg
pg
coe
cm
sg
coe
cm
100 µm

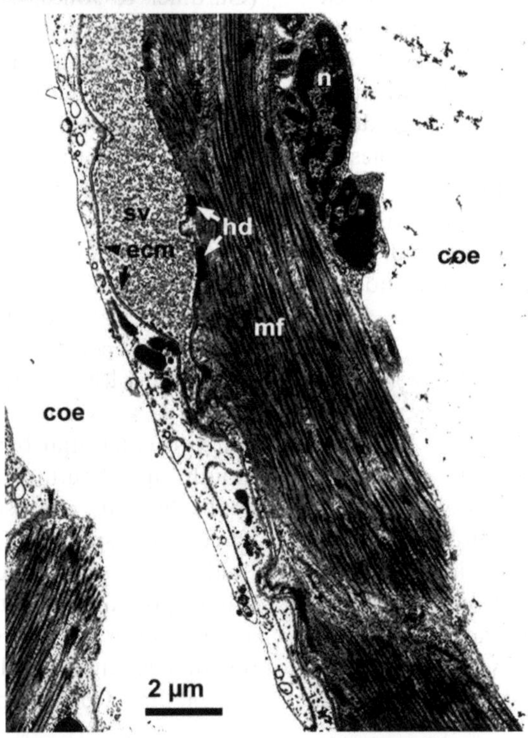
B
n
sv
hd
ecm
coe
mf
coe
2 µm

Figure 10. Opisthosomal septa in the vestimentiferan *Ridgeia piscesae*, associated glands and septal vessels. (A) horizontal section, LM, showing pyriform and septal glands. (B) TEM, opisthosomal septum, anterior is to the left. cm – circular muscle; coe – coelom; ecm – extracellular matrix; hd – hemidesmosomes; mf – myofilaments; n – nucleus; pg – pyriform glands; sg – septal glands; sv – septal vessel.

earlier, took 3 days to develop from cleavage to a larva with prototroch (at 9 °C, 1, 50 and 100 atm.) and survived at least 3 weeks (Young et al., 1996). *Lamellibrachia satsuma* larvae studied by Miura et al. (1997) had both prototroch and telotroch. Eggs removed directly from the oviduct of this species began cleavage without insemination and developed to a trochophore-like larva in three days (Fukunaga et al., 2000) suggesting that they had already been fertilized in the oviduct. A similar observation of cleavage starting *in vitro* without the addition of sperm was noted in the frenulate *Siboglinum fiordicum* (Bakke, 1976).

The larvae of vestimentiferans may have to survive for some weeks, without feeding, in fully oxygenated conditions until they can settle in a suitable environment for the adult to develop. The question of whether the symbionts are transmitted via the egg, or whether they are definitely acquired from the environment is still debatable. Cary et al. (1993) could not detect any symbionts in the oocytes of *Riftia pachyptila*, and none have been found in early settlement stages of vestimentiferans or frenulates, but slightly later stages, having a temporary mouth and alimentary canal, do have bacteria in some of their endodermal cells and this seems a logical way to take them up from the environment (Southward, 1988; Jones & Gardiner, 1988, 1989; Callsen-Cencic & Flügel, 1995). Schmaljohann & Flügel (1987) found that the symbionts in the trophosome of *Siboglinum poseidoni* and methanotrophic bacteria isolated from its environment were similar in ultrastructure and size. Other evidence for environmental origin of the bacteria rests on the discovery that vestimentiferans of different genera, but from the same locality, host identical symbionts, demonstrated by molecular techniques (Feldman et al., 1997). Further detailed analysis and review (McMullin et al., 2003) supports environmental acquisition of symbionts and absence of co-speciation between vestimentiferans and their symbionts. However, free-living bacteria identifiable by new molecular

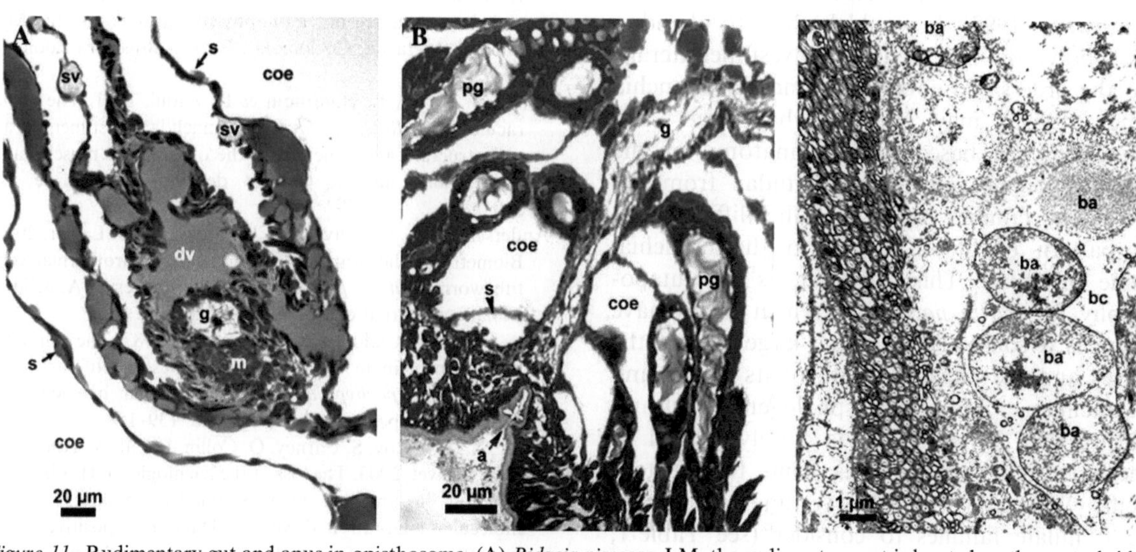

Figure 11. Rudimentary gut and anus in opisthosome. (A) *Ridgeia piscesae*, LM, the rudimentary gut is located on the ventral side of the dorsal vessel, enclosed in muscular tissue, ciliated gut lumen appears dark. (B) *Tevnia jerichonana*, LM, hindgut and anus, lumen of gut lies in a different plane. (C) *Ridgeia piscesae*, TEM, ciliated lumen of rudimentary gut and bacteriocytes with endosymbionts in close proximity; a – anus; ba – bacteria; bc – bacteriocyte; coe – coelom; dv – dorsal vessel; g – gut; m – muscle; pg – pyriform glands; s – septa; sv – septal vessel.

techniques as the vestimentiferan symbionts have not yet been reported, to our knowledge, from the natural environment near vents or seeps.

Systematic implications

Aspects of the anatomy of pogonophores have been discussed here with the aim of understanding their way of life, and understanding the differences between frenulates, vestimentiferans and moniliferans/sclerolinids. Halanych (2005) has reviewed publications on their molecular phylogeny, concluding that vestimentiferans are monophyletic, *Sclerolinum* is sister to the vestimentiferan clade and frenulates form a monophyletic clade sister to the vestimentiferan/moniliferan clade. This fits the anatomical differences discussed here, though the position of *Sclerolinum* seems a little uncertain.

Relationships within the groups have not been discussed in this paper, but we agree with Halanych that a new investigation of frenulate phylogeny would be worthwhile. Ivanov's anatomical work resulted in a division into two orders, Athecanephria and Thecanephria, on the basis of differences in the excretory anatomy, the arrangement of papillae and glands on the postannular trunk, and the shape of spermatophores, none of which seem as clear cut now (Southward, 1993). It would be best to abandon this subdivision. The division of vestimentiferans into the orders Axonobranchia and Basibranchia (Jones, 1985a) may also be abandoned. The distinction was based on the anatomy of the branchial plume, separating Riftiidae from the rest of the families because of the difference in the pattern of blood supply to the branchial plume of *Riftia*. This, however, is an autapomorphy for *Riftia pachyptila*, which might have arisen as a consequence of the large size of the species and the unusual length of its plume and obturaculum. Molecular phylogenies clearly indicate that *R. pachyptila* is closely related to other vent-inhabiting species from the Eastern Pacific. We then have seven vestimentiferan and five frenulate families to consider (see Table 1, p. 226). This is not the place to revise them, but the phylogenetic trees of genera produced by Halanych (2005) and Rouse (2001) should provide guidance, and have already indicated that some of the original descriptions are inadequate, so there is scope for new descriptive and anatomical work, as well as new molecular work if suitable material can be collected.

Acknowledgements

We thank the organizers of the Osnabrück Symposium for the opportunity to prepare this review. Eve Southward is grateful to the late Brij L. Gupta for encouragement in the past and for the gift of some of his electron micrographs, to Mrs J. Gupta for permission to use his unpublished work and Dr A.V. Grimstone of Cambridge University for providing negatives. Alan Southward improved the manuscript and helped with the preparation of the figures. Prof. Verena Tunnicliffe supplied specimens, photographs and encouragement to Eve Southward and Anja Schulze. Dr K. M. Halanych made useful changes to the manuscript. Laura Berner, a former undergraduate student at Bryn Mawr College, assisted with obturacular sectioning and analysis.

References

Almeida, W. de O., M. L. Christoffersen, D. de S. Amorim, A. R. S. Garraffoni & G. S. Silva, 2003. Polychaeta, Annelida, and Articulata are not monophyletic: articulating the Metameria (Metazoa, Coelomata). Revista Brasileira Zoologia 20: 23–57.

Andersen, A. C., L. Hamraoui & D. Zaoui, 2001. The obturaculum of *Riftia pachyptila* (Annelida, Vestimentifera): ultrastructure and function of the obturacular muscles and extracellular matrix. Cahiers de Biologie Marine 42: 219–237.

Andersen, A. C., S. Jollivet, S. Claudinot & F. H. Lallier, 2002. Biometry of the branchial plume in the hydrothermal vent tubeworm *Riftia pachyptila* (Vestimentifera; Annelida). Canadian Journal of Zoology 80: 320–332.

Arp, A. J., J. J. Childress & R. D. Vetter, 1987. The sulphide-binding protein in the blood of the vestimentiferan tube-worm *Riftia pachyptila* is the extracellular haemoglobin. Journal of Experimental Biology 128: 139–158.

Bailly, X., R. Leroy, S. Carney, O. Collin, F. Zal, A. Toulmond & D. Jollivet, 2003. The loss of the haemoglobin H_2S binding function in annelids from sulfide-free habitats reveals molecular adaptation driven by Darwinian positive selection. Proceedings of the National Academy of Sciences of the USA. 100: 5885–5890.

Bakke, T., 1976. The early embryos of *Siboglinum fiordicum* Webb (Pogonophora) reared in the laboratory. Sarsia 60: 1–12.

Bartolomaeus, T., 1995. Structure and formation of the uncini in *Pectinaria koreni*, *Pectinaria auricoma* (Terebellida) and *Spirorbis spirorbis* (Sabellida): implications for annelid phylogeny and the position of the Pogonophora. Zoomorphology 115: 161–177.

Bartolomaeus, T. and K. Meyer, 1997. Development and phylogenetic significance of hooked setae in Arenicolidae. Invertebrate Biology 116: 227–242.

Blackwell, J., K. D. Parker & K. M. Rudall, 1965. Chitin in pogonophore tubes. Journal of Marine Biological Association of the UK 45: 51–54.

Bosch, C. & P. P. Grassé, 1984a. Cycle partiel des bactéries chimioautotrophes symbiotiques et leurs rapports avec les bacteriocytes chez *Riftia pachyptila* Jones (Pogonophore Vestimentifère). I. Le trophosome et les bacteriocytes. Compte Rendu Lebdomaire des séances de l'Académie des Sciences, Paris (Ser. III), 299: 371–376.

Bosch, C. & P. P. Grassé, 1984b. Cycle partiel des bactéries chimioautotrophes symbiotiques et leurs rapports avec les bacteriocytes chez *Riftia pachyptila* Jones (Pogonophore Vestimentifère). II. L'evolution des bactéries symbiotiques et des bacteriocytes. Compte Rendu Lebdomaire des séances de l'Académie des Sciences, Paris (Ser. III), 299: 413–419.

Breton-Gorius, J., 1963. Etude au microscope électronique des cellules chloragogènes d'*Arenicola marina* L. Leur rôle dans la synthèse de l'hémoglobine. Annales des Sciences Naturelles 5: 211–271.

Bright, M. & A. Sorgo, 2003. Ultrastructural reinvestigation of the trophosome in adults of *Riftia pachhyptila* (Annelida, Siboglinidae). Invertebrate Biology 122: 347–368.

Bright, M., H. Keckeis & C. R. Fisher, 2000. An autoradiographic examination of carbon fixation, transfer, and utilization in the *Riftia pachyptila* symbiosis. Marine Biology 136: 621–632.

Brunet, P. J. C. & D. B. Carlisle, 1958. Chitin in Pogonophora. Nature 182: 1689.

Callsen-Cencic, P. & H. J. Flügel, 1995. Larval development and the formation of the gut of *Siboglinum poseidoni* Flügel & Langhof (Pogonophora, Perviata), evidence of protostomian affinity. Sarsia 80: 73–89.

Carlisle, D. B., 1964. Chitin in a Cambrian fossil, *Hyolithellus*. Biochemical Journal 90: 1963–1964.

Cary, S. C., W. Warren, W. E. Anderson & S. J. Giovannoni, 1993. Identification and localization of bacterial endosymbionts in hydrothermal vent taxa with symbiont-specific polymerase chain reaction amplification and *in situ* amplification techniques. Molecular Marine Biology and Biotechnology 2: 51–62.

Caullery, M., 1914. Sur les Siboglinidae, type nouveau d'invértebrés recuelli par l'expédition du Siboga. Compte Rendu Lebdomaire des séances de l'Académie des Sciences, Paris 158: 2014–2017.

Cavanaugh, C. M., S. L. Gardiner, M. L. Jones, H. W. Jannasch & J. B. Waterbury, 1981. Prokaryotic cells in the hydrothermal vent tube worm *Riftia pachyptila* Jones: Possible chemoautotrophic symbionts. Science 213: 340–342.

Chamoy, L., M. Nicolai, B., Quennedey, F. Gaill & J. Delachambre, 2000. Characterization of a cDNA encoding RP43, a CUB-domain-containing protein from the tube of *Riftia pachyptila* (Vestimentifera), and distribution of its transcript. Biochemical Journal 350: 421–427.

Chamoy, L., M. Nicolai, J. Ravaux, B. Quennedey, F Gaill & J. Delachambre, 2001. A novel chitin-binding protein from the vestimentiferan *Riftia pachyptila* interacts specifically with beta-chitin: Cloning, expression, and characterization. Journal of Biological Chemistry 276: 8051–8058.

Childress, J. J. & C. R. Fisher, 1992. The biology of hydrothermal vent animals: physiology, biochemistry and autotrophic symbiosis. Oceanography and Marine Biology Annual Review 30: 337–441.

Dales, R. P. & J. P. Pell, 1970. Cytological aspects of haemoglobin and chlorocruorin synthesis in polychaete annelids. Zeitschrift für Zellforschung 109: 20–32.

Felbeck, H., J. J. Childress & G. N. Somero, 1981. Calvin-Benson cycle and sulphide oxidation enzymes in animals from sulphide-rich habitats. Nature, London 298: 291–293.

Feldman, R. A., M. B. Black, C. S. Cary, R. A. Lutz & R. C. Vrijenhoek, 1997. Molecular phylogenetics of bacterial endosymbionts and their vestimentiferan hosts. Molecular Marine Biology and Biotechnology 6: 268–277.

Flügel, H. J., 1978. Dense core tubuli – a new structural element of the spermatophores of Pogonophora. 9th International Congress of Electron Microsc. Toronto 1978 2: 574–575.

Foucart, M. F., S. Bricteux-Grégoire & C. Jeuniaux, 1965. Composition chimique du tube d'un pogonophore (*Siboglinum* sp.) et des formations squelettiques de deux pterobranches. Sarsia, 20: 36–41.

Freytag, J. K., P. R. Girguis, D. K. Bergquist, J. P. Andras, J. J. Childress & C. R. Fisher, 2001. A paradox resolved: sulfide acquisition by roots of seep tubeworms sustains net chemoautotrophy. Proceedings of the National Academy of Sciences of the United States of America 98: 13408–13413.

Fukunaga, K., T. Miura & J. Tsukahara, 2000. Structural studies of gonads and developmental larvae of *Lamellibrachia satsuma*. Zoological Science (Tokyo) 17 Supplement: 59.

Gaill, F. & S. Hunt, 1986. Tubes of deep sea hydrothermal vent worms *Riftia pachyptila* (Vestimentifera) and *Alvinella pompejana* (Annelida). Marine Ecology Progress Series 34: 267–274.

Gaill, F., L. Hamraoui, F. X. Sicot & R. Timpl, 1994. Immunological properties and tissue localization of two different collagen types in annelid and vestimentiferan species. European Journal of Cell Biology 65: 392–401.

Gaill, F., K. Mann, H. Wiedemann, J. Engel & R. Timpl, 1995. Structural comparison of cuticle and interstitial collagens from annelids living in shallow sea-water and at deep-sea hydrothermal vents. Journal of Molecular Biology 246: 284–294.

Gaill, F., B. Shillito, F. Ménard, G. Goffinet & J. J. Childress, 1997. Rate and process of tube production by the deep-sea hydrothermal vent tubeworm *Riftia pachyptila*. Marine Ecology Progress Series 148: 135–143.

Gardiner, S. L. & M. L. Jones, 1985. Ultrastructure of spermiogenesis in the vestimentiferan tube worm *Riftia pachyptila* (Pogonophora, Obturata). Transactions of the American Microscopical Society 104: 19–44.

Gardiner, S. L. & M. L. Jones, 1993. Vestimentifera. In Harrison, F. W. & M. E. Rice (eds), Microscopic Anatomy of Invertebrates, 12, Onychophora, Chilopoda, and lesser Protostomata, Wiley-Liss, New York: 371–460.

Gardiner, S. L., E. McMullin & C. R. Fisher. 2001. *Seepiophila jonesi*, a new genus and species of vestimentiferan tube worm (Annelida: Pogonophora) from hydrocarbon seep communities in the Gulf of Mexico. Proceedings of the Biological Society of Washington 114: 694–707.

George, J. D. & E. C. Southward, 1973. A comparative study of the setae of Pogonophora and polychaetous annelids. Journal of the Marine Biological Association of the United Kingdom 53: 403–424.

Girguis, P. R., R. W. Lee, N. Desaulniers, J. J. Childress, M. Pospesel, H. Felbeck & F. Zal, 2000. Fate of nitrate acquired by the tubeworm *Riftia pachyptila*. Applied and Environmental Microbiology 66: 2783–2790.

Goffredi, S.K., J. J. Childress, F. H. Lallier & N. T. Desaulniers, 1998. How to be the perfect host: CO_2 and HS^- accumulation and H^+ elimination in the hydrothermal vent tube-worm *Riftia pachyptila*. Cahiers de Biologie Marine 39: 297–300.

Green, A. W., T. Gotoh, T. Suzuki, F. Zal, F. H. Lallier, A. Toulmond & S. N. Vinogradov, 2001. Observation of large, non-covalent globin subassemblies in the appr 3600 kDa hexagonal bilayer hemoglobins by electrospray ionization time-of-flight spectrometry. Journal of Molecular Biology 309: 553–560.

Gupta, B. L. & C. Little, 1969. Studies on Pogonophora. II. Ultrastructure of the tentacular crown of *Siphonobrachia*. Journal of the Marine Biological Association of the United Kingdom 49: 717–741.

Gupta, B. L. & C. Little, 1970. Studies on Pogonophora. 4. Fine structure of the cuticle and epidermis. Tissue and Cell 2: 637–696.

Gupta, B. L. & C. Little, 1975. Ultrastructure, phylogeny and Pogonophora. Zeitschrift für Zoologische Systematik und Evolutionsforschung Sonderheft 1975: 43–63.

Halanych, K. M., 2005. Molecular phylogeny of siboglinid annelids (a.k.a. pogonophorans): a review. Hydrobiologia 535/536 (Dev. Hydrobiol. 179): 295–305.

Ivanov, A. V., 1951. On including the genus *Siboglinum* in the class Pogonophora. Doklady Akademii Nauk. SSSR 76: 739–742 (in Russian).

Ivanov, A. V., 1957. Material on the embryological development of the Pogonophora. Zoologicheskii Zhurnal 36: 1127–1144 (in Russian).

Ivanov, A. V., 1963. Pogonophora. Academic Press, London.

Ivanov, A. V., 1975. Embryonalentwicklung der Pogonophora und ihre systematische Stellung. Zeitschrift für Zoologische Systematick und Evolutionsforschung Sonderheft 1975: 10–44.

Ivanov, A. V., 1988. Analysis of the embryonic development of Pogonophora in connection with the problems of phylogenetics. Zeitschrift für Zoologische Systematick und Evolutionsforschung 26: 161–185.

Ivanov, A. V., 1989. On the morphological nature of the obturacula in Pogonophora. Doklady Akademii Naukii SSSR 308: 758–759.

Ivanov, A. V., 1991. Monilifera, a new subclass of Pogonophora. Doklady Akademii Naukii SSSR 312: 505–507.

Ivanov, A. V., 1994. On the systematic position of the Vestimentifera. Zoologische Jahrbücher, Abteilung für Systematik, Ökologie und Geographie der Tiere 121: 409–456.

Jensen, H. & R. Myklebust. 1975. Ultrastructure of muscle cells in *Siboglinum fiordicum* (Pogonophora). Cell and Tissue Research 163: 185–197.

Jeuniaux, C., 1963. Chitin et Chitinolyse. Masson, Paris: 1–181.

Johansson, K. E., 1937. Über *Lamellisabella zachsi* und ihre systematische Stellung. Zoologischer Anzeiger 117: 23–26.

Johansson, K. E., 1939. *Lamellisabella zachsi* Uschakow, ein Vertreter einer neuen Tierklasse Pogonophora. Zoologiska Bidrag från Uppsala 18: 253–268.

Jones, M. L. 1981a. *Riftia pachyptila*, new genus, new species, the vestimentiferan worm from the Galápagos Rift geothermal vents (Pogonophora). Proceedings of the Biological Society of Washington 93: 1295–1313.

Jones, M. L. 1981b. *Riftia pachyptila* Jones: observations on the vestimentiferan worm from the Galápagos Rift. Science 213: 333–336.

Jones, M. L. 1985a. On the Vestimentifera, new phylum: six new species, and other taxa, from hydrothermal vents and elsewhere. Bulletin of the Biological Society of Washington 6: 117–185.

Jones, M. L., 1985b. Vestimentiferan pogonophores: Their biology and affinities. In Conway-Morris S., J. D. George, R. Gibson & H. M. Platt (eds), The Origins and Relationships of Lower Invertebrates. Systematics Association Special Volume 28, Clarendon Press, Oxford: 327–342.

Jones, M. L. & S. L. Gardiner, 1988. Evidence for a transient digestive tract in Vestimentifera. Proceedings of the Biological Society of Washington 101: 423–433.

Jones, M. L. & S. L. Gardiner, 1989. On the early development of the vestimentiferan tube worm *Ridgeia* sp. and observations on the nervous system and trophosome of *Ridgeia* sp. and *Riftia pachyptila*. Biological Bulletin 177: 254–276.

Jouan, L., J. C. Taveau, S. Marco, F. H. Lallier & J. N. Lamy, 2001. Occurrence of two architectural types of hexagonal bilayer hemoglobin in annelids: comparison of 3D reconstruction volumes of *Arenicola marina* and *Lumbricus terrestris* hemoglobins. Journal of Molecular Biology 305: 757–771.

Jouin, C. & F. Gaill, 1990. Gills of hydrothermal vent annelids: structure, ultrastructure and functional implications in two alvinellid species. Progress in Oceanography 24: 59–69.

Jouin-Toulmond, C., D. Augustin, D. Desbruyères & A. Toulmond, 1996. The gas transfer system in alvinellids (Annelida Polychaeta, Terebellida). Anatomy and ultrastructure of the anterior circulatory system and characterization of a coelomic, intracellular, haemoglobin. Cahiers de Biologie Marine 37: 135–151.

Julian, D., F. Gaill, A. J. Arp & C. R. Fisher, 1999. Roots as a site of hydrogen sulfide uptake in the hydrocarbon seep vestimentiferan *Lamellibrachia* sp. Journal of Experimental Biology 202: 2245–2257.

Kennedy, G. Y. & R. P. Dales, 1958. The function of the heart-body in polychaetes. Journal of the Marine Biological Association of the United Kingdom 37: 15–31.

Knight-Jones, P. 1981. Behaviour, setal inversion and phylogeny of Sabellida (Polychaeta). Zoologica Scripta 10: 184–202.

Knight-Jones, P. & M. R. Fordy, 1979. Setal structure, functions and interrelationships in Spirorbidae (Polychaeta, Sedentaria). Zoologica Scripta 8: 119–138.

Lanzavecchia, G., M. de Eguileor & R. Valvassori. 1988. Muscles. In Westheide, W. & C.O. Hermans (eds), The Ultrastructure of Polychaeta. Gustav Fischer Verlag, Stuttgart: 71–88.

Little, C. & B. L. Gupta, 1968. Pogonophora: uptake of dissolved nutrients. Nature 873–874.

Little, C. & B. L. Gupta, 1969. Studies on Pogonophora III. Uptake of nutrients. Journal of Experimental Biology 51: 759–773.

MacDonald, I., V. Tunnicliffe & E. C. Southward, 2002. Detection of sperm transfer and synchronous spawning in *Ridgeia piscesae* Jones at Endeavour Segment, Juan de Fuca Ridge. Cahiers de Biologie Marine 43: 395–398.

Malakhov, V. V. & S. V. Galkin, 1998. Vestimentifera, Gutless Invertebrates of the Sea Floor. KMK Scientific Press, Moscow: 206 pp. (In Russian, with English summary).

Malakhov, V. V., I. S. Popelyaev & S. V. Galkin, 1996a. Microscopic anatomy of *Ridgeia phaeophiale* Jones, 1985 (Pogonophora, Vestimentifera) and the problem of the position of Vestimentifera in the system of the animal kingdom: I. General anatomy, obturacula and tentacles. Russian Journal of Marine Biology 22: 63–74.

Malakhov, V. V., I. S. Popelyaev & S. V. Galkin, 1996b. Microscopic anatomy of *Ridgeia phaeophiale* Jones, 1985 (Pogonophora, Vestimentifera) and the problem of the position of Vestimentifera in the system of the animal kingdom: III. Rudimentary digestive system, trophosome, and blood vascular system. Russian Journal of Marine Biology 22: 189–198.

Malakhov, V. V., I. S. Popelyaev & S. V. Galkin, 1996c. Microscopic anatomy of *Ridgeia phaeophiale* Jones, 1985 (Pogonophora, Vestimentifera) and the problem of the position of Vestimentifera in the system of the animal kingdom: V. Position of Vestimentifera and Pogonophora in the system of the Animal Kingdom. Russian Journal of Marine Biology 22: 307–313.

Mangum, C. P. & R. P. Dales, 1965. Products of haem synthesis in polychaetes. Comparitive Biochemistry and Physiology 15: 237–257.

Mann, K., D. E. Mechling, H. P. Bächinger, C. Eckerskorn, F. Gaill & R. Timpl, 1996. Glycosylated threonine but not 4-hydroxyproline dominates the triple helix stabilizing positions in the sequence of a hydrothermal vent worm cuticle collagen. Journal of Molecular Biology 261: 255–266.

Marsh, A. G., L. S. Mullineaux, C. M. Young & D. T. Manahan, 2001. Larval dispersal potential of the tubeworm *Riftia pachyptila* at deep-sea hydrothermal vents. Nature 411: 77–80.

Matsuno, A. & Y. Sasayama, 2002. A comparative study of body wall structures of a pogonophore and an annelid from a phylogenetic viewpoint. Zoological Science (Tokyo) 19: 695–701.

McHugh, D., 1997. Molecular evidence that echiurans and pogonophorans are derived annelids. Proceedings of the National Academy of Sciences of the United States of America 94: 8006–8009.

McMullin, E. R., S. Hourdez, S. W. Schaeffer & C. R. Fisher, 2003. Phylogeny and biogeography of deep sea vestimentiferan tubeworms and their symbionts. Symbiosis 34: 1–41.

Meyer, K. & T. Bartolomaeus, 1996. Ultrastructure and formation of the hooked setae in *Owenia fusiformis* delle Chiaje, 1842: implications for annelid phylogeny. Canadian Journal of Zoology 74: 2143–2153.

Miura, T., M. Nedachi & J. Hashimoto, 2002. Sulphur sources for chemoautotrophic nutrition of shallow water vestimentiferan tubeworms in Kagoshima Bay. Journal of the Marine Biological Association of the United Kingdom 82: 537–540.

Miura, T., J. Tsukahara & J. Hashimoto, 1997. *Lamellibrachia satsuma*, a new species of vestimentiferan worms (Annelida: Pogonophora) from a shallow hydrothermal vent in Kagoshima Bay, Japan. Proceedings of the Biological Society of Washington 110: 447–456.

Negrisolo, E., A. Pallavicini, R. Barbato, S. Dewilde, A. Ghiretti-Magaldi, L. Moens & G. Lanfranchi, 2001. The evolution of extracellular hemoglobins of annelids, vestimentiferans, and pogonophorans. Journal of Biological Chemistry 276: 26391–26397.

Nørrevang, A., 1970. The position of Pogonophora in the phylogenetic system. Zeitschrift für Zoologische Systematik und Evolutionsforschung 8: 161–172.

Orrhage, L. & M. C. M. Müller, 2005. Morphology of the nervous system of Polychaeta (Annelida). Hydrobiologia 535/536 (Dev. Hydrobiol. 179): 79–111.

Rau, G. H., 1981. Hydrothermal vent clam and tube worm $^{13}C/^{12}C$: further evidence of nonphotosynthetic food sources. Science 213: 338–340.

Ravaux, J., L. Gay, M. F Voss-Foucart & F. Gaill, 1998. Tube growth process in the deep-sea hydrothermal vent tubeworm *Riftia pachyptila* (Vestimentifera): Synthesis and degradation of chitin. Cahiers de Biologie Marine 39: 99–107.

Rouse, G. 2001. A cladistic analysis of Siboglinidae Caullery, 1914 (Polychaeta, Annelida): formerly the phyla Pogonophora and Vestimentifera. Zoological Journal of the Linnean Society 132: 55–80.

Rouse, G. & K. Fauchald, 1995. The articulation of annelids. Zoologica Scripta 24: 71–138.

Rouse, G. & K. Fauchald, 1997. Cladistics and polychaetes. Zoologica Scripta 26: 139–204.

Rudall, K. M. & W. Kenchington, 1973. The chitin system. Biological Reviews 48: 597–636.

Ruppert, E. E. & R. S. Smith, 1988. The functional organization of filtration nephridia. Biological Reviews 63: 231–258.

Salvini-Plawen, L. v., 2000. What is convergent/homoplastic in Pogonophora? Journal of Zoological Systematic and Evolutionary Research 38: 133–147.

Schmaljohann, R. & H. J. Flügel, 1987. Methane oxidizing bacteria in Pogonophora. Sarsia 72: 91–98.

Schulze, A. 2001a. Ultrastructure of opisthosomal chaetae in Vestimentifera (Pogonophora, Obturata) and implications for phylogeny. Acta Zoologica 82: 127–135.

Schulze, A. 2001b. Comparative anatomy of excretory organs in vestimentiferan tube worms (Pogonophora, Obturata). Journal of Morphology 250: 1–11.

Schulze, A. 2003. Phylogeny of Vestimentifera (Siboglinidae, Annelida) inferred from morphology. Zoologica Scripta 32: 321–342.

Schulze, A. 2002. Histological and ultrastructural characterization of the intravasal body in Vestimentifera (Siboglinidae, Polychaeta, Annelida). Cahiers de Biologie Marine 43: 355–358.

Schulze, A. & K. M. Halanych, 2003. Siboglinid evolution shaped by habitat preference and sulfide tolerance. Hydrobiologia 496 (Dev. Hydrobiol. 170): 199–205.

Shillito, B., J. P. Lechaire & F. Gaill, 1993. Microvilli-like structures secreting chitin crystallites. Journal of Structural Biology 111: 59–67.

Shillito, B., J. P. Lechaire, G. Goffinet & F. Gaill, 1995a. Composition and morphogenesis of the tubes of vestimentiferan worms. Geological Society Special Publication 87: 295–302.

Shillito, B., B. Lübbering, J. P. Lechaire, J. J. Childress & F. Gaill, 1995b. Chitin localization in the secretory system of a repressurized deep-sea tube worm. Journal of Structural Biology 114: 67–75.

Shillito, B., J. P. Lechaire, J. J. Childress & F. Gaill, 1997. Diffraction contrast imaging of extracellular matrix components using zero-loss filtering. Journal of Structural Biology 120: 85–92.

Smirnov, R. V., 2000. Two new species of Pogonophora from the arctic mud volcano off northwestern Norway. Sarsia 85: 141–150.

Smith, P. R. & E. E. Ruppert, 1988. Nephridia. In Westheide, W. & C. O. Hermans (eds), Ultrastructure of the Polychaeta. Gustav Fischer, Stuttgart: 231–262.

Sorgo, A., F. Gaill, J. P. Lechaire, C. Arndt & M. Bright, 2002. Glycogen storage in the *Riftia pachyptila* trophosome: contribution of host and symbiont. Marine Ecology Progress Series 231: 115–120.

Southward, A. J. & E. C. Southward, 1966. A preliminary account of the general and enzyme histochemistry of *Siboglinum atlanticum* and other Pogonophora. Journal of Marine Biological Association of the United Kingdom 46: 579–616.

Southward, A. J. & E. C. Southward, 1970. Observations on the role of dissolved organic compounds in the nutrition of benthic invertebrates. Sarsia 45: 69–96.

Southward, A. J. & E. C. Southward, 1981. Dissolved organic matter and the nutrition of Pogonophora: a reassessment based on recent studies of their morphology and biology. Kieler Meeresforschung 5: 445–453.

Southward, A. J., E. C. Southward, P. R. Dando, R. B. Barrett & R. L. Ling, 1986. Chemoautotrophic function of bacterial symbionts in small Pogonophora. Journal of Marine Biology Association U.K. 66: 415–437.

Southward, E. C., 1961. Pogonophora. Siboga Expedition Monograph 25, 3: 1–22.

Southward, E. C., 1972. On some Pogonophora from the Caribbean and the Gulf of Mexico. Bulletin of Marine Science 22: 739–776.

Southward, E. C. 1973. The distribution of glycogen in the tissues of *Siboglinum atlanticum* (Pogonophora). Journal of Marine Biological Association of the United Kingdom 53: 665–671.

Southward, E. C. 1975a. A study of the structure of the opisthosoma of *Siboglinum fiordicum*. Zeitschrift für Zoologische Systematik und Evolutionsforschung Sonderheft 1975: 64–76.

Southward, E. C., 1975b. Fine structure and phylogeny of the Pogonophora. Symposia of the Zoological Society of London 36: 235–251.

Southward, E. C., 1982. Bacterial symbionts in Pogonophora. Journal of the Marine Biological Association of the United Kingdom 62: 889–906.

Southward, E. C., 1984. Pogonophora. In Bereiter-Hahn, J., A. G. Matoltsy & K. S. Richards (eds), Biology of the Integument, 1. Invertebrates. Springer-Verlag, Berlin: 376–388.

Southward, E. C., 1988. Development of the gut and segmentation of newly settled stages of *Ridgeia* (Vestimentifera): implications for the relationship between Vestimentifera and Pogonophora. Journal of the Marine Biological Association of the United Kingdom 68: 465–487.

Southward, E. C., 1991. Three new species of Pogonophora, including two vestimentiferans, from hydrothermal sites in the Lau Back-arc Basin (Southwest Pacific). Journal of Natural History 25: 859–881.

Southward, E. C., 1993. Pogonophora. In Harrison, F. W. & M. E. Rice (eds), Microscopic Anatomy of Invertebrates 12, Onychophora, Chilopoda, and lesser Protostomata. Wiley-Liss., New York: 327–369.

Southward, E. C., 1999. Development of Perviata and Vestimentifera (Pogonophora). Hydrobiologia 402: 185–202.

Southward, E. C., 2000. Class Pogonophora. In Beesley, P. L., G. J. B. Ross & C. J. Glasby (eds), Polychaetes and Allies: The Southern Synthesis, Fauna of Australia 4A. CSIRO Publishing, Melbourne: 331–351.

Southward, E. C. & K. A. Coates, 1989. Sperm masses and sperm transfer in a vestimentiferan, *Ridgeia piscesae* Jones 1985 (Pogonophora Obturata). Canadian Journal of Zoology 67: 2776–2781.

Southward, E. C. & S. V. Galkin, 1997. A new vestimentiferan (Pogonophora: Obturata) from the hydrothermal vent fields in the Manus Back-arc Basin (Bismarck Sea, Papua New Guinea, Southwest Pacific Ocean). Journal of Natural History 31: 43–55.

Southward, E. C. & A. J. Southward, 1963. Notes on the biology of some Pogonophora. Journal of the Marine Biolocial Association of the United Kingdom 43: 57–64.

Southward, E. C. & A. J. Southward, 1967. The distribution of Pogonophora in the Atlantic Ocean. Symposia of the Zoological Society of London 19: 145–158.

Southward, E. C., A. Schulze & V. Tunnicliffe, 2002. Vestimentiferans (Pogonophora) in the Pacific and Indian Oceans: a new genus from Lihir Island (Papua New Guinea) and the Java Trench, with the first report of *Arcovestia ivanovi* from the North Fiji Basin. Journal of Natural History 36: 1179–1197.

Spies, R. B., 1973. The blood system of the flabelligerid polychaete *Flabelliderma commensalis*. Journal of Morphology 139: 465–490.

Storch, V. & G. Alberti, 1978. Ultrastructural observations on the gills of polychaetes. Helgoländer Wissenschaftliche Meeresuntersuchungen 31: 169–179.

Terwilliger, R. C., N. B. Terwilliger, G. M. Hughes, A. J. Southward & E. C. Southward, 1987. Studies on the haemoglobins of the small Pogonophora. Journal of the

Marine Biological Association of the United Kingdom 67: 219–234.

Truchet, M., C. Ballan-Dufrancais, A.-Y. Jeantet, J.-P. Lechaire & R. Cosson, 1998. The trophosome of the Vestimentifera *Riftia pachyptila* and *Tevnia jerichonana*: Cahiers de Biologie Marine 39: 129–141.

Tunnicliffe, V., A. G. MacArthur & D. McHugh, 1998. A biogeographical perspective of the deep-sea hydrothermal vent fauna. Advances in Marine Biology 34: 353–442.

Uschakow, P., 1933. Eine neue Form aus der Familie Sabellidae (Polychaeta). Zoologischer Anzeiger 104: 205–208.

Van der Land, J. & A. Nørrevang, 1977. Structure and relationships of *Lamellibrachia* (Annelida, Vestimentifera). Det Kongelige Danske Videnkabernes Selskab Biologiske Skrifter. 21, 3:1–102.

Van Dover, C. L., 1994. *In situ* spawning of hydrothermal vent tubeworms (*Riftia pachyptila*). Biological Bulletin 186: 134–135.

Vovelle, J., 1982. Structures et materiaux squelletogenes organiques et mineraux chez les annelids et les mollusques. Bulletin de la Société Zoologique Française 107: 401–417.

Webb, M., 1964. The posterior extremity of *Siboglinum fiordicum* (Pogonophora). Sarsia 15: 33–36.

Webb, M., 1965. Additional notes on the adult & larva of *Siboglinum fiordicum* and on the possible mode of tube formation. Sarsia, 20: 21–34

Webb, M., 1969. *Lamellibrachia barhami* gen. nov. sp. nov., (Pogonophora) from the northeast Pacific. Bulletin of Marine Science 19: 18–47.

Webb, M., 1971. The morphology and formation of the pogonophoran tube and its value in systematics. Zeitschrift für Zoologische Systematik und Evolutionsforschung 9, 3: 179–181.

Young, C. M., E. Vaquez, A. Metaxas & P. A. Tyler, 1996. Embryology of vestimentiferan tube worms from deep-sea methane/sulphide seeps. Nature 381: 514–516.

Yuasa, H. J., B. N. Green, T. Takagi, N. Suzuki, S. N. Vinogradov & T. Suzuki, 1996. Electrospray ionization mass spectrometric composition of the 400 kDa hemoglobin from the pogonophoran *Oligobrachia mashikoi* and the primary structures of three major globin chains. Biochimica Biophysica Acta 1296: 235–244.

Zal, F., F. H. Lallier, B. N. Green, S. N. Vinogradov & A. Toulmond, 1996. The multi-hemoglobin system of the hydrothermal vent tube worm *Riftia pachyptila*: II. Complete polypeptide chain composition investigated by maximum entropy analysis of mass spectra. Journal of Biological Chemistry 271: 8875–8881.

Zal, F., T. Suzuki, Y. Kawasaki, J. J. Childress, F. H. Lallier & A. Toulmond, 1997. Primary structure of the common polypeptide chain *b* from the multi-hemoglobin system of the hydrothermal vent tube worm *Riftia pachyptila*: an insight on the sulfide binding-site. Proteins 29: 562–574.

Zal, F., E. Leize, F. H. Lallier, A. Toulmond, A. Van Dorsselaer & J. J. Childress, 1998. S-sulfohemoglobin and disulfide exchange: the mechanisms of sulfide binding by *Riftia pachyptila* hemoglobins. Proceedings of the National Academy of Sciences of the United States of America 95: 8997–9002.

Hydrobiologia (2005) **535/536:** 253–275
T. Bartolomaeus & G. Purschke (eds), Morphology, Molecules, Evolution and Phylogeny in Polychaeta and Related Taxa

Myzostomida: A review of the phylogeny and ultrastructure

Igor Eeckhaut* & Déborah Lanterbecq
Marine Biology Laboratory, Natural Sciences Building, University of Mons-Hainaut, Mons, Belgium
*(*Author for correspondence: Tel.: +32-65-373439; Fax: +32-65-373434; E-mail: Igor.Eeckhaut@umh.ac.be)*

Key words: Myzostomida, Annelida, phylogeny, ultrastructure

Abstract

Myzostomids are minute, soft-bodied, marine worms associated with echinoderms since the Carboniferous. Due to their long history as host-specific symbionts, they have acquired a highly derived body plan that obscures their phylogenetic affinities to other metazoans. Because certain organs are serially arranged a closer relationship between polychaetes and myzostomids has repeatedly been discussion. We presented here a review on the ultrastructure of myzostomids with the most recent analyses that concern their phylogenetic position. The ultrastructure of the integument, digestive system, excretory system and nervous system are summarized. Unpublished information on the gametogenesis and reproductive systems of myzostomids are also exposed with a view on their reproductive process.

Introduction

Myzostomids, or myzostomes, are minute, soft-bodied, marine worms that are all associated with echinoderms (but see Grygier, 2000 for possible exceptions). They are found in all oceans from subtidal to a depth of over 3000 m. Most of them are ectocommensals of crinoids but some species that are parasites of crinoids, asteroids, or ophiuroids infest the gonads, coelom, integument or digestive system. The association between myzostomids and echinoderms is very old: signs of parasitic activities, similar to those induced by extant gallicolous myzostomids (i.e., those deforming echinoderm stereom), are found on fossilized crinoid skeletons dating back to the Carboniferous (Warn, 1974; Meyer & Ausich, 1983; Eeckhaut, 1998). Some pits found on Ordovician crinoid fossils could also have been induced by parasitic myzostomids (Eeckhaut, 1998). Due to their long history as host-specific symbionts, myzostomids have acquired a unique, highly derived anatomy that obscures their phylogenetic affinities to other metazoans (see Fig. 1). The body plan of most myzostomids is indeed singular inasmuch as they are incompletely segmented, parenchymous, acoelomate organisms with chaetae (see Grygier, 2000 for a review of the myzostomid body plans).

For most myzostomids, the body consists of an anterior cylindrical introvert (also called proboscis) and a flat, oval or disk-like trunk (Fig. 1). The introvert is extended when the individuals feed but it is retracted into an antero-ventral pouch of the trunk most of the time. The trunk ranges from a few millimeters to three centimeters long for the largest species. Five pairs of parapodia are located latero-ventrally in two rows, each parapodium containing a protrusible hook, some replacement hooks, and a support rod (or acicula). Most species have four pairs of slit- or disk-like latero-ventral sense organs, commonly named lateral organs, and the trunk margin often bears flexible needle-like cirri (more than one hundred in some species). Hump-like or pointed cirri also occur at the base of each parapodium of ca. 20 species. Two male gonopores are located at the level of the third pair of parapodia and the female gonopore opens close to the anus, postero-ventrally.

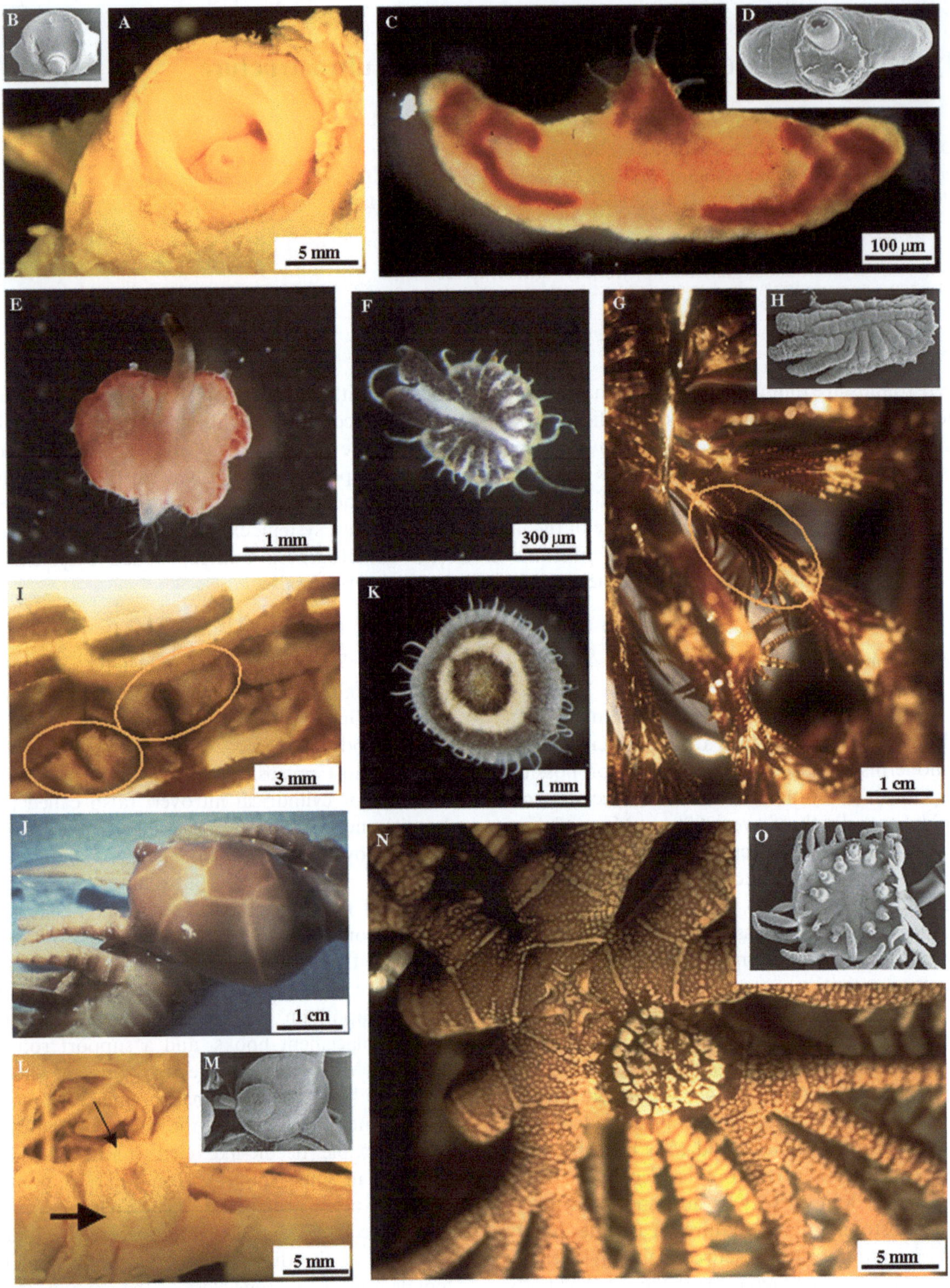
B
A
5 mm
C
D
100 μm
E
1 mm
F
300 μm
G
H
1 cm
I
3 mm
K
1 mm
J
1 cm
N
O
L
M
5 mm
5 mm

Figure 1. Illustrations of different species of myzostomids (OM and SEM views) and of the deformities they induce on crinoids. (A, B) The female of *Pulvinomyzostomum pulvinar* in the anterior part of the digestive system of its Mediterranean crinoid host, *Leptometra phalangium*; the myzostomid is seen after partial dissection of the host (OM) and after total extraction from the host (SEM). (C, D) *Contramyzostoma bialatum* (OM view of the lateral side and SEM view of the dorsal side), a parasite of the integument of the Singaporean crinoid *Comaster gracilis*. (E) A parasite of the integument, *Myzostoma toliarense*, dorsal view, with its introvert everted. (F) A young ectocommensal, *Myzostoma pseudocuniculus* (characterized by one pair of caudal appendages), observed dorsally. The two species in E and F live on crinoids from Madagascar. (G, H) *Myzostoma fissum*, a species infesting preferentially crinoids of the family Mariametridae; the individual is seen in life on its host (within the oval) and with SEM; the body of the individual observed is very similar in shape and coloration to some parts of the crinoid arms. (I, J) Two cysts induced by *Contramyzostoma sphaera* and a gall created by *Endomyzostoma tenuispinum*; all are induced on crinoid arms; the cysts are made of soft tissue and the wall of the gall is made of deformed original ossicles and newly formed ossicles induced by the presence of the parasite. (K) Dorsal side of *Myzostoma polycyclus*, a common Indo-West Pacific species mainly observed on *Comanthus* crinoids. (L, M) A small male of *Myzostoma alatum* (small arrow) attached on the dorsal side of a big female (large arrow), both located close to the mouth of their Mediterranean crinoid host, *Leptometra phalangium*; they are observed in life and with SEM. (N, O) *Myzostoma mortenseni*, a large ectocommensal observed on the aboral side of a *Clarkcomanthus albinotus* from Papua New Guinea, seen in life and with SEM; the SEM view of the ventral side shows the everted introvert, parapodia, and lateral organs.

The body of parasites is often highly modified (Fig. 1). The introvert, external appendages and sensory organs are usually reduced or have disappeared. According to the location of myzostomids in the host, their trunk will be folded up dorsally (e.g., *Pulvinomyzostomum pulvinar*; Fig. 1A and B), very much longer than wide (e.g., Protomyzostomatidae and Mesomyzostomatidae), very much wider than long (*Contramyzostoma* species; Fig. 1C and D), mushroom-shaped (e.g., *Mycomyzostoma calcidicola*), or totally irregular (unnamed species described by Heinzeller et al., 1995).

Leuckart (1827, 1830, 1836) was the first to observe myzostomids and he made the first diagnosis on a new genus of unknown organisms that he named *Myzostoma* (Leuckart, 1836). At present, ca. 170 species are known from the scientific literature, with 80% of them being ectocommensals of crinoids. Most of the ultrastructural analyses of myzostomids have been made on a single species, the European myzostomid *Myzostoma cirriferum*. Phylogenetic analyses are extremely recent. They are based on morphological data (Haszprunar, 1996; Rouse & Fauchald, 1997), molecular data (Eeckhaut et al., 2000), or both (Zrzavý et al., 2001). This work is a review of myzostomid ultrastructure and phylogeny. Only analyses of adult ultrastructure are mentioned, the fine structure of myzostomid larvae having been investigated recently (Eeckhaut et al., 2003). Some unpublished data from Master and Ph.D. theses are included. An overview of the anatomy, taxonomy, and ecology of myzostomids can be found in Grygier (2000).

Phylogenetic position within the Metazoa

The phylogenetic position of myzostomids within metazoa has been a subject of controversy since their discovery. Myzostomids have been first considered as Trematoda (Leuckart, 1827), Crustacea (Semper, 1858) or Stelechopoda (i.e., a taxon grouping myzostomids with Tardigrada and Pentastomida; Graff, 1877). Benham (1896) was the first to suggest them as a separate class of annelids (rather than derived polychaete annelids), a position supported later by other investigators (e.g., Fedotov, 1929; Kato, 1952). Jägersten (1940) grouped the Myzostomida and the Annelida (as two separate classes) into a coelomate protostome clade called Chaetophora. More recently, Mattei & Marchand (1987), based on ultrastructural similarities between the spermatozoa of Myzostomida and Acanthocephala, considered these two taxa as sister groups defining a phylum they called Procoelomata. Because myzostomids exhibit characters such as parapodia with chaetae and acicula, a trochophora-type larva, and a segmentation (though incomplete), they are classified in all textbooks and encyclopaedias as a family or an order of Polychaeta or as a class of Annelida. This suggestion is also based on the very intuitive idea that the body plan of ectocommensal myzostomids evolved from that of an errant polychaete ancestor

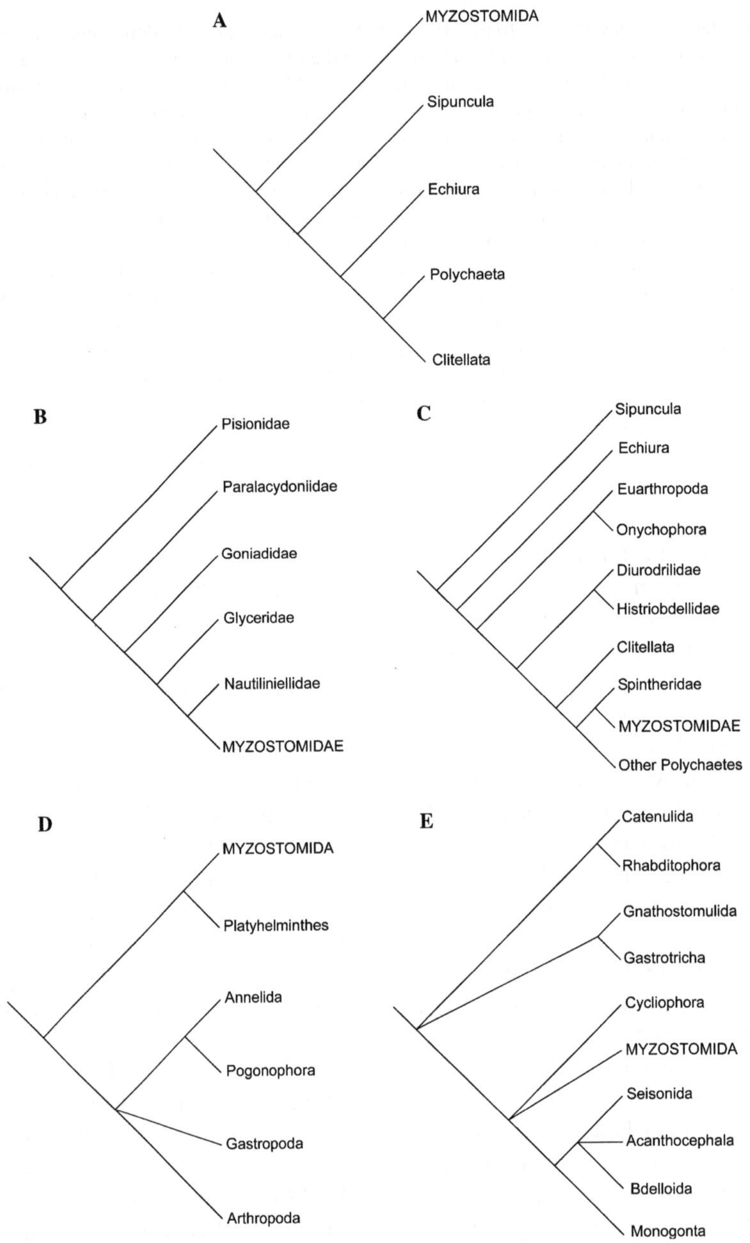

Figure 2. Dendrograms illustrating the results of the analyses concerning the phylogenetic position of the Myzostomida within the Metazoa. The analyses are those of Haszprunar (1996) (A), Rouse and Fauchald (1997) (B and C), Eeckhaut et al. (2000) (D) and Zrzavý et al. (2001) (E).

that lost the notopodial chaetal system, with a migration of the neuropodia onto the ventral side of the body. The overdevelopment of the ventral side of myzostomids and the regression of locomotory and sense organs could have happened once or several times and led to the body plan(s) of parasitic myzostomids.

During the last seven years, eight phylogenetic analyses based on phenotypical and/or molecular characters have included data from the Myzostomida. The first two analyses were based on morphological data. Haszprunar (1996) made a cladistic analysis to point out molluscan affinities with other Bilateria. His analyses suggested that the Myzostomida are the sister group of a clade including the Sipuncula, Echiura, and Annelida (Fig. 2A). Rouse & Fauchald (1997) investigated polychaete systematics by means of cladistic analyses including myzostomids as a family (i.e., the Myzostomatidae). The different analyses supported the placement of the myzostomids as a family nested within the Polychaeta; the Myzostomatidae were either associated to the Spintheridae (a family including ectoparasitic species of sponges) or they clustered with polychaetes with a hypertrophied axial pharynx (Fig. 2B and C).

Chenuil et al. (1997) were the first to obtain a DNA sequence from a myzostomid: they sequenced a small part of the large ribosomal subunit RNA gene (LSU hereafter) of *Myzostoma* sp. They compared the secondary structure of this segment with that of other metazoans including polychaetes and oligochaetes and pointed out the difference existing between myzostomids and annelids. Zrzavý et al. (1998) used phenotypical (morphological, ultrastructural, developmental and ecological characters) and molecular data (DNA sequences coding for the RNA included in the small ribosomal subunit, SSU hereafter) in their cladistic analyses. No SSU sequences were available for myzostomids and the analyses comprised only phenotypical characters for this group. The analyses suggested that the Myzostomida are the sister group of a clade including Echiura, Pogonophora, and annelids.

Eeckhaut et al. (2000) inferred the phylogenetic position of the Myzostomida within the Metazoa by analysing the DNA sequences of two slowly-evolving nuclear genes: the SSU and the elongation factor-1α (EF-1α hereafter). Eeckhaut et al. (2000) sequenced five species of myzostomids (*Endomyzostoma clarki*, *Myzostoma cirriferum*, *M. fissum*, *Notopharyngoides aruense*, and *Contramyzostoma sphaera*) for SSU sequences and two species (*Myzostoma alatum* and *Pulvinomyzostomum pulvinar*) for EF-1α sequences. All analyses yielded best trees with the Myzostomida not nested within the Annelida and suggested that myzostomids are closer to flatworms than they are to annelids (Fig. 2D).

Zrzavý et al.'s (2001) analyses comprised SSU and LSU sequences of *Myzostoma glabrum*, together with phenotypical characters. The analyses showed myzostomids as the sister group of Cycliophora, closely related to the rotifer-acanthocephalan clade (=Syndermata) (Fig. 2E). The myzostomid-cycliophoran-syndermate clade, accommodated within the Platyzoa, was strongly supported by most analyses. Zrzavý et al. (2001) proposed the new name Prosomastigozoa for this group, due to the presence of highly derived spermatozoa with an anteriorly directed flagellum, at least present in myzostomids and syndermates (cycliophoran sperm ultrastructure was insufficiently known).

Recently, two works inferred the phylogenetic position of non-myzostomidan clades but used myzostomidan sequences in the ingroup. Littlewood et al. (2001) used EF-1α amino-acid sequences to estimate the position of the Acoela within the Metazoa. Arthropods, vertebrates, molluscs, myzostomids, and annelids plus pogonophorans were each represented as monophyletic groups. Platyhelminthes appeared to be a polyphyletic group and Myzostomida were found outside the Annelida. Rota et al. (2001) inferred the phylogenetic position of the only two truly terrestrial non-clitellate annelids, *Parergodrilus heideri* and *Hrabeiella periglandulata*, using a data set of new 18S rDNA sequences as well as homologous sequences already available for 18 polychaetes including one species of Myzostomatidae, *Myzostoma* sp. The results included Mollusca, Sipuncula, Echiura, and Pogonophora, as well as *Myzostomum* sp. within the Polychaeta in all of the most parsimonious trees but without bootstrap support for the position of the myzostomids.

In summary, three possibilities exist concerning the phylogenetic position of the Myzostomida within the Metazoa: (1) phylogenetic analyses considering the Myzostomida as a group outside the Annelida are wrong and similarities between the two groups are true synapomorphies; these analyses are right and the similarities between the two groups are either (2) convergences or (3) plesiomorphies.

Integument

The ultrastructure of the regular (i.e., non-sensory) integument of myzostomids is known in four species: *Myzostoma cirriferum* (Eeckhaut & Jangoux, 1993), *Contramyzostoma bialatum* (Eeckhaut

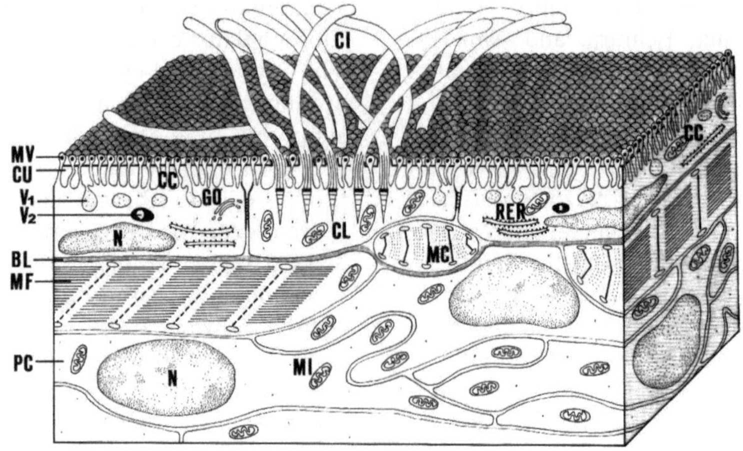

Figure 3. Myzostoma cirriferum. Schematic drawing of a portion of a longitudinal section through a non-sensory area of the integument (Eeckhaut & Jangoux, 1993). Abbreviations: BL – basal lamina; CC – covering cell; CI – cilium; CL – ciliated cell; CU – cuticle; GO – Golgi apparatus; MC – myoepithelial cell; MI – mitochondrion; MF – muscular fibre; MV – microvillus; N – nucleus; PC – parenchymal cell; RER – rough endoplasmic reticulum; V1 and 2 – vesicles of 1st and 2nd types.

& Jangoux, 1995), *Pulvinomyzostomum pulvinar* (Kronenberg, 1997), and *Myzostoma alatum* (Kronenberg, 1997). In all of them, the integument consists of an epidermis with cuticle and a parenchymo-muscular layer that extends between the internal systems (i.e., the digestive, nervous, excretory and genital systems). A mesothelium lining a coelom is not evident in myzostomids, although the female genital tract and ovaries are sometimes considered as coeloms. There is no mineralized skeleton and the only hard parts are the chitinous hooks and aciculae of parapodia that have not yet been studied ultrastructurally except in larvae (Eeckhaut et al., 2003).

Non-ciliated and ciliated cells form most of the myzostomid epidermis (Fig. 3). They have been observed in the four species cited here above and in *Contramyzostoma sphaera*, *Myzostoma cuniculus*, *M. laingense*, *M. horologium*, and *Notopharyngoides aruense* (unpublished observations). The cuticle of all these species is ca.1 μm thick and consists of several superposed fibrillar layers, the outer ones being more electron-dense. The cuticle is crossed by numerous microvilli that end in bulges (Fig. 3). Non-ciliated cells, also called covering cells, are flattened to cylindrical, and mainly characterised by having two types of vesicles (Fig. 3). Vesicles of the first type are spherical to ovoid in shape and contain electron-dense material that forms a dark margin. Vesicles of the second type are generally ovoid and are located in the most apical part of the cytoplasm, just under the cuticle. They are either empty or full of a granular material and are thought to participate in the formation of the cuticle, while the role of the vesicles of the first type is unknown. Ciliated cells are of similar shape and size to covering cells and bear from 20 to 50 cilia (Fig. 3). The cilia are 10–20 μm long, have the classical microtubular arrangement ($9 \times 2 + 2$ microtubules), and are each prolonged by a ciliary rootlet. Most species have usual, tube-like cilia but paddle-like cilia have been observed in *M. jagersteni* (Eeckhaut et al., 1994), *M. fissum*, and *M. ambiguum* (Lanterbecq, 2000). The proportions of covering and ciliated cells vary according to the myzostomid species considered and this seems to be related to the type of association existing between the myzostomid and the host: in ectocommensals, ciliated cells are sparse with a ratio of ciliated cells to non-ciliated cells of about 1:5. At the opposite extreme, the trunk of intradigestive myzostomids such as *P. pulvinar* and *N. aruense* is almost totally covered by cilia (CC/NC ratio of about 1:1). Ciliary beating induces a water current at the surface of the individuals and surely facilitates the intake of both oxygen and dissolved nutrients through the integument. The overdevelopment of epidermal cilia in intradigestive myzostomids could be an adaptation to their symbiotic way of life.

Two types of gland cell have been observed in the myzostomid epidermis. The first type is mainly found in parapodia, in cirri, and in the folds that surround lateral organs. It has been observed in *M. cirriferum* (Eeckhaut & Jangoux, 1993), *P. pulvinar* (Kronenberg, 1997), *M. alatum* (Kronenberg, 1997), and in *N. aruense*, *M. laingense*, and *M. horologium* (unpublished observations). These cells are cylindrical in shape and their cytoplasm is full of ovoid vesicles including Alcian Blue-positive granular material. These gland cells most likely release mucous for protecting places where contact with the substratum or the host is frequent (Eeckhaut & Jangoux, 1993). The second type of gland cell has been observed in the villous and ciliated central part of the lateral organs of *M. cirriferum* (Fig. 5). This type has a large cell body resting under the other epidermal cells, and an elongated apical process running to the apex of the epidermis. The cytoplasm is full of spherical vesicles filled with a homogenous, finely granular, Alcian Blue-positive material. The vesicle content has various electron densities and some vesicles appear empty.

Muscle fibre cells, called myoepithelial cells by Eeckhaut & Jangoux (1993), were observed in the epidermis of *M. cirriferum* (Eeckhaut & Jangoux, 1993) and *C. bialatum* (Eeckhaut & Jangoux, 1995), but they were not found in *M. alatum* nor in *P. pulvinar* (Kronenberg, 1997). They are thin, elongated muscle cells lying under the covering and ciliated cells (they never contact the cuticle) in such a way that their longitudinal axis is perpendicular to the antero-posterior axis of the myzostomid (Fig. 3). They are particularly abundant at the level of the introvert (Eeckhaut & Jangoux, 1993), where their main action is to reduce the introvert diameter and its length while inducing deep pleats when they contract.

The parenchymo-muscular layer includes muscle cells and parenchymal cells. Dermal muscles are conspicious ventrally as well as in the introvert and parapodia; they are less developed dorsally. They consist of longitudinal, circular, and transverse muscle cells of the double-obliquely striated type. In addition, dorso-ventral muscles, which form septa, occur in the trunk. The thickness of the parenchyma varies according to the development of the gonads (it is the thinnest when gonads are mature). The parenchyma is particularly well developed in the dorsal part of the myzostomid trunk where parenchymal cells occur in contact with developing female gametes. In the dorsal side of the trunk of *Myzostoma gopalai*, the parenchyma is completed by a wall of collagen fibres that has only been analysed by histochemical methods (Rao & Sowbhagyavathi, 1972). Parenchymal cells have a highly variable size and shape. Their cytoplasm includes very few organelles. The external matrix that surrounds parenchymal cells is generally poorly developed and is made of a granular material.

Nervous system and epidermal sensory regions

The anatomy of the nervous system has been described most thoroughly for several species of

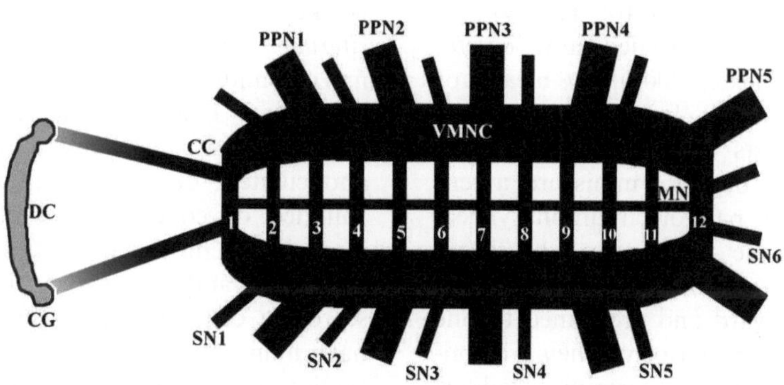

Figure 4. Myzostoma cirriferum. Schematic drawing of the central nervous system observed in the trunk of individuals (drawn from the pictures of Müller & Westheide, 2000). Abbreviations: CC – circumpharyngeal connective; CG – cerebral ganglia; DC – dorsal commissure; MN – median nerve; PPN 1–5 – parapodial nerves one to five; SN 1–6 – side nerves one to six; VMNC – ventral main nerve cord. The numbers 1–12 indicate the twelve commissures that connect the two ventral main nerve cords. Black and grey structures are ventral and dorsal, respectively.

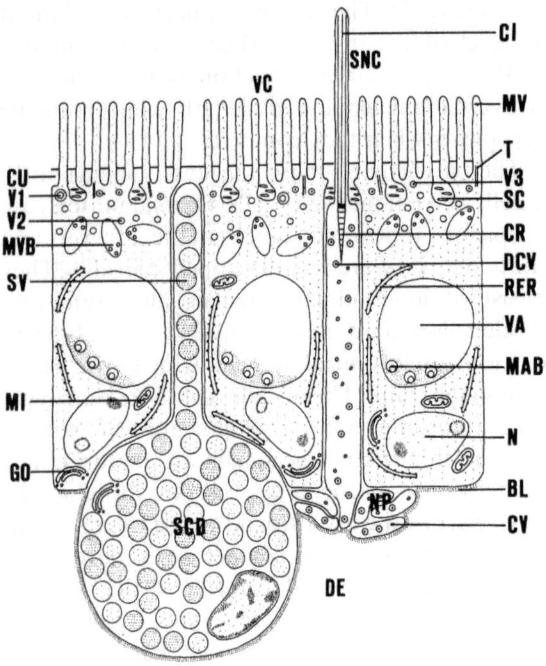

Figure 5. Myzostoma cirriferum. Schematic drawing of a section through the epidermis of the lateral organ dome (not to scale) (Eeckhaut & Jangoux, 1993). Abbreviations: BL – basal lamina; CI – cilium; CR – ciliary rootlet; CU – cuticle; CV – clear vesicle; DCV – dense cored vesicle; DE – dermis; GO – golgi apparatus; MAB – multiannular body; MI – mitochondrion; MV – microvillus; MVB – multivesicular body; N – nucleus; NP – nerve process; RER – rough endoplasmic reticulum; SC – subcuticular chamber; SCD – secretory cell of the dome; SNC – sensory cell; SV – secretory vesicle; T – tonofilaments; V1, V2 and V3 – vesicles of the 1st, 2nd and 3rd type; VA – vacuole; VC – vacuolar cell.

Myzostoma (see Grygier 2000 for review). Its ultrastructure has only been studied in *M. cirriferum* (Müller & Westheide, 2000), by means of confocal laser microscopy. In this species, the nervous system consists of two small cerebral ganglia connected by a dorsal commissure, a ventral nerve mass, and a pair of circumpharyngeal connectives joining the former to the latter (Fig. 4). The two neuropil cords within the ventral nerve mass curve outward and are joined to one another anteriorly and posteriorly. They are connected by twelve commissures forming a ladder-like system (Fig. 4). A single median nerve runs along the midventral axis (Fig. 4). In addition to the circumpharyngeal connectives, eleven peripheral nerves arise from each cord (Fig. 4). The first innervates the anterior body region. The others form five groups of two nerves each, the first and thicker nerve of which is the parapodial nerve, innervating the parapodium and two corresponding cirri (Fig. 4). Except for those of the most posterior group, the second nerves innervate the lateral organs and the trunk margin. One pair of dorsolateral longitudinal nerves was visualized by tubulin staining. The arrangement of the peripheral nerves and commissures strongly suggests that the myzostomid body is made of six segments (Müller & Westheide, 2000).

Sensory regions of the epidermis either show small structural variations from the regular epidermis or markedly differ from it (Eeckhaut & Jangoux, 1993). Small variations occur in the cirri, buccal papillae that surround the mouth of some species, body margin and parapodia, where ciliated sensory cells insinuate between epidermal cells. These are supposed to be mechanoreceptor sites that give information on the structural variations of the host integument; they could also participate in self-recognition of individuals (Eeckhaut & Jangoux, 1993). Ciliated sensory cells are dendritic processes of nervous cells, the cell bodies of which lie mostly in the ventral nerve cord. They usually run either singly or in pairs from bundles of basiepidermal nerve processes and reach the apex of the epidermis. Each sensory process bears up to five small cilia that cross the cuticle and have the usual microtubular arrangement ($9 \times 2 + 2$). Their basal body is prolonged by a ciliary rootlet. The sensory epidermis in the four pairs of lateral organs differs markedly from the regular epidermis. In *M. cirriferum* (Eeckhaut & Jangoux, 1993) and other myzostomids, there are almost always four pairs of ventral lateral organs alternating with the parapodia. A lateral organ consists of a villous and ciliated, dome-like central part that is surrounded or covered by a peripheral fold. The dome-like central part is the sensitive region of the organ and consists in ciliated sensory cells, secretory cells, and complex vacuolar cells that have numerous long microvilli and multivesicular bodies (Fig. 5). Lateral organs are presumed to allow the myzostomids to recognize the host integument and prevent them from becoming displaced onto the surrounding inhospitable substratum (Eeckhaut & Jangoux, 1993).

Digestive system

Ectocommensals of crinoids and many parasitic myzostomids that induce cysts or galls on crinoids feed on particles that they divert thanks to their introvert (i.e., most species of the genus *Myzostoma*; all the species of *Contramyzostoma* and *Endomyzostoma*). *Mesomyzostoma*, *Mycomyzostoma,* and *Protomyzostoma* species feed on host tissues, which is probably also the case for the *Asteromyzostomum* species. *Pulvinomyzostomum pulvinar*, *Asteriomyzostomum asteriae*, and the *Notopharyngoides* species live in the digestive lumen of their hosts where they ingest particles. No indication about the feeding behaviour of *Asteriomyzostomum fisheri* and *Stelechopus hyocrini* exists.

The anatomy of the digestive system of the Myzostomida is very similar in almost all species. It consists of a pharynx which is included in the introvert, a stomach with usually two or three pairs of blind, branching caeca, and an intestine. The only exceptions to this general scheme are: (1) there are no digestive diverticula in the female of *Mycomyzostoma calcidicola* and in *Stelechopus hyocrini*, (2) there are one right and one left U-shaped digestive diverticula in *Contramyzostoma bialatum*, (3) the male of *Mycomyzostoma calcidicola* has no digestive system; (4) the introvert is absent in the representatives of the Pharyngidea. The ultrastructure of the myzostomid digestive system is only known for *Myzostoma cirriferum* (Eeckhaut et al., 1995). All the cell types described in the digestive system of this species have, however, also been observed in *Contramyzostoma sphaera*, *Myzostoma capitocutis*, *Myzostoma cuniculus*, and *M. horologium* (unpublished observations).

The myzostomid pharyngeal epithelium is covered by a cuticle similar to the one that covers the epidermis. The structure of the epithelium differs on the lip (i.e., the part that surrounds the mouth) and in the pharynx *sensu stricto* (Fig. 6A and B). The lip is made of ciliated sensory cells, supporting cells, and salivary gland cells (Fig. 6A). Supporting cells are goblet-shaped cells made of an upper part, contacting the cuticle, and an inner part where the nucleus lies, both being connected by a thin cell process. The upper part of the cell is filled with Alcian blue-positive, ovoid vesicles. Salivary gland cells are common in myzostomids; they have been described in representatives of all genera except in *Mycomyzostoma calcidicola* (Eeckhaut, 1998) and *Pulvinomyzostomum pulvinar* (Jägersten, 1940). They lie in the parenchyma at the junction between the pharynx and the stomach. In *M. cirriferum*, they are about 20 in number, each being made of a large cell body from which starts a long, narrow cell process that ends on the lip. The cytoplasm is full of vesicles filled with electron-dense material that is assumed to be released in the pharyngeal lumen to digest food particles (Eeckhaut et al., 1995). Behind the lip, the pharyngeal lumen is only bordered by secretory cells that look very similar to epidermal covering cells but have many more apical, Alcian blue-positive, electron-dense vesicles (Fig. 6B). Myoepithelial cells, similar to those of the epidermis, have been observed in the pharynx of *M. cirriferum* (Eeckhaut & Jangoux, 1993).

Eeckhaut et al. (1995) proposed a model explaining how food particles are swallowed and carried into the stomach in ectocommensal species. When in search of food, the introvert continuously retracts and protrudes until it is applied at an appropriate place. Retraction results from the contraction of the longitudinal muscle cells that extend through the parenchyma of the introvert. The mechanism for protrusion appears to be more indirect because the introvert does not contain any antagonistic muscular or fibrillar (e.g., collagen fibre) system. The introvert of myzostomids has no internal cavity and changes in volume are due to the fact that muscles and parenchymal cells move from the introvert to the trunk and *vice versa*. The initiator of protrusion is probably located outside the introvert and could be the dorso-ventral muscle cells of the trunk; protrusion would depend on the forcing action of these muscle cells, which would push the pharynx and parenchyma into the relaxed introvert. Once the introvert is extended into an ambulacral groove of the host, food particles are swallowed due to the action of the muscle cells of the lip first, then those of the pharynx. The contraction of radial muscle cells of the lip increases the mouth diameter, thus sucking up food particles from the host's groove. The muscle sheet of the pharynx, made mainly of alternating radial and circular muscle cells, then starts to work. The

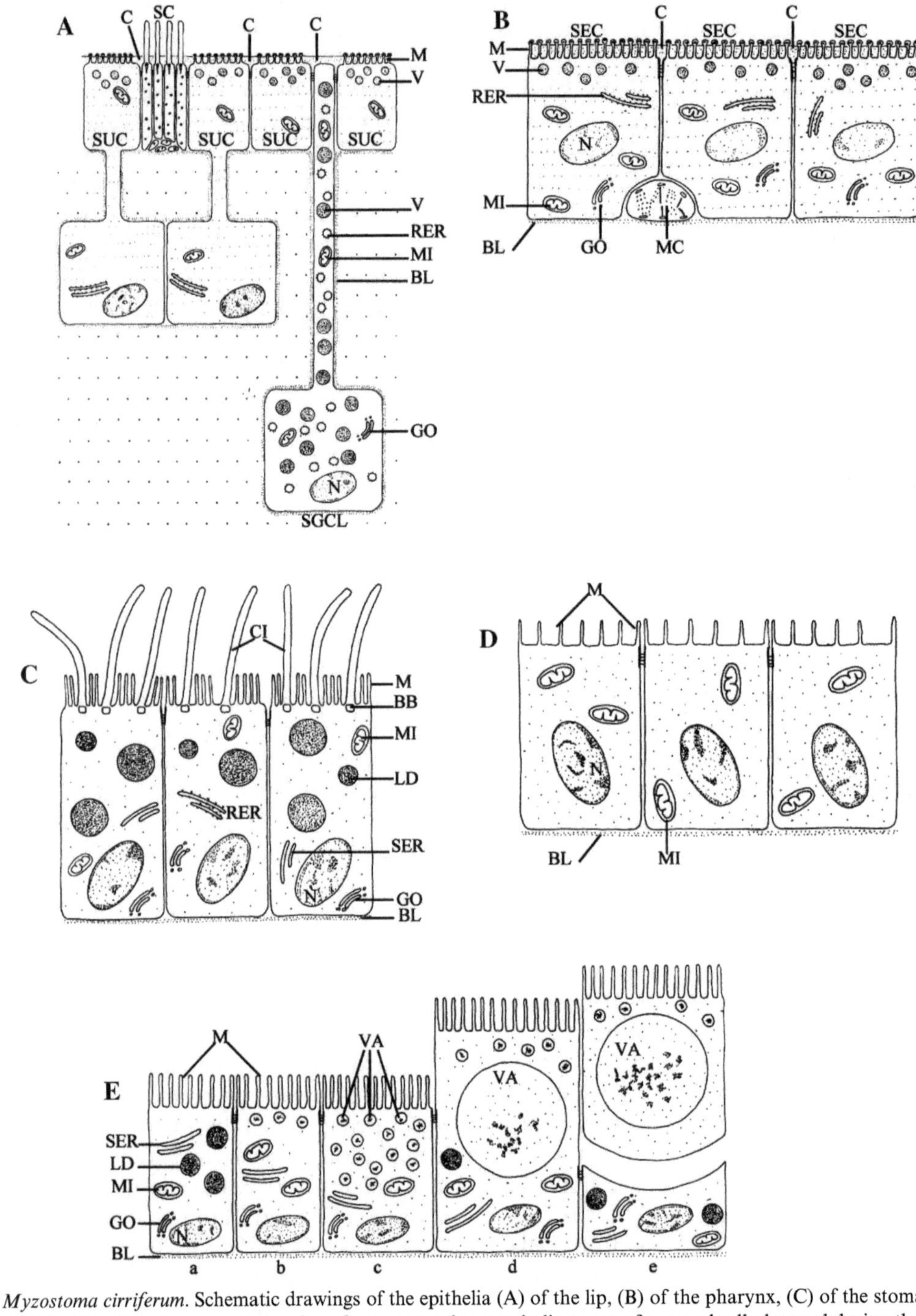

Figure 6. Myzostoma cirriferum. Schematic drawings of the epithelia (A) of the lip, (B) of the pharynx, (C) of the stomach, (D) of the intestine and (E) of the digestive caeca (a,b,c,d,e refer to successive metabolic stages of a caecal cell observed during the intradigestive process). Abbreviations: BB – basal body; BL – basal lamina; C – cuticle; CI – cilium; GO – Golgi apparatus; LD – lipidic droplet; M – microvillus; MC – myoepithelial cell; MI – mitochondria; N – nucleus; RER – rough endoplasmic reticulum; SC – sensory cell; SEC – secretory cell; SER – smooth endoplasmic reticulum; SGCL – salivary gland cells of the lip; SUC – supporting cell; V – vesicle; VA – vacuole.

contraction of radial cells enlarges the diameter of the pharyngeal lumen (sucking up the food particles) while the contraction of circular muscle cells reduces the diameter of the pharyngeal lumen (pushing down the food particles). The resulting peristaltic wave carries food particles into the inner organs of the digestive system, namely the stomach, the digestive caeca, and the intestine, each possessing a non-cuticularized epithelium surrounded by circular muscle cells.

The epithelium of the myzostomid stomach is made of a single cell type that is ciliated and cylindrical (Fig. 6C). In *M. cirriferum*, these cells bear many microvilli and several cilia that have the usual 9 × 2 + 2 microtubular arrangement and no rootlet below their basal body. Their cytoplasm includes numerous lipidic droplets that disappear when individuals are starving. Food particles carried by the pharyngeal peristalsis are conveyed towards the digestive caeca due to the action of the stomachal ciliature. Once in the caeca, particles are endocyted by caecal cells, the single cell type occurring in the caecal epithelium, and transferred into a single large vacuole – *viz.* a phagolysosome – in which they are digested. That vacuole regularly increases in size due to its fusion with additional phagosomes (Fig. 6E). When it has reached a size roughly corresponding to half the caecal cell volume, the vacuole, together with a fringe of cytoplasm that surrounds it, is expelled into the caecal lumen by an apocrine process. Detached cell fragments are forced out of the caecal lumen to the stomachal lumen due to a contraction of the caecal musculature. The cell fragments progressively gather together in the stomachal lumen, being embedded in an alcian blue-positive agglutinating matrix that is supposedly produced as a secretion of the pharyngeal secretory cells. A spindle-shaped faecal mass is finally formed, transferred to the intestine, and expelled to the outside by the contraction of the stomachal and intestinal musculatures. The intestinal epithelium is made of a single flattened cell type that harbours many microvilli but lacks cilia (Fig. 6D).

Excretory system

The excretory system of myzostomids consists of protonephridia whose ultrastructure has been described in *M. cirriferum* for both adults (Pietsch & Westheide, 1987) and larvae (Eeckhaut et al., 2003). The presence of protonephridia has not yet been investigated in any other species. In adults of *M. cirriferum*, five pairs of protonephridia were first described in the trunk (Pietsch & Westheide, 1987), but a closer examination using confocal laser scanning microscopy revealed six pairs of 90 μm long, S-shaped protonephridia in this species (Müller & Westheide, 2000). Nephridiopores are located ventrally in two rows of six that are parallel to the sagittal plane. Each protonephridium comprises three terminal cells and one duct cell (Fig. 7). Each terminal cell has six to nine flagella (without rootlets and with a 9 × 2 + 2 microtubular arrangement), which are each surrounded by rod-like cell processes. The duct cell bears microvilli and cilia. Weir-like fenestrations in the peripheral wall of the terminal cells make up the connection between the central lumina and the parenchymal extracellular space.

Usually, a pair of ducts connecting the uterus to the intestine occur in myzostomids and these have long and erroneously been named metanephridia. Spermatozoa have been observed in the lumen of these ducts and the only obvious function that can actually be assigned to them is to expel the excess of free spermatozoa, thus possibly preventing polyspermy during fertilization.

Oogenesis and female genital system

Most myzostomids are hermaphrodites. They often are functional simultaneous hermaphrodites even though the male genital system develops a bit earlier than the female genital system during organogenesis. In some other species, a male and a female are found and are often interpreted as the two stages of a protandrous hermaphroditic species, the dwarf male being supposed to transform into a female once it lives alone. Only *Mycomyzostoma calcidicola* is considered to be dioecious (Eeckhaut, 1998). The breeding period is only known for the European species *M. cirriferum* in which it extends throughout the year (Eeckhaut & Jangoux, 1997).

The female genital system of adult myzostomids consists of a branched duct and a diffuse

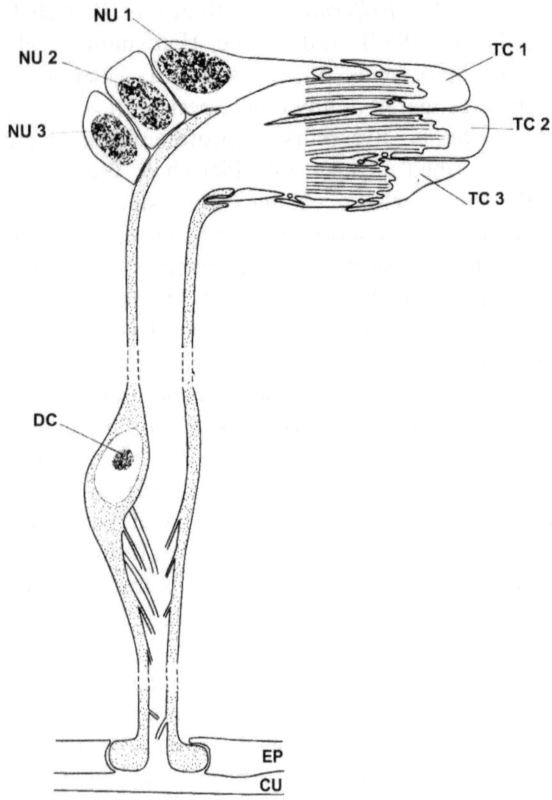

Figure 7. Myzostoma cirriferum. Schematic drawing of a protonephridium (Pietsch & Westheide, 1987). Abbreviations: CU – cuticle; DC – duct cell; EP – epidermal cell; NU 1–3 – nuclei of terminal cells; TC 1–3 – terminal cells.

ovary (i.e., female germinal cells growing in between parenchymal cells), both located in the dorsal part of the trunk in most species. The female genital system is found ventrally or has an ovary developed both ventrally and dorsally in some other species, however, usually, the branched genital duct follows the course of the digestive organs of the trunk (i.e., intestine, stomach, and digestive caeca). The ovary lies above it in such a way that a section through the dorsal part of most individuals will show the epidermis, then the parenchyma with developing female germinal cells, below that the genital duct and finally the digestive system. In fully mature specimens, germinal cells often invade the whole dorsal part of the trunk, with parenchymal cells scattered between them. The ultrastructure of the developing germinal cells and that of the genital duct have only been described for *Myzotoma cirriferum* (Eeckhaut, 1995). In this species, the genital duct consists of a uterus and uterine diverticula. The uterus is a sagittal cavity that lies above the stomach and intestine and opens to the outside through a postero-ventral gonopore. The uterine diverticula branch off from the anterior part of the uterus, dichotomise and end at the body margin. According to the state of maturity of individuals, the uterus and uterine diverticula may be flat and empty or thick and full of fertilised eggs. The lumen of both organs is bordered by an epithelium surrounded by a fine basal lamina 50 nm thick. The epithelium consists of ciliated cells and microvillous cells (Figs 8 and 9). The former cell type forms the dorsal wall of the uterine diverticula and the latter is observed in the ventral wall of the uterine diverticula and borders the uterus lumen both ventrally and dorsally. The ciliated cells are flattened, ca. 10 μm long and 3 μm thick at most, their periphery being extremely thin (Figs 8 and 9A). They are joined together by septate junctions and zonula adhaerentes and bear numerous cilia that have a $9 \times 2 + 2$ microtubular arrangement. Each cilium has a basal body and one ciliary rootlet with two branches that form an angle of ca. 150° (Fig. 9A). As a result of the current created by the ciliary beating, the ciliated cells drive fertilised eggs towards the uterus. The microvillous cells are ca. 10 μm long and 3 μm thick (Fig. 9A–C). In the uterine diverticula, they are separated from the caecal cells of the digestive system by the basal lamina; in the uterus, they are underlain by circular muscle cells whose contractions expel the fertilised eggs through the gonopore. They have numerous microvilli and may be storage cells; their cytoplasm sometimes includes lipidic droplets and possesses a well developed smooth endoplasmic reticulum (Fig. 9B).

The developing germinal cells are not separated by a basal lamina from the surrounding parenchymal cells, nor are they enclosed by an epithelium. Oogonia, previtellogenic oocytes, and vitellogenic oocytes occur in the ovary of adult myzostomids (Figs 8 and 10A–C). The older they are, the closer to the uterine diverticula they come, and vitellogenic oocytes pierce the epithelium of the diverticula and fall into the lumen where they are fertilised (Fig. 8). The oogonia are ovoid cells of 5–10 μm in diameter (Figs 8 and 10A). They have a voluminous, nucleolated nucleus where heterochromatin is condensed into

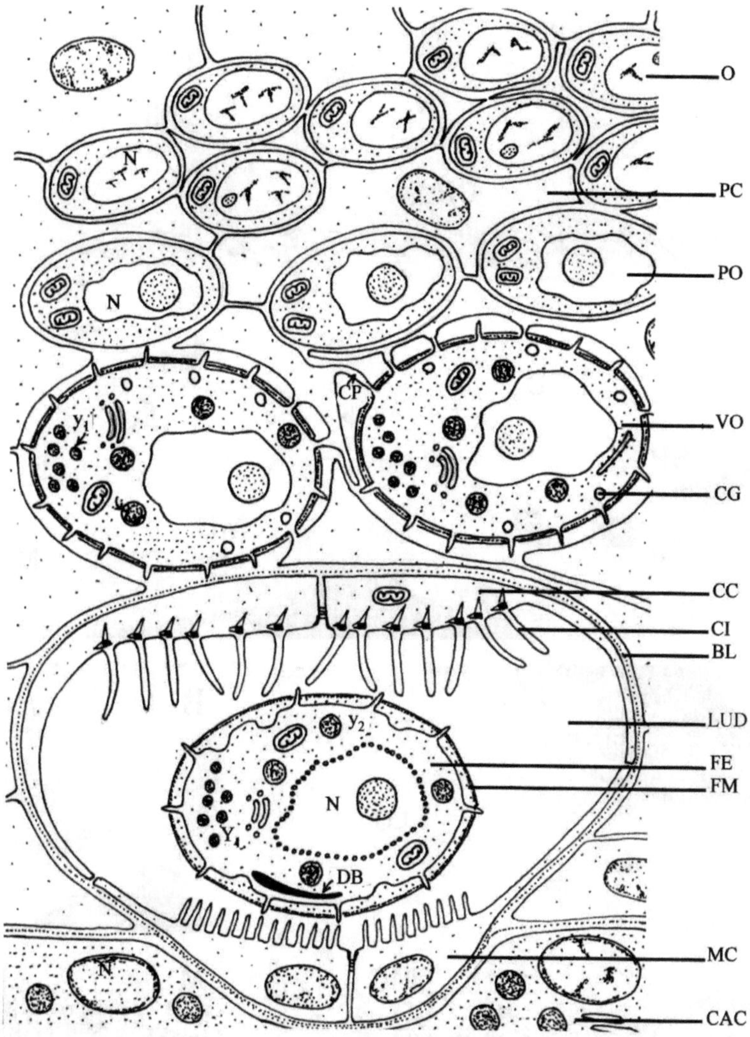

Figure 8. Myzostoma cirriferum. Schematic drawing of a portion of the ovary and of a uterine diverticulum (sectioned transversally). Abbreviations: BL – basal lamina; CAC – caecal cell; CC – ciliated cell; CG – cortical granule; CI – cilium; CP – cytoplasmic process; DB – dense body; FE – fertilised egg; FM – fertilising membrane; LUD – lumen of an uterine diverticulum; MC – microvillous cell; N – nucleus; PC – parenchymal cell; PO – previtellogenic oocyte; O – oogonium; VO – vitellogenic oocyte; $Y_{1,2}$ – yolk granules of types 1 and 2.

chromosomes. The cytoplasm includes some scattered mitochondria. Previtellogenic oocytes are cells of 10–20 μm in diameter with an irregular-shaped nucleus (Figs 8 and 10B). The cytoplasm includes mitochondria often concentrated at one side of the cell (Fig. 10B), a well developed rough endoplasmic reticulum and Golgi apparati. The cell membranes of both the oogonia and previtellogenic oocytes are smooth and deprived of microvilli. The vitellogenic oocytes are cells of 20–30 μm in diameter, with a nucleus similar in shape to that of the previtellogenic oocytes (Figs 8 and 10C). Mitochondria are scattered throughout the cytoplasm, which includes a highly developed rough endoplasmic reticulum (Fig. 10E). Two types of yolk granules occur within the cytoplasm: some are small granules of 100–500 nm in diameter, concentrated at the vegetal pole of the cell (Fig. 10F); the others consist of large granules of 1–3 μm in diameter scattered in the cytoplasm (Fig. 10G). Cortical granules of 0.5–1 μm in diameter are found at the

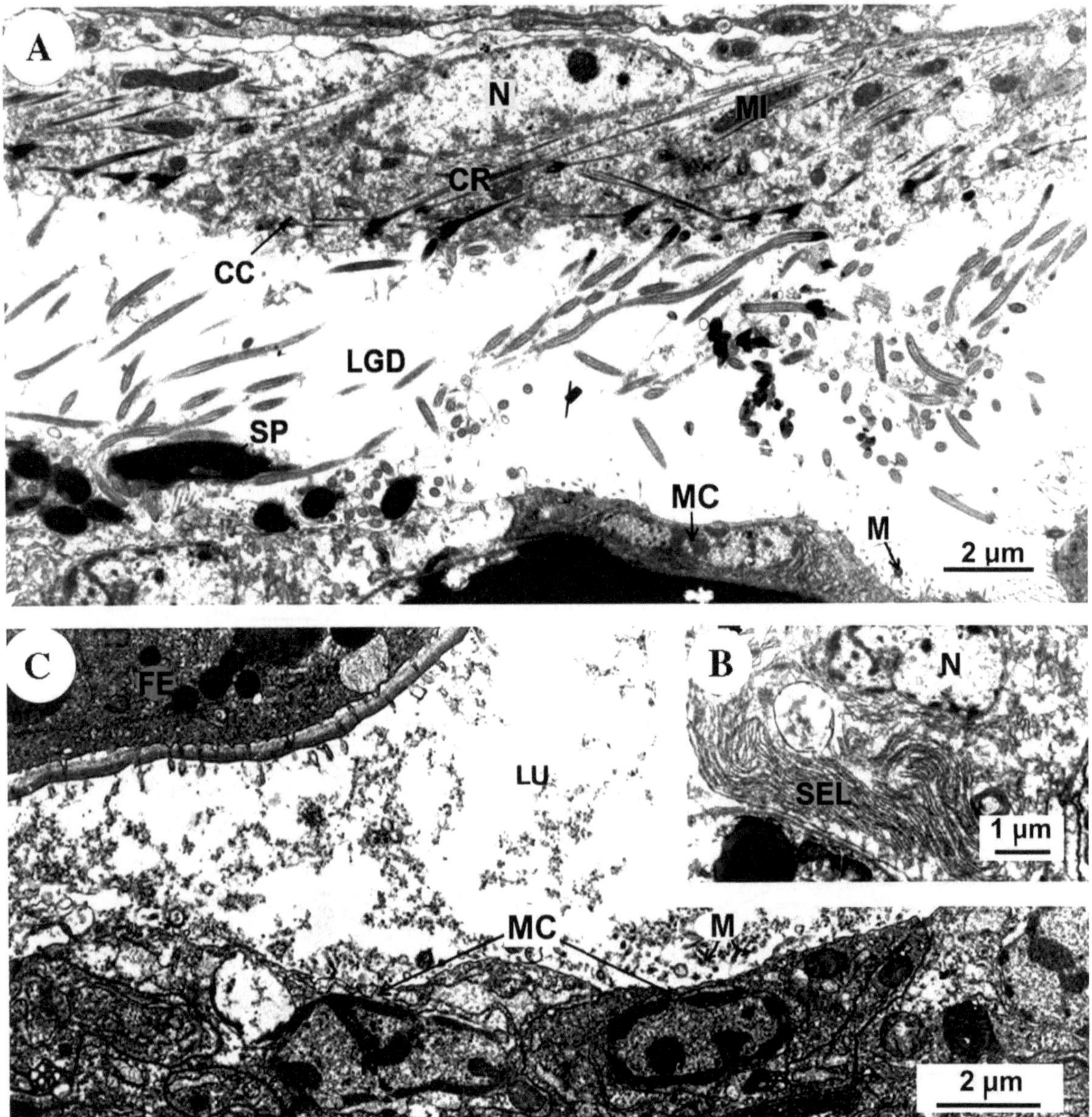

Figure 9. Myzostoma cirriferum. Female genital duct (TEM views). (A) transversal section of an uterine diverticula, (B) basal part of a microvillous cell and (C) sagittal section in the mid ventral part of the uterus. Abbreviations: CC – ciliated cell; CR – ciliary root; FE – fertilised egg; LGD – lumen of a genital duct; LU – lumen of the uterus; M – microvillus; MC – microvillous cell; MI – mitochondria; N – nucleus; SEL – smooth endoplasmic reticulum; SP – spermatozoon.

periphery of the cell just under the cell membrane, which is outlined by a thin vitelline envelope 100 nm thick (Fig. 10H and I). The vitelline envelope is crossed by microvilli, some of which forming cytoplasmic bridges between germinal cells and follicle cells (Fig. 10I and J). Cytoplasmic bridges are small cell processes that are sometimes filled with small vesicles (Fig. 10I). Except for the presence of these intercellular bridges, the ultrastructure of follicle cells does not differ from that of parenchymal cells: they do not show any sign of a particular metabolic activity. Nurse cells have been observed in some Myzostomatidae (Eckelbarger, 1992). They are abortive germ cells that maintain cytoplasmic continuity with the developing oocytes through intercellular bridges. According to Wheeler (1896), nurse cells are absorbed by the developing oocytes, but Jägersten (1939) suggested that nurse cells transform into follicle cells and eventually are disposed in several layers around the oocytes. Eeckhaut (1995) did not observe such nurse cells in *M. cirriferum*.

Fertilised eggs only occur in the lumen of the uterine diverticula and uterus (Figs 8 and 10D). They are 30 μm in diameter and have both types of yolk granules, mitochondria scattered throughout the cytoplasm, and cisternae of a highly developed rough endoplasmic reticulum and Golgi apparatus. The cytoplasm includes an electron-dense body that is supposed to come from the fertilising spermatozoon (Fig. 10D and N). Cortical granules are empty, their membrane having fused with the oolema (Fig. 10L). The fertilising envelope, 200 nm thick, is formed of the upper, old vitelline envelope and of a new, inner layer made of the material secreted by cortical granules (Fig. 10D and M). Microvilli cross the fertilising envelope. The nuclear membrane is fragmented into numerous small vesicles of 10 nm in diameter (Fig. 10K). In some eggs where the fragmentation is advanced, the vesicles are scattered into the nucleo-cytoplasm (Fig. 10O).

Spermatogenesis and male genital system

The male genital system of myzostomids is basically paired: there are usually two male genital apparati that are ventral and each separated from the other by the nerve cord. Each male genital apparatus consists of one or two diffuse testes (according to the species) and one genital duct, the latter consisting of a penis, a seminal vesicle, one or two vasa deferentia, and numerous vasa efferentia. Except for the vasa efferentia that are lined by a matrix, the lumen of all male organs is bordered by an epithelium, itself surrounded by muscles. Epithelial cells are joined together by zonula adhaerentes and septate junctions. Muscles form a continuous (at the level of the seminal vesicles) or discontinuous (at the level of the penises and vasa deferentia) sheath of circular muscle cells. The wall of the vasa efferentia is a thick 300 nm basal lamina (also named the tunica propria) that is continuous with the 50 nm thick basal lamina of the vasa deferentia.

Structural variability in the male genital system of myzostomids includes the presence of an epithelium lining at least part of the vasa efferentia in some species and the absence of seminal vesicles or their division into a narrow proximal duct and a sac-like distal portion in some others (Jägersten, 1939). Compact rather than diffuse testes occur in many Endomyzostomatidae and in *Stelechopus hyocrini* (Jägersten, 1939, see also Grygier, 2000).

The ultrastructure of myzostomidan spermatozoa and spermatogenesis has been extensively studied (Bargalló, 1977; Afzelius, 1983, 1984; Mattei & Marchand, 1987, 1988) but the fine structure of male genital ducts is only known for *M. cirriferum* (Eeckhaut, 1995). In this species, the lumen of the penis is bordered by non-ciliated and ciliated cells similar to those found in the epidermis (Fig. 11A–C). Their shape varies according to whether the penis is retracted or extended and will be cylindrical or flat, respectively. These cells are covered by a cuticle while all the other epithelial cells of the male genital system are not. Vacuolar and spumous gland cells form the epithelium of the seminal vesicles (Fig. 12A, C and E). The first are located at the base of each seminal vesicles, close to the penis (Fig. 12A and C). They are cylindrical cells, 20 μm high with a basal nucleus and cytoplasm full of large vacuoles of 5–10 μm in diameter. They include an electron-lucent fibrillar material. These vacuoles are expelled with the spermatophores that are formed in the seminal vesicles. They will form the base of it and will participate in the lysis of the integument of the receiver individual during the intradermic penetration of the germinal cells (described below). The rest of the seminal vesicle is lined by spumous gland cells that are flat cells 7 μm high (Fig. 12E). They contain vacuoles whose content is assumed to be secreted to form the matrix that coats the spermatophore contents. The vasa deferentia (Fig. 12B) are lined by vesicular gland cells. These are cylindrical cells 15 μm high with a basal nucleus; they have well-developed smooth endoplasmic reticulum and possess rod-shaped, electron-dense vesicles apically (Fig. 12F and G). These rod-shaped vesicles are secreted during the

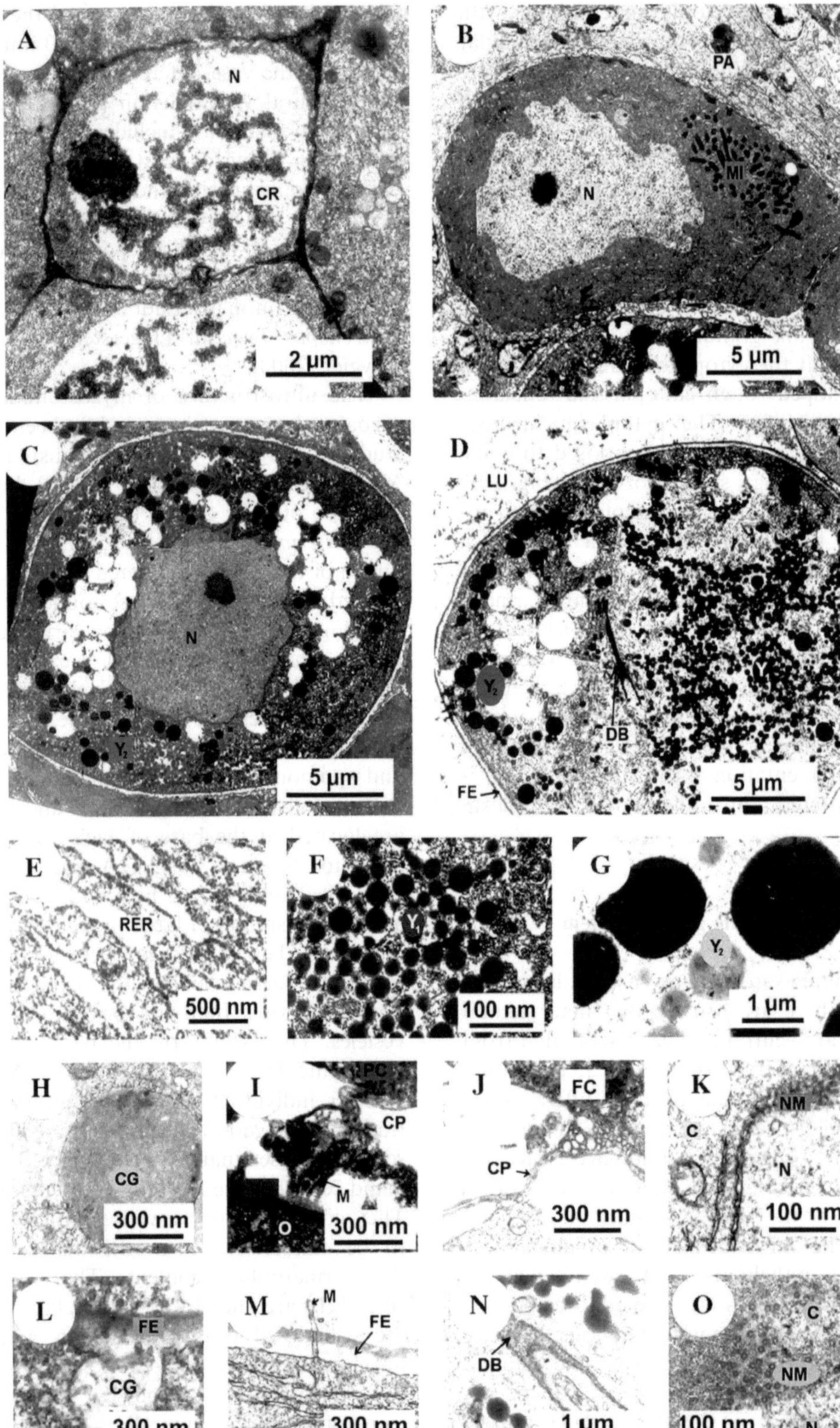
A
N
CR
2 µm
B
PA
MI
N
5 µm
C
N
Y2
5 µm
D
LU
Y2
DB
FE
5 µm
E
RER
500 nm
F
Y1
100 nm
G
Y2
1 µm
H
CG
300 nm
I
PC
CP
M
O
300 nm
J
FC
CP
300 nm
K
NM
C
N
100 nm
L
FE
CG
300 nm
M
M
FE
300 nm
N
DB
1 µm
O
C
NM
100 nm
N

Figure 10. Myzostoma cirriferum. Female germinal cells (TEM views). General views of (A) an oogonia, (B) of a previtellogenic oocyte, (C) of a vitellogenic oocyte and (D) of a fertilised egg. Details of a vitellogenic oocyte illustrating (E) the rough endoplasmic reticulum, (F) yolk granule of first and (G) second type, (H) a cortical granule, (I) vitelline envelope and microvilli, and (J) a cytoplasmic process between an oocyte and a parenchymal cell. Details of a fertilised egg illustrating (L) a cortical granule, (M) the fertilising envelope and microvilli, (N) the intracytoplasmic dense body and fragmentation of the nuclear membrane into numerous small vesicles (beginning and advanced stages of fragmentation: K and O, respectively). Abbreviations: C – cytoplasm; CG – cortical granule; CP – cytoplasmic process; CR – chromatin; DB – dense body; FC – follicle cell; FE – fertilised egg; LU – lumen of uterus; M – microvillus; MI – mitochondria; N – nucleus; NM – nuclear membrane; NU – nucleole; O – oocyte; PA – parenchyma; RER – rough endoplasmic reticulum; VE – vitelline envelope; $Y_{1,2}$ – yolk granules of types 1 and 2.

advance of the germinal cells to the seminal vesicles. They are also found in the lumen of the seminal vesicles, and at the base of the spermatophores between the vacuolar gland cells and the male germinal cells that came from testes. The vasa efferentia are dichotomously branched ducts connected to the testes; they convey the germinal cells to the vasa deferentia. The testes comprise many follicles scattered in the parenchyma. The follicles include many cysts, also called

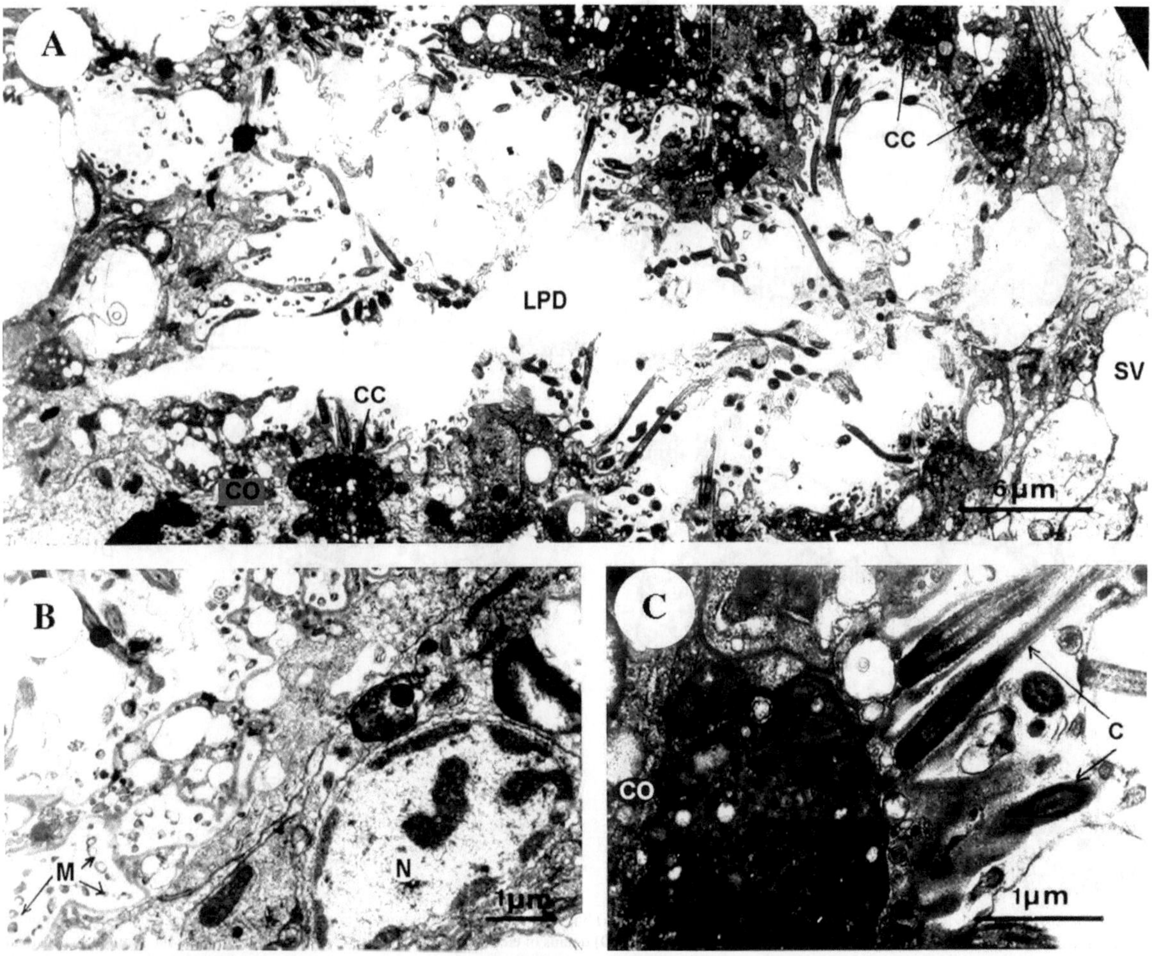

Figure 11. Myzostoma cirriferum. Male genital duct (TEM views). (A) General view of the epithelium bordering the retracted penial duct and details of (B) covering and (C) ciliated cells. Abbreviations: C – cilium; CC – ciliated cell; CO – covering cell; LPD – lumen of penial duct; M – microvillus; N – nucleus; SV – seminal vesicle.

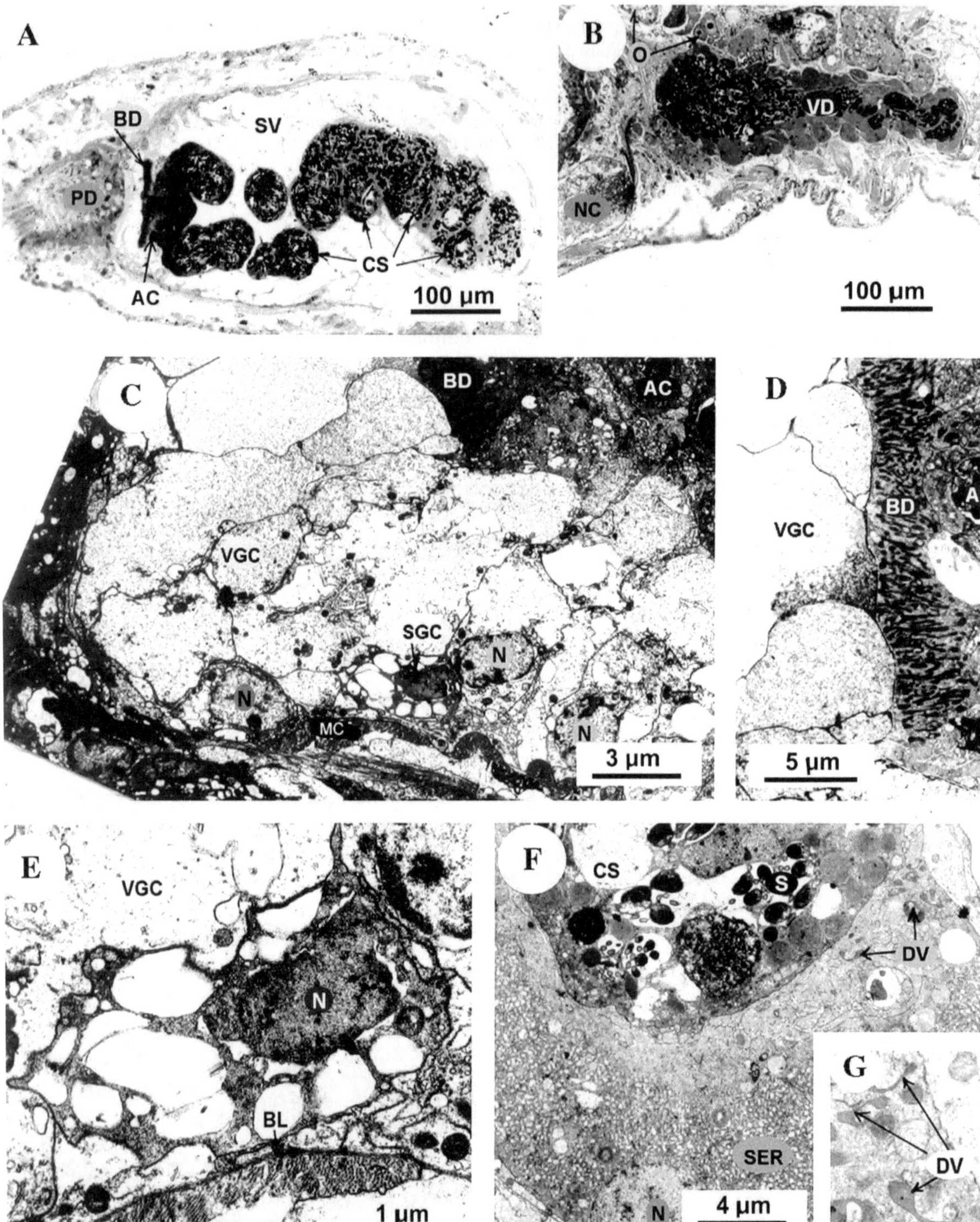

Figure 12. Myzostoma cirriferum. Male genital duct (OM and TEM views). (A) Sagittal sections through a penis and a seminal vesicle; (B) longitudinal section of a vas deferens; (C) aspect of the basal part of a seminal vesicle (near the penis); (D) details of electron-dense vesicles lying on the apex of vacuolar gland cells; (E) details of a spumous cell; (F) transverse section through a portion of a vas deferens with the lumen filled with spermatocysts and (G) details of their electron-dense, apical vesicles. Abbreviations: AC – abortive cyst; BD – basal disc with electron-dense vesicles; BL – basal lamina; CS – cyst containing spermatozoa; DV – dense vesicle; N – nucleus; NC – nervous chain; MC – muscle cell; O – ovary ; PA – parenchyma; PD – penial duct; S – spermatozoon; SER – smooth endoplasmic reticulum; SGC – spumous gland cell; SV – seminal vesicle; VD – vasa deferentia; VGC – vacuolar gland cell.

spermatocysts, each consisting of one cyst cell that encloses developing germinal cells. Cyst cells are ovoid cells of 20–40 μm in diameter. They contain many mitochondria, well-developed rough endoplasmic reticulum and numerous polysaccharide-rich vesicles. The number of germinal cells surrounded by a cyst cell varies according to the stage of development of the former (from 1 to 64 in *M. cirriferum*). Germinal cells that are surrounded by the same cyst cell are all at the same stage. Nothing is known about the cyst structure in young myzostomids but it is probable that, initially, a single germinal cell becomes surrounded by a cyst cell and divides later. The youngest cyst cells observed in mature myzostomids are those containing spermatogonia, which are ovoid cells of 5 μm in diameter (Fig. 13A). Spermatogonia have a large, nucleolated nucleus where the chromatin condenses into chromosomes. Young spermatids are still ovoid cells of 5 μm in diameter (Fig. 13B). In the periphery of the nucleus, intranuclear, electron-dense spheres form. These have been considered either as heterochromatin (Afzelius, 1984; Eeckhaut & Jangoux, 1991) or protein granules (Mattéi & Marchand, 1988). Spermatids gradually lengthen and acquire a flagellum with a $9 \times 2 + 0$ axoneme (Fig. 13C). At the end of spermiogenesis, the spermatids separate from the cytoplasmic remnants and transform into spermatozoa (Fig. 13D and E). Myzostomid spermatozoa are elongated cells about 30 μm long and 1 μm in diameter. They are provided with a long flagellum whose length is almost twice that of the spermatozoon body. The flagellum arises from one extremity of the cell; it bends as soon as it leaves the cell body and borders the latter along its whole length, being attached to the cell membrane through extracellular processes. It ends in a 40 μm long free portion that extends out opposite to the flagellar pole of the spermatozoon and that is supposed to beat forwards according to Mattei & Marchand (1988) . The spermatozoon nucleus is highly elongated and typically includes one row of 40–50 dense spheres. According to Mattei & Marchand (1987), the nuclear membrane is open in the spermatozoa which allows the nucleoplasm and cytoplasm to come into contact. One or two enlarged mitochondria and a manchette of 16–22 microtubules extend from one pole of the spermatozoon to the other. The microtubules are located between the nucleus and the cell membrane in the cytoplasm facing the attached part of the flagellum. On the opposite side, a myelin-like sheath presumably derived from the Golgi apparatus, caps the nucleus over its whole length. An acrosome is not obvious in *Myzostoma* spermatozoon.

Cysts that contain spermatogonia are located in the outermost ends of the follicles, close to the myzostomid body margin. With the division of spermatogonia, cysts occupy more space and push the rear cysts into the vasa efferentia first, then into the vasa deferentia. Only cysts that include spermatozoa enter vasa deferentia. There, contractions of circular muscle cells force the cysts into the seminal vesicles, where spermatophores form.

Spermatophores and reproduction

Reproduction in myzostomids is realised by the emission of spermatophores followed by the intradermic penetration of sperm cells. Emissions of spermatophores have been observed in *M. ambiguum* (Kato, 1952), *M. cirriferum* (Eeckhaut & Jangoux, 1991), *M. alatum* (Eeckhaut & Jangoux, 1992), and *M. capitocutis, M. nigromaculatum*, *M. polycyclus*, and *Contramyzostoma sphaer*a (unpublished observations). All are ectocommensal species except the last one, whose individuals live singly in cysts that they induce on the arms of their crinoid hosts. In ectocommensals, matings involve two mature individuals that contact each other, one of them ejecting one spermatophore that attaches to the integument of the other individual. The spermatophore in the other seminal vesicle does not participate in this process. Contacts between the two myzostomids are very brief and the two separate after mating. Spermatophores are generally attached to the back of the receiver but they can be emitted successfully to any part of the receiver's body. In most cases, receiver individuals have one attached spermatophore, sometimes two. In *C. sphaera*, gamete exchange occurs when an individual extends a very long penis to contact another individual lying in an adjacent cyst (unpublished observation).

Emitted spermatophore are white V-shaped, club-shaped, or ball-shaped baskets according to

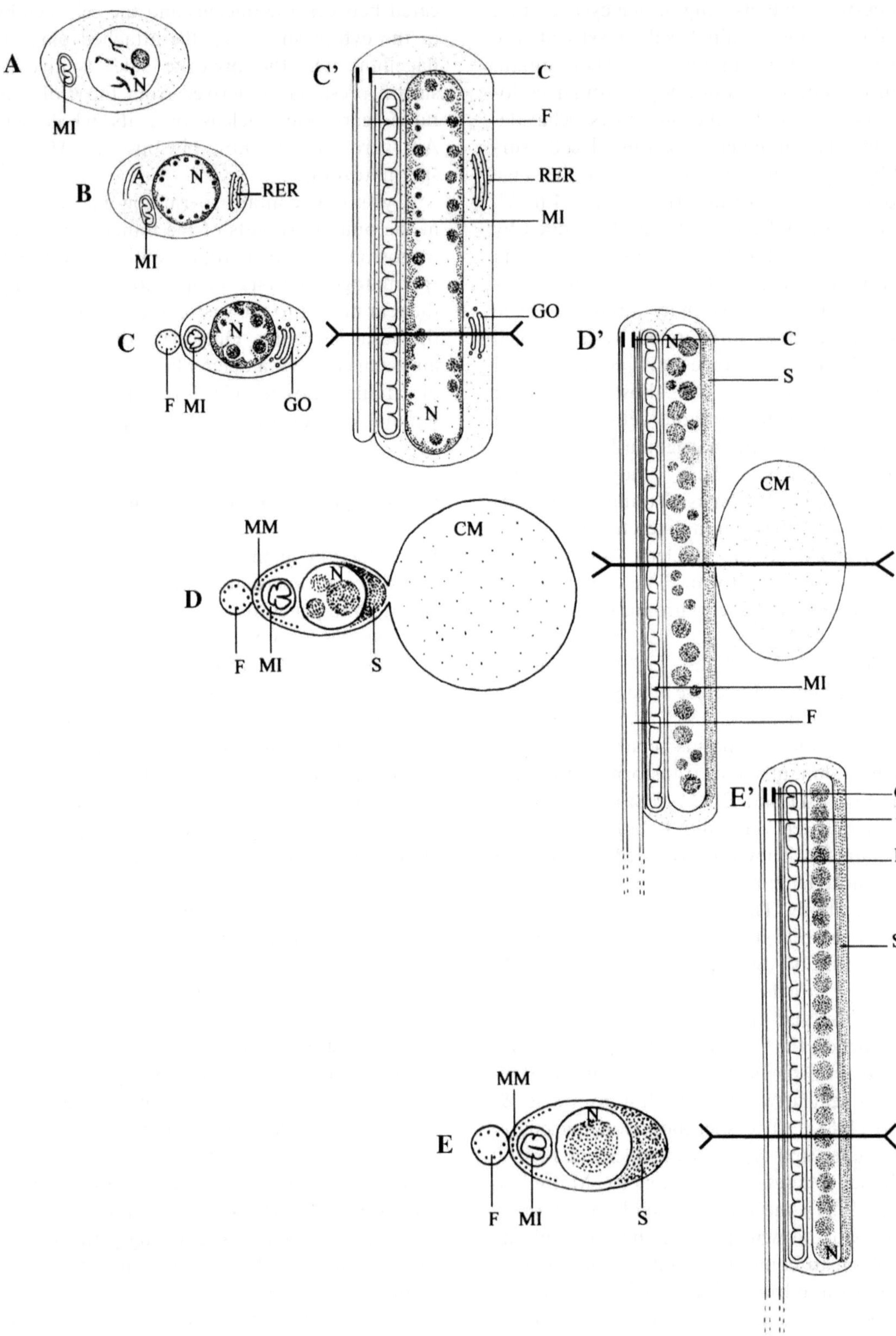
A
N
MI
B
A
N
RER
MI
C
N
F
MI
GO
C'
C
F
RER
MI
GO
N
D'
N
C
S
CM
MM
D
N
CM
F
MI
S
MI
F
E'
C
F
MI
S
MM
E
N
F
MI
S
N

Figure 13. Myzostoma cirriferum. Schematic drawings of male germinal cells undergoing spermatogenesis. (A) spermatogony; (B) young spermatid; (C) a more advanced spermatid sectioned transversely and (C′) sagitally; (D and D′) old spermatid; (E and E′) spermatozoon. Abbreviations: A – axoneme; C – centriole; CM – cytoplasmic mass; F – flagellum; GO – Golgi apparatus; MI – mitochondria; MM – manchette of microtubules; N – nucleus; RER – rough endoplasmic reticulum; S – myelin-like sheet.

the species considered. They are 50–500 μm long and are formed by a translucent extracellular matrix containing cysts. The cysts are packed close together and tend to form numerous sinuous chains which interlace each other. In *M. cirriferum*, spermatophores consist of three regions: the body with the curved horns, the foot, and the basal disc (Fig. 14A). The body-horns region forms the upper two third of the spermatophore and includes spermiocysts each of which contains ca. 64 spermatozoa (Fig. 14B). The foot extends below the body-horns region and includes cysts with abortive germinal cells (Fig. 14C). These cysts are assumed to prevent the spermiocysts from being attacked by the lytic enzymes that create a hole in the receiver's integument. The basal disc of the spermatophore attaches to the receiver's cuticle. It is composed of upper,

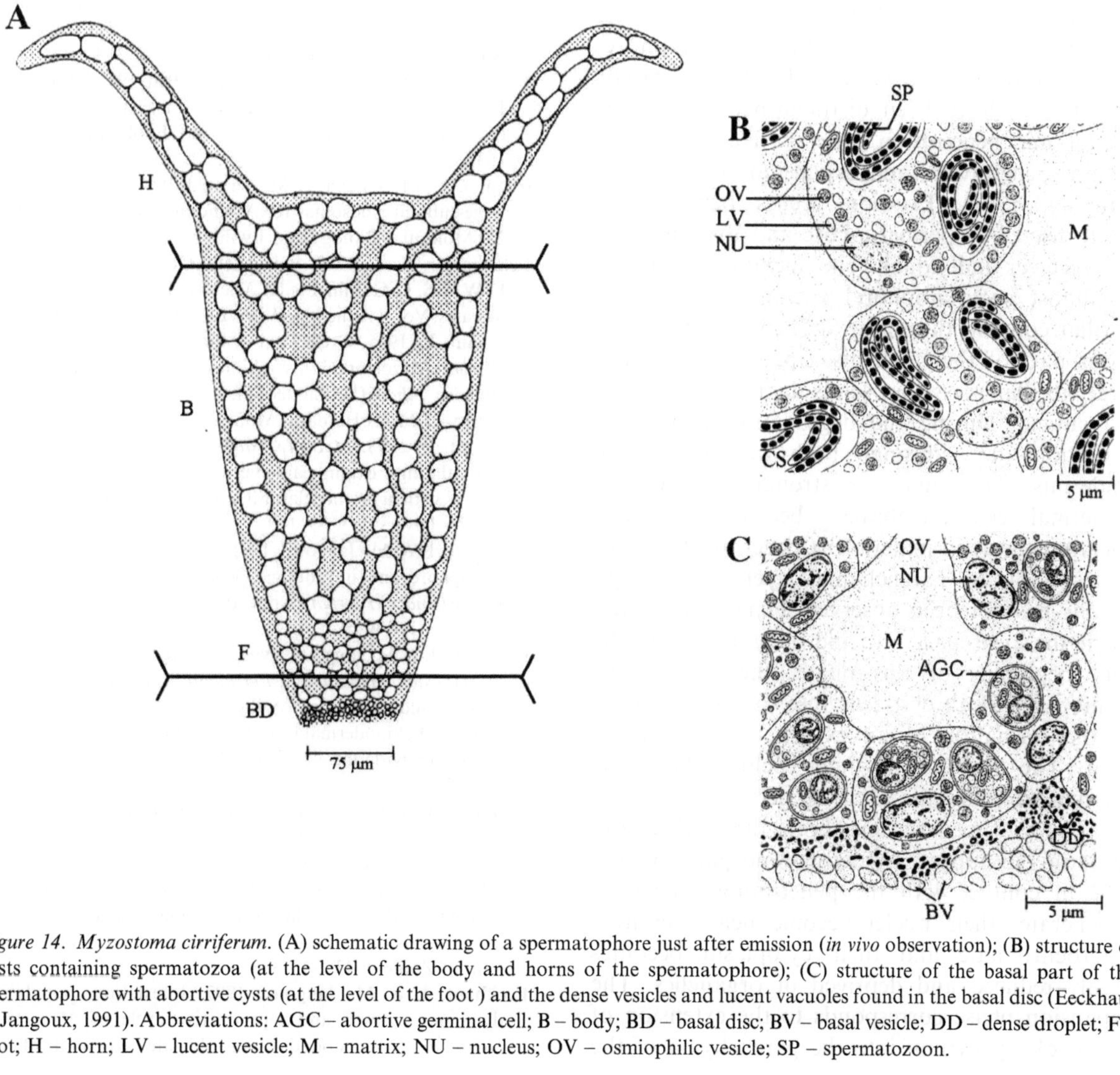

Figure 14. Myzostoma cirriferum. (A) schematic drawing of a spermatophore just after emission (*in vivo* observation); (B) structure of cysts containing spermatozoa (at the level of the body and horns of the spermatophore); (C) structure of the basal part of the spermatophore with abortive cysts (at the level of the foot) and the dense vesicles and lucent vacuoles found in the basal disc (Eeckhaut & Jangoux, 1991). Abbreviations: AGC – abortive germinal cell; B – body; BD – basal disc; BV – basal vesicle; DD – dense droplet; F – foot; H – horn; LV – lucent vesicle; M – matrix; NU – nucleus; OV – osmiophilic vesicle; SP – spermatozoon.

electron-dense, rod-shaped vesicles of unknown function and lower vacuoles filled with a fibrillar material that is responsible of the degradation of the integument (Fig. 14C). The two kinds of vesicles are respectively synthesised by the vesicular gland cells of the vasa deferentia and the vacuolar gland cells of the seminal vesicles.

After attachment, the spermatophores pierce the integument and release all the cysts through it. Intradermic penetration has been observed in *M. ambiguum* (Kato, 1952), *M. cirriferum* (Eeckhaut & Jangoux, 1991) and *M. alatum* (Eeckhaut & Jangoux, 1992). Penetration can be observed *in vivo* thanks to the presence of white trails representing the spermatophore contents that are introduced into the translucent body of the receiver. These trails appear 10–30 min after attachment and after 1–5 h, the spermatophores are reduced to matrix only. In *M. cirriferum*, four phases have been distinguished during the intradermic penetration process: fixation, degradation, penetration, and expansion (Eeckhaut & Jangoux, 1991). Fixation occurs between the basal disc of a spermatophore and the cuticle of the receiver individual. The membranes of the basalmost spermatophoral vacuoles (those with fibrillar content) disappear and, consequently, the lower part of the basal disc appears as a single large fibrillar mass that sends digitations through the cuticle. During the degradation phase, the digitations flow through both the cuticle and epidermis. The cuticle is strongly altered and epidermal cell membranes become degraded, allowing the lytic material to penetrate the cells. Invasion of the parenchymal layer by the digitations has not been observed. Penetration starts when the cysts pass into the integument. At the very beginning of this phase, the cytoplasmic membranes of all cyst-cells fuse together, leading to the formation of an extremely large syncytium. The whole syncytium encloses both the spermatozoa and the abortive germinal cells. At the point of penetration, both the cuticle and epidermis disappear. Epidermal and parenchymal cells in contact with the penetrating syncytium degenerate: their nuclei become nearly entirely heterochromatic and their cytoplasm becomes highly reduced and deprived of organelles. The expansion phase corresponds to the extension of the whole syncytium into the parenchyma of the receiver. Once at the level of the uterine diverticula, the syncytium breaks up and the spermatozoa are released free into the uterine lumen where they fertilise the vitellogenic oocytes.

References

Afzelius, B. A., 1983. The spermatozoon of *Myzostomum cirriferum* (Annelida, Myzostomida). Journal of Ultrastructure Research 83: 58–68.

Afzelius, B. A., 1984. Spermiogenesis in *Myzostomum cirriferum* (Annelida; Myzostomida). Vidensk. Meddr Dansk Naturh. Foren. 145: 11–21.

Bargalló, R., 1977. Polimorfisme dels gàmetes masculins. III. Consideracions espermatològiques a propòsit dels Myzostòmids. Bull. Soc. Catalana Biol. 2: 33–39.

Benham, W. B., 1896. Archiannelida, Polychaeta, Myzostomidaria. In Farmer, S. F. & A. E. Shipley (eds), The Cambridge Natural History. Macmillan, London: 241–334.

Berney, C., J. Pawloski & L. Zaninetti, 2000. Elongation factor 1-alpha sequences do not support an early divergence of the Acoela. Molecular Biology and Evolution 17: 1032–1039.

Chenuil, A., M. Solignac & M. Bernard, 1997. Evolution of the large-subunit ribosomal RNA binding site for protein L23/25. Molecular Biology and Evolution 14: 578–588.

Eckelbarger, K. J., 1992. Polychaeta: Oogenesis. In Harrison, F. W. & S. L. Gardiner (eds), Microscopic Anatomy of Invertebrates, Vol. 7 Annelida. Wiley-Liss, New York: 109–127.

Eeckhaut, I., 1995. Cycle vital et biologie de *Myzostoma cirriferum* (Myzostomida), symbiote obligatoire de la comatule *Antedon bifida* (Echinodermata). Ph.D. dissertation, University of Mons-Hainaut, Mons, Belgium: 106 pp.

Eeckhaut, I., 1998. *Mycomyzostoma calcidicola* gen. nov., sp. nov., the first extant parasitic myzostome infesting crinoid stalks, with a nomenclatural appendix by M. J. Grygier. Species Diversity 3: 89–103.

Eeckhaut, I. & M. Jangoux, 1991. Fine structure of the spermatophore and intradermic penetration of sperm cells in *Myzostoma cirriferum* (Annelida, Myzostomida). Zoomorphology 111: 49–58.

Eeckhaut, I. & M. Jangoux, 1992. Development and behaviour of *Myzostoma alatum* and *Pulvinomyzostomum pulvinar*, two myzostomid symbiotes of the comatulid *Leptometra phalangium* (Echinodermata). In Scalera-Liaci, L. & C. Canicatti (eds), Echinoderms Through Time. Balkema, Rotterdam: 229–236.

Eeckhaut, I. & M. Jangoux, 1993. Integument and epidermal sensory structures of *Myzostoma cirriferum* (Myzostomida). Zoomorphology 113: 33–46.

Eeckhaut, I. & M. Jangoux, 1995. *Contramyzostoma bialatum* (Annelida: Myzostomida), a new genus and species of parasitic myzostome infesting comatulid crinoids. Raffles Bulletin of Zoology 43: 343–353.

Eeckhaut, I. & M. Jangoux, 1997. Infestation, population dynamics, growth and reproductive cycle of *Myzostoma cirriferum* (Myzostomida), an obligate symbiont of the comatulid crinoid *Antedon bifida* (Crinoidea, Echinodermata). Cahiers De Biologie Marine 38: 7–18.

Eeckhaut, I., B. Dochy & M. Jangoux, 1995. Feeding behaviour and functional morphology of the introvert and digestive system of *Myzostoma cirriferum* (Myzostomida). Acta Zoologica 76(4): 307–315.

Eeckhaut, I., L. Fievez & M. C. Müller, 2003. Larval development of *Myzostoma cirriferum* (Myzostomida). J. Morphol. 258: 269–283.

Eeckhaut, I., D. VandenSpiegel & M. J. Grygier (1994). Myzostomida (Annelida) from Singapore with related Indo-Pacific distribution records and description of three new species. Raffles Bulletin of Zoology 42: 669–688.

Eeckhaut, I., D. McHugh, P. Mardulyn, R. Tiedemann, D. Monteyne, M. Jangoux & M. C. Milinkovitch, 2000. Myzostomida: a link between trochozoans and flatworms? Proceedings of the Royal Society of London Series B 267: 1383–1392.

Fedotov, D., 1929. Beiträge zur Kenntnis der Morphologie der Myzostomiden. Z. Morphol. Ökol. Tiere 15: 156–191.

Graff, L., 1877. Das Genus *Myzostoma* (F.S. Leuckart). Wilhelm Engelmann, Leipzig, 82 pp.

Grygier, M. J., 2000. Class Myzostomida. In Beesley, P. L., G. J. B. Ross & C. J. Glasby (eds), Polychaetes and Allies: The Southern Synthesis. Fauna of Australia, Vol. 4A Polychaeta, Myzostomida, Pogonophora, Echiura, Sipuncula. CSIRO Publishing, Melbourne: 297–330.

Haszprunar, G., 1996. The Mollusca: coelomate turbellarians or mesenchymate annelids? In Taylor, J. D. (ed.), Origin and Evolutionary Radiation of the Mollusca. Oxford University Press, Oxford: 3–28.

Heinzeller, T., B. Aschauer, A. Lange & U. Welsch, 1995. A myzostomid invading the connective tissue of its host *Comanthus parvicirrus* (Crinoidea). In Emson, R. H., A. B. Smith & A. C. Campbell (eds), Echinoderm Research 1995. Balkema, Rotterdam: 3–8.

Jägersten, G., 1939. Über die Morphologie und Physiologie des Geschlechtsapparats und den Kopulationsmechanismus der Myzostomiden. Zool. Bidr., Uppsala 18: 163–242.

Jägersten, G., 1940. Zur Kenntnis der Morphologie, Entwicklung und Taxonomie der Myzostomida. Nova Acta Reg. Soc. Sci. Ups. (4)11(8): 1–84.

Kato, K., 1952. On the development of myzostome. Sci. Rep. Saitama Univ. (B)1: 1–16.

Kronenberg, K., 1997. Ultrastructureller Vergleich der Epidermisstrukturen von *Pulvinomyzostomum pulvinar* (Graff) und *Myzostoma alatum* Graff (Myzostomida, Annelida). Master thesis, University of Munich, 55 pp.

Lanterbecq, D., 2000. Phylogenèse et évolution des plans de structure chez les myzostomides (Spiralia: Myzostomida). Master thesis, University of Mons-Hainaut, Mons, Belgium: 48 pp.

Leuckart, F. S., 1827. Versuch einer naturgemässen Eintheilung der Helminthen. Neue Akademische Buchhandlung von Karl Gross, Heidelberg, Leipzig, 88 pp.

Leuckart, F. S., 1830. Untitled paragraph no 92. Isis von Oken 23: 612–613.

Leuckart, F. S., 1836. In Beziehung auf den Haarstern (*Comatula*) und *Pentacrinus europaeus*, so wie auf das Schmarotzerthier auf *Comatula*. Notizen aus dem Gebiete der Natur- und Heilkunde gesammelt und mitgetheilt von Dr. L.G.V. Froriep 59(9): 129–131.

Littlewood, D. T. J., P. D. Olson, M. J. Telford, E. A. Herniou & M. Riutort, 2001. Elongation Factor 1-Alpha sequences alone do not assist in resolving the position of the Acoela within the Metazoa. Molecular Biology and Evolution 18: 437–442.

Mattei, X. & B. Marchand, 1987. Les spermatozoïdes des Acanthocephales et des Myzostomides. Ressemblances et conséquences phylétiques. C. R. Acad. Sci., Série III, Sci. Vie, Paris 305: 525–529.

Mattei, X. & B. Marchand, 1988. La spermiogénèse de *Myzostomum* sp. (Procoelomata, Myzostomida). Journal of Ultrastructure and Molecular Structure Research 100: 75–85.

Meyer, D. L. & W. Ausich, 1983. Biotic interactions among recent and among fossil crinoids. In Tevesz, M. J. S. & P. L. McCall (eds), Biotic Interactions in Recent and Fossil Benthic Communities. Plenum Publishing Corporation, New York: 377–427.

Müller, M. C. & W. Westheide, 2000. Structure of the nervous system of *Myzostoma cirriferum* (Annelida) as revealed by Immunohistochemistry and cLSM Analyses. Journal of Morphol. 245: 87–98.

Pietsch, A. & W. Westheide, 1987. Protonephridial organs in *Myzostoma cirriferum* (Myzostomida). Acta Zoologica 68: 195–203.

Rao, K. H. & R. Sowbhagyavathi, 1972. Observations on the associates of crinoids at Waltair Coast with special reference to myzostomes. Proc. Indian natn. Sci. Acad. 38 B: 360–366.

Rota, E., P. Martin & C. Erséus, 2001. Soil-dwelling polychaetes: enigmatic as ever? Some hints on their phylogenetic relationships as suggested by a maximum parsimony analysis of 18S rDNA gene sequences. Contributions to Zoology 70: 127–138.

Rouse, G. W. & K. Fauchald, 1997. Cladistics and polychaetes. Zoolologica Scripta 26: 139–204.

Semper, C., 1858. Zur Anatomie und Entwickelungsgeschichte der Gattung *Myzostoma* Leuckart. Z. Wiss. Zool. 9: 48–65.

Warn, J. M., 1974. Presumed myzostomid infestation of an Ordovician crinoid. Journal of Paleontology 48: 506–513.

Wheeler, W. M., 1896. The sexual phases of *Myzostoma*. Mitth. Zool. Stat. Neapel 12: 227–302.

Zrzavỳ, J., S. Mihulka, P. Kepka, A. Bezdek & D. Tietz, 1998. Phylogeny of the Metazoa based on morphological and 18S ribosomal DNA evidence. Cladistics 14: 249–285.

Zrzavỳ, J., V. Hypsa & D. Tietz, 2001. Myzostomida are not annelids: molecular and morphological support for a clade of animals with anteriorsperm flagella. Cladistics 17: 1–29.

Hydrobiologia (2005) **535/536:** 277–296 © Springer 2005
T. Bartolomaeus & G. Purschke (eds), Morphology, Molecules, Evolution and Phylogeny in Polychaeta and Related Taxa

Reconstructing the phylogeny of the Sipuncula

Anja Schulze[1,*], Edward B. Cutler[1] & Gonzalo Giribet[1,2]
[1] *Department of Invertebrate Zoology, Museum of Comparative Zoology, 26 Oxford Street, Harvard University, Cambridge, MA 02138, USA*
[2]*Department of Organismic and Evolutionary Biology, Harvard University, 16 Divinity Avenue, Cambridge, MA 02138, USA*
*(*Author for correspondence [current address]: Smithsonian Marine Station, 701 Seaway Drive, Fort Pierce, FL 34949, USA; Tel: 772-465-6630 ext. 105, Fax: 772-461-8154, E-mail: anjaschulze@bigfoot.com)*

Key words: Sipuncula, phylogeny, 18S rRNA, 28S rRNA, histone H3, Spiralia, cladistic analysis

Abstract

Sipunculans are marine spiralian worms with possible close affinities to the Mollusca or Annelida. Currently 147 species, 17 genera, 6 families, 4 orders and 2 classes are recognized. In this paper we review sipunculan morphology, anatomy, paleontological data and historical affiliations. We have conducted cladistic analyses for two data sets to elucidate the phylogenetic relationships among sipunculan species. We first analyzed the relationships among the 45 species of Phascolosomatidea with representatives of the Sipunculidea as outgroups, using 35 morphological characters. The resulting consensus tree has low resolution and branch support is low for most branches. The second analysis was based on DNA sequence data from two nuclear ribosomal genes (18S rRNA and 28S rRNA) and one nuclear protein-coding gene, histone H3. Outgroups were chosen among representative spiralians. In a third analysis, the molecular data were combined with the morphological data. Data were analyzed using parsimony as the optimality criterion and branch support evaluated with jackknifing and Bremer support values. Branch support for outgroup relationships is low but the monophyly of the Sipuncula is well supported. Within Sipuncula, the monophyly of the two major groups, Phascolosomatidea and Sipunculidea is not confirmed. Of the currently recognized families, only Themistidae appears monophyletic. The Aspidosiphonidae, Phascolosomatidae and Golfingiidae would be monophyletic with some adjustments in their definition. The Sipunculidae is clearly polyphyletic, with *Sipunculus nudus* as the sister group to the remaining Sipuncula, *Siphonosoma cumanense* the sister group to a clade containing *Siphonosoma vastum* and the Phascolosomatidea, and *Phascolopsis gouldi* grouping within the Golfingiiformes, as suggested previously by some authors. Of the genera with multiple representatives, only *Phascolosoma* and *Themiste* are monophyletic as currently defined. We are aiming to expand our current dataset with more species in our molecular database and more detailed morphological studies.

Introduction

The Sipuncula are in several respects an ideal group for systematic studies: 1. The taxon has only 147 recognized species (see Cutler, 1994 for the current taxonomy), theoretically enabling researchers to include every single one instead of exemplars, 2. The majority of species (ca. 90%) are relatively large (i.e. >5 mm), facilitating examination, 3. Approximately 64% of the species are, at least in some locations, shallow-water inhabitants (<20 m) and easy to collect. On the other hand, morphological uniformity within the Sipuncula restricts the number of phylogenetically informative characters for cladistic analyses.

Keferstein (1863, 1865a,b, 1866, 1867), Selenka (1875, 1885, 1888, 1897) and Selenka et al. (1883) laid the groundwork for the understanding

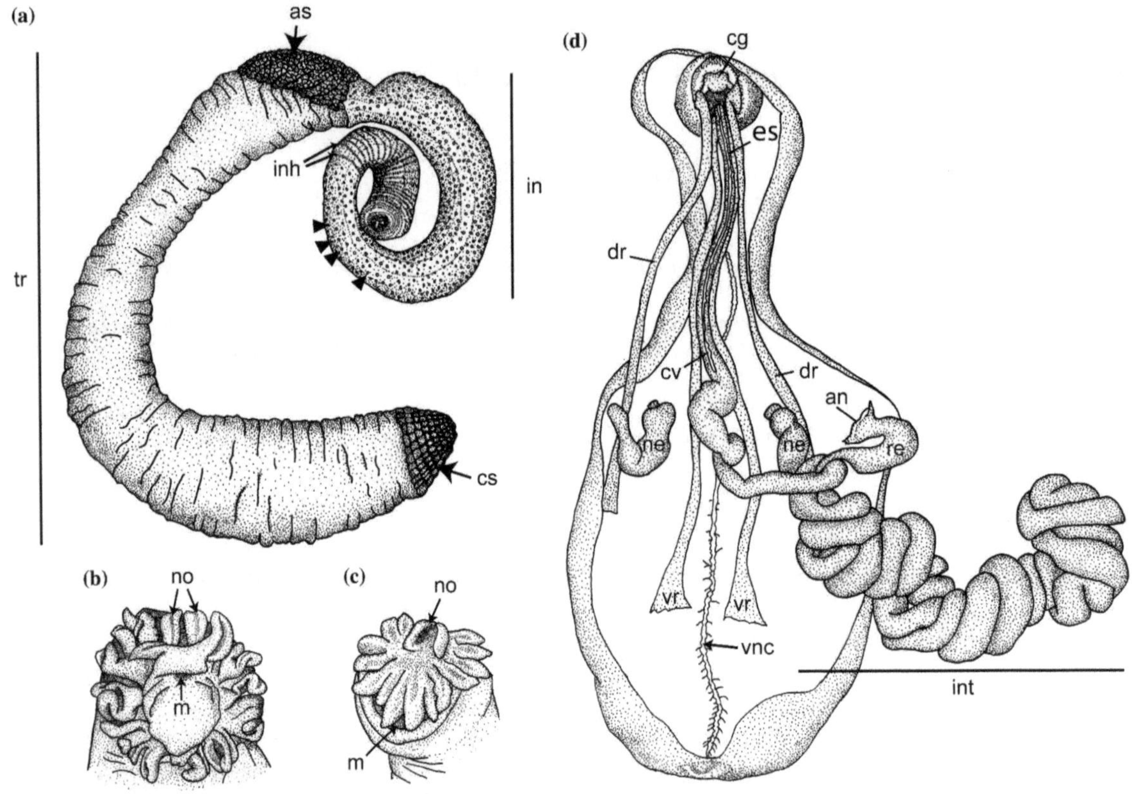

Figure 1. Morphology and anatomy of representatives of the Sipuncula. (a) External morphology of *Aspidosiphon fischeri*, lateral view. Note introvert retractor muscles and esophagus shining through body wall of introvert; tentacles visible at tip of introvert; arrowheads: introvert papillae. (b) Tentacular arrangement in *Golfingia margaritacea*, representative of the Sipunculidea. (c) Tentacular arrangement in *Phascolosoma nigrescens*, representative of the Phascolosomatidea. (d) Anatomy of *Golfingia margaritacea*. an – anus; as – anal shield; cg – cerebral ganglion; cs – caudal shield; cv – contractile vessel; dr – dorsal introvert retractor muscle; in – introvert; inh – introvert hooks; int – intestine; m – mouth; ne – nephridium; no – nuchal organ; es – esophagus; re – rectum; tr – trunk; vnc – ventral nerve cord; vr – ventral introvert retractor muscle.

of the morphology and internal anatomy of sipunculans. Several detailed accounts have been published in the past decade (see Rice, 1993; Cutler, 1994; Edmonds, 2000). We will therefore keep our review of morphology and anatomy short.

The sipunculan body is divided into trunk and retractable introvert (Fig. 1a). The ratio between introvert and trunk length varies among species and also depends on the state of relaxation of a specimen. Introvert length is only meaningful when measuring specimens with a fully extended introvert. The mouth, at the anterior end of the introvert, is surrounded by an array of tentacles in the Sipunculidea (Fig. 1b). In the Phascolosomatidea, the tentacles are arranged in an arc around the nuchal organ, also located at the tip of the introvert (Fig. 1c). The anus lies dorsally, usually at the anterior end of the trunk, except in *Onchnesoma* and four *Phascolion* species where it is shifted anteriorly onto the introvert. The nephridiopores lie ventrolaterally, typically at the level of the anus.

Hooks are often present on the distal part of the introvert. These are proteinaceous, non-chitinous specializations of the epidermis (Voss-Foucart et al., 1977) which are either arranged in rings or scattered. They are usually curved posteriorly and can have a variety of shapes and internal structures (Fig. 2). In *Aspidosiphon*, *Lithacrosiphon* and *Cloeosiphon* the epidermis forms specializations in the form of an anal shield (Fig. 1a). In *Aspidosiphon* and *Lithacrosiphon* the anal shield is restricted to the dorsal side, causing

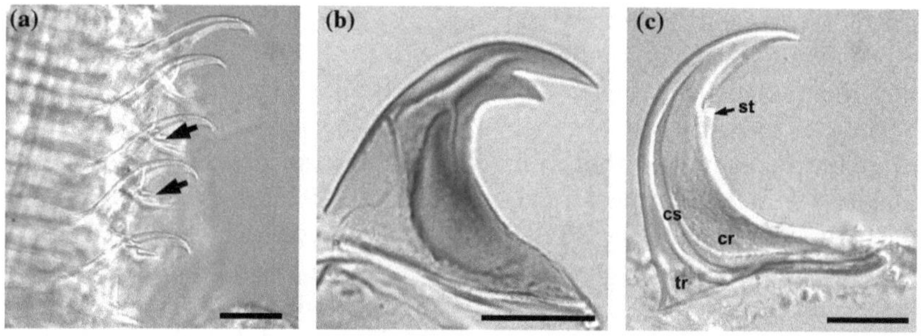

Figure 2. Shape and internal structure of sipunculan hooks, LM. (a) Partial ring of hooks in *Apionsoma pectinatum*; arrows point to basal spinelets. (b) *Aspidosiphon steenstrupi*, typical bidentate hook. (c) *Phascolosoma stephensoni*. cl – clear streak; cr – posterior crescent; st – secondary tooth; tr – clear anterior clear triangle; scale bars 20 μm in all cases.

the introvert to emerge at an angle, whereas it surrounds the anterior trunk in *Cloeosiphon* with the introvert emerging from its center. In *Aspidosiphon* the shield is a hardened, horny structure; in *Lithacrosiphon* it is a calcareous cone; in *Cloeosiphon* it is composed of separate plates. At the posterior end, a hardened caudal shield is sometimes present in *Aspidosiphon*.

Numerous papillae, associated with epidermal organs (combination of glandular and sensory organ) may be present on the trunk and introvert. *Phascolion* has specialized holdfast papillae. These are large papillae on the mid to posterior parts of the trunk with a hardened margin, shaped as a letter U, V or broken O (Åkesson, 1958; Cutler, 1994).

The body wall consists of a non-ciliated epidermis, overlain by a cuticle, an outer layer of circular and an inner layer of longitudinal musculature. In some larger species, oblique muscle fibers may be present between the longitudinal and circular muscle layers (Rice, 1993). A coelomic cavity fills most of the body and encloses the interior organs. Dybas (1981) distinguishes five types of coelomic cells: haemocytes, granulocytes, large multinuclear cells, ciliated urns and immature cells.

The alimentary canal starts with the esophagus, located between the introvert retractor muscles. In the trunk the intestine runs posteriorly, forms a loop and turns anteriorly again. The downward and upward sections of the gut are coiled around each other, forming a double helix (Fig. 1d). At the anterior end of the gut coil the rectum emerges and ends in the anus. A rectal caecum, present in most species, is a blind ending sac at the transition between intestine and rectum with unknown function.

Apart from the body wall musculature, the two other muscle systems are the introvert retractor muscles and intestinal fasteners. There are usually two pairs of introvert retractor muscles which may be fused to various degrees. Some species have a single pair. The spindle muscle inserts anteriorly near the anus, runs through the gut coil and is posteriorly attached either to the body wall near the posterior end or inside the gut coil. In addition, thin filamentous muscles often attach the esophagus and the anterior intestine to the body wall.

A pair of metanephridia is usually present, except in *Phascolion* and *Onchnesoma* which have only a single nephridium. The nephridia have the shape of elongated sacs and are orange or light brown in live or freshly fixed material. A ciliated funnel, or nephrostome, opens into the coelomic cavity at the anterior end, close to the nephridiopore (Ocharan, 1974; Rice, 1993).

The central nervous system consists of an anterior cerebral ganglion with a circumesophageal connective and a ventral nerve cord. The cerebral organ is a non-ciliated structure at the anterior margin of the cerebral ganglion (Purschke et al., 1997) and has been interpreted as a larval vestige (Åkesson, 1958). Purschke et al. (1997), however, suggested a sensory function because the epithelium contains bipolar sensory cells. The nuchal organ, located posterior to the cerebral organ, has fewer sensory cells. Both organs may function as a unit for chemoreception. Based on ultrastructural evidence the nuchal organ is probably not homologous to the nuchal organ in polychaetes (Purschke et al., 1997).

Sipuncula lack a blood vascular system. Fluid transport and gas exchange are instead accomplished by the coelom and the tentacular system. The latter connects the tentacles to a ring canal at their base, from which a contractile vessel that runs along the esophagus and ends blindly posteriorly arises (Edmonds, 2000). Pilger & Rice (1987) found evidence that the contractile vessel might also serve as a site for ultrafiltration.

Evolutionary origin of the Sipuncula

The phylogenetic position of the Sipuncula has long been subject to controversial opinions. While they were regarded as close relatives to holothurians in the early 1800s (Lamarck, 1816; Cuvier, 1830), Quatrefages (1847) erected the group Gephyrea, which he regarded as an intermediate between 'worms' and holothurians and which also contained echiurans, sternaspids and priapulids. Nichols (1967) revived the idea of echinoderm affinities, but found little acceptance in the scientific community. Today, there is general agreement that Sipuncula are protostomes and belong in the Lophotrochozoa, with affinities to annelids and/or molluscs (Cutler, 1994; Winnepenninckx et al., 1995; Zrzavý et al., 1998; Giribet et al., 2000), although their precise position remains unresolved.

The only unambiguous fossil record of sipunculans has recently been revealed by Huang et al. (2004) who describe three species from the Lower Cambrian Maotianshan Shale in southwest China, suggesting that sipunculan morphology has changed little over the past 520 million years. *Ottoia* has been proposed as another fossil sipunculan (Banta & Rice, 1976), but is now presumed to be a primitive 'aschelminth' or Priapulida (Conway Morris, 1989). The paleozoic *Hyolitha* has a mix of attributes of sipunculans and molluscs, suggesting a close phylogenetic relationship with both (Runnegar et al., 1975; Conway Morris, 1998; Marti Mus & Bergstrom, 2001). Fossilized burrows thought to be created by sipunculans in soft sediments are known from early and mid-Paleozoic times (Pemberton et al., 1980; Brett et al., 1983), but the inhabitants of such burrows are hard to determine. More recent Mesozoic and Cenozoic fossil burrows have also been attributed to sipunculan worms (Wetzel & Werner, 1981; Frey et al., 1984; Romero-Wetzel, 1987; McBride & Picard, 1991). Other sipunculans appear to have lived in association with corals and in vacated mollusc shells since the mid-Paleozoic, throughout the Mesozoic, and Cenozoic (Hyman, 1959; Gill & Coates, 1977; Brett & Cottrell, 1982; Pisera, 1987).

Embryological evidence for the origin of the Sipuncula is ambiguous. A trochophore larva in sipunculan development confirms affinities with Mollusca and Annelida. Scheltema (1993) regards the presence of a molluscan cross during cleavage as an indication to place Sipuncula as the sister taxon to the Mollusca. However, in recent years cell lineage studies have shown that the concept of the molluscan cross vs. the annelidan cross is oversimplified and of limited phylogenetic significance (Guralnick & Lindberg, 2001; Guralnick, 2002). Scheltema (1993) further suggests homologies in the head regions of sipunculan and mollusc larvae. She argues that the sipunculan lip shows similarities to the molluscan foot, the sipunculan lip glands to the molluscan pedal glands and the sipunculan buccal organ to the molluscan radular sac.

Rice (1985), on the other hand, lists certain similarities between sipunculan and annelid development. She notes the resemblance of the prototrochal and metatrochal bands in the larvae of both taxa, but Nielsen (1987) suspects that the posterior ciliary band in the sipunculan pelagosphera is an accessory ciliary band and not a true metatroch. Rice (1985) further mentions the retention of the egg envelope in some species to form the larval cuticle. She notes that, although in most sipunculan species examined to date, the nerve cord develops as a single tract, there are two known exceptions that may point to annelidan affinities: in *Phascolosoma agassizii* the nerve cord is double in early larval stages and in *Nephasoma pellucidum* it is partially split in the young pelagosphera larva. A notable difference between polychaete and sipunculan larvae is that there is no opposed-band feeding in sipunculans.

Even less conclusive with regard to phylogenetic affinities is evidence from comparative biochemistry, such as carbonic anhydrase activity (Henry, 1987), actin/myosin control of muscle contraction (Lehman & Szent-Györgyi, 1975), chromatin subunit structure in erythroid cells (Wilheim & Wilheim, 1978), the presence/absence of different pyruvate oxireductases (Livingstone et al., 1983), phospholipids (Kostetskii, 1984), hemerythrin biochemistry (Florkin, 1975) and properties of the immune system (Ionescu-Varo & Tufesco, 1982). The study of ultrastructural evidence such as septate junctions (Green & Bergquist, 1982), ciliary bands (Nielsen, 1987) and sperm (Klepal, 1987) gives no clear answer either.

Previous cladistic analyses of morphological, molecular and, recently, gene order data (Boore & Staton, 2002) have rendered a number of hypotheses relating Sipuncula: to an unresolved clade containing the Mollusca, Echiura, and a clade grouping the annelid taxa with the Arthropoda (Brusca & Brusca, 1990; Backeljau et al., 1993); sister group to Echiura (Meglitsch & Schram, 1991); sister group to Annelida (Erber et al., 1998); sister group to Mollusca (Brusca & Brusca, 2003); derived Annelida (Boore & Staton, 2002); sister group to Echiura + Annelida (Eernisse et al., 1992); sister group to an unresolved clade containing Mollusca, Annelida and the Panarthropoda (Nielsen et al., 1996), or sister group to a clade as follows: (Echiura (Mollusca (Annelida (Onychophora (Tardigrada + Arthropoda))))) (Sørensen et al., 2000); sister group to Mollusca (Zrzavý et al., 1998; Giribet et al., 2000); sister group to Mollusca + Annelida (Peterson & Eernisse, 2001); or within an unresolved clade also containing Mollusca, Annelida and Echiura (Zrzavý et al., 2001). The number of molecular or combined morphological/molecular hypotheses is even greater since the monophyly of Mollusca or Annelida is often not recovered. In summary, little agreement is found about the exact position of Sipuncula within the protostome worms, perhaps due to the lack of structure for resolving some of those animal phyla (see Giribet, 2002).

The Sipuncula constitutes the first protostome phylum in which the ParaHox cluster has been fully characterized, presenting the three expected genes (Gsx, Xlox, Cdx) in two sipunculan species (Ferrier & Holland, 2001). This confirms the hypothesis that protostomes and deuterostomes shared a common ancestor whose ProtoHox cluster duplicated into ParaHox and Hox clusters that were conserved in both bilaterian lineages (Holland, 1998).

Sipuncula have been ranked as a family, order, sub-class or class at various times until Sedgwick (1898) proposed the name Sipunculoidea for the group, which he considered a phylum. However, this ranking did not gain much acceptance until Hyman (1959) proposed the spelling Sipunculida as she 'obliterated' the biologically meaningless construct Gephyrea. The present name, Sipuncula, and the use of 'sipunculan' for the vernacular name (not sipunculid) was proposed by Stephen (1964) and restated by Stephen & Edmonds (1972). Prior to this latter work there had been only two 20th century informal proposals regarding the arrangement of genera into unnamed family-like sets (Pickford, 1947; Åkesson, 1958). This void of intermediate taxa was partially filled when Stephen & Edmonds (1972) erected four families. Cutler & Gibbs (1985) set forth a more complete arrangement of the 17 genera into two classes, four orders, and six families. This arrangement has been followed by subsequent authors.

Phylogeny of sipunculan species

The first attempts to reconstruct the internal phylogeny of the Sipuncula were made by Cutler & Gibbs (1985), Gibbs & Cutler (1987) and Cutler (1994) (Fig. 3). The three studies relied on the same character set, however the polarities of several characters were changed since they were linked to the newly revised version of the hypothetical ancestral sipunculan (RHAS) in Cutler (1994). These analyses did provide some forward momentum, but currently they fall short of more rigorous standards for phylogenetic analyses. There are too few characters and too many unresolved branch points. Polarizing characters using a hypothetical ancestor can be accepted as an act of creative synthesis, or rejected as something less than objective science.

Some of the elements used in Cutler (1994) cannot be incorporated into a strict phylogenetic analysis based on a matrix of character states due to the incompleteness of the data set and thus some potentially meaningful information gets lost. In the present case this includes the number and shape of chromosomes, type of epidermal glands, or which of the four types of developmental patterns is exhibited. Additionally, interpretations of current patterns of distribution set against the

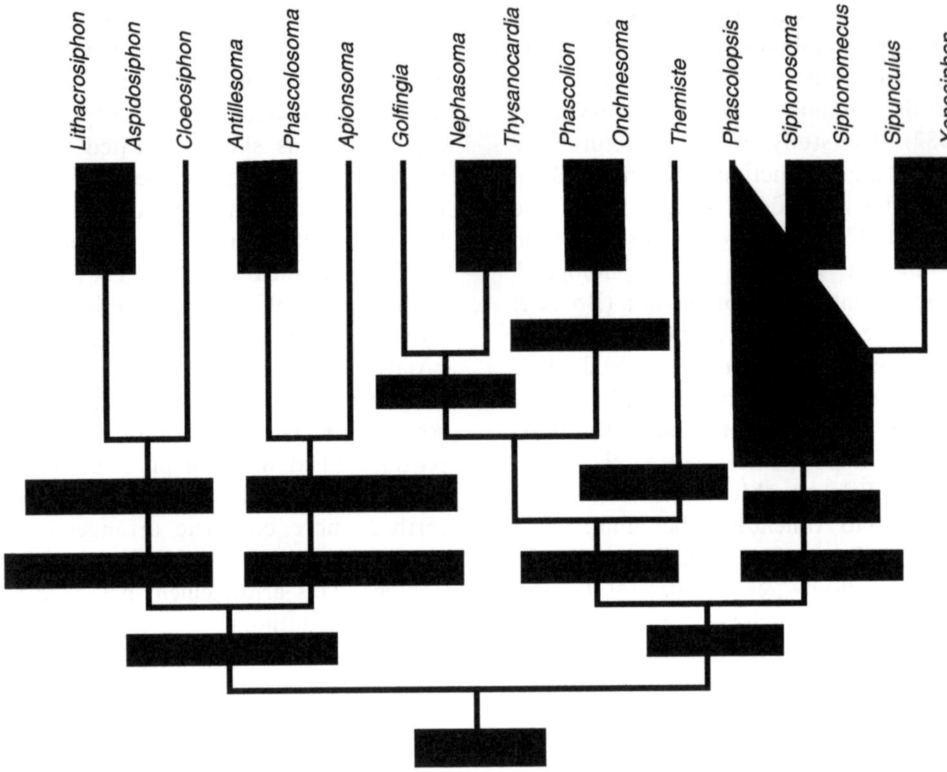

Figure 3. Phylogenetic relationships among sipunculan genera as proposed by Cutler & Gibbs (1985).

background of paleo-oceanography used by Cutler (1994: 360–374) in the construction of his evolutionary scenario, are difficult to incorporate into more restrictive analytical approaches.

Maxmen et al. (2003) recently tested Cutler's (1994) phylogenetic hypothesis. They analyzed molecular sequence data of the nuclear ribosomal genes 18S rRNA and 28S rRNA and the nuclear protein-coding gene histone H3 for 24 sipunculan species distributed in 13 genera, using a direct optimization approach with parsimony as the optimality criterion. This study showed polyphyly of the family Sipunculidae, with the genus *Sipunculus* being the sister taxon to the remaining sipunculans, *Siphonosoma* grouping with the Phascolosomatidea and *Phascolopsis* with the Golfingiidae (*Xenosiphon* and *Siphonomecus* were not sampled). Apart from the polyphyly of the Sipunculidae, the phylogenetic scheme mostly agreed with Cutler's (1994) system, with the exception of the problematic genus *Apionsoma* for which only partial sequences for the 18S rRNA were obtained (see Maxmen et al., 2003: their Fig. 3).*

Cladistic analysis of Phascolosomatidea based on morphology

To study the relationships among members of the Phascolosomatidea we assembled a morphological data matrix for the 45 species currently classified in the group, with representatives of all of the families in the Sipunculidea as outgroups (Appendix A). Thirty-five characters (31 parsimony informative) were included in the analyses. Data were analyzed using parsimony as an optimality criterion in

*Note added in proof: Another analysis of sipunculan phylogenetic relationships has been published recently (Staton, J.L. 2003, Inv. Biol. 122: 252–264), based on cytochrome *c* oxidase subunit I and encompassing thirteen sipunculan genera. This analysis is largely congruent with Cutler's 1994 analysis, except for the position of *Phascolopsis*. Sipuncula were found to be most closely related to annelids.

PAUP* 4.0b10 (Swofford, 2000). All characters were treated as unordered and no differential weighting was applied. Searches for the shortest trees were performed with the heuristic search option, for 1000 replicates of random addition sequence. Tree-bisection-reconnection (tbr) was chosen for branch-swapping, saving no more than 10 trees at each step per replicate. All trees were unrooted. Bremer support indices were calculated in AutoDecay 4.0.2 (Eriksson, 1998) in conjunction with PAUP. For each of the constraint trees generated by AutoDecay, ten random addition replicates were performed.

After the 1000 replicates, 980 equally most parsimonious trees of length 132 were retained. The strict consensus of all most parsimonious cladograms is presented in Figure 4. Resolution in the strict consensus is low and the ingroup does not appear as monophyletic.

Due to the relatively small size of the phylum a cladistic analysis including all species would be possible. However, when including all Sipuncula in the analysis, rooting becomes impracticable when relying on morphology alone. Other than in the molecular analyses, the inclusion of representatives of a number of other phyla as outgroups is problematic because morphological homologies are unclear and most characters of Sipuncula are inapplicable to other taxa. Therefore we preferred to explore the resolution of morphological features within the Sipuncula by selecting a subset of sipunculan taxa (the Phascolosomatidea) as ingroup and using other representatives of the phylum to polarize the phascolosomatidean tree. Morphologically, the Phascolosomatidea are well characterized by the presence of nuchal tentacles.

The low number of phylogenetically informative characters is obviously a problem and one would not expect full resolution with a character/taxon ratio of 0.58 (it is said that a ratio of three is desirable to obtain well resolved nodes, if there are no contradictory characters). Yet another problem seems to be the proportional representation of the various organ systems. Of the 31 phylogenetically informative characters included, 15 refer to hooks, leading to an unproportionately high weight of hook-associated characters in the analysis. The first 10 taxa in the tree in Figure 4 have no hooks and the presence of hooks appears as a synapomorphy for the clade including the remaining species. In light of all our previous data on sipunculan phylogeny, considering both morphological and molecular studies, we find this result questionable. It is possible that the morphological dataset contains more homoplasy than phylogenetic signal. This could be tested by combining the morphological data with a molecular data matrix. This has not been attempted here due to the lack of molecular data for a large proportion of the morphologically represented species.

Analysis of molecular data

For the molecular sequence analyses we used the sipunculan and outgroup sequences generated by Maxmen et al. (2003). Additional sequences were obtained for the following species:

Siphonosoma vastum (Selenka & Bülow, 1883): Bath, Barbados; June 24, 2002; Schulze, Saiz-Salinas & Cutler; MCZ DNA100625

Siphonosoma cumanense (Keferstein, 1867): Bath, Barbados; June 24, 2002; Schulze, Saiz-Salinas & Cutler; MCZ DNA100622

Aspidosiphon (Paraspidosiphon) fischeri ten Broeke, 1925: Martin's Bay, Barbados, June 21, 2002; Schulze, Saiz-Salinas, Cutler; MCZ DNA 100620

Lithacrosiphon cristatus (Sluiter, 1902): Bank Reef, Barbados, June 25, 2002; Schulze & Saiz-Salinas, MCZ DNA100623

Apionsoma (Edmondsius) pectinatum (Gibbs & Cutler, 1987): Six Mens Bay, Barbados, June 27, 2002; Schulze, Saiz-Salinas & Cutler; MCZ DNA100624

Phascolosoma nigrescens Baird, 1868: Six Mens Bay, Barbados, June 27, 2002, Schulze, Saiz-Salinas & Cutler; MCZ 100622

DNA sequences were deposited in GenBank (accession numbers in Table 1). Outgroup representatives were chosen among the spiralian phyla Nemertea, Mollusca, Entoprocta and Annelida (Table 1).

Methods for DNA extraction, amplification and sequencing are outlined in Maxmen et al. (2003). DNA electropherograms were edited in Sequencher^{TM} 4.0. Complete sequences were

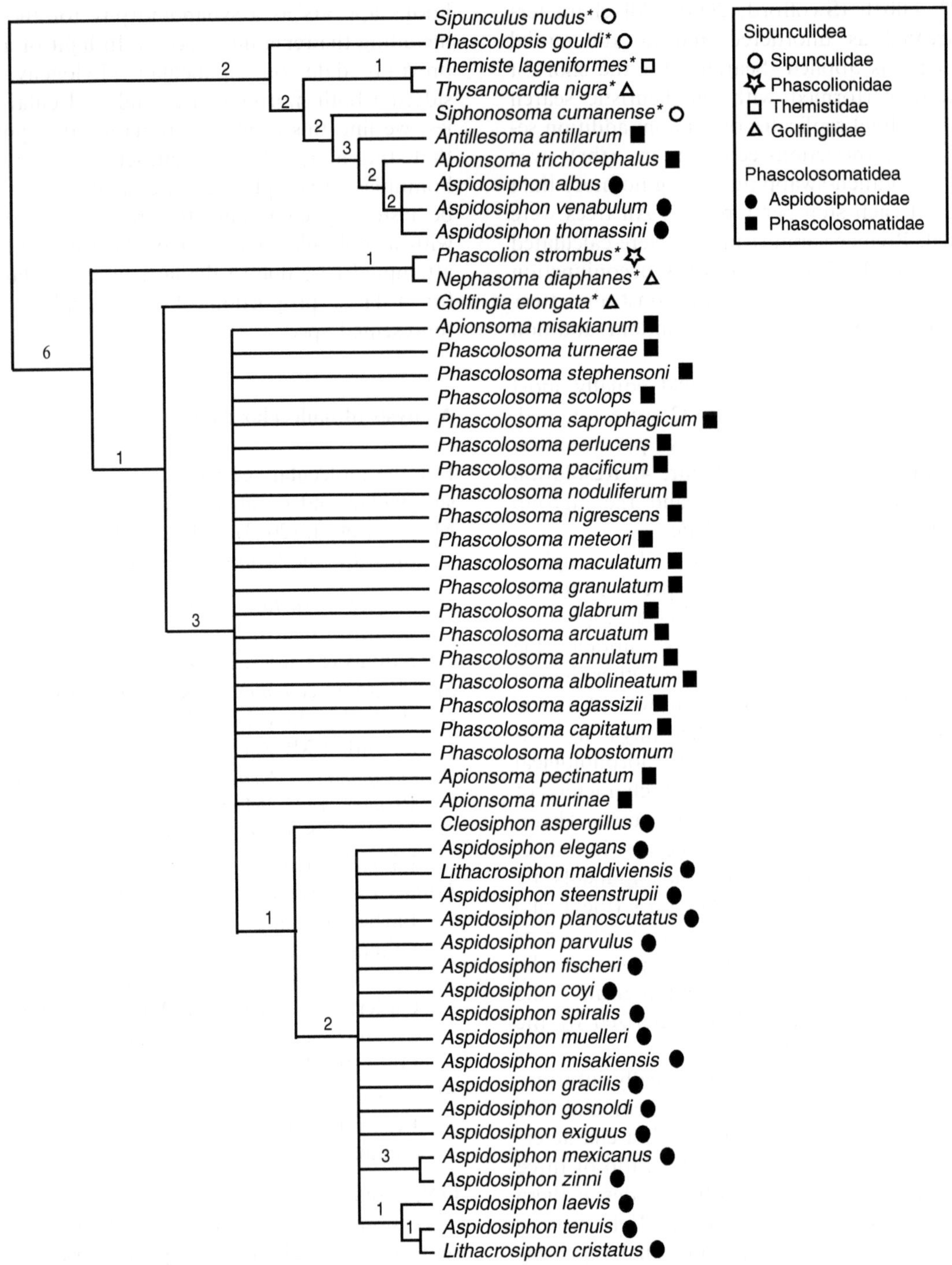

Figure 4. Strict consensus tree of 980 most parsimonious trees, generated in PAUP* using the morphological character matrix in Appendix A. Numbers above branches indicate Bremer support values. Asterisks indicate outgroup taxa.

edited in Genetic Data Environment (GDE) (Smith et al., 1994). All sequence data were analyzed simultaneously, using the program POY (Wheeler et al., 2002) assuming equal weights for all transformations (indels and base substitutions), as this was the parameter set that minimized

Table 1. Taxon sampling and accession codes to GenBank for the loci used in the analyses

		18S rRNA	28S rRNA	Histone H3
Phylum Nemertea				
Lineus sp.		X79878		
Argonemertes australiensis		AF519235	AF519264	AF519293
Amphiporus sp.		AF119077	AF519265	AF519294
Phylum Mollusca				
Lepidopleurus cajetanus		AF120502	AF120565	AY070142
Rhabdus rectius		AF120523	AF120580	AY070144
Haliotis tuberculata		AF120511	AF120570	AY070145
Yoldia limatula		AF120528	AF120585	AY070149
Phylum Entoprocta				
Barentsia hildegardae		AJ001734		
Pedicellina cernua		U36273		
Phylum Annelida				
Polyophthalmus pictus		AF519236		AF185259
Paralepidonotus ampulliferus		AF519237	AF519266	AF185247
Lumbrineris latreilli		AF519238	AF519267	AF185253
Chaetopterus variopedatus		U67324*		U96764
Lamellibrachia spp.		AF168742		AF185235
Urechis caupo		AF119076	AF519268	X58895
Lumbricus terrestris		AJ272183*		AF185262
Phylum Sipuncula Sipunculidae				
Sipunculus nudus	DNA100245	AF519239	AF519269	
Sipunculus nudus	DNA100246	AF519240[a]	AF519270	AF519295
Siphonosoma cumanense	DNA100235	AF519241	AF519271	AF519296
*Siphonosoma cumanense**	DNA100622	AY326291	AY445139	AY326296
*Siphonosoma vastum**	DNA100625		AY445137	AY326297
Phascolopsis gouldi	DNA100199	AF123306	AF519272	AF519297
Golfingiidae				
Golfingia elongata	DNA100466	AF519242[b]		AF519298
Golfingia vulgaris	DNA100207	AF519244[c]	AF519273	
Nephasoma flagriferum	DNA100439	AF519243		AF519299
Nephasoma diaphanes	DNA100443	AF519245[d]		
Nephasoma diaphanes	DNA100445	AF519246[e]		
Thysanocardia nigra	DNA100606	AF519247[f]	AF519274	AF519300
Phascolionidae				
Phascolion strombus	DNA100101	AF519248	AF519275	AF519301
Themistidae				
Themiste lageniformis	DNA100229	AF519249[g]	AF519276	AF519302
Themiste minor	DNA100210	AF519250[h]	AF519277	AF519303
Phascolosomatidae				
Phascolosoma albolineatum	DNA100396	AF519251[i]	AF519278	
Phascolosoma granulatum	DNA100201	AF519252[j]	AF519279	AF519304
Phascolosoma granulatum		X79874		
*Phascolosoma nigrescens**	DNA100622	AY326292	AY445140	AY326299
Phascolosoma noduliferum	DNA100208	AF519253[k]	AF519280	AF519305
Phascolosoma perlucens	DNA100228	AF519254	AF519281	AF519306
Phascolosoma scolops	DNA100373	AF519255[l]	AF519282	AF519309

Continued on p. 286

Table 1. (Continued)

		18S rRNA	28S rRNA	Histone H3
Phascolosoma stephensoni	DNA100469	AF519256	AF519283	AF519310
Phascolosoma stephensoni	DNA100209	AF519257[m]	AF519284	AF519307
Phascolosoma stephensoni	DNA100485	AF519258[n]	AF519285	AF519308
Antillesoma antillarum	DNA100390	AF519259	AF519286	AF519311
Apionsoma (A.) misakianum	DNA100231	AF519260[o]	AF519287	
*Apionsoma (E.) pectinatum**	DNA100624	AY326293[p]	AY445142	AY326300
Aspidosiphonidae				
*Aspidosiphon (P.) fischeri**	DNA100620	AY326294		AY326301
Aspidosiphon (A.) misakiensis	DNA100205	AF119090	AF519288	AF519312
Aspidosiphon (P.) laevis	DNA100467	AF519261[q]	AF519289	AF519313
Aspidosiphon (P.) parvulus	DNA100202	AF119075	AF519290	AF519314
Aspidosiphon (P.) steenstrupii	DNA100232	AF519262	AF519291	AF519315
Cloeosiphon aspergillus	DNA100393	AF519263	AF519292	AF519316
Lithacrosiphon cristatus	DNA100623*	AY326295	AY445141	AY326302

*Asterisks after species names indicate new sequences for this study. For incomplete 18S rRNA sequences, the number of bp sequenced (excluding primers) is indicated.
[a] 376 bp sequenced; [b] 381 bp; [c] 943 bp; [d] 383 bp; [e] 383 bp; [f] 988 bp; [g] 460 bp; [h] 923 bp; [i] 394 bp; [j] 1282 bp; [k] 1429 bp; [l] 1064 bp; [m] 394 bp; [n] 602 bp; [o] 378 bp; [p] 1432 bp; [q] 1389 bp.

overall incongruence (Maxmen et al., 2003). As there is little conflict among the three genes (Maxmen et al., 2003), separate analyses of the three datasets were not attempted here.

Due to length differences, fixed homology statements (alignments) were not implied for the two ribosomal genes which were analyzed using the direct optimization method (Wheeler, 1996). For the protein-coding gene histone H3 no insertions/deletions had to be inferred and the sequences were treated as prealigned. Hundred replicates of spr, tbr and tree fusing and tree drifting were performed.

Combined analysis of molecular and morphological data

The morphological data matrix used in the analysis of the Phascolosomatidea was combined with the molecular data but only those taxa for which sequence data were available were included. The analysis was performed with the same parameter set as in the analysis of the molecular data alone. All data files, batch file and results can be downloaded from the following website: http://www.mcz.harvard.edu/Departments/InvertZoo/giribet_data.htm.

The results of the molecular data only (Fig. 5) and of the combined analysis (Fig. 6) were very similar. We will therefore base our discussion on the results of the combined analysis and mention discrepancies wherever appropriate.

While showing support for the monophyly of the Sipuncula, there is low branch support for the relationships among outgroup taxa, and the sister group to the Sipuncula could not be determined. With respect to the ingroup, *Sipunculus* is the sister group to the remaining genera (*Xenosiphon*, *Siphonomecus* and *Onchnesoma* not studied). This second clade includes two main groups: the first one comprises *Themiste*, *Phascolopsis*, *Golfingia*, *Thysanocardia*, *Nephasoma*, and *Phascolion*; the second one includes *Siphonosoma cumanense* as the sister group to a clade containing *Siphonosoma vastum* and the Phascolosomatiformes.

Overall, our results are very similar to those of Maxmen et al. (2003). The main difference between both analyses is the placement of *Apionsoma misakianum*: in our combined analysis its position is unresolved whereas in the previous analysis it is the sister taxon to a clade comprising the remaining Phascolosomatidea and *Siphonosoma cumanense*. This placement is also

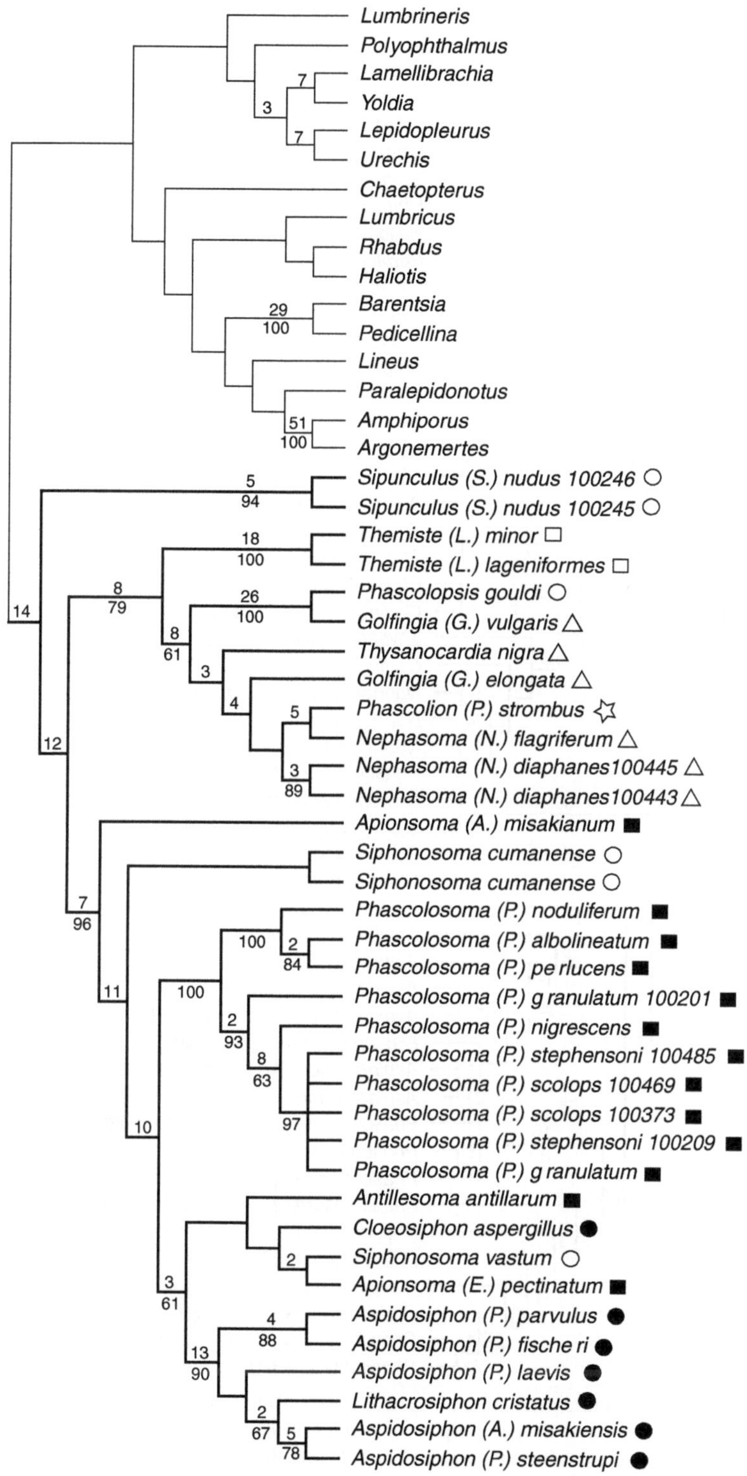

Figure 5. Strict consenus tree of five most parsimonious trees of length 4098, based on the analysis of DNA sequence data (18S rRNA, 28S rRNA and histone H3) alone. Branch support: jackknife proportions (36% deletion) underneath branches, Bremer support values on top of branches. Symbols after taxon names as in Figure 4.

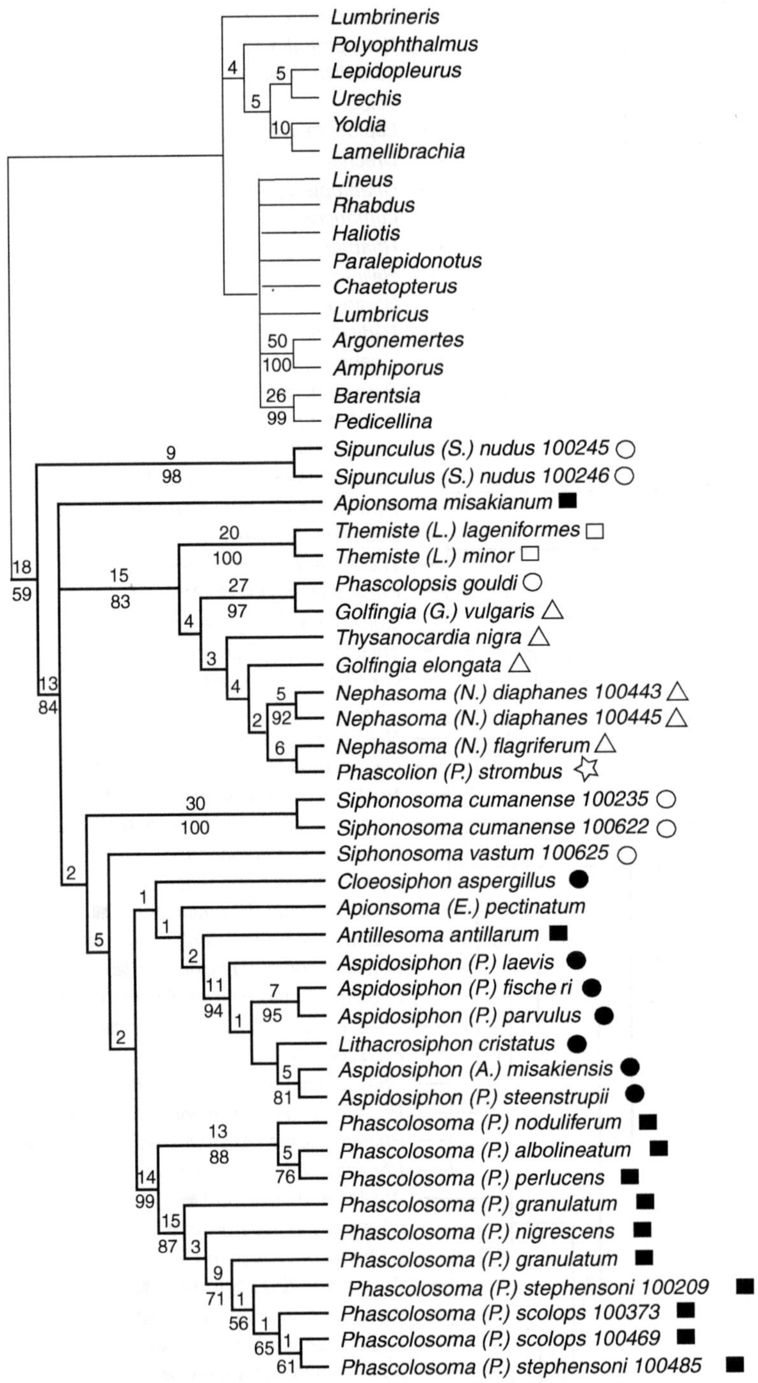

Figure 6. Strict consensus tree of three most parsimonious trees of length 4186, based on the combined analysis of 18S RNA, 28S rRNA, histone H3 and morphological data. Branch support: jackknife proportions (36% deletion) underneath branches, Bremer support values on top of branches. Symbols after taxon names as in Figure 4.

supported in our analysis of the molecular data alone (Fig. 5) and in one of the three shortest trees in the combined analysis. These results apparently indicate that nuchal tentacles may have evolved more than once in the Sipuncula. It should be noted, however, that the placement of

Apionsoma misakianum might be an artifact. The 18S rRNA sequence is currently incomplete for this species and histone H3 has not been successfully sequenced. Once full sequences are available, the phylogenetic affinities of *Apionsoma misakianum* will probably become less ambiguous.

Of the currently recognized families, only the Themistidae is monophyletic according to our analysis, but only 2 out of 10 species have been analyzed, both belonging to the same subgenus. Morphologically, Themistidae are characterized by stem-like extensions of the oral disk which bear the tentacles (Cutler, 1994).

The Phascolosomatidae and Aspidosiphonidae could be re-defined with slight modifications in their diagnosis: if *Antillesoma antillarum* and *Apionsoma pectinatum* were excluded from the Phascolosomatidae and moved to the Aspidosiphonidae, both families would be monophyletic. This would result in a monogeneric Phascolosomatidae, morphologically defined by two pairs of retractor muscles and laterally compressed, posteriorly directed hooks, arranged in rings and with a characteristic clear streak (Fig. 2c). Within *Phascolosoma*, species for which several representatives were included do not necessarily form clades. This is the case for *P. granulatum* and for *P. stephensoni*. One of the sequences for *P. granulatum* is from GenBank and we cannot rule out a misidentification. Species within *Phascolosoma* show similar morphology and intraspecific variability. A more comprehensive study of the genus, including multiple representatives of each species from a variety of locations would shed more light on species boundaries and intrageneric phylogeny. Morphological synapomorphies for the Aspidosiphonidae would be less obvious.

The tree topology within the Aspidosiphonidae differs between the combined analysis and the analysis of molecular data alone. In the combined analysis *Cloeosiphon aspergillus* is the sister taxon to a clade comprising the *Aspidosiphon* species, *Antillesoma antillarum*, *Apionsoma pectinatum* and *Lithacrosiphon cristatus*. The same result, with the exception of *L. cristatus* and *Apionsoma pectinatum* which were not sampled, was obtained by Maxmen et al. (2003). In our analysis of molecular data alone, the Aspidosiphonidae comprise two clades (Fig. 5). One of them contains *Antillesoma antillarum*, *Cloeosiphon aspergillus*, *Siphonosoma vastum* and *Apionsoma pectinatum*. With the exception of Cloeosiphon, none of these species has previously been assiociated with the Aspidosiphonidae. The 18S rRNA sequences are currently lacking for *Siphonosoma vastum* and its placement might be preliminary. The grouping of *Antillesoma antillarum* and *Apionsoma pectinatum* with the Aspidosiphonidae in both the molecular and the combined analysis is also surprising because both species do not have an anal or caudal shield as typical for the Aspidosiphonidae. This could imply that the anal shield in *Cloeosiphon* is not homologous to the anal shield in *Aspidosiphon* and *Lithacrosiphon*. Considering that the anal shield shows quite a different morphology among these genera (see section on external morphology) this interpretation could be plausible.

The Golfingiidae would be monophyletic if *Phascolion strombus* and *Phascolopsis gouldi* were incorporated. Again, morphological synapomorphies are not obvious. More drastic re-arrangements would be required to accommodate the taxa Sipunculidea, Sipunculiformes, and Sipunculidae. The Sipunculidae are clearly polyphyletic with *Sipunculus* as the sister group to the remaining Sipuncula, *Siphonosoma cumanense* at the base of the Phascolosomatidea clade and *Phascolopsis gouldi* the sister group to *Golfingia vulgaris*. *Siphonosoma* does not appear as monophyletic but, again, this might be due to the lack of 18S rRNA data in *Siphonosoma vastum*.

Maxmen et al. (2003) showed that the rooting between *Sipunculus* and the remaining Sipuncula is not dependent on the choice of outgroups, and therefore the results obtained here are interpreted the same way. It would be interesting to include more species of *Sipunculus* as well as *Siphonomecus* and *Xenosiphon* in the analysis. *Xenosiphon* might group with *Sipunculus* whereas *Siphonomecus* might group with *Siphonosoma* (Fig. 3). *Phascolopsis gouldi* has had a confusing nomenclatural history, but prior to Stephen (1964) who moved it into its own monotypic genus and Cutler & Gibbs (1985) who shifted the genus into a different family, it was associated with species that are currently members of the genus *Golfingia*.

Conclusions

Our molecular database currently includes 28 out of the 147 recognized sipunculan species, a rough 20% of the known diversity for the entire phylum. These cover 14 out of the 17 genera, except *Xenosiphon*, *Siphonomecus* and *Onchnesoma*. For most genera, except *Phascolion* and *Thysanocardia* we have more than one representative, enabling us to conduct initial tests of monophyly, although in a few cases we were not able to obtain samples of the different subgenera. Some genera, especially *Nephasoma* and *Onchnesoma* are difficult to obtain because most or all of the species occur beyond Scuba diving depth and often only in waters deeper than 500 m. Scientific cruises aimed for benthic deep-sea fauna are rare today and even rarer are dredging activities, making it difficult to obtain fresh and appropriately fixed material for molecular studies.

The main difficulties we are facing with regard to morphological data are the problems in determining the sister taxon of the Sipuncula and the paucity of phylogenetically informative characters. We plan to conduct more detailed morphological studies in the future, employing histological, ultrastructural and immunohistochemical techniques, to address these problems. For example, the arrangement of the body wall musculature might provide more characters than we are currently using. Another potential source of characters are developmental data which need to be explored thoroughly in a phylogenetic context.

Eventually our results should lead to a revision of sipunculan taxonomy and systematics. This would include elimination of taxon names that are nested within others, for example *Lithacrosiphon*, which appears to be nested in *Aspidosiphon* or *Phascolopsis gouldi*, nested within *Golfingia*. Most of the family names could be retained as long as a few species are moved from one family to another with slight changes in family definitions as required.

Acknowledgements

We thank Günter Purschke und Thomas Bartolomaeus for organizing an outstanding symposium and for inviting us to present our research. Pat Hutchings, J. Hylleberg, Damhnait McHugh, Michele Nishiguchi, Teruaki Nishikawa, Cruz Palacín, Mary Rice, Iñaki Saiz-Salinas, and Xavier Turon provided samples for this study. Iñaki Saiz-Salinas and Harlan K. Dean assisted with specimen work. Amy Maxmen and Burnett King generated most of the sequence data used in this study. Funding was provided by the MarCraig foundation and by the Fundamental Biology Program of NASA. Damhnait McHugh and two anonymous reviewers provided constructive criticism and valuable suggestions which improved the quality of this manuscript.

Morphological characters

Tentacles

Nuchal tentacles (nuc): tentacles arranged in an arc around the nuchal organ (Fig. 1c); 0 = absent; 1 = present.

Peripheral tentacles (per): tentacles arranged in a circle around the mouth (Fig. 1b); 0 = absent, 1 = present.

Branched tentacles (bra): tentacles arising from stemlike outgrowths; 0 = absent, 1 = present.

Nephridia

Number of nephridia (npn): 0 = paired, 1 = single.

Nephridial shape (nps): 0 = unilobed, 1 = bilobed.

Nephridial attachment (nat): 0 = unattached, 1 = (partially) attached to body wall.

Body wall

Coelomic extensions in body wall (coe): in several sipunculan genera the coelom extends into the body wall in the form of coelomic canals or sacs. The canals either run longitudinally between the longitudinal muscle bands (*Sipunculus*), or diagonally as short, subcutaneous canals (*Xenosiphon*). In *Siphonosoma* and *Siphonomecus* the extensions are more sac-like (Ruppert & Rice, 1995). 0 = absent; 1 = present.

Longitudinal muscles (lmu): 0 = continuous; 1 = in bands.

Anal shield

All currently recognized Aspidosiphonidae are characterized by a hardened shield at the anterior end of the trunk. However, its chemical composition, extend and morphology are variable among the species.

Dorsal anal shield (dsh): calcareous or horny protein shield at anterior end of trunk; 0 = absent, 1 = present.

Shape of dorsal anal shield (sha): 0 = ± flat, 1 = cone-shaped.

Pineapple shield (psh): shield at anterior end of trunk, composed of calcareous plates; 0 = absent, 1 = present.

Grooves in anal shield (gro): 0 = absent, 1 = present.

Spindle muscle

Spindle muscle (spm): slender, thread-like muscle running through the intestinal coil; 0 = absent, 1 = present.

Attachment of spindle muscle (att): 0 = attached at posterior end of trunk, 1 = ends in gut coil.

Hooks

Hooks on introvert (hoo): 0 = absent, 1 = present.

Hooks in rings (rin): 0 = absent, 1 = present.

Number of rings of hooks (nrh): 0 = < 50; 1 = > 50.

Scattered hooks (sca): hooks not arranged in rings; 0 = absent, 1 = present.

Basal spinelets (spi): small pointed units found at the base of introvert hooks (Fig. 2a); 0 = absent, 1 = present.

Bidentate hooks (bid): hooks with two pointed teeth (Fig. 2B); 0 = absent, 1 = present.

Secondary tooth (sec): accessory tooth on posterior concave side of hooks (Fig. 2c): 0 = absent, 1 = present.

Shape of secondary tooth (sst): 0 = blunt, 1 = pointed.

Anterior clear triangle (tri): clear space in anterior basal position in hook (Fig. 2c): 0 = absent, 1 = present.

Clear streak (cls): tube-like hollow space extending from the base toward the tip of the hook (Fig. 2c); 0 = approximately uniform diameter, 1 = with distinct bulge.

Crescent (cre): crescent-shaped clear space posterior to clear streak; 0 = absent, 1 = present

Posterior basal structures (pbs): structures at posterior basal edge of hooks (see Cutler, 1994, Fig. 45); 0 = absent, 1 = present, Type of posterior basal structures (tbs): 0 = warts, 1 = long processes, 2 = warts.

Angle of hooks (ang): angle of hook tip relative to main axis; 0 = < 90, 1 > 90.

Pyramidal hooks (pyr): hooks with triangular bases; 0 = absent, 1 = present.

Conical hooks (con): hooks with a nearly circular cross section; 0 = absent, 1 = present.

Other

Pigmented introvert bands (pib): 0 = absent, 1 = present.

Contractile vessel villi (cvv): The contractile vessel is part of the tentacular coelomic system. It runs dorsally along the esophagus, has a coelomic lining. It contains hemocytes and is considered an analogue to a blood vascular system. In some species, in particular of the genus *Themiste*, digitiform villi are present along the length of the vessel (Rice, 1993); 0 = absent, 1 = present. Type of contractile vessel villi (tvv) (see Cutler 1994, Fig. 36): 0 = villi, 1 = tubules.

Introvert retractor muscles (irm): This set of strong muscles insert anteriorly near the brain and are posteriorly attached to the body wall (Fig. 1d); 0 = 2 pairs, 1 = 1 pair.

Introvert/trunk length (itl): 0 = introvert < 75% trunk length, 1 = 75–200%, 2 = > 200%.

References

Åkesson, B., 1958. A study of the nervous system of the Sipunculoideae, with some remarks on the development of the two species *Phascolion strombi* Montagu and *Golfingia minuta* Keferstein. Undersokningar over Oresund 38: 1–249.

Backeljau, T., B. Winnepenninckx & L. De Bruyn, 1993. Cladistic analysis of metazoan relationships: a reappraisal. Cladistics 9: 167–181.

Banta, W. C. & M. E. Rice, 1976. A restudy of the Middle Cambrian Burgess Shale fossil worm, *Ottoia prolifica*. In Rice, M. E. & M. Todorovic (eds), Proceedings of the International Symposioum on the Biology of the Sipuncula and Echiura. Nauçno Delo Press, Belgrade: 79–90.

Boore, J. L. & J. L. Staton, 2002. The mitochondrial genome of the sipunculid *Phascolopsis gouldii* supports its association with Annelida rather than Mollusca. Molecular Biology and Evolution 19: 127–137.

Brett, C. E. & J. F. Cottrell, 1982. Substrate specificity in the Devonian tabulate coral *Pleurodictyum*. Lethaia 15: 247–262.

Brett, C. E., W. D. Liddell & K. L. Derstler, 1983. Late Cambrian hard substrate communities from Montana/Wyoming: the oldest known hardground encrusters. Lethaia 16: 281–289.

Brusca, R. C. & G. J. Brusca, 1990. Invertebrates. Sinauer Associates, Sunderland.

Brusca, R. C. & G. J. Brusca, 2003. Invertebrates. Second edition Sinauer Associates, Sunderland.

Conway Morris, S., 1989. Burgess Shale faunas and the Cambrian explosion. Science 246: 339–346.

Conway Morris, S., 1998. The Crucible of Creation. The Burgess Shale and the Rise of Animals. Oxford University Press, Oxford, 242 pp.

Cutler, E. B., 1994. The Sipuncula. Their Systematics, Biology and Evolution. Cornell University Press, Ithaca, NY, 453 pp.

Cutler, E. B. & P. E. Gibbs, 1985. A phylogenetic analysis of higher taxa in the phylum Sipuncula. Systematic Zoology 34: 162–173.

Cuvier, B., 1830. Le règne animal distribué d'après son oranisation, pour servir de base a l'histoire naturelle des animaux et d'introduction à l'anatomie comparèe, Ser. 2, Vol. 3. Deterville, Paris: 1–244.

Dybas, L., 1981. Sipunculans and echiuroids. In Ratcliff, N. A. & A. F. Rowley (eds), Invertebrate Blood Cells. Academic Press, New York: 161–188.

Edmonds, S. J., 2000. Phylum Sipuncula. In Beesley, P. L., G. J. B. Ross & C. J. Glasby (eds), Fauna of Australia 4, Polychaetes & Allies: The Southern Synthesis. CSIRO Publishing, Melbourne: 375–400.

Eernisse D. J., J. S. Albert & F. E. Anderson, 1992. Annelida and Arthropoda are not sister taxa: a phylogenetic analysis of spiralian metazoan morphology. Systematic Biology 41: 305–330.

Erber, A., D. Riemer, M. Bovenschulte & K. Weber, 1998. Molecular phylogeny of metazoan intermediate filament proteins. Journal of Molecular Evolution 47: 751–762.

Eriksson, T., 1998. AutoDecay. Stockholm, Distributed by the Author. Department of Botany, Stockholm University.

Ferrier, D. E. & P. W. Holland, 2001. Sipunculan ParaHox genes. Evolution & Development 3: 263–270.

Florkin, M., 1975. Biochemical evidence for the phylogenetic relationships of the Sipuncula. In Rice, M. E. & M. Todorovic (eds), Proceedings of the International Symposium on the biology of the Sipuncula and Echiura. Naucno Delo Press, Belgrade: 95–110.

Frey, R. W., G. Pemberton & J. A. Fagerstrom, 1984. Morphological, ethological and environmental significance of the ichnogenera *Scoyenia* and *Ancorichnus*. Journal of Paleontology 58: 511–528.

Gibbs, P. E. & E. B. Cutler, 1987. A classification of the phylum Sipuncula. Bulletin of the British Museum of Natural History (Zoology) 52: 43–58.

Gill, G. A. & A. G. Coates, 1977. Mobility, growth patterns and substrate in some fossil and Recent corals. Lethaia 10: 119–134.

Giribet, G., 2002. Current advances in the phylogenetic reconstruction of metazoan evolution. A new paradigm for the Cambrian explosion? Molecular Phylogenetics and Evolution 24: 345–357.

Giribet, G., D. L. Distel, M. Polz, W. Sterrer & W. C. Wheeler, 2000. Triploblastic relationships with emphasis on the acoelomates and the position of Gnathostomulida, Cycliophora, Platyhelminthes, and Chaetognatha: a combined approach of 18S rDNA sequences and morphology. Systematic Biology 49: 539–562.

Green, C. R. & P. R. Berquist, 1982. Phylogenetic relationships within the Invertebrata in relation to the structure of septate junctions and the development of occluding junctional types. Journal of Cell Science 53: 279–306.

Guralnick, R. 2002. A recapitulation of the rise and fall of the cell lineage research program: the evolutionary developmental relationship of cleavage to homology, body plans and life history. Journal of History of Biology 35: 527–567.

Guralnick, R. P. & Lindberg, D. R. 2001. Reconnecting cell and animal lineages: what do cell lineages tell us about the evolution and development of Spiralia? Evolution 55: 1501–1519.

Henry, R. P., 1987. Invertebrate red blood cell carbonic anhydrase. Journal of Experimental Zoology 242: 113–116.

Holland, P. W. H., 1998. Major transitions in animal evolution: A developmental genetic perspective. American Zoology 38: 878–887.

Huang, D, J. Chen, J. Vannier & J. I. Saiz Salinas, 2004. Early Cambrian sipunculan worms from southwest China. Proceedings of the Royal Society of London, Series B 271: 1671–1676.

Hyman, L. H., 1959. The Invertebrates 5, Smaller Coelomate Groups. McGraw-Hill, New York, 783 pp.

Ionescu-Varo, M. & M. Tufescu, 1982. Quantitative remarks on immune evolution in the animal series. Rev. Roumaine Biol. série Biol. Animale 27: 29–39.

Keferstein, W., 1863. Beiträge sur Kenntnis der Gattung *Phascolosoma* F.S. Leuck. Untersuchungen über niedere Seethiere. Zeitschrift für Wissenschaftliche Zoologie 12: 35–51.

Keferstein, W., 1865a. Beiträge zur anatomischen und systematischen Kenntnis der Sipunculiden. Nachrichten der Gesellschaft der Wissenschaften, Göttingen 1865: 189–209.

Keferstein, W., 1865b. Beiträge zur anatomischen und systematischen Kenntniss der Sipunculiden. Zeitschrift für Wissenschaftliche Zoologie 15: 404–445.

Keferstein, W., 1866. Untersuchungen uber einige amerikanische Sipunculiden. Nachrichten von der Königlichen Gesellschaft der Wissenschaften und der Georg-August-Universität zu Göttingen 14: 215–228.

Keferstein, W., 1867. Untersuchungen über einige amerikanische Sipunculiden. Zeitschrift Fur Wissenschaftliche Zoologie 17: 44–55.

Klepal, W., 1987. The anchoring fiber apparatus in the sperm of *Aspidosiphon* sp. (Sipunculida). European Journal of Cell Biology Suppl. 0: 18.

Kotetskii, E. Y., 1984. The phospholipid composition of Spongia, Coelenterata and free living Platyhelminthes, Nemertini, Annelida, Sipunculida and Echiurida. Biologiya Morya 0: 46–53.

Lake, J. A., 1990. Origin of the Metazoa. Proceedings of the National Academy of Sciences of the United State of America 87: 763–766.

Lamarck, J. P. B. A. de M., 1816. Histoire naturelle des animaux sans vertébrés. Vol. 3. Verdière: Paris, 683 pp.

Lehman, W. & A. G. Szent-Györgyi, 1975. Regulation of muscular contraction: distribution of actin control and myosin control in the animal kingdom. Journal of General Physiology 66: 1–30.

Livingstone, D. R., A. Dezwaan, M. Leopold & E. Marteijn, 1983. Studies on the phylogenetic distribution of pyruvate oxidoreductases. Biochem. Syst. Ecol. 1: 415–425.

Maddison, D. R. & W. P. Maddison, 2001. MacClade. Sinauer Associates, Sunderland, MA.

Mangum, C. P., 1990. The role of physiology and biochemistry in understanding animal phylogeny. Proceedings of the Biological Society of Washington 103: 235–247.

Marti Mus, M. & J. Bergstrom, 2001. The skeleton-muscular system of hyolithids. American Zoologist 41: 1514.

Maxmen, A. B., B. F. King, E. B. Cutler & G. Giribet, 2003. Evolutionary relationships within the protostome phylum Sipuncula: a molecular analysis of ribosomal genes and histone H3 sequence data. Molecular Phylogenetics and Evolution 27: 489–503.

McBride, E. F. & M. D. Picard, 1991. Facies implications of Trichichnus and Chondrites in turbidites and hemipelagites, Marnoso-arenacea formation (Miocene), northern Apennine, Italy. Palaios 6: 281–290.

Meglitsch, P. A. & F. R. Schram, 1991. Invertebrate Zoology. Oxford University Press, Oxford.

Nichols, D. 1967. The origin of echinoderms. Symposia of the Zoological Society of London 20: 209–229.

Nielsen, C., 1987. Structure and function of metazoan ciliary bands and their phylogenetic significance. Acta Zoologica 68.

Nielsen, C., N. Scharff & D. Eibye-Jacobsen, 1996. Cladistic analyses of the animal kingdom. Biological Journal of the Linnean Society 57, 385–410.

Ocharan, F. J., 1974. Sobre los nefridios de *Phascolosoma granulatum* (Sipuncula). Separata de la revista de la "Facultad de Ciencias" 15: 21–40.

Pemberton, S. G., D. R. Kobluk, R. K. Yeo & M. J. Risk, 1980. Boring Trypanites at the Silurian-Devonian disconformity in southern Ontario, Canada. Journal of Paleontology 54: 1258–1266.

Peterson, K. & D. J. Eernisse, 2001. Animal phylogeny and the ancestry of bilaterians: inferences from morphology and 18S rDNA gene sequences. Evolution & Development 3: 170–205.

Pickford, G. E., 1947. Sipunculida. In Encyclopedia Britannica, Vol. 20. University of Chicago, Chicago: 717–718b.

Pilger, J. F. & M. E. Rice, 1987. Ultrastructural evidence for the contractile vessel of sipunculans as a possible ultrafiltration site. American Zoology. 27: 810a.

Pisera, A., 1987. Boring and nestling organisms from Upper Jurassic coral colonies from Northern Poland. Acta Palaeontologia Polonica 32: 83–104.

Purschke, G., F. Wolfrath & W. Westheide, 1997. Ultrastructure of the nuchal organ and cerebral organ in *Onchnesoma squamatum* (Sipuncula, Phascolionidae). Zoomorphology 117: 23–31.

Qautrefages, A. de, 1847. Mémoires sur l'échine de Gaertner. Annales des Sciences Naturelles, Zoologie, Série 3: 307.

Rice, M. E., 1985. Sipuncula: developmental evidence for phylogenetic inference. In Morris, S. C., J. D. George, R. Gibson & H. M. Platt (eds), The Origins and Relationships of Lower Invertebrates. Oxford University Press, Oxford: 274–296.

Rice, M. E., 1993. Sipuncula. In Harrison, F. W. & M. E. Rice (eds), Microscopic Anatomy of Invertebrates: Onychophora, Chilopoda, and lesser Protostomata. Wiley-Liss, New York: 237–325.

Romero-Wetzel, M. B., 1987. Sipunculans as inhabitants of very deep narrow burrows in deep-sea sediments. Marine Biology 96: 87–92.

Runnegar, B., B. Pojeta, N. J. Morris, J. D. Taylor, M. E. Taylor & G. McClung, 1975. Biology of the *Hyolitha*. Lethaia 8: 181–192.

Ruppert, E. E. & M. E. Rice, 1995. A functional organization of dermal coelomic canals in *Sipunculus nudus* (Sipuncula) with a discussion of respiratory designs in sipunculans. Invertebrate Zoology 114: 51–63.

Scheltema, A. H., 1993. Aplacophora as progenetic aculiferans and the coelomate origin of mollusks as the sister taxon of Sipuncula. Biological Bulletin 184: 57–78.

Sedgwick, A., 1898. A student's textbook of zoology. S. Sonnenschein, London; Macmillan, New York, 783 pp.

Selenka, E., 1875. Eifurchung und Larvenbildung von *Phascolosoma elongatum*. Zeitschrift für Wissenschaftliche Zoologie 25: 442–450.

Selenka, E., 1885. Report on the Gephyrea collected by H. M. S. Challenger during 1873–76. Report of scientific results of the Voyage of Challenger Zoology 13: 1–25.

Selenka, E., 1888. On the Gephyrea of the Mergui Archipelago collected for the Trustees of the Indian Museum. J. Linn. Soc. London, Zool. 21: 220–222.

Selenka, E., 1897. Die Sipunculiden-Gattung *Phymosoma*. Zoologischer Anzeiger 20: 460.

Selenka, E., J. G. de Man & C. Bülow, 1883. Die Sipunculiden, eine systematische Monographie. Semper Reisen im Archipel der Phillippinen II 4: 1–131.

Smith, S. W., R. Overbeek, C. R. Woese, W. Gilbert & P. M. Gillevet, 1994. The Genetic Data Environment: an expandable GUI for multiple sequence analysis. Computer Applications in the Biosciences 10: 671–675.

Sørensen, M. V., P. Funch, E. Willerslev, A. J. Hansen & J. Olesen, 2000. On the phylogeny of Metazoa in the light of Cycliophora and Micrognathozoa. Zoologischer Anzeiger 239: 297–318.

Stephen, A. C., 1964. A revision of the classification of the phylum Sipuncula. Annals and Magazine of Natural History 7: 457–462.

Stephen, A. C. & S. J. Edmonds, 1972. The phyla Sipuncula and Echiura. Trustees British Mus. (Nat. Hist.), London, 528 pp.

Swofford, D. L., 2000. PAUP*: Phylogenetic Analysis Using Parsimony (and Other Methods). Sinauer Associates, Sunderland, MA.

Voss-Foucart, M. F., S. Barzin, C. Jeuniaux & J. C. Bussers, 1977. Étude comparée de la composition chimique des régions souples et durcies de la cuticule de quatre espèces de sipunculiens. Cahiers de Biologie Marine 18: 135–145.

Wetzel, A. & F. Werner, 1981. Morphology and ecological significance of Zoophycos in deep-sea sediments off north-west Africa. Palaeogeogr. Palaeoclimatology Palaeoecology 32: 185–212.

Wheeler, W. C., 1996. Optimization alignment: the end of multiple sequence alignment in phylogenetics? Cladistics 12: 1–9.

Wheeler, W. C., D. Gladstein & J. DeLaet, 2002. POY: the Optimization of Alignment Characters, version 3.0. Program and documentaition available from ftp.amnh.org/pub/molecular.

Wilheim, M. L. & F. X. Wilheim, 1978. Subunit structure of erythroid cells chromatin. Studia Biophysica 67: 105–106.

Winnepenninckx, B., T. Backeljau & R. De Wachter, 1995. Phylogeny of protostome worms derived from 18S rRNA sequences. Molecular Biology and Evolution 12: 641–649.

Zrzavý, J., V. Hypša & D. F. Tietz, 2001. Myzostomida are not annelids: molecular and morphological support for a clade of animals with anterior sperm flagella. Cladistics 17: 170–198.

Zrzavý, J., S. Mihulka, P. Kepka, A. Bezdek, D. Tietz, 1998. Phylogeny of the Metazoa based on morphological and 18S ribosomal DNA evidence. Cladistics 14: 249–285.

Appendix A

Morphological dataset and characters for cladistic analysis of Phascolosomatidea; ? = unknown state; x napplicable.

	nuc	per	bra	npn	nps	nat	coe	dsh	sha	gro	psh	spm	att	hoo	rin	nrh	sca	Spi	bid	sec	ssh	tri	cls	cre	pbs	tbs	ang	pyr	con	pib	lmu	cvv	tvv	irm	itl
Antillesoma antillarum	1	0	x	0	0	1	0	0	x	x	0	1	0	0	x	X	x	x	x	x	x	x	x	x	x	x	x	0	0	0	1	1	0	0	0
Apionsoma misakianum	1	0	x	0	1	?	0	0	x	x	0	1	0	1	1	1	0	1	0	0	x	0	0	0	0	x	1	0	0	0	0	0	x	0	2
Apionsoma murinae	1	0	x	0	0,1	0	0	0	x	x	0	?	0	1	1	0	0	1	0	0	x	0	0	0	0	x	0	0	0	0	0	0	x	0	2
Apionsoma trichocephalus	1	0	x	0	1	?	0	0	x	x	0	1	0	0	x	x	x	x	x	x	x	x	x	x	x	x	x	0	0	0	0	0	x	0	2
Apionsoma pectinatum	1	0	x	0	1	1	0	0	x	x	0	1	1	1	1	0	0	1	0	0	x	0	0	0	0	x	1	0	0	0	1	0	x	0	2
Aspidosiphon albus	0	0	x	0	0	?	0	1	0	0	0	1	0	0	x	x	x	x	x	x	x	x	x	x	x	x	x	0	0	0	0	0	x	1	2
Aspidosiphon mexicanus	0	0	x	0	0	?	0	1	0	0	0	1	0	1	0	x	1	0	0	0	x	0	0	0	0	x	0	0	0	0	0	0	x	1	2
Aspidosiphon thomassini	0	0	x	0	0	1	0	1	0	0	0	1	0	0	x	x	x	x	x	x	x	x	x	x	x	x	x	0	0	0	0	0	x	1	2
Aspidosiphon venabulum	?	0	x	0	0	1	0	1	0	0	0	1	0	0	x	x	x	x	x	x	x	x	x	x	x	x	x	0	0	0	0	0	x	1	2
Aspidosiphon zinni	1	0	x	0	0	?	0	1	0	0	0	1	0	1	0	x	1	0	0	0	x	0	0	0	0	x	0	0	0	0	0	0	x	1	1
Aspidosiphon elegans	1	0	x	0	0	1	0	1	0	0	0	1	0	1	1	1	1	0	1	0	x	0	0	0	0	x	0	0	1	0	0	0	x	1	2
Aspidosiphon exiguus	1	0	x	0	0	?	0	1	0	0	0	1	0	1	1	0	1	0	1	0	x	0	0	0	0	x	0	0	0	0	0	0	x	1	2
Aspidosiphon gosnoldi	1	0	x	0	0	1	0	1	0	0	0	1	0	1	1	0	1	0	1	0	x	0	0	0	0	x	0	1	0	0	0	0	x	1	1,2
Aspidosiphon gracilis	1	0	x	0	0	1	0	1	0	0	0	1	0	1	1	0	1	0	0	0	x	0	0	0	0	x	0	1	0	0	0	0	x	1	1
Aspidosiphon misakiensis	1	0	x	0	0	1	0	1	0	0	0	1	0	1	1	0	1	0	1	0	x	0	0	0	0	x	0	0	0	0	0	0	x	1	1
Aspidosiphon muelleri	1	0	x	0	0	1	0	1	0	1	0	1	0	1	1	0	1	0	1	0	x	0	0	0	0	x	0	1	0	0	0	0	x	1	1
Aspidosiphon spiralis	1	0	x	0	0	1	0	1	0	0	0	1	0	1	1	0	0	0	1	0	x	0	0	0	0	x	0	0	0	0	0	0	x	1	1
Aspidosiphon coyi	1	0	x	0	0	?	0	1	0	1	0	1	0	1	1	0,1	0	0	1	0	x	0	0	0	0	x	0	1	0	0	1	0	x	1	1
Aspidosiphon fischeri	1	0	x	0	0	1	0	1	0	0	0	1	0	1	1	0	1	0	1	0	x	0	0	0	0	x	0	1	0	0	1	0	x	1	1
Aspidosiphon laevis	1	0	x	0	0	1	0	1	0	1	0	1	0	1	1	0,1	1	0	0	0	x	0	0	0	0	x	0	0	0	0	1	0	x	1	1
Aspidosiphon parvulus	1	0	x	0	0	1	0	1	0	0	0	1	0	1	1	0	1	0	1	0	x	0	0	0	0	x	0	1	0	0	1	0	x	1	1
Aspidosiphon planoscutatus	1	0	x	0	0	?	0	1	0	0	0	1	0	1	1	0	0	0	0	0	x	0	0	0	0	x	0	0	0	0	1	0	x	1	1
Aspidosiphon steenstrupii	1	0	x	0	0	1	0	1	0	0	0	1	0	1	1	0	1	0	1	0	x	0	0	0	0	x	0	1	0	0	1	0	x	1	1
Aspidosiphon tenuis	1	0	x	0	0	0	0	1	0	0	0	1	0	1	1	0	1	0	1	0	x	0	0	0	0	x	0	0	0	0	1	0	x	1	1
Cloesiphon aspergillus	1	0	x	0	0	1	0	0	x	x	1	1	0	1	1	0	0	0	1	0	x	0	0	0	0	x	1	0	0	0	0	0	x	1	1
Lithacrosiphon cristatus	1	0	x	0	0	0	0	1	1	1	0	1	0	1	1	0	1	0	1	0	x	0	0	0	0	x	0	0	0	0	1	0	x	1	1
Lithacrosiphon maldiviensis	1	0	x	0	0	1	0	1	1	0	0	1	0	1	1	?	0	0	1	0	x	0	0	0	0	x	0	0	0	0	1	0	x	1	1
Phascolosoma lobostomum	1	0	x	0	0	0	0	0	x	x	0	1	0	1	1	0	0	0	0	1	1	0	0	0	0	x	0	0	0	?	0	0	x	0	0
Phascolosoma capitatum	1	0	x	0	0,1	1	0	0	x	x	0	1	0	1	1	0	0	0	0	0	x	0	0	0	0	0	0	0	0	0	0	0	x	0	1
Phascolosoma agassizii	1	0	x	0	0	1	0	0	x	x	0	1	0	1	1	0	0	0	0	0,1	0	1	0	0	1	0	0	0	0	1	1	0	x	0	1
Phascolosoma albolineatum	1	0	x	0	0	1	0	0	x	x	0	1	0	1	1	0	0	0	0	1	0	1	0	0	1	0	1	0	0	1	1	0	x	0	1
Phascolosoma annulatum	1	0	x	0	0	1	0	0	x	x	0	1	0	1	1	0	0	0	0	0	x	1	0	0	1	0	0	0	0	1	1	0	x	0	1
Phascolosoma arcuatum	1	0	x	0	0	1	0	0	x	x	0	1	0	1	1	1	0	0	0	0	x	0	1	0	1	0	0	0	0	0	1	0	x	0	1
Phascolosoma glabrum	1	0	x	0	0	1	0	0	x	x	0	1	0	1	1	0	0	0	0	1	0	0	0	0	1	2	1	0	0	1	1	0	x	0	1

Continued on p. 296

Appendix A. (Continued).

	nuc	per	bra	npn	nps	nat	coe	dsh	sha	gro	psh	spm	att	hoo	rin	nrh	sca	Spi	bid	sec	ssh	tri	cls	cre	pbs	tbs	ang	pyr	con	pib	lmu	cvv	tvv	irm	itl
Phascolosoma granulatum	1	0	x	0	0	1	0	0	x	x	0	1	0	1	1	1	0	0	0	0,1	0	0	0	0	1	0	0	0	0	0	1	0	x	0	1
Phascolosoma maculatum	1	0	x	0	0	1	0	0	x	x	0	1	0	1	1	0	0	0	0	1	1	1	0	0	1	0	0	0	0	1	1	0	x	0	1
Phascolosoma meteori	1	0	x	0	0	1	0	0	x	x	0	1	0	1	1	0	0	0	0	0	x	0	0	0	?	?	0	0	0	?	1	0	x	0	0
Phascolosoma nigrescens	1	0	x	0	0	1	0	0	x	x	0	1	0	1	1	1	0	0	0	0,1	0	0	1	0	1	0	0	0	0	1	1	0	x	0	1
Phascolosoma noduliferum	1	0	x	0	0	1	0	0	x	x	0	1	0	1	1	1	0	0	0	0	x	0	0	0	0	x	0	0	0	0	1	0	x	0	1
Phascolosoma pacificum	1	0	x	0	0	1	0	0	x	x	0	1	0	1	1	1	0	0	0	1	0	1	0	0	1	0	0	0	0	0	1	0	x	0	1
Phascolosoma perlucens	1	0	x	0	0	1	0	0	x	x	0	1	0	1	1	0	0	0	0	1	0	1	0	0	1	0	0	0	0	1	1	0	x	0	1
Ph. saprophagicum	1	0	x	0	0	1	0	0	x	x	0	1	0	1	1	0	0	0	0	1	0	0	0	0	0	x	1	0	0	0	1	0	x	0	1
Phascolosoma scolops	1	0	x	0	0	1	0	0	x	x	0	1	0	1	1	0	0	0	0	0,1	0	1	0	0	1	0	0	0	0	1	1	0	x	0	1
Phascolosoma stephensoni	1	0	x	0	0	1	0	0	x	x	0	1	0	1	1	1	0	0	0	1	1	1	0	1	1	0	0	0	0	1	1	0	x	0	1
Phascolosoma turnerae	1	0	x	0	0	1	0	0	x	x	0	1	0	1	1	0,1	0	0	0	0	x	0	0	0	1	1	0,1	0	0	0	1	0	x	0	1
Sipunculus nudus	0	1	0	0	0	1	2	0	x	x	x	1	1	0	x	x	0	x	x	x	x	x	x	x	0	x	x	0	0	0	1	0	x	0	0
Golfingia elongata	0	1	0	0	0	0	0	0	x	x	x	1	1	1	1	0	0	0	0	0	0	0	0	0	0	x	0	0	0	0	0	0	x	0	0
Phascolion strombus	0	1	0	1	0	1	0	0	x	x	x	0	x	1	0	x	1	0	0	0	0	0	0	0	0	x	0	0	0	0	0	0	x	0	1
Themiste lageniformes	0	1	1	0	0	0	0	0	x	x	x	1	1	0	x	x	x	x	x	x	x	x	x	x	x	x	x	0	0	0	0	1	0	1	0
Siphonosoma cumanense	0	1	0	0	0	0	1	0	x	x	x	1	0	0	x	x	x	x	x	x	x	x	x	x	x	x	x	0	0	0	0	1	0	0	0
Phascolopsis gouldi	0	1	0	0	0	0	0	0	x	x	x	1	1	0	x	x	x	x	x	x	x	x	x	x	x	x	x	0	0	0	0	0	x	0	0
Nephasoma diaphanes	0	1	0	0	0	0	0	0	x	x	x	1	1	1	0	x	1	0	0	0	x	0	0	0	0	x	0	1	0	0	0	0	x	1	0
Thysanocardia nigra	1	1	0	0	0	0	0	0	x	x	x	1	1	0	x	x	x	x	x	x	x	x	x	x	x	x	x	0	0	0	0	1	0	1	0

Hydrobiologia (2005) **535/536:** 297–307
T. Bartolomaeus & G. Purschke (eds), Morphology, Molecules, Evolution and Phylogeny in Polychaeta and Related Taxa

Molecular phylogeny of siboglinid annelids (a.k.a. pogonophorans): a review

Kenneth M. Halanych
Department of Biological Sciences, Auburn University, 101 Life Science Building, AL 36849, USA
(Tel.: +1-334-844-3222, Fax: +1-334-844-2333, E-mail: ken@auburn.edu)

Key words: Phylogeny, Pogonophora, Siboglinidae, Vestimentifera, Frenulata

Abstract

Siboglinid, or pogonophoran, annelids are tubicolous worms that rely on chemoautotrophic endosymbionts for nutrition. Three clades within the siboglinids are recognized: Frenulata, Vestimentifera, and Monilifera. As a group, these worms have received considerable attention from molecular phylogenetists. Most studies have focused either on the evolutionary origins of the group or on the relationships within vestimentiferans, which live at hydrocarbon seeps and hydrothermal vents. Here I review the literature to date on siboglinid molecular phylogeny and summarize the clade's evolution. The vestimentiferans have been well studied, especially in the eastern Pacific. The seep taxon *Lamellibrachia* is basal in the clade with vent species being more derived. Recent studies of seeps are finding new species and suggest that habitat depth can be correlated with species boundaries. In contrast to the vestimentiferans, frenulate evolution has been poorly studied. Despite their greater apparent diversity, frenulate specimens have not been sampled so extensively, and thus little is known about their evolution. *Sclerolinum*, also referred to as Monilifera, is a recognized genus of siboglinids that forms the sister group to Vestimentifera. Like the frenulates, little is known about the history of this group. Our present understanding of siboglinid phylogeny has, in large part, been dictated by insufficient sampling effort.

Introduction

The reliance on chemoautotrophic endosymbionts in siboglinid (a.k.a., pogonophoran) tubeworms has prompted considerable interest in their evolutionary history. Typically the group has been split into two major lineages, frenulates, and vestimentiferans. A third lesser-known group, *Sclerolinum*, also referred to as moniliferans, is important from the evolutionary perspective. Vestimentiferans are the best-known members of the clade because of their association with, and dominance at, hydrothermal vents and cool seeps. In particular, the high levels of productivity due to the endosymbionts, has drawn considerable attention from a wide range of biologists. Vestimentiferans tend to have considerably larger bodies than frenulates, which are found in numerous different types of sedimented habitats including fiords, continental slopes, and deep sea trenches. Whereas frenulates comprise 19 nominal genera with ~136 nominal species, vestimentiferans contain about 10 genera with 15 nominal species. Thus, many vestimentiferan genera are monospecific. The seven nominal species of *Sclerolinum* live either in reducing sediments or on rotting organic material.

The focus of this manuscript is to review and summarize the molecular data relevant to the evolutionary history of siboglinids. Where do siboglinids fit into animal phylogeny? What are the clades within the siboglinids and how are they related? What are the relationships within those clades? There are several related topics that will not be discussed because they are covered elsewhere. In the case of morphology, considerable information bears on the evolution of the group.

This topic has been addressed in this volume by Southward et al. (2005) who discuss similarities and differences between the groups. Readers interested in molecular evolutionary analyses of endosymbionts are referred to Feldman et al. (1997), Di Meo et al. (2000), Nelson & Fisher, (2000), and McMullin et al. (2003) for vestimentiferans. No published analyses of frenulate endosymbionts exist, but data from *Siboglinum fiordicum* endosymbionts places these organisms in the γ-proteobacteria (Halanych et al., unpublished). Also nomenclature and higher-level taxonomy have been hotly debated and confusingly inconsistent. Recent papers cover this topic in detail (e.g., Rouse & Fauchald, 1995; Halanych et al., 2002; Southward et al., 2005) and it will not be reiterated here. The definition of the three siboglinid clades (Vestimentifera, Frenulata, and Monilifera) follows Halanych et al. (2001). As a prelude to the discussion below, Table 1 summarizes the relevant molecular studies dealing with siboglinids.

Are siboglinids annelids?

Yes! Abundant morphological (e.g., Rouse & Fauchald, 1995, 1997), embryological (e.g., Young et al., 1996; Southward, 1999) and molecular data (Kojima et al., 1993; McHugh, 1997; Black et al., 1997; Kojima, 1998; Halanych et al., 1998, 2001) place siboglinids within the annelid radiation. Much of the confusion on the status of this group is due to proposals that Pogonophora (a.k.a., Frenulata, Perviata) and Vestimentifera both be elevated to the rank of phylum. I have chosen to use the term siboglinid rather than pogonophoran to emphasize its placement within the annelid radiation. However others (e.g., Southward et al., 2005) retain the term 'Pogonophora' because of its familiarity.

Field et al. (1988) was the first molecular phylogenetic study addressing possible placement of the siboglinids within animal phylogeny. However, 18S nuclear ribosomal gene data in that study, and subsequent work by Winnepenninckx et al. (1995), failed to provide convincing support for the placement of siboglinids. Whereas Field et al. found a vestimentiferan (*Riftia pachyptila*) and a brachiopod between two annelids, Winnepenninckx et al. found siboglinids to be close to echiurans. More recent 18S studies (e.g., Halanych et al., 1998, 2001) have also suggested an annelid affinity for siboglinids, but again support for internal nodes is weak.

Molecular support showing an annelid affinity for siboglinids came from Elongation Factor-1α data by Kojima et al. in 1993. Subsequent EF-1α papers (McHugh, 1997; Kojima, 1998) included more diverse taxon sampling and provided evidence that siboglinids are within the annelid radiation. Interestingly, McHugh also found echiurans to be within annelids, suggesting a new interpretation of the Winnepenninckx et al.'s (1995) finding. Amino acid coding sequence from the Cytochrome Oxidase c subunit I gene is also consistent with the annelid placement of siboglinids (Black et al., 1997). Similarly, hemoglobin genes in siboglinids look very similar to other annelids (Suzuki et al., 1989, 1993; Zal et al., 1997; Bailly et al., 2002, 2003) and place siboglinids within annelids (Negrisolo et al., 2001). Hemoglobin studies have received less attention for phylogenetic inference due in part to taxon sample sizes, but mainly because the papers focus on the evolution of hemoglobin proteins and not the organisms. Bailly et al. (2003) did use globin sequences to build a neighbor joining tree. However, their results and conclusions must be critically examined as they did not have the taxon sampling to justify their conclusion concerning the evolution of sulfide tolerance and their tree is contra all other published phylogenetic analyses. Schulze and Halanych (2003) come to the opposite conclusion the same topic.

Two other sources of molecular data also bear on the issue. Available mitochondrial gene rearrangement data (frenulate – Boore & Brown, 2000, vestimentiferan – Jennings & Halanych, unpublished) is identical between siboglinids and clitellate annelids. Combined 28S and 18S data finds weak support for a monophyletic Annelida, including echiurids and siboglinids (Passamaneck & Halanych, submitted). In contrast to these six sources of data, molecular phylogenetic analyses reporting siboglinids to be outside the annelid radiation are lacking (except Winnepenninckx et al., 1995).

Where siboglinids fit within the annelids is not clear. Uschakov (1933) described *Lamellisabella* as a type of sabellid polychaete. Although an evolutionary link with the Sabellida is suggested

Table 1. Molecular studies of siboglinid (a.k.a. pogonophoran) evolution

References	Clade of interest					Molecular marker									
	Siboglinid origins	Within Siboglinids	Within Frenulates	Within Vestimentiferans	Intra-specific	EF-1α	Haemoglobin	Mitochondrial genome	Nuclear 18S rDNA	Nuclear 28S rDNA	Mitochondrial 16S rDNA	CO1	AFLPs	RFLPs (ITS region)	Allozymes
Bailly et al. (2002)	X						X								
Boore & Brown (2000)	X							X							
Field et al. (1988)	X								X						
Jennings & Halanych (unpub.)	X							X							
Kojima (1998)	X					X									
Kojima et al. (1993)	X					X									
McHugh (1997)	X					X									
Negrisolo et al. (2001)	X						X								
Passamaneck & Halanych (submitted for publication)	X								X	X					
Suzuki et al. (1988)	X						X								
Suzuki et al. (1989)	X						X								
Suzuki et al. (1993)	X						X								
Winnepenninckx et al. (1995)	X								X						
Zal et al. (1997)	X						X								
Halanych et al. (1998)		X	X						X						
Halanych et al. (2001)		X	X						X		X				
Bailly et al. (2003)		X					X								
Black et al. (1997)	(X)			X								X			
McMullin et al. (2003)				X								X			
Kojima et al. (2000)				X								X			
Kojima et al. (2001)				X	X							X			
Kojima et al. (2002)				X	X							X			
Kojima (2003)				X								X			
Williams et al. (1993)				X						X					
Black et al. (1994)					X										X
Black et al. (1998)					X							X			X
Bucklin (1988)					X										X
Carney et al. (2002)					X										X
Shank & Halanych (submitted for publication)					X								X		
Southward et al. (1996)					X									X	X
Southward et al. (1995)					X										X

by morphological cladistic analysis (Rouse & Fauchald, 1997), recent Elongation Factor-1α data does not support this view (McHugh, pers. com.).

What are the major clades within Siboglinidae?

Webb (1969) was the first to formally recognize the evolutionary kinship between vestimentiferans and frenulates. Since then, several authors have acknowledged the close evolutionary relationship between the groups (e.g., Webb, 1969; Jones, 1981; Southward, 1988; Ivanov, 1994). As for molecular papers, at least one frenulate and one vestimentiferan were included with enough other taxa to demonstrate monophyly in studies on hemoglobin (Zal et al., 1997; Negrisolo et al., 2001; Bailly et al., 2002), CO1 (Black et al., 1997) and 18S (Halanych et al., 1998, 2001).

Major relationships within siboglinids are poorly understood because frenulate sampling has been insufficient. Specifically, because many frenulates occur in deep water, they are difficult to collect and are typically taken as bycatch whose ultimate fate is to end up in a jar of formalin limiting its utility for molecular studies. To my knowledge, Halanych et al. (1998, 2001) are the only molecular analyses to include more than two non-vestimentiferan siboglinids. Both 16S mitochondrial ribosomal data and 18S nuclear ribosomal data suggest: (1) vestimentiferans are monophyletic, (2) *Sclerolinum* (a.k.a., Monilifera *sensu* Ivanov, 1991) is sister to the vestimentiferan clade, and (3) frenulates form a monophyletic clade sister to the vestimentiferan/moniliferan clade.

Points 1 and 2 are also corroborated by CO1 data (Black et al., 1997). However, a caveat must be added to point 3. These molecular analyses (Halanych et al., 1998, 2001) include only five frenulate species representing a group with ~136 nominal species. In contrast, vestimentiferans have been well represented (see below). From the molecular perspective, examination of more taxa is required to determine if frenulates are monophyletic or form a paraphyletic grade at the base of the Siboglinidae.

Morphological cladistic analysis (Rouse, 2001) is consistent with the molecular data on all three points. Synapomorphies (Rouse, 2001; Schulze, 2003) for the clade Vestimentifera include the presence of an obturaculum, vestimentum, and multicellular pinnules. The vestimentiferan/moniliferan clade is supported by the presence of posterior chaetae in rows on the opisthosomal segments, and a tube with the posterior end closed. The frenulate clade is supported by the presence of sparse posterior peg-like chaetae, spermatophores, and a tube with the posterior end open.

What are the relationships within the major clades?

Frenulata

As mentioned above, a maximum of five frenulate species have been included in molecular phylogenetic studies of the group. Nonetheless two interesting results have been observed in both the 16S and 18S data (Halanych et al., 2001). *Siboglinum* (*S. ekmani* and *S. fiordicum*) appears not to be monophyletic, consistent with recent morphological observations (Southward, pers. com.). Additionally, *Polybrachia* sp. and *Galathealinum brachiosum* look very similar to each other (genetic distance 0.01 for 18S and 0.02 for 16S data) calling into question the validity of separate generic status. At present the only working hypotheses of frenulate evolutionary history are based on morphology (e.g., Ivanov, 1963; Webb, 1964; Rouse, 2001). Within siboglinids, we clearly have the most to learn about frenulates.

Despite this lack of molecular knowledge about frenulates, we can look to two sources for an understanding of their evolutionary history. Given that taxonomists have grouped siboglinids in to 'genera' and 'families' based on morphological similarity, we can use the existing taxonomic framework as a working hypothesis of frenulate evolutionary history (Fig. 1); however, this framework is untested. The one morphological cladistic study that includes a diversity of frenulates is Rouse (2001). This study finds the Oligobrachiidae and Athecanephria to be paraphyletic grades. Also *Choanophorus indicus*, is placed separately from the rest of the Polybrachiidae. *Cyclobrachia auriculata* seems closer to Lamellisabellidae than Polybrachiidae (also confirmed by Southward, pers. com.). More data from both morphological and molecular sources need to be gathered to further evaluate the history of frenulates.

Monilifera

This group is represented by *Sclerolinum* which has 7 nominal species. To date, only two individuals have been sequenced. *Sclerolinum brattstromi* for 16S and 18S (Halanych et al., 2001) and an unidentified *Sclerolinum* species from the Loihi seamount for CO1 (Black et al., 1998). Because there has never been a phylogenetic analysis of the group, little can be said about their evolutionary history.

Vestimentifera

In contrast to other siboglinids, vestimentiferans have been well studied with several molecular papers focusing on within clade relationships (Table 1). Prior to the first molecular studies, Jones (1988) had provided a taxonomic framework for the group (Fig. 2) which, to some degree, implied a hypothesized phylogeny. The first molecular work to test this framework used 28S nuclear ribosomal gene data and placed *Lamellibrachia*, not *Riftia*, basal in Vestimentifera (Williams et al., 1993). This conclusion was later corroborated with mitochondrial Cytochrome Oxidase c subunit I (CO1) gene data (Black et al., 1997) and 18S rDNA data (Halanych et al., 1998, 2001). (Note the mitochondrial 16S rDNA data places *Riftia* basal but this is not well supported; Halanych et al., 2001; also see McMullin et al., 2003). Thus, neither the Basibranchia nor the Lamellibrachiidae as proposed by Jones appear to be real entities, a result supported by morphological cladistic analyses (Rouse, 2001; Schulze, 2003).

Additionally, all molecular studies mentioned above revealed surprisingly limited diversity across vestimentiferan lineages; certainly far less than expected for a group that was purported to be a distinct phylum (see Halanych et al., 1998). The possibilities to explain this observation is that the clade is young or that there has been a slowdown in the rate of genetic change in the lineage. The available rDNA data show that vestimentiferan genes evolved more slowly than frenulate copies (Halanych et al., 2001), but even taking the slowdown into account, diversity is limited, suggesting the clade is young.

Determining if hydrothermal vent endemic species form a monophyletic clade has also been of interest to biologists. Of the six known vent-en-

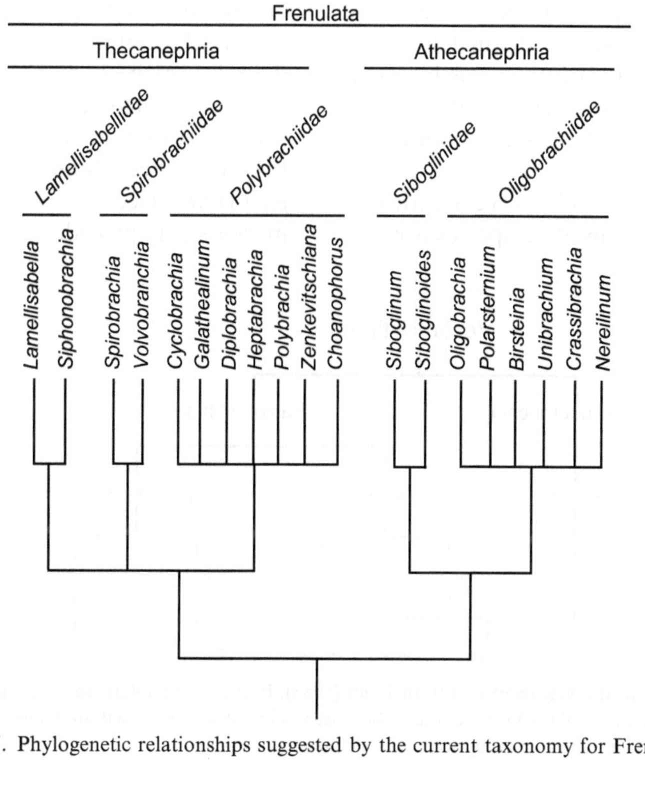

Figure 1. Phylogenetic relationships suggested by the current taxonomy for Frenulata.

demic species, four occur in the Eastern Pacific, a region that is well understood geologically (Tunnicliffe, 1988). The other species, *Arcovestia ivanovi* and *Alaysia spiralis*, occur in the western Pacific. Whereas CO1 data provides the most resolution within the vestimentiferan clade and provides marginal support for a monophyletic vent clade (Black et al., 1998, but *A. ivanovi* and *A. spiralis* were not included), other molecular data (28S – Williams et al., 1993, and 16S – Halanych et al., 2001) did not recover a vent clade. In these latter cases, support for internal branches was weak. CO1 is one of the slowest evolving genes in the mitochondrion, and perhaps a gene with a higher rate of nucleotide substitution will yield more phylogenetic signal for vestimentiferan relationships.

Even so, available data tell us something about vestimentiferan evolution in the eastern Pacific. Approximately 37 million years ago, the Juan de Fuca, Explorer and Gorda ridges (where *Ridgeia piscesae* occurs) were contiguous with the East Pacific Rise (EPR; where *Oasisia alvinae, Tevnia jerichonana*, and *Riftia pachyptila* occur). A clade consisting of all three EPR species is not found with any molecular data sets suggesting that either two separate vestimentiferan lineages have colonized EPR vents or at least two distinct lineages existed on one ridge system before it was bisected by Baja California. Given the genetic diversity of the group, the former hypothesis seems more tenable.

Despite the attention paid to vent-endemic vestimentiferans, the seep dwelling species appear to have greater phylogenetic diversity. The systematic work on seep vestimentiferans has focused on the western Pacific led by the efforts of Kojima and co-workers (Kojima et al., 1997, 2000, 2001, 2002, 2003). They have used CO1 data to explore the diversity and relationships of *Escarpia, Lamellibrachia* and *Arcovestia* populations. Although Kojima and workers were interested in the relationships between species, much of this work focuses on delineating species and population boundaries through a phylogeographic approach.

A couple of key generalizations can be drawn from this body of work. First, within a region species appear to be stratified according to depth rather than geographic proximity. In the case of *Lamellibrachia*, species 'L1' (characterized mtDNA haplotypes) was only found between 300 and 1450 m and 'L2' was restricted to below 2000 m even though they both occurred in the same geographic locality (e.g., Nankai Trough; Kojima et al., 2001). A similar story was found in *Escarpia* with one species found at 300 m and the other between 1100 and 1650 m (Kojima et al., 2002). This type of biogeographic pattern has been reported in other deep-sea animals (e.g., Etter & Rex, 1990; Etter et al., 1999). A second important finding by Kojima and co-workers was limited genetic diversity within species. Population bottlenecks are perhaps the most common explanation for limited intraspecific diversity. However, using a conservative genetic marker can produce the same effect. Similar to the situation in *Riftia pachyptila* (see below), more rapidly evolving markers may provide better insight on the patterns

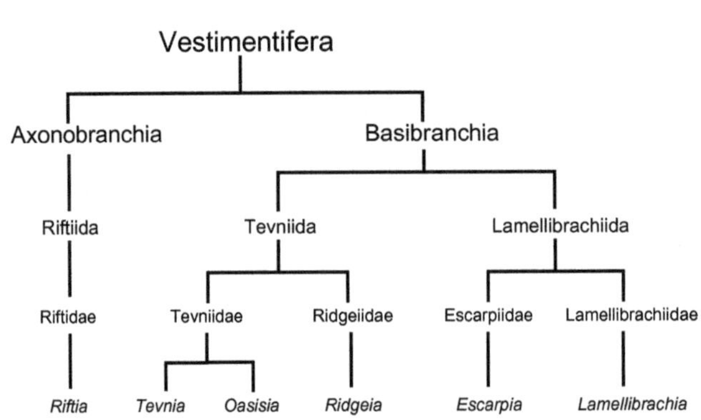

Figure 2. Taxonomic scheme for the Vestimentifera from Jones (1988). In the original diagram Jones provided a rank designation for each taxon name starting with 'phylum' for Vestimentifera. Because vestimentiferans are within Annelida, and because of the problems associated with ranks, Jones' hierarchical ranking were not included.

of genetic diversity in seep dwelling vestimentiferans.

Are populations within species genetically structured?

Although population genetic issues have never been dealt with in frenulates or moniliferans, several vestimentiferans have been studied. *Riftia pachyptila* has been the best studied with perhaps the most colorful history. Bucklin et al. (1988) conducted the first intraspecific study of *R. pachyptila*. They examined two eastern Pacific populations using 13 allozyme loci and concluded that levels of genetic diversity were low, but that there was a significant difference between the populations. In contrast, Black et al. (1994) examined six populations of *R. pachyptila* and found that levels of diversity and gene flow were high, concluding that localities were not genetically isolated. Interestingly, they also report a slight 'isolation by distance' trend. Note that these studies only had three loci in common.

Although allozymes have been useful as an inexpensive way to screen many individuals, numerous studies (e.g., Johnson et al., 1977; Sites & Davis, 1989; Reeb & Avise, 1990; Steinger et al., 1996) have repeatedly demonstrated that allozyme data often substantially underestimate genetic variation. This problem is particularly relevant when only a few polymorphic loci are used. These problems motivated Shank & Halanych (submitted) to revisit *R. pachyptila* genetics using amplified fragment length polymorphism (AFLP) analysis sampling hundreds of loci across the nuclear genome. They found genetic structure and some (but not total) genetic isolation between recognized *R. pachyptila* populations. Multidisciplinary research (Marsh et al., 2001) examined larval energetics and possible retention mechanisms that would account for along-axis retention of *R. pachyptila* larvae. Their model is compatable with the genetic observations of Shank & Halanych but more data is needed.

Black et al. (1998) examined *Oasisia alvinae, Tevnia jerichonana*, and *Ridgeia piscesae* in the eastern Pacific. The sampling of *O. alvinae* and *T. jerichonana* was limited and allozymes were mainly used to identify small juveniles. For both of these species, Black et al. report estimated rates of gene flow that were sufficient to prevent isolation between populations, but the variation observed was suggestive of a stepping-stone model of dispersal. *R. piscesae* has a very plastic morphology, leading earlier workers to suspect it was several closely related species (Jones, 1985), but subsequent allozyme, CO1 sequence data, and AFLP data revealed these different morphologies belonged to a common gene pool (Southward et al., 1995; Black et al., 1998; Carney et al., 2002). This work was further expanded (Southward et al., 1996) with additional allozyme data and restriction fragment length polymorphism (RFLP) data from the internal transcribed spacer (ITS) region of the nuclear ribosomal repeat. These data showed that *R. piscesae* experienced some disruption to gene flow between the Juan de Fuca and Gorda Ridges, but within a ridge segment, larvae were able to disperse considerable distances. Southward et al. (1996) also used these molecular tools to confirm the presence of *Lamellibrachia barhami* at sedimented vent localities.

What is our current view of siboglinid evolution?

Figure 3 shows our current understanding of siboglinid evolution based on several sources. This tree was not reconstructed based on any explicit reconstruction program, but is a summary of the relevant available data on the siboglinid phylogeny as understood by the author. Thus, this 'meta-tree', merely designed stimulate future research by highlighting unresolved nodes, should not be used as a definitive phylogenetic framework for other studies. The vestimentiferan/moniliferan clade is based on molecular works listed in Table 1. In contrast, the frenulate clade is based on a composite of morphology (Rouse, 2001) and taxonomic nomenclature. *Polybrachia* and *Galathealinum* are placed together based on molecular data (Halanych et al., 2001). Furthermore, some taxa (*Alaysia, Arcovestia, Polarsternium,* and *Volvobrachia*) have not been included in phylogenetic analyses that allow their position to be determined. The placement of these taxa in this diagram merely represents a conservative 'best guess' by the author. Nonetheless, this diagram is intended to be a useful tool representing our current knowledge of siboglinid phylogeny.

What needs to be done?

Siboglinids, or pogonophorans, are highly derived annelids whose evolution has been shaped by the environment they live in and their dependence on endosymbiotic bacteria (Schulze & Halanych, 2003). As expected, our understanding of the group has been shaped by sample availability. Because hydrothermal vents have received much attention, we know the most about the vent-endemic vestimentiferans. However, if we really want to understand the evolution of the clade and the origins of endosymbiosis in these worms, we need to examine frenulate and moniliferan evolution. At present delineating the frenulate clades and determining relationships between them is the biggest gap in our knowledge of siboglinid evolution. The limiting factor is access to specimens that can be used in molecular phylogenetic analysis. Because many siboglinids live in environments that are not easily accessible, understanding the larger picture of siboglinid evolution will require a concerted sampling effort from multiple researchers.

Note added in proof

A novel type of siboglinid called *Osedax* was reported while this work was in press (Rouse et al. 2004). *Osedax* is a unique lineage basal to the Vestimentifera/Monilifera clade that makes a living by using heterotrophic bacterial symbionts to live on whale bone.

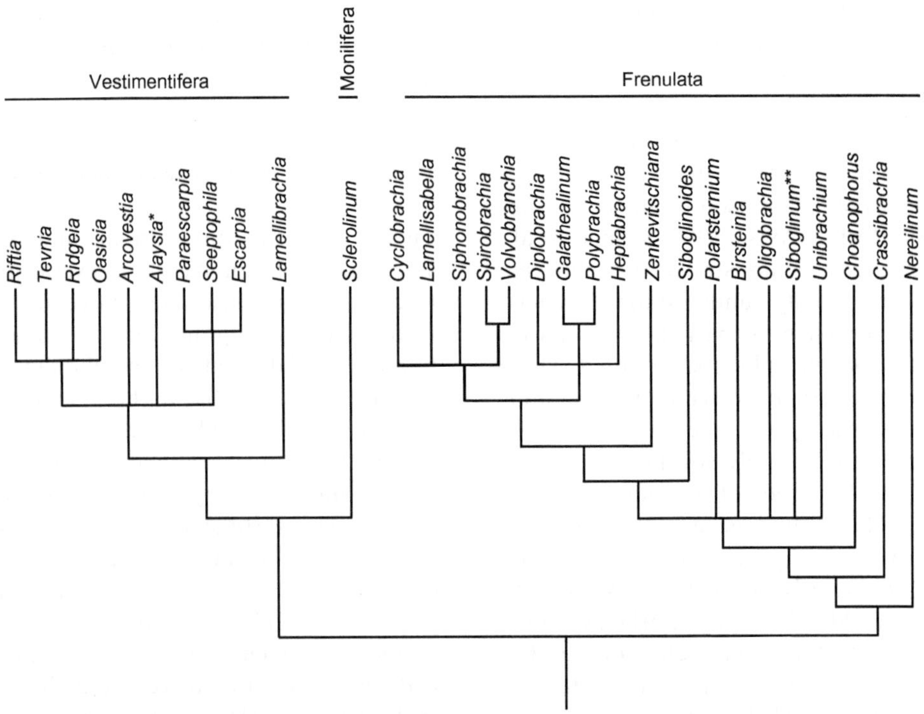

Figure 3. Current understanding of siboglinid relationships. This diagram was not produced with an explicit reconstruction algorithm, but summarizes results from several different sources. The Vestimentifera topology is a composite of the works of Kojima, Halanych, and their collaborators (see Table 1). The polytomies are shown to be conservative as many of the internal branches within the vestimentiferans are short and variously resolved in different studies (e.g., the *Riftia*/*Tevnia*/*Ridgeia*/*Oasisia* clade). The placement of *Sclerolinum* is based on Halanych et al. (2001). The Frenulata topology is based on the morphological work of Rouse (2001). The grouping of *Galathealinum* and *Polybrachia* and the unsolved position of *Siboglinum* relative to *Unibrachium* (see Rouse, 2001) is based on molecular data (Halanych et al., 2001). Data on the placement of some taxa (*Alaysia*, *Arcovestia*, *Polarsternium*, and *Volvobrachia*) are lacking, and their placement here are merely represents a conservative 'best guess' by the author based on their descriptions. *Currently, the validity of *Alaysia* is under debate (Southward, pers. com.) and it inclusion here is for consistency with the published literature. ***Siboglinum* is probably polyphyletic.

Acknowledgements

The efforts of Günter Purschke and Thomas Bartolomaeus in organizing and hosting such a wonderful symposium are gratefully acknowledged. Additionally Wilfred Westheide's lifetime of work and achievement has set an impressive standard for younger polychaete researchers. Helpful comments were provided by Eve Southward, Tim Shank, and two anonymous reviewers. Support from the National Science Foundation to KMH (DEB-0075618, EAR-0120646) is gratefully acknowledged.

References

Bailly, X., D. Jollivet, S. Vanin, J. Deutsch, F. Zal, F. Lallier & A. Toulmond, 2002. Evolution of the sulfide-binding function within the globin multigenic family of the deep-sea hydrothermal vent tubeworm *Riftia pachyptila*. Molecular Biology and Evolution 19: 1421–1433.

Bailly, X., R. Leroy, S. Carney, O. Collin, F. Zal, A. Toulmond & D. Jollivet, 2003. The loss of the hemoglobin H_2S-binding function in annelids from sulfide-free habitats reveals molecular adaptation driven by Darwinian positive selection. Proceedings of the Natural Academy of Sciences 100: 5885–5890.

Black, M. B., K. M. Halanych, P. A. Y. Maas, W. R. Hoeh, J. Hashimoto, D. Desbruyères, R. A. Lutz & R. C. Vrijenhoek, 1997. Molecular systematics of vestimentiferan tubeworms from hydrothermal vents and cold-water seeps. Marine Biology 130: 141–149.

Black, M. B., R. A. Lutz & R. C. Vrijenhoek, 1994. Gene flow among vestimentiferan tube worm (*Riftia pachyptila*) populations from hydrothermal vents of the eastern Pacific. Marine Biology 120: 33–39.

Black, M. B., A. Trivedi, P. A. Y. Maas, R. A. Lutz & R. C. Vrijenhoek, 1998. Population genetics and biogeography of vestimentiferan tube worms. Deep-Sea Research II 45: 365–382.

Boore, J. L. & W. M. Brown, 2000. Mitochondrial genomes of *Galathealinum*, *Helobdella*, and *Platynereis*: sequence and gene arrangement comparisons indicate that Pogonophora is not a phylum and Annelida and Arthropoda are not sister taxa. Molecular Biology and Evolution 17: 87–106.

Bucklin, A., 1988. Allozymic variability of *Riftia pachyptila* populations from the Galapagos Rift and 21° N hydrothermal vents. Deep-Sea Research 35: 1759–1768.

Carney, S. L., J. R. Peoples, C. R. Fisher & S. W. Schaeffer, 2002. AFLP analyses of genomic DNA reveal no differentiation between two phenotypes of the vestimentiferan tubeworm, *Ridgeia piscesae*. Cahiers De Biologie Marine 43: 363–366.

Di Meo, C. A., A. E. Wilbur, W. E. Holben, R. A. Feldman, R. C. Vrijenhoek & S. C. Cary, 2000. Genetic variation among endosymbionts of widely distributed vestimentiferan tubeworms. Applied and Environmental Microbiology 66: 651–658.

Etter, R. J. & M. A. Rex, 1990. Population differentiation decreases with depth in deep-sea gastropods. Deep-Sea Research 37: 1251–1261.

Etter, R. J., M. A. Rex, M. C. Chase & J. M. Quattro, 1999. A genetic dimension to deep-sea biodiversity. Deep-Sea Research Part I 46: 1095–1099.

Feldman, R. A., M. B. Black, C. S. Cary, R. A. Lutz & R. C. Vrijenhoek, 1997. Molecular phylogenetics of bacterial endosymbionts and their vestimentiferan hosts. Molecular Marine Biology and Biotechnology 6: 268–277.

Field, K. G., G. J. Olsen, D. J. Lane, S. J. Giovannoni, M. T. Ghiselin, E. C. Raff, N. R. Pace & R. A. Raff, 1988. Molecular phylogeny of the animal kingdom. Science 239: 748–753.

Halanych, K. M., T. G. Dahlgren & D. McHugh, 2002. Evidence that some lesser-known "phyla" are annelids. American Zoologist 42: 678–684.

Halanych, K. M., R. A. Feldman & R. C. Vrijenhoek, 2001. Molecular evidence that *Sclerolinum brattstromi* is closely related to vestimentiferans, not frenulate pogonophorans (Siboglinidae, Annelida). Biological Bulletin 201: 65–75.

Halanych, K. M., R. A. Lutz & R. C. Vrijenhoek, 1998. Evolutionary origins and age of vestimentiferan tube-worms. Cahiers De Biologie Marine 39: 355–358.

Ivanov, A. V., 1963. Pogonophora. Academic Press, London.

Ivanov, A. V., 1991. Monilifera – a new subclass of Pogonophora. Doklady Akademii Nauk SSR 319: 505–507.

Ivanov, A. V., 1994. On the systematic position of Vestimentifera. Zoologische Jahrbücher Systematik 121: 409–456.

Johnson, G. B., 1977. Assessing electrophoretic similarity: the problem of hidden heterogeneity. Annual Review of Ecology Systematics 8: 309–328.

Jones, M. L., 1981. *Riftia pachyptila* Jones: Observations on the vestimentiferan worm from the Galapagos Rift. Science 213: 333–336.

Jones, M. L., 1985. On the vestimentifera, new phylum: six new species, and other taxa, from hydrothermal vents and elsewhere. Biol. Soc. Wash. Bull. 6: 117–158.

Jones, M. L., 1988. The Vestimentifera, their biology, systematic and evolutionary patterns. In Laubier, L. (ed.), Actes du Colloque "Les Sources Hydrothermales de la Ride du Pacifique Oriental. Biologie et Ecologie." Institut Océanographique, Paris, November 4–7, 1985. Gauthier-Villars, Montrouge: 69–82.

Kojima, S., 1998. Paraphyletic status of Polychaeta suggested by phylogenetic analysis based on the amino acid sequences of elongation factor-1-alpha. Molecular Phylogenetics and Evolution 9: 255–261.

Kojima, S., T. Hashimoto, M. Hasegawa, S. Murata, S. Ohta, H. Seki & N. Okada, 1993. Close phylogenetic relationship between Vestimentifera (tube worms) and Annelida revealed by amino acid sequence of elongation factor-1a. Journal of Molecular Evolution 37: 66–70.

Kojima, S., S. Ohta, T. Miura, Y. Fujiwara & J. Hashimoto, 2000. Molecular phylogenetic study of chemoautosynthesis-based communities in the Manus Basin. JAMSTEC Journal of Deep Sea Biology 16: 7–13.

Kojima, S., S. Ohta, T. Yamamoto, T. Miura, Y. Fujiwara, K. Fujikura & J. Hashimoto, 2002. Molecular taxonomy of vestimentiferans of the western Pacific and their phylogenetic relationship to species of the eastern Pacific: II. Families Escarpiidae and Arcovestiidae. Marine Biology 141: 57–64.

Kojima, S., S. Ohta, T. Yamamoto, T. Yamaguchi, T. Miura, Y. Fujiwara, K. Fujikura & J., 2003. Hashimoto Molecular taxonomy of vestimentiferans of the western Pacific, and their phylogenetic relationship to species of the eastern Pacific III. Alaysia-like vestimentiferans and relationships among families. Marine Biology 142: 625–635.

Kojima, S., S. Ohta, T. Yamamoto, T. Miura, Y. Fujiwara & J. Hashimoto, 2001. Molecular taxonomy of vestimentiferans of the western Pacific and their phylogenetic relationship to species of the eastern Pacific. 1. Family Lamellibrachiidae. Marine Biology 139: 211–219.

Kojima, S., R. Segawa, J. Hashimoto & S. Ohta, 1997. Molecular phylogeny of vestimentiferans collected around Japan, revealed by the nucleotide sequences of mitochondrial DNA. Marine Biology 127: 507–513.

Marsh, A. G., L. S. Mullineaux, C. M. Young & D. T. Manahan, 2001. Larval dispersal potential of the tubeworm *Riftia pachyptila* at deep-sea hydrothermal vents. Nature 411: 77–80.

McHugh, D., 1997. Molecular evidence that echiurans and pogonophorans are derived annelids. Proceedings of the National Academy of Sciences of the United States of America 94: 8006–8009.

McMullin, E. R., S. Hourdez, S. W. Schaeffer & C. R. Fisher, 2003. Phylogenetics and biogeography of deep sea vestimentiferan tubeworms and their bacterial symbionts. Symbiosis 34: 1–41.

Negrisolo, E., A. Pallavicini, R. Barbato, S. Dewilde, A. Ghiretti-Magaldi, L. Moens & G. Lanfranchi, 2001. The evolution of extracellular hemoglobins of annelids, vestimentiferans, and pogonophorans. Journal of Biological Chemistry 276: 26391–26397.

Nelson, K. & C. R. Fisher, 2000. Absence of Cospeciation in Deep-Sea Vestimentiferan Tube Worms and Their Bacterial Endosymbionts. Symbiosis 28: 1–15.

Reeb, C. A. & J. C. Avise, 1990. A genetic discontinuity in a continuously distributed species: mitochondrial DNA in the American Oyster, *Crassostrea virginica*. Genetics 124: 397–406.

Rouse, G. W., 2001. A cladistic analysis of Siboglinidae Caullery, 1914 (Polychaeta, Annelida): formerly the phyla Pogonophora and Vestimentifera. Zoological Journal of the Linnean Society 132: 55–80.

Rouse, G. W. & K. Fauchald, 1995. The articulation of annelids. Zoologica Scripta 24: 269–301.

Rouse, G. W. & K. Fauchald, 1997. Cladistics and polychaetes. Zoological Scripta 26: 139–204.

Rouse, G. W., S. K. Goffredi & R. C. Vrijenhoek, 2004. *Osedax*: Bone-eating marine worms with drawf males. Science 305: 668–671.

Schulze, A., 2003. Phylogeny of Vestimentifera (Siboglinidae, Annelida) inferred from morphology. Zoologica Scripta 32: 321–342.

Schulze, A. & K. M. Halanych, 2003. Siboglinid evolution shaped by habitat preference and sulfide tolerance. Hydrobiologia 496: 199–205.

Sites, J. W. & S. K. Davis, 1989. Phylogenetic relationships and molecular variablility within and among six chromosome races of *Sceloporus grammicus* (Sauria, Iguanidae), based on nuclear and mitochondial markers. Evolution 43: 296–317.

Southward, E. C., 1988. Development of the gut and segmentation of newly settled stages of *Ridgeia* (Vestimentifera): implications for relationship between Vestimentifera and Pogonophora. Journal of the Marine Biological Association of the United Kingdom 68: 465–487.

Southward, E. C., 1999. Development of Perviata and Vestimentifera (Pogonophora). Hydrobiologia 402: 185–202.

Southward, E. C., V. Tunnicliffe & M. Black, 1995. Revision of the species of *Ridgeia* from northeast Pacific hydrothermal vents, with a redescription of *Ridgeia piscesae* Jones (Pogonophora: Obturata = Vestimentifera). Canadian Journal of Zoology 73: 282–295.

Southward, E. C., V. Tunnicliffe, M. B. Black, D. R. Dixon & L. R. Dixon. 1996. Ocean-ridge Segmentation and Vent Tubeworms (Vestimentifera) in the NE Pacific. In MacLeod, C. J., P. A. Tyler & C. L. Walker (eds), Tectonic, Magmatic, Hydrothermal and Biological Segmentation of Mid-Ocean Ridges, Geological Society Special Publication. Geological Society, pp. 211–224.

Southward, E. C., A. Schulze, S. L. Gardiner, 2005. Pogonophora (Annelida): form and function. Hydrobiologia 535/536 (Dev. Hydrobiol. 179): 225–249.

Steinger, T. S., C. Körner & B. Schid, 1996. Long-term persistence in a changing climate: DNA analysis suggests very old ages of clones of alpine *Carex curvula*. Oecologia 105: 94–99.

Suzuki, T., T. Takagi & S. Ohta, 1993. N-terminal amino acid sequences of 440 kDa hemoglobins of the deep-sea tubeworms, *Lamellibrachia* sp. 1, *Lamellibrachia* sp. 2, and slender Vestimentifera gen. sp. 1. Evolutionary relationship with annelid hemoglobins. Zoological Science 10: 141–147.

Suzuki, T., T. Takagi, K. Okuda, T. Furukohri & S. Ohta, 1989. The deep-sea tubeworm hemoglobin: subunit structure and phylogenetic relationship with annelid haemoglobin. Zoological Science 6: 915–926.

Suzuki, T., T. Takagi & O. S., 1988. N-terminal amino acid sequence of the deep-sea tubeworm haemoglobin remarkably resembles that of annelid haemoglobin. Biochemical Journal 255: 541–545.

Tunnicliffe, V., 1988. Biogeography and evolution of hydrothermal-vent fauna in the eastern Pacific Ocean. Proceedings of the Royal Society Series B – Biological Sciences, London 233: 347–366.

Uschakov, P., 1933. Eine neue Form aus der Familie Sabellidae (Polychaeta). Zoological Anzeiger 104: 205–208.

Webb, M., 1969. *Lamellibrachia barhami*, gen. nov. sp. nov. (Pogonophora), from the northeast Pacific. Bull. Mar. Sci. 19: 18–47.

Williams, N. A., D. R. Dixon, E. C. Southward & P. W. H. Holland, 1993. Molecular evolution and diversification of the vestimentiferan tube worms. Journal of the Marine Biological Association of the United Kingdom 73: 437–452.

Winnepenninckx, B., T. Backeljau & R. De Wachter, 1995. Phylogeny of protostome worms derived from 18S rRNA sequences. Molecular Biology and Evolution 12: 641–649.

Young, C. M., E. Vázquez, A. Metaxas & P. A. Tyler, 1996. Embryology of vestimentiferan tube worms from deep-sea methane/sulphide seeps. Nature 381: 514–516.

Zal, F., Y. Kawasaki, J. J. Childress, F. H. Lallier & A. Toulmond, 1997. Primary structure of the common polypeptide chain b from the multi-haemoglobin system of the hydrothermal vent tube worm *Riftia pachyptila*: an insight on the sulphide binding-site. Proteins 29: 562–574.

Hydrobiologia (2005) **535/536:** 309–318
T. Bartolomaeus & G. Purschke (eds), Morphology, Molecules, Evolution and Phylogeny in Polychaeta and Related Taxa

Molecular systematics of polychaetes (Annelida)

Damhnait McHugh
Biology Department, Colgate University, Hamilton, NY 13346, USA
E-mail: dmchugh@mail.colgate.edu

Key words: Annelida, Polychaeta, phylogeny, 18S rRNA, root

Abstract

Some progress has been made in the field of molecular systematics of polychaetes over the past couple of years. In particular, phylogenetic analyses of sequence data from the 18S rRNA gene have included increasing numbers of taxa, and explicit hypothesis testing of sister-group relationships is being incorporated into the most recent studies. An increasing number of analyses of relationships within polychaete groups are being undertaken, with specific inferences being drawn regarding the evolution of characters such as reproductive mode. Despite this progress, the unanswered questions regarding annelid relationships outlined by McHugh (2000, p. 1881) remain: "what are the relationships among the polychaete annelids, what group is sister to the Clitellata, what extant group is most basal on the annelid tree, and what group is sister to Annelida?" Continued expansion of taxon sampling and further combined investigation of conserved nuclear coding genes, in conjunction with rRNA genes, may help to resolve some of these issues. Furthermore, only by expanding molecular systematic studies of polychaetes to analyses of nuclear coding genes for comprehensive taxon samples will it become clear whether the lack of basal-node resolution observed in analyses of 18S rRNA reflects a rapid radiation of the group, or is a feature associated with the 18S rRNA gene itself. Genomic-level data (e.g., mitochondrial gene order) may also be informative, and the cautious use of gene copies in phylogenetic analyses may point to a root of the annelid tree.

Introduction

Molecular phylogenetic analyses indicate that "Polychaeta" includes not only the polychaetes as we generally recognize them; derived positions of Clitellata, the siboglinids, and likely also the echiurids within a paraphyletic polychaete grade are supported to varying degrees (see McHugh, 2000; Martin, 2001; Rota et al., 2001; Siddall et al., 2001; Mallatt & Winchell, 2002; Struck et al., 2002a,b; Bleidorn et al., 2003). The most comprehensive phylogenetic analyses of the polychaetes based on morphological characters support a monophyletic "Polychaeta" that includes Siboglinidae (Rouse & Fauchald, 1997; Rouse, 1999; Rouse & Pleijel, 2001). However, the secondary absence of morphological characters may account for the exclusion from the polychaete clade of the clitellates (e.g., nuchal organs secondarily absent: see Westheide et al., 1999; Purschke et al., 2000; Purschke, 2002) and the echiurids (segmentation secondarily absent: see Purschke et al., 2000; Hessling, 2002; Hessling & Westheide, 2002; Purschke, 2002) in those analyses.

Accepting paraphyly of "Polychaeta" on the basis of molecular phylogenetic analyses, the task then becomes deciphering the relationships among polychaete groups and identifying the basal-most extant annelid groups. Here, the focus is on the results of molecular analyses of polychaete relationships that have appeared since a recent synopsis of the molecular phylogeny of annelids (McHugh, 2000). Since that time, several broad studies of polychaete relationships based on 18S rRNA gene sequences have appeared, and some within-group analyses for a few polychaete families have been published. Despite this progress, lack of resolution of basal nodes is a recurring issue in molecular studies of polychaete systematics, and rooting of the annelid tree remains an outstanding problem. Some avenues for future research that might help address these difficulties are suggested in the hopes that a more complete understanding of annelid evolution will soon be possible.

Polychaete phylogeny

A few wide-ranging analyses of polychaete relationships based on molecular sequence data have been published in the past couple of years (e.g., Rota et al., 2001; Struck et al., 2002a,b; Bleidorn et al., 2003). They are similar to most previous analyses of annelid phylogeny in that they use sequences of the 18S ribosomal RNA (18S rRNA) gene (e.g., Winnepenninckx et al., 1995, 1998; Kim et al., 1996; Moon et al., 1996; Eernisse, 1997). They differ, however, in that the taxon sampling has been expanded and the approach of testing hypotheses of specific sister-group relationships has been established (e.g., Struck et al., 2002b).

Rota et al. (2001) undertook a parsimony analysis of 18S rRNA sequences from 46 taxa (including 27 polychaetes, 11 clitellates, and an echiurid) to examine the relationships of the soil-dwelling, non-clitellate annelids, *Parergodrilus heideri* and *Hrabeiella periglandulata*. The parsimony analyses presented by Rota et al. (2001) were based on two different alignments of the 18S rRNA sequences (done using DCSE (De Rijk & De Wachter, 1993) or ClustalW (Thompson et al., 1994)). As with other analyses of 18S rRNA sequences for annelids, few relationships beyond sister groupings of terminal taxa are strongly supported in this study (Rota et al., 2001). Monophyly of Annelida (including polychaetes, Clitellata, and echiurids) is not supported, with three molluscan taxa and a sipunculan falling within the group. Furthermore, some expected groupings based on Rouse & Fauchald (1997) or Rouse (1999) are not recovered. While this study was relatively comprehensive in its sampling of taxa, only 20 of the approximately 80 family-designated groups of polychaetes were represented, and Rota et al. (2001) see this uneven representation as a possible explanation for the poor resolution of basal nodes.

The analyses did support a sister relationship between *Parergodrilus heideri* and *Stygocapitella subterranea* (Bootstrap proportion (BP) >78), thereby affirming monophyly of Parerogodrilidae; however, the position of Parergodrilidae remains unresolved (Rota et al., 2001). *Hrabeiella periglandulata* apparently represents an independent case of the evolution of terrestriality in a polychaete; it is sister to the meiofaunal freshwater group, Aeolsomatidae, in both analyses although there is only weak support for this relationship (Rota et al., 2001).

In a similar study, Struck et al. (2002a) examined the phylogenetic position of Parerogodrilidae (*Stygocapitella subterranea*) and Aeolosomatidae (*Aeolosoma* sp.), using 18S rRNA sequences from 49 annelids (including 40 polychaetes and 9 clitellates), and designating molluscs and arthropods as outgroups. Unlike Rota et al. (2001), Struck et al. (2002a) explored various weighting schemes based on detailed analyses of the alignment prior to tree construction using maximum parsimony, distance, and maximum likelihood methods, but none resolved the position of these clitellate-like groups. However, neither *Stygocapitella subterranea* nor *Aeolosoma* sp. appeared within or as sister to the monophyletic clitellate clade on any of the trees. It would have been interesting to see some specific hypothesis testing in the studies by both Rota et al. (2001) and Struck et al. (2002a), e.g., Shimodaira–Hasegawa tests of significance for Bunke's 1967 hypothesis of Parerogodrilidae and Aeolosomatidae as sister taxa.

In both Rota et al. (2001) and Struck et al. (2002a), a sister relationship between Questidae and Orbiniidae was strongly supported (BP = 100), which argues against an early hypothesis of a questid–clitellate relationship. Monophyly of Clitellata, Hirudinea, Dinophilidae and Spionidae was also well supported, and a monophyletic Eunicida (*sensu* Rouse & Fauchald, 1997) was weakly supported in one maximum parsimony analysis (Struck et al., 2002a). On the other hand, some well-supported unexpected groupings indicated by the analyses suggest the need to confirm the identification of sequences deposited in public databases. For example, the strongly supported grouping of *Aphrodita aculeata* with *Neanthes virens*, to the exclusion of other species of *Neanthes* and other representatives of Aphroditidae warrants caution (Martin, 2001; Rota et al., 2001; Struck et al., 2002a; Bleidorn et al., 2003).

Struck et al. (2002b) focused on Eunicida and addressed a specific question regarding the evolution of small body size and simple body plans in some dorvilleids and in Dinophilidae: Did progenesis evolve more than once independently in these groups? Some authors consider Dinophilidae as a separate family (e.g., Orensanz, 1990), while others include them within Dorvilleidae (e.g., Rouse & Pleijel, 2001). Struck et al. (2002b) used maximum parsimony, maximum likelihood and minimum evolution tree-building methods to analyze 18S rRNA sequences from 43 taxa (including 31 polychaeta and 9 clitellates). The jawless species, *Parapodrilus psammophilus* fell within a dorvilleid clade (BP > 84%), indicating that it arose by progenesis from a dorvilleid ancestor. Kishino–Hasegawa tests rejected the hypothesis of a common origin for *P. psammophilus* and the small, simple-bodied Dinophilidae ($p < 0.05$); however, Templeton tests did not ($p > 0.05$). Another dorvilleid species, *Pettiboneia uriciensis*, was more closely related to Lumbrineridae than Dorvilleidae in all analyses, but non-parametric tests did not support rejection of dorvilleid monophyly on this basis (Struck et al., 2002b). Paraphyly of Eunicidae, with the inclusion of onuphid species in the eunicid clade, is perhaps not an unexpected result of the analysis (Struck et al., 2002b), given the sharing of jaw asymmetry and aragonite mineralization by the two groups (see Rouse & Pliejel, 2001). Interestingly, no clear morphological synapomorphy has been identified for either the Eunicidae or the Dorvilleidae (Struck et al., 2002b). Monophyly of Dinophilidae was strongly supported (BP = 100), but its position relative to the eunicidan groups in the analysis was unresolved.

Bleidorn et al. (2003) conducted a phylogenetic analysis of relationships among sedentary polychaeta using 18S rRNA gene sequences from 70 taxa (including 47 polychaeta and four clitellates) (Fig. 1). As with other 18S rRNA analyses, monophyly of several well-established polychaete groups for which two or more taxa were included was supported (Cirratulidae, Opheliidae, Orbiniidae, Spionidae, Siboglinidae), irrespective of the tree-building method used in the analysis (maximum parsimony or maximum likelihood). As with Rota et al. (2001) and Struck et al. (2002a), a close relationship between questids and orbiniids is well supported in the Bleidorn et al. (2003) study. Furthermore, a sister relationship between a capitellid and the two echiurids (BP = 91) indicates additional support for a derived position of the echiurids within a polychaete grade (McHugh, 1997, 2000). While the study by Bleidorn et al. (2003) includes the most annelid taxa in a molecular analysis to date, the taxon sampling is very uneven. For example, 10 siboglinids are included, but only one terebellid. The authors acknowledge the need to increase taxon sampling and suggest that it may help to resolve polychaete relationships. Alternatively, as has been done by several others (see McHugh, 2000; Martin, 2001; Rota et al., 2001; Struck et al., 2002a,b), Bleidorn et al. (2003) suggest that a rapid radiation may explain the lack of resolution of basal annelid nodes in phylogenetic analyses of 18S rRNA sequences. This explanation will be supported if multiple independent gene sequences for extensive taxonomic samples also yield poor resolution in phylogenetic analyses.

Phylogenetic relationships within polychaete groups

In addition to broad scale analyses of annelid relationships, several recent studies have focused on relationships within and among particular polychaete groups. In several cases, combined analyses of both molecular and morphological data are used to examine these relationships. For example, Rousset et al. (2003) used sequences of 28S rRNA (D1 domain) and 52 morphological characters to examine the phylogenetic position of Alvinellidae, a group of polychaetes known only from hydrothermal vents, and originally classified as a subfamily of Ampharetidae (Desbruyères & Laubier, 1980). With a maximum parsimony analysis of just 13 taxa in their molecular data set, Rousset et al. (2003) found a trichobranchid as sister to one of the two alvinellids in the analysis; a combined analysis of 16 taxa (3 lacking molecular data) showed weak support for a sister group relationship between Alvinellidae and the trichobranchid, as well as monophyly of Terebellidae and Ampharetidae. However, the very limited taxon sampling leaves the issue of alvinellid relationships unresolved.

Focussing on the same taxonomic groups, Colgan et al. (2001) analyzed the phylogenetic relationships of terebellomorph polychaetes using a combined data set based on three nuclear genes

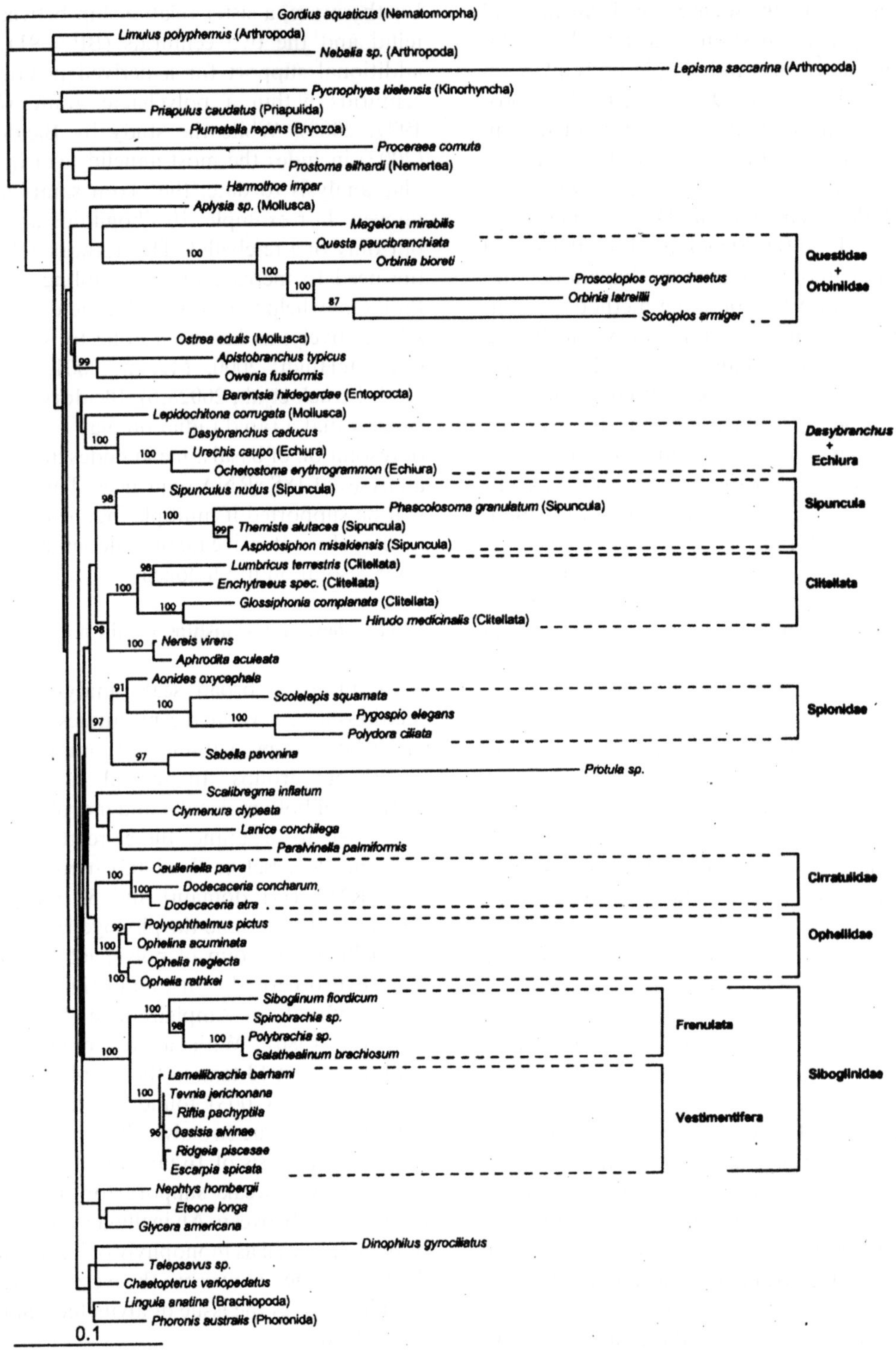

Figure 1. Results of a maximum likelihood analysis of 18S rRNA sequences from 70 taxa (from Bleidorn et al., 2003). Numbers above the nodes represent posterior probabilities from a Bayesian analysis of selected groups. Monophyly of the Clitellata, Spionidae, Cirratulidae, Opheliidae, Orbiniidae, and Siboglinidae is supported; a sister-group relationship between Questidae and Orbiniidae is also supported. Polychaetes do not form a monophyletic group and none of the orders of Fauchald (1977) or the clades proposed by Rouse & Fauchald (1997) are supported.

(Histone H3, U2 snRNA, and two regions of 28S rRNA) and the mitochondrial gene, cytochrome oxidase I (COI). Using a single clitellate outgroup to root the tree, Colgan et al. (2001) restricted their study of the data to maximum parsimony analysis of equally weighted characters. The topology resulting from analysis of all characters available for 25 taxa supported monophyly of Cirratulidae, but showed several other well-established groups to be polyphyletic (Terebellidae, Alvinellidae, Ampharetidae, Trichobranchidae). Despite the analysis of sequence data from multiple genes, no nodes on the tree below terminal-taxon sister relationships are supported by bootstrap proportions greater than 50%. In addition to the limited taxon sampling, these results likely also reflect the fact that some of the genes chosen lack the phylogenetic signal needed to resolve the relationships among annelid groups (e.g., Histone H3 (see McHugh, 2000) and COI (Nylander et al., 1999).

In a recent study of another group of polychaeta, Nereidiformia, Dahlgren et al. (2000) undertook a parsimony analysis of COI sequences for nine taxa representing Hesionidae, Pilargidae, Nereididae, Chrysopetalidae, and Pisionidae. All but one node on the resulting topology was weakly supported (BP < 50%). Combining their molecular data with previously analyzed morphological data for 13 taxa, Dahlgren et al. (2000) hypothesized a sister relationship between the two chrysoptelids and the nereid, albeit with weak support (BP = 52%); the Hesionidae was polyphyletic. Again, this study is so limited in the scope of taxonomic sampling that few conclusions can be drawn from the results.

The body of molecular phylogenetic studies of polychaete relationships that move beyond tree-building to infer the evolution of features such as reproductive mode, feeding habit, larval development mode, heterochrony, etc. is building slowly. One example of such a study focuses on the relationships of species of *Ophryotrocha*, a group of small worms that displays a full range of reproductive modes from sequential hermaphroditism to gonochorism (Dahlgren et al., 2001). Using 16S rRNA gene sequence data, Dahlgren et al. (2001) carried out parsimony, distance, and maximum likelihood analyses to reconstruct relationships among 22 taxa, including 18 assigned to *Ophryotrocha*. While the basal relationships among the ingroup taxa were not resolved, the resulting trees allowed the inference that the reproductive mode of *Ophryotrocha* changed once and that simultaneous hermaphroditism is the reproductive mode of the immediate ancestor to the sequential hermaphroditic *Ophryotrocha* clade (Dahlgren et al., 2001).

Schulze et al. (2000) were also interested in the evolution of reproductive modes in a polychaete group. In this case, the authors used COI sequences to hypothesize relationships among populations of *Streblospio*, a polychaete group known to exhibit poecilogony, i.e., the presence of more than one developmental mode within a species. Maximum parsimony and distance analyses of 88 sequences from individuals of *S. benedicti* and *S. gynobranchiata* along the east and west coasts of North America supports paraphyly of *S. benedicti* with respect to *S. gynobranchiata*, and also corroborates poecilogony in this group (Schulze et al., 2000). Using molecular clock estimates of divergence times, Schulze et al. (2000) proposed that divergence times among clades of *Streblospio* are recent and thus the evolutionary changes in larval developmental modes have been rapid.

In a recent study of syllid polychaetes, Nygren & Sundberg (2003) analyzed 16S rRNA and 18S rRNA gene sequences for 47 taxa to reconstruct relationships and infer patterns of change in epitokous reproduction in the group. Irrespective of the tree-building method they used (maximum parsimony, maximum likelihood, or Bayesian inference), character reconstruction on the resultant trees supported epigamy as the ancestral reproductive mode in Syllidae, with the independent evolution of schizogamy in Syllinae and Autolytinae (Nygren & Sundberg, in press). Because the relationships among three clades of the syllid *Autolytus* were unresolved, the evolution of reproductive modes within Autolytinae remains ambiguous.

Rooting the Annelida

In phylogenetic reconstruction of any organismal group, an outgroup is usually used to root the tree. The outgroup is often the sister group of the ingroup, or multiple outgroups may be used. Unfortunately, there is no undisputed sister group known for Annelida, and inclusion of multiple presumed outgroups has resulted in their inclusion within Annelida in the molecular analyses undertaken so far. Rouse & Pleijel (2001) highlight the rooting of the polychaete tree as a major problem in annelid systematics, and discuss the alternative hypotheses that have been proposed and their implications for the body form of the basal annelids. One possible solution to the lack of an obvious sister group to root the Annelida tree comes from the molecular phenomenon of gene duplication.

If a gene duplicated prior to the divergence of a group of extant organisms, then a combined analysis of both gene copies for the same taxa will produce an unrooted network of two subtrees that basically mirror each other (Page & Holmes, 1998; Mathews & Donoghue, 1999) (Fig. 2a). By placing the root of this network at the point at which the fewest gene duplications are required to explain the data, the root of each subtree will be identified (Fig. 2b, c). Thus, no outgroup is required. This approach has been used to identify the root of the tree of life (Doolittle & Brown, 1995) and the root of the angiosperms (Mathews & Donoghue, 1999). It is possible that the root of annelid subtrees for copies of a duplicated gene can also be identified. The first task is to recognize a gene duplication event in the line leading to the Annelida. Some potential candidate genes include members of the actin (e.g., Carlini et al., 2000), tyrosine kinase (e.g., Miyata & Suga, 2001), and Delta/Serrate/LAG-2 (e.g., Lissemore & Starmer, 1999) gene families, among others. Extensive preliminary work would be required to determine which, if any genes, are appropriate. There are many possible pitfalls to rooting trees using duplicated genes that must be recognized and avoided if it is to be successful. For example, gene conversion, recombination between the two gene copies, and very unequal evolutionary rates of the two gene copies could yield misleading data. However, given the lack of alternatives, it is worth pursuing this method, albeit cautiously.

Other future directions

Several recent molecular studies of polychaete relationships have used maximum parsimony to

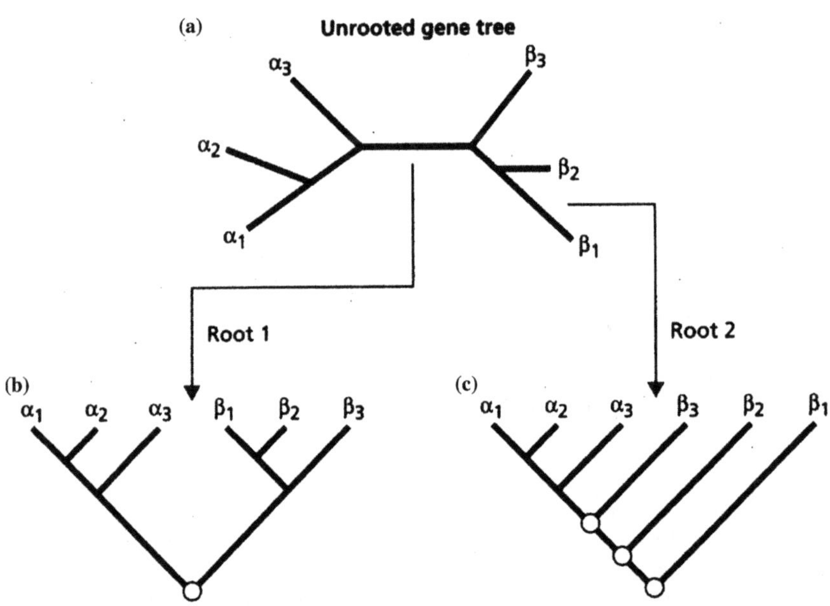

Figure 2. Using gene duplications to root a tree (from Page & Holmes, 1998). By minimizing the number of gene duplications that must be invoked to explain the relationships among sequences from two copies of a gene for the same taxa, the root of the tree for those taxa can be identified. (a) Simultaneous analysis of two gene copies (α and β) from three species (1, 2, and 3) results in an unrooted network of the $\alpha 1$, $\alpha 2$, $\alpha 3$, $\beta 1$, $\beta 2$, and $\beta 3$ sequences. (b) Placing the root of the tree on the branch that separates the α and β gene sequences requires one duplication event (denoted by open circle). (c) Placing the root anywhere else requires more than one duplication event (denoted by three open circles in this case).

analyze equally weighted characters (e.g., Colgan et al., 2001; Dahlgren et al., 2001; Rota et al., 2001). This restriction is unfortunate and difficult to justify, because it overlooks the great deal that is known about molecular evolution, which can be incorporated as explicit substitution models in maximum likelihood analyses or as weighting schemes in maximum parsimony analyses, for example. Substitution models and weighting schemes are based on evidence for different rates of substitution along a molecule (e.g., stems versus loops in rRNA genes), and for different rates of accrual of distinct character changes (e.g., transversions versus transitions (Struck et al., 2002a; Bleidorn et al., 2003) (see Page & Holmes, 1998). Future molecular systematic studies of polychaete relationships should fully explore the data by integrating such models and schemes.

Beyond full analyses of molecular data, critical assessment of tree topology is another important step in any systematic study. Critical assessment of node support on trees of polychaete relationships is usually done by getting bootstrap proportions for each node. Basically, bootstrap proportions represent the percentage of 1000 analyses of pseudoreplicates of the original data matrix that support nodes also found in the tree based on the analysis of the original data matrix. These proportions are easy to evaluate; they indicate how well the nodes on a tree are supported by the data matrix. Bootstrap proportions of greater than 70 are associated with well-supported nodes in simulations (Hillis & Bull, 1993). Another measure of node support has been used in some of the most recent studies of polychaete relationships, i.e., posterior probabilities from Bayesian inference (e.g., Bleidorn et al., 2003). However, as mentioned in Bleidorn et al. (2003), a number of authors have found that posterior probabilities do not correlate well with bootstrap proportions (e.g., in Leaché & Reeder, 2002, posterior probabilities of 95% were found for nodes with bootstrap proportions less than 50). Thus, high Bayesian posterior probability values must be interpreted with caution (see Huelsenbeck et al., 2002; Suzuki et al., 2002 for detailed discussion).

As of August 2002, there were only 2103 Annelida sequences available on GenBank, of which approximately 540 were for polychaetes, and almost half of these were ribosomal gene sequences. The heavy reliance on 18S rRNA sequences for molecular analyses of polychaete relationships continues, because it allows authors to build on the richest molecular database thus far for the annelids. A few sister group relationships are supported in the recent 18S rRNA analyses, e.g., Orbinidae + Questidae (Rota et al., 2001; Struck et al., 2002a; Bleidorn et al., in press), but beyond that analyses based on these data have been inconclusive or contradictory regarding polychaete relationships and monophyly of Annelida.

Further exploration of the phylogenetic usefulness of 28S rRNA, a large-subunit rRNA gene, especially in combination with the small-subunit 18S rRNA gene, for resolution of annelid relationships is warranted. Mallatt & Winchell (2002) recently presented combined analyses of these two genes for protostomes and showed that together they provide higher support for Ecdysozoa and Lophotrochozoa than 18S rRNA alone. In addition, the combined data set supports a sister relationship between the polychaete and the echiurid in the analyses, and monophyly of Annelida. While these analyses included very few taxa (16), they nonetheless illustrate the benefits of combining 28S rRNA and 18S rRNA sequences, i.e., 28S rRNA can be easily sequenced and addition of 28S rRNA adds phylogenetic signal (Mallatt & Winchell, 2002).

Both 18S rRNA and 28S rRNA, as structural genes, present some difficulties for alignment of sequences from diverse taxa. In most cases, authors have used the secondary structure model of 18S rRNA to align annelid sequences and/or they have "manually edited" the alignments; ambiguous regions in the alignments are usually excluded from the analyses (e.g., Colgan et al., 2001; Struck et al., 2002b). It is difficult to avoid the subjectivity involved in doing this, and the analyses cannot be replicated. These alignments (and those used in any published analysis) should be made immediately and publicly available to all readers for further investigation. "Alignments available from the author" is not satisfactory and the deposition of all published alignments in the EMBL ALIGN (as done by Rota et al., 2001) or TREEBASE databases is urged. Given the likelihood that many annelid relationships will remain unresolved until combined analyses of multiple genes are undertaken, it is particularly important to facilitate the use of all published data in future studies.

Several highly conserved nuclear protein-coding genes have been used in previous analyses of annelid relationships. These genes have the advantage of being easily aligned, but in some cases they have not been very useful. For example, given the highly conserved nature of the Histone H3 gene, it does not provide many parsimony informative sites for analysis of polychaete relationships (McHugh, 2000). Previous studies that include multiple data sets include those by Brown et al. (1999) and Colgan et al. (2001), both of which use the nuclear genes U2 snRNA, Histone H3, and 28S, and neither of which yielded resolved, well-supported basal nodes. There was also a lack of support for basal nodes in a combined analysis of elongation factor-1α, U2 snRNA, Histone H3, and 28S sequences from terebellidan, sabellidan, and spionidan polychaetes (McHugh, 2001). However, all of these studies were limited to 25 or fewer polychaeta.

Additional genes that are potentially informative for polychaete systematic studies include enolase, Na^+, K^+-ATPase (Friedlander et al., 1994), and myosin heavy chain type II. The latter of these has recently been used to hypothesize a basal bilaterian position of the small and simple acoel and nemertodermatid flatworms (Ruiz-Trillo et al., 2002). The polychaete, two clitellates, and echiurid included in that analysis formed a weakly supported monophyletic annelid clade; the availability of these sequences allows the design of annelid-specific primers for the myosin heavy chain type II gene. RNA polymerase II and elongation factor-2, as well as elongation factor-1α are other nuclear coding genes that are currently being sequenced for more than 100 polychaetes as part of an ongoing collaborative project on annelid evolution (seehttps://www.fastlane.nsf.gov/servlet/showaward?award=012064). This project also includes analysis of mitochondrial gene order and the complete mitochondrial genome sequences for the same taxa. Mitochondrial genomic-level data support inclusion of the siboglinids within Annelida (Boore & Brown, 2000), and indicate that Annelida and Sipuncula form a monophyletic clade to the exclusion of the Mollusca and Brachiopoda (Boore & Staton, 2002).

From the relatively few molecular studies of polychaete systematics, it appears that a combination of sequence data from several conserved genes, genomic-level data, and morphological characters for a sample of taxa that fully represents the great diversity of polychaetes and other annelids may be necessary for a robust, stable phylogenetic hypothesis for Annelida. While this may seem insurmountable, phylogenetic analyses at lower taxonomic levels can provide a good start (Bleidorn et al., in press), provided they draw on the same molecular data that can ultimately be combined for a complete analysis of annelid relationships.

Acknowledgements

I am very grateful to Günter Purschke (Universität Osnabrück) and Thomas Bartolomaeus (Freie Universität Berlin) for inviting me to participate in the symposium on "Morphology, Molecules, Evolution and Phylogeny in Polychaeta and Related Taxa" at Haus Ohrbeck in Osnabrück, Germany. I appreciate the helpful comments of two anonymous reviewers of this manuscript. I also thank Barbara Hoopes (Colgate University) for making it possible for me to attend the symposium, Nancy Schult (Colgate University) for her comments on the manuscript, and Laura Katz (Smith College) for discussion of gene duplication. I gratefully acknowledge the financial support of the US National Science Foundation (Grant No. 0120646 EAR BIOCOMPLEXITY).

References

Bleidorn, C., L. Vogt & T. Bartolomaeus, 2003. A contribution to sedentary polychaete phylogeny using 18S rRNA sequence data. Journal of Zoological Systematics and Evolutionary Research 41: 186–195.

Boore, J. L. & W. M. Brown, 2000. Mitochondrial genomes of *Galathealinum, Helobdella*, and *Platynereis*: sequence and gene arrangement comparisons indicate the Pogonophora is not a phylum and Annelida and Arthropoda are not sister taxa. Molecular Biology and Evolution 17: 87–106.

Boore, J. L. & J. L. Staton, 2002. The mitochondrial genome of the sipunculid *Phascolopsis gouldii* supports its association with Annelida rather than Mollusca. Molecular Biology and Evolution 19: 127–137.

Brown, S., G. Rouse, P. Hutchings & D. Colgan, 1999. Assessing the usefulness of histone H3, U2 snRNA and 28S

rDNA in analyses of polychaete relationships. Australian Journal of Zoology 47: 499–516.
Carlini, D. B., K. S. Reece & J. E. Graves, 2000. Actin gene family evolution and the phylogeny of coleoid cephalopods (Mollusca: Cephalopoda). Molecular Biology and Evolution 17: 1353–1379.
Colgan, D. J., P. A. Hutchings & S. Brown. 2001. Phylogenetic relationships within the Terebellomorpha. Journal of the Marine Biological Association of the United Kingdom 81: 765–773.
Dahlgren, T. G., B. Akesson, C. Schander, K. M. Halanych & P. Sundberg, 2001. Molecular phylogeny of the model annelid *Ophryotrocha*. The Biological Bulletin 201: 193–203.
Dahlgren, T. G., J. Lundberg, F. Pleijel & P. Sundberg, 2000. Morphological and molecular evidence of the phylogeny of Nereidiform polychaetes (Annelida). Journal of Zoological Systematics and Evolutionary Research 38: 249–253.
De Rijk, P. & R. De Wachter, 1993. DCSE, an interactive tool for sequence alignment and secondary structure research. Computational Applications in Bioscience 9: 735–740.
Desbruyères, D. & L. Laubier, 1980. *Alvinella pompejana* g. et sp. nov., Ampharetidae aberrant des sources hydrothermales de la ride Est-Pacifique. Oceanologica Acta 3: 267–274.
Doolittle, W. R. & J. R. Brown, 1995. Root of the universal tree of life based on ancient aminoacyl-transfer-RNA synthetase gene duplications. Proceedings of the National Academy of Sciences of the United States of America 92: 2441–2445.
Eernisse, D. J., 1997. Arthropod and annelid relationships re-examined. In Fortey, R. A. & R. H. Thomas (eds), Arthropod Relationships, Systematics. Association Special Volume Series 55. Chapman and Hall, London: 43–56.
Fauchald, K., 1977. The polychaete worms. Definitions and keys to the orders, families and genera. Natural History Museum Los Angeles County, Science Series 28: 1–188.
Friedlander, T. P., J. C. Regier & C. Mitter, 1994. Phylogenetic information content of five nuclear gene sequences in animals: initial assessment of character sets from concordance and divergence studies. Systematic Biology 43: 511–525.
Hessling, R., 2002. Metameric organisation of the nervous system in developmental stages of *Urechis caupo* (Echiura) and its phylogenetic implications. Zoomorphology 121: 221–234.
Hessling, R. & W. Westheide, 2002. Are Echiura derived from a segmented ancestor? Immunohistochemical analysis of the nervous system in developmental stages of *Bonellia viridis*. Journal of Morphology 252: 100–113.
Huelsenbeck, J. P., B. Larget, R. E. Miller & F. Ronquist, 2002. Potential applications and pitfalls of Bayesian inference of phylogeny. Systematic Biology 51: 673–688.
Kim, C. B., S. -Y. Moon, S. Gelder & W. Kim, 1996. Phylogenetic relationships of annelids, molluscs, and arthropods evidenced from molecules and morphology. Journal of Molecular Evolution 43: 207–215.
Leaché, A. D. & T. W. Reeder, 2002. Molecular systematics of the Eastern Fence lizard (*Sceloporus undulates*): a comparison of parsimony, likelihood, and Bayesian approaches. Systematic Biology 51: 44–68.
Lissemore, J. L. & W. T. Starmer, 1999. Phylogenetic analysis of vertebrate and invertebrate Delta/Serrate/LAG-2 (DSL) proteins. Molecular Phylogenetics and Evolution 11: 308–319.
Mallatt, J. & C. J. Winchell, 2002. Testing the new animal phylogeny: first use of combined large-subunit and small-subunit rRNA gene sequences to classify the protostomes. Molecular Biology and Evolution 19: 289–301.
Martin, P., 2001. On the origin of the Hirudinea and the demise of the Oligochaeta. Proceedings of the Royal Society of London B 268: 1089–1098.
Mathews, S. & M. J. Donoghue, 1999. The root of angiosperm phylogeny inferred from duplicate phytochrome genes. Science 286: 947–950.
McHugh, D., 1997. Molecular evidence that echiurans and pogonophorans are derived annelids. Proceedings of the National Academy of the Sciences of the United States of America 94: 8006–8009.
McHugh, D., 2000. Molecular phylogeny of the Annelida. Canadian Journal of Zoology 78: 1873–1884.
McHugh, D., 2001. Molecular phylogenetic analyses indicate a rapid radiation of polychaete annelids. American Zoologist 41: 1520–1521.
Miyata, T. & H. Suga, 2001. Divergence pattern of animal gene families and relationship with the Cambrian explosion. BioEssays 23: 1018–1027.
Moon, S. Y., C. B. Kim, S. R. Gelder & W. Kim, 1996. Phylogenetic positions of the aberrant branchiobdellidans and aphanoneurans within the Annelida as derived from 18S ribosomal RNA gene sequences. Hydrobiologia 324: 229–236
Nygren, A. & P. Sundberg, 2003. Phylogeny and evolution of reproductive modes in Autolytinae (Syllidae, Annelida). Molecular Phylogenetics and Evolution 29: 235–249.
Nylander, J. A. A., C. Erseus & M. Kallersjo, 1999. A test of monophyly of the gutless Phallodrilinae (Oligochaeta, Tubificidae) and the use of a 573-bp region of the mitochondrial cytochrome oxidase I gene in analysis of annelid phylogeny. Zoologica Scripta 28: 305–313.
Orensanz, J. M., 1990. The eunicemorph polychaete annelids from Antarctic and Subantarctic Seas. Antarctic Research Series 52: 1–183.
Page, R. D. M. & E. C. Holmes, 1998. Molecular Evolution. A Phylogenetic Approach. Blackwell Science, Oxford, 346 pp.
Purschke, G., 2002. On the ground pattern of Annelida. Organisms, Diversity and Evolution 2: 181–196.
Purschke, G., R. Hessling & W. Westheide, 2000. The phylogenetic position of the Clitellata and the Echiura – on the problematic assessment of absent characters. Journal of Zoological Systematics and Evolutionary Research 38: 165–173.
Rota, E., P. Martin & C. Erséus, 2001. Soil-dwelling polychaetes: enigmatic as ever? Some hints on the phylogenetic relationships as suggested by a maximum parsimony analysis of 18S rRNA gene sequences. Contributions to Zoology 70: 127–138.
Rouse, G. W., 1999. Trochophore concepts: ciliary bands and the evolution of larvae in spiralian Metazoa. Biological Journal of the Linnean Society 66: 411–464.
Rouse G. W. & K. Fauchald, 1997. Cladistics and polychaetes. Zoologica Scripta 26: 139–204.

Rouse, G. W. & F. Pleijel. 2001. Polychaetes. Oxford University Press, New York, 354 pp.

Rousset, V., G. W. Rouse, J.-P. Féral, D. Desbruyères & F. Pleijel, 2003. Molecular and morphological evidence of Alvinellidae relationships (Terebelliformia, Polychaeta, Annelida). Zoologica Scripta 32: 185–197.

Ruiz-Trillo, I., J. Paps, M. Loukota, C. Ribera, U. Jondelius, J. Baguña & M. Riutort, 2002. A phylogenetic analysis of myosin heavy chain type II sequences corroborates that Acoela and Nemertodermatida are basal bilaterians. Proceedings of the National Academy of Sciences of the United States of America 99: 11246–11251.

Schulze, S. R., S. A. Rice, J. L. Simon & S. A. Karl, 2000. Evolution of poecilogony and the biogeography of North American populations of the polychaete *Streblospio*. Evolution 54: 1247–1259.

Siddall, M. E., K. Apakupakul, E. M. Burreson, K. A. Coates, C. Erséus, S. R. Gelder, M. Källersjo & H. Trapido-Rosenthal, 2001. Validating Livanow: molecular data agree that leeches, branchiobdellidans, and *Acanthobdella peledina* form a monophyletic group of oligochaetes. Molecular Phylogenetics and Evolution 21: 346–351.

Struck, T. H., R. Hessling & G. Purschke, 2002a. The phylogenetic position of the Aeolosomatidae and Parergodrilidae, two enigmatic oligochaete-like taxa of the "Polychaeta", based on molecular data from 18S rDNA sequences. Journal of Zoological Systematics and Evolutionary Research 40: 155–163.

Struck, T. H., W. Westheide & G. Purschke, 2002b. Progenesis in Eunicida ("Polychaeta," Annelida) – separate evolutionary events? Evidence from molecular data. Molecular Phylogenetics and Evolution 25: 190–199.

Suzuki, Y., G. V. Glazko & M. Nei, 2002. Overcredibility of molecular phylogenies obtained by Bayesian phylogenetics. Proceedings of the National Academy of Sciences of the United States of America 99: 16138–16143.

Thompson, J. D., D. G. Higgins & T. J. Gibson, 1994. Clustal W: improving the sensitivity of progressive multiple sequence alignment through sequence weighting, positions-specific gap penalties and weight matrix choice. Nucleic Acids Research 22: 4673–4680.

Westheide, W., D. McHugh, G. Purschke & G. Rouse, 1999. Systematization of the Annelida: different approaches. Hydrobiologia 402: 291–307.

Winnepenninckx, B., T. Backeljau & R. De Wachter, 1995. Phylogeny of protostome worms derived from 18S rRNA sequences. Molecular Biology and Evolution 12: 641–649.

Winnepenninckx, B. M. H., Y. Van de Peer & T. Backeljau, 1998. Metazoan relationships on the basis of 18S rRNA sequences: A few years later... American Zoologist 38: 888–906.

Hydrobiologia (2005) **535/536**: 319–340
T. Bartolomaeus & G. Purschke (eds), Morphology, Molecules, Evolution and Phylogeny in Polychaeta and Related Taxa

Evolution of interstitial Polychaeta (Annelida)

Katrine Worsaae* & Reinhardt Møbjerg Kristensen
Zoological Museum, University of Copenhagen, Universitetsparken 15, 2100 Copenhagen Ø
*(*Author for correspondence: E-mail: kworsaae@zmuc.ku.dk, rmkristensen@zmuc.ku.dk)*

Key words: Polychaeta, Annelida, interstitial, meiofauna, systematics, evolution, progenesis

Abstract

An update of the systematics is given for the eight most important interstitial polychaete families: Diurodrilidae, Nerillidae, Protodrilidae, Protodriloididae, Saccocirridae, Parergodrilidae, Polygordidae and Psammodrilidae. Additional information and new observations are presented for the Diurodrilidae, Nerillidae and Psammodrilidae. Three new supplementary evolutionary hypotheses for these families are here suggested: (I) basal position of Diurodrilidae in Polychaeta, (2) evolution of Nerillidae in mud, and (3) evolution from meio- to macrofaunal forms of Psammodrilidae.

Introduction

Interstitial fauna is a term used for animals capable of moving between sediment particles, quartz grains, and pieces of broken shells or corals, with minimum disturbance of the constituent particles. Mud-dwelling animals are therefore excluded from this classification (Swedmark, 1964; Higgins & Thiel, 1988). The space factor, or size of the interstices, is of great importance for the interstitial species, which are generally small sized (0.3–3 mm), but not necessarily of meiofaunal dimensions (Swedmark, 1964; Higgins & Thiel, 1988). According to Coull & Bell (1979) it is found that in those meiofaunal taxa that have both interstitial and burrowing representatives, the sand fauna tends to be long and slender, whereas the mud fauna is not restricted to a particular morphology and is generally larger. The meiofaunal as well as the interstitial polychaetes are polyphyletic groups of species, which have evolved by many independent lines from various macrofaunal forms (Swedmark, 1964; Westheide, 1971, 1985, 1987, 1988). Many of these are highly derived species evolved by gradual transition from larger to smaller size or by progenesis (maturation in juvenile/larval stage) (Westheide, 1987).

Most of the sediment-inhabiting meiofaunal polychaetes are representatives from macrofauna families, including Acrocirridae, Dorvilleidae, Goniadidae, Hesionidae, Orbiniidae, Paraonidae, Pholoidae, Phyllodocidae, Pisionidae, Questidae, Sabellidae, Sphaerodoridae, Spionidae, and Syllidae. However, several exclusively meiofaunal or interstitial families exist, of which only the eight most important will be treated here. These include the Diurodrilidae, Nerillidae, Parergodrilidae, Polygordidae, Protodrilidae, Protodriloididae, Psammodrilidae and Saccocirridae. Based on a cladistic analysis of morphological data the interstitial taxon Dinophilidae has been included in Dorvilleidae (Eibye-Jacobsen & Kristensen, 1994). It here represents the most reduced forms in a progenetic evolution series leading from large dorvilleids with parapodia, cirri and jaws to small dinophilids without parapodia or jaws. In fact, it is the first time that progenesis has been shown to be the driving force in the annelid evolution. However, a recent cladistic analysis based on molecular data has cast doubt on the monophyly of the Dorvilleidae (Struck et al., 2002b). Yet, additional information is necessary to then clarify the position of the dinophilids and explain the altered evolutionary scenario. Because the dinophilids still belong to the macrofaunal Dorvilleidae, these are not treated further here. Most of the eight exclusively interstitial families, mentioned above, were formerly classified as archiannelids (defined by Hatschek, 1878), which is now considered an invalid taxon (Westheide, 1985). These families are now believed be highly derived and most of them to have evolved independently (Fauchald, 1977; Westheide 1985, 1987). However, their phylogenetic position and origin is controversial and the present paper will seek to update the systematics of these families.

Several characteristics have been mentioned for interstitial and former archiannelid polychaetes: small size; complete homonymy; weak segmentation; long, spionid-like primary tentacles (palps); few or no appendages, parapodia, or chaetae; larval characteristics like ventral ciliation (gastrotroch), dorsal ciliary bands and protonephridia; incomplete mesenteries and mesenterial channels; no circular musculature, muscle cells of myoepithelial origin; simple blood vascular system; simple ciliated intestine; simple structure of brain; brain apical; obvious subesophagous ganglion missing; nerve system closely linked to epidermis; ventral nerve cord not segmented; direct development; and novel characters like specialized pharyngeal apparatus; peculiar spermatozoa; special reproductive modes (e.g. internal fertilization) and larvae; and adhesive organs (Hempelmann, 1931; Clark, 1969; Swedmark, 1964; Westheide, 1971, 1984, 1985, 1987, 1990; Bubko, 1973; Orrhage, 1974). Some of these characters, like e.g. protonephridia, which were formerly interpreted as plesiomorphic or larval states, have now been demonstrated to be secondarily reduced or specialized states (Bartolomaeus, 1999).

Many of the interstitial polychaetes have evolved by genetically fixed progenesis (Westheide, 1971, 1987). It is now the general belief that the morphological resemblance of the interstitial polychaetes does not reflect a common ancestor, but convergent adaptations to the environment, small size and progenetic origin (Swedmark, 1964; Westheide, 1971, 1985, 1988). According to Westheide (1987) a progenetic origin is most compelling, when the taxon has great similarity with larval or juvenile stages of a macrofaunal taxon. However, the evolution should not be viewed as totally regressive, as it will often also involve strongly specialized and novel characters, which are crucial for a successful existence in the interstitial environment (Westheide, 1987). The many apomorphies and problems with scoring absent or reduced characters have made it very difficult to analyze the phylogenetic position of the interstitial families in Polychaeta (Purschke et al., 2000).

The theory of progenesis has been linked rather closely to the size-specific niche of the interstitial environment, indicating that the hard adaptational demands for entering this environment require a one-step specialization to smaller size (Westheide, 1987). Mud-dwelling meiofauna have never been proposed to evolve by progenesis. The four nerillids, *Meganerilla swedmarki* Boaden, 1961, *Nerilla australis* Willis, 1951, *Paranerilla cilioscutata* Worsaae & Kristensen, 2003 and *P. limicola* Jouin & Swedmark, 1965, are among the only examples of mud-dwelling meiofauna polychaetes (Willis, 1951; Jouin & Swedmark, 1965; Saphonov & Tzetlin, 1997; Worsaae & Kristensen, 2003), but mud habitats are not as thoroughly examined for this fauna as the interstitial environment. The possibility that meiofauna polychaetes like Nerillidae could have evolved in mud, either by progenesis or gradual transition to smaller size (miniaturization), is discussed here.

One of the arguments for a progenetic evolution as an alternative to miniaturization is that only few examples of gradual transition of middle-sized species to true small interstitial forms are known (Westheide, 1987). Some of these examples can even be interpreted such that the evolutionary pathway has the opposite direction, from interstitial meiofauna to non-interstitial macrofauna (Westheide, 1987). *Microphthalmus hamosus* Westheide, 1982 is an example of a commensal macrofaunal species, which has evolved from an otherwise interstitial genus (Westheide, 1982). Speciation has led to an increase in body size and number of segments. However, chaetae lost in the previous adaptation of the genus to the interstitial environment were not regained in this species (Westheide, 1982). It was argued that genetic information has been lost and that Dollo's law may affect the evolution from smaller to larger forms (Westheide, 1982). The irreversibility of character reduction as well as the phylogenetic interpretation of macrofaunal taxa with 'simple' characters has wide implications for our understanding of relationships within Annelida. Evolution from macro- to meiofaunal forms (Swedmark, 1955, 1958; Kristensen & Nørrevang, 1982) and back to larger forms is here discussed in relation to the various coelomic conditions found in especially Psammodrilidae (Fransen, 1980; this paper). Finally, it is discussed whether the family Diurodrilidae may not be a secondarily derived 'pseudocoelomate taxon' of progenetic origin, but instead may occupy a more basal position within the Polychaeta or Annelida.

Results

An update of the systematics of the eight most important interstitial polychaete families is presented in alphabetic order. The dinophilids are not treated here because they belong to the macrofaunal family Dorvilleidae (see Introduction). The genera and number of species (in parenthesis) is summarized and an update on current hypotheses on their phylogenetic position within Polychaeta is presented. Additional information summarizing previous and new results on morphology and distribution is provided for the Diurodrilidae, Nerillidae and Psammodrilidae. This is provided in relevance to the remarks on their evolution, which may not be so simple that they all have evolved from macrofaunal forms by progenesis. Detailed reviews and supplementary information on the interstitial polychaete families can be found in Westheide (1988), Westheide (1990), Beesley et al. (2000), and Rouse & Pleijel (2001).

Diurodrilidae Kristensen & Niilonen, 1982

The family consists of six described species belonging to *Diurodrilus* Remane, 1925 (250–500 μm long). Two undescribed species are known from New Zealand (Riser, 1984) and an undescribed species from Queensland, Australia with dorsal 'cuticular plates' is illustrated by Paxton (2000, fig. 1.64). A detailed description and morphological comparison of the six described species was made by Kristensen & Niilonen (1982) and Villora-Moreno (1996).

Systematics

The genus *Diurodrilus* was originally placed in the Dinophilidae by Remane (1925). Kristensen & Niilonen (1982) demonstrated that this genus deserved family status and erected the Diurodrilidae. The diurodrilids have been proposed to have a resemblance to gnathostomulids (Kristensen & Niilonen, 1982) and it has been hypothesized that psammobiontic 'worms', primarily the Gnathostomulida, are related to Annelida (e.g. Nielsen, 1995). However, Kristensen & Eibye-Jacobsen (1995) totally refuted the idea that Diurodrilidae and Gnathostomulida could be related, but stated that the phylogenetic relationship of the family within the Polychaeta remains unclear. A comparison between Gnathifera, especially *Limnognathia maerski* Kristensen & Funch, 2000 (Micrognathozoa) and Diurodrilidae was made by Kristensen & Funch (2000). Although several striking similarities were found, the overall conclusion was that these were either superficial or have evolved by convergence. Recently, the Diurodrilidae was included in Dorvilleidae (Rouse & Pleijel, 2001) but as these authors stated, they based this alone on superficial similarities of *Apodotrocha* Westheide & Riser, 1983 and *Dinophilus* Schmidt, 1848.

Additional information

Diurodrilids possess a tripartite prostomium, a metastomium (peristomium), 5 trunk segments (weakly segmented) and a pygidium with forked toes (Fig. 1A–D) provided with duo-glands system (adhesive glands) and muscles. Adhesive glands are also found on the prostomium. The oval mouth opening is located ventrally on the metastomium, and surrounded by a cuticular ring. Labial lobes may be present in all species but are so far only observed in the here presented SEM illustrations of *D. minimus* (Fig. 1C). Evidence for the monophyly of the family is the single epidermal cells (ciliophores) located ventrally on head and trunk. The ciliophores carry regularly arranged cilia with very long rootlets (Fig. 2A–B). The prostomium and metastomium (peristomium) are provided with ovoid ciliophores, while the trunk has rectangular ciliophores, forming a discontinuous, midventral band (Figs 1A–B, 2A–B). The trunk ciliophores are the locomotory organs, while the prostomial

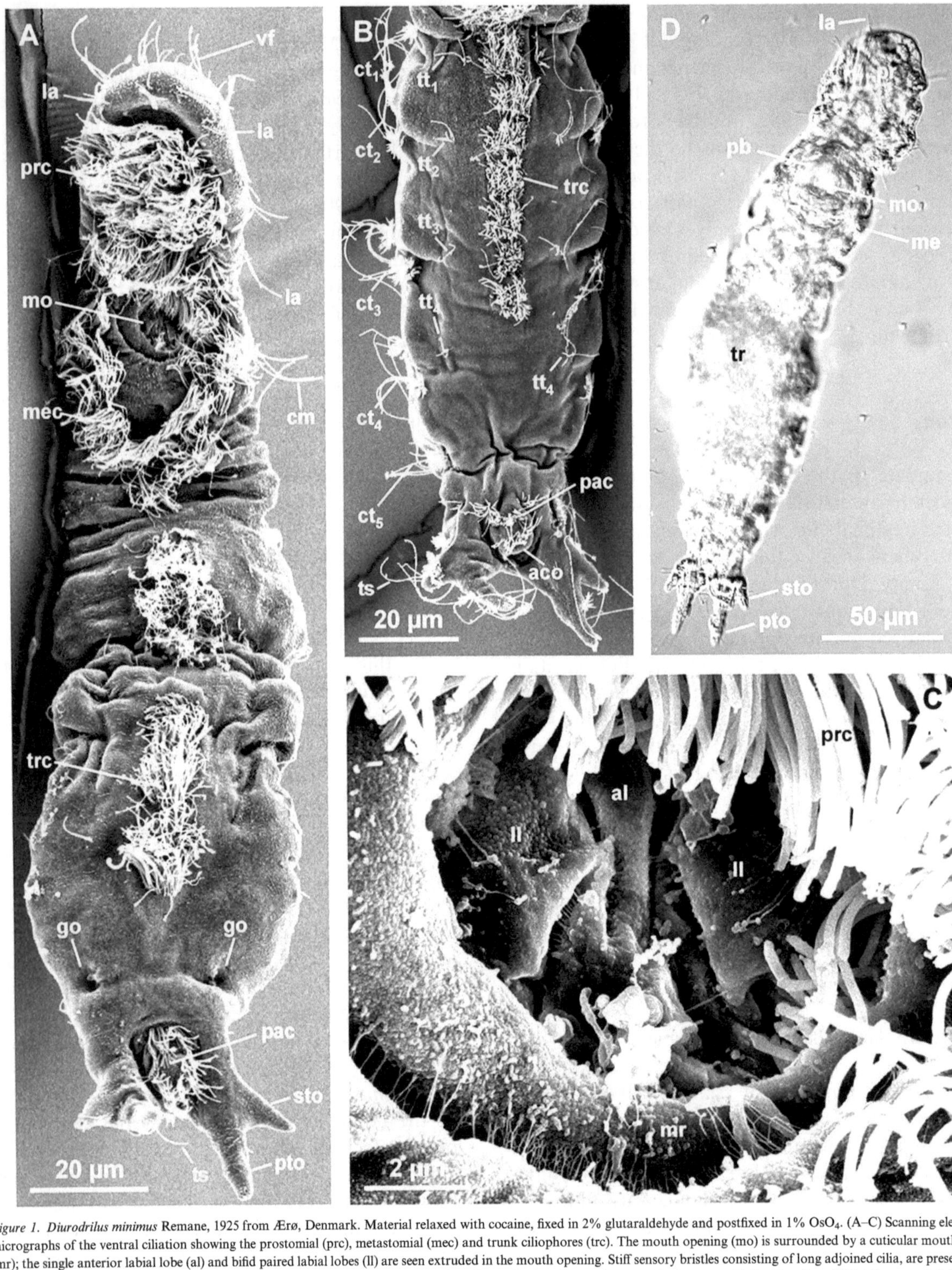

Figure 1. Diurodrilus minimus Remane, 1925 from Ærø, Denmark. Material relaxed with cocaine, fixed in 2% glutaraldehyde and postfixed in 1% OsO_4. (A–C) Scanning electron micrographs of the ventral ciliation showing the prostomial (prc), metastomial (mec) and trunk ciliophores (trc). The mouth opening (mo) is surrounded by a cuticular mouth ring (mr); the single anterior labial lobe (al) and bifid paired labial lobes (ll) are seen extruded in the mouth opening. Stiff sensory bristles consisting of long adjoined cilia, are present on the prostomium (vf, la), metastomium (cm) and the five trunk segments (ct_{1-5}). The primary (pto) and secondary toes (sto) are provided with a few stiff bristles as well (ts). Not previously observed, ciliary tufts of the trunk (tt_{1-4}) are located ventrally on the first four trunk segments. The preanal cilia field (pac) is found around the short anal cone (aco). Paired male gonopores (go) are located on the fifth trunk segment. (D) Differential interference contrast micrograph of live animal (relaxed with cocaine), showing the prostomium (pr), metastomium (me) with the pharyngeal bulb (pb), and the very weak segmentation of the trunk (tr).

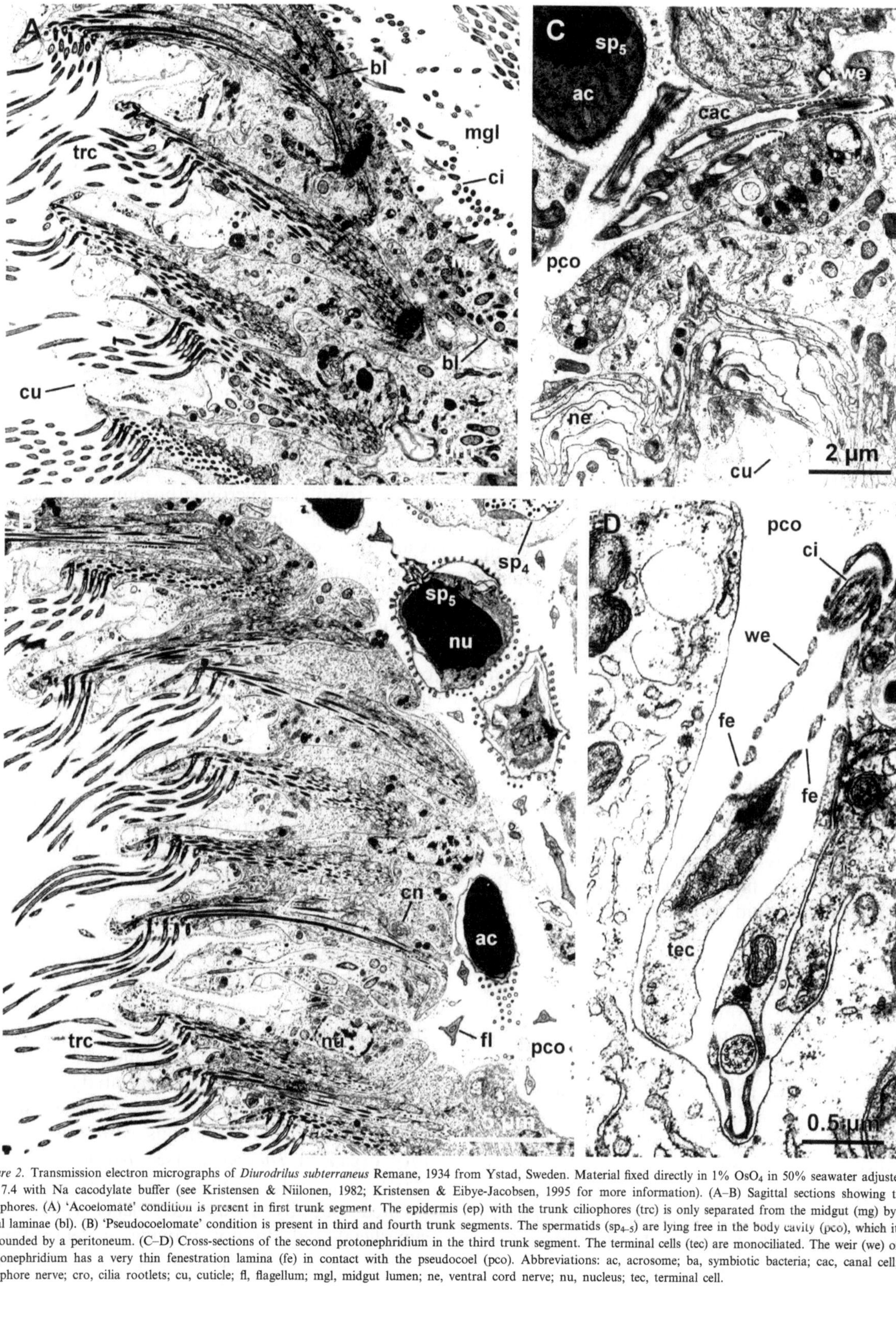

Figure 2. Transmission electron micrographs of *Diurodrilus subterraneus* Remane, 1934 from Ystad, Sweden. Material fixed directly in 1% OsO_4 in 50% seawater adjusted to pH 7.4 with Na cacodylate buffer (see Kristensen & Niilonen, 1982; Kristensen & Eibye-Jacobsen, 1995 for more information). (A–B) Sagittal sections showing trunk ciliophores. (A) 'Acoelomate' condition is present in first trunk segment. The epidermis (ep) with the trunk ciliophores (trc) is only separated from the midgut (mg) by two basal laminae (bl). (B) 'Pseudocoelomate' condition is present in third and fourth trunk segments. The spermatids (sp_{4-5}) are lying free in the body cavity (pco), which it not surrounded by a peritoneum. (C–D) Cross-sections of the second protonephridium in the third trunk segment. The terminal cells (tec) are monociliated. The weir (we) of the protonephridium has a very thin fenestration lamina (fe) in contact with the pseudocoel (pco). Abbreviations: ac, acrosome; ba, symbiotic bacteria; cac, canal cell; cn, ciliophore nerve; cro, cilia rootlets; cu, cuticle; fl, flagellum; mgl, midgut lumen; ne, ventral cord nerve; nu, nucleus; tec, terminal cell.

ciliophores may be used as a broom to collect food particles. Although the trunk ciliophores may be functional homologous with the midventral ciliary band found in other meiofaunal polychaetes, the ciliary structure and discontinuous distribution in small transverse bands clearly discriminate them (Figs 1A–B, 2A–B). Long sensory bristles (compound cilia) (vf, la) are found dorsally and latero-ventrally on the prostomium (Fig. 1A), and latero-segmentally positioned (cm) on the metastomium (Fig. 1A) and the trunk (ct_{1-5}, Fig. 1B). The forked toes are also provided with tactile bristles (Fig. 1B, ts). Transverse ciliary bands seem to be lacking, except for in the new species from Australia, but the ventral ciliary tufts of the trunk observed on *D. minimus* (Fig. 1B, tt_{1-4}) may be homologous with such ciliary bands. There are no signs of parapodial appendages or chaetae.

Diurodrilids are gonochoristic. The females have paired ovaries that contain one or two large oocytes. The trunk of the males is often found filled with spermatids or spermatozoa lying free in a single large cavity ('pseudocoelom') in the trunk. The spermatozoa are highly specialized with a giant acrosome (Fig. 2B) and the middle piece covered with modified microvilli (mushroom-shaped bodies, see Kristensen & Eibye-Jacobsen, 1995). Males have paired gonopores (go) located on the fifth trunk segment (Fig. 1A).

Two pairs of protonephridia are observed in *D. westheidei* Kristensen & Niilonen, 1982 and *D. subterraneus*. The terminal cell is monociliated and has a weir without basal lamina (Fig. 2C, D), and the very tiny fenestration lamina (fe) functions as the only filter for ultrafiltration of the pseudocoelomic fluid into the protonephridium (Fig. 2D). The cilium from the terminal cell is spiraled several times in the canal cell (Fig. 2C), and it seems that the protonephridia lie in close contact with the germinal cells. Nephridiopores were not observed. Basal laminae are weakly developed or totally lacking between the protonephridia and the germinal cells. The blood vascular system seems to be absent, and the somatic muscles are very tiny. The coelomic condition is unsatisfactorily understood. In the first trunk segment of *D. subterraneus* the ciliophores are in close contact with the cells of the midgut ('acoelomate condition', Fig. 2A). In the other four trunk segments the ciliophores are separated from the midgut by a large cavity where the spermatozoa and the protonephridia are located ('pseudocoelomate' condition, Fig. 2B).

Remarks on evolution

The Diurodrilidae do not possess any larval characters (besides small size) and have no close resemblance to any present macrofaunal families, which are two criteria for recognizing a progenetic origin. The ventral band of trunk ciliophores is not interpreted as a larval feature but as a unique specialization. The 'acoelomate/pseudocoelomate' condition and various morphological specializations cannot be used as evidence for a secondary miniaturization or that this exclusive meiofaunal family should have evolved from a macrofaunal form by progenesis. Progenetic origins of other meiofaunal taxa like the gnathostomulids, micrognathozoans and rotifers from juvenile eucoelomate Spiralia during the 'Cambrian explosion' is even an improbable explanation for their 'acoelomate/pseudocoelomate' condition. Although no phylogenetic justification exists, it has been well argued by Bartolomaeus & Ax (1992) that 'acoelomate' condition and lack of larval state are plesiomorphic character states in the Bilateria. However, the variability of coelomic conditions throughout the animal kingdom (see Bartolomaeus, 1993) indicates that the 'acoelomate', 'pseudocoelomate' or 'eucoelomate' condition should be used with caution analyzing phylogenetic relationships. Ciliophores are found in both *Neotenotrocha sterreri* Eibye-Jacobsen & Kristensen, 1994 (Dorvilleidae) and *Limnognathia maerski* (Micrognathozoa), but they differ in several aspects and are more developed in *Diurodrilus* (see Eibye-Jacobsen & Kristensen, 1994; Kristensen & Funch, 2000). Nuchal organs, characteristic for polychaetes, have never been found in Diurodrilidae. The findings of gonopores above the pygidium in this study (Fig. 1A) contradicts the idea that the anal cone in diurodrilids should be homologous with the copulatory organ in *Dinophilus* (see Rouse & Pleijel, 2001). The other similarities with dorvilleids pointed out by Rouse & Pleijel (2001) are superficial and do not provide evidence for a relationship within the Dorvilleidae. It is therefore at present not possible to eliminate that the Diurodrilidae could have branched off very early within the Polychaeta. The new data on the protonephridia and the body cavity condition of *Diurodrilus subterraneus* (Fig. 2) may indicate that diurodrilids are in fact primary small interstitial annelids, but it remains an unproven alternative to the current comprehension of a progenetic origin of the family. Additional morphological studies in cLSM and TEM, and in particular molecular data are strongly needed for small aberrant families like the Diurodrilidae.

Nerillidae Levinsen, 1883

Nerillidae is the largest meiofauna family in Polychaeta with 48 species in 17 genera (generally, 300 μm–2 mm in length): *Afronerilla* Faubel, 1978 (1 species), *Akessoniella* Tzetlin & Larionov, 1988 (1), *Aristonerilla* Müller, 2002 (1), *Bathychaetus* Faubel, 1978 (1), *Leptonerilla* Westheide & Purschke, 1996 (2), *Meganerilla* Boaden, 1961 (3), *Mesonerilla* Remane, 1949 (9), *Micronerilla* Jouin, 1970 (1), *Nerilla* Schmidt, 1848 (12), *Nerillidium* Remane, 1925 (9), *Nerillidopsis* Jouin, 1966 (1), *Paranerilla* Jouin & Swedmark, 1965 (2), *Psammoriedlia* Kirsteuer, 1966 (1), *Thalassochaetus* Ax, 1954 (1), *Trochonerilla* Tzetlin & Saphonov, 1992 (1), *Troglochaetus Delachaux*, 1921 (1), and *Xenonerilla Müller*, Bernhard & Jouin, 2001 (1). Previous studies on the systematics and morphology of Nerillidae have been made by Swedmark (1959), Jouin (1967, 1968, 1970a, 1971), Schmidt & Westheide (1977), Westheide (1990) and Müller et al. (2001).

Systematics

In the earlier literature the nerillids have been placed along with nereids and syllids (e.g. Schmidt, 1848; Quatrefages, 1866). The family was later on classified as archiannelids, and various relationships within this group have been proposed (e.g. Beauchamp, 1910; Goodrich, 1912). The pharyngeal apparatus has been shown to have some similarities with that of protodrilids (Purschke, 1985). However, great structural differences exist as well and the resemblances are most likely a matter of convergence (Purschke & Jouin, 1988). Rouse & Fauchald (1997) place them in Aciculata as incertae sedis, which is followed by Rouse & Pleijel (2001). Their superficial resemblance to juvenile onuphids, studied by Hsieh & Simon (1987) is pointed out by Westheide & Purschke (1996). This resemblance, as well as their otherwise meiofaunal characteristics, may point to a progenetic origin of the family (Westheide, 1990; Westheide & Purschke, 1996). Based on morphological studies, Rouse & Pleijel (2001) suggest a close affinity to *Aberranta* Hartman, 1965.

Westheide & Purschke (1996) proposed a regressive evolutionary pathway within Nerillidae, with *Leptonerilla* positioned most basally in the family. *Leptonerilla diplocirrata* Westheide & Purschke, 1996, *L. prospera* Sterrer & Iliffe, 1982 and *Mesonerilla diatomeophaga* Núñez, 1997 in Núñez et al. 1997 all have the following combination of character states: compound chaetae and maximum number of segments (9) and appendages among nerillids (3 antennae, 2 palps, 2 pygidial cirri, double parapodial cirri). *Mesonerilla diatomeophaga* should according to the definitions of *Leptonerilla* by Westheide & Purschke (1996) be reassigned to this genus. The presented character states were believed to represent the plesiomorphic conditions of the family (Westheide & Purschke, 1996). However, the hypothesis was not based on cladistic analyses, which are necessary to verify the basal position of *Leptonerilla* within the family, as well as the evolution of characters. Some of the character states presented may also prove to behave more homoplastic than predicted by Westheide & Purschke (1996).

Additional information

Nerillids are nearly all marine and distributed worldwide from the intertidal to abyssal depths (3660 m – see Worsaae & Kristensen, 2003). Nerillidae have generally been characterized as an interstitial polychaete family as several species are described from either the interstitial sandy habitat or shell gravel. The gravel (with or without shells) containing nerillids is sometimes 'dirty', holding mud, although the content of silt in coarse interstitial habitats is generally not reported. When the very large interstices of the gravel contain deposits of silt it may in fact not represent a true interstitial environment. Many nerillid species are actually found outside the interstitial environment: *Nerilla australis*, *Paranerilla cilioscutata* and *P. limicola* are described from mud bottoms (Willis, 1951; Jouin & Swedmark, 1965; Worsaae & Kristensen, 2003) and *Meganerilla swedmarki* has occasionally been found in mud (Saphonov & Tzetlin, 1997; personal observations of animals at 100–250 m depth in mud at Iqpik, Disko, West Greenland); *Leptonerilla prospera* is found in caves with fine silt (Sterrer & Iliffe, 1982); *Bathychaetus heptapous* Faubel, 1978 in mud with sand (Faubel, 1978); *Nerilla* spp. in detritus sand, organic debris, green algae and macrophytes (for review see Gelder, 1974); *Xenonerilla bactericola* Müller, Bernhard & Jouin-Toulmond, 2001 in bacterial mats of the Santa Barbara Basin (Müller et al., 2001); *Troglochaetus beranecki* Delachaux, 1921 in freshwater caves and pebbles in rivers (see review in Morselli et al., 1995); *Leptonerilla diatomeophaga* (Núñez, 1997 in Núñez et al. 1997) in caves with diatom carpets on lapilli (Núñez et al., 1997); and as mentioned several species have been found in 'dirty' gravel with or without shells e.g. *Aristonerilla brevis* (Saphonov & Tzetlin, 1997), *Meganerilla swedmarki*, *Mesonerilla armoricana* Swedmark, 1959, *M. fagei* Swedmark, 1959, *M. roscovita* Lévi, 1953, *Nerilla* spp., *Nerillidium troglochaetoides* Remane, 1925, *Thalassochaetus palpifoliaceus* Ax, 1954 and *Trochonerilla mobilis* Tzetlin & Saphonov, 1992 (Ax, 1954; Boaden, 1961; Gelder, 1974; Saphonov & Tzetlin, 1997; and personal observations on habitats of *A. brevis*, *Mesonerilla* spp., *N. troglochaetoides*. and *T. mobilis*). Many nerillid descriptions only give a poor habitat report, and the few observations of live animals generally do not describe behavior from the natural habitat but from sieved or decanted material. However, if a well-supported phylogeny of the family Nerillidae existed, it would be interesting to map the different nerillid habitats on the tree and then trace and analyze the evolutionary scenario.

Most nerillids have direct development, sometimes including brooding. They would only be able to spread from one locality to another and from the interstitial environment to other habitats by migration, dispersal of the sediment or continental drift. However, the obligate mud-dwelling *Paranerilla limicola* is found to have indirect development with a pelagic trochophore larva, which is more easily spread by currents over larger distances to different habitats (Jouin & Swedmark, 1965). It would be interesting to know whether the plesiomorphic condition in nerillids is indirect development, which could indicate a non-interstitial origin of the family.

The freshwater nerillid *Troglochaetus beranecki* Delachaux, 1921 has been reported from many localities around central Europe (Germany, France, Italy) as well as from the Colorado Rocky Mountains (e.g. Delachaux, 1921; Pennak, 1971). According to Pennak (1971) the species most likely originated before the continental drift was well under way, which would explain its existence on continents on each side of the Atlantic, and he predicts that the species will be found in many other parts of North America. Nerillids are known from all continents except the Antarctic, and this wide geographical distribution as well as the diversity in habitats may very well reflect an old history of the family.

Remarks on evolution

Nerillids may have evolved in the interstitial habitat and secondarily spread to non-interstitial habitats. However, the opposite evolutionary history cannot be rejected based on current knowledge of their distribution.

Westheide (1987) considered different evolutionary pathways of meiofauna in the interstitial environment and the selective forces supporting them. The induction of progenesis was linked closely to the interstitial habitat, which possessed so extraordinary adaptational demands, that a one-step adaptation in size would be necessary to enter this habitat (Westheide, 1987). However, some of these evolutionary pathways, including progenesis, may also be applicable to the muddy environment.

The highest concentration of meiofauna in mud has generally been found in the upper one centimeter of the bottom (Coull & Bell, 1979). *Paranerilla cilioscutata*, *P. limicola*, *Meganerilla swedmarki* from Disko, West Greenland, *M. swedmarki* from the White Sea and *Nerilla australis*, all have been found in mud (Willis, 1951; Saphonov & Tzetlin, 1997; Worsaae & Kristensen, 2003, unpublished observations). The three first of which were more specifically found in the sediment–water interface, which may be a size-specific niche for meiofauna organisms. This flocculent layer contains a higher concentration of small sized food particles compared to the underlying layers of the mud due to suspension and resuspension of less heavy particles. Furthermore, the layer is well oxygenated, thereby allowing respiratory exchanges of animals living there and facilitating a higher production of bacteria and algae. The meiofauna has therefore access to higher concentrations of oxygen and appropriate food items in this layer. In the interface, small organisms would be less exposed than larger motile organisms to selective predation. In some areas, the lower part of the sediment may be anoxic thereby limiting the meiofauna to the upper oxygen-rich layers. The uppermost part of the sediment–water interface may also be so loose that only small organisms using ciliary motion could move around without costly use of the musculature in actual swimming.

The flocculent layer of the sediment–water interface of muddy habitats may therefore function as a size-related niche, comparable to the size-specific niche of the interstices in the interstitial environment. The specific size limitations are dependent on the depth and density of the muddy flocculent layer. These factors are defined by local current conditions and the composition of sediment particles, which are the same abiotic parameters that define the space available in the interstitial environment (Swedmark, 1964). In lower parts of the mud bottom with less organic material and oxygen, the expenditure/use analysis for motile deposit-feeding macrofauna may be negative. The same may apply to existence above the bottom, where the current is higher. Energetically expensive burrowing or swimming is then only worthwhile when a great deal of food is available and small distances have to be covered. Thus, a selection pressure may exist in or above mud bottoms for a decrease in size down to dimensions that allow an exploration of the sediment–water interface.

In a very thin or loose flocculent layer a progenetic evolution would be advantageous in providing a one-step speciation to a small size. If the interface is more extensive or particle dense, middle-sized animals may be able to explore some of the same advantages of this environment, and a miniaturization by gradual decrease in size is also possible.

Besides a reduction in size a well-developed ciliary covering may be particularly advantageous when entering the niche of the sediment–water interface of the muddy environment. Adhesive organs, as found on many interstitial forms would not be useful. It is more important to have the ability to move on or in the flocculent layer as well as to sink into the mud during current and other turbulent influence to avoid dispersal to the water column. *Paranerilla cilioscutata* and *P. limicola* possess a distinct dense dorsal ciliation (Fig. 3A, C) making them capable of not only gliding 'epibenthically' on top of the flocculent layer (by help of the ventral ciliation) but also entering the mud. The animal burrows into the flocculent layer by transporting mud particles across the dorsal surface by help of the very dense dorsal ciliation (Jouin & Swedmark, 1965; Worsaae & Kristensen, 2003). No information is given on the motility of *Nerilla australis*; however, Willis (1951) describes it with thin dorsal transverse ciliary bands. Although far from as dense in their distribution, they may have some resemblance to the dorsal ciliation found in *Paranerilla*. *Meganerilla swedmarki* almost lacks a dorsal ciliation (Fig. 3B) and cannot burrow, but it is capable of gliding on the sediment surface. When doing so, it generates a thick mucus-string from the posterior end of the midventral ciliary band, attaching it to the uppermost flocculent layer (personal observations). Indirect development, which is found in the obligate mud-dwelling *P. limicola* (Jouin & Swedmark, 1965), would provide a possible way of spreading and of decreasing intraspecific competition, when living outside the interstitial environment. Jouin & Swedmark (1965) argued that the large number of relatively small eggs found in the facultative mud-dwelling *Meganerilla swedmarki* by Boaden (1961) might also indicate an indirect development.

Meganerilla swedmarki is a facultative mud-dweller (Saphonov & Tzetlin, 1997; personal observations of animals from West Greenland) and does not seem as well adapted morphologically to the muddy habitat as the obligate mud-dwelling *Paranerilla* species. It therefore seems possible that more nerillids may be facultative mud-dwellers. Although *Nerilla*, *Paranerilla* and *Meganerilla* have a very different body shape,

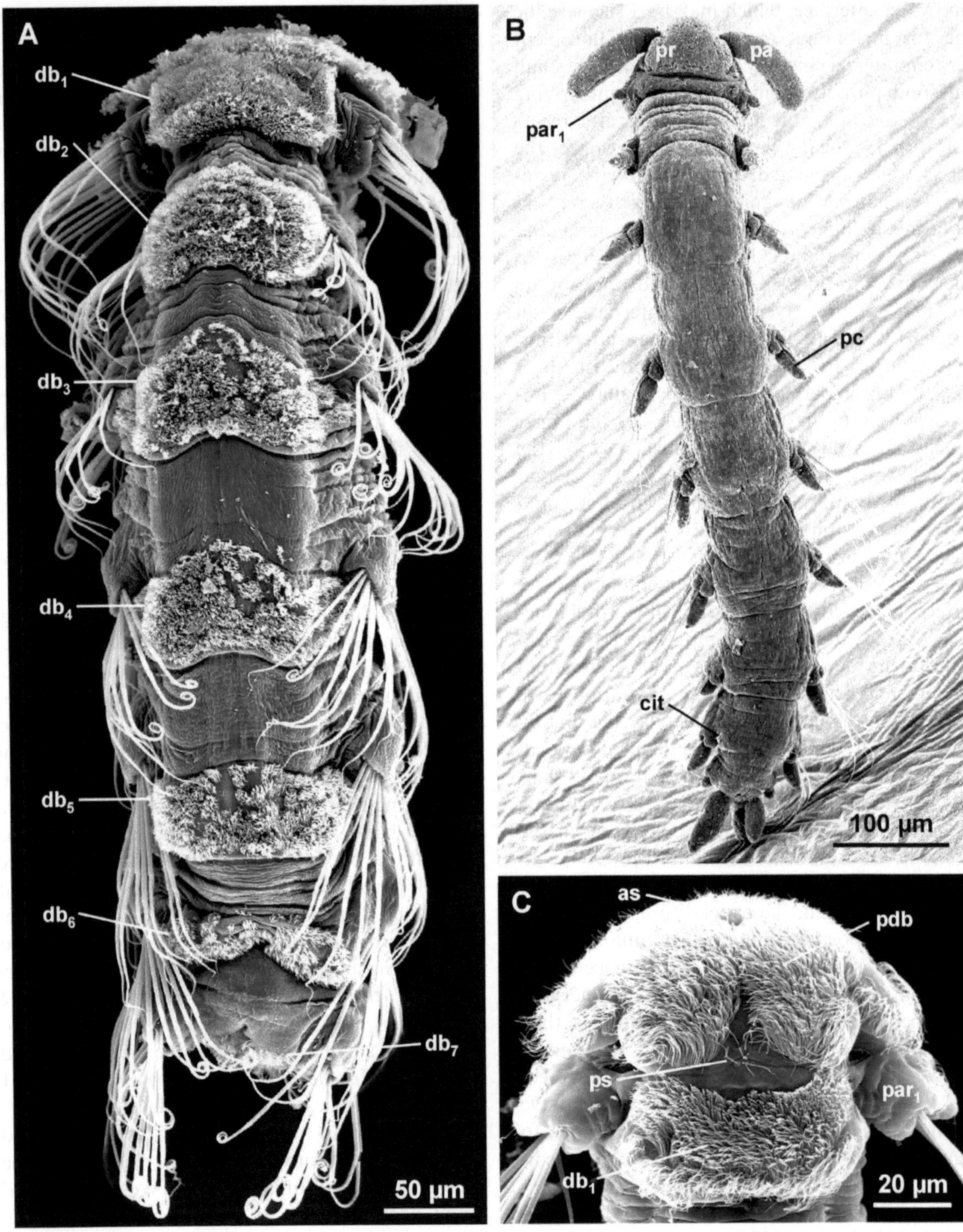

Figure 3. Scanning electron micrographs of *Paranerilla cilioscutata* Worsaae & Kristensen, 2003, *P. limicola* Jouin & Swedmark, 1965 and *Meganerilla* cf. *swedmarki* Boaden, 1961 (see Worsaae & Kristensen, 2003 for details on fixation). (A) Dorsal view of *Paranerilla cilioscutata* from North East Greenland, showing dorsal ciliary plates on segment 1–7 (db_{1-7}). (C) Close-up of *Paranerilla limicola* from Kristineberg, Sweden, showing dense ciliation on prostomium with prostomial ciliary plate (pdb) and on segment 1 with dorsal ciliary plate (db_1). Parapodia of segment 1 (par_1) and two groups of sensory cilia are present: anterior sensory cilia (as) and posterior sensory cilia (ps). (B) Dorsal view of *Meganerilla* cf. *swedmarki* from Disko, West Greenland, showing sparsely ciliated dorsal surface only with parapodial ciliary tufts (cit). Abbreviations: pr, prostomium; pa, palp; pc, parapodial cirri.

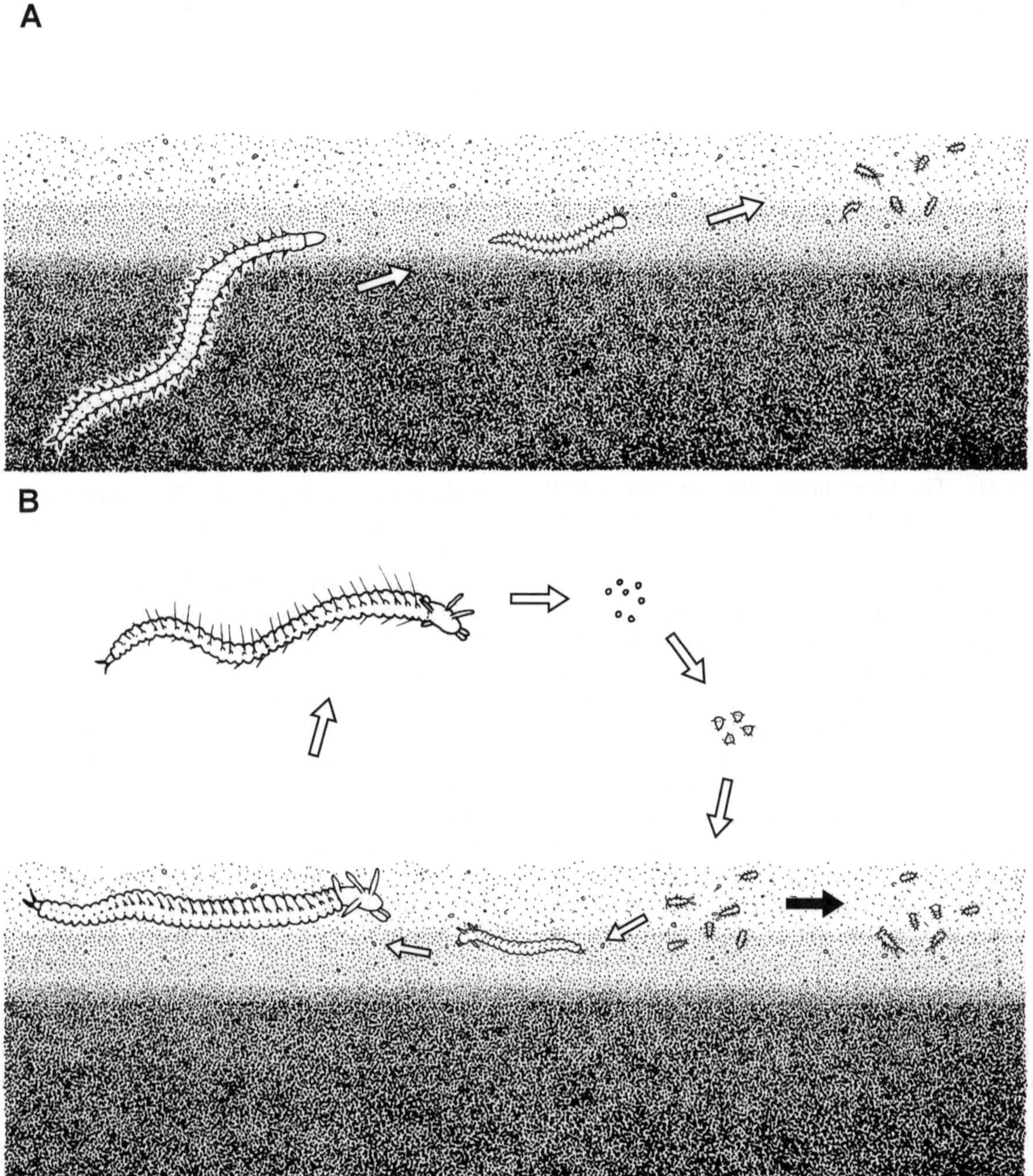

Figure 4. Hypothetical evolutionary pathways in muddy habitats (following the schemes presented by Westheide, 1987). (A) Regressive evolution, by gradual decrease in size from endopsammic macrofaunal form (in lower mud layers) to meiofaunal form in the mud interface (flocculent layer). (B) Example of progenetic origin (black arrow) of present meiofauna in the mud interface (flocculent layer) from a temporary meiofaunal juvenile stage of an epibenthic macrofauna organism with a pelago-benthic lifecycle.

they are all relatively large and robust compared to the interstitial nerillids (e.g. most species of *Nerillidium*). This is in accordance with the generally larger size of the mud-dwelling meiofauna (Coull & Bell, 1979). According to Coull & Bell (1979) the sand fauna tends to be long and slender, whereas the mud fauna is not restricted to a particular morphology. Compared to most other interstitial polychaetes, the nerillids possess many appendages and long chaetae and are not particularly slender, however, they are relatively small.

Based on these arguments, it is at present not possible to reject the idea that the nerillids could have evolved in mud – either by a gradual transition in size (Fig. 4A) or by progenesis (Fig. 4B). However, more sampling in muddy habitats as well as phylogenetic analyses are needed to clarify their evolution.

Parergodrilidae Reisinger, 1925

Two species in two genera are described in the family (0.7–2.6 mm long): *Parergodrilus* Reisinger, 1925 (1) and *Stygocapitella* Knöllner, 1934 (1). The latter taxon may contain several species according to Schmidt & Westheide (2000).

Systematics

The family was described as 'archiannelids' (Reisinger, 1925), and has since then been characterized as oligochaetes (Meyer, 1927), capitellids (in the description of *Stygocapitella* Knöllner, 1934), related to nerillids (Reisinger, 1960), to ctenodrilids (Fauchald, 1977), and to *Hrabeiella* Pizl & Chalupsky, 1984. Rouse & Fauchald (1997) placed the family as incertae sedis in Polychaeta. Although the family shares many similarities with clitellates and the terrestrial *Hrabeiella*, most of these have been explained by convergent evolution and recent morphological studies regard them as polychaetes (Rota, 1998; Purschke, 1999; Purschke et al., 2000). Two cladistic analyses based on molecular data of 18S rDNA support the independent evolution of Parergodrilidae and Clitellata (Rota et al., 2001; Struck et al., 2002a). Struck et al. (2002a) also found a relationship of *Stygocapitella subterranea* Knöllner, 1934 (although weakly supported) with a cluster comprising the polychaetes *Scoloplos armiger* (O.F. Müller, 1776), *Questa paucibranchiata* Giere & Erseus, 1998 and sometimes *Magelona mirabilis* (Johnston, 1865).

Polygordiidae Czerniavsky, 1881

The family consists of one genus, *Polygordius* Schneider, 1868 with 15 nominal species (1–10 cm long). The validity of *Chaetogordius canaliculatus* Moore, 1904 described from fragmentary material was questioned by Hermans (1969) and the species (and genus) was considered invalid by Westheide (1990). A revision of *Polygordius* was made by Rota & Carchini (1999).

Systematics

Early literature placed the family near the Opheliidae (e.g. McIntosh, 1875), and later on in the 'archiannelids' together with *Protodrilus* Hatschek, 1880 and closely related to *Saccocirrus* Bobretzky, 1871 (e.g. Marion & Bobretzky, 1875; Hatschek, 1878, 1893). Hatschek (1893) suggested that opheliids (classified with the Spiomorpha) were derived from saccocirrids that in their turn had evolved from polygordiids. This hypothesis has been rejected, and the family is generally believed to be highly derived within the polychaetes. According to Westheide (1990), Polygordius is a rather isolated genus with no close affinity to Protodrilida. However, the relationship to Spiomorpha was indirectly supported by Rouse & Fauchald (1997), who placed the family as incertae sedis in Canalipalpata, partly by assuming that the palps are grooved, which they are not. A sister group within the Opheliidae has been reconsidered, based on similarities in the structure of cuticle, musculature and locomotion as well as by rejecting the interpretation of grooved palps (Hermans, 1969; Rouse & Pleijel, 2001). According to Westheide (1990), these unproven similarities with Opheliidae might as well reflect convergent adaptations to a life in coarse gravel sediments. In a recent cladistic analysis based on molecular data (18S rDNA) a species of *Polygordius* tends to clade with *Saccocirrus papillocercus* Bobretzky, 1871 (Struck et al., 2002a). Although more genes and taxa need to be examined, molecular systematics seems an obvious way to gather more information on the phylogenetic position of this morphologically indistinct family.

Protodrilidae Czerniavsky, 1881

The family consists of 31 species in two genera (2–15 mm long): *Parenterodrilus* Jouin, 1992 (1 species) and *Protodrilus* Hatschek, 1880 (30). Detailed morphological studies and revisions of *Protodrilus* have been published by Pierantoni (1908), Jägersten (1952), Jouin (1970b), and von Nordheim (1989) and of *Parenterodrilus* by Jouin (1979, 1992).

Systematics

The Protodrilidae were originally considered to be archiannelids by Hatschek (1893). Morphological studies by Purschke & Jouin (1988) supported the definition of three families (Protodrilidae, Protodriloididae and Saccocirridae) and their inclusion in the monophyletic order Protodrilida. Within this order, the Saccocirridae was suggested to form a clade with the Protodrilidae, and with the Protodriloididae, as their closest relative. The order Protodrilida may be a sister group to Spionida (Purschke & Jouin, 1988). This view supported the close relationship to Spionida proposed by Orrhage (1974), who found similarities in the nervous system and the development of the palps, especially those of Apistobranchidae. A progenetic origin of these families was indicated by Westheide (1985). The three families were placed as incertae sedis in Canalipalpata by Rouse & Fauchald (1997) and Rouse & Pleijel (2001); in the latter they are grouped as an order.

Additional information

The ultrastructure of protodrilids is generally well investigated – especially the gutless species *Parenterodrilus taenioides* (Jouin, 1979) (see Jouin, 1992). A cross section of *Protodrilus* sp. from Portugal will serve to illustrate the main characters and coelomic conditions for an anterior trunk segment (Fig. 5). The epidermal cells are covered with a well-developed cuticle. Single stiff sensory cilia (se) surrounded by microvilli penetrate the cuticle. The continuous median ventral band of locomotory cilia is always present. It is formed by two clusters of epidermal cells. The cilia rootlets of the locomotory cilia are relatively short, however, an offshoot of the cell with ciliary rootlets extends into a neighboring cell (icr). The two intraepidermal nerve cords are in close contact with the median ventral ciliary band. True ganglia were not observed, but the nerve cord neuropil may be surrounded by clusters of glial cells. Protodrilids are true coelomates with a body cavity along each side (Fig. 5, eco) crossed by oblique muscles, but the peritoneal layer is partially incomplete, e.g. the coelomic space is lined by the peritoneum and by large longitudinal muscles (lmu) arranged in distinct segmental blocks separated from the epidermis by a thin basal lamina. The oblique muscles lack basal laminae. Circular muscles seem to be absent. The ventral blood vessel (vbv) is surrounded by the midventral basal laminae. The midgut cells are densely ciliated.

The Protodrilidae is an example of a clearly 'eucoelomate' interstitial family, which may have evolved by progenesis. A progenetic origin and/or interstitial habits do therefore not necessarily involve a reduction of the coelom.

Protodriloididae Purschke & Jouin, 1988

The family was erected based on species formerly assigned to Protodrilidae and consists of two species in the genus *Protodriloides* Jouin, 1966 (up to 13 mm long). Detailed morphological studies were made by Jouin (1966) and Purschke & Jouin (1988).

Systematics

The family differs morphologically from Protodrilidae, by, e.g., the presence of tentacles appearing as anterior extensions of the prostomium and lacking a central canal; external fertilization, aflagellate spermatozoa, large yolky eggs laid in 'cocoons', direct development of benthic larvae; and presence of chaetae in *Protodriloides chaetifer* (Remane, 1926) (Jouin, 1966, 1978–1979; Westheide, 1990). For the phylogenetic position in Polychaeta, see above (Protodrilidae, *Systematics*).

Saccocirridae Czerniavsky, 1881

The family consists of 18 nominal species belonging to Saccocirrus Bobretzky, 1871 (length up to 3 cm). Detailed morphological studies were provided by Jouin & Rao (1987) and Purschke & Jouin (1988), and the most recent revision is given by Brown (1981).

Systematics

The family is thought to be closely related to Protodrilidae (Purschke & Jouin, 1988). It shows some resemblance to Protodrilidae in, e.g., the tentacle morphology and pharyngeal apparatus.

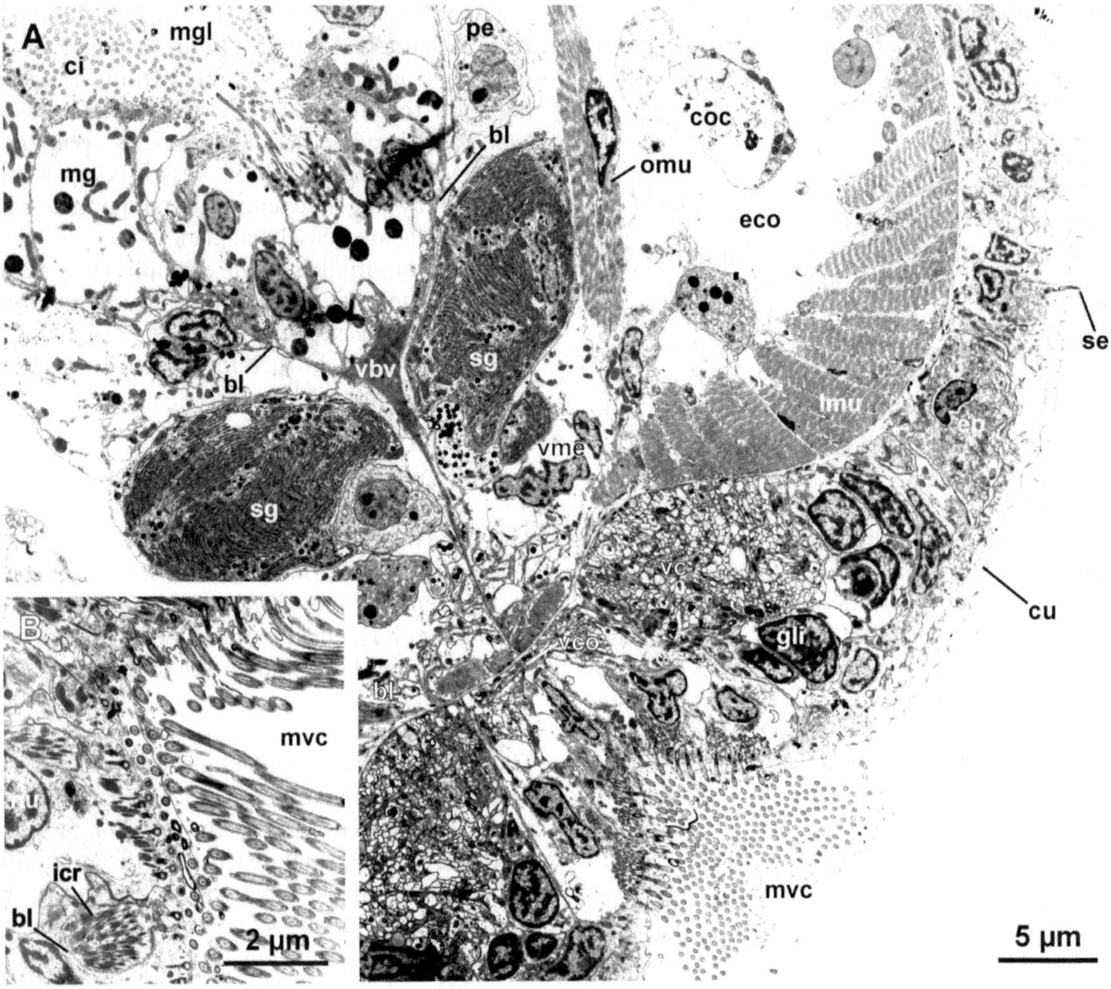

Figure 5. Transmission electron micrographs of *Protodrilus* sp. from Valle de Lama, Portugal. Material fixed in modified trialdehyde and postfixed in 1% OsO_4 (see Lake, 1973 about fixation). Ultrathin sections of Epon-embedded material stained in 3.5% uranyl acetate at 60 °C and in lead citrate at 20 °C. (A) Ventral cross section of anterior trunk segment showing 'eucoelomate' condition (eco), ciliated midgut cells (mg), salivary glands (sg), and the midventral ciliary band (mvc) covered by the intraepithelial ventral nerve cords (vc) with a ventral commissure (vco). (B) Close-up of the midventral ciliary band. The cells that contain ciliary rootlets (icr) are surrounded by a basal lamina (bl). Abbreviations: ci, cilia in midgut; coc, coelomocyte; cu, cuticle; ep, epidermis; gli, glial cell; lmu, longitudinal muscle; mgl, midgut lumen; nu, nucleus; omu, oblique muscle; pe, peritoneal cell; se, sensory cell with stiff cilium; vbv, ventral blood vessel; vme, ventral mesentery.

Other characteristics of the family are, e.g., the presence of various types of chaetae, parapodial stumps, and a complicated reproductive system. For the phylogenetic position of the family, see above (Protodrilidae, *Systematics*).

Psammodrilidae Swedmark, 1952

The family consists of three described species in two genera (1–8 mm long): *Psammodrilus Swedmark*, 1952 (2 species) and *Psammodriloides* Swedmark, 1958 (1).

Systematics

Whereas there is little doubt that Psammodrilidae must be placed systematically within the Polychaeta, their relationship to any particular polychaete family is much more doubtful. The only clue so far seems to be the fine structure of the abdominal uncini. As already pointed out by Swedmark

(1958), uncini with barbules are present only in Arenicolidae and Maldanidae. Recently, the family was regarded as the sister group of Arenicolidae and Maldanidae, based on detailed ultrastructural studies of these uncini (Meyer & Bartolomaeus, 1997). Rouse & Fauchald (1997) placed it as incertae sedis in Polychaeta.

Additional information

Only *Psammodriloides fauveli* Swedmark, 1958 is a true interstitial species (up to 1 mm long), and it is possible that this species evolved by progenesis from a macrofaunal psammodrilid. However, several morphological data presented here point towards an alternative evolutionary scenario.

TEM-investigations of both *Psammodrilus balanoglossoides* and *P. aedificator* have shown unique cavities in the prostomium (the paired lateral and paired median cavities, see Swedmark, 1958), which are true coeloms with a peritoneum (Fig. 6). Furthermore, the peristomium (buccal region) of these species has a distinct segment-like collar region. The prostomial coeloms and the collar region are not found in juveniles, in the interstitial *P. fauveli,* or any other polychaetes. However, it seems that all psammodrilids have the two 'diaphragma sacs' (coeloms) in the anterior region described by Swedmark (1958, fig. 8). The ultrastructure of the collar region is unique within the polychaetes. The epidermis of the collar consists of hexagonally arranged microvillar cells, and totally lack cilia and cuticle (Fig. 7). The epidermal cells interdigitate and form ducts in the intercellular space. Especially the collar cells of *P. balanoglossoides* are basally strongly infolded, as is often seen in osmoregulatory cells (Fig. 7B). Kristensen & Nørrevang (1982) suggested that the collar cells are secretory, but they did not mention that the collar might secrete the mucous that glues the sand grains together in the tube or 'house' of the two species of *Psammodrilus*. Alternatively, the abundant and extensive microvilli of the collar region may represent an absorptive surface of small dissolved molecules in seawater.

Below the epidermal cells, the collar is bordered by two muscular diaphragms. These are believed to function as a suctorial pharynx (Westheide, 1990). In *P. balanoglossoides* the epidermal cells are separated from the muscular diaphragms by a very thick basement membrane with collagen fibres (Fig. 7B). In *P. aedificator* the muscular diaphragms are compact and 'acoelomate' (Fig. 7A, see also Kristensen & Nørrevang, 1982). The border between the epidermal cells and the muscle cells consists only of a sandwich-like structure (two thin basal laminae), and the muscle cells attach directly to the basal lamina. Interstitial and juvenile psammodrilids lacking a collar region also totally lack muscular diaphragms.

The coelomic situation in the thorax and abdomen is very different in the three described species. In *P. balanoglossoides* the longitudinal muscles (lmu, Fig. 8A) form large sacs (coeloms) in the thorax. The animal is a true 'eucoelomate'. In *P. aedificator* the coelomic situation of the thorax is quite different. The animal is 'acoelomate', as the giant endoderm gut cells and the coelenchyme cells (Fig. 8B) fill up most of the thorax and abdomen and a true coelom, i.e. a body cavity lined by a continuous mesodermal epithelium, seems to be absent (see also Kristensen & Nørrevang, 1982, fig. 13). The longitudinal muscles and other muscles are diminutive, except from the muscles moving the three first pair of aciculae. Especially in the abdomen, the muscles are very tiny, the epidermis is very thin, and in the mid-dorsal sulcus the large midgut cells are in near contact with the environment (see Kristensen & Nørrevang, 1982, fig. 13). The endoderm gut cells may have a chordoid function. Most of the cytoplasm consists of a homogeneous granular matrix. The same kind of parenchymatic cells (endodermal) with a chordoid function are observed in plathyhelminthes and in gnathostomulids, e.g. *Rastrognathia* (see fig. 6 in Kristensen & Nørrevang, 1977).

The metanephridial system of *P. balanoglossoides* is well-developed with a ciliated and microvillar funnel (mef) and a metanephridial duct (med) passing through the longitudinal muscles as a pore is observed in the sixth segment using TEM (Fig. 8A), but additional metanephridia may be present in other abdominal segments. The metanephridial funnel opens into a true coelom lined with peritoneum. The metanephridia of *P. aedificator* (Fig. 8B) contain only 7–9 cilia and the whole metanephridial system seems to be less developed.

Very little is known about the coelomic condition in the interstitial *Psammodriloides fauveli*, but

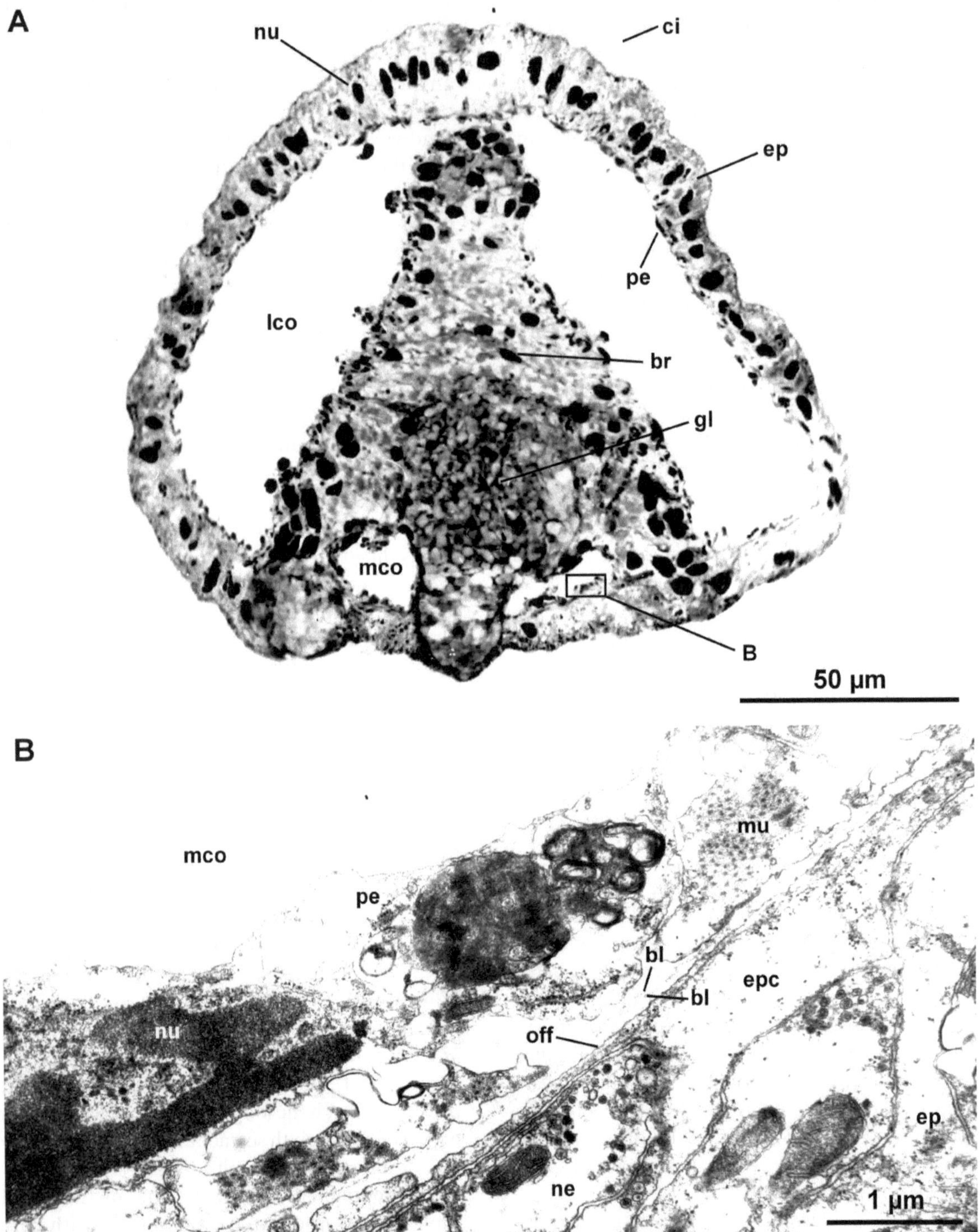

Figure 6. Psammodrilus balanoglossoides Swedmark, 1952 from Helsingør, Denmark. Material fixed in modified trialdehyde and postfixed in 1% OsO_4. (see Kristensen & Nørrevang, 1982). (A) Light microscope micrograph of semi-thin cross section (1 μm) stained with toluidine blue and examined with DIC-technique. The section is located through the prostomium with the two large lateral coeloms (lco) and the two smaller median coeloms (mco). The very thin peritoneum (pe) covers the cavities of both coeloms. (B) Transmission electron micrograph of median coelom (mco). The coelom cavity is covered by peritoneal cells (pe) and some myoepithelial cells (mu). The epidermal cell (epc) has many cellular offshoots (off) in close contact with the epidermal basal lamina. Abbreviations: br, brain; ci, cilia covering the entire surface of the prostomium; ep, epidermis; gl, gland in close contact with the brain; ne, nerve process; nu, nucleus.

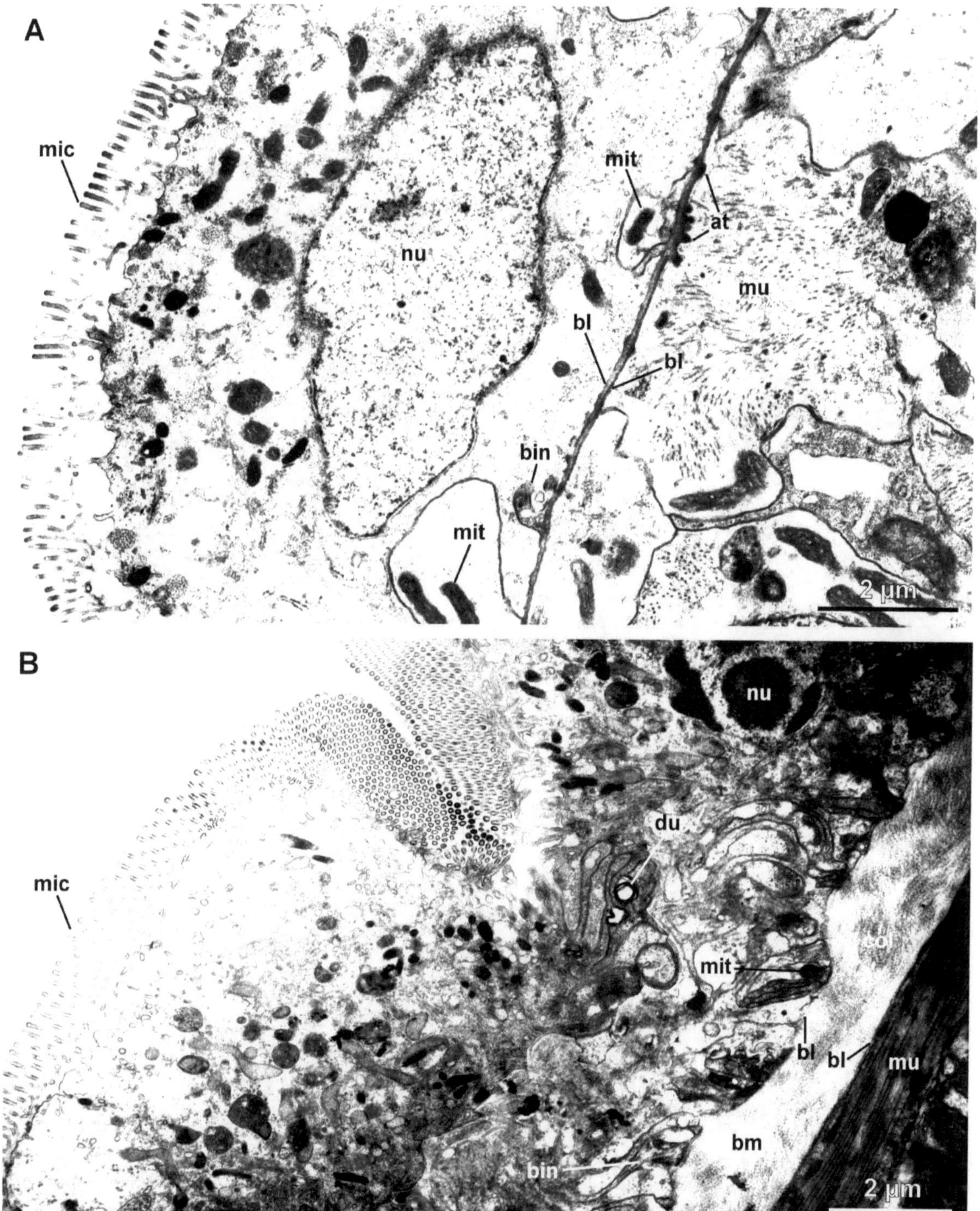

Figure 7. Transmission electron micrographs of the collar region of Psammodrilidae. For information on fixation and techniques see Kristensen & Nørrevang (1982). (A) *Psammodrilus aedificator* Kristensen & Nørrevang, 1982 from Disko Island, Greenland fixed directly in 1% OsO_4. Two closely opposed basal laminae (bl) present and few basal infoldings (bin) of the epidermal cells. No coelomic cavity present. (B) *Psammodrilus balanoglossoides* from Denmark. The two basal laminae form a true basement membrane (bm) with collagen fibers (col). The basal infoldings of the epidermal cells form a labyrinth with ducts (du). Abbreviations: at, muscle attachment; mi, microvilli of collar cell; mit, mitochondrion; mu, muscle; nu, nucleus.

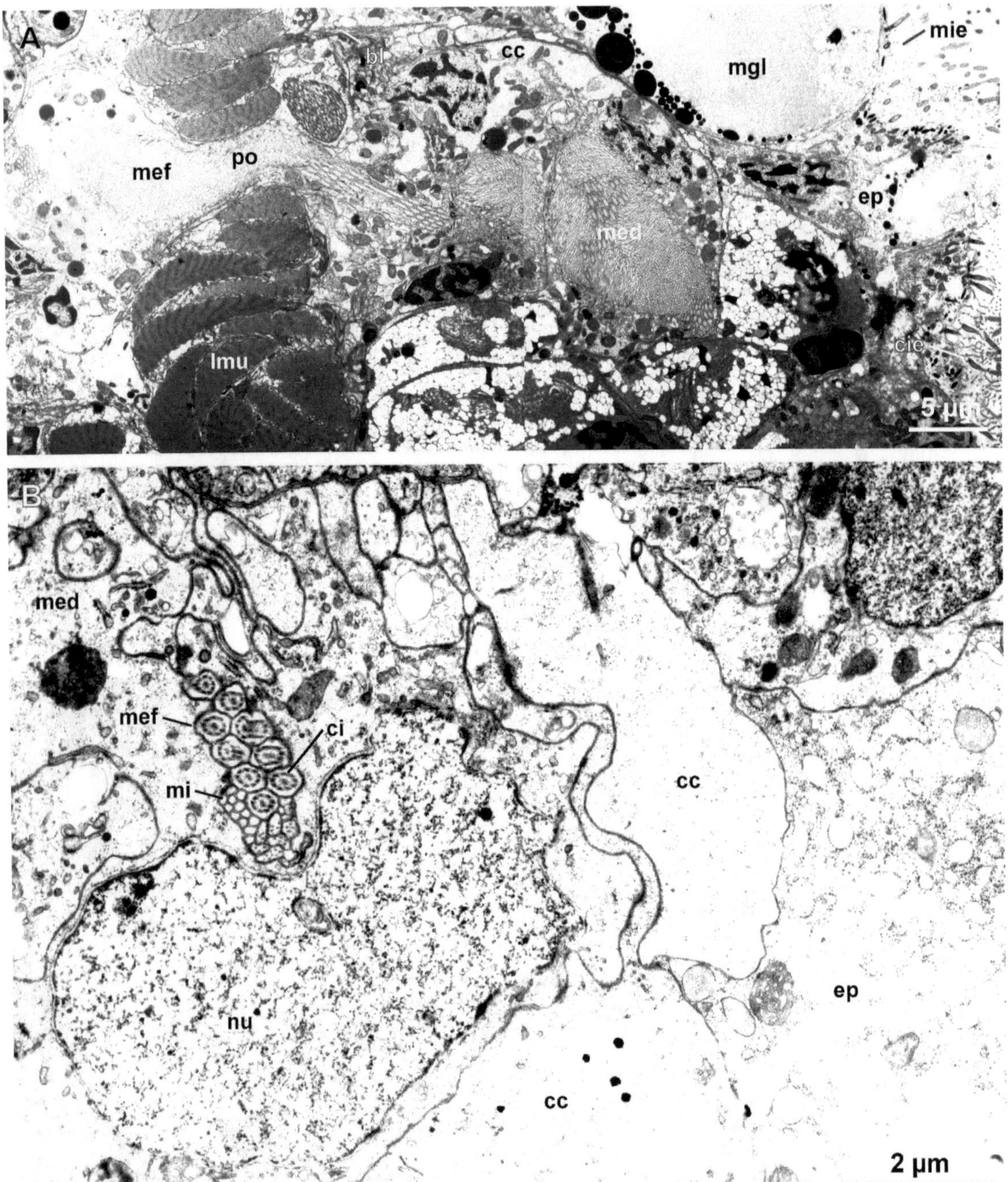

Figure 8. Transmission electron micrographs of the metanephridial system of Psammodrilidae. (A) *Psammodrilus balanoglossoides* from Denmark. A well-developed metanephridial system from the thorax region with a ciliated and microvillar funnel (mef) and a metanephridial duct (med) passing through longitudinal muscles (lmu) as a pore (po). (B) *Psammodrilus aedificator* from Greenland. A weakly developed metanephridial system from the thorax region with few cilia (ci) and microvilli (mi) surrounded by coelenchyme cell (cc). Abbreviations: bl, basal lamina; cie, cilia of epidermis; ep, epidermis; mie, microvilli of epidermis; mgl, mucous gland of epidermis; nu, nucleus.

Fransen (1980) published some information on an undescribed species of *Psammodriloides* from USA. The animal is 'acoelomate' and the metanephridia are weakly developed. What is very interesting is that the compactness of the trunk is formed by large coelenchyme cells and not by the endoderm cells as in *P. aedificator*.

Remarks on evolution

The conclusion of the TEM-investigations of the psammodrilids is that the 'acoelomate' *P. aedificator* is not an intermediate form between *P. balanoglossoides* and *Psammodriloides fauveli*, where *P. fauveli* should have evolved by progenesis from a larger form. *Psammodrilus aedificator* may instead have evolved from a line of interstitial psammodrilids and secondarily have become larger. This may explain the nearly 'acoelomate' condition, less complex metanephridia, and the minor length of the 6 pairs of cirri with aciculae in this species. Except for the lack of cilia on the collar (Fig. 7) and in the middorsal sulcus of the trunk, all of the Psammodrilidae have an almost uniform and completely ciliated epidermis which counterpart is found only among the Lobatocerebrida (see Rieger, 1980); however, the latter taxon has a true cuticle, a character lacking in Psammodrilidae. The well-developed ciliation (see Fig. 6A) of psammodrilids (including the larger species) may indicate that the whole family has evolved from an interstitial ancestor. Further phylogenetic analyses, preferably with additional new taxa, are needed to explain the evolution within the Psammodrilidae and prove whether *Psammodrilus* is a paraphyletic genus or *Psammodrilus balanoglossoides* descended from an interstitial ancestor as well. If the last scenario is followed to its extreme, the whole family may have evolved by progenesis.

Conclusion

The many examples of meiofauna representatives from macrofauna polychaete families can most easily be explained by progenetic evolution. These examples are found in all major clades in the phylogeny of polychaetes, thus it must be assumed that progenesis is a common evolutionary pathway. This pathway may also apply to many of the exclusively meiofaunal families. However, it is very difficult to prove a progenetic origin phylogenetically (and reject a primary meiofaunal form) by morphological data alone, since the lack of characters and many specializations make it difficult to compare these taxa with macrofaunal polychaetes. Although several hypotheses on the evolution of meiofauna polychaetes have been presented here, cladistic analyses have generally failed to substantiate the phylogenetic position of meiofaunal families within Polychaeta and explain their evolution. Techniques like SEM, TEM and cLSM can still provide important comparative information on the external and especially internal morphology (see, e.g., Müller & Westheide, 2002; this paper). Time consuming studies of animals in culture could provide comparative information on, e.g., ontogeny, reproduction and common parasites (see, e.g., Åkesson, 1977). However, molecular studies may seem the most obvious solution for gathering additional comparative characters of the 'simple' looking meiofauna taxa. However, as in morphological studies, the genes used at present in molecular systematics seldom provide convincing information on the more basal splits in the evolution of polychaetes and new genes may have to be considered (see, e.g., Struck et al., 2002a).

Acknowledgements

The authors wish to thank T. Bartolomaeus and G. Purschke for inviting us to the symposium 'Morphology, Molecules, Evolution and Phylogeny in the Polychaeta and Related Taxa, September 23–27, 2002, Osnabrück'. We thank D. Eibye-Jacobsen and two anonymous reviewers for valuable comments on the manuscript and S. Elle for making the final version of Figure 4. The kind assistance of D. Abrahamsen in collecting *D. minimus* from Ærø, Denmark is greatly appreciated. We dedicate this paper to W. Westheide in honor of his exceptional, wide-ranging work with interstitial polychaetes.

References

Ax, P., 1954. *Thalassochaetus palpifoliaccus* nov. gen., nov. spec. (Archiannelida, Nerillidae) ein marine Verwandtes von *Troglochaetus beranecki* Delachaux. Zool. Anz. 153: 64–75.

Åkesson, B., 1977. Parasite-host relationships and phylogenetic systematics. The taxonomic position of dinophilids. Microfauna Meersboden 61: 19–28.

Bartolomaeus, T., 1993. Die Leibeshöhlenverhältnisse und Nierenorgane der Bilateria – Ultrastruktur, Entwicklung und Evolution. Habilitationsschrift, Universität Göttingen, 298 pp.

Bartolomaeus, T., 1999. Structure, function and development of segmental organs in Annelida. In Dorresteijn, A. W. C. & W. Westheide (eds), Reproductive Strategies and Developmental Patterns in Annelids. Hydrobiologia 402: 21–37.

Beauchamp, P. de, 1910. Sur l'organisation de la *Nerilla*. Bull. biol. Fr. Belg. 44: 11–22.

Beesley, P. L., G. J. B. Ross & C. J. Glasby (eds), 2000. Polychaeta, Myzostomida, Pogonophora, Echiura, Sipuncula, Polychaetes & Allies: The Southern Synthesis. Fauna of Australia. Vol. 4, CSIRO Publishing, Melbourne, xii + 465 pp.

Boaden, P. J. S., 1961. *Meganerilla swedmarki,* nov. gen., nov. spec., an archiannelid of the family Nerillidae. Arkiv för Zoologi 13: 553–559.

Brown, R., 1981. Saccocirridae (Annelida: Polychaeta) from the central coast of New South Wales. Australian Journal of Marine and Fresh Water Research 32: 439–456.

Bubko, O. V., 1973. On the systematic position of Oweniidae and Archiannelida (Annelida). Zoologichesky Zhurnal 52: 1286–1296.

Clark, R. B., 1969. Systematics and phylogeny: Annelida, Echiura, Sipuncula. In Florkin, M. & B. T. Scheer (eds), Chemical Zoology, Annelida, Echiura, Sipuncula, Vol. 4 Academic Press, New York: 1–68.

Coull, B. C. & S. S. Bell, 1979. Perspectives of marine meiofaunal ecology. In Livingston, R. J. (ed.), Ecological Processes in Coastal and Marine Ecosystems. New York Plenum Publishing Company: 189–216.

Delachaux, Th., 1921. Un polychète d'eau douce cavernicole *Troglochaetus beranecki* nov. gen. nov. spec. Bull. Soc. Neuchâtel. Sc. Nat. 45: 1–11.

Eibye-Jacobsen, D. & R. M. Kristensen, 1994. A new genus and species of Dorvilleidae (Annelida, Polychaeta) from Bermuda, with a phylogenetic analysis of Dorvilleidae, Iphitimidae and Dinophilidae. Zoologica Scripta 23: 107–131.

Faubel, A., 1978. Neue Nerillidae (Archiannelida) aus dem sublitoral der Nordsee und des Mittelatlantik (Nordwest Afrika). Zoological Scripta 7: 257–262.

Fauchald, K., 1977. The polychaete worms: definition and keys to the orders, families and genera. Natural History Museum of Los Angeles County, Science Series 28, 188 pp.

Fransen, M. E., 1980. Ultrastructure of coelomic organization in annelids. I. Archiannelids and other small polychaetes. Zoomorph. 95: 235–249.

Gelder, S. R., 1974. A review of the zoogeography and habitat data of the genus *Nerilla* Schmidt, 1848 (Annelida, Archiannelida). Journal of Natural History 8: 631–643.

Goodrich, E. S., 1912. *Nerilla* an archiannelid. Quarterly Journal of Microscopical Science 57: 397–425.

Hatschek, B., 1878. Studien über Entwicklungsgeschichte der Anneliden. Ein Beitrag zur Morphologie der Bilaterien. Arb. Zool. Inst. Univ. Wien 1: 277–404.

Hatschek, B., 1893. System der Anneliden, ein vorläufiger Bericht. Lotos 13: 123–126.

Hempelmann, F., 1931. Erste und zweite Klasse der Vermes Polymera (Annelida). Archiannelida und Polychaeta. In Kükentahal, W. & T. Krumbach (eds), Handbuch der Zoologie, Band 2, Teil 2, Leif 12u. 13. Berlin: pp. 1–112.

Hermans, C. O., 1969. The systematic position of the Archiannelida. Systematic Zoology 18: 85–102.

Higgins, R. P. & T. Thiel, 1988. Prospectus. In Higgins, R. P. & H. Thiel (eds), An Introduction to the Study of Meiofauna. Smithsonian Institution Press, Washington: 11–13.

Hsieh, H. L. & J. L. Simon, 1987. Larval development of *Kinbergonuphis simoni,* with a summary of development patterns in the family Onuphidae (Polychaeta). Bulletin Biological Society of Washington 7: 194–210.

Jägersten, G., 1952. Studies on the morphology, larval development and biology of *Protodrilus*. Zool. Bidr. Uppsala 29: 425–512.

Jouin, C., 1966. Morphologie et anatomie comparée de *Protodrilus chaetifer* Remane et *Protodrilus symbioticus* Giard; création du nouveau genre *Protodriloides* (Archiannélides). Cahiers de Biologie Marine 7: 139–155.

Jouin, C., 1967. Etude morphologique et anatomique de *Nerillidopsis hyalina* Jouin et de quelques *Nerillidium* Remane (Archiannélides Nerillidae). Arch. Zool. Exp. Gén. 108 (1): 97–110.

Jouin, C., 1968. Sexualité et biologie de la reproduction chez *Mesonerilla* Remane et *Meganerilla* Boaden. Cahiers de Biologie Marine 9: 31–52.

Jouin, C., 1970a. Archiannélides interstitielles de Nouvelle-Calédonie. Expedition Française sur le Récifs Coralliens de la Nouvelle-Calédonie 4: 149–167.

Jouin C., 1970b. Recherches sur le Protodrilidae (Archiannélides). I. Etude morphologique et systématique du genre *Protodrilus*. Cahiers de Biologie Marine 11: 367–434.

Jouin C., 1971. Status of the knowledge of the systematics and ecology of Archiannelida. In Hulings, N. C. (ed.), Proceedings of the First International Conference on Meiofauna. Smith. Contrib. Zool. 76: 46–56.

Jouin, C., 1978–1979. Spermatozoide non flagellé et fécondation externe chez *Protodriloides symbioticus* (Giard) (Annélides Polychètes, Archiannélides). Vie Milieu, Sér. A Biolog. Mar. 28–29: 473–487.

Jouin, C., 1979. Description of a free-living polychaete without gut: *Astomus taenioides* n. gen., n. sp. (Protodrilidae, Archiannelida). Canalian Journal of Zoology 57: 2448–2456.

Jouin C., 1992. The ultrastructure of a gutless annelid, *Parenterodrilus* gen. nov. *taenoides (= Astomus taenioides)* (Polychaeta, Protodrilidae). Canadian Journal of Zoology 70: 1833–1848.

Jouin, C. & G. C. Rao, 1987. Morphological studies on some Polygordiidae and Saccocirridae (Polychaeta) from the Indian Ocean. Cahiers de Biologie Marine 28: 389–401.

Jouin, C., & B. Swedmark, 1965. *Paranerilla limicola* n.g., n. sp., Archiannélide Nerillidae du benthos vaseux marine. Cahiers de Biologie Marine 6: 201–218.

Lake, P. S., 1973. Trialdehyde fixation of crustacean tissue for electron microscopy. Crustaceana 24: 244–246.

Knöllner, F., 1934. *Stygocapitella subterranea,* nov. gen. nov. spec. Schr. Naturw. Ver. Schleswig-Holst. 20: 468–472.

Kristensen, R. M. & D. Eibye-Jacobsen, 1995. Ultrastructure of spermiogenesis and spermatozoa in *Diurodrilus subterraneus* (Polychaeta, Diurodrilidae). Zoomorphology 115: 117–132.

Kristensen, R. M. & P. Funch, 2000. Micrognathozoa: a new class with complicated jaws like those of Rotifera and Gnathostomulida. Journal of Morphology 246: 1–49.

Kristensen, R. M. & T. Niilonen, 1982. Structural studies on *Diurodrilus* Remane (Diurodrilidae fam. n.), with description of *Diurodrilus westheidei* sp. n. from the Arctic interstitial fauna of Disko Island, W. Greenland. Zoologica Scripta 11: 1–12.

Kristensen, R. M. & A. Nørrevang, 1977. On the fine structure of *Rastrognathia macrostoma* gen. et sp. n. placed in Rastrognathidae fam. n. (Gnathostomulida). Zoologica Scripta 6: 27–41.

Kristensen, R. M. & A. Nørrevang, 1982. Description of *Psammodrilus aedificator* sp. n. (Polychaeta) with notes on the Arctic interstitial fauna of Disko Island, W. Greenland. Zoologica Scripta. 11: 265–279.

Marion, A. F. & N. Bobretzky, 1875. Etude des Annélides du golfe de Marseille. Ann. Sci. Nat. Paris, Ser. 6, 2: 1–106.

McIntosh, W. C., 1875. On a new example of the Opheliidae, *Linotrypane apogon* from Shetland. Proceeding of the Royal Society Edinburgh 8: 386–390.

Meyer, A., 1927. Ist *Parergodrilus heideri* (Reisinger) ein Archiannelide? Zoologischer Anzeiger. 72: 19–35.

Meyer, R. & T. Bartolomaeus, 1997. Ultrastruktur und Morphogenese der Hakenborsten bei *Psammodrilus balanoglossoides* – Bedeutung für die Stellung der Psammodrilida (Annelida). Microfauna Mar.11: 87–113.

Morselli, I., M. Mari & M. Sarto, 1995. First record of the stygobiont "archiannelid" *Troglochaetus beranecki* Delachaux from Italy. Bollenttino di Zoologia 62: 287–290.

Müller, M. C. M., 2002. *Aristonerilla*: a new nerillid genus (Annelida: Polychaeta) with description of *Aristonerilla (Micronerilla) brevis* comb. nov. from a seawater aquarium. Cahiers de Biologie Marine 43: 131–139.

Müller, M. C., J. M. Bernhard & C. Jouin-Toulmond, 2001. A new member of Nerillidae (Annelida: Polychaeta), *Xenonerilla bactericola* gen. et sp. nov., collected off California, USA. Cahiers de Biologie Marine 42: 203–217.

Müller, M. C. & W. Westheide, 2002. Comparative analysis of the nervous systems in presumptive progenetic dinophilid and dorvilleid polychaetes (Annelida) by immunohistochemistry and cLSM. Acta Zoologica 83(1): 33–48.

Nielsen, C., 1995. Animal Evolution: Interrelationships of the Living Phyla. Oxford University Press, Oxford.

Núñez, J., O. Ocaña & M. del C. Brito., 1997. Two new species (Polychaeta: Fauveliopsidae and Nerillidae) and other polychaetes from the marine lagoon cave of Jameos del Agua, Lanzarote (Canary Islands). Bulletin of Marine Science 60(2): 252–260.

Orrhage, L., 1974. Über die Anatomie, Histologie und Verwandtschaft der Apistobranchidae (Polychaeta, Sedentaria) nebst Bemerkungen über die systematische Stellung der Archianneliden. Zeitschrift für Morphologie der Tiere 79: 1–45.

Paxton, H., 2000. Family Diurodrilidae. In Beesley, P. L., G. J. B. Ross & C. J. Glasby (eds), Polychaetes & Allies: The Southern Synthesis. Fauna of Australia. Polychaeta, Myzostomida, Pogonophora, Echiura, Sipuncula, Vol. 4A CSIRO Publishing, Melbourne: 104–105.

Pennak, R. W., 1971. A fresh-water archiannelid from the Colorado Rocky Mountains. Transcations of the American Microscopical Society 90 (3): 372–375.

Pierantoni, U., 1908. *Protodrilus*. Fauna Florades Golfes von Neapel 31: 1–226.

Pizl, V. & J. Chalupsky, 1984. *Hrabeiella periglandulata* gen. et sp. n. (Annelida): a curious worm from Czechoslovakia. Acta Societatis Zoologicae Bohemoslovenicae. Věstn. Československ. Spol. Zool. 48: 291–295.

Purschke, G., 1985. Anatomy and ultrastructure of ventral pharyngeal organs and their phylogenetic importance in Polychaeta (Annelida). II. The pharynx of the Nerillidae. Microfauna Mar. 2: 23–60.

Purschke, G., 1999. Terrestrial polychaetes – models for the evolution of the Clitellata (Annelida)? Hydrobiologia 406: 87–99.

Purschke, G., 2000. Sense organs and central nervous system in an enigmatic terrestrial polychaete, *Hrabeiella periglandulata* (Annelida) – implications for annelid evolution. Invertebrate Biology 119: 329–341.

Purschke, G. & C. Jouin, 1988. Anatomy and ultrastructure of the ventral pharyngeal organs of *Saccocirrus* and *Protodriloides* with remarks on the phylogenetic relationships within the Protodrilida (Annelida, Polychaeta). Journal of Zoology (London) 215: 405–432.

Purschke, G., R. Hessling & W. Westheide, 2000. The phylogenetic position of the Clitellata and the Echiura – on the problematic assessment of absent characters. Journal of Zoological Systematics and Evolutionary Research 38: 165–173.

Quatrefages, A. M. de, 1866. Annélides et géphyriens. In: Histoire naturelle des annélides, marine et d'eau douce. Librairie Encyclopédique Roret, Paris 2: 67–70.

Reisinger, E., 1925. Ein landbewohnender Archiannelide. Zugleich ein Beitrag zur Systematik der Archianneliden. Z. Morph. Ökol. Tiere 13: 197–254.

Reisinger, E., 1960. Die Lösung des *Parergodrilus*-Problems. Z. Morph. Ökol. Tiere 48: 517–544.

Rieger, R. M., 1980. A new group of interstial worms, Lobatocerebridae nov. fam. (Annelida) and its significance for metazoan phylogeny. Zoomorphologie 95: 41–84.

Riser, N. W., 1984. General observations on the intertidal interstitial fauna of New Zealand. Tane 30: 239–250.

Riser, N. W., 1988. Morphology of a new species of nerillid polychaete from the north shore of Massachusetts Bay, U.S.A. Transaction of American Microscopical Society 107 (2): 171–179.

Rota, E., 1998. Morphology and adaptations by *Parergodrilus* Reisinger and *Hrabeiella* Pizl and Chalupsky, two enigmatic soil-dwelling annelids. Italian Journal Zoology 65: 75–84.

Rota, E. & G. Carchini, 1999. A new *Polygordius* (Annelida: Polychaeta) from Terra Nova Bay, Ross Sea, Antarctica. Polar Biology 21: 201–213.

Rota, E., P. Martin & C. Erséus, 2001. Soil-dwelling polychaetes: enigmatic as ever? Some hints on their phylogenetic relationship as suggested by a maximum parsimony analysis of 18S rDNA gene sequences. Contributions to Zoology 70: 127–138.

Rouse, G. W. & K. Fauchald, 1997. Cladistics and polychaetes. Zoological Scripta 26(2): 139–204.

Rouse, G. W. & F. Pleijel, 2001. Polychaetes. Oxford University Press: New York, 354 pp.

Saphonov M. V. & A. B. Tzetlin, 1997. Nerillidae (Annelida: Polychaeta) from the White Sea, with description of a new species of *Micronerilla* Jouin. Ophelia, 47(3): 215–226.

Schmidt, E. O., 1848. Neue Beiträge zur Naturgeschichte der Würmer gesammelt auf einer Reise nach den Färör im Frühjahr 1848. F. Manke, Jena, 44 pp.

Schmidt, P. & W. Westheide, 1977. Interstitielle Fauna von Galapagos. XVII. Polygordiidae, Saccocirridae, Protodrilidae, Nerillidae, Dinophilidae (Polychaeta). Mikrofauna Meeresboden 62: 1–38.

Schmidt, H. & W. Westheide, 2000. Are the meiofaunal polychaetes *Hesionides arenaria* and *Stygocapitellla subterranea* true cosmopolitan species? Results of RAPD-PCR investigations. Zoologica Scripta 29: 17–28.

Sterrer, W. & T. M. Iliffe, 1982. *Mesonerilla prospera,* a new archiannelid from marine caves in Bermuda. Proceedings of the Biological Society of Washington 95(3): 509–514.

Struck, T., R. Hessling & G. Purschke, 2002a. The phylogenetic position of Aeolosomatidae and Parergodrilidae, two enigmatic oligochaete-like taxa of the "Polychaeta", based on molecular data from 18S rDNA sequences. Journal of Zoological Systematics and Evolutionary Research 40: 155–163.

Struck, T. H., W. Westheide & G. Purschke, 2002b. Progenesis in Eunicida ("Polychaeta," Annelida) – separate evolutionary events? Evidence from molecular data. Molecular Phylogenetics and Evolution 25: 190–199.

Swedmark, B., 1952. Note preliminaire sur un Polychète sedentaire abérrant, *Psammodrilus balanoglossoides* n. g. n. sp . Arkiv for Zoologi 4: 159–162.

Swedmark, B., 1955. Recherches sur la morphologie, le developpement et la biologie de *Psammodrilus balanoglossoides,* Polychète sedentaire de la microfaune des sables. Archs Zool. Exp. Gén. 92: 141–220.

Swedmark, B., 1958. *Psammodriloides fauveli* n. gen. n. sp. et la famille des Psammodrilidae (Polychaeta Sedentaria). Arkiv for Zoologi 12: 55–64.

Swedmark, B., 1959. Archiannélides Nerillidae des côtes du Finistère. Arch. Zool. Exp. Gen. 98: 26–42.

Swedmark, B., 1964. Interstitial fauna of marine sand. Biological Review. 39: 1–42.

Tzetlin, A. B. & M. W. Saphonov, 1992. *Trochonerilla mobilis* gen. et sp. n., a meiofaunal nerillid (Annelida, Polychaeta) from a marine aquarium in Moscow. Zoologica Scripta 21: 251–254.

Villora-Moreno, S., 1996. Ecology and distribution of the Diurodrilidae (Polychaeta), with redescription of *Diurodrilus benazzi.* Cahiers de Biologie Marine 37: 99–108.

von Nordheim, H. 1989., Six new species of *Protodrilus* (Annelida: Polychaeta) from Europe and New Zealand, with a concise presentation of the genus. Zoologica Scripta 18: 245–268.

Westheide, W., 1971. Interstitial Polychaeta (excluding Archiannelida). In Hulings, N. C. (ed.) Proceedings of the First International Conference on Meiofauna. Smithson. Contributions to Zoolgy 76: 57–70.

Westheide, W., 1982. *Microphthalmus hamosus* sp. n. (Polychaeta, Hesionidae) – an example of evolution leading from the interstitial fauna to a macrofaunal interspecific relationship. Zoologica Scripta 11: 189–193.

Westheide, W., 1984. The concept of reproduction in polychaetes with small body size: adaptation in interstitial species. In Fischer, A. & H-D. Pfannenstiel (eds), Polychaete Reproduction. Fortschr. Zool. 29: 265–287.

Westheide, W., 1985. The systematic position of the Dinophilidae and the archiannelid problem. In Conway Morris, S., J. S. George, R. Gibson & H. M. Platt (eds), The Origin and Relationships of Lower Invertebrates Vol. 397. Clarendon Press, Oxford: 310–326.

Westheide, W., 1987. Progenesis as a principle in meiofauna evolution. Journal of Natural History 21: 843–854.

Westheide, W., 1988. Polychaeta. In Higgins R. P. & H. Thiel (eds), An Introduction to the Study of Meiofauna, pp. 332–344. Smithsonian Institution Press: Washington, 488 pp.

Westheide, W., 1990. Polychaetes: Interstitial families. Synopsis of the British Fauna (new series): 1–152. Universal Book Services/Dr W. Backhuys, Oegstgeest, 152 pp.

Westheide, W. & G. Purschke, 1996. *Leptonerilla diplocirrata,* a new genus and species of interstitial polychaetes from the island of Hainan, south China (Nerillidae). Procedings at the Biological Society Washington 109(3): 586–590.

Wilke, U., 1953. *Mesonerilla intermedia* nov. sp., ein neuer Archiannelide aus dem Golf von Neapel. Zoologischer Anzeiger 150: 211–215.

Willis, A. G., 1951. On *Nerilla australis*, a new Australian archiannelid. Proceedings at the Zoological Society London 120: 731–735.

Worsaae, K. & R. M. Kristensen, 2003. A new species of *Paranerilla* (Polychaeta: Nerillidae) from Northeast Greenland Waters, Arctic Ocean. Cahiers de Biologie Marine 44: 23–39.

Hydrobiologia (2005) **535/536:** 341–356
T. Bartolomaeus & G. Purschke (eds), Morphology, Molecules, Evolution and Phylogeny in Polychaeta and Related Taxa

Polychaete phylogeny based on morphological data – a comparison of current attempts

Thomas Bartolomaeus[1,*], Günter Purschke[2] & Harald Hausen[1]
[1]*Animal Systematics and Evolution, Free University Berlin, Koenigin-Luise-Str. 1-3, 14195 Berlin, Germany*
[2]*Systematic Zoology, Department of Biology, University of Osnabrueck, 49069 Osnabrueck, Germany*
*(*Author for Correspondence: Tel: 0049-30-838-56288; Fax: 0049-30-838-54869; E-mail:tbartol@zoosyst-berlin.de)*

Key words: Annelida, ultrastructure, phylogeny, Echiura, Clitellata, Pogonophora

Abstract

Annelid phylogeny is one of the largest unresolved problems within the Metazoa. This is due to the enormous age of this taxon and also strongly influenced by the current discussion on the position of the Arthropoda, which traditionally is hypothesized to be the annelid sister taxon. Within the framework of recent discussions on the position of the Annelida, the ground pattern of this taxon is either a clitellate-like, parapodia-less dwelling organism or an organisms that resembles errant polychaetes in having parapodia and gills and probably being a predator. To solve this problem different attempts have been made in the past, cladistic analysis, scenario based plausibility considerations and a successive search for sister taxa base on isolated characters. These attempts are presented and critically discussed. There is at least strong support for the Annelida as wells as for several of its taxa above the level of traditional families; the monophyly of the Polychaeta, however, remains questionable.

Introduction

The phylogenetic relationships among Annelida is still one of the largest unresolved problems in metazoan systematics and is most controversely discussed (Rouse & Fauchald, 1995, 1997; Eibye-Jacobson & Nielsen, 1997; Westheide, 1997; Westheide et al., 1999; Rouse & Pleijel, 2001; Purschke, 2002). Central problems concern the monophyly of Annelida and, presumed they are a monophyletic group, the organization of the annelid stem species as well as the interrelationships between the different annelid taxa. Annelida are multisegmented organisms with a multiple repetition of identically organized segments. The first and the last section differ from this organization. The anteriormost section, called prostomium, contains the cerebral ganglia, and the caudalmost section, called pygidium, contains a terminally or dorsally situated anus. The mouth is situated ventrally behind the prostomium. A growth zone of continuous mitotic activity which gives rise to additional segments lies anteriorly to the anus. Each segment contains a pair of ganglia, a pair of coelomic cavities, a pair of metanephridia and at least paired ventral and dorsal groups of chaetae. Most of these characters vary within the Annelida, and the mentioned generalized description is much more a result of an idealized 'body plan' than of a precise phylogenetic analysis (Nielsen, 2001; Ruppert et al., 2003).

Annelida were recognized as a taxon[1] of segmented soft bodied worms by Lamarck (1801). Cuvier (1812) placed Annelida and Arthropoda

[1] The term taxon is used here in the sense of group of things that share certain characteristics. Biological taxa are not necessarily monophyletic, although many of them turned out to be. In terms of phylogenetic systematics taxa should be monophyletic.

Table 1. Characters present in the ground pattern of the Annelida. Hypothesis 1: Polychaeta and Clitellata monophyletic; hypothesis 2: Polychaeta paraphyletic, Clitellata related to subordinate polychaete taxon only. Autapomorphies bold, plesiomorphies normal font. Characters marked by asterisks will become synapomorphies with Arthropoda if the Articulata-hypothesis is followed. From Purschke (2002)

Hypothesis 1	Hypothesis 2
Biphasic life cycle with planktonic larva and benthic adult	Biphasic life cycle with planktonic larva and benthic adult
Collageneous cuticle	Collageneous cuticle
epidermis without kinocilia	Epidermis with kinocilia
Coelom and blood vessels	Coelom and blood vessels
Nephridia	Nephridia
Foregut with dorsolateral ciliated folds	Foregut with dorsolateral ciliated folds
Microphageous	Microphageous
Endobenthic	Epibenthic
Burrowing	Crawling
Rhabdomeric photoreceptors in pigment cup ocelli	Rhabdomeric photoreceptors in pigment cup ocelli
Gut straight tube	Gut straight tube
Homonomous segmentation*	**Homonous segmentation***
Longitudinal muscle bands*	**Longitudinal muscle bands***
Capillary chaetae (ß-chitin) in four groups	**Complex chaetae (ß-chitin) in four groups**
No parapodia	**Biramous parapodia***
Small prostomium*	**Large prostomium***
No prostomial appendages	**Palps and antennae**
Pygidium*	**Pygidium***
No pygidial cirri	**Pygidial cirri**
Praepygidial proliferation zone*	**Praepygidial proliferation zone***
No nuchal organs	**Nuchal organs**
Dorsal brain and ventral nerve cord within Orthogonal NS*	**Dorsal brain and ventral nerve cord** within Orthogonal NS*

into a taxon Articulata, a view that is held until today and regarded as the best explanation for the distribution of morphological characters in both taxa (Wägele et al., 1999; Wägele & Misof, 2001; Scholtz, 2002). Since 1997 this view is opposed by the results of molecular analyses that argue for a close relationship of Arthropoda and some taxa of Nemathelminthes (see Aguinaldo et al., 1997; Schmidt-Rhaesa et al., 1998). This taxon has been called Ecdysozoa. If Ecdysozoa should turn out to represent a monophyletic taxon, Annelida were one of the best supported monophyletic taxa (Table 1) within Bilateria belonging to a clade named Lophotrochozoa. The situation is completely different under the perspective of the Articulata-hypothesis: Segmental organisation, generation of additional segments in a caudal preterminal growth zone, paired coelomic cavities in the embryo and metanephridia are synapomorphies of Annelida and Arthropoda then (Bartolomaeus, 1998; Bartolomaeus & Ruhberg, 1999). Whether other characters, like nuchal organs, head appendages and parapodia can be assumed to represent autapomorphies of Annelida depends on the relationships within Annelida (Table 1; for review see Purschke, 2002). The only character that presently supports the hypothesis of monophyletic Annelida is a paired dorsal and ventral group of chaetae in each segment.

Unravelling the relationships among Annelida is the first step to find further evidence that could support Annelida. This paper wants to summarize and discuss the recent attempts to cast light on annelid and especially polychaete evolution. It will concentrate on morphological analyses, which are proposed by Rouse & Fauchald (1995, 1997), Westheide (1997) and our own studies (see Bartolomaeus, 1998; Hausen, 2001; Purschke, 2002). By doing this we also want to substantiate hypotheses on monophyletic taxa within Annelida.

We will show that some of them are supported by a single character only. In these cases, however, this single withstands every outgroup comparison.

Taxa of the Annelida

Traditionally, annelids are subdivided in three systematic entities, i.e. polychaetes, oligochaetes and hirudineans. Only one of them, i.e. the Hirudinea, can clearly regarded as monophyletic. This is sustained by a large number of autapomorphies which include the fixed number of 32 segments plus prostomium and peristomium, a posterior sucker consisting of seven posteriormost segments, spermatophores, a strong development of the oblique musculature, the reduction of the mesenteries and coelomic extensions, as well as separation of the nephrostome from the nephridial duct (s. Purschke et al., 1993).

Certain oligochaetes like Branchiobdellida are discussed to represent the sister taxon of Hirudinea (Purschke et al., 1993), and not a single autapomorphy is found until now, which supports any hypothesis on the monophyletic origin of the oligochaetes (see Erséus, 2005). Despite of this, Hirudinea and all those annelids which traditionally are classified as oligochaetes must have shared a common ancestor. This ancestor is characterized by several autapomorphies, like hermaphroditism, lecithotrophic eggs and the clitellum, a spatially restricted epidermal region where secretory gland cells produce a cocoon, the eggs are shed into, modification of the spiral cleavage and specific development (Purschke et al., 1993) and the special sperm ultrastructure (see Ferraguti, 1984; Ferraguti & Erséus, 1999). The entire taxon has been termed Clitellata and represents a very well supported monophyletic taxon within Annelida.

Initially Echiura and Sipuncula were included into Annelida (see Fauchald & Rouse, 1997; for review), later both taxa were removed. At least for Echiura this assumption has recently been revived. Studies on the development of the echiuran nervous system provided traces of a metamerical organisation (Hessling & Westheide, 2002; Hessling, 2002). These investigations clearly support the view that Echiura are derived from a segmented ancestor and the lack of segmentation is secondary. Moreover, it follows that the echiurid trunk is made up of numerous fused segments. This view receives independent support from molecular analyses (Halanych et al., 2002, Bleidorn et al., 2003a,b). Despite their possible position, Echiura clearly are monophyletic taxon due to their anal sacs (see Bartolomaeus & Quast, 2005). This character is unique among bilaterians and withstands any outgroup comparison.

The most diverse group within Annelida is Polychaeta. Grube (1850) introduced the name Polychaeta to distinguish them from Oligochaeta. While a taxon Sèdentaires (Sedentaria) had already been introduced by Lamarck (1818), the name Annèlides errantes (Errantia) was at first mentioned by Andouin & Milne Edwards (1834). Perrier (1897) resurrected these terms and erected two large taxa, one consisting of mostly vagile and free living polychaetes, the Errantia, and one consisting of hemisessile, sessile and mostly tubicolous polychaetes, the Sedentaria. Fauvel (1923), Uschakov (1955) and others (Hartmann-Schröder, 1971) adopted them for polychaete taxonomy. Parallel to these terms smaller entities were established by Hatschek (1893), especially Archiannelida, Spiomorpha, Serpulimorpha, Terebellomorpha and Drilomorpha as well as Amphinomorpha and Nereimorpha. These taxon names were in part adopted by Benham (1896), Fauvel (1923), Uschakov (1955) and Hartmann-Schröder (1971). In the first half of the twentieth century a taxon Polychaeta consisting of Archiannelida, Errantia and Sedentaria was widely accepted. This system of polychaetes was used until the seventies, although there were doubts, whether errant and sedentary polychaetes were actually reliable entities (Dales, 1962; Day, 1967). Archiannelida was recognized as an artificial assemblage of small annelids that invaded the mesopsammon in different lineages (Jouin, 1971; Westheide, 1985, 1987). However, due to their apparent simple organization, their relationships are largely unresolved till now. The remaining groups, i.e. Errantia and Sedentaria were subsequently eliminated end of the seventies, when Fauchald (1977) presented a system of 17 more or less isolated taxa, which are equally ranked as orders in a more typologically orientated system (Fauchald, 1977; George & Hartmann-Schröder, 1985). All entities above these orders, i.e. taxa like

Terebellomorpha, Spiomorpha or Errantia were eliminated. The old sequence with the errant taxa listed at first, followed by the sedentary ones was changed by Fauchald (1977). By doing this he obviously followed morpho-functional considerations proposed by Clark (1964, 1977), according to which the annelid stem species had an earthworm-like organization without parapodia and head appendages (further discussion see Westheide, 1997). The number of orders was extended by Pettibone (1982) and Hartmann-Schröder (1996).

But, even most of the orders established by Fauchald (1977) were hard to be kept in the view of phylogenetic systematics. In most keys the families became those entities which were easily to characterize (George & Hartmann-Schröder, 1985) and, more recently, by Rouse & Pleijel (2001). The first cladistic analysis of polychaetes by Rouse & Fauchald (1997) was accordingly performed on the family level and a tree consisting of 53 out of about 80 polychaete families was presented. The remaining taxa were excluded from this restricted analysis. This analysis provided the basis of the classification in Rouse & Pleijel (2001). The monophyly of several families, however, was still uncertain (Fauchald & Rouse, 1997).

Prior to any critical evaluation, it seem essential to draw a few conclusions from this extremely short summary of history of polychaete taxonomy (for details see Fauchald & Rouse, 1997; Westheide et al., 1999; Rouse & Pleijel, 2001): Continuous subdividing of the polychaetes, application of such high ranks as orders (Fauchald, 1977) reflects the tremendous structural diversity found within polychaetes. This is certainly a result of the enormous age of this group; their oldest known representatives have been found among middle Kambrian fossils (Conway-Morris, 1979). This fossil record already represents a surprisingly high diversity of body forms. It also reflects the enormous adaptability to different ecological niches, so that polychaetes are found in nearly all marine environments often playing a major role in certain marine systems, as for instance arenicolid species in sandy intertidal or siboglinid (pogonophoran) species in the hydrothermal vent community. However, even a few limnetic, ground water and terrestrial species are known (Purschke, 2002). A second conclusion that can be drawn from the above is that morpho-functional considerations influenced systematization especially if the system also should reflect evolution. This lasts on until the most recent contributions (Westheide et al., 1999; Rouse & Pleijel, 2001; Purschke, 2002).

Cladistic analysis – analysing all characters simultaneously

As there is no doubt that all available characters must be used for any attempt to unravel phylogeny, cladistic methods are clearly those which have to be used (see Westheide et al., 1999; Rouse & Pleijel, 2001). An enormous progress has been made in this respect in the last years; these methods rationalize discussions, they allow to test *a priori* homology hypotheses and trees as results of other analysis and to include further characters. A crucial problem in all cladistic analyses, which often is underestimated in subsequent discussions, is character coding and the resulting data matrices (see below).

Rouse & Fauchald (1997) presented the first cladistic analysis based on data, which were available for a large number of taxa to almost the same extend. Many of these data were extracted from old literature, so that their analysis also made these data available again. With their analysis they provided the first evaluation of morphological data without any *a priori* hypothesis on the evolution of the Annelida. One of the exiting results was that Pogonophora, which thus far were regarded as a separate phylum, have to be included into Annelida. They integrated Pogonophora into Annelida as the taxon Siboglinidae. We want to repeat and comment the most important results of the Rouse & Fauchald (1997) analysis at first, summarize the autapomorphies of the taxa above the family level and relate their names to the older taxonomies (Fig. 1). By doing this we want to focus on such characters that are autapomorphies with respect to any outgroup comparison within annelids. We also will hint at some problems along with analysis.

Siboglinidae (Pogonophora) are a member of Sabellida (sensu Rouse & Fauchald, 1997), which also contains Oweniidae, Serpulidae, Sabellidae and Sabellariidae. The latter three represent the Hatschek's (1893) Serpulimorpha. Rouse & Fauchald's (1997) results indirectly corroborate

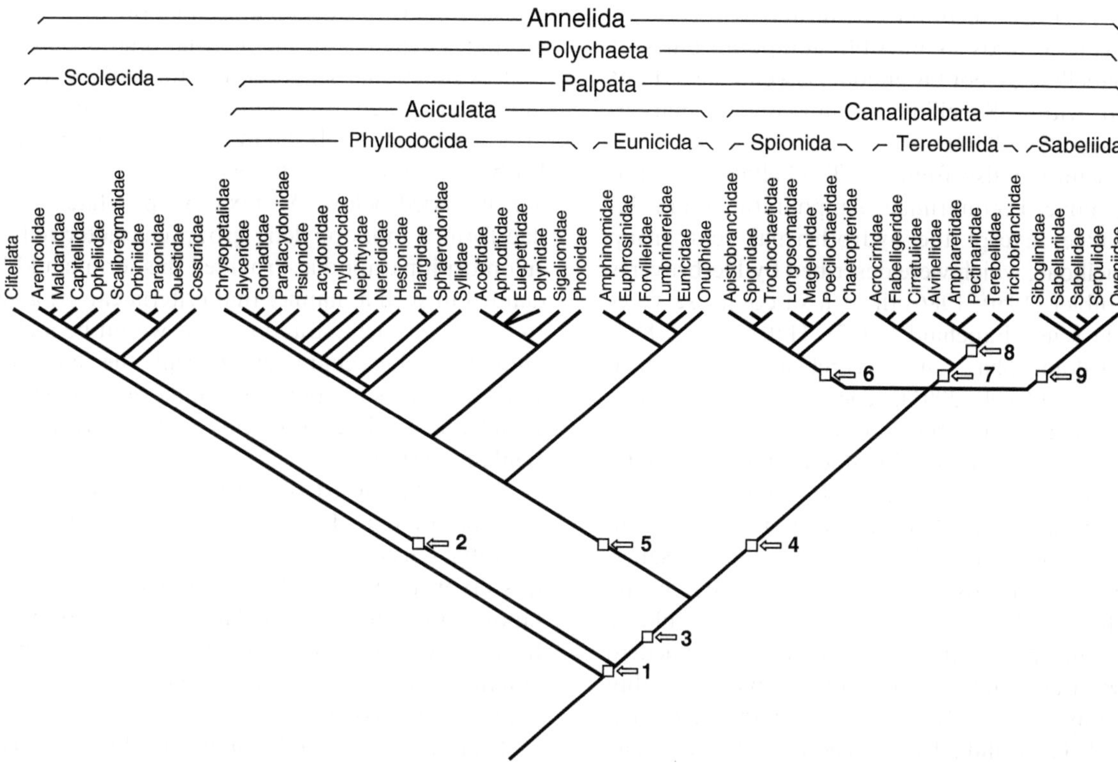

Figure 1. Phylogenetic relationships among the Annelida according to Rouse & Fauchald (1997). The shown tree has been used by the authors to discuss character distribution. It is one of three trees that result from successive weighting of a reduced, a priorily weighted data set. Autapomorphies of selected taxa are: **1** Mixonephridia, nuchal organs as pits or grooves, parapodia. **2** Parapodia with similar rami, two or more pairs of of pygidial cirri. **3** Peristomium limited to lips, palps. **4** Peristomial and grooved palps. **5** Prostomial and (sensory) palps, lateral and medial prostomial antennae, dorsal cirriform cirri, ventral cirri, one pair of pygidial cirri, nephridia and coelomoducts in most segments, acicula. **6** Paired peristomial palps, anterior nephridia and posterior gonoducts, nuchal organs form posterior projections. **7** heart body, absence of chaetae and appendages in the first segment, gular membrane. **8** dorsal branchiae in the first segments, buccal tentacles, coiled gut. **9** limited fusion of prostomium and peristomium, peristomium not limited to lips any longer.

the hypothesis of a sister group relationship between Serpulidae + Sabellidae and Siboglinidae (Bartolomaeus, 1995). The monophyly of Sabellida (sensu Rouse & Fauchald, 1997) is weakly supported by only two characters, i.e. a limited fusion of prostomium and peristomium and a peristomium that is not limited to lips any longer (9 in Fig. 1). However, the first character is homoplasious, the second one is a reduction of an earlier evolved character, an interpretation resulting from the general character distribution. At least the sister group relationship between Sabellidae and Serpulidae is supported by three unique character complexes, (1) their branchial crown is remnant of the prostomium and consists of radioles with paired series of pinnules (Rouse & Fauchald, 1997), (2) the nuchal organ is hidden within a dorsal pit that lies ventral to the nephridiopore or to the distal section of the nephridioduct, respectively (Orrhage, 1980: 119–124; Purschke 1997), and (3) nervous system with inversion of the dorsal and ventral root of the circumoesophageal connectives (Orrhage, 1980). Sabellariidae is regarded as their sister taxon. This hypothesis is substantiated by palp nerve roots 1–3 (Hausen, 2001; Orrhage & Müller, 2005) the chaetal inversion which exchanges the position of the neuro- and notopodial chaetae between thorax and abdomen (Rouse & Fauchald, 1997, but see Bartolomaeus, 2002). Siboglinidae (Pogonophora) share a single pair of excretory organs draining the second segment with the fore-mentioned taxa. The metanephridial duct forms a large caudally extending U; its pori are dorsal. Rouse & Fauchald (1997) also mention uncini and reduction of dorsolateral folds as possible autapomorphies of the Sabellida except Oweniidae. Oweniidae did not group within Sabellida in subsequent analyses (Rouse, 1999, 2000).

Uncini are also found in Terebellidae, Trichobranchidae, Pectinariidae, Ampharetidae and Alvinellidae. Except the hydrothermal vent group Alvinellidae, which were described by Desbruyères & Laubier (1980, 1986) these taxa comprise Terebellida sensu Fauchald (1977), Hatschek's (1893) Terebellomorpha, and Terebelliformia sensu Rouse & Pleijel (2001). The monophyly of this taxon is supported by (1) dorsal branchiae in the first segments, (2) buccal tentacles (Orrhage, 2001) and (3) a coiled gut (Rouse & Fauchald, 1997). At least the latter is unique within polychaetes (8 in Fig. 1). Buccal tentacles are also found in Sabellariidae; their possible homology awaits testing (Orrhage, 1978, 2001). Dorsal branchiae could be autapomorphic if this character were specified to allow discrimination from other branchiae within the polychaetes. Rouse & Fauchald (1997) extended Terebellida by inclusion of Acrocirridae, Cirratulidae and Flabelligeridae. Rouse & Pleijel (2001) classified the latter three plus further taxa like Fauveliopsidae, Ctenodrilidae and *Sternaspis* as Cirriformia within Terebellida. Terebellida (sensu Rouse & Fauchald, 1997) is supported by a heart body, absence of chaetae and appendages in the first segment and a gular membrane (7 in Fig. 1). All characters are homoplasies; their evaluation as autapomorphies depends on the internal relationships within Annelida.

Spionida sensu Rouse & Fauchald (1997) consist of Apistobranchidae, Spionidae, Trochochaetidae, Longosomatidae, Magelonidae, Poecilochaetidae and Chaetopteridae. Their monophyly is supported by paired peristomial palps, anterior nephridia and posterior gonoducts and the fact that nuchal organs form posterior projections (6 in Fig. 1). All species of *Magelona* lack nuchal organs and knowledge on nuchal organs of Chaetopteridae is sparse. In Apistobranchidae the presumed nuchal organs are associated to projections lateral to the palps. This is in contrast to the situation found in Spionidae, Trochochaetidae and Poecilochaetidae. The other characters mentioned in favour for the monophyly of the Spionida sensu Rouse and Fauchald (1997) are homoplasies and their evaluation as autapomorphies depends on the internal relationships as well. A functional separation of the metanephridia into anterior nephridia and posterior gonoducts is also found in terebellidan and sabellidan taxa. Problems along with the evaluation of the palps are discussed below. All taxa except Chaetopteridae form a monophyletic group characterized by spiomorph parapodia. Provided this term could be defined in such a manner that criteria can be given to recognize such parapodia, this would be a clear autapomorphic character. Except the later described Longosomatidae, Spionida sensu Rouse & Fauchald (1997) corresponds to Spioniformia of Benham (1896).

Aciculata consist of mostly vagile polychaetes with parapodia. They were formerly known as Errantia sensu Perrier (1897). The monophyly of this taxon is supported by a number of clear autapomorphies like (1) aciculae inside the parapodia to stabilize them, (2) lateral and medial prostomial antennae and (3) ventral cirri (4 in Fig. 1). The palps are discussed below.

Scolecida are weakly supported by two characters: parapodia with similar rami and two or more pairs of pygidial cirri (2 in Fig. 1). Both characters are parallelisms and not uniform within the group. They depend on the internal relationships among the polychaetes. Being aware of this Rouse & Pleijel (2001) mention that further cladistic analyses might reveal the paraphyly of this taxon. This group is the most basal in their tree.

Three characters support the monophyly hypothesis of Polychaeta, i.e. parapodia, nuchal organs as pits and grooves and mixonephridia (Rouse & Fauchald, 1997) (1 in Fig. 1). The parapodia are unspecified, a problem that results from absence/presence coding of structure and substructures that are logically not independent. It will be outlined below. Using the term mixonephridium Rouse & Fauchald (1997) adopted the terminology of Goodrich (1895). These terms describe two different features of the nephridia, i.e. their specific structure and their function as both, organs to release genital products and to eliminate metabolic wastes from the coelomic cavity. If Rouse & Fauchald (1997) used the term in the latter sense, this must represent a primary feature of metanephridia, because these organs fulfil the same function in Echiura and Sipuncula. As such, mixonephridia can hardly be an autapomorphy of the polychaetes, because at least sipunculids are outgroup to the annelids. Recent results on the structure and formation of nephridia, however, do not allow maintaining the terminology of Goodrich (1895, 1945) as no empirical data support his ideas (see Bartolomaeus & Quast, 2005). During maturity the ciliated funnel of certain metanephridia enlarges. This enlargement has been explained by Goodrich (1945) in the light of the gonocoel theory (Hatschek, 1878; Meyer, 1890; Goodrich, 1895) as witness for a merge between a gonoduct and a metanephridium. In accordance with the gonocoel theory Goodrich (1945) assumed that the gonoduct was a derivative of the coelothelium while the metanephrida had a different origin. Due to the presumed extend of fusion he chose different terms for the presumed fused metanephridia (see Bartolomaeus & Hausam, 2005). In none of the species subsequently studied has such enlargement of the funnel by coelothelially derived cells been observed (see Bartolomaeus, 1999). Thus, Goodrich's terms only inform about enlargement of the metanephridial funnel (metanephromixium) or lack of such an enlargement (mixonephridium). These different terms imply higher information content than they actually have. Some of the polychaetes possess protonephridia during their entire life time (see Table 1 in Bartolomaeus & Quast, 2005). During maturity these organs are used to discharge the genital product from the coelom in some species and acquire a funnel. Proliferation of the proximal duct cells generates this funnel which degenerates at the end of the reproductive period (Stecher, 1968). Again this temporarily restricted phenomenon was interpreted by Goodrich (1945) in the light of the gonocoel theory as remnant of the ancestral gonoduct. He termed the organ protonephromixium, but this term describes merely a modification of the protonephridium during the period of reproduction. In some polychaetes with protonephridia such a modification cannot be observed and the genital products are released by rupture. Goodrich also termed them protonephromixia, because he interpreted some ciliated structure in the coelomic wall as remnant of a funnel. Thus, adopting Goodrich's terms as characters inevitably results in coding a hypothetical process, but not the organ itself. Moreover, recent studies revealed that protonephridia can be found in a variety of different taxa, and not all of them are changed during maturity (Bartolomaeus & Quast, 2005). There is some evidence that all of them are not necessarily homologous among the polychaeta, while certain types, like those with solenocytes seem to be.

Adopting terms when coding data from the older literature causes conflicts with new data, mostly because ancient assumptions on the evolution of organs are also adopted with these terms. This can easily be seen in Rouse & Fauchald (1997) when they code "nephridia and coelomoducts in most segments" for most phyllodocidans. Like shown for the nephridia, new data on the structure of the chaetae can also not be integrated into the Rouse & Fauchald (1997) tree without larger conflicts. Recent studies into the formation and structure of chaetae provided strong evidence that uncini and certain hooded hooks are homologous (Hausen, 2005). While these data in part corroborated the cladistic analysis of Rouse & Fauchald (1997) some of the results indicate a complete different position of different groups of Scolecida, namely Arenicolidae, Maldanidae and Capitellidae. We will outline this in more detail later (Fig. 4).

Beside the central problem of some recycling of literature (Jenner, 2001), character coding itself is a wide field and different character concepts have been developed. Meanwhile it became quite clear that absence/presence coding in parsimony analysis is very problematic (see Jenner & Schramm, 1999). Handling of complex characters with a large number of substructures needs extreme care, because these substructures might be logically not independent from each other. Logical correlation has already been discussed by Sokal & Sneath (1963: 66). Presence of haemoglobin and redness of blood, they mention as an example are logically not independent characters, if blood's redness is strictly a consequence of the presence of haemoglobin (see Fristrup, 2001: 20). If logical independence of the coded substructures is not guaranteed absence/presence coding can cause the reconstruction of non-sense ground patterns (in contrast to Pleijel, 1995 and in accordance with Meier, 1994). Because of their shorter tree length the corresponding non-sense trees can be preferred

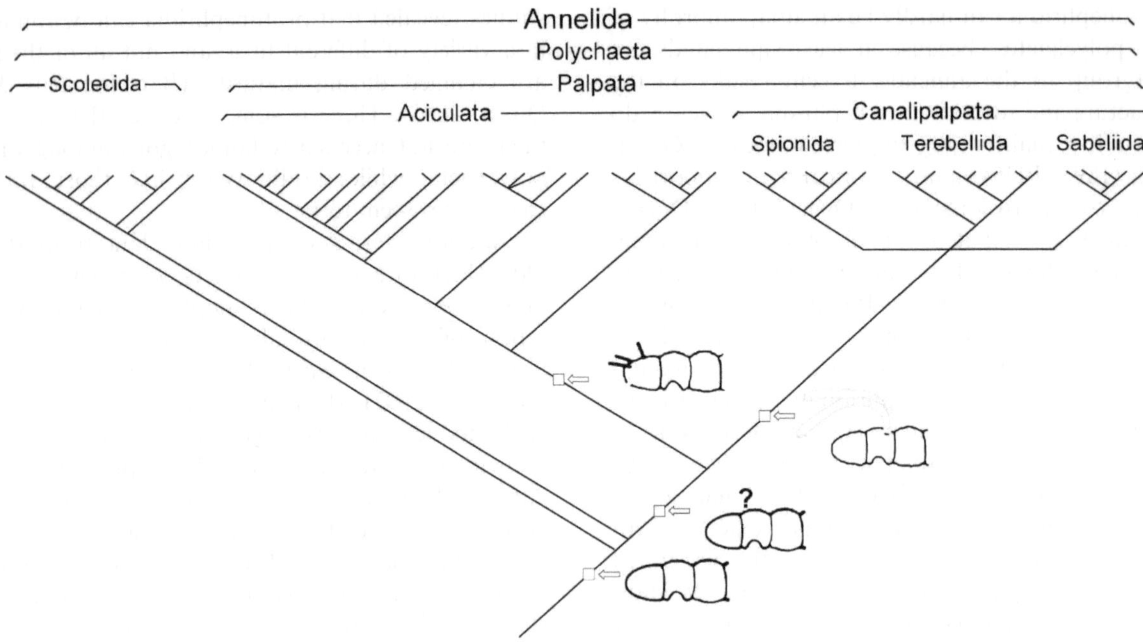

Figure 2. The palp problem. Ignoring logical dependence of palps and their substructures during coding causes non-sense character complexes. Insertion and function of the palps in the ground pattern of Palpata are not specified. Both ways of insertion and both functions are used as autapomorphies above the first node. This clearly results from coding palp substructures (insertion, structure, function) as logically independent characters.

instead of trees that show ground patterns which are at least logically plausible. We want to exemplify this for the basal radiations for the Polychaeta in Rouse & Fauchald (1997) tree (Fig. 2). Palpata are supported by two characters, i.e. (1) a peristomium that is limited to lips and (2) palps (3 in Fig. 1). The highest ranking taxa within the Palpata, Canalipalpata and Aciculata are characterized by peristomial, grooved palps (Canalipalpata) (5 in Fig. 1) or, among other characters by prostomial, sensory palps (Aciculata) (4 in Fig. 1). Considering the stem species of the Palpata one inevitably asks for the quality of the palps. Which position and which structure had the palps that evolved in the Palpata stem lineage? At least one of the two conditions assumed to have evolved in either of the subsequent lineages must be plesiomorphic. Thus, the character composition of the Palpata stem species is incomplete and structural integrity of the stem species is not given. This happened because the palp substructures were handled as logicially independent structures (Fig. 2). A comparable problem occurs when coding parapodia as a character being logically independent of their different morphologies (Rouse & Fauchald, 1997).

We chose this example to show the necessity to remember that as far as this is possible the morpho-functional integrity of the stem species must be guaranteed within a tree. The essential role of such considerations has been outlined by Westheide (1997) and Westheide et al. (1999), but in contrast to these contributions, we are convinced that such considerations cannot result in *a posteriori* assumptions. They can merely be used to estimate whether the character composition generated for a stem species is possible at all – at least from the fact that the functional integrity of a stem species must be maintained.

Evolutionary scenarios – trying to establish the annelid ground pattern

All statements on phylogeny are hypothetical. Some of the hypotheses are less corroborated than others and unravelling phylogeny is a process of continuous corroboration and rejection of hypotheses. While phylogenetic analyses usually

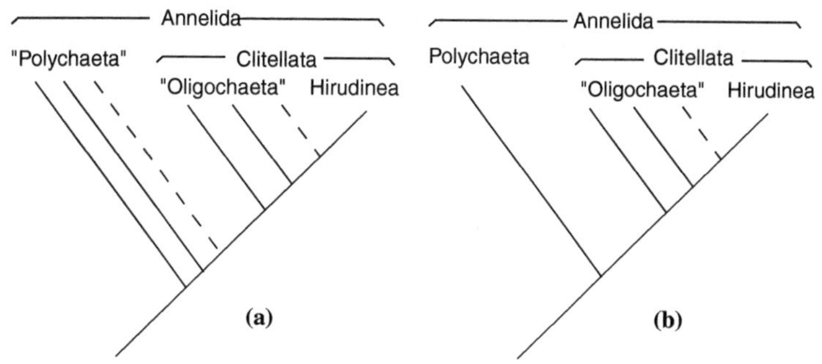

Figure 3. Annelid relationships according to Westheide (1987). The annelid stem species was either an infaunal, clitellate-like organism or a vagile, predatory organism with parapodia used for locomotion. In the latter case, the Polychaeta are not monophyletic. For details in character distribution (see Tables 1 and 2).

start with comparing characters on a lower level (primary homology) there is a long tradition in approaches to unravel phylogenies by starting the reconstruction of the ground pattern or basic character composition of a larger entity (see Ax, 1987, 1996, 2000, 2003; Westheide, 1997). This procedure generally focuses on preselected characters and on the structural and morpho-functional integrity of the assumed stem species. This is not an arbitrary procedure, but character selection is generally made intuitively, often based on an enormous background knowledge concerning the quality and information content of characters. In most cases criteria for character election are hard to test intersubjectively. This does not necessarily result in wrong trees, but tends to overemphasize considerations on character evolution compared to character comparison – and can only be done with a restricted number of characters. While the Rouse & Fauchald analyses (see Rouse & Fauchald, 1997; Westheide et al., 1999, Rouse & Pleijel, 2001) tried to include all available information on polychaete structure that has been gathered for a number of polychaete taxa large enough to allow comparison, Westheide (1997), Westheide et al. (1999) start with morpho-functional consideration to analyse the basal split in Annelida and, thus, start at a rather high taxonomic level and step down to lower levels (Fig. 3).

The general problem is to find a decision between two alternative hypotheses on the organisation of the annelid stem species. Assuming that annelids are monophyletic the stem species was either an epibenthic organism with a prostomium and prostomial appendages, and several segments with parapodia and a strong and diverse chaetation (Fig. 3A), or an endobenthic organism with a small prostomium without appendages and several segments without parapodia and with a weak chaetation (Fig. 3B). The organization of the latter largely corresponds modern clitellate annelids, while that of the assumed epibenthic organism resembles an errant polychaete. A comparison with Arthropoda does not help to decide between both hypotheses, but there is some evidence from the molecular data that Clitellata are embedded within the polychaetes (e.g., Struck et al. 2002). The state of these considerations has recently been summarized by Purschke (2002).

Westheide (1997) proposed the idea that septa evolved to connect the blood lacunae surrounding the gut to the peripheral blood vessels. This is a very interesting functional explanation that gives rise to some assumptions on possible evolutionary pathways. If this were true, enlargement of the surface would allow a better gas exchange. As the volume/surface ratio decreases if an animal increases body size, in aquatic environments one would expect that annelids that have a large body size also possess gills as long as they live in an aquatic environment. Higher oxygen content of the air allows sufficient oxygen supply without special structure that enlarges the body surface. Actually, the large terrestrial clitellates lack such parapodia despite of their large body size. Westheide's (1997) assumption does not inevitably imply that blood vessels were necessarily associated with external gills, as long as the animals did not exceed a certain diameter. It does not necessarily mean that the annelid ancestor had parapodia with gills. How-

Table 2. Autapomorphies of Polychaeta and Clitellata with respect to the conflicting hypotheses. Hypotheses 1: Polychaeta and Clitellata monophyletic; hypothesis 2. Polychaeta paraphyletic and Clitellata related to subordinate taxon of the former. Apomorphies bold, plesiomorphies normal font. From Purschke (2002)

Hypothesis 1	Hypothesis 2
Polychaeta	
Nuchal organs	Nuchal organs
Parapodia	Parapodia
Pygidial cirri	Pygidial cirri
Clitellata	
Epidermis without kinocilia	**Epidermis without kinocilia**
Chaetae simple spines	**Chaetae simple spines**
No parapodia	**No parapodia**
Small prostomium	**Small prostomium**
No prostomial appendages	**No prostomial appendages**
No pygidial cirri	**No pygidial cirri**
No nuchal organs	**No nuchal organs**
Brain situated behind prostomium	**Brain situated behind prostomium**
Simple circumoesophageal connectives	**Simple circumoesophageal connectives**
Burrowing	Burrowing
Phaosomes	**Phaosomes**
Ciliary cerebral sense organs	Ciliary cerebral sense organs
Hermaphroditism	**Hermaphroditism**
Gonads in specific segments	**Gonads in specific segments**
Specific type of spermatozoon	**Specific type of spermatozoon**
Spermathecae outside female organs	**Spermathecae outside female organs**
Cocoons formed by the **clitellum**, a girdle of at least Two types of gland cells	**Cocoons** formed by the **clitellum**, a girdle of at least Two types of gland cells
External fertilization within the cocoon	**External fertilization within the cocoon**
Ectoteloblasts	**Ectoteloblasts**
No larva	**No larva**
Dorsal pharynx	**Dorsal pharynx**

ever, as the annelids are primarily marine and not terrestrial, the stem species must have had a general organisation that resembles errant polychaetes rather then terrestrial clitellates – provided the stem species was large. Thus, those characters all those characters that could support the monophyly of the Polychaeta are plesiomorphies of all polychaetous annelids – and the polychaetes are not monophyletc (Fig. 3A, Tables 1 and 2).

Nuchal organs are characteristic for marine polychaetes; in terrestrial polychaete species they are either modified or lacking (Purschke, 1997, 1999, 2000). In Clitellata they are lacking. From the data available the lack of nuchal organs is functionally related to a terrestrial habitat and to the posterior displacement of the brain (Westheide et al., 1999). The problem now is, to estimate the direction of the evolution that caused this condition. No such organs are found in any outgroup (Purschke et al., 1997), so that the question concentrates on whether the lack in clitellates is primary or secondary. Plausibility considerations seem to allow a decision, but plausibility is a necessary, but extremely weak criterion. The general problem along with the lack of characters is that the hypothesis of a complete reduction cannot be tested by the character in question. There is no chance to find out whether a structure was initially there and has subsequently been reduced, as no testable observation could directly falsify the hypothesis of a complete reduction. Any hypothesis of reduction of a structure instead can be justified indirectly by congruence with other character transformations within a cladistic

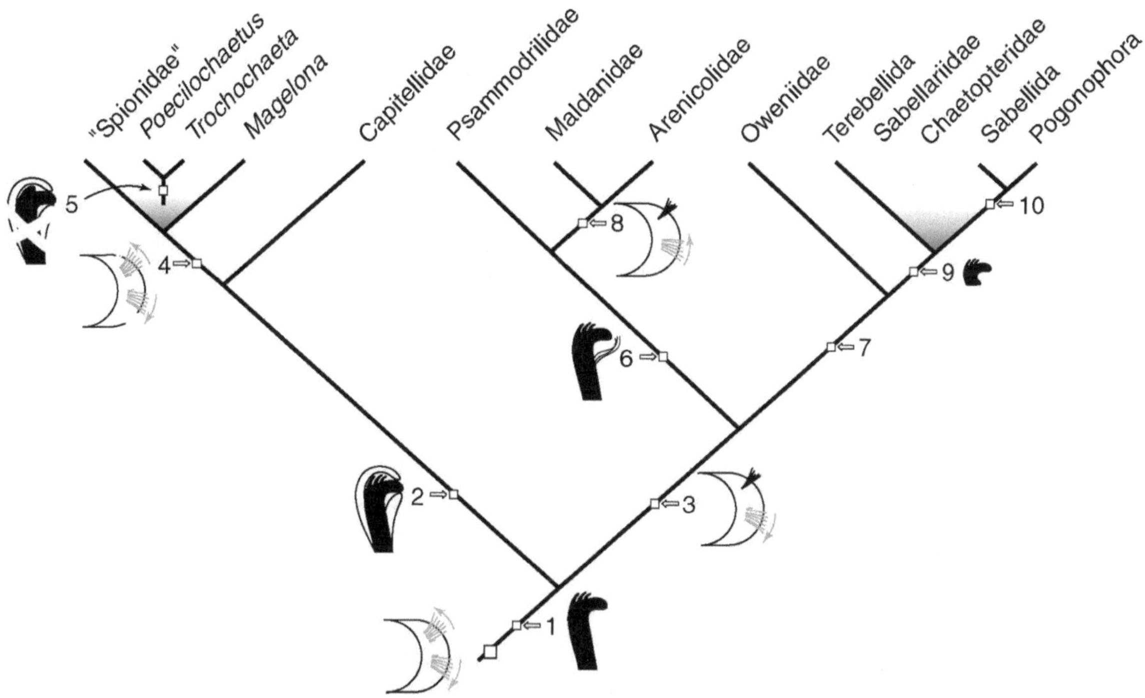

Figure 4. Phylogenetic relationships inferred from comparative analysis of chaeta and sense structures according to Bartolomaeus (1998) and Hausen (2001) – present state of knowledge. The key innovations are indicated by small sketches. **1** Hooked chaeta with a special mode of formation evolved in the stem lineage a taxon consisting of the spionidans (sensu Rouse & Fauchald, 1997) "Spionidae", Magelonidae, Poecilochaetidae, Trochochaetidae and Chaetopteridae, the scolecidans (sensu Rouse & Fauchald, 1997) Capitellidae, Maldanidae and Arenicolidae, the Psammodrilidae, the Terebellida except the Cirratuliformia (sensu Rouse & Pleijel, 2001) and the Sabellida (sensu Rouse & Fauchald, 1997). **2** Hood surrounding the apex of uncinus. Hood consists of two separately formed sheaths. **3** Rows of notopodial chaetae modified into bundles. **4** Additional transversal row of chaetae in neuro- and notopodia (see Hausen, this volume), nuchal organs shifted posteriorly, pave stone microvilli in nuchal organs, special photoreceptors (see Hausen 2001). **5** Reduction of hooded hooks. **6** Uncini with beard, lecithrotrophic development. **7** reduction of neuropodial hooks in chaetiger 1, **8** inverted formative side, sediment feeding. **9** reduction of the manubrial length, replacement of actin filaments at the end of chaetogenesis to attach chaetae; tube by secretions of the anterior ventral epidermis. **10** Nephridia in segment 2, dorsal nephridiopores.

analysis. However, in cladistic analyses loss of a given structure can only be detected if the group under consideration is supported by a sufficient number of autapomorphies (e.g. loss of nuchal organs in Pisionidae rather than primary absence). If the group considered is characterized by a large number of possible secondary absences, the systematic position may be incorrectly inferred (Rouse in Westheide et al., 1999; Purschke et al., 2000). This may account for the exclusion of Clitellata and Echiura from the polychaete clade in the Rouse & Fauchald (1997) tree (see McHugh, 2005).

Compared to the situation found in terrestrial polychaetes, the assumed reduction of nuchal organs implicates a terrestrial stem lineage of the Clitellata (Purschke, 2003). Accordingly freshwater habitats must have been invaded secondarily by Clitellata. If, as implied by molecular studies (see Erséus, 2005) freshwater habitats should be the primary environment of Clitellata, the loss of nuchal organs must have other reasons than terrestrialisation. Interestingly, freshwater polychaetes belonging to Nerillidae and Aeolosomatidae clearly possess these organs (Purschke, 1997; Hessling & Purschke, 2000). However, no final conclusion can be drawn yet, without taking a look at lower taxonomic levels.

Homology hypothesis of isolated structures – starting sister taxon search within the Polychaeta

Different to aiming at a complete cladistic analysis of polychaetes we started to fill the gaps in our knowledge of certain polychaete groups and erected trees using a few isolated structures (Bartolomaeus, 1995, 1998). Starting point had to be a taxon within the polychaetes that was considered to be monophyletic with respect to any outgroup comparison. We chose Sabellida sensu Fitzhugh (1989) and focussed at first on the ultrastructure and development of their neuropodial chaetae, the uncini. Although we are aware of the possible risk of this procedure (excluding possibly informative characters, studying localized maxima only) it allows to evaluate characters within the framework of selected other characters (Fig. 4).

Annelid chaetae are formed within an ectodermal invagination, the chaetal sac. Spatially and temporarily modulation of the apical microvilli pattern of its basalmost cell, i.e. the chaetoblast, determinates the structure of the chaeta (O'Clair & Cloney, 1974). Realizing this as well as the fact that structure and arrangement of chaetae are highly specific for polychaete species and higher taxonomic entities, we assume that the underlying information which guarantees formation of a certain kind of chaetae is rather conservative (see Hausen, 2005). Under this assumption, any hypothesis of the homology of chaetae could, thus, be tested, as an identical formation process was expected for presumed homologous chaetae. Studies of chaetogenesis of the uncini and hooded hooks of certain "sedentary" Polychaeta, revealed that the structure of these chaetae results from a uniform chaetogenesis (Arenicolidae and Maldanidae: Bartolomaeus & Meyer, 1997; Bobin, 1949; unpubl. data; Psammodrilidae and Oweniidae: Meyer & Bartolomaeus, 1996, 1997; Pectinariidae; Amphitritinae, Serpulidae, Sabellidae, Pogonophora (Siboglinidae): Bartolomaeus, 1995, 1998, 2002; Schulze, 2001; further unpubl. data; for summary see Hausen, 2005). One of the major conclusions inferred from theses studies says that several substructures and the course of development support the hypothesis of a homology of the hooked chaetae and uncini. This homology hypothesis has been extended for Capitellidae (Schweigkofler et al., 1998) and Spionidae (Hausen & Bartolomaeus, 1997). The studies allowed inclusion of Pogonophora into Polychaeta in a similar position as recovered in the cladistic analyses (Rouse & Fauchald, 1997, Rouse, 1999, 2000). Pogonophora[2] (Siboglinidae) are sister taxon to Sabellida consisting of Serpulidae and Sabellidae. This sister group relationship is supported by a reduction of the nephridia to a single pair in the second segment and the dorsal position of the nephridiopore. This sister group relationship implies a homology of the sabellid branchial crown and the dorsal tentacle of pogonophorans, which could explain their comparable organization comprising one blood vessel, one nerve and one coelomic cavity per tentacle (B in Fig. 4); other character concerning the reproduction and spermiogenesis are questionable (Bartolomaeus, 1998). Due to the aberrant structure of the central nervous system in Pogonophora, no conclusion can be drawn from their innervation pattern whether these tentacles actually represent palps in *Sclerolinum brattstromi* and *Siboglinum fjordicum* (Purschke, unpubl. obs.). Terebellida (sensu Rouse & Fauchald, 1997, excluding Acrocirridae, Flabelligeridae and Cirratulidae) are the sister group to that taxon, because a reduction of the length of the uncini shaft and replacement of the actin filaments by intermediate filaments to adhere the chaeta to the chaetoblast at the end of chaetogenesis represents the autapmorphy of this taxon. Formation of a tube by secretions of anterior-ventral glands in the anterior body region possibly is a further synapomorphy. Chaetopteridae and Sabellariidae also possess uncini and should belong to this group, although their position is uncertain yet (Fig. 4). The taxon consisting of the latter two and of Sabellida, Pogonophora and Terebellida has been termed Uncinifera (Bartolomaeus, 1998). Oweniidae are regarded as their sister taxon, substantatiated by the lack of hooked chaetae in the first setiger, by reduction of the beard and probably a completely incrusted tube (A in Fig. 4). Sister taxon to Uncinifera plus Oweniidae are Maldanomorpha and Psammodrilidae. Chaetation of the common ancestor of these taxa consists of dorsal capillary chaetae and ventral rows of uncini. Thus, some taxa of Scolecida (sensu Rouse & Fauchald, 1997) are included into the lineage of

[2] In terms of phylogenetic systematics there is no empirical justification to apply caterories to the taxa. Consequently, certain endings of taxa names introduced to indicate a hierachical level should also be neglected. There is accordingly no need to replace the taxon name Pogonophora by Siboglinidae. Figure 4, thus, uses taxa names irrespective of their endings.

those polychaetes with uncini. These studies were expanded for a possible inclusion of Spionida (sensu Rouse & Fauchald, 1997) and studies into the arrangement of chaetae (see Hausen, 2005) as well as the ultrastructure of unpigmented photoreceptor-like sense organs were included (Hausen, 2001). These studies supported the hypothesis that the spionid taxa except Apistobranchidae and Chaetopteridae are monophyletic and represent the sister taxon of Capitellidae. Presently, we assume that their sister taxon is a group which includes all species with uncini. We want to emphasize that these hypotheses have been concluded from isolated but intensely studied characters and we know that inclusion of further characters may either support this view or may lead to contradicting phylogenies. Provided that chaetal structure is as informative as we believe, palps must have been reduced several times within the Polychaeta (Fig. 4). Besides this, the proposed relationships change the composition of the superfamiliar taxa of Rouse & Pleijel (2001), but also recover a part of them.

Conclusions

Like most other comparative morphological studies we also end up claiming for further morphological investigations as prerequisite to unravel polychaete phylogeny. Out of the different attempts presented here to resolve polychaete phylogeny, cladistic analyses are the most decisive ones. They require a complete matrix containing substantiated homology hypotheses. Presently, however, cladistic analyses of polychaetes suffer from incomplete data sets, ambiguous character coding and provide conflicting trees. Such conflicting trees always indicate that at least one of the homology hypotheses coded in the matrix is not valid – and that these homology hypotheses need a re-evaluation. As long as there are large gaps in our knowledge of polychaete morphology, polychaete evolution can hardly be unravelled. Because any attempt to resolve higher level taxonomy (see Ecdysozoa vs. Articulata discussion: Giribet, 2003; Schmidt-Rhaesa, 2003) needs information on the annelid ground pattern, different attempts tried to infer the ground pattern from the known data and from evolutionary scenarios. This, however, does not succeed in any decisive result (see Purschke 2002), but provides some characters that must belong to annelid ground pattern (see Tables 1 and 2). On the other hand, some progress has been made by comparative studies of selected characters within the polychaetes. During this attempt stepwise search for sister group relationships produced trees while establishing homology hypothesis. They allow to present preliminary results of comparative studies and find questions for tightly focussed studies.

Much progress has been made during the last two decades by resolving the relationships within families using species or genera as lowest category (e.g. Fitzhugh, 1989, 1991; Rouse & Fitzhugh, 1994 for Sabellidae; Bellan et al., 1990 for Opheliidae; Pleijel, 1991; Orrhage & Eibye-Jacobson, 1998 for Phyllodocidae, 1998 for Hesionidae; Licher & Westheide, 1994 for Pilargidae; Pleijel & Dahlgren, 1998 for Chrysopetalidae and Hesionidae; Bartolomaeus & Meyer, 1999 for Arenicolidae; Blake & Arnofsky, 1999 for Spionida; Nygren, 1999 for Syllidae; Blake, 2000 for Orbiniidae; Rouse, 2001 for Siboglinidae/Pogonophora). These attempts allow describing the ground pattern of the families used in cladistic analyses on this level and are extremely reliable, because the taxa they use are clearly monophyletic and possess a uniform character distribution. We are sure that both ways, i.e. stepwise resolving the phylogeny on a low taxonomic level (species or genera level) and gathering further data to complete the morphological data base, will finally provide a sound picture of polychaete evolution.

Acknowledgements

Our thanks are due to Dr Lars Vogt. His and a further referees suggestions improved the manuscript. This study was supported by the Deutsche Forschungsgemeinschaft (Ba 1520/2, Ba 1520/4).

References

Aguinaldo, A. M. A., J. M. Turbeville, L. S. Linford, M. C. Rivera, J. R. Garey, R. A. Raff & J. A. Lake, 1997. Evidence for a clade of nematodes, arthropods and other moulting animals. Nature 387: 489–493.

Andouin, J. V. & H. Milne Edwards, 1834. Recherches pour servir a l'histoire naturelle du littoral de la France, ou Recueil de mémoires sur l'anatomie, la physiologie, la classification et les mœurs des animaux de nos côtes; ouvrage accompagné de planches faites d'après nature, 2, Annélides Pt. 1. Crochard, Paris.

Ax, P., 1987. The Phylogenetic System. Wiley, Chichester: 340 pp.

Ax, P., 1996. Multicellular animals. A New Approach to the Phylogenetic Order in Nature, I. Springer, Berlin: 225 pp.

Ax, P., 2000. Multicelluar animals. A New Approach to the Phylogenetic Order in Nature, II. Springer, Berlin: 396 pp.

Ax, P., 2003. Multicellular animals. A New Approach to the Phylogenetic Order in Nature, III. Springer, Berlin: 317 pp.

Bartolomaeus, T., 1999. Ultrastructure, development and function of segmental organs in Annelida. Hydrobiologia 402: 21–37

Bartolomaeus, T., 1995. Structure and formation of the uncini in *Pectinaria koreni*, *Pectinaria auricoma* (Terebellida) and *Spirorbis spirorbis* (Sabellida): implications for annelid phylogeny and the position of the Pogonophora. Zoomorphology 115: 161–177.

Bartolomaeus, T., 1998. Chaetogenesis in polychaetous Annelida – Significance for annelid systematics and the position of the Pogonophora. Zoology (ZACS) 100: 348–364.

Bartolomaeus, T., 2002. Structure and formation of the uncini in *Fabricia stellaris* (Ehrenberg, 1836) – Implication for the evolution of Sabellida (Annelida). Zoologischer Anzeiger 241: 1–17.

Bartolomaeus, T. & B. Quast (2005). Structure and development of nephridia in Annelida and related taxa. Hydrobiologia 535/536 (Dev. Hydrobiol. 179): 139–164.

Bartolomaeus, T. & H. Ruhberg, 1999. Ultrastructure of the body cavity lining in embryos of *Epiperipatus biolleyi* (Peripatidae, Onychophora). Invertebrate Biology 118: 165–174.

Bartolomaeus, T. & K. Meyer, 1997. Development and phylogenetic significance of hooked setae in Arenicolidae (Polychaeta, Annelida). Invertebrate Biology 116: 227–242.

Bartolomaeus, T. & K. Meyer, 1999. Zur Phylogenie der Arenicolidae. Senckenberg Courier 215: 23–29.

Bellan, G., D. Bellan-Santini & J. C. Dauvin (1991) Phénetique et phylogénie des Opheliidae (Annélides, Polychètes). Comptes rendus de l'Academic Science, Paris, Series III 310: 175–181.

Benham, W. B., 1896. The Archiannelida, Polychaeta, Myzostomaria. In Harmer, S. F. & A. E. Shipley (eds), The Cambridge Natural History, 2. MacMillan, London: 241–344.

Blake, J. A. & P. L. Arnofsky 1999. Reproduction and larval development of the spioniform Polychaeta with application to systematics and phylogeny. Hydrobiologia 402: 57–106.

Blake, J. A., 2000. A new genus and species of polychaete worm (familiy Orbiniidae) from methane seeps in the Gulf of Mexico, with a review of the systematics and phylogeentic interrelationships of the genera of Orbiniidae. Cahiers de Biologie Marine 41: 435–449.

Bleidorn, C., L. Vogt & T. Bartolomaeus, 2003a. New insights into polychaete phylogeny (Annelida) inferred from 18S rDNA sequences. Molecular phylogenetics and evolution 29: 279–288

Bleidorn, C., L. Vogt & T. Bartolomaeus, 2003b. A contribution to sedentary polychaete phylogeny using 18S rRNA sequence data. Journal of Zoological Systematics and Evolutionary Research 41: 186–195.

Bobin, G., 1944. Morphogénèse des soies chez les annélides polychètes. Annales des Institute Oceanographique, Monaco 22: 1–106.

Clark, R. B., 1964. Dynamics in Metazoan Evolution. The Origin of the Coelom and Segments. Clarendon Press, Oxford.

Clark, R. B., 1977. Phylogenetic reconstruction. Verhandlungen der deutschen Zoologischen Gesellschaft 70 175–183.

Conway-Morris, S., 1979. Middle Cambrian polychaetes from the Burgess shale of British Columbia. Philosophical Transactions of the Royal Society B 285: 227–274.

Cuvier, G. 1812. Sur un nouveau rapprochement à établir entre les classes qui composant le Règne Animal. Annales de Muséum Histoire 19: 73–84.

Dales, R. P., 1962. The polychaete stomatodeum and the inter–relationships of the families of Polychaeta. Proceedings of the Zoological Society, London 139: 389–428.

Day, J. H., 1967. A monograph of the Polychaeta of Southern Africa. British Museum (Natural history). Publication 656. British Museum (Natural History), London: 878 pp.

Desbruyéres, D. & L. Laubier, 1980. *Alvinella pompejana* gen.-sp.nov., Ampharetidae aberrant des sources hydrothermales de la ride Est-Pacifique. Oceanologica Acta 3: 267–274.

Desbruyéres, D. & L. Laubier, 1986. Les Alvinellidae, une famille nouvelle d'annélides polychètes inféodées aux sources hydrothermales sous-marines: systématique, biologie et écologie. Canadian Journal of Zoology 64: 2227–2245.

Eibye-Jacobson, D. & C. Nielsen, 1997. The rearticulation of annelids. Zoologica Scripta 25: 275–282.

Erséus, C., 2005. Phylogeny of oligochaetous Clitellata. Hydrobiologia 535/536 (Dev. Hydrobiol. 179): 355–370.

Fauchald, K. & G. W. Rouse, 1997. Polychaete systematics: Past and present. Zoologica Scripta 26: 71–138.

Fauchald, K., 1977. The polychaete worms: definitions and keys to the orders, families and genera. Natural History Museum of Los Angeles County, Sciences Series 28: 190 pp.

Fauvel, P., 1923. Polychétes errantes, 5 (Faune de France). Paul Lechevalier, Paris: 488 pp.

Ferraguti, M. & C. Erseus (1999). Sperm types and their use for a phylogenetic analysis of aquatic clitellates. Hydrobiologia 402: 225–237.

Ferraguti, M., 1984. The comparative ultrastructure of sperm flagella central sheath in Clitellata reveals a new autapomorphy of the group. Zoologica Scripta 13: 201–207.

Fitzhugh, K., 1989. A systematic revision of the Sabellidae–Caobangiidae–Sabellongidae complex (Annelida: Polychaeta). Bulletin of the American Museum of National History 192: 1–104

Fitzhugh, K., 1991. Further revisions of the Sabellidae subfamilies and cladistic relationships among the Fabricinae (Annelida: Polychaeta). Zoological Journal of the Linnean Society 102: 305–332.

Fristrup, K. M., 2001. A history of character concepts in evolutionary biology. In Wagner, G. P. (ed.), The Character Concept in Evolutionary Biology. Academic Press, San Diego: 13–35.

George, J. D. & G. Hartmann-Schröder, 1985. Polychaetes: British Amphinomida, Sphinterida and Eunicida. In Kermack, D. M. & R. S. K. Barnes (eds), Synopses of the British Fauna. Brill Backhuys, Leiden.

Giribeth, G., 2003. Molecules, development and fossils in the study of metazoan evolution; Articulata versus Ecdysozoa revisited. Zoology 106: 303–326.

Goodrich, E. S., 1895. On the coelom, genital ducts, and nephridia. Quarterly Journal of microscopical Science 37: 477–510.

Goodrich, E. S., 1945. The study of nephridia and genital ducts since 1895. Quarterly Journal of microscopical Science 86: 113–392.

Grube, A. E., 1850. Die Familien der Anneliden. Archiv für Naturgeschichte 16: 249–364.

Halanych, K., T. Dahlgren & D. McHugh 2002. Unsegmented annelids? Possible origin of four lophotrochozoan worm taxa. Integrative Comparative Biology 42: 678–684.

Hartmann-Schroeder, G., 1971. Annelida, Borstenwürmer, Polychaeta. In Dahl, M. & F. Peus (eds), Die Tierwelt Deutschlands, 58. G. Fischer, Jena.

Hartmann-Schröder, G., 1996. Annelida, Borstenwürmer, Polychaeta. Zweite Auflage. In In Dahl, M. & F. Peus (eds), Die Tierwelt Deutschlands, 58. G. Fischer, Jena

Hatschek, B., 1878. Studien über die Entwicklungsgeschichte der Anneliden. Ein Beitrag zur Morphologie der Bilaterien. Arbeiten aus dem Zoologischen Institut der Universität Wien und der Zoologischen Station Triest 1: 277–404.

Hatschek, B., 1893. System der Anneliden, ein vorläufiger Bericht. Lotos 13: 123–126.

Hausen, H. & T. Bartolomaeus, 1997. Setal structure and chaetogenesis in *Scolelepis squamata* and *Malacoceros fuliginosus* (Spionidae, Annelida). Acta Zoologica (Stockholm) 79: 179–141.

Hausen, H., 2005. Chaetae and chaetogenesis in polychaetes (Annelida). Hydrobiologia 535/536 (Dev. Hydrobiol. 179): 37–52.

Hausen, H., 2001. Untersuchungen zur Phylogenie "spiomorpher" Polychaeten (Annelida). Logos, Berlin: 145 pp.

Hessling, R. & G. Purschke, 2000. Immunohistochemical (cLSM) and ultrastructural analysis of the central nervous system and sense organs in *Aeolosoma hemprichi* (Annelida, Aeolosomatidae). Zoomorphology 120: 65–78.

Hessling, R. & W. Westheide, 2002. Are Echiura derived from a segmented ancestor? Immunhistochemical analysis of the nervous system in developmental stages of *Bonellia viridis*. Journal of Morphology 252: 100–113.

Hessling, R., 2002. Metameric organisation of the nervous system in developmental stages of *Urechis caupo* (Echiura) and its phylogenetic implications. Zoomorphology 121: 221–234.

Jenner, R. A. & F. R. Schram, 1999. The grand game of metazoan phylogeny – rules and strategies. Biology Reviews 74: 121–142.

Jenner, R. A., 2001. Bilaterian phylogeny and uncritical recycling of morphological data sets. Systematic Biology 50: 730–742.

Jouin, C., 1971. Status of the knowledge of the systematics and ecology of Archiannelida. In Hulings, N. C. (ed.), Proceedings of the First Interantional Conference on Meiofauna. Smithsonian Contribution Zoology 67: 47–56.

Lamarck, J.-B. de, 1801. Système des Animaux sans Vertèbres, ou Tableau général des classes, des ordres et des genres de ces animaux; Présentant leurs caractères essentiels et leur distribution, d'après la considération de leurs apports naturelles et da leur organisation, et suivant l'arrangement établi dans les galeries du Muséum d'Histoire Naturelle, parmi leurs dépouilles conservées ; Précédé du discours d'ouverture du Cours de Zoologie, donné dans la Muséum National d'Histoire Naturelle l'an 8 de la Républiques. Chez l'auteur Déterville, Paris.

Lamarck, J.-B. de, 1818. Histoire naturelle des Animaux sans Vertèbres, présentant les caractères généraux et particuliers de ces animaux, leur distribution, leur classes, leur familles, leur genres, et la citation des principales espèces, qui s'y rapportent ; précédée d'une Introduction offrant la Détermination des caractères essentiels de l'Animal, sa distinction du Végétal et des autres corps naturels, enfin, l'Exposition des Principes fondamentaux de la Zoologie, 5. Déterville and Verdière, Paris.

Licher, F. & W. Westheide, 1994. The phylogenetic position of the Pilargidae with a cladistic analysis of the taxon – facts and ideas. Mémoirs du Muséum national d'Histoire Naturelle, Paris 162: 223–235.

McHugh, D., 2005. Molecular systematics of polychaetes. Hydrobiologia 535/536 (Dev. Hydrobiol. 179): 307–316.

Meier, R., 1994. On the inappropriateness of presence/absence recoding for non-additive multistate characters in computerized cladistic analyses. Zoologischer Anzeiger 232: 201–212.

Meyer, E., 1890. Die Abstammung der Anneliden. Der Ursprung der Metamerie und die Bedeutung des Mesoderms. Biologisches Centralblatt 10: 296–308.

Meyer, K. & T. Bartolomaeus, 1996. Ultrastructure and formation of the hooked setae in *Owenia fusiformis* delle Chiaje, 1842: Implications for annelid phylogeny. Canadian Journal of Zoology 74: 2143–2153.

Meyer, R. & T. Bartolomaeus, 1997. Ultrastruktur und Morphogenese der Hakenborsten bei *Psammodrilus balanoglossoides* – Bedeutung für die Stellung der Psammodrilida (Annelida). Microfauna marina 11: 87–113

Nielsen, C., 2001. Animal Evolution. Interrelationships of the Living Phyla. Oxford University Press, Oxford: 563 pp.

Nygren, A., 1999. Phylogeny and reproduction in Syllidae (Polychaeta). Zoological Journal of the Linnean Society 126: 365–386.

O'Clair, R. M. & R. A. Cloney, 1974. Patterns of morphogenesis mediated by dynamic microvilli: Chaetogenesis in *Nereis vexillosa*. Cell Tissue Research 151: 141–157.

Orrhage, L., 1978. On the structure and evolution of the anterior end of the Sabellariidae (Polychaeta, Sedentaria. With some remarks on the general organisation of the polychaete brain. Zoologische Jahrbücher für Anatomie und Ontogenie der Tiere 100: 343–374.

Orrhage, L., 1980. On the structure and homologues of the anterior end of the polychaete families Sabellidae and Serpulidae. Zoomorphology 96: 113–168.

Orrhage, L., 2001. On the anatomy of the central nervous system and the morphological value of the anteriorend appendages pf Ampharetidae, Pectinariidae and Terebellidae (Polychaeta). Acta Zoologica (Stockholm) 82: 57–71.

Orrhage, L. & D. Eibye-Jabobsen, 1998. On the anatomy of the central nervous system of Phyllodocida (Polychaeta) and the

phylogeny of phyllodocid genera: a new alternative. Acta Zoologica (Stockholm) 79: 215–234.
Orrhage, L. & M. C. M. Müller, 2005. Morphology of the Nervous system of Polychaeta (Annelida). Hydrobiologia 535/536 (Dev. Hydrobiol. 179): 79–111.
Perrier, E., 1897. Traité de Zoologie, Fascicule IV. Vers, Mollusques, Tuniciers. Masson et Cie, Paris.
Pettibone, M. H., 1982. Annelida. In Parker, S. P. (ed.), Synopsis and Classification of Living Organisms, 2. McGraw Hill, New York: 1–43.
Pleijel, F., 1991. Phylogeny and classification of the Phyllodocidae (Polychaeta). Zoological Screening. 20: 225–261.
Pleijel, F., 1995. On character coding for phylogeny reconstruction. Cladistics 11: 309–315
Pleijel, F. & T. Dahlgren (1998). Position and delineation of Chrysopetalidae and Hesionidae (Annelida, Polychaeta, Phyllodocida). Cladistics 14: 129–150.
Purschke, G., 1997. Ultrastructure of the nuchal organs in polychaets (Annelida) – new results and review. Acta Zoologica (Stockholm) 78: 123–143.
Purschke, G., 1999. Terrestrial polychaetes – models for the evolution of the Clitellata (Annelida)? Hydrobiologia 406: 87–99.
Purschke, G., 2000. Sense organs and central nervous system in an enigmatic terrestrial polychaete, *Hrabeiella periglandulata* (Annelida). Invertebrate Biology 119: 329–341.
Purschke, G., 2002. On the ground pattern of Annelida. Organisms Diversity and Evolution 2: 181–196.
Purschke, G., 2003. Is *Hrabeiella periglandulata* (Annelida, "Polychaeta") the sitser group of Clitellata? Evidence from an ultrastructural analysis of the dorsal pharynx in *H. periglandulata* and *Enchytraeus minutus* (Annelida, Clitellata). Zoomorphology 122: 55–66.
Purschke, G., F. Wolfrath & W. Westheide, 1997. Ultrastructure of the nuchal organ and cerebral organ in *Onchnesomum squamatum* (Sipuncula, Phascolionidae). Zoomorphology 117: 23–31.
Purschke, G., R. Hessling, W. Westheide, 2000. The phylogenetic position of the Clitellata and the Echiura – on the problematic assessment of absent characters. Journal of Zoological Systematics and Evolutionary Research 38: 165–173.
Purschke, G., W. Westheide, D. Rhode & R.O. Brinkhurst, 1993. Morphological reinvestigation and phylogenetic relationship of *Acanthobdella peledina* (Annelida, Clitellata). Zoomorphology 113: 91–101.
Rouse, G. W., 1999. Trochophore conepts: ciliary bands and the evolution of larvae in spiralian Metazoa. Biological Journal of the Linnean Society 66: 411–464.
Rouse, G. W., 2000. Bias? What bias? Gain and loss of downstream larval-feeding in animals. Zoologica Scripta 29: 213–236.
Rouse, G. W., 2001. A cladistic analysis of Siboglinidae Caullery, 1914 (Polychaeta, Annelida): formerly the phyla Pogonophora and Vestimentifera. Zoological Journal of the Linnean Society 132: 55–80
Rouse, G. W. & K. Fauchald, 1995. The articulation of annelids. Zoologica Scripta 24: 269–301.
Rouse, G. W. & K. Fauchald, 1997. Cladistics and polychaetes. Zoologica Scripta 26: 139–204.
Rouse, G. W. & K. Fitzhugh, 1994. Broadcasting fables: is external fertilization really primitive? Sex, size and larvae in sabellid polychaetes. Zoologica Scripta 23: 271–312.
Rouse, G. W. & F. Pleijel, 2001. Polychaetes. Oxford University Press, New York: 354 pp.
Ruppert, E. E., R. S. Fox & R. D. Barnes, 2003. Invertebrate Zoology. A functional evolutionary approach. Brooks/Cole-Thomson learning, Belmont, CA.
Schmidt-Rhaesa, A., 2003. Old trees, new trees – is there any progress? Zoology 106: 291–301.
Schmidt-Rhaesa, A., T. Bartolomaeus, C. Lemburg, U. Ehlers & J. R. Garey, 1998. The position of the Arthropoda in the phylogenetic system. Journal of Morphology 238: 263–285.
Scholtz, G., 2002. The Articulata hypothesis – or what is a segment? Organisms Diversity and Evolution 2: 197–215.
Schulze, A. 2001. Ultrastructure of opisthosomal chaetae in Vestimentifera (Pogonophora, Obturata) and implications for phylogeny. Acta Zoologica (Stockholm) 82: 127–135.
Schweigkofler, M., T. Bartolomaeus & L. von Salvini-Plawen, 1998. Ultrastructure and formation of hooded hooks in *Capitella capitata* (Annelida, Capitelida). Zoomorphology 118: 117–128.
Sokal, R. R. & P. H. A. Sneath, 1963. Principles of Numerical Taxonomy. WH Freeman, San Francisco.
Stecher, H.-J., 1968. Zur Organisation und Fortpflanzung von *Pisione remota* (Southern) (Polychaeta Pisionidae). Zeitschrift für Morphologie der Tiere 61: 347–410.
Struck, T. H., R. Hessling & G. Purschke, 2002a. The phylogenetic position of the Aeolosomatidae and Parergodrilidae, two enigmatic oligochaete-like taxa of the "Polychaeta", based on molecular data from 18S rDNA sequences. Journal of Zoological Systematics and Evolutionary Research 40: 155–163.
Uschakov, P. V. 1955. Polychaeta of the far eastern seas of the U.S.S.R. Keys to the fauna of the USS 56: 1–445.
Wägele, J.-W. & B. Misof, 2001. On the quality of evidence in phylogeny reconstruction: a reply to Zrzavy's defence of the, ecdysozoa' hypothesis. Journal of Zoological Systematics and Evolutionary Research 39: 165–176.
Wägele, J.-W., T. Erikson, P. Lockhart & B. Misof, 1999. The Ecdysozoa: artifact or monophylum. Journal of Zoological Systematics and Evolutionary Research 37: 211–223.
Westheide, W., 1985. The systematic position of the Dinophilidae and the archiannelid problem. In Conways Morris, S., D. George, R. Gibson & H. M. Platt (eds), The Origins and Relationships of Lower Invertebrates. Oxford Universsity Press, Oxford: 310–326.
Westheide, W., 1987. Progenesis as a principle in meiofaunal evolution. Journal of Natural History 21: 843–854.
Westheide, W., 1997. the direction of the evolution within the Polychaeta. Journal of Natural History 31: 1–15.
Westheide, W., D. McHugh, G. Purschke & G. W. Rouse, 1999. Systematization of the Annelida: different approaches. Hydrobiologia 401: 291–307.

Hydrobiologia (2005) **535/536:** 357–372
T. Bartolomaeus & G. Purschke (eds), Morphology, Molecules, Evolution and Phylogeny in Polychaeta and Related Taxa

Phylogeny of oligochaetous Clitellata

Christer Erséus
Department of Invertebrate Zoology, Swedish Museum of Natural History, Stockholm, Sweden
Present address: Department of Zoology, Göteborg University, P.O. Box 463, SE-405 30 Göteborg, Sweden
Tel.: +46-31-773-3645, Fax: +46-31-416-729, E-mail: christer.erseus@zool.gu.se

Key words: Clitellata, Oligochaeta, phylogeny, cladistics, molecular systematics

Abstract

Clitellata, with more than one third of all annelid species described, is briefly introduced, and an overview of the hypotheses of phylogenetic relationships among the groups traditionally referred to as oligochaetes is given. The presentation is placed in a historical context and describes the trend to move from intuitive, narrative approaches to more formal analyses of character patterns. Monophyly of the earthworms (the megadriles, or Metagynophora *sensu* Jamieson), or at least a major part of them (Crassiclitellata *sensu* Jamieson), and paraphyly of the 'microdrile' largely aquatic, groups are supported by both morphological and molecular data. Further, DNA sequences as well as spermatozoal ultrastructure corroborate that all leech-like taxa (Hirudinida, Acanthobdellida and Branchiobdellida) constitute a clade derived within 'Oligochaeta', closely related to the family Lumbriculidae. Molecular systematic studies also support relationships already identified on the basis of morphological data, e.g., the position of Naididae within Tubificidae, the position of Phreodrilidae close to, but outside, the same family, and the putative sister-group relationship between the newly discovered Capilloventridae and the rest of Clitellata. A recent study using 18S rDNA suggests that Enchytraeidae is closely related to Metagynophora, and that these two taxa, which contain all terrestrial oligochaetous clitellates, form a clade derived from aquatic 'microdriles' This refutes a recent hypothesis proposing that the ancestor of Clitellata was terrestrial. To a great extent, however, the basal resolution of the oligochaetous clitellates remains unclear.

Introduction

Clitellata comprises more than one third of the totally about 15 000 species of segmented worms (Annelida) described and currently considered as valid (estimate based on Reynolds & Cook, 1993; Rouse & Pleijel, 2001; and M. J. Wetzel, information available at http://www.inhs.uiuc.edu/). The taxon name refers to a glandular, epidermal structure, the clitellum (girdle), which develops in the shape of a ring or saddle around a specific part of the body when the worm reaches sexual maturity, and which is involved in the formation of cocoons at egg-laying. Clitellates are primarily hermaphroditic, but vegetative or parthenogenetic reproduction occurs in some species. Several morphological characters, including the clitellum and its tight association with hermaphrodism, support monophyly of the group (Purschke et al., 1993; Westheide, 1997; McHugh, 2000; Purschke, 2002). In a recent paper, Almeida et al. (2003) argue that the polychaete taxon Questidae is likely to be the sister group of Clitellata; questid females have a body wall structure reminiscent of the clitellate clitellum. However, other morphological as well as molecular data do not support this relationship (Giere & Riser, 1981; Jamieson & Webb, 1984; Westheide, 1997; Giere & Erséus, 1998; Erséus et al., 2000; Rota et al., 2001; Struck et al., 2002).

Traditionally, Clitellata has been divided into two main groups, 'Oligochaeta' (earthworms, sludge worms, etc.) and Hirudinea *sensu lato* (the 'true' leeches plus the Arctic salmon parasite, *Acanthobdella peledina*), with Branchiobdellida (leech-like commensals or parasites on crayfish) either as a taxon within the oligochaetes, or as a third group by itself. The rank of Hirudinea equal to that of 'Oligochaeta' was not formally challenged until recently, when analyses of DNA sequences corroborated that both leeches and branchiobdellidans are members of the oligochaete clade (Martin, 2001; Siddall et al., 2001). Thus, Clitellata has become synonymous to 'Oligochaeta', and it is a matter of taste whether one should use the one name or the other for the group as a whole. As pointed out by Martin (2001), Clitellata may be preferred as it was originally meant to be more inclusive than 'Oligochaeta'. Siddall et al. (2001) proposed the leech-like forms to be classified as 'orders', Branchiobdellida, Hirudinida, and Acanthobdellida, within Clitellata.

The aim of this paper is to give an overview of the hypotheses of phylogenetic relationships among the groups traditionally referred to as oligochaetes, a paraphyletic assemblage now better termed oligochaetous Clitellata, i.e., the clitellates bearing [few] chaetae, as opposed to branchiobdellidans and hirudinidans. I will summarize what we know about clitellate phylogeny at this date, and by putting this into a historical context, discuss the recent trend to move from intuitive phylogenies to more formal analyses of character patterns, including those of DNA sequences.

Higher level classification of oligochaetous Clitellata

About 100 years ago and throughout a great part of the 20th century, oligochaetes *sensu stricto* were often classified in two main groups, largely corresponding to two rather distinct size classes and habitat preferences among the worms: the smaller forms generally being associated with water, the larger forms with soil. Benham (1890) coined the names Microdrili and Megadrili, respectively, for these groups. Beddard (1895) summarized the classification of 'Oligochaeta' of his time, using the basic framework of Microdrili/Megadrili laid out by Benham, but redefined Microdrili to include also the 'family Naidomorpha' (=Naididae in more recent terminology), a group that Benham had regarded as a subclass separate from all other oligochaetes.

Beddard (1895) recognized five families of microdriles and seven of megadriles, but today at least 38 different taxa have been referred to as 'families' within Oligochaeta *sensu stricto*, i.e., counting the families listed by Reynolds & Cook (1993) plus Tumakidae Righi, 1995, and Parvidrilidae (see below). Some authors, Jamieson (1988) and Omodeo (1998) in particular, have suggested a reduction of this number by regarding several former families as lower-level taxa within others. There is general consensus that the about 15 microdrile families do not comprise a monophylum. On the other hand, the assemblage of earthworm-like taxa, sometimes still being called Megadrili, is seen by most workers as a natural group. In a reclassification of the oligochaetes following a cladistic analysis, Jamieson (1988) proposed a new name for this group, Metagynophora, in recognition of the inferred loss of an anterior pair of ovaries. He also established a less inclusive taxon, Crassiclitellata, for all those about 3000 earthworm species that have a multilayered clitellum (with epidermis composed of several cell layers), a most likely apomorphic condition. All other oligochaetes, including about 120 metagynophoran species (largely 'families' Moniligastridae and Alluroididae) outside Crassiclitellata, have a single-layered clitellum.

Morphological (Jamieson, 1988) as well as molecular data (Jamieson et al., 2002) support monophyly of crassiclitellates, while the monophyly of Metagynophora (also supported by Jamieson, 1988) has not been assessed in any molecular study. However, at least two authors in the past (Cekanovskaya, 1962: fig. 67; Jamieson, 1974: 199) considered a possible polyphyly of the groups now referred to as Crassiclitellata. More recently, Omodeo (1998, 2000) presented a similar view, suggesting that the earthworm taxon Lumbricoidea originated from a haplotaxid-like microdrile stock, while two other earthworm groups, Megascolecoidea and Eudriloidea, have one or two different alluroidid-like (i.e., non-crassiclitellate) ancestors; implying that also Metagynophora is polyphyletic. [In his 1998 paper, Omodeo attributed a former suggestion of a relationship between Alluroididae and primitive megascolecines (ocnerodrilids) to Stephenson (1930), but this was a lapsus; as corrected by Omodeo himself (in litt.), the correct reference here is Jamieson (1974).]

Thus, there is still some disagreement as to whether a multi-layered clitellum has evolved only once, and how the various earthworm groups are related to each other, but for a review focusing on the basal resolution of the clitellate family tree, all earthworms can be treated as a single clade, whereas some microdrile (or rather, non-metagynophoran) families deserve more specific attention. It can be estimated that of a total of about 5000 valid species of oligochaetous Clitellata, about 1600 (30%) are aquatic or at least semiaquatic, and about 600 of the aquatic forms are marine. At least 450 of the 3400 terrestrial species belong to Enchytraeidae (see below), the rest are (metagynophoran) earthworms.

Various systems of ordinal division of 'Oligochaeta' have been proposed (a recent one by Jamieson, 1988). However, due to the on-going phylogenetic re-assessments of Clitellata using molecular methods, oligochaete 'orders' will not be treated any further in this account.

Some selected families

The most speciose aquatic family is Tubificidae (Fig. 1: Tub), with about 800 described species worldwide, a majority of which are marine. The worms vary greatly in size (2–185 mm) but are typically 4–20 mm long, an average marine species being smaller than a typical freshwater one. Tubificids either burrow in fine sediments, or live interstitially among sand grains. Some freshwater forms, such as the ubiquitous *Tubifex tubifex*, *Limnodrilus hoffmeisteri* and *Branchiura sowerbyi*, are highly tolerant to organic pollution, but the majority of tubificids are adapted to more specific habitats. Two marine genera, *Inanidrilus* and *Olavius*, are unique among clitellates in lacking a normal alimentary system, but instead thriving in association with endosymbiotic chemoautotrophic bacteria (see, e.g., Giere, 1981; Richards et al., 1982; Erséus, 1984a; Dubilier et al., 1999). In most cases, tubificids reproduce sexually. They are

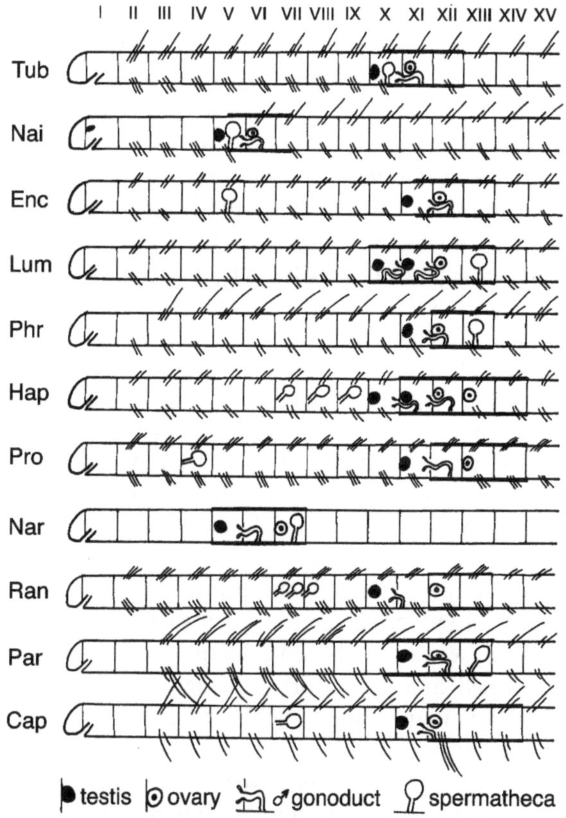

Figure 1. Schematic outline and comparison of chaetae, clitellum (indicated by thickened lines in body wall), gonads, gonoducts and spermathecae in families of oligochaetous Clitellata treated in the text: Tubificidae (Tub), 'Naididae' (Nai), Enchytraeidae (Enc), Lumbriculidae (Lum), Haplotaxidae (Hap), Phreodrilidae (Phr), Propappidae (Pro), Narapidae (Nar), Randiellidae (Ran), Parvidrilidae (Par), and Capilloventridae (Cap). Chaetal patterns, and position of clitellum and genitalia, vary to some extent within most families. Eyes occur in some species of 'Naididae' only. Segment numbers are indicated by Roman numerals.

tetragonadal, with one pair of testes in segment X, and one pair of ovaries in segment XI. A pair of spermathecae (organs receiving sperm at copulation) are typically located in the testicular segment while male gonoducts develop in close proximity of the ovaries in segment XI (see Gustavsson & Erséus, 1997, 1999). These ducts tend to be elaborate, with glandular and muscular details useful for identifying taxa at lower levels. Virtually all tubificids have bifid chaetae (i.e., chaetae with bifurcate tips), and the dorsal bifid chaetae are sometimes associated with capillary (hair-like) chaetae.

'Naididae' (Fig. 1: Nai) is another aquatic (largely limnic) taxon, with about 180 species. Traditionally, it has been treated as a family, but Erséus & Gustavsson (2002) recently proposed that it should be regarded as a subfamily, Naidinae, within Tubificidae. Naidines are similar to other tubificids with regard to the general morphology of chaetae and sexual organs, but they are more active and live a less benthic life, preferring superficial sediment layers (e.g., those covering rocks and sunken wood), or the surfaces of aquatic plants. They are small, some barely over 1 mm long. Some species can swim and/or have eyes, a few others (*Chaetogaster* spp.) are predators feeding on other microinvertebrates. All naidines reproduce asexually by paratomic fission (including regeneration of a fixed number of anterior segments in secondary individuals), providing great potential for population growth and propagation. However, at some stage or season they may become sexual, with reproductive organs being developed more anteriorly (in segments IV–V to VII–VIII) than in the typical tubificid.

Enchytraeidae (Fig. 1: Enc) is often thought of as a terrestrial taxon, as so many of its about 650 valid species are found in soil, but it is probably the most ubiquitous of all clitellate families. Enchytraeids dwell also in a wide range of limnic (lakes including the profundal zone, rivers and streams including the riparian zone, bogs, springs, caves, etc.), and marine/brackish-water habitats (salt marshes, wrack beds, intertidal and subtidal sands, filamentous algae or macroalgal holdfasts on hard bottoms, brackish-water, deep-sea sediments), and sometimes in extreme alpine situations such as the snow of glaciers. The species are either burrowing or interstitial, some being only 1–2 mm, others up to 170 mm long (Eisen, 1904; see also Rota, 2001), but the typical size is about 5–20 mm. Enchytraeids have simple-pointed, stout chaetae, similar to those of earthworms, and their cuticle is often thick, making the worms more sluggish than their aquatic relatives. A thick layer of collagen fibres in the cuticle, noted for both enchytraeids and earthworms, appears to be correlated with a terrestrial mode of life (Richards, 1977; Jamieson, 1981). Virtually all enchytraeids reproduce sexually, and they are tetragonadal with testes and ovaries in adjacent segments (generally in XI–XII). Their male gonoducts differ from those of Tubificidae in lacking atrial and (tubificid-like) prostate structures, but instead they have peculiar, glandular sperm funnels and copulatory apparati. Moreover, the spermathecae open to the exterior in segment V, and even when extending posteriorly through a number of segments, they are always well separated from the rest of the genitalia.

Lumbriculidae (Fig. 1: Lum) is a freshwater family, with some species found also in wet soils. It is primarily restricted to the Holarctic, but a few species, such as the opportunistic *Lumbriculus variegatus*, have been introduced to southern latitudes. Most lumbriculids are cold-stenothermal, preferring deep lakes, high altitude streams, and phreatic waters. In their general morphology and size, they resemble freshwater tubificids. For instance, they have atria (often bearing prostates) associated with their male gonoducts. However, although bifid chaetae are common, hair chaetae are always lacking. Moreover, lumbriculids have a different and more complex gonadal arrangement, with a basic pattern of two pairs of testes, in consecutive segments (somewhere in VII–XIII), followed by one pair of ovaries in the next segment (see Brinkhurst, 1989).

Phreodrilidae (Fig. 1: Phr) is a tetragonadal aquatic taxon with about 50 described species, largely confined to the Southern Hemisphere (Pinder & Brinkhurst, 1997a), and with a few also known from Sri Lanka and Northern Africa (Giani et al., 1995). Most phreodrilids live in freshwater sediments, some are commensal on freshwater crayfish, and there are a few records from estuarine or intertidal habitats (Pinder & Erséus, 2000). They are of about the same size as tubificids and often bear bifid and capillary chaetae, but they also have chaetal types (e.g., 'support chaetae'; see Pinder & Brinkhurst, 1997a) found nowhere else within Clitellata. Their male ducts bear atria, possibly homologous with those of Tubificidae. The genital organs are usually located in segments XI–XII, i.e., one segment posterior to the position typical of most Tubificidae.

Haplotaxidae (Fig. 1: Hap) is a rare, but widely distributed group. It contains about 30 freshwater forms, often associated with phreatic or wet soil habitats. The worms are somewhat earthworm-like, but more slender than a typical megadrile. They have single-pointed or bifid chaetae, and short, simple male gonoducts devoid of atria and prostates. The gonads are most often eight, with two pairs of testes and two pairs of ovaries in segments X–XIII or thereabouts, and Brinkhurst (1982, 1984a, 1984b, 1994, 1999) suggested that the gonadal pattern of all other oligochaetes has been derived from this arrangement. He regarded Haplotaxidae as a paraphyletic, ancestral assemblage, and a basal position of this family has been widely accepted over the years, at least with reference to haplotaxids being close to the ancestor of megadriles (Beddard, 1895; Michaelsen, 1930; Stephenson, 1930; Čekanovskaya, 1962; Brinkhurst & Jamieson, 1971; Timm, 1981; Kasprzak, 1984a; Erséus, 1987; Jamieson, 1988; Omodeo, 1987, 1998, 2000).

In recent years, five new aquatic families have been established, none of which with more than a few known species. First, *Propappus* (Fig. 1: Pro) with its three Palearctic freshwater species and originally regarded as a member of Enchytraeidae, was placed in a family of its own, Propappidae, by Coates (1986). The same author suggested that this taxon is phylogenetically near both the Enchytraeidae and some species of Haplotaxidae, based on a preliminary cladistic analysis of all these taxa (Coates, 1986, 1987), i.e., her studies supported the paraphyly of Haplotaxidae. Propappids have bifid chaetae, one pair of testes in segment XI and one pair of ovaries in segment XIII, and their male gonoducts are simple, lacking atria and prostates.

The phylogenetic affinities of the South-American Narapidae (with one freshwater species; Righi & Varela, 1983; Fig. 1: Nar), and Randiellidae (with four marine species in the Atlantic and Pacific Oceans; Erséus & Strehlow, 1986; Erséus, 1997; Fig. 1: Ran) are uncertain, while the North-American/European Parvidrilidae (with two phreatic freshwater species described; Fig. 1: Par) may be a Northern Hemisphere sistergroup to Phreodrilidae (Erséus, 1999; Martínez-Ansemil et al., 2002). Finally, Capilloventridae (Fig. 1: Cap), so far known only from the Southern Hemisphere, is unique among the new families in that it contains both marine and freshwater species (Harman & Loden, 1984; Erséus, 1993a; Pinder & Brinkhurst, 1997b). It has a lateral displacement of all its chaetal bundles, reminiscent of the arrangement seen in polychaetous annelids (Erséus, 1993a). Parvidrilidae and Capilloventridae are the only clitellate families with capillary chaetae ventrally as well as dorsally.

From scenario to formal analysis

As in all other systematic biology following the establishment of evolutionary theory, assessments of oligochaete phylogeny for a long time were characterized by a intuitive, narrative approach. The phylogenetic hypotheses were dependent on the author's subjective evaluation of facts at hand. An early example of this (Fig. 2) is the scenario of oligochaete relationships presented by Beddard (1895), but similar approaches were seen throughout most of the last century (e. g., Michaelsen, 1929; Stephenson, 1930; Brinkhurst & Jamieson, 1971; Timm, 1981; Erséus, 1984b; Omodeo, 1998, 2000). These intuitive phylogenies are often characterized by the following features:

(1) They reflect the author's great knowledge of the taxa involved.
(2) They involve descriptions of hypothetical ancestors or intermediate forms.
(3) They put emphasis on rather few character systems, and especially those related to reproductive organs. This includes the frequent association of particular criteria to different ranks in the classification. For instance, for a long time certain kinds of features were conceived as familial, others as subfamilial, yet others as generic 'criteria' (see, e.g., Erséus, 1980; Kasprzak, 1984b).
(4) The narratives leading to the intuitive hypotheses sometimes stress the adaptive value of the assumed transformation of various character states.

Nothing critical should be said about the first point. Rather, it is important that also proponents of modern cladistic approaches (see below) carry on the good tradition to spend considerable time on acquiring detailed information about many species, and then critically evaluate primary homologies; i.e., to test whether a superficial similarity will remain after a critical scrutiny of the details in the feature in question. The second point, however, using hypothetical ancestors or intermediate forms is risky. Such forms are not based on actual observations, which increases the number of *ad hoc* hypotheses. Similarly, selecting a

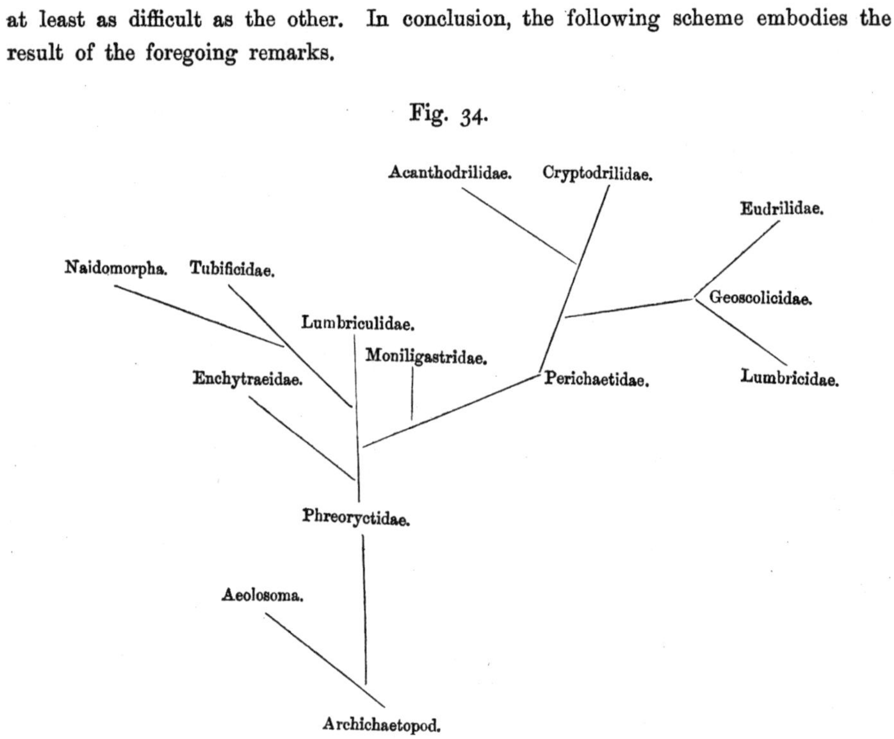
at least as difficult as the other. In conclusion, the following scheme embodies the result of the foregoing remarks.

Fig. 34.

Figure 2. Hypothesis of oligochaete relationships first published by Beddard (1895: 173). Phreoryctidae is an abandoned name, corresponding to today's Haplotaxidae. The seven families on the right branch are all megadriles (Metagynophora *sensu* Jamieson, 1988), and all of them except Moniligastridae are today classified in Crassiclitellata. Note the final sentence in the original author's (long) discussion leading to this diagram.

few characters as more important than others (third point) introduces subjective bias. Finally, although authors could be given credit for considering evolutionary processes in their phylogenetic assessment, *ad hoc* assumptions about the adaptability of different character states (fourth point) concern historical events that are hardly testable.

Brinkhurst (1982, 1984a, 1984b) developed logical interpretations of the patterns of gonads and gonoducts in various oligochaete families, and used them as a basis for evolutionary schemes (an example shown in Fig. 3). A common theme in his phylogenetic reconstructions is the idea of an octogonadal ancestor of all oligochaetes; the hexa- or tetragonadal patterns seen in a majority of recent oligochaetes are then interpreted as the results of reduction of various pairs of gonads in a basic octogonadal form. He thus also used the octogonadal pattern in Haplotaxidae as evidence for the ancestral and paraphyletic status of this family. It is noteworthy, however, that the derivation of tetragonadal forms was outlined as four convergent events (Fig. 3; Brinkhurst, 1984a), and on the whole, Brinkhurst's scheme is not parsimonious. In corresponding scenarios, Brinkhurst (1984b: Figs 4A, 5A, 6A) superimposed his haplotaxid ancestor theory on the evolution of chaetal characters and habitat preferences.

Needless to say, the intuitively inferred oligochaete phylogenies have their merits. As they have been based on a considerably amount of data and treated with a lot of common sense, they have laid a useful framework for subsequent testing using more objective methods.

Along with the development of more formalised methods of phylogenetic systematics (Hennig, 1950), also referred to as cladistics, the approaches to the evolutionary history of clitellates drastically changed about 20 years ago. The first study applying Hennigian methodology was that by Jamieson (1980), and it was soon followed by a number of publications using computer-aided cladistic methods based on the principle of

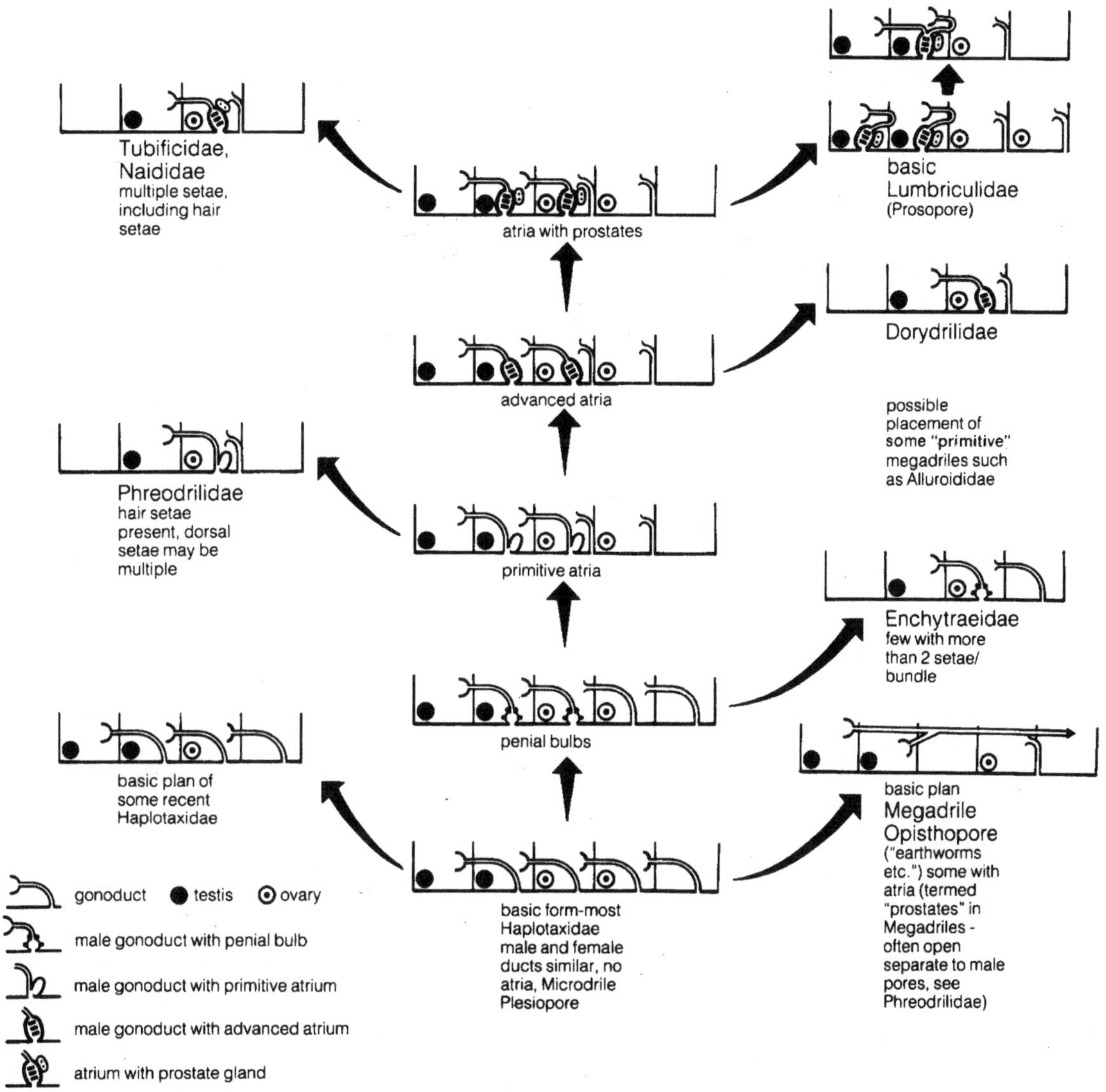

Figure 3. Evolutionary scheme of oligochaete families, based on patterns of gonads and gonoducts, first published as a "possible evolutionary pathway of microdrile oligochaetes assuming that atria evolved only once" (Brinkhurst, 1984a: Fig. 1). [A part of this figure was later corrected by Brinkhurst & Nemec (1987: Fig. 9), to account for the fact that prostate glands are associated with the atria in Dorydrilidae, a family however not considered in the present account.]

parsimony. Parsimony methods search for hypotheses that maximize congruence among characters, which also provides a powerful test of homology (Patterson, 1982). Some of these studies dealt with relationships between closely related species or genera (Erséus, 1984a, 1990a, 1991, 1993b, 1994; Erséus & Milligan, 1993), others with subfamilial or intra-familial phylogenies (Brinkhurst, 1988, 1989; 1991, 1994; Erséus, 1987, 1990b; Brinkhurst & Nemec, 1987; Nemec & Brinkhurst, 1987; Coates 1987, 1989; Jamieson, 1988; Gelder & Brinkhurst, 1990). In general, they were based on a limited number of conventional morphological characters.

Some of these studies identified cases of putative paraphyly in the current classification. One conclusion repeatedly made was that Naididae is a derived group within Tubificidae, on the basis of the fact that naidids have modified genital chaetae, diffuse prostate glands, and numerous granulated coelomocytes, features also typical of the (probably paraphyletic) tubificid subfamily Rhyacodrilinae (Erséus, 1987, 1990b; Brinkhurst, 1994).

However, most of these early cladistic studies still suffered from some of the same subjectivity as that characterizing the intuitive phylogenies. In particular, they often used generalized descriptions of taxa, rather than real representatives (species) of them, and rooted with hypothetical ancestors. As a rule also, monophyly of a particular taxon was not properly tested prior to using it as an operational unit in a higher level phylogenetic analysis. In those days of non-molecular systematics, an excuse for using hypothetical ancestors was that, if using real outgroups, many morphological characters, although informative within the ingroup, would not be present in (and not applicable for) the outgroup. Then it would be impossible to establish character polarities (plesiomorphic versus apomorphic states). In molecular systematics, one always uses real outgroup taxa, as homologous genes are normally recognized in the outgroup as well as the ingroup.

Despite the shortcomings of some of the early cladistic analyses of clitellate phylogeny, a formalised analysis of congruence among characters, today often associated with measurements of support (bootstrap, jackknife, etc.), will reduce subjectivity and better fulfil the requirements of scientific testing, and thus is to be preferred to intuitive scenarios of evolution. Still, however, a careful *a priori* assessment of primary homologies (such as studying ultrastructural detail and ontogenetic processes; see below) is of fundamental importance for the quality of any cladistic study.

Relationships revealed by morphological data

By and large, morphological characters alone have not been sufficient for resolving a number of fundamental issues concerning clitellate origin and relationships. For instance, no one has yet established which other annelids are the sistergroup to Clitellata (Purschke et al., 1993; Westheide, 1997; Rouse & Fauchald, 1997; McHugh, 2000; Purschke, 2002). Aeolosomatidae, which once was considered as a part of "Oligochaeta" (see Fig. 2; Beddard, 1895), appeared to be intermediate between polychaetes and clitellates, but as concluded by Timm (1981) and Purschke et al. (2000), and now also corroborated by DNA studies (Rota et al., 2001; Struck et al., 2002), this family does not belong to Clitellata. Together with a similar taxon, Potamodrilidae, it is today regarded as a separate group, Aphanoneura, with an uncertain position in Annelida (Timm, 1981; Purschke & Hessling, 2002). On the basis of morphology, two recent studies propose *Hrabeiella periglandulata* (a terrestrial polychaete) and Questidae (another polychaete taxon), respectively, as putative sister-groups to Clitellata (Almeida et al., 2003; Purschke, 2003), but neither of these relationships are supported by molecular data (Erséus et al., 2000; Rota et al., 2001; Struck et al., 2002).

As already mentioned, morphological evidence has resulted in a rather broad (but usually not so explicit) consensus about the monophyly of Metagynophora, and the ancestral position of Haplotaxidae (at least vis-à-vis Metagynophora), but the systematic position of Enchytraeidae has not been settled, and although several aquatic taxa, such as Lumbriculidae, Branchiobdellida and Phreodrilidae, have atrial structures similar to those of Tubificidae, the exact relationships among these groups, or whether or not the atria are homologous, have remained unresolved. Nor is there any consensus as to in which habitat the clitellates evolved. Timm (1981, Fig. 2) starts his 'pedigree of Oligochaeta' in the 'fresh inland waters', while a coastal lagoon interstitial environment was the cradle of Clitellata according to Omodeo (1998). Westheide (1997), Purschke (1999, 2003), and Purschke et al. (2000), on the other hand, argue that the first clitellates were terrestrial.

Recent achievements regarding clitellate morphology have broadened the basis for phylogenetic analyses. First, a particular focus on spermatozoal ultrastructure and 'spermiocladistics' (Jamieson, 1987a) during the last 20 years has proved beneficial for inferring clitellate relationships. Clitellate spermatozoa show several synapomorphic features (see, e.g., Jamieson, 1986; Ferraguti & Erséus, 1999; Ferraguti, 2000), but also a great variation that has been explored in numerous papers (e.g., Garavaglia et al., 1974; Jamieson, 1978, 1982, 1983, 1984, 1987a, 1987b; Braidotti & Ferraguti, 1982; Jamieson et al., 1983, 1987; Ferraguti, 1984, 2000; Ferraguti & Jamieson, 1987; Ferraguti et al.,

1989, 1994, 1999; Ferraguti & Gelder, 1991; Westheide et al., 1991; Erséus & Ferraguti, 1995; Westheide & Purschke, 1996; Ferraguti & Erséus, 1999; Cardini et al., 2000; Gelder & Ferraguti, 2001; Marotta et al., 2003). These studies demonstrate that sperm often carry a phylogenetic signal that is congruent with that of other morphological data. For instance, sperm ultrastructure does not only support monophyly of a group consisting of leeches, *Acanthobdella*, and Branchiobdellida, but also that these taxa are all derived oligochaetes (Ferraguti & Erséus, 1999). Further, Ferraguti et al. (1996) showed that the sperm of Capilloventridae have plesiomorphic traits, supporting the hypothesis, proposed on the basis of other morphological data (Erséus, 1993a), that this family represents an early offshoot within Clitellata.

Second, recent work on the ontogenetic development of the male gonoducts in Tubificidae (Gustavsson & Erséus, 1997, 1999; Gustavsson, 2002) has revealed features of relevance for understanding the phylogeny of this and other aquatic families. An important discovery in these studies is that tubificid prostate glands have at least two different origins; one kind is derived from ectodermal, another from mesodermal tissues. This means that the various prostates, previously assumed to be all derived from one kind of gland present in a common ancestor (see, e.g., Erséus, 1987; 1990b; Brinkhurst, 1994), instead should be regarded as at least two separate characters.

Morphological features will undoubtedly remain important for the assessment of clitellate phylogeny, particularly if combined with the molecular data now becoming available at an accelerating rate.

Molecular phylogenies – a progress report

Since the mid-1990s, new efficient methods for sequencing and analysing DNA have revolutionized systematics, or at least drastically improved its potential. To date, molecular systematics has not given any final answers to many of the questions regarding clitellate phylogeny, but we now have the perspective to review some first steps towards this end.

Table 1 lists publications involving clitellate worms and molecular phylogenies. It was confirmed early on that Clitellata is monophyletic; both the 18S rDNA and elongation factor -1α genes support this (Kim et al., 1996; Moon et al., 1996; McHugh, 1997; Kojima, 1998; Winnepenninckx et al., 1998). Sequence data now also show that Clitellata is nested somewhere among the polychaetous annelids, although the exact position is yet unknown (Martin, 2001; Rota et al., 2001; Struck et al., 2002).

A few molecular studies have dealt with particular relationships within leeches (Siddall & Burreson, 1998; Apakupakul et al., 1999; Trontelj et al., 1999) and branchiobdellidans (Gelder & Siddall, 2001). Further, using different genes, Christensen & Theisen (1998) and Erséus et al. (2000) obtained the first molecular evidence for a position of Naididae within Tubificidae (see above). A preliminary study of partial COI mtDNA sequences (Nylander et al., 1999) corroborated that the gutless tubificid genera *Inanidrilus* and *Olavius* (in the study represented by six species) comprise a monophyletic group, but for establishing other relationships within the oligochaetous Clitellata, the gene proved too variable. Using another highly variable mitochondrial gene, 16S rDNA, Sturmbacher et al. (1999) and Beauchamp et al. (2001) showed that various populations of '*Tubifex tubifex*' in North America and Europe represent what in effect appears to be a number of distinct species.

A DNA-based suggestion that leeches and their allies are derived oligochaetes was first discussed by Martin et al. (2000). Using a larger dataset, with 83 clitellate species and 18S and COI sequences combined, Siddall et al. (2001) presented a better supported hypothesis of the oligochaete-leech relationships (Fig. 4). According to this study, *Acanthobdella*, Branchiobdellida and Hirudinida indeed comprise a monophyletic group within Clitellata, with Lumbriculidae as its most likely sister taxon. In a more or less simultaneous publication, but based on fewer taxa and 18S data only, Martin (2001) arrived at the same conclusion. Moreover, the study by Siddall et al. (2001) indicated a sistergroup relationship between the crassiclitellates (represented by three earthworm species) and Enchytraeidae (also represented by three species).

In another investigation, Erséus et al. (2002) analysed 18S data for 52 species of Tubificidae,

Table 1. Published molecular studies involving clitellate taxa

Reference	Gene	Taxa include	Main focus/result
Moon et al. (1996)	18S	3 clitellates	Clitellata monophyletic
Kim et al. (1996)	18S	6 clitellates	Clitellata monophyletic
McHugh (1997)	EF-1α	3 clitellates	Clitellata monophyletic
Winnepenninckx et al. (1998)	18S	9 clitellates	Clitellata monophyletic
Kojima (1998)	EF-1α	4 clitellates	Clitellata monophyletic
Christensen & Theisen (1998)	23S/COI	11 clitellates	Naididae a part of Tubificidae
Siddall & Burreson (1998)	COI	27 clitellates	Leech phylogeny
Nylander et al. (1999)	COI	35 clitellates	Gutless group monophyletic
Apakupakul et al. (1999)	18S/COI	39 clitellates	Leech phylogeny
Sturmbauer et al. (1999)	16S	intra-specific	Emphasis on *Tubifex tubifex*
Trontelj et al. (1999)	12S/18S	15 clitellates	Leech relationships
Martin et al. (2000)	18S/COI	13 clitellates	Leech-like taxa in Oligochaeta
Erséus et al. (2000)	18S	23 clitellates	Naididae a part of Tubificidae
Jamieson (2000)	12S/16S/28S	44 clitellates	Crassiclitellate relationships
Beauchamp et al. (2001)	16S	intra-specific	Emphasis on *Tubifex tubifex*
Gelder & Siddall (2001)	18S/COI	27 clitellates	Branchiobdellidan phylogeny
Martin (2001)	18S	21 clitellates	Leech-lika taxa in Oligochaeta
Siddall et al. (2001)	18S/COI	83 clitellates	Leech-like taxa in Oligochaeta
Erséus et al. (2002)	18S	67 clitellates	Tubificid/naidid relationships
Jamieson et al. (2002)	12S/16S/28S	59 clitellates	Crassiclitellate relationships
Erséus & Källersjö (2004)	18S	39 clitellates	Clitellate relationships

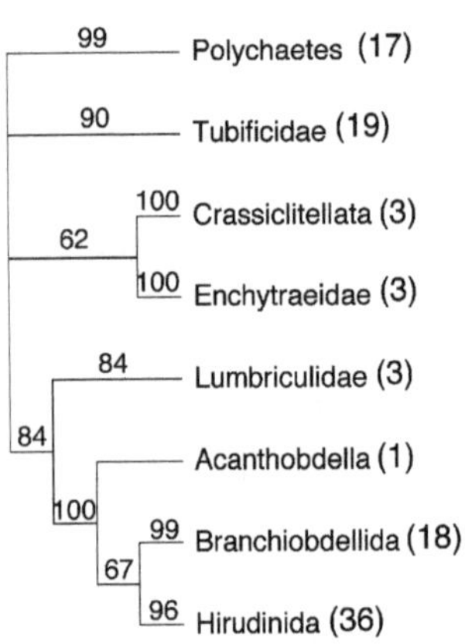

Figure 4. Phylogenetic tree of Clitellata based on analysis of combined 18S and COI gene sequences of 83 ingroup clitellates and 17 outgroup polychaetes, with focus on the position of Hirudinida, Branchiobdellida and *Acanthobdella peledina* (simplified after Siddall et. al., 2001: Fig. 1). Individual species (nos. in parentheses) are conglomerated into higher-level taxa. Jackknife support values are shown for all nodes. The ingroup node (with support 99) is rooted among the polychaetes.

and seven species of Naididae. It supported the earlier notions (see above) that naidids are derived tubificids, and that the gutless species are monophyletic, otherwise it failed to resolve much of the subfamilial relationships within Tubificidae. It did provide, however, the first molecular evidence that Phreodrilidae is closely related to Tubificidae.

In yet another recent work, with focus on megascolecid earthworms, Jamieson (2000) and Jamieson et al. (2002) analysed data of three different genes (28S, 12S and 16S rDNA). In their analysis of 28S only, Jamieson et al. (2002) got strong support for the monophyly of Crassiclitellata, as well as modest support for the sistergroup relationship between Lumbriculidae and the hirudinidan/branchiobdellidan clade. The combined data set of the three genes (op. cit.) yielded a topology within Megascolecidae that is not congruent with the present internal classification of this family (Jamieson, 1971; Gates, 1972).

Finally, Erséus & Källersjö (2004) revisited some of the Erséus et al. (2002) data, and by adding 11 new sequences (some representing families not previously sampled) were able to present a first DNA-based hypothesis regarding the more basal relationships among Clitellata.

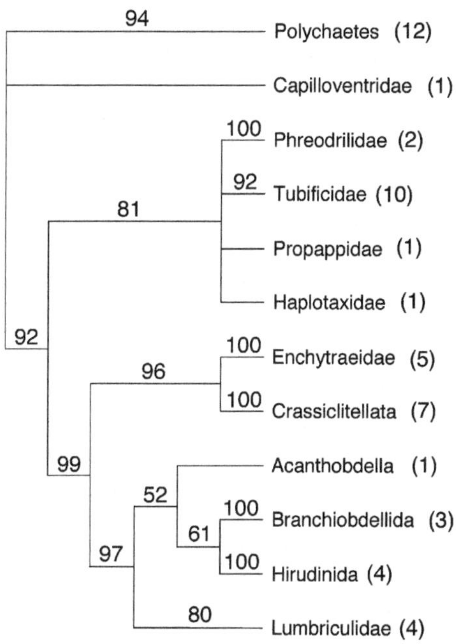

Figure 5. Phylogenetic tree of Clitellata based on an analysis of 18S rDNA gene sequences of 39 ingroup clitellates and 12 outgroup polychaetes, using an aligned dataset in which gaps were coded as a fifth character state (simplified after Erséus & Källersjö, 2004: Fig. 4). Individual species (nos. in parentheses) are conglomerated into higher-level taxa. Jackknife support values are shown for all nodes. The ingroup node (with support 94) is rooted among the polychaetes.

Using one particular set of aligned sequences and treating gaps in the alignment as a fifth character state, a tree shown in a simplified form here (Fig. 5) was obtained. It suggests that Capilloventridae is the sistergroup to all other clitellates included in the study, a position already supported by the ancestral traits in its general morphology (Erséus, 1993a) and sperm ultrastructure (Ferraguti et al., 1996). There is a basal dichotomy among the remaining clitellates, which puts Enchytraeidae, Crassiclitellata, Lumbriculidae and all leech-like taxa in one clade, and Propappidae, Haplotaxidae, Phreodrilidae and Tubificidae/Naididae in another. This is unexpected, as it places Haplotaxidae well separated from Crassiclitellata (Metagynophora). However, Haplotaxidae is represented by only a single species in the study, and as indicated several times by Brinkhurst (e.g., 1988, 1992, 1999), this family may be a highly paraphyletic assemblage of survivors, today scattered along several lines of evolution in Clitellata. The issue can only be resolved with extended taxon sampling.

Another noteworthy aspect of the Erséus & Källersjö 18S study (Fig. 5) is the close relationship between Enchytraeidae and all earthworms (Crassiclitellata), as was already indicated in the Siddall et al. (2001) tree (Fig. 4). A great majority of the enchytraeids are terrestrial, as are most crassiclitellates. This means that all non-aquatic oligochaetous clitellates appear to be restricted to a clade, which is not at the base of the proposed clitellate family tree. The analysis thus refutes the proposal that the ancestor of Clitellata was adapted to terrestrial life (Westheide, 1997; Purschke, 1999, 2003; Purschke et al., 2000), and instead supports the traditional view that the first clitellates were aquatic.

Summary and conclusions

The following points summarize the present knowledge regarding the phylogeny of oligochaetous Clitellata.

(1) There is strong support for monophyly of Clitellata and its position somewhere among the polychaetous annelids. However, its sistergroup is still unknown.
(2) Hirudinida, Branchiobdellida and *Acanthobdella* (Acanthobdellida) comprise a clade, the ancestor of which was an oligochaetous clitellate closely related to Lumbriculidae.
(3) So far, molecular data corroborate that earthworms with a multilayered clitellum (Crassiclitellata) are monophyletic, but additional taxa of this group and other metagynophorans need to be analysed.
(4) Capilloventridae is a putative sistergroup to all other Clitellata.
(5) 18S rDNA gives some support for the hypothesis that Enchytraeidae is closely related to earthworms, but this needs to be tested with additional molecular data.
(6) The systematic position of Haplotaxidae (often considered as an ancestral group) is still somewhat unclear, and this can only be settled with extending taxon sampling.
(7) Phreodrilidae appears to be closely related to Tubificidae and the new, enigmatic taxon Parvidrilidae, although the relationship with

the latter has so far been inferred from morphology only.

(8) 'Naididae' is positioned within Tubificidae and is now suggested to be called Naidinae (Erséus & Gustavsson, 2002), but the relationships between the tubificid subfamilies are still largely unresolved.

The phylogeny of the oligochaetous Clitellata still involves many challenges for future work. Analyses of molecular data have given promising results and will remain and develop as an important source of phylogenetic information. However, gene sequence patterns are loaded with homoplasy and entail great problems of primary homology assessment, in particular, when alignment concerns sequences of different length. Moreover, different genes evolve at different rates, making each gene informative only at certain taxonomic levels (and this varies between different kinds of organisms). Thus we can still anticipate a lot of trial and error approaches, with the goal of finding combinations of genes that work satisfactorily for each particular group of taxa.

It is also important to emphasize the need for additional characters, including those pertaining to conventional morphology, ontogeny, and ultrastructure. Here it is particularly important to encourage a continuous critical analysis of statements regarding primary homologies, so that, as far as possible, convergence can be revealed and accounted for (re-scored) prior to the phylogenetic analysis. In my opinion, a formal analysis of congruence among characters actually observed (and combined from as many sources as possible) is to be preferred to intuitive phylogenies whether or not they are based on discussions about the adaptive value of possible character states in hypothetical ancestors. Well corroborated phylogenetic hypotheses will then instead show the most likely character states of the ancestors inferred from them.

A final remark is that increased taxon sampling will be likely to improve the quality of phylogenetic hypotheses, at least in a longer perspective. Considering the limited number of morphological characters that have been available for work on clitellate phylogeny in past times, just adding taxa to old datasets may not improve resolution, but with the virtually unlimited molecular information soon available, as well as the great potential in new ultrastructural and ontogenetic data, it can be anticipated that the more 'densely' the taxa are sampled, the more precise the estimated phylogeny will be.

Acknowledgements

I am indebted to Dr Günter Purschke and Dr Thomas Bartolomeus, for inviting me to the Symposium, 'Morphology, Molecules, Evolution and Phylogeny in the Polychaeta and Related Taxa', in Osnabrück, in September 2002; to Dr Emilia Rota and Prof Pietro Omodeo, for consultance in matters relating to earthworm systematics; to Ms. Christine Hammar, for drawing Figure 1; to Dr Emilia Rota, and Prof Barrie G. M. Jamieson, for their constructive reviews of the manuscript; and to the Swedish Research Council for financial support (Contract No. 621-2001-2788).

References

Almeida, W. O., M. L. Christoffersen, D. S. Amorin, A. R. S. Garraffoni & G. S. Silva, 2003. Polychaeta, Annelida, and Articulata are not monophyletic: articulating the Metameria (Metazoa: Coelomata). Revista Brasileira de Zoologia 20: 23–57.

Apakupakul, K., M. E. Siddall & E. M. Burreson, 1999. Higher level relationships of leeches (Annelida: Clitellata: Euhirudinea) based on morphology and gene sequences. Molecular Phylogenetics and Evolution 12: 350–359.

Beauchamp, K. A., R. D. Kathman, T. S. McDowell & R. P. Hedrick, 2001. Molecular phylogeny of tubificid oligochaetes with special emphasis on *Tubifex tubifex* (Tubificidae). Molecular Phylogenetics and Evolution 19: 216–224.

Beddard, F. E., 1895. A Monograph of the Order Oligochaeta. Clarendon Press, Oxford, 769 pp.

Benham, W. B., 1890. An attempt to classify earthworms. Quarterly Journal of Microscopical Science (new series) 31: 201–315.

Braidotti, P. & M. Ferraguti, 1982. Two sperm lines in the spermatozeugmata of *Tubifex tubifex* (Annelida, Oligochaeta). Journal of Morphology 171: 123–136.

Brinkhurst, R. O., 1982. Evolution in the Annelida. Canadian Journal of Zoology 60: 1043–1059.

Brinkhurst, R. O., 1984a. Comments on the evolution of the Annelida. Hydrobiologia 109: 189–191.

Brinkhurst, R. O., 1984b. The position of the Haplotaxidae in the evolution of oligochaete annelids. Hydrobiologia 115: 25–36.

Brinkhurst, R. O., 1988. A taxonomic analysis of the Haplotaxidae (Annelida, Oligochaeta). Canadian Journal of Zoology 66: 2243–2252.

Brinkhurst, R. O., 1989. A phylogenetic analysis of the Lumbriculidae (Annelida, Oligochaeta). Canadian Journal of Zoology 67: 2731–2739.

Brinkhurst, R. O., 1991. A phylogenetic analysis of the Tubificinae (Oligochaeta, Tubificidae). Canadian Journal of Zoology 69: 392–397.

Brinkhurst, R. O., 1992. Evolutionary relationships within the Clitellata. Soil Biology and Biochemistry 24: 1201–1205.

Brinkhurst, R. O., 1994. Evolutionary relationships within the Clitellata: an update. Megadrilogica 5: 109–116.

Brinkhurst, R. O., 1999. Retrospect and prospect: reflections on forty years of study of aquatic oligochaetes. Hydrobiologia 406: 9–19.

Brinkhurst, R. O. & B. G. M. Jamieson, 1971. Aquatic Oligochaeta of the World. Oliver & Boyd, Edinburgh, 860 pp.

Brinkhurst, R. O. & A. F. L. Nemec, 1987. A comparison of phenetic and phylogenetic methods applied to the systematics of Oligochaeta. Hydrobiologia 155: 65–74.

Cardini, A., M. Ferraguti & S. R. Gelder, 2000. A phylogenetic assessment of the branchiobdellidan family Branchiobdellidae (Annelida, Clitellata) using spermatological and somatic characters. Zoologica Scripta 29: 347–366.

Čekanovskaya, O. V., 1962. Aquatic Oligochaeta of the U. S. S. R. Akademia Nauk S. S. S. R., Moskva-Leningrad (in Russian), 411 pp.

Christensen, B. & B. F. Theisen, 1998. Phylogenetic status of the family Naididae (Oligochaeta, Annelida) as inferred from DNA analyses. Journal of Zoological Systematics and Evolutionary Research 36: 169–172.

Coates, K. A., 1986. Redescription of the oligochaete genus Propappus, and diagnosis of the new family Propappidae (Annelida: Oligochaeta). Proceedings of the Biological Society of Washington 99: 417–428.

Coates, K. A., 1987. Phylogenetics of some Enchytraeidae (Annelida: Oligochaeta): a preliminary investigation of relationships to the Haplotaxidae. Hydrobiologia 155: 91–106.

Coates, K. A., 1989. Phylogeny and origins of Enchytraeidae. Hydrobiologia 180: 17–33.

Dubilier, N., R. Amann, C. Erséus, G. Muyzer, S. Y. Park, O. Giere & C. M. Cavanaugh, 1999. Phylogenetic diversity of bacterial endosymbionts in the gutless marine oligochaete Olavius loisae (Annelida). Marine Ecology Progress Series 178: 271–280.

Eisen, G., 1904. Enchytraeidae of the West Coast of North America. Harriman Alaska Expedition 12: 1–166.

Erséus, C., 1980. Specific and generic criteria in marine Oligochaeta, with special emphasis on Tubificidae. In Brinkhurst, R. O. & D. G. Cook (eds), Aquatic Oligochaete Biology. Plenum Publishing Corporation, New York: 9–24.

Erséus, C., 1984a. Taxonomy and phylogeny of the gutless Phallodrilinae (Oligochaeta, Tubificidae), with descriptions of one new genus and twenty-two new species. Zoologica Scripta 13: 239–272.

Erséus, C., 1984b. Aspects of the phylogeny of the marine Tubificidae. Hydrobiologia 115: 37–44.

Erséus, C., 1987. Phylogenetic analysis of the aquatic Oligochaeta under the principle of parsimony. Hydrobiologia 155: 75–277.

Erséus, C., 1990a. The marine Tubificidae (Oligochaeta) of the barrier reef ecosystems at Carrie Bow Cay, Belize, and other parts of the Caribbean Sea, with descriptions of twenty-seven new species and revision of Heterodrilus, Thalassodrilides and Smithsonidrilus. Zoologica Scripta 19: 243–303.

Erséus, C., 1990b. Cladistic analysis of the subfamilies within the Tubificidae. Zoologica Scripta 19: 57–63.

Erséus, C., 1991. Two new species and a phylogenetic analysis of the genus Tectidrilus (Oligochaeta, Tubificidae). Zoologica Scripta 20: 333–338.

Erséus, C., 1993a. Taxonomy of Capilloventer (Capilloventridae), a little-known group of aquatic Oligochaeta, with descriptions of two new species. Journal of Natural History 27: 1029–1040.

Erséus, C., 1993b. A new marine species of Smithsonidrilus (Oligochaeta: Tubificidae) from the Florida Keys. Proceedings of the Biological Society of Washington 106: 587–590.

Erséus, C., 1994. Marine Tubificidae (Oligochaeta) of Antarctica, with descriptions of three new species of Phallodrilinae. Zoologica Scripta 23: 217–224.

Erséus, C., 1997. A record of Randiella from New Caledonia, the first known occurrence of the marine interstitial family Randiellidae (Annelida; Oligochaeta) in the South Pacific Ocean. Journal of Natural History 31: 1745–1750.

Erséus, C. 1999. Parvidrilus strayeri, new genus and new species, an enigmatic interstitial clitellate from underground waters in Alabama. Proceedings of the Biological Society of Washington 112: 327–337.

Erséus, C. & M. Ferraguti, 1995. The use of spermatozoal ultrastructural in phylogenetic studies of Tubificidae. In Jamieson, B. G. M., J. Ausio & L. L. Justine (eds), Advances in Spermatozoal Phylogeny and Taxonomy. Mémoires du Muséum National d'Histoire Naturelle, Paris: 189–201.

Erséus, C. & L. Gustavsson, 2002. A proposal to regard the former family Naididae as a subfamily within Tubificidae (Annelida, Clitellata). Hydrobiologia 485: 253–256.

Erséus, C. & M. Källersjö, 2004. 18S rDNA phylogeny of Clitellata (Annelida). Zoologica Scripta 33: 187–196.

Erséus, C., M. Källersjö, M. Ekman & R. Hovmöller, 2002. 18S rDNA phylogeny of the Tubificidae (Clitellata) and its constituent taxa: Dismissal of the Naididae. Molecular Phylogenetics and Evolution 22: 414–422.

Erséus, C. & Milligan, M.R. 1993. A new species of Uniporodrilus (Oligochaeta, Tubificidae) from the Gulf of Mexico coast of Florida, and a phylogenetic analysis of the genus. Proceedings of the Biological Society of Washington 106: 243–250.

Erséus, C., T. Prestegaard & M. Källersjö, 2000. Phylogenetic analysis of Tubificidae (Annelida, Clitellata) based on 18S rDNA sequences. Molecular Phylogenetics and Evolution 15: 381–389.

Erséus, C. & D. R. Strehlow, 1986. Four new interstitial species of marine Oligochaeta representing a new family. Zoologica Scripta 15: 53–60.

Ferraguti, M., 1984. The comparative ultrastructure of sperm flagella central sheath in Clitellata reveals a new autapomorphy of the group. Zoologica Scripta 13: 201–207.

Ferraguti, M., 2000. Euclitellata. In Adyiodi, K. G. & R. G. Adyiodi (series eds), Reproductive Biology of Invertebrates,

Vol. IX, Part B: Progress in Male Gamete Ultrastructure and Phylogeny (Jamieson, B. G. M., volume ed.). John Wiley, Chichester: 165–222.

Ferraguti, M. & C. Erséus, 1999. Sperm types and their use for a phylogenetic analysis of aquatic clitellates. Hydrobiologia 402: 225–237.

Ferraguti, M., C. Erséus, I. Kaygorodova & P. Martin, 1999. New sperm types in Naididae and Lumbriculidae (Annelida: Oligochaeta) and their possible phylogenetic implications. Hydrobiologia 406: 213–222.

Ferraguti, M., C. Erséus & A. Pinder, 1996. The spermatozoon of Capilloventer australis and the systematic position of the Capilloventridae (Annelida: Oligochaeta). Australian Journal of Zoology 44: 469–478.

Ferraguti, M. & S. R. Gelder, 1991. The comparative ultrastructure of spermatzoa from five branchiobdellidans (Annelida: Clitellata). Canadian Journal of Zoology 69: 1945–1956.

Ferraguti, M., G. Grassi & C. Erséus, 1989. Different models of tubificid spermatozeugmata. Hydrobiologia 180: 73–82.

Ferraguti, M. & B. G. M. Jamieson, 1987. Spermiogenesis in Bythonomus lemani and the phylogenetic position of Lumbriculidae. Hydrobiologia 155: 123–134.

Ferraguti, M., D. Ruprecht, C. Erséus & O. Giere, 1994. An ultrastructural overview of tubificid spermatozoa. Hydrobiologia 278: 165–178.

Garavaglia, C., C. Lora Lamia Donin & G. Lanzavecchia, 1974. Ultrastructural morphology of spermatozoa of Hirudinea. Journal of Submicroscopical Cytology 6: 229–244.

Gates, G. E., 1972. Burmese earthworms. An introduction to the systematics and biology of megadrile oligochaetes with special reference to southeast Asia. Transactions of the American Philosophical Society, new series, 62: 1–326.

Gelder, S. R. & R. O. Brinkhurst, 1990. An assessment of the phylogeny of the Branchiobdellida (Annelida: Clitellata), using PAUP. Canadian Journal of Zoology 68: 1318–1326.

Gelder, S. R. & M. Ferraguti, 2001. Diversity of spermatozoan morphology in two families of Branchiobdellida (Annelida. Clitellata) from North America. Canadian Journal of Zoology 79: 1380–1393.

Gelder, S. R. & M. E. Siddall, 2001. Phylogenetic assessment of the Branchiobdellidae (Annelida, Clitellata) using 18S rDNA, mitochondrial cytochrome c oxidase subunit I and morphological characters. Zoologica Scripta 30: 215–222.

Giani, N., P. Martin & J. Juget, 1995. A new species of Phreodrilidae (Oligochaeta), Astacopsidrilus naceri sp nov, from Morocco (North Africa), with notes on the biogeography of the family. Canadian Journal of Zoology 73: 2375–2381.

Giere, O., 1981. The gutless marine oligochaete Phallodrilus leukodermatus. Structural studies on an aberrant tubificid associated with bacteria. Marine Ecology Progress Series 5: 353–357.

Giere, O. & C. Erséus, 1998. A systematic account of the Questidae (Annelida, Polychaeta), with description of new taxa. Zoologica Scripta 27: 345–360.

Giere, O. & N. W. Riser, 1981. Questidae – polychaetes with oligochaetoid morphology and development. Zoologica Scripta 10: 95–103.

Gustavsson. L., 2002. The development of the genital ducts in Limnodriloidinae (Tubificidae, Clitellata). Journal of Zoology, London 257: 83–91.

Gustavsson, L. & C. Erséus, 1997. Morphogenesis of the genital ducts and spermathecae in Clitellio arenarius, Heterochaeta costata, Tubificoides benedii (Tubificidae) and Stylaria lacustris (Naididae) (Annelida, Oligochaeta). Acta zoologica 78: 9–31.

Gustavsson, L. M. & C. Erséus, 1999. Development of the genital ducts and spermathecae in the rhyacodrilines Rhyacodrilus coccineus and Monopylephorus rubroniveus (Oligochaeta, Tubificidae). Journal of Morphology 242: 141–156.

Harman, W. J. & M. S. Loden, 1984. Capilloventer atlanticus gen. et sp. n., a member of a new family of marine Oligochaeta from Brazil. Hydrobiologia 115: 51–54.

Hennig, W., 1950. Grundzüge einer Theorie der phylogenetischen Systematik. Deutscher Zentralverlag Berlin, Berlin, 370 pp.

Jamieson, B. G. M., 1971. A review of the megascolecoid earthworm genera (Oligochaeta) of Australia. Part I. Reclassification and checklist of the megascolecoid genera of the world. Proceedings of the Royal Society of Queensland 82: 75–86.

Jamieson, B. G. M., 1978. A comparison of spermiogenesis and spermatozoal ultrastructure in megascolecid and lumbricid earthworms (Oligochaeta: Annelida). Australian Journal of Zoology 26: 225–240.

Jamieson, B. G. M., 1980. Preliminary discussion of an Hennigian analysis of the phylogeny and systematics of opistoporous oligochaetes. Revue dÉcologie et de Biologie du Sol 17: 261–275.

Jamieson, B. G. M., 1981. The Ultrastructure of the Oligochaeta. Academic Press, London, 462 pp.

Jamieson, B. G. M., 1982. The ultrastructure of the spermatozoon of Haplotaxis ornamentus (Annelida, Oligochaeta, Haplotaxidae) and its phylogenetic significance. Zoomorphology 100: 177–188.

Jamieson, B. G. M., 1983. Spermatozoal ultrastructure: evolution and congruence with a holomorphological phylogeny of the Oligochaeta (Annelida). Zoologica Scripta 107–114.

Jamieson, B. G. M., 1984. A phenetic and cladistic study of spermatozoal ultrastructure in the Oligochaeta (Annelida). Hydrobiologia 115: 3–13.

Jamieson, B. G. M., 1987a. The Ultrastructure and Phylogeny of Insect Spermatozoa. Cambridge University Press, Cambridge, 320 pp.

Jamieson, B. G. M., 1987b. A biological classification of sperm types, with special reference to annelids and molluscs, and an example of spermiocladistics. In Mohri. H. (ed.), New Horizon in Sperm Cell Research. Japan Science Society, Tokyo/Gordon, and Breach, New York: 311–332.

Jamieson, B. G. M., 1988. On the phylogeny and higher classification of the Oligochaeta. Cladistics 4: 367–410.

Jamieson, B. G. M., 2000. Native Earthworms of Australia (Megascolecidae, Megascolecinae). Science Publishers, Enfield, New Hampshire, U.S.A., unpaginated CD ROM.

Jamieson, B. G. M., C. Erséus & M. Ferraguti, 1987. Parsimony analysis of the phylogeny of some Oligochaeta

(Annelida) using spermatozoal ultrastructure. Cladistics 3: 145–155.

Jamieson, B. G. M., K. S. Richards, T. P. Fleming & C. Erséus, 1983. Comparative morphometrics of oligochaete spermatozoa and egg-acrosome correlation. Gamete Research 8: 149–169.

Jamieson, B. G. M., S. Tillier, A. Tillier, J. -L. Justine, E. Ling, S. James, K. McDonald & A. F. Hugall, 2002. Phylogeny of the Megascolecidae and Crassiclitellata (Oligochaeta, Annelida): combined versus partitioned analysis using nuclear (28S) and mitochondrial (12S, 16S) rDNA. Zoosystema 24: 707–734.

Jamieson, B. G. M. & R. I. Webb, 1984. The morphology, spermatozoal ultrastructure and phylogenetic affinities of new species of questid (Polychaeta; Annelida). In Hutchings, P. A. (ed.), Proceedings of the 1st International Polychaete Conference, Sydney, Australia. Linnean Society of New South Wales, Sydney, Australia: 21–34.

Kasprzak, K., 1984a. The previous and contemporary conceptions on phylogeny and systematic classifications of Oligochaeta (Annelida). Annales zoologici, Warszawa 38: 205–223.

Kasprzak, K., 1984b. Generic criteria in Enchytraeidae (Oligochaeta) family. Biólogia (Bratislava) 39: 163–172.

Kim, C. B., S. Y. Moon, S. R. Gelder & W. Kim, 1996. Phylogenetic relationships of annelids, molluscs, and arthropods evidenced from molecules and morphology. Journal of Molecular Evolution 43: 207–215.

Kojima, S., 1998. Paraphyletic status of Polychaeta suggested by phylogenetic analysis based on the amino acid sequences of elongation factor-1α. Molecular Phylogenetics and Evolution 9: 255–261.

Marotta, R., M. Ferraguti & C. Erséus, 2003. A phylogenetic analysis of Tubificinae and Limnodriloidinae (Annelida, Clitellata, Tubificidae) using sperm and somatic characters. Zoologica Scripta 32: 255–278.

Martin, P., 2001. On the origin of the Hirudinea and the demise of the Oligochaeta. Proceedings of the Royal Society, London (B) 268: 1089–1098.

Martin, P., I. Kaygorodova, D. Yu. Sherbakov & E. Verheyen, 2000. Rapidly evolving lineages impede the resolution of phylogenetic relationships among Clitellata (Annelida). Molecular Phylogenetics and Evolution 15: 355–368.

Martínez-Ansemil, E., B. Sambugar & N. Giani, 2002. First record of Parvidrilidae (Annelida, Oligochaeta) in Europe with a description of a new species (*Parvidrilus spelaeus* sp. nov.) and comments on the family and its phyletic relationships. Journal of Zoology, London 256: 495–503.

McHugh, D., 1997. Molecular evidence that echiurans and pogonophorans are derived annelids. Proceedings of the National Academy of Sciences of the United States of America 94: 8006–8009.

McHugh, D., 2000. Molecular phylogeny of the Annelida. Canadian Journal of Zoology 78: 1873–1884.

Michaelsen, W., 1929. Zur Stammesgeschichte der Oligochäten. Zeitschrift für wissenschaftliche Zoologie 134: 693–716.

Moon, S. Y., C. B. Kim, S. R. Gelder & W. Kim, 1996. Phylogenetic positions of the aberrant branchiobdellidans and aphanoneurans within the Annelida as derived from 18S ribosomal RNA gene sequences. Hydrobiologia 324: 229–236.

Nemec, A. F. L. & R. O. Brinkhurst, 1987. A comparison of methodological approaches to the subfamilial classification of the Naididae (Oligochaeta). Canadian Journal of Zoology 65: 691–707.

Nylander, J. A. A., C. Erséus & M. Källersjö, 1999. A test of monophyly of the gutless Phallodrilinae (Oligochaeta, Tubificidae) and the use of a 573-bp region of the mitochondrial cytochrome oxidase I gene in analysis of annelid phylogeny. Zoologica Scripta 28: 305–313.

Omodeo, P., 1987. Some new species of Haplotaxidae (Oligochaeta) from Guinea and remarks on the history of the family. Hydrobiologia 155: 1–13.

Omodeo, P., 1998. History of Clitellata. Italian Journal of Zoology 65: 51–73.

Omodeo, P., 2000. Evolution and biogeography of megadriles (Annelida, Clitellata). Italian Journal of Zoology 67: 179–201.

Patterson, C., 1982. Morphological characters and homology. In Joysey, K. A. & A. E. Friday (eds), Problems of Phylogenetic Reconstruction. Academic Press, London and New York: 21–74.

Pinder, A. M. & R. O. Brinkhurst, 1997a. Review of the Phreodrilidae (Annelida: Oligocheata: Tubificida) of Australia. Invertebrate Taxonomy 11: 443–523.

Pinder, A. M. & R. O. Brinkhurst, 1997b. The family Capilloventridae (Annelida: Clitellata) in Australia, with descriptions of two new species of Capilloventer. Zoologica Scripta 26: 255–265.

Pinder, A. M. & C. Erséus, 2000. New Phreodrilidae (Annelida: Clitellata) from Tasmanian estuaries. Papers and Proceedings of the Royal Society of Tasmania 134: 29–33.

Purschke, G., 1999. Terrestrial polychaetes – models for the evolution of the Clitellata (Annelida)? Hydrobiologia 406: 87–99.

Purschke, G., 2002. On the ground pattern of Annelida. Organisms, Diversity and Evolution 2: 181–196.

Purschke, G., 2003. Is *Hrabeiella periglandulata* (Annelida, "Polychaeta") the sister group of Clitellata? Evidence from an ultrastructural analysis of the dorsal pharynx in *H. periglandulata* and *Enchytraeus minutus* (Annelida, Clitellata). Zoomorphology 122: 55–66.

Purschke, G. & R. Hessling, 2002. Analysis of the central nervous system and sense organs in *Potamodrilus fluviatilis* (Annelida: Potamodrilidae). Zoologischer Anzeiger 241: 19–35.

Purschke, G., R. Hessling & W. Westheide, 2000. The phylogenetic position of the Clitellata and Echiura – on the problematic assessment of absent characters. Journal of Zoological Systematics and Evolutionary Research 38: 165–173.

Purschke, G., W. Westheide, D. Rodhe & R. O. Brinkhurst, 1993. Morphological reinvestigation and phylogenetic relationship of *Acanthobdella peledina* (Annelida, Clitellata). Zoomorphology 113: 91–101.

Reynolds, J. W. & D. G. Cook, 1993. Nomenclatura oligochaetologica. Supplementum tertium. A catalogue of names, descriptions and type specimens of the Oligochaeta. New

Brunswick Museum Monographic Series (Natural Science) 9: i-vi, 1–33.

Richards, K. S., 1977. Structure and function in the oligochaete epidermis (Annelida). In Spearman, R.C. (ed.), Comparative Biology of Skin. Academic Press, New York: 171–193.

Richards, K. S., T. P. Fleming & B. G. M. Jamieson, 1982. An ultrastructural study of the distal epidermis and the occurrence of subcuticular bacteria in the gutless tubificid *Phallodrilus albidus* (Oligochaeta: Annelida). Australian Journal of Zoology 30: 327–336.

Righi, G., 1995. Colombian earthworms. Studies on tropical Andean ecosystems/Estudios de ecosystemas tropandinos. Cramer, Berlin-Stuttgart, 4: 485–607.

Righi, G. & M. E. Varela, 1983. Narapa bonettoi, gen. nov. sp. nov. (Oligochaeta, Narapidae, fam. Nov.) de aqua doce de Argentina. Revista de la Asociacion de Ciencias Naturales de Litoral 14: 7–15.

Rota, E., 2001. Oversized enchytraeids (Annelida, Clitellata): a comparative study, with a revised description of *Lumbricillus maximus* (Michaelsen). Organisms, Diversity and Evolution 1: 225–238.

Rota, E., P. Martin & C. Erséus, 2001. Soil-dwelling polychaetes: enigmatic as ever? Some hints on their phylogenetic relationships as suggested by a maximum parsimony analysis of 18S rRNA gene sequences. Contributions to Zoology 70: 127–138.

Rouse, G. W. & K. Fauchald, 1997. Cladistics and polychaetes. Zoologica Scripta 26: 139–204.

Rouse, G. W. & F. Pleijel, 2001. Polychaetes. Oxford University Press, Oxford, 354 pp.

Siddall, M. E. & E. M. Burreson, 1998. Phylogeny of leeches (Hirudinea) based on mitochondrial cytochrome c oxidase subunit I. Molecular Phylogenetics and Evolution 9: 156–162.

Siddall, M. E., K. Apakupakul, E. M. Burreson, K. A. Coates, C. Erséus, M. Källersjö, S. R. Gelder & H. Trapido-Rosenthal, 2001. Validating Livanow: molecular data agree that leeches, branchiobdellidans and *Acanthobdella peledina* are a monophyletic group of oligochaetes. Molecular Phylogenetics and Evolution 21: 346–351.

Stephenson, J., 1930. The Oligochaeta. Clarendon Press, Oxford, 978 pp.

Struck, T., R. Hessling & G. Purschke, 2002. The phylogenetic position of the Aeolosomatidae and Parergodrilidae, two enigmatic oligochaete-like taxa of the 'Polychaeta', based on molecular data from 18S rDNA sequences. Journal of Zoological Systematics and Evolutionary Research 40: 155–163.

Sturmbauer, C., G. B. Opadiya, H. Niederstätter, A. Riedmann & R. Dallinger, 1999. Mitochondrial DNA reveals cryptic oligochaete species differing in cadmium resistence. Molecular Biology and Evolution 16: 967–974.

Timm, T., 1981. On the origin and evolution of aqautic Oligochaeta. Eesti NSV Teaduste Akadeemia Toimetised (Bioloogia) 30: 174–181.

Trontelj, P., B. Sket & G. Steinbrück, 1999. Molecular phylogeny of leeches: congruence of nuclear and mitochondrial rDNA data sets and origin of bloodsucking. Journal of Zoological Systematics and Evolutionary Research 37: 141–147.

Westheide, W., 1997. The direction of evolution within the Polychaeta. Journal of Natural History 31: 1–15.

Westheide, W. & G. Purschke, 1996. Proacrosome and acrosome of the spermatozoon in *Acanthobdella peledina* (Annelida: Clitellata). Invertebrate Reproduction and Development 29: 223–230.

Westheide, W., G. Purschke & K. Middendorf, 1991. Spermatozoal ultrastructure of the taxon Enchytraeus (Annelida, Oligochaeta) and its significance for species discrimination and identification. Zeitschrift für Zoologische Systematik und Evolutionsforschung 29: 323–342.

Winnepenninckx, B. M. H., Y. Van de Peer & T. Backeljau, 1998. Metazoan relationships on the basis of 18S rRNA sequences: a few years later. American Zoologist 38: 888–906.

Hydrobiologia (2005) **535/536:** 373–379
T. Bartolomaeus & G. Purschke (eds), Morphology, Molecules, Evolution and Phylogeny in Polychaeta and Related Taxa

Subject index

Figures are in boldface and tables are in italics

D

R

S

Hydrobiologia (2005) **535/536:** 381–387
T. Bartolomaeus & G. Purschke (eds), Morphology, Molecules, Evolution and Phylogeny in Polychaeta and Related Taxa

Species/generic index

Figures are in boldface and tables are in italics

MIX
Papier aus verantwortungsvollen Quellen
Paper from responsible sources
FSC® C105338

If you have any concerns about our products,
you can contact us on
ProductSafety@springernature.com

In case Publisher is established outside the EU,
the EU authorized representative is:
Springer Nature Customer Service Center GmbH
Europaplatz 3, 69115 Heidelberg, Germany

Printed by Libri Plureos GmbH
in Hamburg, Germany